ENCYCLOPEDIA OF MATHEMATICS AND ITS APPLICATIONS

EDITED BY G.-C. ROTA

Volume 32

Iterative Functional Equations

ENCYCLOPEDIA OF MATHEMATICS AND ITS APPLICATIONS

Iterative Functional Equations

MAREK KUCZMA

Institute of Mathematics, Silesian University, Katowice, Poland

BOGDAN CHOCZEWSKI

Institute of Mathematics, Academy of Mining and Metallurgy, Kraków, Poland

ROMAN GER

Institute of Mathematics, Silesian University, Katowice, Poland

The right of the
University of Cambridge
to print and sell
all manner of books
was granted by
Henry VIII in 1534.
The University has printed
and published continuously
since 1584.

CAMBRIDGE UNIVERSITY PRESS

Cambridge

New York   Port Chester

Melbourne   Sydney

Published by the Press Syndicate of the University of Cambridge
The Pitt Building, Trumpington Street, Cambridge CB2 1RP
40 West 20th Street, New York, NY 10011, USA
10 Stamford Road, Oakleigh, Melbourne 3166, Australia

© Cambridge University Press 1990

First published 1990

Printed in Great Britain at The Bath Press, Avon

British Library cataloguing in publication data
Kuczma, Marek
Iterative functional equations.
1. Functional equations
I. Title   II. Choczewski, Bogdan
III. Ger, Roman   IV. Series
515.7

Library of Congress cataloguing in publication data
Kuczma, Marek
Iterative functional equations / Marek Kuczma, Bogdan Choczewski, Roman Ger.
p.   cm. – (Encyclopedia of mathematics and its applications; v. 32)
Includes bibliographies and index.
ISBN 0 521 35561 3
1. Functional equations.   2. Functions of real variables.
I. Choczewski, Bogdan.   II. Ger, Roman.   III. Title.   IV. Series.
QA431.K792   1990
515′.62 – dc19   88-27536 CIP

ISBN 0 521 35561 3

MP

BST
REf

CONTENTS

99.50
599

PREFACE

The present book contains an outline of the modern theory of iterative functional equations. The expression *functional equations* is here understood in a narrow sense (*equations of finite kind*; see Kuczma [20]). It does not include equations in which infinitesimal operations are performed on the unknown functions. So, e.g., differential equations with transformed argument do not fall under this notion.

Nowadays, mainly owing to various activities of Professor J. Aczél, functional equations have grown to be a large, independent branch of mathematics, with its own methods, rich in results and abounding in applications. In such a large area further subdivisions are indispensable. The main line of division runs between *equations in several variables* in which at least one unknown function depends on fewer variables than the number of independent variables actually occurring in the equation (see Aczél [2], Kuczma [12]) and *equations in a single variable*, which can be written using one independent variable only.

Functional equations containing several variables are dealt with in another Encyclopedia volume written by J. Aczél and J. Dhombres [1]. The reader interested in the history of functional equations can consult Dhombres [4], [2] and also Aczél [3], Aczél–Dhombres [1].

'*Iterative functional equations*' is just another name for functional equations in a single variable (such equations are also referred to as *equations of rank 1*). Thus the subject matter of this book is approximately the same as that of Kuczma [26]. However (because of recent developments in the subject), most results now have a more general form and, of course, a number of results dating from after 1968 are found here. Moreover, we present more examples and applications. On the other hand, we do not consider some topics discussed in Kuczma [26]. In particular these include the problems of continuous iteration (embedding functions in flows and semiflows) and construction of general solutions.

Embedding problems are now at the centre of attention of many researchers and they on their own form a very large topic. Here we were only able to report on some aspects of the theory of continuous iteration; see Section 1.7. Construction of general solutions seems less important for applications; we have decided to concentrate on problems of the uniqueness and existence of solutions, justly considered as fundamental for equations of all kinds.

It would have been impossible to present everything about iterative functional equations. In order to keep this book to a reasonable size we had to make a selection, which is always the case with every book. Such a selection undoubtedly reflects the authors' personal preference. Note that at the end of each chapter there is a special section (Notes) in which some further results are briefly discussed and suitable references are given.

Although finite difference equations belong to the general type discussed in this book (and so some theorems can be applied, in particular, to difference equations), they are not dealt with here separately. Finite differences form an independent branch of mathematics and have their special problems and methods, which are not characteristic of more general iterative functional equations. There exist many books devoted entirely to difference equations and the interested reader is referred, e.g., to Nörlund [1] or Gelfand [1].

On the other hand, there are only a few books in which more general iterative functional equations are treated. We may mention here Ghermănescu [2], Gumowski–Mira [1], Kuczma [26], Maier–Kiesewetter [1], Montel [1], Pelyukh–Šarkovskiĭ [3], Targoński [8]. Every one of these books presents a different approach and contains also results not included in the other, nor in the present work. See also the booklets Neuman [12], Smítal [1].

The list of references at the end of this book is by no means complete. We have listed only those items that are referred to in the text. A fairly complete bibliography on iterative functional equations (except, however, finite differences) up to 1967 can be found in Kuczma [26]. But in the last two decades more publications concerning iterative functional equations have appeared than during the previous two centuries.

The story of this book requires a few words of explanation. Originally the book was written by M. Kuczma; the manuscript was ready by the end of 1976 and early in 1977 was sent to the publisher. However, it was felt that its style did not conform with that of the Encyclopedia series and major revisions were suggested. But in 1978 and 1980 the author suffered two strokes, which have resulted in an almost complete paralysis, and so the task of rewriting (and also updating, for several years had already passed since the original manuscript had been completed) the book had to be commissioned to somebody else. The author's colleagues and co-workers B. Choczewski and R. Ger were chosen for this job. The present Chapters 1,

7 and 11 are the work of R. Ger, the remaining chapters have been written by B. Choczewski. R. Ger also compiled the references. M. Kuczma has read everything and approved the changes. During the work the three authors were in constant contact and so the final text may be regarded as a result of their joint effort.

We want to conclude this preface by saying our thanks to those who have helped us at various stages of the work on this book.

We owe very much to Professor J. Aczél, on whose initiative this book has been written and who, all the time, was always ready to aid us with helpful advice and to ease our way through the meanders of formal matters. Our warmest thanks go to him also for his patience and understanding when the work on this book was getting delayed.

It is a pleasant duty to express here our gratitude to Dr K. Baron, Dr D. Brydak, Dr J. Drewniak, Dr H.-H. Kairies, Dr M. Sablik, Dr A. Smajdor and Dr M. C. Zdun, who have provided us with many useful pieces of information and bibliographical hints. Especially appreciated is the help of Dr K. Baron and Dr M. Sablik.

We entrust our book to the readers with the hope that it will be a help for those who want to apply iterative functional equations, and it will become a key to the enchanted world of iterative functional equations for those more interested in theory

THE AUTHORS

SYMBOLS AND CONVENTIONS

There are no preliminaries necessary to this book. However, the reader is assumed to have a basic knowledge of the undergraduate mathematics at a good university.

Throughout the book some particular sets are denoted by special symbols. $\mathbb{P}$ is the set of all prime numbers. $\mathbb{N}$ denotes the set of positive integers, $\mathbb{N}_0 = \mathbb{N} \cup \{0\}$ the set of nonnegative integers. $\mathbb{Z}$ is the set of all integers. $\mathbb{Q}$ is the set of rational numbers. $\mathbb{R}$ is the set of reals, $\bar{\mathbb{R}} = [-\infty, \infty]$ the set of extended reals, $\mathbb{R}^+ = [0, \infty)$ the set of nonnegative real numbers. $\mathbb{C}$ denotes the set of complex numbers – the complex plane. $\mathbb{K}$ stands for $\mathbb{R}$ or $\mathbb{C}$. For $n \in \mathbb{N}$ the symbol $\mathbb{K}^n$ denotes the set of all n-tuples $(\xi_1, \ldots, \xi_n)$ with $\xi_i \in \mathbb{K}$, and for $m, n \in \mathbb{N}$ the symbol $\mathbb{K}^{m \times n}$ denotes the set of all $m \times n$ matrices with entries from $\mathbb{K}$.

For $x = (\xi_1, \ldots, \xi_n) \in \mathbb{K}^n$ the symbol $|x|$ denotes the usual Euclidean norm: $|x| = (\sum_{i=1}^n |\xi_i|^2)^{1/2}$, and for $A \in \mathbb{K}^{m \times n}$ the norm $\|A\|$ is defined as the norm of the corresponding linear operator: $\|A\| = \sup_{|x|=1} |Ax|$. With the addition and multiplication by scalars from $\mathbb{K}$ defined in the usual manner, the sets $\mathbb{K}^n$ and $\mathbb{K}^{m \times n}$ endowed with the respective norms are Banach spaces over $\mathbb{K}$.

If $X \subset \mathbb{K}^n$, and $f : X \to \mathbb{K}^m$, $f = (f_1, \ldots, f_m)$, is a differentiable function, then f' denotes the matrix $(\partial f_i / \partial \xi_j) \in \mathbb{K}^{m \times n}$. Thus f' is a function, $f' : X \to \mathbb{K}^{m \times n}$. The symbol $\|\cdot\|$ stands for the norm also in any normed space other than $\mathbb{K}^n$.

The symbol 0 may have three different meanings. It may denote the number zero, or the origin in $\mathbb{K}^n$ (or in a more general linear space), or the function whose value is zero at every point of its domain. In every instance it is clear from the context which meaning we have in mind.

Somewhat similar is Landau's O-symbol (Oh) occurring in asymptotic conditions.

The symbol id denotes the identity function, i.e. the function whose value at every point of its domain is equal to that point. If we want to bring the domain (say, X) into evidence, we write id_X. Thus id_X is the function $id_X: X \to X$ such that $id_X(x) = x$ for every $x \in X$.

Open and closed real intervals are written, as usual, with round and square brackets, respectively. If it is not essential whether or not an end belongs to the interval in question, we write simply a vertical line. Thus, for instance, $X = [0, a|$ means that X may be either $[0, a)$ or $[0, a]$. But $|a, \infty|$ always means $|a, \infty)$.

The words increasing, decreasing, monotonic are used in the broader sense. Thus a function f is *increasing* iff ($=$ if and only if) $x_1 < x_2$ implies $f(x_1) \leq f(x_2)$. If $x_1 < x_2$ implies $f(x_1) < f(x_2)$, f is said to be *strictly increasing*; and similarly for decreasing, monotonic.

The word *vicinity* (of a point ξ in a space endowed with a topology) denotes a neighbourhood of ξ from which the point ξ has been removed. So, if U is a neighbourhood of ξ, then $U \setminus \{\xi\}$ is a vicinity of ξ. Neighbourhood is always understood relatively to the underlying set in which the equation in question is considered. The closure of a set A is denoted by cl A, and its interior by int A.

A sequence with terms a_n will be denoted by $(a_n)_{n \in \mathbb{N}}$ or simply (a_n). If X is a topological space, Y is a metric space, and we are given a sequence of functions $f_n: X \to Y$, $n \in \mathbb{N}$, we say that the sequence $(f_n)_{n \in \mathbb{N}}$ converges to a function $f: X \to Y$ *almost uniformly* (abbreviated to a.u.) in X iff it converges to f uniformly on every compact subset of X.

The abbreviation a.e. denotes, of course, almost everywhere (i.e., with the exception of a set of measure zero).

The symbol $\mathscr{F}(X, Y)$ stands for the family of all mappings from X into Y; $\mathscr{F}(X) := \mathscr{F}(X, X)$ whereas $C^r(X, Y)$, $r \in \mathbb{N}_0$, consists of all r times continuously differentiable mappings from $\mathscr{F}(X, Y)$; $C^r(X) := C^r(X, X)$, $r \in \mathbb{N}$; $C(X, Y) := C^0(X, Y)$, $C(X) := C^0(X)$.

Concerning sums and products, we use the convention that

$$\sum_{i=k}^{m} a_i = 0 \quad \text{and} \quad \prod_{i=k}^{m} a_i = 1 \quad \text{whenever } m < k.$$

The present book is divided into 12 chapters and the introduction (Chapter 0). Each chapter is divided into a number of sections, and most sections are further divided into subsections. Within a section we use separate numerations for formulae, theorems, lemmas, corollaries, remarks and definitions.

The end of each proof is marked by the sign ■.

0

Introduction

0.0 Preliminaries

0.0A. Types of equations considered

We will not attempt to define what is a functional equation. Such a definition can be found, e.g., in Aczél [2] or Kuczma [26] (see also Pelyukh–Sarkovskiĭ [3]). However, for our use this definition is too complicated and general. Therefore we shall settle for the indication that (with a few exceptions mentioned below) in the present book we shall study functional equations of the general form

$$F(x, \varphi(x), \varphi(f_1(x)), \ldots, \varphi(f_n(x))) = 0, \tag{0.0.1}$$

where φ is the unknown function and the remaining functions are considered as given. Only in Chapter 11 do we deal with the case where the functions f_i may depend again on the unknown function φ, whereas in Chapter 12 we touch upon the theory of *functional inequalities*, i.e., the case where instead of the equality sign in (0.0.1) we have the inequality $\leqslant$ or $\geqslant$.

The number n occurring in (1) is referred to as the *order* of the equation. Thus the equation of order 1 becomes

$$F(x, \varphi(x), \varphi(f(x))) = 0. \tag{0.0.2}$$

Equations of order zero, i.e. those of implicit functions $F(x, \varphi(x)) = 0$, will not be dealt with in the present book.

We consider also equations of infinite order, of the form

$$F(x, \varphi(x), \varphi \circ f(\cdot, x)) = 0, \tag{0.0.3}$$

where $(f(s, \cdot))_{s \in S}$ is a given family of functions, S being a nonvoid set. As a matter of fact, equation (0.0.3) contains equations (0.0.1) and (0.0.2) as particular cases. Equation (0.0.3) reduces to (0.0.2) when S consists of a single element, to (0.0.1) when S has n elements, and to the equation

$$F(x, \varphi(x), \varphi(f_1(x)), \varphi(f_2(x)), \ldots) = 0$$

when S is countable $(S = \mathbb{N})$. If the cardinality of S is the continuum, (0.0.3) may be regarded as an equation of order continuum.

However, in our book we shall confine ourselves mainly to equations of finite orders. Equations of infinite orders will occur only in Chapter 7.

We shall always consider equation (0.0.1) in the form solved with respect to $\varphi(x)$ –

$$\varphi(x) = h(x, \varphi(f_1(x)), \ldots, \varphi(f_n(x))) \qquad (0.0.4)$$

– or with respect to $\varphi(f_n(x))$ –

$$\varphi(f_n(x)) = g(x, \varphi(x), \varphi(f_1(x)), \ldots, \varphi(f_n(x))). \qquad (0.0.5)$$

It can be shown that (0.0.1) always is equivalent to a system of equations of form (0.0.4) and (0.0.5); see Reghiş–Vuc [1], Balint [1], [2].

The main attention will be paid to the case where x is a real variable and the values of the functions occurring are real. A few sections are devoted to the case of a complex variable, and in some cases we consider equations in more general spaces. There are two main reasons for this fact. Firstly, in the overwhelming majority of applications we meet equations in real or complex variables. Secondly, the uniqueness theorems which form the bulk of the present work require rather rich structures in the domain and in the range of the unknown function φ.

0.0B. Problems of uniqueness

In general, equations of form (0.0.1) have too many solutions. The situation is very well exemplified in the case of the equation

$$\varphi(x + 1) = \varphi(x), \qquad (0.0.6)$$

which is a particular case of (0.0.2) with $f(x) = x + 1$ and $F(x, y, z) = z - y$. Here every periodic function of period 1 is a solution. Such a function can be prescribed arbitrarily on any segment of length 1, open on one side, and thus the general solution φ of equation (0.0.6) depends on an arbitrary function.

In practice such a situation is disadvantageous. If we are led to a functional equation by a practical problem, we rather need a single well-defined solution which could be taken to represent the solution of the original problem. Therefore we need uniqueness theorems yielding a single solution characterized by a particular additional property.

The conditions which may yield the uniqueness of the solution depend very heavily on the form of the equation considered, on the properties of the given functions, and even on the numerical value of some parameters connected with the equation. For instance, in the case of equation (0.0.6) the situation seems to be hopeless. The general continuous or differentiable solution of equation (0.0.6) still depends on an arbitrary function. Nor does the analyticity condition guarantee the uniqueness of the solution. But if we assume that φ is monotonic then necessarily $\varphi = $ const. This is also the family of solutions of (0.0.6) that approach a finite limit at infinity.

Other equations may behave quite differently. Thus the equation

$$\varphi(2x) = \varphi(x) \qquad (0.0.7)$$

has the constant functions as the only continuous solutions $\varphi: \mathbb{R} \to \mathbb{R}$. And in the case of the equation (for a fixed $r \in \mathbb{N}$)

$$\varphi(2x) = 2^r \varphi(x) \qquad (0.0.8)$$

the constant functions are the only solutions $\varphi: \mathbb{R} \to \mathbb{R}$ which are of class C^r, whereas in the class C^{r-1} the solution depends on an arbitrary function. On the other hand, the monotonicity condition does not yield uniqueness of solution for equation (0.0.8).

One of the conclusions resulting from the above examples is that there is no universal uniqueness theorem. We have many uniqueness theorems, which are not comparable; none is more general than the others. The more uniqueness theorems we have at our disposal, the bigger is the chance of finding a suitable one for applications. And it is for this reason that theorems of this kind are our main concern in this book.

Of course, we shall deal also with the existence of solutions, in particular with existence-and-uniqueness theorems. However, occasionally there will appear existence theorems that do not guarantee uniqueness, and vice versa.

0.0C. Fixed points

There is also another important conclusion resulting from the analysis of equations (0.0.6), (0.0.7), (0.0.8). In the case of equation (0.0.6) the requirement of the existence of a limit of φ at infinity cannot be replaced by that of the existence of a limit at any other point. In the case of equations (0.0.7) or (0.0.8) the assumption of the continuity or differentiability on the whole real line may be weakened to the condition of the continuity or differentiability at zero. But again zero cannot be replaced by any other point.

The point $\xi = 0$ for the function $f(x) = 2x$, or $\xi = \infty$ for $f(x) = x + 1$ is distinguished as the fixed point of f, $f(\xi) = \xi$. Thus we are led to the very just observation that the conditions yielding the uniqueness of the solution of an iterative functional equation should always be postulated at, or in a neighbourhood of, the fixed point ξ of the function f. A simple change of variables allows us to shift this fixed point to the origin. Therefore, without loss of generality, we will usually assume that the function f has a fixed point at zero.

0.0D. General solution

We leave out from this book the problem of the construction of the general solution φ (without any further conditions) of (0.0.1) or (0.0.2). In this matter the reader is referred to Kuczma [4], [18], [26, Ch. I], Reghiş–Vuc [1], Balint [1], [2], Pelyukh–Šarkovskiĭ [2], [3]. Sometimes, however, we shall wish to describe the general solution (depending on an arbitrary function) of an equation of form (0.0.2) in a particular class of functions. This is essentially possible in two ways which we explain for the example of the equation

$$\varphi(x + 1) = x\varphi(x), \qquad x \in (0, \infty). \tag{0.0.9}$$

The general continuous solution $\varphi: (0, \infty) \to \mathbb{R}$ of (0.0.9) can be constructed as follows. Fix arbitrarily an $x_0 \in (0, \infty)$. Then every continuous function $\varphi_0: [x_0, x_0 + 1] \to \mathbb{R}$ fulfilling the boundary condition $\varphi_0(x_0 + 1) = x_0\varphi_0(x_0)$ can be uniquely extended onto $(0, \infty)$ to a continuous solution $\varphi: (0, \infty) \to \mathbb{R}$ of equation (0.0.9).

The other method consists in expressing the general continuous solution $\varphi: (0, \infty) \to \mathbb{R}$ of equation (0.0.9) by the formula

$$\varphi(x) = \hat{\phi}(x)\Gamma(x),$$

where Γ denotes Euler's gamma function (see Section 10.4), and $\hat{\phi}: (0, \infty) \to \mathbb{R}$ is an arbitrary continuous periodic function of period 1. Here the general continuous solution of (0.0.9) has been expressed with the aid of the general continuous solution of (the simpler) equation (0.0.6).

It depends on the particular situation which of the two methods is preferable. Essentially the first method offers more information about the degree of arbitrariness to which the solution is determined. The second one usually yields more elegant, closed formulae, more convenient to handle. The first approach is represented in Kuczma [26], whereas the second can be found in Pelyukh–Šarkovskiĭ [3]. It may be observed that the first method is not possible in the case of analytic solutions.

0.0E. Solution depending on an arbitrary function

Whereas the intuitive sense of this expression is rather clear, it is difficult to give it a precise meaning (see Kuczma [48]). In the present book we shall follow the definition (in spite of all its deficiencies) given in Choczewski–Kuczma [1] and used also in Kuczma [26, p. 45].

Definition 0.0.1. Let $\mathbf{\Phi}$ be a class of functions and let $\mathbf{\Phi}[A]$ denote the class of the restrictions of the functions in $\mathbf{\Phi}$ to the set A. Let X be a subset of a metric space. We say that a functional equation has in the class $\mathbf{\Phi}[X]$ a *solution depending on an arbitrary function* iff there exists an open set $A \subset X$ such that every function in $\mathbf{\Phi}[A]$ can be extended (not necessarily uniquely) to a solution in $\mathbf{\Phi}[X]$ of the equation in question.

0.1 Special equations

0.1A. Change of variables

This is an important procedure which facilitates solving functional equations. An ingenious change of variables may result in a considerable simplification of the equation considered.

Suppose that we are studying the equation

$$F(x, \varphi(x), \varphi(f(x))) = 0, \tag{0.1.1}$$

for functions $\varphi \colon X \to Y$, where X and Y are arbitrary sets, and 0 denotes an arbitrary, but fixed, element of Y. Further suppose we are given two bijections $\sigma \colon X \to S$ and $\tau \colon Y \to T$ (where S, T are again arbitrary sets), which satisfy the functional equations

$$\sigma(f(x)) = g(\sigma(x)), \tag{0.1.2}$$

$$\tau(F(x, y, z)) = L(x, \tau(y), \tau(z)) \tag{0.1.3}$$

(τ may possibly depend on y). Then equation (0.1.1) goes over to

$$L(\sigma^{-1}(t), \psi(t), \psi(g(t))) = \tau(0), \tag{0.1.4}$$

where $t = \sigma(x)$ and the new unknown $\psi \colon S \to T$ is connected with φ by the formula $\psi = \tau \circ \varphi \circ \sigma^{-1}$. Equation (0.1.4) is again of form (0.1.1), but now it may happen that the new known functions g, L are considerably simpler than f, F. For instance, we may be able to get equation (0.1.2) and (0.1.3) with linear g and L. Then we speak about a *linearization*.

0.1B. Schröder's, Abel's and Böttcher's equations

The most important equations of linearization are those of Schröder and Abel; see Chapter 8. The equation

$$\sigma(f(x)) = s\sigma(x) \tag{0.1.5}$$

will be referred to as the *Schröder equation* (Schröder [1]). Here s may be a scalar factor, as in the classical Schröder equation, or a linear operator on the range of σ.

The *Abel equation* (Abel [1])

$$\alpha(f(x)) = \alpha(x) + A, \tag{0.1.6}$$

where $A \neq 0$ is a fixed element in the range of α (usually one takes $A = 1$) is more subtle than Schröder's and may often be used when equation (0.1.5) fails.

If the linearization is not possible, one may try to reduce the function f in (0.1.1) to the next simplest form. The reduction to a power function may be realized by means of the *Böttcher equation* (Böttcher [1], [2])

$$\beta(f(x)) = [\beta(x)]^p. \tag{0.1.7}$$

The occurrence in the book of the Abel and Schröder equations and of their applications is indicated in Section 9.0.

0.2 Applications

Functional equations we are dealing with have, first of all, many interesting applications in other branches of mathematics. We indicate in Table 0.1 the sections in which such applications may be found.

Table 0.1

Topic	Section(s)
Characterization of functions	6.1A, 10.1–10.5, 11.8
Dynamical systems	8.2
Ergodic theory	3.7A, 6.2A, 6.9
Functional analysis	5.5BC, 9.5
Functional equations in several variables	9.3, 9.6
Functional inequalities	12.2–12.7, 12.9
Geometry	3.5D, 11.9
Iteration theory	1.7, 8.5, 8.7, 11.1–11.7
Ordinary differential equations	3.7C, 9.4, 9.6
Partial differential equations	4.8B
Probability theory	1.4, 3.7B, 6.1B, 6.9
Stochastic processes	2.1, 2.6

A few examples of applications in behavioural and natural sciences are given in Subsections 0.2A–C.

0.2A. Synthesizing judgements

In group decision-making procedure one meets the problem of synthesizing n quantifiable judgements $x_1, x_2, \ldots, x_n$ into the quantity $F(x_1, x_2, \ldots, x_n)$ (see Aczél–Saaty [1] and also Aczél [4]). The values x_k may be obtained, e.g., by ratio estimation which is made by comparing an object with another according to a given criterion.

The synthesizing function $F: P^n \to \mathbb{R}$, where $P :=]\bar{a}, \bar{b}]$, $0 < \bar{a} < \bar{b} \leqslant \infty$, should meet several reasonable requirements. Among them are the following:

– separability of individual judgements,

$$F(x_1, x_2, \ldots, x_n) = g(x_1) * g(x_2) * \cdots * g(x_n), \qquad x_k \in P, \quad (0.2.1)$$

where '$*$' is a continuous, associative and cancellative operator on $Q := g(P)$, and Q is assumed to be a proper interval;

– consensus condition (if all judgements are x, then the synthesized judgement should be x too),

$$F(x, x, \ldots, x) = x, \qquad x \in P; \tag{0.2.2}$$

– reciprocity property (if second object is compared to first rather than first to second, both individual and synthesized ratios are replaced by their reciprocals),

$$F\left(\frac{1}{x_1}, \frac{1}{x_2}, \ldots, \frac{1}{x_n}\right) = \frac{1}{F(x_1, \ldots, x_n)}. \tag{0.2.3}$$

By a result of J. Aczél from 1949, all continuous associative and cancellative operators on an interval Q (which is then necessarily open or half-open) are given by the formula

$$y * z = \varphi^{-1}(\varphi(y) + \varphi(z)), \qquad y, z \in Q,$$

where $\varphi: Q \to \mathbb{R}$ is an arbitrary continuous, strictly monotonic function, and

$$\varphi(Q) = \mathbb{R} \quad \text{or} \quad (-\infty, d] \quad \text{or} \quad [e, \infty), \qquad d \leqslant 0 \leqslant e \quad (0.2.4)$$

(see Aczél [2, pp. 254–67]). Therefore, by (0.2.1),

$$F(x_1, \ldots, x_n) = \varphi^{-1}\left(\sum_{k=1}^{n} \varphi(g(x_k))\right), \qquad x_k \in P. \qquad (0.2.5)$$

This, when used in (0.2.2), yields

$$\varphi(x) = n\varphi(g(x)), \qquad x \in P, \qquad (0.2.6)$$

so that (0.2.5) becomes the quasiarithmetic mean:

$$F(x_1, \ldots, x_n) = \varphi^{-1}\left(\frac{1}{n}\sum_{k=1}^{n} \varphi(x_k)\right), \qquad x_k \in P. \qquad (0.2.7)$$

For the reciprocity property we need P to contain with each element its reciprocal. Thus 1 is in P and 0 is not, and P cannot be half-closed. We claim that P is open. For from (0.2.6) we see that both functions $g: P \to Q$ and $\varphi: Q \to \varphi(Q)$ are homeomorphisms. By this and (0.2.4), both $\varphi(Q)$ and P are either open or half-closed, and we have shown that the latter is not the case.

Combining (0.2.3) and (0.2.7) we get the functional equation for φ:

$$\varphi^{-1}\left(\frac{1}{n}\sum_{k=1}^{n} \varphi\left(\frac{1}{x_k}\right)\right) = 1 / \varphi^{-1}\left(\frac{1}{n}\sum_{k=1}^{n} \varphi(x_k)\right) \qquad (0.2.8)$$

for $x_k \in P$ which is now an open (finite or infinite) interval of positive numbers, $1 \in P$. After we execute the transformation

$$t_k = \log x_k, \qquad \psi(t) = \varphi(\exp t),$$

equation (0.2.8) goes over into

$$\psi^{-1}\left(\frac{1}{n}\sum_{k=1}^{n} \psi(-t_k)\right) = -\psi^{-1}\left(\frac{1}{n}\sum_{k=1}^{n} \psi(t_k)\right), \qquad t_k \in \log P, \quad (0.2.9)$$

and $\log P$ is an open interval symmetric with respect to the origin.

Obviously, every continuous strictly monotonic odd function ψ satisfies (0.2.9). We are going to show that the general strictly monotonic continuous solution of (0.2.9) (for fixed $n > 1$) on an interval $P' = (-c, c)$, $0 < c \leqslant \infty$, is given by

$$\psi(t) = \alpha\psi_0(t) + \beta, \qquad \alpha \neq 0, \qquad (0.2.10)$$

where ψ_0 is an arbitrary continuous strictly monotonic odd mapping from P' into $\mathbb{R}$.

Indeed, if ψ_0 satisfies (0.2.9), then so does every $\alpha\psi_0 + \beta$ with arbitrary α and β, $\alpha \neq 0$. Now assume that a ψ satisfies (0.2.9) on P' and rewrite the equation in the form

$$\tilde{\psi}^{-1}\left(\frac{1}{n}\sum_{k=1}^{n}\tilde{\psi}(t_k)\right) = \psi^{-1}\left(\frac{1}{n}\sum_{k=1}^{n}\psi(t_k)\right), \qquad t_k \in P', \qquad (0.2.11)$$

where we have put $\tilde{\psi}(t) = \psi(-t)$, so that $\tilde{\psi}^{-1}(x) = -\psi^{-1}(x)$.

It is known by a result of K. Knopp from 1929 (see Hardy–Littlewood–Pólya [1, pp. 66–7]) that (0.2.11) implies

$$\tilde{\psi}(t) = a\psi(t) + b$$

for some constants a and b. We have arrived at the linear iterative functional equation

$$\psi(-t) = a\psi(t) + b, \qquad t \in P' \qquad (0.2.12)$$

which is easily solved by a single iteration

$$\psi(t) = \psi(-(-t)) = a\psi(-t) + b = a^2\psi(t) + (a+1)b.$$

Since ψ cannot be constant, we get from this $a^2 = 1$ and $(a+1)b = 0$. If $a = 1$, then $b = 0$ and (0.2.12) shows that ψ is an even function, so it cannot be strictly monotonic. In the case $a = -1$ the function $\psi_0(t) = \psi(t) - b/2$ is odd, by (0.2.12). Thus (0.2.10) holds with $\alpha = 1$ and $\beta = b/2$.

Since $\log P$ may serve as P', and equation (0.2.9) goes over into (0.2.8), we get all the quasiarithmetic means (0.2.7) having the reciprocal property (0.2.3):

$$F(x_1, \ldots, x_n) = \exp\left[\psi_0^{-1}\left(\frac{1}{n}\sum_{k=1}^{n}\psi_0(\log x_k)\right)\right], \qquad x_k \in P,$$

where ψ_0 is an arbitrary continuous strictly monotonic odd function.

If the variables $x_1, \ldots, x_n$ are measures rather than their ratios, then the reciprocity property is less natural. In this case one may wish to replace (0.2.3) by the condition

$$F(f(x_1), f(x_2), \ldots, f(x_n)) = f(F(x_1, x_2, \ldots, x_n)), \qquad (0.2.13)$$

where $f: P \to P$ is a given continuous strictly monotonic function, and ask for functions (0.2.7) satisfying (0.2.13). Combining (0.2.7) and (0.2.13) we see that the resulting equation for φ is just (0.2.11) with ψ replaced by φ and $\tilde{\psi}$ by $\varphi \circ f$, respectively. By the same argument as previously we get, in place of (0.2.12), the equation

$$\varphi(f(x)) = a\varphi(x) + b, \qquad x \in P. \qquad (0.2.14)$$

This is Abel's equation (2.6) when $a = 1$, whereas if $a \neq 1$ (0.2.14) may be transformed to Schröder's equation

$$\sigma(f(x)) = a\sigma(x),$$

where $\sigma(x) := \varphi(x) + b/(1-a)$.

In the case where $f(x) = x^p$, $p \neq -1, 0, 1$, the general continuous strictly monotonic solution to (0.2.14) is described in Aczél–Alsina [1]. The solution depends on an arbitrary function, except for some cases where it does not exist at all.

In the above presentation we have followed the lines of Aczél–Saaty [1]. More information on the multiattribute approach to decision-making may be found, e.g., in the book by T. L. Saaty [1].

Monotonic solutions, however, not necessarily strict, of equations like (0.2.14) are discussed in Section 2.3.

0.2B. Clock-graduation and the concept of chronon

Let A be an observer making observations of a particle B by means of light-signals (see Crum [1]). The light-signal sent from A at a time t returns to A, after having been reflected by B, at a time $g(t)$ by A's clock. If we assume that at the time $t = 0$ the particle B leaves A and remains in motion in the time-interval $(0, \xi)$, returning to A at $t = \xi \leqslant \infty$, then the function g fulfils the following conditions:

$$g: [0, \xi] \to [0, \xi] \text{ is continuous, } g(0) = 0, \; g(\xi) = \xi, \; g(t) > t \text{ in } (0, \xi).$$

Moreover, if B has a continuous positive velocity in the time-interval considered, g will be of class C^1 in $[0, \xi]$ with the positive derivative, whence g is strictly increasing.

Now suppose that there is another observer at B, making observations of the light-signals sent by A. Suppose that the light-signal sent by A at the time t (by A's clock) reaches B at a time $\chi(t)$ (by B's clock). If both clocks are congruent, then the light-signal sent (or reflected) by B at the time t returns to A at the time $\chi(t)$. Hence

$$\chi(\chi(t)) = g(t). \qquad (0.2.15)$$

Consulting, in particular, Theorem 11.1.1 and Lemma 11.2.2, we see that χ must have similar properties to g. Write $f = g^{-1}$, $\varphi = \chi^{-1}$. Then equation (0.2.15) is equivalent to

$$\varphi(\varphi(t)) = f(t). \qquad (0.2.16)$$

There is an infinity of continuous solutions $\varphi: [0, \xi] \to [0, \xi]$ of equation (0.2.16), and all of them fulfil the condition $f(t) < \varphi(t) < t$ in $(0, \xi)$; see Section 11.2. But if we add the condition

$$|f'(t) - f'(0)| \leqslant Ct^\delta \qquad \text{for small } t > 0$$

with some positive constants C and δ (it is equivalent to an analogous one for g) then, by Theorem 11.3.2, equation (0.2.15) has a unique solution φ, a continuous self-mapping of $[0, \xi]$, which is of class C^1 in $[0, \xi)$. Thus the knowledge of g allows us to determine χ uniquely. Note that in general the derivative $\chi'(\xi)$ will not exist; see Section 11.4.

A functional equation similar to (0.2.16) appears also in a different context (Targoński [7]). Let X be the set of all possible states of the universe, and let $f: X \to X$ describe the state evolution in a unit time-interval. Thus if $x \in X$ is the state of the universe at a time t, then $f(x)$ is the state at the time $t + 1$. Now, if $\varphi: X \to X$ describes the state evolution in the time-interval $1/N$, then clearly

$$\varphi^N(x) = f(x), \qquad (0.2.17)$$

where φ^N denotes the Nth functional iterate of φ. If, for a certain $N \in \mathbb{N}$, equation (0.2.17) has no solution (this actually can happen; see Lemma 11.1.1 and the examples in Sections 11.4 and 11.10), then there is no time-interval of length $1/N$. If N is the largest positive integer for which (0.2.17) has a solution, then $\tau = 1/N$ represents the *chronon*, the smallest, indivisible, nonzero time-interval, the quantum of time, as suggested by the analogous quantum of energy in the quantum theory.

0.2C. Sensation scale and Fechner's law

It is well known that very small changes of the stimulus do not result in changes of the sensory experience. For instance, two sounds differing very little in frequency or amplitude (of sound waves) are felt as equally high or loud. The stimulus can be measured in well-defined physical units (like frequency or amplitude). A measurement of sensory experience, that is a derivation of a sensation scale, is a research topic in psychology (see Luce–Edwards [1]).

Suppose that the stimulus is represented by a number x from an interval $X \subset \mathbb{R}^+$, and let $\alpha(x)$ be a measure of the sensory experience caused by this stimulus. There is a Weber function $w: X \to \mathbb{R}^+$ such that a stimulus magnitude $y \geqslant x$ is detected as larger than x if $y \geqslant x + w(x)$, whereas for $x \leqslant y < x + w(x)$ the stimuli x and y are indistinguishable. (The function w is determined statistically from experiments). Write

$$g(x) = x + w(x). \qquad (0.2.18)$$

Then the distance $w(x)$ from x to $g(x)$ is the *just noticeable difference* (JND) in the stimulus scale. Fechner's condition (which can be assumed as a definition) says that all JNDs on the sensation scale are equivalent to each other. Analytically this means that

$$\alpha(g(x)) - \alpha(x) = \text{const}.$$

Assuming the constant to be 1, i.e. taking the JND as the unit on the sensation scale, we see that the function α must satisfy the Abel equation

$$\alpha(g(x)) - \alpha(x) = 1. \qquad (0.2.19)$$

Let $X = (0, \infty)$ and assume that the function $w: X \to X$ is continuous and strictly increasing. Then so are $g: X \to X$ given by (0.2.18) and $f := g^{-1}$.

Further suppose that $\lim_{x \to 0} w(x) = 0$. Then $g(x) > x$ in X and $0 < f(x) < x$ in X. Putting $\varphi = -\alpha$ we see that equation (0.2.19) is equivalent to

$$\varphi(f(x)) - \varphi(x) = 1. \tag{0.2.20}$$

Equation (0.2.20) has infinitely many solutions $\varphi : X \to \mathbb{R}$ of the same regularity as f (see Theorems 3.1.1 and 3.4.1); actually, the solution depends on an arbitrary function. Every such solution yields a measure of the subjective experience generated by physical stimuli.

Assume that there exists the limit $r := \lim_{x \to 0}(w(x)/x)$. It seems reasonable to take as the best and most representative solution of (0.2.20) its principal solution (see Section 9.1) in the case where $r = 0$, and if $r > 0$ to choose $\varphi(x) = \log \sigma(x)/\log s$, where σ is the principal solution of the Schröder equation (0.1.5) and $s = 1/(1 + r)$. (Both principal solutions do exist under further mild assumptions on w.)

The most important Weber function $w(x) = kx$, $k > 0$, results from Weber's law stating that the value of stimulus of the JND divided by the stimulus size is constant. Starting from Weber's law, Fechner wrote (0.2.19) as $\Delta \alpha = 1$, consequently $\Delta x = g(x) - x = w(x)$, whence $\Delta \alpha / \Delta x = 1/w$. Replacing Δs by differentials he deduced hence (incorrectly) that α must be a solution of the differential equation

$$\alpha' = 1/w. \tag{0.2.21}$$

In general the solutions of (0.2.19) and (0.2.21) do not agree. However, for the Weber function $w(x) = kx$ the solution of (0.2.21) is $\alpha(x) = k^{-1} \log x$, and for (0.2.20) we have $\sigma(x) = cx$ and $\alpha(x) = \log x/\log(k + 1)$ (each of them up to an additive constant). Thus these solutions generate the same sensation scale up to the unit and the origin of the scale. They give rise to Fechner's law that the sensation caused by a stimulus is proportional to the logarithm of the value of the stimulus.

The derivation of sensation scales and Fechner's law as presented above follows the lines of Luce–Edwards [1], where also further details can be found.

Remark 0.2.1. Other applications, of the kind we discuss in this section, come via the theory of branching processes; see Chapter 2. For some applications in astronomy see Lundmark [1], and in geology Gersevanov [1].

0.3 Iterative functional equations

The title of this book requires a comment. The name *iterative functional equations*, adopted here instead of 'functional equations in a single variable' (see the Preface), has been used previously (Reghiş–Vuc [1], Balint [1], [2]). It is certainly short and convenient to use, and we believe that it is rather

imaginative. However, it may arouse objections[1]; in fact, equations like
(0.0.1)–(0.0.3) do not contain any iterates (of order greater than 1) at all.
But iteration is the fundamental technique for solving such functional
equations, and iterates usually appear in the formulae for solutions. Many
results we present here may be interpreted in both ways: either as theorems
about the behaviour of iterates, or as theorems about the solutions of
functional equations. A part of the theory of iteration can be deduced from
theorems on functional equations (see, for instance, Targoński [8] and Zdun
[13]). On the other hand, iteration is the basic tool in the investigation of
functional equations of the general type (0.0.1). And therefore we begin this
book with a chapter on iteration.

[1] See S. Gołąb's review of Balint [1] in *Mathematical Reviews* **49** (1975), # 9464.

1

Iteration

1.0 Introduction

This chapter is intended to give a brief review of the basic notions in the theory of iteration which (with one exception) will be needed in what follows. It is one of our leading ideas, expressed also in the title of the book, to emphasize the fruitful and intriguing interplay between iteration and the theory of functional equations in a single variable. We feel that any attempt to divorce iteration from functional equation investigations would be an extremely, indeed totally, fruitless task. In the whole book we shall endeavour to make the reader believe this.

Some readers may find some serious omissions in this chapter but we have attempted to minimize the contents of weighty terminology and, on the other hand, we are very far from an aspiration to any kind of completeness. Basic to this chapter are the following questions: iteration sequences (splinters) and orbits, cycles, attractive fixed points and the domain of attraction. Fixed points play a distinguished role both in functional equations and in iteration theory. Some fixed-point theorems which will be useful in the sequel are also included in the present chapter. One may consider them as various generalizations of the Banach contraction principle in a complete metric space.

1.1 Basic notions and some substantial facts

1.1A. Iterates, orbits and fixed points

The operation of composition is the only natural inner operation which can be defined in the family $\mathscr{F}(X)$ of all self-mappings of a given set X. It is intrinsically connected with the notion of function and does not require any additional structures or properties of X. Since the operation '$\circ$' of *composition* is associative, the system $(\mathcal{F}(X), \circ)$ naturally forms a (noncommutative, in general) semigroup with identity id_X.

The powers f^n, $n \in \mathbb{N}_0$, of an element $f \in \mathcal{F}(X)$ in this semigroup are called *iterates* of f, i.e.

$$f^0 = \mathrm{id}_X, \qquad f^{n+1} = f \circ f^n, \qquad n \in \mathbb{N}_0.$$

Noteworthy is the fact that any two iterates of a function $f \in \mathcal{F}(X)$ commute:

$$f^n \circ f^m = f^m \circ f^n = f^{n+m}, \qquad n, m \in \mathbb{N}_0. \tag{1.1.1}$$

The iteration sequence $(f^n(x))_{n \in \mathbb{N}}$ will also be called the *splinter* (of x) (as proposed by Gy. Targoński [8] after J. S. Ullian [1] who has introduced the term in the theory of recursive functions).

In the case of an invertible function $f \colon X \to X$ one may define the iterates for negative integers, too, by putting

$$f^{-k-1} := f^{-1} \circ f^{-k}, \qquad k \in \mathbb{N}_0,$$

where f^{-1} is the usual inverse function to f. Now, f^{-k} is the inverse function to f^k for $k \in \mathbb{N}_0$. Unfortunately, unless f maps X onto X, the iterates f^{-k} for $k \in \mathbb{N}$ are not elements of $\mathcal{F}(X)$ since they are defined in a subset $f^k(X) \subset X$ only. However, relation (1.1.1) remains true for arbitrary $n, m \in \mathbb{Z}$ on a suitably restricted domain (depending on n and m).

In the sequel we shall usually have also some algebraic structure on X. Then, in order to distinguish the iterates from the powers with respect to multiplication in X, the latter will be denoted by $(f)^n$, the upper indices after a parenthesis. Thus, e.g., $f^2(x)$ denotes $f[f(x)]$, whereas $[f(x)]^2$ denotes $f(x)f(x)$.

In 1924, K. Kuratowski in a remark added to the paper of R. Tambs Lyche [1] introduced the following equivalence relation connected with a given self-mapping f of X: we say that points $x, y \in X$ are *equivalent under iteration* of f, and write $x \sim_f y$, if and only if $f^n(x) = f^m(y)$ for some $n, m \in \mathbb{N}_0$. The equivalence classes under $\sim_f$ are called *orbits*.

As pointed out in the book of Gy. Targoński [8] (to our knowledge the first monograph completely devoted to iteration theory) it seems that this notion went unnoticed at the time. In 1941 G. T. Whyburn [1], [2] rediscovered Kuratowski's relation, named the equivalence classes orbits and began their systematic study.

Without doubt the notion of orbit is fundamental for the whole of iteration theory. The structure of orbits plays, among others, an important role in the construction of the general solution of functional equations (see, e.g., Reghiş–Vuc [1], Kuczma [18], [26, Ch. I]), but since we do not develop this line in the present book, we will not discuss the details here.

It should come as no surprise that fixed points play a distinguished role in iteration as well as functional equations theory. To fix the terminology, let us introduce the following definition.

If, for an $x_0 \in X$ and a $k \in \mathbb{N}$, we have

$$f^k(x_0) = x_0 \quad \text{and} \quad f^i(x_0) \neq x_0 \quad \text{for } i \in \{1, \ldots, k-1\},$$

then x_0 is called a *fixed point of order k* of f. Fixed points of order 1 are called, for short, *fixed points* and they are characterized by the property: $f(x_0) = x_0$. Obviously, every fixed point of order k of f is a fixed point of f^k.

Clearly, if x_0 is a fixed point of order k of f, then so are the points $f^n(x_0)$, $n \in \mathbb{N}_0$, and there are exactly k distinct points in this sequence. These pairwise distinct points form a set called a *cycle of order k* (k-cycle).

The following simple result turns out to be extremely useful.

Theorem 1.1.1. *Let X be a Hausdorff topological space and let $f: X \to X$ be a sequentially continuous function. If, for an $x \in X$, the sequence $(f^n(x))_{n \in \mathbb{N}}$ converges to an $x_0 \in X$, then x_0 is a fixed point of f.*

This results immediately from the fact that $f^{n+1}(x) = f(f^n(x))$, $n \in \mathbb{N}$, by passing to the limit as $n \to \infty$.

1.1B. Limit points of the sequence of iterates

Let $L_f(x)$ denote the set of all limit points of the sequence $(f^n(x))_{n \in \mathbb{N}}$, $x \in X$. In the case where X is a compact real interval B. Barna [1] generalized Theorem 1.1.1 proving that for any point $x \in X$ the set $L_f(x)$ having exactly k distinct points yields a cycle of order k. Barna's result has been extended to more general spaces (Kuczma [49] and also Graw–Kuczma [1]). The proof we present below is after Kuczma [49].

Theorem 1.1.2. *Let X be a locally compact Hausdorff topological space satisfying the first axiom of countability. Let $f: X \to X$ and $x \in X$ be such that $L_f(x) \neq \varnothing$ and let f be continuous at each point of $L_f(x)$. Then we have the following.*

(a) *If $L_f(x)$ is finite, then $L_f(x)$ is a cycle under f.*
(b) *If $L_f(x)$ is infinite, but has no limit points, then $L_f(x)$ contains no cycle under f.*

Proof. Part (a). To prove that a nonempty finite set $L \subset X$ is a cycle under f it suffices to show that $f(L) \subset L$ and for any proper subset A of L one has $f(A) \not\subset A$. The inclusion $f(L_f(x)) \subset L_f(x)$ follows from the continuity of f on $L_f(x)$. Let A be a proper subset of $L_f(x)$ and let G be an open set containing A whose closure cl G is compact and such that $(L_f(x) \setminus A) \cap$ cl $G = \varnothing$; such a choice of G is possible in view of the finiteness of $L_f(x)$ and the local compactness of the Hausdorff space X. Since the open sets G and $X \setminus$ cl G both contain points of $L_f(x)$, each of them must contain infinitely many points of the set $\{f^n(x): n \in \mathbb{N}\}$. Consequently, there exists a sub-sequence $(f^{n_m}(x))_{m \in \mathbb{N}}$ of $(f^n(x))_{n \in \mathbb{N}}$ such that

$$\{f^{n_m}(x): m \in \mathbb{N}\} \subset G \qquad \text{and} \qquad \{f^{n_m+1}(x): m \in \mathbb{N}\} \subset X \setminus G.$$

Since cl G is sequentially compact, we may assume that the sequence $(f^{n_m}(x))_{m \in \mathbb{N}}$ converges to a point p; obviously, $p \in A$. By the continuity of f we get

$$f(p) = \lim_{m \to \infty} f(f^{n_m}(x)) = \lim_{m \to \infty} f^{n_m+1}(x) \in \mathrm{cl}(X \setminus G) = X \setminus G$$

and hence $f(p) \notin A \subset G$.

Part (b). Suppose, for the indirect proof, that $L_f(x)$ contains a cycle L. Then L, being finite, is a proper subset of $L_f(x)$ and $(L_f(x) \setminus L) \cap \mathrm{cl}\, G = \varnothing$ for some nonempty open set G whose closure is compact since $L_f(x)$ has no limit points. Applying the method used in the first part, one obtains $f(L) \not\subset L$ which contradicts the fact that L is a cycle. ■

Of course, the story of limit sets of iteration sequences does not end here since, in general, these sets may be infinite and have a rather complicated structure (see Barna [1], Šarkovskiĭ [4], [5]). The investigation of iteration sequences has been quite fruitful. Not entering into details let us only mention here the works of Šarkovskiĭ [1], I. N. Baker [3]–[5], Fatou [1], [5], Krüppel [1], Drewniak–Mrózek–Ulewicz [1], Zdun [9], [10], Targoński [8], Collet–Eckmann [1]; see also the references in the last two books as well as in Kuczma [26, p. 17]). The increased interest in research on topological dynamics including the up-to-date theory of chaos entailed further eventful development of investigations of splinters and, more generally, of orbit structures.

On the other hand, again owing to Theorem 1.1.1, the well-known classical method of *successive approximations* plays a fundamental role in solving equations of various types (see Hamilton [2]) and it is nothing other than dealing with some special iteration sequences. Plainly, we confine ourselves to those properties of iteration sequences which will be useful in our further study of functional equations.

1.1C. Theorem of Šarkovskiĭ

Like iteration, the search for fixed points of a given self-mapping is also one of the central components of the theory of equations of various types and of functional equations in particular. Some aspects of these investigations will be discussed in Section 1.5. Here we quote, without proof, an interesting result belonging to both domains – fixed-point and iteration theories – namely, the following theorem of A. N. Šarkovskiĭ [2].

Theorem 1.1.3. *Consider the following linear ordering in* $\mathbb{N}$:

$$3 \prec 5 \prec 7 \prec \cdots \prec 2 \cdot 3 \prec 2 \cdot 5 \prec 2 \cdot 7 \prec \cdots \prec 2^n \cdot 3 \prec 2^n \cdot 5 \prec 2^n \cdot 7 \prec \cdots$$
$$\prec 2^m \prec 2^{m-1} \prec \cdots \prec 2 \prec 1.$$

If a continuous self-mapping f on an interval in $\mathbb{R}$ (bounded or not) has a fixed point of order p, then f has also fixed points of all orders q such that $p \prec q$.

In particular, if f has a fixed point of order 3, then it has fixed points of all orders, and, if f has a fixed point of any order $p \in \mathbb{N}$, then it has to have a fixed point. This intriguing result may also be found in a celebrated paper due to T. Y. Li–J. A. Yorke [1], published under the very suggestive title 'Period 3 implies chaos'. Šarkovskiĭ's original proof of Theorem 1.1.3 is pretty long and involved. Nowadays, shorter proofs are at the reader's disposal: see Block–Guckenheimer–Misiurewicz–Young [1], Burkart [1], Štefan [1] and Gaweł [1]. For further details, references and some new results in this direction the reader is referred to Targoński [8].

1.1D. Attractive fixed points

Let X be a topological space and $f: X \to X$ be an arbitrary function. Let $x_0 \in X$ be a fixed point of f. The set

$$A_f(x_0) := \left\{ x \in X : \lim_{n \to \infty} f^n(x) = x_0 \right\}$$

is called the *domain of attraction* of x_0. A fixed point x_0 of f is called *attractive* provided $x_0 \in \text{int } A_f(x_0)$. Figuratively speaking, an attractive fixed point attracts towards itself the iterates of all points from a neighbourhood of itself.

We have the following (Fatou [1], Barna [1]).

Theorem 1.1.4. *Let f be a continuous self-mapping of a topological space X and let $x_0 \in X$ be a fixed point of f. Then*

(a) $f(A_f(x_0)) \subset A_f(x_0)$

and

(b) *$A_f(x_0)$ is open provided x_0 is attractive.*

Proof. Part (a) is obvious. To prove (b) assume that U is a neighbourhood of x_0 contained in $A_f(x_0)$ and fix an $x \in A_f(x_0)$; then there exists an $N \in \mathbb{N}$ such that $f^n(x) \in U$ for all $n \geq N$. On account of the continuity of f^N at x one may find a neighbourhood V of the point x such that $f^N(V) \subset U \ni f^N(x)$. Now, for any $z \in V$, we get $f^N(z) \in U \subset A_f(x_0)$ and, consequently,

$$f^{n+N}(z) = f^n(f^N(z)) \to x_0 \qquad \text{as } n \to \infty.$$

This shows that $V \subset A_f(x_0)$. ∎

Remark 1.1.1. The assertion (b) fails to hold if we drop out the assumption that x_0 is attractive. For instance, in the case where $X = [0, 1]$ and f is a hat function

$$f(x) := \begin{cases} 2x & \text{for } x \in [0, \frac{1}{2}], \\ 2 - 2x & \text{for } x \in [\frac{1}{2}, 1], \end{cases}$$

one gets easily that the domain of attraction of the point $0 = f(0)$ is equal to

$$A_f(0) = \{k/2^m : k = 0, \dots, 2^m, m \in \mathbb{N}_0\}$$

and coincides with the orbit of 0 (see Baayen–Kuyk–Maurice [1], Sablik [1], Zdun [9], [10]).

More generally, if (X, ρ) is a metric space and $x_0 \in X$ is a fixed point of a transformation $f: X \to X$ such that there exists a $\Theta > 1$ with

$$\rho(f(x), x_0) \geq \Theta \rho(x, x_0)$$

for all x from some neighbourhood of x_0, then the only points in X which generate iteration sequences convergent to x_0 are those belonging to the orbit of x_0.

On the other hand, if there exists a $\Theta < 1$ such that

$$\rho(f(x), x_0) \leq \Theta \rho(x, x_0)$$

for all x in a neighbourhood U of x_0, then x_0 is an attractive fixed point. In fact, for any $x \in U$, one has $\rho(f^n(x), x_0) \leq \Theta^n \rho(x, x_0) \to 0$ as n tends to infinity. Hence we deduce the following simple result.

Theorem 1.1.5. *Let* $X \subset \mathbb{R}$ *be an interval and let* $f: X \to X$ *be differentiable at a point* $x_0 = f(x_0) \in \text{int } X$. *If* $|f'(x_0)| < 1$, *then* x_0 *is an attractive fixed point of* f.

Proof. There exist positive constants $\vartheta < 1$ and δ such that

$$\left| \frac{f(x) - x_0}{x - x_0} \right| < 1 - \vartheta$$

for all $x \in (x_0 - \delta, x_0) \cup (x_0, x_0 + \delta)$ whence, taking $\Theta := 1 - \vartheta$, we get

$$|f(x) - x_0| \leq \Theta |x - x_0|$$

for all $x \in (x_0 - \delta, x_0 + \delta)$. ∎

More generally, we have the following.

Theorem 1.1.6. *Let* X *be a subset of* $\mathbb{R}^n$ *or* $\mathbb{C}^n$ *such that* $0 \in \text{int } X$ *and let* $f: X \to X$ *be a map differentiable at the point* $0 = f(0)$. *If all the characteristic roots of the matrix* $f'(0)$ *lie inside the unit circle, then* 0 *is an attractive fixed point of* f.

To prove this theorem we need A. Ostrowski's lemma (Ostrowski [2]; see also Kordylewski [3]), which we state here without proof.

Lemma 1.1.1. *Let A be a real or complex $n \times n$ matrix and let $\{\lambda_1, \ldots, \lambda_n\}$ be the set of its characteristic roots. Then for any $\varepsilon > 0$ there exists a nonsingular $n \times n$ matrix T such that*

$$-\varepsilon + \min_{i \in \{1,\ldots,n\}} |\lambda_i| \leq \|TAT^{-1}\| \leq \varepsilon + \max_{i \in \{1,\ldots,n\}} |\lambda_i|.$$

Now we are able to present the following.

Proof of Theorem 1.1.6. From Lemma 1.1.1 and from our assumption on the matrix $f'(0)$ we may find a nonsingular matrix T such that

$$\|Tf'(0)T^{-1}\| < 1.$$

Write $t(x) := T \cdot x$ and $F(x) := T \cdot f(T^{-1} \cdot x)$ so that $t : X \to t(X)$ is a homeomorphism and $F : t(X) \to t(X)$ has the properties $F(0) = 0$, $F'(0) = T \cdot f'(0) \cdot T^{-1}$. Since $\lim_{x \to 0}(1/|x|)(F(x) - F'(0)x) = 0$ we have $|F(x)| \leq \Theta|x|$ for some $\Theta < 1$ and all x from some neighbourhood U of the origin. Thus 0 is an attractive fixed point of F, i.e. $\lim_{n \to \infty} F^n(x) = 0$ for all $x \in U$. Consequently,

$$f^n(x) = (t^{-1} \circ F^n \circ t)(x) = T^{-1} \cdot F^n(T \cdot x) \to 0 \qquad \text{as } n \to \infty$$

for all $x \in t^{-1}(U)$. ■

Remark 1.1.2. The fact that the fixed point of f is placed at zero is obviously meaningless; a simple change of variables allows us to shift a given fixed point to the origin.

The condition for the attractive character of the fixed point $x = 0$ of f contained in Theorem 1.1.6 is sufficient, but not necessary. In the case $n = 1$, where X is an interval and f is continuous, J. Drewniak–J. Kalinowski–B. Ulewicz [1] have given a complete characterization of attractive fixed points of f. We quote here their result without proof (the proof is long and involved).

Theorem 1.1.7. *Let $X \subset \mathbb{R}$ be an interval, let $0 \in \text{int } X$, and suppose that $f : X \to X$ is a continuous function, $f(0) = 0$. Then the following conditions are equivalent:*

(a) *0 is an attractive fixed point of f.*
(b) *There exists a neighbourhood $U \subset X$ of 0 such that for $x \in U$ we have*

$$f^2(x) > x \text{ for } x < 0; \qquad f^2(x) < x \text{ for } x > 0.$$

(c) *There exists a neighbourhood $U \subset X$ of 0 and a decreasing function $g : U \to U$ such that $g(0) = 0$, $g^2(x) = x$ in U, and for $x \in U$ we have*

$$g(x) > f(x) > x \text{ for } x < 0; \qquad g(x) < f(x) < x \text{ for } x > 0 \quad (1.1.2)$$

(d) *There exist a neighbourhood* $U \subset X$ *of* 0, $f(U) \subset U$, *and an increasing homeomorphism* $\varphi: U \to U$ *such that* $\varphi(0) = 0$, $\varphi(U) = U$ *and for* $x \in U \setminus \{0\}$ *we have*

$$\left| \varphi[f(\varphi^{-1}(x))] \right| < |x|. \tag{1.1.3}$$

Actually, condition (d) is not contained in the paper quoted above; it has been given by B. Ulewicz (unpublished) and therefore we show here that it is actually equivalent to the remaining conditions.

If condition (d) is fulfilled, then by Theorem 1.2.5 the origin is an attractive fixed point of the function $\varphi \circ f \circ \varphi^{-1}$, and hence also of f (see the proof of Theorem 1.1.6). Now assume that 0 is an attractive fixed point of f, and hence condition (c) is fulfilled. By Theorems 11.2.2 and 11.1.1 the function g is continuous and $g(U) = U$, whence, by (1.1.2), $f(U) \subset U$. Write $U = (a, b)$, $a < 0 < b$, and let $\varphi_0: (a, 0] \to (a, 0]$ be an arbitrary increasing homeomorphism. Define φ by

$$\varphi(x) = \begin{cases} \varphi_0(x) & \text{for } x \in (a, 0], \\ -\varphi_0[g(x)] & \text{for } x \in [0, b). \end{cases}$$

Then φ is an increasing homeomorphism of U onto U and $g(x) = \varphi^{-1}[-\varphi(x)]$, $x \in U$. Now (1.1.3) results from (1.1.2).

1.2 Maximal domains of attraction

Often more important than recognizing the attractive character of a fixed point is recognizing its domain of attraction. Below we give a number of theorems which state that under certain conditions the whole set X coincides with the domain of attraction of a fixed point x_0 of a mapping $f: X \to X$. If that is the case, one has simply

$$\lim_{x \to \infty} f^n(x) = x_0 = f(x_0) \qquad \text{for all } x \in X.$$

Here, again, the question arises as to whether a convergence better than pointwise should not be considered. As we shall see, in many instances, the convergence considered is almost uniform.

1.2A. Convergence of splinters

The results presented here are not too far-reaching; most of them concern the case where X is an interval on the real line with the origin as the left point of its boundary, i.e.

$$X = |0, a|, \qquad 0 < a \leq \infty. \tag{1.2.1}$$

We begin with the following.

Theorem 1.2.1. *Assume* (1.2.1) *and consider a right upper semicontinuous function* $f: X \to X$. *If*

$$f(x) < x \qquad \text{for all } x \in X \setminus \{0\}, \tag{1.2.2}$$

then, for any $x \in X$, *the splinter* $(f^n(x))_{n \in \mathbb{N}}$ *is decreasing and*

$$\lim_{n \to \infty} f^n(x) = 0. \tag{1.2.3}$$

If, moreover, $0 < f(x) < x$ *for* $x \in X \setminus \{0\}$, *then for every* $x \in X \setminus \{0\}$ *the splinter* $(f_n(x))_{n \in \mathbb{N}}$ *is strictly decreasing.*

Proof. The right upper continuity of f together with (1.2.2) implies $f(0) = 0$ whenever $0 \in X$. Thus $f(x) \leq x$ for all $x \in X$ whence $f^{n+1}(x) = f(f^n(x)) \leq f^n(x)$, $x \in X$. If we had $l := \lim_{n \to \infty} f^n(x) > 0$ for some $x \in X$, then

$$\lim_{n \to \infty} f^{n+1}(x) = \lim_{n \to \infty} f(f^n(x)) \leq f(l) < l,$$

a contradiction. Hence (1.2.3) follows. The latest assertion is obvious. ∎

Theorem 1.2.2. *Assume* (1.2.1) *and consider a right continuous and increasing mapping* $f: X \to X$. *If condition* (1.2.2) *is satisfied then the convergence* (1.2.3) *is almost uniform in* X.

Proof. Take an arbitrary compact $C \subset X$; then $C \subset [0, x_0]$ for some $x_0 \in X$ and $0 \leq f^n(x) \leq f^n(x_0)$ for all $x \in C$ and $n \in \mathbb{N}_0$. Now, the theorem follows in virtue of (1.2.3). ∎

Theorem 1.2.3. *Assume* (1.2.1) *and consider a metric space* (T, ρ). *Suppose that a map* $f: X \times T \to X$ *is continuous and* $f(x, t) < x$ *for all* $(x, t) \in (X \setminus \{0\}) \times T$. *Put* $g_t(x) := f(x, t)$, $(x, t) \in X \times T$. *Then the splinters* $(g_t^n(x))_{n \in \mathbb{N}}$ *tend to zero almost uniformly with respect to* $(x, t) \in X \times T$.

Proof. Take an arbitrary compact $C \subset X \times T$ and consider its projections C_X and C_T onto X and T, respectively. Write

$$F(x) := \sup\{f(u, t): (u, t) \in (X \cap [0, x]) \times C_T\}, \qquad x \in X.$$

Evidently, F is an increasing function. Moreover, F is right continuous on X (as a matter of fact, F is simply continuous, but this fact is irrelevant in the sequel). To prove that, fix an $x \in X$ and take any decreasing sequence $(y_n)_{n \in \mathbb{N}}$ of elements of X such that $y_n \to x$ as $n \to \infty$. Since f is continuous on the compact set $K := [0, y_1] \times C_T$ we infer that $F(x) \leq F(y_n) = f(u_n, t_n)$ for some $(u_n, t_n) \in K$, $n \in \mathbb{N}$. Consequently, there exists a convergent subsequence $(u_{n_k}, t_{n_k})_{k \in \mathbb{N}}$ of the sequence just obtained, say, $(u_{n_k}, t_{n_k}) \to (u_0, t_0) \in K$ as $k \to \infty$. Now, $F(x) \leq F(y_{n_k}) = f(u_{n_k}, t_{n_k})_{k \to \infty} \to f(u_0, t_0) \leq F(x)$, because, obviously, $u_n \leq y_n$ for all $n \in \mathbb{N}$ whence $u_0 \leq x$. Therefore, the decreasing sequence $(F(y_n))_{n \in \mathbb{N}}$ having a subsequence $(F(y_{n_k}))_{k \in \mathbb{N}}$ convergent to $F(x)$, must tend to $F(x)$ itself.

Finally, observe that $F(x) < x$ for $x \in X \setminus \{0\}$. Actually, if necessary, extend f continuously onto $(X \cup \{0\}) \times T$ by putting $f(0, t) := 0$, $t \in T$; then

$$F(x) = \max_{(u,t) \in [0,x] \times C_T} f(u,t) = f(\bar{u}, \bar{t}) \qquad \text{for some } (\bar{u}, \bar{t}) \in [0, x] \times C_T$$

and either $\bar{u} = 0$ which implies $F(x) = 0 < x$, or $\bar{u} > 0$ whence $F(x) = f(\bar{u}, \bar{t}) < \bar{u} \leq x$, again. Thus, all the assumptions of Theorem 1.2.2 are satisfied whence, in particular, the uniform convergence of $(F^n|C_X)_{n \in \mathbb{N}}$ to zero results. Now, our assertion results from the inequalities

$$0 \leq g_t^n(x) \leq F^n(x)$$

satisfied for all $n \in \mathbb{N}$ and all $(x, t) \in C_X \times C_T \supset C$. ∎

Taking as T a set consisting of a single point and suppressing the irrelevant dependence of f on t, we come to the following variant of Theorem 1.2.2.

Theorem 1.2.4. *Assume* (1.2.1) *and consider a continuous mapping* $f: X \to X$ *fulfilling condition* (1.2.2). *Then the convergence* (1.2.3) *is almost uniform in* X.

For higher-dimensional sets X we have the following

Theorem 1.2.5. *Let* X *be a closed subset of* $\mathbb{K}^N$ *containing the origin. Consider a continuous mapping* $f: X \to X$ *such that* $|f(x)| < |x|$ *for all* $x \in X \setminus \{0\}$. *Then the convergence* (1.2.3) *is almost uniform in* X.

Proof. Put $a := \sup\{|x|: x \in X\}$ and $X_0 = [0, a]$ or $[0, a)$ depending on whether the supremum is attained or not. Write

$$F(r) := \sup\{|f(x)|: x \in X, |x| \leq r\}, \qquad r \in X_0. \tag{1.2.4}$$

Plainly, F is increasing and, since the intersection $X \cap \{x: |x| \leq r\}$ is compact, the sup sign may, in fact, be replaced by max, from which the inequality $F(r) < r$ follows for all $r \in X_0 \setminus \{0\}$. Moreover, F is right continuous; this may be proved in a manner similar to that used in the proof of Theorem 1.2.3.

Take any compact $C \subset X$; then $C \subset \{x: |x| \leq r_0\}$ for some $r_0 \in X_0$ and a straightforward induction procedure shows that $|f^n(x)| \leq F^n(r_0)$ for all $x \in C$ and all $n \in \mathbb{N}_0$. It remains to apply Theorem 1.2.2. ∎

Remark 1.2.1. A careful inspection of the proof allows one to replace the assumption that X is closed by a slightly less restrictive one; namely, it suffices to assume that for every $x_0 \in X$ the set $\{x \in X: |x| \leq |x_0|\}$ is compact.

1.2B. Analytic mappings

We conclude this section with a theorem concerning analytic self-mappings of a complex domain.

Theorem 1.2.6. *Let $X \subset \mathbb{C}$ be an open and connected neighbourhood of the origin such that the boundary of X contains at least two finite points. Consider an analytic mapping $f: X \to X$ such that $f(0) = 0$ and $|f'(0)| < 1$. Then the convergence (1.2.3) is almost uniform in X.*

Proof. Since $f^n(X) \subset X$ for $n \in \mathbb{N}_0$ the set $\{f^n(x): x \in X, n \in \mathbb{N}\}$ omits the two finite boundary points of X and, by means of the classical Montel Theorem, the family $\{f^n: n \in \mathbb{N}\}$ is normal in X. In view of the fact that $|f(x)| < |x|$ for all $x \in D := \{z \in \mathbb{C} \setminus \{0\}: |z| \leqslant r\} \subset X$ for sufficiently small $r > 0$, relation (1.2.3) holds uniformly in D. ∎

In connection with the above result let us mention here the so-called *Julia set* of a transcendental entire function $f: \mathbb{C} \to \mathbb{C}$, namely the set $J(f)$ of all points $x \in \mathbb{C}$ for which the family $\{f^n: n \in \mathbb{N}\}$ fails to be normal (in the sense of Montel) in any neighbourhood of x (see Fatou [5], Blanchard [1] and Devaney [1]). P. Fatou has proved, among others, that $J(f)$ is closed in $\mathbb{C} \cup \{\infty\}$ and is contained in the closure of the set of all fixed points (of any order) of f but $J(f)$ does not contain any strongly attractive fixed point x_0 (i.e. such that $|f'(x_0)| < 1$).

The Julia set $J(f)$ always contains a finite point. Fatou himself asked for an example of an entire function f with $J(f) = \mathbb{C}$ (Fatou [5]). He conjectured that $f = \exp$ might serve for such an example. The question had remained unanswered for almost 60 years. In 1970 I. N. Baker [11] proved that there exists a constant $k > e^2$ such that for $f(z) = kze^z$, $z \in \mathbb{C}$, one has $J(f) = \mathbb{C}$. In 1981 W. Ogińska [1] gave a number of suitable examples including entire functions which were, in a sense, 'closer' to the exponential function (see also Matkowski–Ogińska [1] and Ogińska [2]). In the same year Fatou's conjecture was definitely answered in the affirmative by M. Misiurewicz [1] (see also Devaney [1]).

Now, we know much more: the one-parameter family $\{\lambda \exp: \lambda \in (0, \infty)\}$ of entire transcendental functions has the following intriguing property: the Julia sets $J(\lambda \exp)$ explode from a nowhere dense subset of the plane to cover the entire plane. The value $\lambda_0 := 1/e$ is the critical point: $J(\lambda \exp)$ is small for $\lambda < \lambda_0$ and $J(\lambda \exp) = \mathbb{C}$ for $\lambda > \lambda_0$ (see Devaney [1]).

1.3 The speed of convergence of iteration sequences

Knowing that an iteration sequence is convergent we may ask how fast is this convergence. In this section, following J. Drewniak [2] we present some results of this kind in the case where

  (i) f is a self-mapping of the real interval $X = [0, a|$ with $0 < a \leqq \infty$,
  (ii) $f(x) = x(s + p(x))$, $x \in X$, with $s \in [0, 1]$, $0 < p(x) + s < 1$, $x \in X$, and $\lim_{x \to 0} p(x) = 0$.

(Some results concerning the case where f is a self-mapping of a domain in a complex plane have also been obtained by J. Drewniak [2]).

1.3A. Some lemmas

We start with the following.

Lemma 1.3.1. *Let* $(x_n)_{n\in\mathbb{N}_0}$ *and* $(y_n)_{n\in\mathbb{N}_0}$ *be two positive sequences and let* $s\in(0,1)$ *be such that both of the sequences with the terms*

$$p_n:=\frac{x_{n+1}}{x_n}-s, \qquad q_n:=\frac{y_{n+1}}{y_n}-s, \qquad n\in\mathbb{N}_0,$$

tend to zero as $n\to\infty$. *If, moreover,* $p_n, q_n\in(-s,1-s)$, $n\in\mathbb{N}_0$, *and*

$$\sum_{n=1}^{\infty}|p_n-q_n|<\infty \tag{1.3.1}$$

then

$$\lim_{n\to\infty}\frac{x_n}{y_n} \text{ exists and belongs to } (0,\infty). \tag{1.3.2}$$

If, moreover, the differences p_n-q_n, $n\in\mathbb{N}_0$, *have a constant sign, then* (1.3.2) *implies* (1.3.1).

Proof. For any $n\in\mathbb{N}$ we have

$$\frac{x_n}{y_n}=\frac{x_{n-1}}{y_{n-1}}\left(1+\frac{p_{n-1}-q_{n-1}}{s+q_{n-1}}\right)=\frac{x_0}{y_0}\prod_{i=0}^{n-1}\left(1+\frac{p_i-q_i}{s+q_i}\right). \tag{1.3.3}$$

Condition (1.3.1) implies the convergence of the product in (1.3.3) to a finite and positive limit, and the converse implication holds under the additional assumption that the terms p_n-q_n, $n\in\mathbb{N}_0$, have a constant sign. ∎

For further considerations we need a family of 'test' functions.

Definition 1.3.1. We denote by $\mathscr{R}$ the family of all measurable functions $r:X\to\mathbb{R}^+$ such that

$$\int_0^{\delta}\frac{r(x)}{x}\,dx<\infty \qquad \text{for } \delta\in(0,a)$$

and for every $\alpha\in(0,1)$ there exists a $\beta\in(1,\infty)$ such that either

$$r(y)\leq\beta r(x) \qquad \text{for all } y\in X\setminus\{0\} \text{ and } x\in[\alpha y,y) \tag{1.3.4}$$

or

$$r(x)\leq\beta r(y) \qquad \text{for all } y\in X\setminus\{0\} \text{ and } x\in[\alpha y,y). \tag{1.3.5}$$

Condition (1.3.4) (resp. (1.3.6)) is obviously fulfilled provided r is decreasing (resp. increasing) but it is definitely less restrictive than monotonicity. The following lemma describes the fundamental property of the members of $\mathscr{R}$.

Lemma 1.3.2. *Let* $(x_n)_{n \in \mathbb{N}_0}$ *be a sequence of positive elements of* X *and let* $s \in (0, 1)$ *be such that*

$$p_n := \frac{x_{n+1}}{x_n} - s \to 0 \qquad \text{as } n \to \infty.$$

If, moreover, $p_n \in (-s, 1-s)$, $n \in \mathbb{N}_0$, *then*

$$\sum_{n=0}^{\infty} r(x_n) < \infty$$

for any $r \in \mathcal{R}$.

Proof. Fix an $r \in \mathcal{R}$ and assume that (1.3.4) is satisfied (in case (1.3.5) the argument is similar). Take a $d \in (s, 1)$. Since $s + p_n < d$ for almost all $n \in \mathbb{N}$ we have $x_{n+1} < dx_n$ for those n. Moreover, for every $k \in \mathbb{N}_0$,

$$x_k \in (0, x_0] = \bigcup_{n=0}^{\infty} (d^{n+1}x_0, d^n x_0]$$

whence $x_k \in (d^{n_k+1}x_0, d^{n_k}x_0]$ for some $n_k \in \mathbb{N}_0$; consequently $d^{n_k+1}x_0 \in [dx_k, x_k)$ and the sequence $(n_k)_{k \in \mathbb{N}_0}$ is strictly increasing for large k. Taking $\alpha := d^{n_k+1}x_0$ and $y := x_k$ in (1.3.4), we get

$$0 \le r(x_k) \le \beta r(d^{n_k+1}x_0), \qquad k \in \mathbb{N}_0. \tag{1.3.6}$$

Write

$$h(x) := \int_{n_0}^{x} r(d^t x_0) \, dt, \qquad x \in [n_0, \infty).$$

We have

$$h(k+1) - h(k) = \int_{k}^{k+1} r(d^t x_0) \, dt \ge \inf_{t \in [k, k+1]} r(d^t x_0) \qquad \text{for all } k \ge n_0.$$

Fix a $t \in (k, k+1]$ and put $\alpha := d$, $x := d^t x_0$ and $y := d^k x_0$ in (1.3.4). Then

$$r(d^k x_0) \le \beta r(d^t x_0), \qquad t \in [k, k+1]$$

since for $t = k$ the inequality holds trivially. Hence

$$r(d^k x_0) \le \beta \inf_{t \in [k, k+1]} r(d^t x_0) \le \beta[h(k+1) - h(k)], \qquad k \ge n_0.$$

Now, by means of (1.3.6),

$$\sum_{k=n_0}^{\infty} r(x_k) \le \beta \sum_{k=n_0}^{\infty} r(d^{n_k+1}x_0) \le \beta \sum_{k=n_0}^{\infty} r(d^k x_0) \le \beta^2 \sum_{k=n_0}^{\infty} [h(k+1) - h(k)]$$

$$= \beta^2 \int_{n_0}^{\infty} r(d^t x_0) \, dt = -\frac{\beta^2}{\log d} \int_{0}^{x_0 d^{n_0}} \frac{r(x)}{x} \, dx < \infty. \qquad \blacksquare$$

The reverse procedure is mainly an exercise and, jointly with Lemma 1.3.2, leads immediately to the following.

Lemma 1.3.3. *Let* $(x_n)_{n \in \mathbb{N}_0}$ *be a sequence of positive elements of* X *and let* $s \in (0, 1)$ *be such that*

$$p_n := \frac{x_{n+1}}{x_n} - s \to 0 \qquad \text{as } n \to \infty$$

and $p_n \in (-s, 1-s)$ *for all* $n \in \mathbb{N}_0$. *Then, a measurable function* $r: X \to \mathbb{R}^+$ *fulfilling condition* (1.3.4) *or* (1.3.5) *belongs to the class* $\mathcal{R}$ *if and only if*

$$\sum_{n=0}^{\infty} r(x_n) < \infty.$$

1.3B. Splinters behaving like geometric sequences

In what follows, we use Landau's O-symbols (see de Bruijn [1, pp. 5–10]). As a consequence of Lemmas 1.3.1 and 1.3.2 we obtain the following.

Theorem 1.3.1. *Let hypotheses* (i) *and* (ii) *be fulfilled. If* f *is continuous,* $s \in (0, 1)$ *and* $p(x) = O(r(x))$ *as* $x \to 0$ *for some* $r \in \mathcal{R}$, *then for every* $x \in X \setminus \{0\}$ *the limit*

$$\lim_{n \to \infty} \frac{f^n(x)}{s^n}$$

exists and belongs to $(0, \infty)$.

Proof. It suffices to take $x_n := f^n(x)$ and $y_n := s^n$, $n \in \mathbb{N}_0$, in Lemma 1.3.1. Then $p_n = p(x_n)$, $q_n = 0$ for $n \in \mathbb{N}_0$ and hence $\lim_{n \to \infty} p_n = \lim_{n \to \infty} p(x_n) = 0$ because $x_n \to 0$ by means of Theorem 1.1.1, and because of (ii). On the other hand, there exists an $M > 0$ such that $|p(x_n)| \leq Mr(x_n)$, $n \in \mathbb{N}_0$, whence

$$\sum_{n=0}^{\infty} |p_n - q_n| = \sum_{n=0}^{\infty} |p_n| \leq M \sum_{n=0}^{\infty} r(x_n) < \infty$$

on account of Lemma 1.3.2. It remains to apply Lemma 1.3.1. ∎

From Theorem 1.3.1 we derive the following result of E. Seneta [5].

Theorem 1.3.2. *Let hypotheses* (i) *and* (ii) *be fulfilled. If* f *is continuous,* $s \in (0, 1)$ *and* p *is monotonic, then for every* $x \in X$ *there exists the limit*

$$\varphi(x) := \lim_{n \to \infty} \frac{f^n(x)}{s^n}$$

and either $\varphi = 0$ *or* $\varphi = \infty$ *or* $\varphi(x) \in (0, \infty)$ *for all* $x \in X$. *The last case occurs if and only if* $\int_0^\delta \dfrac{p(x)}{x} \, dx$ *is convergent for* $\delta \in (0, a)$.

Proof. Since p is monotonic and $p(x) \to 0$ as $x \to 0$, we infer that $g(x) := (1/s)f(x)/x = 1 + (1/s)p(x)$ is monotonic, too; moreover, $g(x) - 1$ has a constant sign in $X \setminus \{0\}$. This implies easily that the sequence $(f^n(x)/s^n)_{n \in \mathbb{N}}$ is monotonic and hence convergent in X (the convergence for $x = 0$ is trivial). If the integral $I(\delta)$ occurring in the statement of our theorem is convergent, then $\varphi(x) \in (0, \infty)$ for all $x \in X \setminus \{0\}$ in virtue of Theorem 1.3.1 (take $r := p$). Conversely, if $\varphi(x_0) \in (0, \infty)$ for some $x_0 \in X \setminus \{0\}$ then $\sum_{n=0}^{\infty} |p(f^n(x_0))|$ converges on account of Lemma 1.3.1; now Lemma 1.3.3 implies $|p| \in \mathscr{R}$, i.e. $I(\delta) < \infty$ and hence also $\varphi(x) \in (0, \infty)$ for all $x \in X \setminus \{0\}$. ∎

Remark 1.3.1. The usefulness of results in the spirit of Theorem 1.3.1 is self-evident. Intuitively, Theorem 1.3.1 may be interpreted as follows: the splinter $(f^n(x))_{n \in \mathbb{N}}$ tends to zero (the fixed point of f provided $0 \in X$) with the speed of the convergence of the geometric sequence $(s^n)_{n \in \mathbb{N}}$. In particular, under the assumptions of Theorem 1.3.1 we derive the convergence of the series $\sum_{n=0}^{\infty} f^n(x)$, $x \in X$, which is sometimes asked for (W. Żelazko, oral communication; P. A. Schwarzman [1]). The other comparison sequences considered further on are $(s^n n^t)_{n \in \mathbb{N}}$ and $(1/n^\alpha)_{n \in \mathbb{N}}$ with some positive α. Further results of this kind may be found in de Bruijn [1], Drewniak [1], [2], Diamond [2], Karamata [2], Ostrowski [1], [2], Kuczma [13], [42], Riekstiņš [1], Seneta [5], Thron [1]. See also Hamilton [2], Choczewski [11], Drewniak–Kuczma [1] for related problems.

Remark 1.3.2. It is worth noting that the mapping φ occurring in Theorem 1.3.2 satisfies the functional equation of Schröder [1]

$$\varphi(f(x)) = s\varphi(x), \qquad x \in X.$$

In fact,

$$\varphi(f(x)) = \lim_{n \to \infty} \frac{f^n(f(x))}{s^n} = s \lim_{n \to \infty} \frac{f^{n+1}(x)}{s^{n+1}} = s\varphi(x).$$

Schröder's equation is one of the most important iterative functional equations (see, in particular, Chapters 8 and 9).

1.3C. Slower convergence of splinters

The power function $r(x) = x^t$, $x \in \mathbb{R}^+$, $t > 0$, belongs to $\mathscr{R}$ (see Thron [1]), but the family $\mathscr{R}$ contains also functions tending to zero much more slowly.

Throughout the rest of this section the symbol u has a fixed meaning and denotes always the function $u: [0, 1) \to \mathbb{R}$ given by the formula

$$u(x) := \begin{cases} 1/\log \dfrac{1}{x} & \text{for } x \in (0, 1), \\ 0 & \text{for } x = 0. \end{cases}$$

This function is defined on $[0, 1)$ only, but, since the asymptotic properties of splinters depend essentially on the behaviour of the given mapping near zero only, we may assume without loss of generality that the parameter a in (i) is less than unity.

Observe that $(u)^t \in \mathscr{R}$ for all $t > 1$ and $(u)^t \notin \mathscr{R}$ for $t \leq 1$.

Theorem 1.3.3. *Let hypotheses* (i) *and* (ii) *be fulfilled. If f is continuous, $s \in (0, 1)$ and $p(x) = c \cdot u(x) + v(x)$; $v(x) = O(r(x))$ as $x \to 0$ for some $r \in \mathscr{R}$, then for every $x \in X \backslash \{0\}$ the limit*

$$\lim_{n \to \infty} \frac{1}{s^n n^t} f^n(x)$$

exists and belongs to $(0, \infty)$; *here* $t := (c/s)u(s)$.

Proof. Take an $x_0 \in X \backslash \{0\}$ and write $x_n := f^n(x_0)$, $n \in \mathbb{N}_0$. Since $x_{n+1}/x_n = s + p(x_n)$ tends to s as $n \to \infty$, so does the sequence $(\sqrt[n]{x_n})_{n \in \mathbb{N}}$ whence

$$u(x_n) = \frac{1}{n} u(s)(1 + w_n); \qquad w_n = o(1) \qquad \text{as } n \to \infty. \tag{1.3.7}$$

Now, in view of $\log(1 + x) < x$, $x > 0$, and $u(x_n) = O(r(x_n))$, $n \to \infty$, we get

$$\log \frac{x_n}{s^n} = \log\left(x_0 \prod_{i=0}^{n-1} \left(1 + \frac{p(x_i)}{s}\right)\right) = A + \sum_{i=1}^{n-1} \log\left(1 + \frac{c}{s} u(x_i) + \frac{1}{s} v(x_i)\right)$$

$$= A + \sum_{i=1}^{n-1} \log\left(1 + \frac{c}{s} \frac{u(s)}{i}(1 + w_i) + \frac{1}{s} v(x_i)\right)$$

$$= O\left(\sum_{i=1}^{n-1} \frac{1}{i}\right) + O\left(\sum_{i=1}^{n-1} r(x_i)\right) = O(\log n) + O(1) = O(\log n)$$

on account of Lemma 1.3.2; A stands here for $\log[x_0(1 + p(x_0)/s)]$. This, jointly with (1.3.7), gives

$$\lim_{n \to \infty} u(x_n) \log \frac{x_n}{s^n} = 0.$$

Making use of the identity

$$\frac{u(x)}{u(y)} = 1 + u(y) \log \frac{x}{y} + \frac{[u(x)]^2 \left(\log \dfrac{x}{y}\right)^2}{1 + u(x) \log \dfrac{x}{y}}$$

for $x = x_n$ and $y = s^n$ we get

$$u(x_n) = u(s^n)\left[1 + u(s^n) \log \frac{x_n}{s^n} + O\left(\left(u(x_n) \log \frac{x_n}{s^n}\right)^2\right)\right]$$

$$= \frac{u(s)}{n} + O\left(\frac{\log n}{n}\right)^2 \tag{1.3.8}$$

as $n \to \infty$. Therefore

$$p_n := p(x_n) = cu(x_n) + O(r(x_n)) = \frac{cu(s)}{n} + O\left(\left(\frac{\log n}{n}\right)^2\right) + O(r(x_n)),$$

i.e.

$$p_n = \frac{cu(s)}{n} + l_n, \qquad n \in \mathbb{N},$$

where

$$\sum_{n=1}^{\infty} |l_n| \leq \text{const}\left(\sum_{n=1}^{\infty} r(x_n) + \sum_{n=1}^{\infty} \left(\frac{\log n}{n}\right)^2\right) < \infty$$

in view of Lemma 1.3.2. Similarly, putting $y := s^n n^t$ and $q_n := y_{n+1}/y_n - s$, $n \in \mathbb{N}$, one has

$$q_n = \left[s\left(\frac{n+1}{n}\right)^t - 1\right] = \frac{st}{n} + O\left(\frac{1}{n^2}\right) = \frac{cu(s)}{n} + k_n$$

on account of the definition of t; moreover, $\sum_{n=1}^{\infty} |k_n| < \infty$. Since

$$\sum_{n=1}^{\infty} |p_n - q_n| \leq \sum_{n=1}^{\infty} |l_n - k_n| \leq \sum_{n=1}^{\infty} |l_n| + \sum_{n=1}^{\infty} |k_n| < \infty,$$

the theorem follows from Lemma 1.3.1. ∎

Relation (1.3.7) resulted from the simplest properties of the sequence $(x_n)_{n \in \mathbb{N}}$. Making use of (1.3.7) we could derive the improved estimate (1.3.8). This could now be used in order to get still more precise information about the asymptotic properties of the sequence $(u(x_n))_{n \in \mathbb{N}}$. Iterating this procedure one can arrive at the following result which we quote here without proof (see Drewniak [2]).

Theorem 1.3.4. *Let hypotheses* (i) *and* (ii) *be fulfilled. If f is continuous, $s, t \in (0, 1)$ and, for some $r \in \mathcal{R}$,*

$$p(x) = \sum_{k=1}^{m-1} c_k(u(x))^{kt} + c_m u(x) + O(r(x)), \qquad x \to 0$$

$$\left(\text{resp. } p(x) = c \prod_{i=1}^{k+1} u^i(x) + O(r(x)), \ x \to 0\right),$$

where $m \in \mathbb{N}$ is such that $1/t \in (m-1, m]$, then for every $x \in X \setminus \{0\}$ the limit

$$\lim_{n \to \infty} \left(f^n(x)s^{-n} \exp\left[\sum_{k=1}^{m-1} d_k n^{1-tk} + d_m \log n\right]\right)$$

$$\left(\text{resp. } \lim_{n \to \infty} [f^n(x)s^{-n}(\log^k n)^{cu(s)/s}]\right)$$

exists and belongs to $(0, \infty)$.

1.3D. Special cases

So far we have not discussed the case where $s = 0$ or $s = 1$. It turns out that it may be reduced to the case where $s \in (0, 1)$. To visualize such a technique we shall present a result concerning the case $s = 1$ (Thron [1]).

Theorem 1.3.5. *Let hypotheses* (i) *and* (ii) *be fulfilled. If f is continuous, $s = 1$ and $c := \lim_{x \to 0}(x^{-t}p(x))$ is negative and finite for some $t > 0$, then for every $x \in X \setminus \{0\}$ we have*

$$\lim_{n \to \infty} n^{1/t} f^n(x) = \left(\frac{-1}{ct}\right)^{1/t}.$$

Proof. Write $\varphi(x) := (f(x^{1/t}))^t$ and put $g := u^{-1} \circ \varphi \circ u$. Then we have $f(x) = x(1 + cx^t(1 + o(1))$, $\varphi(x) = x(1 + ctx(1 + o(1))$ and $g(x) = x(e^{ct} + 0(1))$ as $x \to 0$. Take an $x_0 \in X \setminus \{0\}$ and put $x_n := f^n(x_0)$, $y_n := \varphi^n(x_0^t) = x_n^t$, and $z_n := u^{-1}(y_n)$, $n \in \mathbb{N}_0$. Observe that

$$y_n = \varphi^n(y_0) = (u \circ g^n \circ u^{-1})(y_0)$$

whence $z_n = u^{-1}(y_n) = g^n(u^{-1}(y_0))$ for all $n \in \mathbb{N}_0$. Consequently, since $e^{ct} \in (0, 1)$, one may apply relation (1.3.7) getting

$$x_n^t = y_n = u(z_n) = \frac{u(e^{ct})}{n}(1 + o(1)) = -\frac{1}{ctn}(1 + o(1)), \qquad n \to \infty.$$

Finally,

$$n^{1/t} x_n = \left(\frac{-1}{ct}\right)^{1/t}(1 + o(1)), \qquad n \to \infty. \qquad \blacksquare$$

The following lemma is obvious.

Lemma 1.3.4. *Let X fulfil* (i) *and let $f, g: X \to X$ be functions such that $f \leq g$ in X. If either f or g is increasing, then $f^n \leq g^n$ for all $n \in \mathbb{N}_0$.*

This lemma allows one to obtain the following modification of Theorem 1.3.5 (see Ostrowski [1], Thron [1]).

Theorem 1.3.6. *Let hypotheses* (i) *and* (ii) *be fulfilled. If f is continuous, $s = 1$ and*

$$\limsup_{x \to 0}\left(-\frac{1}{x^t}p(x)\right) =: C \in (0, \infty)$$

$$\left(resp. \liminf_{x \to 0}\left(-\frac{1}{x^t}p(x)\right) =: c \in (0, \infty)\right),$$

then for every $d > C$ *(resp.* $0 < d < c$*) and for every* $x \in X \setminus \{0\}$ *we have*

$$f''(x) \geq \left(\frac{1}{dtn}\right)^{1/t}$$

$$\left(resp. \ f''(x) \leq \left(\frac{1}{dtn}\right)^{1/t}\right)$$

for all sufficiently large $n \in \mathbb{N}$. *If, moreover,* f *is increasing then, in the latter case, the inequality* $f''(x) \leq (dtn)^{-1/t}$ *holds uniformly with respect to* $z \in X \cap [0, x]$.

In the class of concave mappings there is a converse of Theorem 1.3.6. The result, whose proof will be omitted here, reads as follows (Drewniak–Drobot [1]).

Theorem 1.3.7. *Let hypotheses* (i) *and* (ii) *be fulfilled. If* f *is concave in* X *and if there exist positive constants* t, A *and* B *such that for each* $x \in X \setminus \{0\}$ *the inequalities*

$$A\left(\frac{1}{n}\right)^{1/t} \leq f''(x) \leq B\left(\frac{1}{n}\right)^{1/t}$$

hold true for all $n \geq N_x$, *then necessarily* $s = 1$ *and there exist positive constants* C *and* D *such that*

$$Cx^t \leq -p(x) \leq Dx^t$$

for all $x \in X$.

Finally, for the case $s = 0$, we have the following.

Theorem 1.3.8. *Let hypotheses* (i) *and* (ii) *be fulfilled. If* f *is continuous,* $s = 0$ *and*

$$p(x) = u^{-1}\left(u(x)\left(\frac{1}{d} + O(r(u(x)))\right)\right) \qquad as \ x \to 0 \qquad (1.3.9)$$

for some $r \in \mathscr{R}$, *then for every* $x \in X \setminus \{0\}$ *the limit*

$$\lim_{n \to \infty} (f''(x))^{(1+d)^{-n}} \qquad (1.3.10)$$

exists and belongs to $(0, 1)$.

Proof. Take an $x_0 \in X \setminus \{0\}$ and write $x_n := f''(x_0)$, $n \in \mathbb{N}_0$. We have $x_{n+1} = x_n \, p(x_n)$, $n \in \mathbb{N}_0$, whence, by (1.3.9),

$$u(x_{n+1}) = u(x_n)\left(\frac{1}{1+d} + l_n\right), \qquad n \in \mathbb{N}_0$$

and $l_n = O(r(u(x_n)))$ as $n \to \infty$. Lemma 1.3.2 implies the convergence of the series $\sum_{n=1}^{\infty} |l_n|$ and Lemma 1.3.1 applied for $y_n := u(x_n)$, $n \in \mathbb{N}_0$ (with $1/(1+d)$

in place of s and $1/(1 + d)^n$ instead of x_n therein), gives the existence of a $c \in (0, \infty)$ such that

$$c_n := (1 + d)^n u(x_n) \to c \qquad \text{as } n \to \infty.$$

Now, $x_n = u^{-1}(c_n/(1 + d)^n) = \exp(-(1 + d)^n/c_n)$ for $n \in \mathbb{N}$, whence

$$x_n^{(1+d)^{-n}} = \exp\left(-\frac{1}{c_n}\right) \xrightarrow[n \to \infty]{} \exp\left(-\frac{1}{c}\right) \in (0, 1). \qquad \blacksquare$$

Remark 1.3.3. Noteworthy is the fact that the limit (1.3.10) considered as a function of $x \in X \setminus \{0\}$ satisfies the functional equation of Böttcher [1], [2] (see Chapter 8)

$$\beta(f(x)) = \beta(x)^p, \qquad x \in X \setminus \{0\},$$

with $p := 1 + d$. In fact,

$$\beta(f(x)) = \lim_{n \to \infty} (f^n(f(x)))^{(1+d)^{-n}} = \lim_{n \to \infty} [(f^{n+1}(x))^{(1+d)^{-(n+1)}}]^{1+d} = \beta(x)^{1+d}.$$

The last result of the present section concerns also the case $s = 0$ and is just a simple consequence of Theorem 1.3.8; that is, we have the following (see Thron [1]).

Theorem 1.3.9. *Let hypotheses* (i) *and* (ii) *be fulfilled. If f is continuous, $s = 0$ and*

$$0 < \liminf_{x \to 0} x^{-d}p(x) \leq \limsup_{x \to 0} x^{-d}p(x) < \infty$$

for some positive d, then limit (1.3.10) *exists for all $x \in X \setminus \{0\}$.*

Proof. One has plainly

$$p(x) = x^d \exp O(1) = \exp\left(-\frac{1}{d^{-1}u(x)} + O(1)\right) = \exp\left(-\frac{1}{d^{-1}u(x) + O([u(x)]^2)}\right)$$

$$= u^{-1}(d^{-1}u(x) + O([u(x)]^2)) = u^{-1}\left(u(x)\left(\frac{1}{d} + O(u(x))\right)\right)$$

as $x \to 0$; this shows that (1.3.9) is satisfied by the identity function r. $\qquad \blacksquare$

1.4 Iteration sequences of random-valued functions

For simplicity, in the whole of this section we take $X = [0, 1]$. As we have seen in Section 1.2, any splinter of a continuous map $f: X \to X$ such that $f(x) < x$, $x \in X \setminus \{0\}$, has to converge to zero; in other words, the domain of attraction of zero coincides with the whole of X. In the present section we try to extend this fundamental (although straightforward) result to the case where we know, not the exact value of f, but only the probability distribution of $f(x)$ (Baron–Kuczma [1]). Exclusively in this section we shall use some more advanced notions and facts from probability theory. However, the results presented here will not be that used in the sequel.

1.4A. Preliminaries

To avoid any ambiguity let us fix some notation. $\mathscr{B}$ denotes the σ-field of all Borel subsets of $X = [0, 1]$ and if two σ-fields Σ_1 and Σ_2 are given then $\Sigma_1 \times \Sigma_2$ stands for their product σ-field, i.e. the smallest σ-field containing all the 'rectangles' $A \times B$ where $(A, B) \in \Sigma_1 \times \Sigma_2$. Finally, if P_1 and P_2 are two probability measures on Σ_1 and Σ_2, respectively, then $P_1 \times P_2$ denotes the product probability measure on $\Sigma_1 \times \Sigma_2$. The product space $(\Omega \times \Omega, \Sigma \times \Sigma, P \times P)$ will be denoted by $(\Omega^2, \Sigma^2, P^2)$ and, more generally, $(\Omega^n, \Sigma^n, P^n)$ will stand for the appropriate product probability space of n copies of (Ω, Σ, P).

Let (Ω, Σ, P) be a probability space. A *random-valued function* (abbreviated to *rv-function* in the sequel) is a map $f: X \times \Omega \to X$ such that the inverse image $f^{-1}(B)$ of any Borel set $B \subset X$ belongs to $\mathscr{B} \times \Sigma$. In other words, an rv-function is a measurable stochastic process for which the state space coincides with the time interval. In particular, for each fixed $x \in X$, the mapping $f(x, \cdot)$ (for brevity, denoted simply by $f(x)$ further on) is a random variable on the space Ω.

Prior to any consideration on iteration sequences for an rv-function one has to define the operation of composition. We shall do it as follows: by the *composition of* any two *rv-functions* f and g with $(\Omega_1, \Sigma_1, P_1)$ and $(\Omega_2, \Sigma_2, P_2)$ as the respectively underlying probability spaces we mean the map $g \circ f: X \times \Omega_1 \times \Omega_2 \to X$ given by the formula

$$(g \circ f)(x, \omega_1, \omega_2) := g(f(x, \omega_1), \omega_2), \qquad (x, \omega_1, \omega_2) \in X \times \Omega_1 \times \Omega_2.$$

Let f be an rv-function and let $\mathscr{F}(x|\cdot)$ be the probability distribution of the random variable $f(x)$, i.e.

$$\mathscr{F}(x, t) = P(\{\omega \in \Omega: f(x, \omega) < t\})$$

for all $(x, t) \in [0, 1] \times \mathbb{R}$. With the help of standard techniques (we omit the details) one may prove the following lemma collecting some elementary properties of rv-functions.

Lemma 1.4.1. *Let f, g be two rv-functions with $(\Omega_1, \Sigma_1, P_1)$ and $(\Omega_2, \Sigma_2, P_2)$ as the respective underlying probability spaces and let $\mathscr{F}(x|\cdot), \mathscr{G}(x|\cdot)$ be the respective probability distributions of the random variables $f(x), g(x), x \in X$. Then the following hold.*

(i) *For each $t \in \mathbb{R}$ the mapping $\mathscr{F}(\cdot|t): X \to X$ is Borel measurable;*
(ii) *$g \circ f$ is an rv-function with $(\Omega_1 \times \Omega_2, \Sigma_1 \times \Sigma_2, P_1 \times P_2)$ as the underlying probability space;*

(iii) *the probability distribution* $(\mathcal{G} \circ \mathcal{F})(x|\cdot)$ *of the random variable* $(g \circ f)(x)$
is given by the formula

$$(\mathcal{G} \circ \mathcal{F})(x|t) = \int_X \mathcal{F}(y|t) \, d_y \mathcal{G}(x|y), \qquad (x, t) \in X \times \mathbb{R}.$$

In particular, the probability distributions $\mathcal{F}_n(x|\cdot)$ *of the random variables*
$f^n(x)$ *are given by the recurrence*

$$\mathcal{F}_{n+1}(x|t) = \int_X \mathcal{F}(y|t) \, d_y \mathcal{F}_n(x|y) = \int_X \mathcal{F}_n(y|t) \, d_y \mathcal{F}(y|t), \qquad (x, t) \in X \times \mathbb{R},$$

where $\mathcal{F}_1(\cdot|\cdot) := \mathcal{F}(\cdot|\cdot)$ *and* $n \in \mathbb{N}$.

Observe that $f^n(x)$ is a random variable on the probability space
$(\Omega^n, \Sigma^n, P^n)$. However, all these product probability spaces can be embedded
in the infinite product $(\Omega^\infty, \Sigma^\infty, P^\infty)$ and the function f^n can be naturally
extended onto $X \times \Omega^\infty$ by putting $f(x, \omega_1, \omega_2, \ldots) := f^n(x, \omega_1, \ldots, \omega_n)$,
$x \in X$, $(\omega_k)_{k \in \mathbb{N}} \in \Omega^\infty$ and $n \in \mathbb{N}$. Thus the sequence $(f^n(x))_{n \in \mathbb{N}}$ may be regarded
as a Markov chain of random variables on the same probability space
$(\Omega^\infty, \Sigma^\infty, P^\infty)$.

Denote by $\mathbb{F}_n$ the σ-field of all sets $A \in \Sigma^\infty$ that are of the form

$$A = A_n \mathop{\times}_{i=n+1}^{\infty} \Omega_i \tag{1.4.1}$$

where $A_n \in \Sigma^n$ and $\Omega_i = \Omega$ for $i = n+1, n+2, \ldots$. Clearly $\mathbb{F}_n \subset \mathbb{F}_{n+1}$, $n \in \mathbb{N}$.
Let $m(x)$ stand for the *mean* (expectation) of the random variable $f(x)$, i.e.

$$m(x) = \int_X t \, d_t \mathcal{F}(x|t), \qquad x \in X. \tag{1.4.2}$$

Lemma 1.4.2. *Fix an* $n \in \mathbb{N}$ *and an* $x \in X$. *The conditional expectation of the*
random variable $f^{n+1}(x)$ *with respect to* $\mathbb{F}_n$ *is given by*

$$E(f^{n+1}(x)|\mathbb{F}_n) = m \circ (f^n(x)).$$

Proof. It follows from the Fubini Theorem that the function $m: X \to \mathbb{R}$ is
Borel measurable whence $m \circ (f^n(x))$ is $\mathbb{F}_n$ measurable. For any set $A \in \mathbb{F}_n$ of
the form (1.4.1) we have

$$\int_A m(f^m(x, \omega_1, \ldots)) dP = \int_{A_n} m(f^n(x, \omega_1, \ldots, \omega_n)) dP^n$$

$$= \int_{A_n \times \Omega} f^{n+1}(x, \omega_1, \ldots, \omega_{n+1}) dP^{n+1}$$

$$= \int_A f^{n+1}(x, \omega_1, \ldots) dP^\infty. \qquad \blacksquare$$

1.4B. Convergence of random splinters

We start with the following.

Theorem 1.4.1. *If f is an rv-function with the mean (1.4.2) and if $m(x) \leq x$ in X then, for every $x \in X$, the splinter $(f^n(x))_{n \in \mathbb{N}}$ converges with probability 1.*

Proof. In view of Lemma 1.4.2 one has

$$E(f^{n+1}(x)|\mathbb{F}_n) \leq f^n(x)$$

which says that the sequence $(f^n(x))_{n \in \mathbb{N}}$ is a supermartingale. Let

$$m_n(x) := \int_X t\, d_t \mathbb{F}_n(x|t)$$

be the mean of the random variable $f^n(x)$, $x \in X$, $n \in \mathbb{N}$. On account of Lemma 1.4.1 we arrive at

$$m_{n+1}(x) = \int_X t\, d_t \mathscr{F}_{n+1}(x|t) = \int_X t\, d_t \int_X \mathscr{F}(y|t) d_y \mathscr{F}_n(x|y)$$

$$= \int_X \left(\int_X t\, d_t \mathscr{F}(y|t) \right) d_y \mathscr{F}_n(x|y) = \int_X m(y) d_y \mathscr{F}_n(x|y),$$

whence, for all $n \in \mathbb{N}$ and $x \in X$,

$$m_{n+1}(x) = m_n(x) - \int_X (y - m(y)) d_y \mathscr{F}_n(x|y) \qquad (1.4.3)$$

and the sequence $(m_n(x))_{n \in \mathbb{N}}$ is decreasing. The assertion follows by the theorem on the convergence of supermartingales (see Loève [1, p. 393]). ∎

One may ask whether the iterative sequences considered tend to zero with probability 1. Observe, however, that even in the simplest deterministic case and even with the sharp inequality $f(x) < x$, $x \in X \setminus \{0\}$, the sequence $(f^n(x))_{n \in \mathbb{N}}$, although convergent, need not tend to zero; so, we have to impose some kind of continuity in order to get the desired effect. To this aim adopt the following definition: an *rv-function* f is called *continuous* at the point $x_0 \in X$ if and only if $f(x)$ tends to $f(x_0)$ in law as $x \to x_0$, i.e. $\mathscr{F}(x|t) \to \mathscr{F}(x_0|t)$ at each continuity point of $\mathscr{F}(x_0|\cdot)$; f is called continuous provided it is continuous at each point of X.

Now, we have the following.

Theorem 1.4.2. *If f is a continuous rv-function with mean (1.4.2) and if $m(x) < x$ for all $x \in X \setminus \{0\}$ then, for every $x \in X$, the splinter $(f^n(x))_{n \in \mathbb{N}}$ converges to zero with probability 1.*

Proof. In view of Theorem 1.4.1 it is enough to prove that $(f^n(x))_{n \in \mathbb{N}}$ converges to zero in law. For $x = 0$ the theorem is trivial: the continuity of m implies $m(0) = 0$ whence $f^n(x) = 0$ with probability 1, $n \in \mathbb{N}$. Take an

$x \in X \setminus \{0\}$. By Helly's Theorem, from every sub-sequence $(\mathscr{F}_{n_k}(x|\cdot))_{k\in\mathbb{N}}$ of $(\mathscr{F}_n(x|\cdot))_{n\in\mathbb{N}}$ we can choose a sub-sub-sequence $(\mathscr{F}_{n_{k_l}}(x|\cdot))_{l\in\mathbb{N}}$ which converges weakly to a probability distribution $\mathscr{F}_0(x|\cdot)$ (which *a priori* may depend on the sub-sequence $(n_{k_l})_{l\in\mathbb{N}}$). As we have seen in the preceding proof, the sequence $(m_n(x))_{n\in\mathbb{N}}$ is convergent. Replacing n by n_{k_l} in (1.4.3) and letting l tend to infinity we obtain

$$\int_X (y - m(y)) \mathrm{d}_y \mathscr{F}_0(x|y) = 0.$$

Since $m(y) < y$ for all $y \in X \setminus \{0\}$, we must have

$$\mathscr{F}_0(x|t) = \begin{cases} 0 & \text{for } t \leq 0, \\ 1 & \text{for } t > 0; \end{cases}$$

in particular, $\mathscr{F}_0$ does not depend on the choice of the sub-sequence $(n_{k_l})_{l\in\mathbb{N}}$. Consequently, the sequence $(\mathscr{F}_n(x|\cdot))_{n\in\mathbb{N}}$ must itself weakly converge to $\mathscr{F}_0(x|\cdot)$. ∎

Concerning related results see Diamond [2] and Kuczma [42]. As in the deterministic case one may ask how fast is the convergence of the random splinter considered in Theorem 1.4.2. In response to that question K. Baron [15] has recently obtained some stochastic analogues of W. J. Thron's [1] results (see Section 1.3) on the speed of convergence of iterative sequences.

There are also other interpretations of the notions and results contained in the present section. The distributions may be interpreted as the transition probability from the state x to the interval $(-\infty, t)$; then f^n become iterates of the transition operator. Finally, the problem of iteration of rv-functions is different from that of stochastic approximation. The latter consists in the investigation of the convergence of sequences of random variables X_n given by the recurrence relation $X_{n+1} = f(X_n)$ where, however, the shape of the function f is known. In our case we do not know the exact form of the function f either.

1.5 Some fixed-point theorems

Existence theorems obtained with the aid of the fixed-point method enjoy a universal popularity among specialists in the field of iterative functional equations. Fixed-point theory lying in the intersection of topology and functional analysis is a subject of continuous and intensive search; the best sources to quote here are the recently published monographs of Dugundji–Granas [1] and Goebel–Reich [1] as well as the references therein. The most widely known results of this permanently expanding theory are its corner stones stemming from functional analysis: the Banach contraction principle and Schauder's fixed-point theorem. For the sake of completeness we quote them here explicitly.

Banach's Theorem. *Let f be a self-mapping of a complete metric space (X, ρ) and let*

$$\rho(f(x), f(y)) \leqq \Theta\rho(x, y), \qquad x, y \in X,$$

hold true with some $\Theta \in (0, 1)$. Then f has exactly one fixed point $x_0 \in X$; moreover, the domain of attraction of x_0 coincides with the whole of X.

Schauder's Theorem. *Let X be a nonempty convex and compact subset of a Banach space. Then every continuous self-mapping of X has a fixed point.*

The identity mapping id_X shows that any kind of uniqueness is excluded in Schauder's Theorem.

Numerous extensions of Banach's Theorem in various directions may be found in the survey article by F. E. Browder [1]. Below we prove a few theorems which generalize Banach's principle and which will be useful in the sequel.

1.5A. Generalizations of the Banach contraction principle

We start with the following result due to J. Matkowski [13] (see also Meir-Keeler [1]).

Theorem 1.5.1. *Let f be a self-mapping of a complete metric space (X, ρ) and let*

$$\rho(f(x), f(y)) < \rho(x, y) \qquad \text{for all } x, y \in X, x \neq y. \tag{1.5.1}$$

If for every $\varepsilon > 0$ there exists a $\delta > 0$ such that

$$\varepsilon < \rho(x, y) < \varepsilon + \delta \quad \text{implies} \quad \rho(f(x), f(y)) \leqq \varepsilon \tag{1.5.2}$$

then there exists exactly one fixed point of f; moreover, its domain of attraction coincides with the whole of X.

Proof. Existence. Take an $x_0 \in X$ and put $x_n := f^n(x_0)$, $n \in \mathbb{N}$. Obviously, we may assume that $x_{n+1} \neq x_n$ for all $n \in \mathbb{N}_0$. Then relation (1.5.1) guarantees that the sequence $(\rho(x_{n+1}, x_n))_{n \in \mathbb{N}}$ is strictly decreasing and hence convergent. If we had $c := \lim_{n \to \infty} \rho(x_{n+1}, x_n) > 0$, then

$$c < \rho(x_{n+1}, x_n) < c + \delta(c) \qquad \text{for large } n \in \mathbb{N}$$

and from (1.5.2) we would get $\rho(x_{n+2}, x_{n+1}) \leqq c$, a contradiction. Thus $c = 0$.

Without loss of generality we may suppose that for a given $\varepsilon > 0$ the number $\delta = \delta(\varepsilon)$ is chosen so that $\delta(\varepsilon) < \varepsilon$. Fix an $\varepsilon > 0$. Then $\rho(x_{k+1}, x_k) < \delta(\varepsilon)$ for some $k \in \mathbb{N}$. Put

$$B := \{x \in X : \rho(x, x_k) < \varepsilon + \delta(\varepsilon)\}$$

and note that $f(B) \subset B$; actually, relation (1.5.1) (resp. (1.5.2)) implies that $\rho(f(x), x_k) < \varepsilon + \delta(\varepsilon)$ provided $\rho(x, x_k) \leq \varepsilon$ (resp. $\rho(x, x_k) \in (\varepsilon, \varepsilon + \delta(\varepsilon)))$. Consequently, $x_n \in B$ for all $n \geq k$. Hence, for all $m, n \geq k$, we get

$$\rho(x_n, x_m) \leq \rho(x_n, x_k) + \rho(x_k, x_m) < 2(\varepsilon + \delta(\varepsilon)) < 4\varepsilon$$

which means that $(x_n)_{n \in \mathbb{N}}$ is a Cauchy sequence. Let $\xi := \lim_{n \to \infty} x_n$; since (1.5.1) forces f to be continuous we have

$$f(\xi) = \lim_{n \to \infty} f(x_n) = \lim_{n \to \infty} x_{n+1} = \xi.$$

Uniqueness. Follows immediately from (1.5.1). ∎

Neither (1.5.1) nor the existence of a fixed point of f in X is implied by condition (1.5.2) alone. To see this, take $X := \{1, 2, 3\}$, $\rho(x, y) := |x - y|$, $x, y \in X$, and $f: X \to X$ defined by $f(1) = 3$, $f(2) = 1$, $f(3) = 2$ and put

$$\delta(\varepsilon) := \begin{cases} 1 - \varepsilon & \text{for } \varepsilon \in (0, 1), \\ 2 - \varepsilon & \text{for } \varepsilon \in [1, 2), \\ 1 & \text{for } \varepsilon \in [2, \infty). \end{cases}$$

It is, however, noteworthy that (1.5.1) may be replaced by the assumption that (X, ρ) is *metrically convex* in the sense of Menger, i.e. that for each pair of different points $x, y \in X$ there exists a $z \in X \setminus \{x, y\}$ such that $\rho(x, y) = \rho(x, z) + \rho(z, y)$ (see Matkowski [13]). In connection with this, let us quote here a result of Matkowski–Węgrzyk [1] (see also Węgrzyk [3]).

Theorem 1.5.2. *Let f be a self-mapping of a complete and metrically convex metric space (X, ρ) and let f satisfy one of the following conditions:* (1.5.2) *or*

there exists a decreasing function $\alpha: \mathbb{R}^+ \to [0, 1]$, $\alpha(t) < 1$ for $t > 0$ and $\hspace{3cm}$ (1.5.3)
$$\rho(f(x), f(y)) \leq \alpha(\rho(x, y))\rho(x, y), \qquad x, y \in X,$$

or

for every $\varepsilon > 0$ there is a $\delta > 0$ such that $\hspace{2cm}$ (1.5.4)
$$\varepsilon \leq \rho(x, y) < \varepsilon + \delta \text{ implies } \rho(f(x), f(y)) < \varepsilon, \qquad x, y \in X,$$

or

there exists a function $\gamma: \mathbb{R}^+ \to \mathbb{R}^+$, $\gamma(t) < t$ for $t > 0$, such that f is a γ-contraction, i.e. $\hspace{2cm}$ (1.5.5)
$$\rho(f(x), f(y)) \leq \gamma(\rho(x, y)), \qquad x, y \in X,$$

then f satisfies all the relations (1.5.2)–(1.5.5) and, additionally, one may require γ to be strictly increasing, concave and continuously differentiable in $\mathbb{R}^+$ and α to be continuous in $\mathbb{R}^+$. Moreover, there exists exactly one fixed point of f and its domain of attraction coincides with the whole of X.

This theorem closes, in a sense, the developments originated by E. Rakotch [1] who proposed condition (1.5.3), D. W. Boyd–J. S. Wong [1] who assumed (1.5.5) with a right upper semicontinuous function γ and A. Meir–E. Keeler [1] who considered (1.5.4). By using Theorem 1.5.2, R. Węgrzyk [3] obtained some fixed-point theorems for multivalued mappings.

The following modification of the above mentioned Boyd–Wong theorem was also proved by J. Matkowski [13].

Theorem 1.5.3. *Let f be a self-mapping of a complete metric space (X, ρ) such that $\rho(f(x), f(y)) \leq \gamma(\rho(x, y))$, $x, y \in X$, for some increasing function $\gamma: \mathbb{R}^+ \to \mathbb{R}^+$ fulfilling the condition $\lim_{n \to \infty} \gamma^n(t) = 0$ for all $t > 0$. Then there exists exactly one fixed point of f and its domain of attraction coincides with the whole of X.*

Proof. First we show that $\gamma(t) < t$ for $t > 0$. Obviously, $\gamma(t) \neq t$ for $t \in (0, \infty)$. If we had $\gamma(t_0) > t_0$ for some $t_0 > 0$, then $\gamma^n(t_0) \geq t_0 > 0$ for all $n \in \mathbb{N}_0$ which contradicts the assumed convergence of the iterative sequence of γ to zero. The proof proceeds further like that of Theorem 1.5.1. ∎

1.5B. Case of product spaces

The following, Matkowski's fixed-point theorem (Matkowski [12], [13]), extends Banach's Theorem to self-mappings of product spaces and reduces to the latter for $N = 1$.

Theorem 1.5.4. *Let $N \in \mathbb{N}$ be fixed. Assume that*

$$(X_i, \rho_i) \text{ are complete metric spaces for } i \in \{1, \ldots, N\} \text{ and} \atop f_i: X_1 \times X_2 \times \cdots \times X_N \to X_i \text{ are mappings such that} \qquad (1.5.6)$$

$$\rho_i(f_i(x_1, \ldots, x_N), f_i(y_1, \ldots, y_N)) \leq \sum_{j=1}^N s_{i,j} \rho_j(x_i, x_j) \qquad (1.5.7)$$

for all $x_i, y_i \in X_i$ and some nonnegative constants $s_{i,j}$, $i, j \in \{1, \ldots, N\}$.
Put $a_{i,i}^0 := 1 - s_{i,i}$, $a_{i,j}^0 := s_{i,j}$ for $i \neq j$, $i, j \in \{1, \ldots, N\}$ and, inductively,

$$a_{i,j}^{k+1} := \begin{cases} a_{1,1}^k a_{i+1,j+1}^k - a_{i+1,1}^k a_{1,j+1}^k & \text{for } i = j, \\ a_{1,1}^k a_{i+1,j+1}^k + a_{i+1,1}^k a_{1,j+1}^k & \text{for } i \neq j, \end{cases} \qquad (1.5.8)$$

$i, j \in \{1, \ldots, N-k-1\}$, $k \in \{0, \ldots, N-2\}$.
If

$$a_{i,i}^k > 0 \qquad \text{for } i \in \{1, \ldots, N-k\}, k \in \{0, \ldots, N-1\}, \qquad (1.5.9)$$

then there exists exactly one fixed point $\xi \in X := X_1 \times \cdots \times X_N$ of the transformation $f := (f_1, \ldots, f_N): X \to X$; moreover, $\lim_{n \to \infty} f^n(x) = \xi$ for all $x \in X$, i.e. the domain of attraction of ξ coincides with the whole of X.

As we shall see later (see Lemma 1.5.3 below), in the case where the matrix $S := (s_{i,j})_{i,j=1,\ldots,N}$ has positive entries, inequalities (1.5.9) are equivalent to the statement that all the characteristic roots of S are less than unity in absolute value. Thus Theorem 1.5.4 may be reformulated as follows.

Theorem 1.5.5. *Assume* (1.5.5) *is satisfied with positive* $s_{i,j}$, $i,j \in \{1, \ldots, N\}$. *If all the characteristic roots of the matrix* $(s_{i,j})_{i,j=1,\ldots,N}$ *are less than unity in absolute value, then the transformation* $f := (f_1, \ldots, f_N)$ *has exactly one fixed point in the space* $X := X_1 \times \cdots \times X_N$; *its domain of attraction coincides with the whole of* X.

Such a result has been known long before Theorem 1.5.4 (see e.g. Krasnoselskiĭ *et al.* [1]). The major positive attribute of Theorem 1.5.4 is its effectiveness. Checking its assumptions involves only the simplest arithmetic operations addition, subtraction and multiplication and hence it is very well suited for easy computerization. The advantage of Theorem 1.5.4 is visible chiefly for large N when the practical ascertainment of the fact that all the characteristic roots lie in the open unit disc may cause essential difficulties especially when some of them are localized close to the unit circle.

Finally, note that the assumption $s_{i,j} > 0$, $i,j \in \{1, \ldots, N\}$, in Theorem 1.5.5 does not bear upon its generality since dealing with sharp inequalities we may always replace $s_{i,j}$ in (1.5.7) by slightly larger numbers and small changes will not spoil the inclusion of the characteristic roots in the open unit disc.

To prove Theorem 1.5.4 we need a technical lemma.

Lemma 1.5.1. *Fix an integer* $N \geq 2$ *and assume the matrix* $(a_{i,j}^0)_{i,j=1,\ldots,N}$ *to have nonnegative entries. Define the numbers* $a_{i,j}^k$, $i,j = 1, \ldots, N-k$, $k = 1, \ldots, N-1$, *according to formulae* (1.5.8). *Then the system of inequalities*

$$\sum_{\substack{j=1 \\ j \neq i}}^{N} a_{i,j}^0 r_j < a_{i,i}^0 r_i, \qquad i \in \{1, \ldots, N\}, \tag{1.5.10}$$

has a positive solution $(r_1, \ldots, r_N)$ *if and only if relation* (1.5.9) *holds true.*

Proof. We proceed by induction. For $N = 2$, if $a_{1,2}^0 = a_{2,1}^0 = 0$ the lemma is trivial; so, assume that e.g. $a_{1,2}^0 \neq 0$. Then (1.5.9) can be written as

$$\frac{a_{2,1}^0}{a_{2,2}^0} < \frac{a_{1,1}^0}{a_{1,2}^0}$$

which is satisfied if and only if there exist positive numbers r_1, r_2 such that

$$\frac{a_{2,1}^0}{a_{2,2}^0} < \frac{r_2}{r_1} < \frac{a_{1,1}^0}{a_{1,2}^0}$$

or, equivalently, if

$$a^0_{1,2} r_2 < a^0_{1,1} r_1 \qquad \text{and} \qquad a^0_{2,1} r_1 < a^0_{2,2} r_2$$

which is just (1.5.10) for $N = 2$. Thus the lemma is true for $N = 2$.

Now, let $N \geq 3$ and assume the validity of our lemma for $N - 1$. Let $(r_1, \ldots, r_N)$ be a positive solution of (1.5.10). For $i = 1$ we obtain

$$\frac{1}{a^0_{1,1}} \sum_{j=2}^{N} a^0_{1,j} r_j < r_1$$

whence, again by (1.5.10), for each $i \in \{2, \ldots, N\}$ one has

$$\frac{a^0_{i,1}}{a^0_{1,1}} \sum_{j=2}^{N} a^0_{1,j} r_j + \sum_{\substack{j=2 \\ j \neq i}}^{N} a^0_{i,j} r_j < a^0_{i,i} r_i,$$

i.e.

$$\sum_{\substack{j=2 \\ j \neq i}}^{N} (a^0_{i,1} a^0_{1,j} + a^0_{1,1} a^0_{i,j}) r_j < (a^0_{1,1} a^0_{i,i} - a^0_{i,1} a^0_{1,i}) r_i$$

which, on account of (1.5.8), may be written as

$$\sum_{\substack{j=1 \\ j \neq i}}^{N} a^1_{i,j} r_{j+1} < a^1_{i,i} r_{i+1}, \qquad i = 1, \ldots, N-1. \tag{1.5.11}$$

The definition of $a^1_{i,j}$ for $i \neq j$ and inequalities (1.5.11) imply that $a^1_{i,i} > 0$ for all $i \in \{1, \ldots, N-1\}$. Applying the induction hypothesis to (1.5.11) we get $a^k_{i,i} > 0$ for all $i \in \{1, \ldots, N-k\}$ and $k \in \{1, \ldots, N-1\}$. For $k = 0$ inequalities (1.5.9) result directly from (1.5.10).

To prove the converse, observe that, by the induction hypothesis, there exist positive numbers $r_2, \ldots, r_N$ satisfying (1.5.11) and it suffices to take

$$r_1 := \sum_{j=2}^{N} a^0_{1,j} r_j + \varepsilon$$

with $\varepsilon > 0$ small enough in order that positive numbers $r_1, \ldots, r_N$ satisfy system (1.5.10). ∎

Now, we are in a position to present the following.

Proof of Theorem 1.5.4. Take an $x_0 = (x_{0,1}, \ldots, x_{0,N}) \in X$ and put $x_n = (x_{n,1}, \ldots, x_{n,N}) := f^n(x_0)$, $n \in \mathbb{N}_0$. In virtue of Lemma 1.5.1, we may find positive numbers $r_1, \ldots, r_N$ such that system (1.5.10) is satisfied and by means of the positive homogeneity of the system, without loss of generality, we may assume that

$$r_i \geq \rho_i(x_{1,i}, x_{0,1}) \qquad \text{for all } i \in \{1, \ldots, N\}. \tag{1.5.12}$$

In view of the definition of the $a^0_{i,j}$ and of the sharp inequality in (1.5.10) one can find an $s \in (0, 1)$ such that

$$\sum_{j=1}^{N} s_{i,j} r_j \leq s r_i, \qquad i \in \{1, \ldots, N\}. \tag{1.5.13}$$

Note that $\rho_i(x_{n+1,i}, x_{n,i}) \leq s^n r_i$ for all $i \in \{1, \ldots, N\}$ and all $n \in \mathbb{N}_0$; in fact, for $n = 0$ this is just (1.5.12) and assuming the validity of these estimates for an $n \in \mathbb{N}_0$ and all $i \in \{1, \ldots, N\}$ we get by (1.5.7)

$$\rho_i(x_{n+2,i}, x_{n+1,i}) \leq \sum_{j=1}^{N} s_{i,j} \rho_j(x_{n+1,j}, x_{n,j}) \leq \sum_{j=1}^{N} s_{i,j} s^n r_j \leq s^{n+1} r_i,$$

$i \in \{1, \ldots, N\}$, in view of (1.5.13). Consequently, for each $i \in \{1, \ldots, N\}$ the sequence $(x_{n,i})_{n \in \mathbb{N}}$ satisfies the Cauchy condition. Putting $\xi_i := \lim_{n \to \infty} x_{n,i}$, $i \in \{1, \ldots, N\}$ and $\xi := (\xi_1, \ldots, \xi_N)$ we have $\xi = \lim_{n \to \infty} f^{n+1}(x_0) = f(\lim_{n \to \infty} f^n(x_0)) = f(\xi)$ because f is a continuous map.

If we had another fixed point $\eta = (\eta_1, \ldots, \eta_N)$ then, along the same lines, we would get $\rho_i(\xi_i, \eta_i) \leq s^n r_i$ for $i \in \{1, \ldots, N\}$, $n \in \mathbb{N}_0$, which on letting n tend to infinity yields $\xi_i = \eta_i$ for $i \in \{1, \ldots, N\}$, i.e. $\xi = \eta$. ∎

Remark 1.5.1. In the case $N = 2$ Theorem 1.5.4 was proved by Păvăloiu [1] and Rus [1] by using different methods. See also Jarczyk [2].

1.5C. Equivalence statement

In order to show the announced equivalence between Theorems 1.5.4 and 1.5.5 we have to prove two Lemmas.

Lemma 1.5.2. Let $(s_{i,j})_{i,j=1,\ldots,N}$ be a real matrix with nonnegative entries and suppose that (1.5.9) holds for the $a_{i,i}^k$ defined by (1.5.8). Then any solution $(r_1, \ldots, r_N)$ of the system

$$\sum_{j=1}^{N} s_{i,j} r_j \geq r_i, \qquad i \in \{1, \ldots, N\}$$

is nonpositive.

Proof. Lemma 1.5.1 guarantees the existence of positive solution $(R_1, \ldots, R_N)$ of system (1.5.10) and henceforth

$$\sum_{j=1}^{N} s_{i,j} R_j \leq s R_i, \qquad i \in \{1, \ldots, N\},$$

with some $s \in (0, 1)$. Making use of the homogeneous character of this system we may assume that $r_i \leq R_i$ for all $i \in \{1, \ldots, N\}$. Applying the induction procedure used in the proof of Theorem 1.5.4 we obtain easily that

$$r_i \leq s^n R_i \qquad \text{for all } i \in \{1, \ldots, N\} \text{ and all } n \in \mathbb{N}_0$$

and letting n tend to infinity we get $r_i \leq 0$, $i \in \{1, \ldots, N\}$, as asserted. ∎

With the use of Lemma 1.5.2 we are able to prove the previously mentioned result.

Lemma 1.5.3. *Let $S := (s_{i,j})_{i,j=1,...,N}$ be a real matrix with positive entries and suppose that the $a_{i,j}^k$ are defined by* (1.5.8). *Then* (1.5.9) *holds if and only if all the characteristic roots of the matrix S are less than unity in absolute value.*

Proof. Sufficiency. Let $\Lambda = \{\lambda_1, \ldots, \lambda_N\}$ be the set of all the characteristic roots of the matrix S. By a theorem of Perron and Frobenius (see Gantmacher [1], for instance) the number $s := \max\{|\lambda_i| : i = 1, \ldots, N\}$ belongs to Λ and the corresponding eigenvector $(r_1, \ldots, r_N)$ is positive. Therefore, the positive numbers $r_1, \ldots, r_N$ satisfy the system

$$\sum_{j=1}^{N} s_{i,j} r_j = s r_i < r_i, \qquad i \in \{1, \ldots, N\},$$

which when translated into the language of $a_{i,j}^k$ reads as (1.5.10) whence (1.5.9) results by means of Lemma 1.5.1.

Necessity. Suppose, for the indirect proof, that $s \geq 1$. Again by the Perron–Frobenius Theorem, $s \in \Lambda$ and the corresponding eigenvector $(r_1, \ldots, r_N)$ satisfies the system

$$\sum_{j=1}^{N} s_{i,j} r_j = s r_i \geq r_i > 0, \qquad i \in \{1, \ldots, N\};$$

this, however, is incompatible with Lemma 1.5.2. ∎

So, as claimed before, Theorems 1.5.4 and 1.5.5 are equivalent which was undoubtedly worth realizing. The practical advantages of Theorem 1.5.4 were enumerated in detail after the statement of Theorem 1.5.5 above. On the other hand, Theorem 1.5.5 throws some light onto the geometric meaning of the analytically involved relations (1.5.9).

1.6 Continuous dependence

A natural question arises whether a convergent sequence of transformations having exactly one fixed point each has the property that the unique fixed point of the limit transformation is just the limit of the sequence of fixed points obtained. In case of positive answer to that question we shall say that the (unique) fixed point of a given transformation depends on it in a continuous manner. Such nice behaviour is, in general, desired from the point of view of applications, especially in the fixed-point theory approach to the existence problems of iterative functional equations. In the language of solutions of functional equations of various types we obtain immediately, in such a case, the continuous dependence of solutions on given functions occurring in the equation.

To make the thing clearer, we shall illustrate this idea by considering the 'continuous dependence' question in connection with Theorems 1.5.1 and 1.5.4. Theorem 1.6.4 below covers the case where the metric space considered is compact. We start with a result of K. Baron–J. Matkowski [2] which reads as follows.

Theorem 1.6.1. *Let* $(f_m)_{m \in \mathbb{N}_0}$ *be a sequence of self-mappings of a complete metric space* (X, ρ) *such that*

$$\rho(f_m(x), f_m(y)) < \rho(x, y) \quad \text{for all } x, y \in X, \ x \neq y, \text{ and all } m \in \mathbb{N}_0.$$

Suppose that for every $\varepsilon > 0$ *there exists a* $\delta > 0$ *such that for all* $m \in \mathbb{N}_0$ *and all* $x, y \in X$

$$\varepsilon < \rho(x, y) < \varepsilon + \delta \quad \text{implies} \quad \rho(f_m(x), f_m(y)) \leqq \varepsilon.$$

If f_0 *is the pointwise limit of the sequence* $(f_m)_{m \in \mathbb{N}}$ *and* ξ_m *is the unique fixed point of* f_m, $m \in \mathbb{N}_0$, *then* $\xi_m \to \xi_0$ *as* $m \to \infty$.

Proof. The existence and uniqueness of ξ_m, $m \in \mathbb{N}_0$, result from Theorem 1.5.1. Fix an $\varepsilon > 0$; obviously, we may assume that the corresponding $\delta = \delta(\varepsilon)$ is less than ε. Since $f_m(\xi_0) \to f_0(\xi_0) = \xi_0$ as $m \to \infty$, we have $\rho(f_m(\xi_0), \xi_0) < \delta$ for sufficiently large $m \in \mathbb{N}$, say for $m \geqq M$. By a method used in the proof of Theorem 1.5.1 we check that, for each $m \geqq M$, the mapping f_m transforms the set $B := \{x \in X : \rho(x, \xi_0) < \varepsilon + \delta(\varepsilon)\}$ into itself. Hence, taking arbitrary $x \in B$ and an $m \geqq M$, we get $f_m^n(x) \in B$ for all $n \in \mathbb{N}_0$, which implies that $\xi_m = \lim_{n \to \infty} f_m^n(x)$ belongs to the closure of B. Thus, $\rho(\xi_m, \xi_0) \leqq \varepsilon + \delta(\varepsilon) < 2\varepsilon$ for all $m \geqq M$. ∎

Inspired by the same idea with regard to Theorem 1.5.4, K. Baron [6] has proved the following.

Theorem 1.6.2. *Let* (X_i, ρ_i), $i = 1, \dots, N$, *be complete metric spaces and let* $X := X_1 \times \cdots \times X_N$. *Consider a sequence* $(f_m)_{m \in \mathbb{N}_0}$ *of self-mapping* $f_m = (f_{m,1}, \dots, f_{m,N})$ *of* X *such that*

$$\rho_i(f_{m,i}(x_1, \dots, x_N), f_{m,i}(y_1, \dots, y_N)) \leqq \sum_{j=1}^{N} s_{i,j} \rho_j(x_j, y_j)$$

for all $x_j, y_j \in X_j$ *and some* $s_{i,j}$ *independent of* m, $i, j \in \{1, \dots, N\}$, $m \in \mathbb{N}_0$. *Suppose further that relation* (1.5.9) *holds with the* $a_{i,j}^k$ *defined by* (1.5.8). *If* f_0 *is a pointwise limit of the sequence* $(f_m)_{m \in \mathbb{N}}$ *and* ξ_m *is the unique fixed point of* f_m, $m \in \mathbb{N}_0$, *then* $\xi_m \to \xi_0$ *as* $m \to \infty$.

Proof. The existence and uniqueness of $\xi_m = (\xi_{m,1}, \dots, \xi_{m,N})$, $m \in \mathbb{N}_0$, result from Theorem 1.5.4. Take an $x_0 = (x_{0,1}, \dots, x_{0,N}) \in X$ and put $x_m^n = (x_{m,1}^n, \dots, x_{m,N}^n) := f_m^n(x_0)$, $m, n \in \mathbb{N}_0$. Theorem 1.5.4 ensures that $x_{m,i}^n \to \xi_{m,i}$ as $n \to \infty$, for all $i \in \{1, \dots, N\}$ and all $m \in \mathbb{N}_0$.

Since the convergent sequences $(x^1_{m,i})_{m \in \mathbb{N}}$, $(x^0_{m,i})_{m \in \mathbb{N}}$ are bounded for all $i \in \{1, \ldots, N\}$, with the help of Lemma 1.5.1, one may find positive numbers $r_1, \ldots, r_N$ and $s \in (0, 1)$ such that $\rho_i(x^1_{m,i}, x^0_{m,i}) \leq r_i$, $i \in \{1, \ldots, N\}$, $m \in \mathbb{N}_0$, and (1.5.13) is satisfied. With the aid of the induction procedure used several times in the preceding section we obtain the relations

$$\rho_i(x^{n+1}_{m,i}, x^n_{m,i}) \leq s^n r_i, \qquad i \in \{1, \ldots, N\}, \; m, n \in \mathbb{N}_0.$$

Consequently

$$\rho_i(x^0_{m,i}, x^{n+1}_{m,i}) \leq \sum_{j=0}^{n} \rho_i(x^j_{m,i}, x^{j+1}_{m,i}) \leq \sum_{j=0}^{n} s^j r_i < \frac{1}{1-s} r_i$$

for all $i \in \{1, \ldots, N\}$ and letting n tend to infinity one obtains $\rho_i(x^0_{m,i}, \xi_{m,i}) \leq (1-s)^{-1} r_i$, $i \in \{1, \ldots, N\}$. Now

$$u_{m,i} := \rho_i(\xi_{m,i}, \xi_{0,i}) \leq \rho_i(\xi_{m,i}, x^0_{m,i}) + \rho_i(x^0_{m,i}, \xi_{0,i}) \leq \frac{1}{1-s} r_i + c_i$$

for some c_i independent of m because $x^0_{m,i} = x_{0,i}$ for all $m \in \mathbb{N}_0$ and $i \in \{1, \ldots, N\}$. Thus, all the sequences $(u_{m,i})_{m \in \mathbb{N}_0}$, $i \in \{1, \ldots, N\}$, are bounded. Therefore, we may assume (increasing r_i, if necessary) that $u_{m,i} \leq r_i$ for all $m \in \mathbb{N}_0$ and all $i \in \{1, \ldots, N\}$. Put

$$v^1_{m,i} := \rho_i(f_{m,i}(\xi_0), \xi_{0,i}) \qquad \text{and} \qquad v^{n+1}_{m,i} := \sum_{j=1}^{N} s_{i,j} v^n_{m,j} + v^1_{m,i}$$

for $i \in \{1, \ldots, N\}$ and $m, n \in \mathbb{N}$. Evidently, $v^1_{m,i} \to 0$ as $m \to \infty$, $i = 1, \ldots, N$ because of the convergence of $(f_m(\xi_0))_{m \in \mathbb{N}}$ to $f(\xi_0) = \xi_0$. An instantaneous induction gives also the convergence of $(v^n_{m,i})_{m \in \mathbb{N}}$ to zero for all $i \in \{1, \ldots, N\}$ and all $n \in \mathbb{N}$. On the other hand,

$$u_{m,i} = \rho_i(\xi_{m,i}, \xi_{0,i}) = \rho_i(f_{m,i}(\xi_m), \xi_{0,i}) \leq \rho_i(f_{m,i}(\xi_m), f_{m,i}(\xi_0)) + \rho_i(f_{m,i}(\xi_0), \xi_{0,i})$$

$$\leq \sum_{j=1}^{N} s_{i,j} \rho_i(\xi_{m,i}, \xi_{0,i}) + v^1_{m,i} = \sum_{j=1}^{N} s_{i,j} u_{m,i} + v^1_{m,i}$$

and a simple induction shows that

$$u_{m,i} \leq s^n r_i + v^n_{m,i}$$

for all $i \in \{1, \ldots, N\}$ and all $m, n \in \mathbb{N}$. Letting here m tend to infinity we get

$$0 \leq \limsup_{m \to \infty} u_{m,i} \leq s^n r_i, \qquad i \in \{1, \ldots, N\}, \; n \in \mathbb{N}.$$

Thus on letting $n \to \infty$ we arrive at $\lim_{m \to \infty} u_{m,i} = 0$ for all $i \in \{1, \ldots, N\}$, which proves that $\xi_m \to \xi_0$ as $m \to \infty$. ∎

In particular, for $N = 1$, we get a 'continuous dependence' type result in the case of Banach's Theorem. This noteworthy result has been known much earlier (see e.g. Dugundji–Granas [1, Exercise (6.4), p. 17]).

Theorem 1.6.3. *Let* $(f_m)_{m \in \mathbb{N}_0}$ *be a sequence of self-mappings of a complete metric space* (X, ρ). *Suppose that there exists a* $\Theta \in (0, 1)$ *such that*

$$\rho(f_m(x), f_m(y)) \leq \Theta \rho(x, y) \qquad \text{for all } x, y \in X \text{ and all } m \in \mathbb{N}_0.$$

If f_0 *is a pointwise limit of the sequence* $(f_m)_{m \in \mathbb{N}}$ *and* ξ_m *is the unique fixed point of* f_m, $m \in \mathbb{N}_0$, *then* $\xi_m \to \xi_0$ *as* $m \to \infty$.

We end this short and definitely incomplete survey with the following simple but useful theorem (see Matkowski [4]).

Theorem 1.6.4. *Let* $(f_m)_{m \in \mathbb{N}_0}$ *be a sequence of continuous self-mappings of a compact metric space* (X, ρ). *If* f_0 *is a uniform limit of the sequence* $(f_m)_{m \in \mathbb{N}}$ *and if for every* $m \in \mathbb{N}_0$ *the mapping* f_m *has exactly one fixed point* ξ_m, *then* $\xi_m \to \xi_0$ *as* $m \to \infty$.

Proof. Suppose, for the indirect proof, that there exists a sub-sequence $(\xi_{m_k})_{k \in \mathbb{N}}$ of the sequence $(\xi_m)_{m \in \mathbb{N}}$ such that $\xi_{m_k} \to \xi \neq \xi_0$ as $k \to \infty$. Then

$$\xi = \lim_{k \to \infty} \xi_{m_k} = \lim_{k \to \infty} f_{m_k}(\xi_{m_k}) = f_0(\xi)$$

which contradicts the uniqueness of ξ_0. ∎

1.7 Notes

1.7.1. It is possible to introduce functions in an axiomatic way with the notion of composition brought out in strong relief; see Menger [1], Schweizer–Sklar [1], [2], and also Schweizer–Sklar [3], Sklar [1], Penner–Schroeder [1] and Baillieul [1] for further developments and related problems.

1.7.2. Most of this chapter is devoted to properties of iterative sequences, i.e. we have confined ourselves to the case where the iteration indices were taken from the additive semigroup $(\mathbb{N}_0, +)$ of all nonnegative integers. As we have pointed out at the beginning of Section 1.1, this is very natural and practically the only procedure allowing us to avoid any additional structure. On the other hand, there is a pretty rich mathematical literature concerning the case where iteration indices are taken from $(\mathbb{R}^+, +)$, $(\mathbb{R}, +)$, $(\mathbb{C}, +)$ or even from an abstract commutative semigroup $(S, +)$ with a unit element e. In such a case, instead of iterates of a given mapping $f: X \to X$, one considers a function family $\{f^s : s \in S\}$ (of iterations) for which the equalities

$$f^s \circ f^t = f^{s+t}, \qquad s, t \in S \tag{1.7.1}$$

and

$$f^e = f \tag{1.7.2}$$

hold true.

By setting $g(x, s) := f^s(x)$, $(x, s) \in X \times S$, relation (1.7.1) becomes nothing else but the celebrated *translation equation*

$$g(g(x, t), s) = g(x, s + t) \qquad (1.7.3)$$

which is undoubtedly one of the most important functional equations in several variables (see Aczél [2]). Equation (1.7.3) occurs naturally in the theory of Lie groups or in the theory of semigroups of operators, for instance. Under some slightly restrictive assumptions the solution of (1.7.3) has the form

$$g(x, s) = \varphi^{-1}(s + \varphi(x)), \qquad (x, s) \in X \times S,$$

where φ is a bijection of X onto S. Then taking $f^s := g(\cdot, s)$, $s \in S$, one has to solve Abel's equation

$$\varphi(f(x)) = e + \varphi(x), \qquad x \in X,$$

in order to get the embedding property (1.7.2).

Solutions of (1.7.3) form the so-called *flows* playing crucial roles in the modern theory of dynamical systems, among others.

Definitely, the most important case is $S = \mathbb{R}^+$. Nevertheless, we may take positive rationals as S, for instance (*rational flows*, see Tabor [1], Weitkämper [1]) and we are very close to the problem of finding iterative roots of a given function which we investigate in Chapter 11. There are some attempts to consider (1.7.1) on a restricted domain of admissible indices; for example, instead of assuming (1.7.1) for all pairs (s, t) from the first quadrant, we assume that (1.7.1) is satisfied only for (s, t) lying on a certain curve C (see Zdun [14] for the case $C = \{(s, t): s = t\}$ and Sablik [4] for some other cases).

The scope of our book prevents us from presenting even references concerning at least substantial results on the problem of existence and uniqueness of iteration flows; the reader interested in these is referred to Targoński [8] to look at an outline of the theory.

1.7.3. We would like to mention here only the article of M. C. Zdun [13] where an exhaustive characterization of *continuous* and *differentiable semigroups* (flows) $\{f^s: s \in (0, \infty)\}$ of a self-mapping f of the real interval X is given. One of the most elegant results contained therein states that any *Lebesgue measurable iteration semigroup* (i.e. such that the mapping φ_x sending t into $f^t(x)$ is measurable for each $x \in X$) has necessarily to be continuous (i.e. φ_x is continuous for each $x \in X$); see also Targoński–Zdun [1], [2].

Of course, in the case of invertible functions a natural extension to *iteration groups* is possible. Recently, M. C. Zdun [17] characterized the iteration groups of homeomorphisms of the unit circle (which, however, may easily be replaced by any Jordan curve on the plane).

1.7.4. The case where the semigroup $(S, +)$ of iteration indices is noncommutative seems to be artificial at first glance but, surprisingly, it finds a nontrivial application in the theory of automata (see Moszner [4]). For a brief survey of some achievements of Z. Moszner and his co-workers in a pretty abstract theory of equation (1.7.1), the reader is referred to Moszner [5]. Finally, iteration semigroups in the class of multivalued mappings are widely investigated in the recently published dissertation of A. Smajdor [9].

1.7.5. The theorem of Barna is evidently related to the celebrated Poincaré–Bendixson Theorem (see Hirsch–Smale [1]) which says that a nonempty compact limit set of a C^1 planar dynamical system which contains no equilibrium point is a closed orbit. However, neither theorem can be derived from the other. Theorem 1.1.2 extends the theorem of Barna to locally compact Hausdorff topological space satisfying the first countability axiom. The assumption of local compactness of the given space X is essential as may be seen from the following example. Take $X = l^\infty$ – the space of all bounded sequences with the usual supremum norm and define $f: l^\infty \to l^\infty$ (with $\alpha := (\alpha_n)_{n \in \mathbb{N}}$) by

$$f(\alpha) := \begin{cases} (0, \alpha_1, \alpha_2, \ldots) & \text{if } \|\alpha\| \leq \tfrac{1}{2}, \\ ((2 - \alpha_r)2^{-r}, \alpha_{r+1}, \ldots) & \text{if } \|\alpha\| \in (\tfrac{1}{2}, 1], \\ (2 - \alpha_1, \alpha_2, \ldots) & \text{otherwise}; \end{cases}$$

here, α_r is the first nonzero term in the sequence α. One may show (Kuczma [49]) that for the point $x := (\tfrac{3}{4}, 0, 0, \ldots)$ we have

$$\{f^n(x): n \in \mathbb{N}_0\} = \{(3 \cdot 2^{-q-1}\delta_n^r)_{n \in \mathbb{N}} : r, q \in \mathbb{N}\} \cup \{((2 - 3 \cdot 2^{-q-2})\delta_n^1)_{n \in \mathbb{N}} : q \in \mathbb{N}\};$$

δ_n^r stands here for the usual Kronecker symbol. The only limit points of the splinter $(f^n(x))_{n \in \mathbb{N}}$ are $p_1 := (0, 0, \ldots)$ and $p_2 := (2, 0, 0, \ldots)$ but these two points do not form a cycle. Our function f is continuous on the set $L_f(x) = \{p_1, p_2\}$ but it is not continuous on the whole of l^∞. The question arises to find a counter-example to the generalized Barna Theorem 1.1.2 in the case where X is an infinite-dimensional Banach space with continuous function $f: X \to X$. This question, among others, is partially answered in Graw [1]: if X admits no points with compact neighbourhood then the generalized Barna theorem is no longer valid.

1.7.6. Fixed points x_0 fulfilling the condition

$$\rho(f(x), x_0) \geq \Theta \rho(x, x_0), \qquad \Theta > 1,$$

for all x from some neighbourhood of x_0 (see Remark 1.1.1) are sometimes called *strongly repulsive* (see Kuczma [26, p. 18], Targoński [8]).

A fixed point x_0 is said to be *repulsive* iff there exists a neighbourhood U of x_0 such that the splinter $(f^n(x))_{n \in \mathbb{N}}$ does not converge to x_0 for any $x \in U$ outside the orbit of x_0 under f. However, this definition may be inconvenient since, for instance, for any constant mapping with x_0 as the only value, the point x_0 is both attractive and repulsive because the entire space X coincides with the orbit of x_0. Although various attempts have been made to classify fixed points (Barna [1], Šarkovskiĭ [3]), so far there exists no quite satisfactory classification, nor even a satisfactory definition of a repulsive fixed point. Anyhow, the most important fixed points are the attractive ones and so we have confined our attention to them.

1.7.7. In Theorem 1.4.2 the condition that $m(x) < x$ for all $x \in X \setminus \{0\}$ may be replaced by

$$\Phi(x|s) > e^{-sx}, \qquad x \in X \setminus \{0\},$$

where $\Phi(x|\cdot)$ is the Laplace transform of the random variable $f(x)$, i.e.

$$\Phi(x|s) = \int_X e^{-st} d_t \mathcal{F}(x|t), \qquad s \geq 0.$$

The proof is analogous (see Baron–Kuczma [1]). The two conditions are independent of each other while dealing with sharp inequalities. However, the relation $\Phi(x|s) \geq e^{-sx}$ forces m to satisfy the inequality $m(x) \leq x$, $x \in X$, since

$$-m(x) = \frac{d}{ds} \Phi(x|s) \bigg|_{s=0} = \lim_{s \to 0} \frac{1}{s} (\Phi(x|s) - 1) \geq \lim_{s \to 0} \frac{1}{s} (e^{-sx} - 1) = -x.$$

1.7.8. Matkowski's fixed-point Theorem 1.5.4 serves well while considering systems of functional equations and that was, of course, the original motivation for proving it. This is also the reason we have devoted so much attention to it. The versatile benefits it gives are well illustrated by the exhaustive treatment of integrable solutions of iterative functional equations due to J. Matkowski [13] himself (see Sections 4.7 and 5.8) and also in the works by K. Baron [9] (implicitly), K. Baron–R. Ger [1], K. Baron–R. Ger–J. Matkowski [1], S. Czerwik [18], Z. Kominek [3], R. Węgrzyk [3] (see also Choczewski [7]). Another version of this theorem has been proved by S. Czerwik [18]. This is the following.

Proposition 1.7.1. *Let $N \in \mathbb{N}$ be fixed. Suppose that assumptions (1.5.6) are satisfied and that there exists a positive solution $(r_1, \ldots, r_N)$ of the system*

$$\sum_{i=1}^{N} s_{i,j} r_i < r_j, \qquad j \in \{1, \ldots, N\}. \tag{1.7.4}$$

Then there exists exactly one fixed point $\xi \in X := X_1 \times \cdots \times X_N$ of the transformation $f = (f_1, \ldots, f_N): X \to X$; moreover, the domain of attraction of ξ coincides with the whole of X.

To see this, write

$$s := \max \left\{ \frac{1}{r_j} \sum_{i=1}^{N} s_{i,j} r_i : j = 1, \ldots, N \right\};$$

then $s \in [0, 1)$ because of (1.7.4) and, consequently, the relations

$$\sum_{i=1}^{N} s_{i,j} r_i \leqq s r_j, \qquad j \in \{1, \dots, N\},$$

hold true. The product space (X, ρ) with the metric ρ defined by the formula

$$\rho((x_1, \dots, x_N), (y_1, \dots, y_N)) := \sum_{i=1}^{N} r_i \rho_i(x_i, y_i)$$

for all $x = (x_1, \dots, x_N)$ and $y = (y_1, \dots, y_N)$ from X is a complete metric space. It remains to check that $\rho(f(x), f(y)) < s\rho(x, y)$.

1.7.9. Theorem 1.5.4 and Proposition 1.7.1 are comparable whenever the matrix $(s_{i,j})_{i,j=1,\dots,N}$ is symmetric. In the light of Lemma 1.5.1, both of these results state essentially the same thing. But again, we must underline that the assumption of the existence of a positive solution of system (1.7.4) is expressed in Matkowski's theorem directly and effectively in the language of the $s_{i,j}$. This fact is of an unquestionable value.

1.7.10. Among various versions of a converse to Banach's Theorem one has, in particular, the following result of P. R. Meyers [1].

Proposition 1.7.2. *Let X be a metrizable topological space and let f be a continuous self-mapping of X. Suppose f possesses a fixed point whose domain of attraction coincides with the whole of X and that there exists a neighbourhood U of x such that for any neighbourhood V of x there is an $n(V) > 0$ such that $f^n \subset V$ for all $n \geqq n(V)$. Then, for each $\Theta \in (0, 1)$ there exists a metric ρ_Θ on X, complete if X admits a complete metric, such that f is a ρ_Θ-contraction with contraction constant Θ.*

Therefore, to settle the question whether a given fixed-point theorem is subtle enough we must look at its practical usefulness and not compare with Banach's Theorem since, generally, we may predict that we deal with a contraction with respect to an appropriate metric. A 'trifling' detail remains: to know the metric explicitly.

2

Linear equations and branching processes

2.0 Introduction

We start the investigation of iterative functional equations with the linear equation

$$\varphi(f(x)) = g(x)\varphi(x) + h(x),$$

where φ is a real-valued unknown function. Among the simplest equations of this form are those of Schröder,

$$\sigma(f(x)) = s\sigma(x), \tag{2.0.1}$$

and Abel,

$$\alpha(f(x)) = \alpha(x) + 1. \tag{2.0.2}$$

These two and also some other linear equations appear in a natural way in the theory of branching processes, in which one extensively uses probability generating functions. The theory of iteration of real functions yields limit theorems for such processes. The theory of functional equations becomes a useful tool to establish generating functions of limit distributions as unique solutions of equations like (2.0.1) and (2.0.2), as well as to determine invariant measures for branching processes.

Since probability generating functions are (by their definition: power series with nonnegative coefficients) absolutely monotonic, results on monotonic or convex solutions of functional equations are particularly significant for the theory of branching processes. The class of regularly varying functions, in Karamata's sense, is also useful there.

In the first section of this chapter we introduce a simple Galton–Watson process without and with immigration and show the role of linear iterative functional equations in its theory. In Sections 2.2, 2.3, 2.4 and 2.7 we present results concerning solutions with a nonnegative pth difference,

$$\Delta_y^0 \varphi(x) := \varphi(x),$$
$$\Delta_y^{p+1} \varphi(x) := \Delta_y^p \varphi(x+y) - \Delta_y^p \varphi(x), \qquad p \in \mathbb{N}_0,$$

which amounts to nonnegativity (for $p = 0$), monotonicity (for $p = 1$) and convexity (for $p = 2$, provided that φ is measurable). Such conditions often ensure the uniqueness of solutions. Regularly varying functions (and solutions) are dealt with in Section 2.5. Having collected the results we use some of them for solving problems we have described in the first section. Applications to characterization of polynomials are discussed in Section 2.7. The concluding section (Notes) contains comments, complements, examples and some bibliographical notes.

2.1 Galton–Watson processes

A typical model of a branching process is the following one.

An object can give rise to X offspring of the same type, where X is a random variable (RV) with the probability distribution

$$P(X = j) = p_j, \qquad j \in \mathbb{N}_0,$$

where $0 \leqslant p_j < 1, j \in \mathbb{N}_0, \sum_{j=0}^{\infty} p_j = 1$. There are many possible interpretations: the objects can be biological creatures giving birth to their progeny, can be electrons in electron multipliers, can be particles occurring in (chemical or nuclear) chain reactions, etc. (see Harris [1], Kuczma [36]).

To get a Galton–Watson process (without immigration) assume that $Z_0 = 1$ (the process descends from a single ancestor) and denote by Z_n the size of the nth generation. The *simple Galton–Watson process* is then the Markov chain

$$Z_n = OFZ_{n-1} = X_1 + \cdots + X_{Z_{n-1}}, \qquad n \in \mathbb{N}, Z_0 = 1,$$

where OFZ_{n-1} is the number of offspring of the objects forming the $(n-1)$st generation. The RVs X_i are assumed to be mutually independent and identically distributed as X.

2.1A. Probability generating functions

Let Z be an RV with the probability distribution $P(Z = j) = a_j, j \in \mathbb{N}_0$. The *probability generating function (PGF)* of Z is defined by

$$A(t) = \sum_{j=0}^{\infty} a_j t^j.$$

The power series above converges at least in $[-1, 1]$, as we have $A(1) = \sum_{j=0}^{\infty} a_j = 1$. The PGF A is related to the Laplace transform Φ of Z by the formula $\Phi(t) = A(e^{-t})$, $t \geqslant 0$.

Suppose that W is an RV independent of Z, with the PGF

$$B(t) = \sum_{j=0}^{\infty} b_j t^j.$$

Let $P(Z + W = j) = c_j, j \in \mathbb{N}_0$. Thus

$$\sum_{j=0}^{\infty} c_j t^j = \sum_{j=0}^{\infty} \left(\sum_{i=0}^{j} a_i b_{j-1} \right) t^j = A(t) B(t).$$

We see that the PGF of the sum of independent RVs is the product of their PGFs.

Let the $(n-1)$st generation consist of the objects $x_1, \ldots, x_{Z_{n-1}}$. Each object x_k gives rise to X_k offspring and X_k are independent RVs identically distributed as X. Denote the PGF of X by

$$F(t) = \sum_{j=0}^{\infty} p_j t^j. \tag{2.1.1}$$

Then the transition probabilities for the Markov chain $(Z_n)_{n \in \mathbb{N}_0}$ are

$$p_{ij} = P(Z_n = j \mid Z_{n-1} = i) = P(X_1 + \cdots + X_i = j)$$

and the PGF of the sum of i independent RVs having the same PGF F equals

$$\sum_{j=0}^{\infty} p_{ij} t^j = (F(t))^i. \tag{2.1.2}$$

Moreover, the PGF F_n of Z_n is given for $n \geqslant 2$ by

$$F_n(t) = \sum_{j=0}^{\infty} P(Z_n = j) t^j = \sum_{j=0}^{\infty} \left[\sum_{i=0}^{\infty} P(Z_n = j \mid Z_{n-1} = i) P(Z_{n-1} = i) \right] t^j$$

$$= \sum_{i=0}^{\infty} P(Z_{n-1} = i) \sum_{j=0}^{\infty} p_{ij} t^j = \sum_{i=0}^{\infty} P(Z_{n-1} = i)(F(t))^i$$

$$= F_{n-1}(F(t)),$$

whence for $n \in \mathbb{N}$ (note that $F_1 = F$ as $Z_1 = X$)

$$F_n(t) = F^n(t) = \sum_{j=0}^{\infty} p_j^n t^j, \qquad p_j^n := P(Z_n = j), \tag{2.1.3}$$

where F^n is the nth functional iterate of the PGF F. The probability distribution of Z_n can be thus determined from (2.1.3).

Note that the function F, as a PGF, is defined, continuous, strictly increasing and convex in $[0, 1]$, and even analytic in $[0, 1)$. Moreover, the expectation (mean) of X equals

$$m := E[X] = \sum_{j=1}^{\infty} j p_j = F'(1)$$

(m may be infinite). Thus the graph of F may look as shown in Fig. 2.1 (provided $p_0 > 0$). Here we have denoted by q the smallest fixed point of F in the interval $[0, 1]$. Thus we have $0 < q \leqslant 1$ if $p_0 > 0$ and, according to

Fig. 2.1

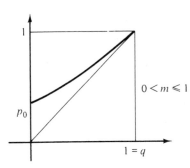 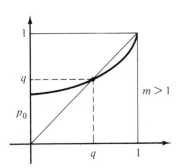

Theorem 1.1.1,

$$\lim_{n \to \infty} F^n(t) = q \qquad \text{for } t \in [0, q], \tag{2.1.4}$$

as the sequence $(F^n(t))_{n \in \mathbb{N}}$ is monotonic and bounded for every $t \in [0, q]$. The following argument shows that the q here is just the probability of extinction in the process (Z_n).

The state $j = 0$ of the process (Z_n) is an absorbing one, i.e., if $Z_n = 0$, then $Z_{n+k} = 0$ for $k \in \mathbb{N}$. Thus the sequences of events '$Z_n = 0$' is increasing and the probability of extinction equals

$$P\left(\bigcup_{n=1}^{\infty} (Z_n = 0) \right) = \lim_{n \to \infty} P(Z_n = 0) = \lim_{n \to \infty} F^n(0) = q,$$

by (2.1.4). Further examination of the Markov chain $(Z_n)_{n \in \mathbb{N}}$ shows that it is unstable, i.e. $\lim_{n \to \infty} P(Z_n = i) = 0$ for $i \in \mathbb{N}$, while $P(\lim_{n \to \infty} Z_n = 0) = q$ (note that (Z_n) is integer valued) and $P(\lim_{n \to \infty} Z_n = \infty) = 1 - q$. Therefore to estimate the size of the nth generation Z_n we should rather turn to conditional limit probabilities for the process $(Z_n)_{n \in \mathbb{N}}$.

2.1B. Limit distributions

This is the first kind of problem which leads to functional equations. Let T be the time of extinction:

$$Z_T = 0 \qquad \text{and} \qquad Z_n \neq 0 \qquad \text{for } n < T.$$

We aim at proving that the conditional limit probabilities

$$a_j := \lim_{n \to \infty} P(Z_n = j \mid n < T < \infty)$$

can be determined through the Schröder equation (2.0.1). To this end we first find, using the Bayes formula, the probabilites

$$P(Z_n = j \mid n < T < \infty) = \frac{P(n < T < \infty \mid Z_n = j) P(Z_n = j)}{\sum_{k=1}^{\infty} P(n < T < \infty \mid Z_n = k) P(Z_n = k)} = \frac{q^j p_j^n}{\sum_{k=1}^{\infty} q^k p_k^n}.$$

Indeed, if $Z_n = j$, then the event '$n < T < \infty$' means extinction in all the j separate Galton–Watson processes starting from single ancestors–members of the nth generation.

Let us form the corresponding PGF:

$$A_n(t) := \sum_{j=1}^{\infty} P(Z_n = j \mid n < T < \infty) t^j = \frac{\sum_{j=0}^{\infty} p_j^n q^j t^j - p_0^n}{\sum_{k=0}^{\infty} p_k^n q^k - p_0^n}$$

$$= \frac{F^n(qt) - F^n(0)}{F^n(q) - F^n(0)} = 1 - \frac{q - F^n(qt)}{q - F^n(0)};$$

see (2.1.3) and note that $F^n(q) = q$. Substituting $x = q(1 - t)$, $f(x) = q - F(q - x)$, we get

$$A_n(t) = 1 - \frac{f^n(x)}{f^n(q)}, \qquad x \in [0, q]. \tag{2.1.5}$$

Assume that the ratio in (2.1.5) converges for $x \in [0, q]$, i.e. there exists a function $\sigma: [0, q] \to \mathbb{R}$,

$$\sigma(x) := \lim_{n \to \infty} \frac{f^n(x)}{f^n(q)}.$$

Then we have ($f^n(q) \to 0$, $f(0) = 0$)

$$\sigma(f(x)) = \lim_{n \to \infty} \left(\frac{f^{n+1}(x)}{f^{n+1}(q)} \cdot \frac{f(f^n(q))}{f^n(q)} \right) = \sigma(x) f'(0),$$

which tells us that σ satisfies the Schröder equation (2.0.1) with $s = f'(0)$. In other words, the limit of the sequence (2.1.5), say A, is a solution of the functional equation

$$1 - A(F(y)/q) = F'(q)(1 - A(y/q)), \qquad y \in [0, q]. \tag{2.1.6}$$

Remark 2.1.1. The convergence of the sequence in question has been shown by using probability theory, e.g., by A. M. Jaglom [1], under the assumptions that $0 < m < 1$ and the variance of the RV X is finite. Results on monotonic and convex solutions of linear functional equations yield this convergence in the case $m \neq 1$ without any additional assumptions; see Sections 2.3 and 2.4.

We turn to the so-called *critical case* $m = 1$. Now necessarily $q = 1$ and $p_0 > 0$.

We shall be concerned with the limit probabilities (for $i, j \in \mathbb{N}$)

$$u_j(i) := \lim_{n \to \infty} u_{ij}^n, \qquad u_{ij}^n := P(Z_n = j \mid T = n + i). \tag{2.1.7}$$

Denote by p_{ij}^n the probability of passing from the state i to state j in exactly n steps (it does not depend on k; see Note 2.8.2):

$$p_{ij}^n = P(Z_{k+n} = j \mid Z_k = i).$$

The corresponding PGF is

$$\sum_{j=0}^{\infty} p_{ij}^n t^j = [F^n(t)]^i. \tag{2.1.8}$$

We proceed as in the case $m \neq 1$ above. Again by the Bayes formula (see (2.1.7))

$$u_{ij}^n = \frac{P(T=n+i\,|\,Z_n=j)P(Z_n=j)}{\sum_{k=1}^{\infty} P(T=n+i\,|\,Z_n=k)P(Z_n=k)} = \frac{(p_{j0}^i - p_{j0}^{i-1})p_j^n}{\sum_{k=1}^{\infty} (p_{k0}^i - p_{k0}^{i-1})p_k^n}. \tag{2.1.9}$$

In fact, '$Z_{n+i}=0$' is the union of the events '$Z_{n+i-1}=0$' and '$T=n+i$' which exclude each other. This implies

$$P(T=n+i\,|\,Z_n=j) = P(Z_{n+i}=0\,|\,Z_n=j) - P(Z_{n+i-1}=0\,|\,Z_n=j) = p_{j0}^i - p_{j0}^{i-1}.$$

We find the corresponding PGF

$$U_i^n(t) := \sum_{j=1}^{\infty} u_{ij}^n t^j = \frac{\sum_{j=1}^{\infty} (p_{j0}^i - p_{j0}^{i-1})p_j^n t^j}{\sum_{k=1}^{\infty} (p_{k0}^i - p_{k0}^{i-1})p_k^n}.$$

In view of (2.1.8), we have $p_{j0}^i = [F^i(0)]^j$. Using this and (2.1.3), we get

$$U_i^n(t) = \frac{F^n(tF^i(0)) - F^n(tF^{i-1}(0))}{F^n(F^i(0)) - F^n(F^{i-1}(0))}.$$

Let $x := 1 - tF^i(0)$, $x' := 1 - tF^{i-1}(0)$, $x_0 := 1 - F^{i-1}(0)$ and use $f(x) = 1 - F(1-x)$ to get

$$U_i^n(t) = \frac{f^n(x) - f^n(x')}{f^{n+1}(x_0) - f^n(x_0)}. \tag{2.1.10}$$

If there exists the limit

$$\alpha(x) := \lim_{n \to \infty} \frac{f^n(x) - f^n(x_0)}{f^{n+1}(x_0) - f^n(x_0)}, \qquad x \in [0, 1], \tag{2.1.11}$$

then it is a solution of the Abel equation (2.0.2); see Section 2.4. The limit PGF is then given by

$$U_i(t) := \lim_{n \to \infty} U_i^n(t) = \alpha(x) - \alpha(x')$$

and the probability distribution $(u_j(i))_{j \in \mathbb{N}_0}$ can be determined.

As will be proved in Section 2.4, the nature of the problem discussed (more precisely the convexity of the PGF F) itself forces the limit (2.1.11) to exist.

2.1C. Stationary measures for processes with immigration

Consider now the case where at each generation of the simple Galton–Watson process there is an additional input (an immigration), from an outside source, of Y_n objects of the same type as Z_n, and Y_n is an RV with the probability distribution

$$P(Y_n = j) = q_j, \qquad j \in \mathbb{N}_0,$$

which is independent of n. We obtain the *Galton–Watson process with immigration* in which the size of the nth generation is given by

$$\bar{Z}_n = OF\bar{Z}_{n-1} + Y_n, \qquad n \in \mathbb{N}, \bar{Z}_0 = 1.$$

Let the PGF of Y_n be

$$G(t) := \sum_{j=0}^{\infty} q_j t^j. \qquad (2.1.12)$$

Following the lines of considerations in Subsection 2.1A we determine the PGF of $\bar{Z}_n$. The transition probabilities are:

$$\bar{p}_{ij} = P(\bar{Z}_n = j \mid \bar{Z}_{n-1} = i) = P(X_1 + \cdots + X_i + Y_n = j),$$

so that (see (2.1.2))

$$\sum_{j=0}^{\infty} \bar{p}_{ij} t^j = G(t)(F(t))^i. \qquad (2.1.13)$$

In place of (2.1.3) we now have, with $\bar{F}_n$ as the PGF of $\bar{Z}_n$,

$$\bar{F}_n(t) = \prod_{i=0}^{n-1} G(F^i(t)), \qquad n \in \mathbb{N}_0. \qquad (2.1.14)$$

Formula (2.1.14) results from $\bar{F}_n(t) = G(t)\bar{F}_{n-1}(F(t))$, $n \in \mathbb{N}$.

Now, we are going to determine stationary measures for the Markov chain $(\bar{Z}_n)_{n \in \mathbb{N}}$. Assume we are given the probability distribution of $\bar{Z}_n$:

$$v_j := P(\bar{Z}_n = j), \qquad j \in \mathbb{N}_0.$$

Then $\bar{Z}_{n+1}$ has the probability distribution

$$P(\bar{Z}_{n+1} = j) = \sum_{i=0}^{\infty} P(\bar{Z}_{n+1} = j \mid \bar{Z}_n = i)P(\bar{Z}_n = i) = \sum_{i=0}^{\infty} \bar{p}_{ij} v_i.$$

The distribution of $\bar{Z}_{n+1}$ can be identical with that of $\bar{Z}_n$. In such a case all $\bar{Z}_{n+k}$, $k \in \mathbb{N}_0$, will be identically distributed as $\bar{Z}_n$. We then say that $(v_j)_{j \in \mathbb{N}_0}$ forms a *stationary distribution* of $\bar{Z}_n$.

Definition 2.1.1. A *stationary (invariant) measure for* $\bar{Z}_n$ is a measure defined on the state space $\mathbb{N}_0$ (i.e., a sequence of nonnegative numbers v_j, not all zero), invariant under the map induced by the transition probabilities $\bar{p}_{ij}$:

$$\sum_{j=0}^{\infty} \bar{p}_{ij} v_i = v_j, \qquad v_j \geq 0, j \in \mathbb{N}_0. \qquad (2.1.15)$$

Unlike the stationary distribution the stationary measure need not be bounded as the condition $\sum_{j=0}^{\infty} v_j = 1$ is dropped.

For a stationary measure $(v_j)_{j \in \mathbb{N}_0}$ we put

$$V(t) = \sum_{j=0}^{\infty} v_j t^j. \qquad (2.1.16)$$

Let $F(t) = p_0 + p_r t^r + \cdots$ with $p_r > 0$ and $r \geqslant 1$. We have (see (2.1.2))

$$(F(t))^i = \sum_{j=0}^{\infty} p_{ij} t^j.$$

Since $(F(t))^i = (p_0 + p_r t^r + \cdots)^i = (p_0)^i + i(p_0)^{i-1} p_r t^r + \cdots$, we have $p_{ir} = i(p_0)^{i-1} p_r$ and (because of (2.1.13) and (2.1.12))

$$\bar{p}_{ir} = \sum_{k=0}^{r} p_{ik} q_{r-k} \geqslant q_0 p_{ir} = i q_0 (p_0)^{i-1} p_r.$$

By (2.1.15), we may estimate

$$\infty > v_r = \sum_{i=0}^{\infty} \bar{p}_{ir} v_i \geqslant q_0 p_r \sum_{i=0}^{\infty} i v_i (p_0)^{i-1}.$$

Assuming that $q_0 > 0$, we get from this the convergence of the derived series of (2.1.16) for $t = p_0$. Thus (2.1.16) converges in $[0, p_0)$. To check that it is convergent also on $[0, q)$ apply (2.1.15), (2.1.16) and (2.1.13) to get

$$V(t) = \sum_{j=0}^{\infty} \sum_{i=0}^{\infty} \bar{p}_{ij} v_i t^j = G(t) \sum_{i=0}^{\infty} v_i (F(t))^i = G(t) V(F(t)).$$

This means that if (2.1.16) converges on $[0, t_0)$, $t_0 > 0$, then it does so on $[0, F(t_0))$. It remains to recall that $[0, q) = \bigcup_{k=0}^{\infty} [0, F^k(p_0))$ (see (2.1.4)). Thus the function V is defined on $[0, q)$ and it satisfies the functional equation

$$V(t) = G(t) V(F(t)). \tag{2.1.17}$$

Conversely, if a function V has the expansion (2.1.16) with $v_j \geqslant 0$, convergent in a neighbourhood of the origin, and V satisfies (2.1.1), then it follows by inserting (2.1.16) into (2.1.17) that $(v_j)_{j \in \mathbb{N}_0}$ forms a stationary measure for $\bar{Z}_n$.

As we shall show in Section 2.6, in the case $m \leqslant 1$ the stationary measure is unique up to a constant factor, whereas if $m > 1$ in order to assure the uniqueness we need the function $x \mapsto V(q - x)$ to be regularly varying at the origin.

2.1D. Restricted stationary measures for simple processes

In the case of simple Galton–Watson processes we have to restrict the state space of the process to the set $\mathbb{N}$. Namely, the stationary measure from Definition 2.1.1 would be the trivial one, concentrated on the state $j = 0$, i.e., $v_0 > 0$, $v_i = 0$ for $i \in \mathbb{N}$ (see e.g. Harris [1]).

Consequently, for the process $(Z_n)_{n \in \mathbb{N}_0}$ without immigration we need to introduce a *restricted stationary measure* (*RSM*) as a sequence of nonnegative numbers u_j, not all zero, fulfilling

$$u_j = \sum_{k=1}^{\infty} p_{kj} u_k, \qquad u_j \geqslant 0, j \in \mathbb{N}. \tag{2.1.18}$$

Now we can see that the search for an RSM leads to the Abel functional equation. For, assume we are given an RSM $(u_j)_{j\in\mathbb{N}}$ and write

$$U(t):=\sum_{j=1}^{\infty} u_j t^j. \tag{2.1.19}$$

Repeating the argument from the preceding subsection we find that the series (2.1.19) converges in $[0, q)$. Moreover, by (2.1.18) and (2.1.2)

$$U(t) = \sum_{j=1}^{\infty}\sum_{k=1}^{\infty} p_{kj} u_k t^j = \sum_{k=1}^{\infty} u_k \sum_{j=1}^{\infty} p_{kj} t^j$$

$$= \sum_{k=1}^{\infty} u_k((F(t))^k - (F(0))^k)$$

$$= U(F(t)) - U(F(0)),$$

so that the function U satisfies the Abel equation

$$U(F(t)) = U(t) + U(F(0)), \qquad t\in[0, q). \tag{2.1.20}$$

Conversely, if a power series (2.1.19) with nonnegative coefficients converges in a neighbourhood of the origin and U satisfies (2.1.20), then U generates an RSM for Z_n. Indeed, inserting (2.1.19) into (2.1.20), applying (2.1.2) and calculating the coefficients of t^j, we get (2.1.18).

Determining RSMs will be discussed in Subsection 2.6C.

Comments. We do not attempt to give a full bibliography of branching processes. There are several survey papers on this subject, in particular by K. B. Athreya–P. E. Ney [1], T. E. Harris [1], E. Seneta [6]. Our presentation here is based on the collection of articles edited by M. Kuczma [36]; see also Seneta [6] and Comments in Section 2.6.

2.2 Nonnegative solutions

We begin with the case $p = 0$, the study of the linear equation of the first order

$$\varphi(f(x)) = g(x)\varphi(x) + h(x) \tag{2.2.1}$$

in classes of specifically growing functions, namely those possessing a nonnegative pth difference; see Section 2.0. Conditions of this kind are elementary as they do not involve any infinitesimal operations performed on the unknown φ. They also ensure, in a number of cases, the uniqueness of solutions to (2.2.1).

We define the sequence $(G_n(x))_{n\in\mathbb{N}_0}$

$$G_n(x) = \prod_{i=0}^{n-1} g(f^i(x)), \qquad n\in\mathbb{N}_0, \tag{2.2.2}$$

which will occur in the majority of formulae determining solutions of (2.2.1).

Results we present here are due to J. Burek [1].

2.2A. Negative g

If $g < 0$ then necessarily $h \geq 0$, otherwise there do not exist nonnegative solutions of equation (2.2.1).

Theorem 2.2.1. *Let X be a real interval, and let $f: X \to X$, $g: X \to \mathbb{R}$ and $h: X \to \mathbb{R}$ be functions such that $g < 0$ and $h \geq 0$. Further assume that*

$$\lim_{n \to \infty} \frac{h(f^n(x))}{G_{n+1}(x)} = 0, \qquad x \in X. \tag{2.2.3}$$

If a nonnegative function $\varphi: X \to \mathbb{R}^+$ satisfies equation (2.2.1), then

$$\varphi(x) = -\sum_{n=0}^{\infty} \frac{h(f^n(x))}{G_{n+1}(x)}, \qquad x \in X. \tag{2.2.4}$$

If, moreover,

$$h(f(x)) + g(f(x))h(x) \leq 0, \qquad x \in X, \tag{2.2.5}$$

then equation (2.2.1) actually has the (unique) solution $\varphi: X \to \mathbb{R}^+$ given by (2.2.4).

Proof. Let $\varphi: X \to \mathbb{R}^+$ satisfy (2.2.1). Induction yields

$$\varphi(x) = \varphi(f^k(x))/G_k(x) - \sum_{n=0}^{k-1} \frac{h(f^n(x))}{G_{n+1}(x)}, \qquad k \in \mathbb{N}. \tag{2.2.6}$$

Hence, since $\varphi \geq 0$, $g < 0$ and

$$G_k(x) = (-1)^k |G_k(x)|, \qquad k \in \mathbb{N}_0, \tag{2.2.7}$$

we get

$$-\sum_{n=0}^{2k} \frac{h(f^n(x))}{G_{n+1}(x)} \leq \varphi(x) \leq -\sum_{n=0}^{2k-1} \frac{h(f^n(x))}{G_{n+1}(x)}, \qquad k \in \mathbb{N}.$$

Thus

$$\left| \varphi(x) + \sum_{n=0}^{k-1} \frac{h(f^n(x))}{G_{n+1}(x)} \right| \leq \frac{h(f^k(x))}{|G_{k+1}(x)|}, \qquad k \in \mathbb{N}.$$

Condition (2.2.3) implies formula (2.2.4).

Having assumed (2.2.5) we get the inequalities

$$\frac{h(f^{n+1}(x))}{|G_{n+2}(x)|} \leq \frac{h(f^n(x))}{|G_{n+1}(x)|}, \qquad n \in \mathbb{N}_0,$$

which together with (2.2.3) imply the convergence of the series (2.2.4) (as an alternating one; see (2.2.7)). The function (2.2.4) is clearly nonnegative and one can readily check that it satisfies equation (2.2.1). ∎

2.2B. Positive g

First of all, we have

Theorem 2.2.2. *Let X be a real interval and let $f: X \to X$, $g: X \to \mathbb{R}$, $h: X \to \mathbb{R}$*

be functions such that $g > 0$ and $h \leqslant 0$. Then the convergence of the series in (2.2.4) is a necessary and sufficient condition for the existence of a nonnegative solution $\varphi: X \to \mathbb{R}^+$ of equation (2.2.1).

Proof. If the series in (2.2.4) converges, then its sum φ is a solution of (2.2.1) in X and $\varphi \geqslant 0$, since $G_{n+1} > 0$ for $n \in \mathbb{N}_0$.

Conversely, assume that $\varphi: X \to \mathbb{R}^+$ satisfies equation (2.2.1). Then we have (2.2.6), whence

$$0 \leqslant \sum_{n=0}^{k-1} -h(f^n(x))/G_{n+1}(x) \leqslant \varphi(x), \qquad k \in \mathbb{N}, \qquad (2.2.8)$$

which proves the convergence of the series occurring in (2.2.4). ∎

Theorem 2.2.3 below says that now the nonnegative solution of (2.2.1) need not be unique. However, it follows from (2.2.8) that if the function (2.2.4) exists then it is the minimal nonnegative solution of (2.2.1) in X.

Theorem 2.2.3. *Assume that $X = |0, a|, 0 < a \leqslant \infty, f: X \to X$ is continuous and strictly increasing on $X, 0 < f(x) < x$ in $X \setminus \{0\}, g: X \to \mathbb{R}$ is positive in $X \setminus \{0\}$, $h: X \to \mathbb{R}$ satisfies $h(x)(x - x_0) \leqslant 0$ in X, $h(x_0) \geqslant 0$, for some $x_0 \in X \setminus \{0\}$. Moreover, if $0 \in X$, then $g(0) \leqslant 1$, and $h(0) = 0$ when $g(0) = 1$.*

For every function $\varphi_0: (f(x_0), x_0] \to \mathbb{R}^+$ there exists an extension $\varphi: X \to \mathbb{R}^+$ of φ_0 satisfying equation (2.2.1). This extension is unique except that $\varphi(0) \in \mathbb{R}^+$ may be arbitrary when $0 \in X$ and $g(0) = 1$.

Proof. Write $X_n := (f^{n+1}(x_0), f^n(x_0)], n \in \mathbb{N}_0$, and define $\varphi_n: X_n \to \mathbb{R}^+$ by

$$\varphi_{n+1}(x) = g(f^{-1}(x))\varphi_n(f^{-1}(x)) + h(f^{-1}(x)), \qquad n \in \mathbb{N}_0.$$

It follows by induction that actually $\varphi_n \geqslant 0$. We put $\varphi(x) = \varphi_n(x)$ for $x \in X_n$, $n \in \mathbb{N}_0$. The function φ is the unique extension of φ_0 onto $(0, x_0] = \bigcup_{n=0}^{\infty} X_n$ to a solution of (2.2.1) and $\varphi \geqslant 0$. In a similar manner we can uniquely extend φ_0 to the right of x_0 to a solution of (2.2.1) on $(x_0, a|$, preserving nonnegativity. If $0 \in X$, the value $\varphi(0)$ is calculated from (2.2.1) on setting $x = 0$ if $g(0) < 1$, and it is taken arbitrary nonnegative if $g(0) = 1$. ∎

Comments. Nonnegative solutions, of equations of higher orders as well, are considered in Czerwik [2], Jankowski–Kwapisz [1], Kornstaedt [1], Kwapisz [3], Kwapisz–Turo [2]–[4], among others.

2.3 Monotonic solutions

2.3A. Homogeneous equation

We start with the equation

$$\varphi(f(x)) = g(x)\varphi(x) \qquad (2.3.1)$$

and make the following assumptions.

(i) $X = (0, a], 0 < a \leqslant \infty$.
(ii) $f: X \rightarrow X$ is continuous, increasing and $0 < f(x) < x$ in X.
(iii) $g: X \rightarrow \mathbb{R}$ is positive in X.

The three theorems we are going to prove tell us in turn about the uniqueness of solutions of (2.3.1) (A. Smajdor [4], [5]; see also John [1]), the form of them (Burek–Kuczma [1], Kuczma [6], [11], A. Smajdor [5]) and their existence (Kuczma [6], A. Smajdor [5]).

We start with the following.

Lemma 2.3.1. *If hypotheses* (i)–(iii) *are fulfilled and* $\varphi: X \rightarrow \mathbb{R}$ *is a monotonic solution of equation* (2.3.1), *then either* $\varphi = 0$ *or* $\varphi > 0$ *or* $\varphi < 0$.

Proof. Let $\varphi(x_0) = 0$ for an $x_0 \in X$. Then by (2.3.1) $\varphi(f^n(x_0)) = 0$ for $n \in \mathbb{N}_0$. Since φ is monotonic, it vanishes on $[0, x_0]$, as $f^n(x_0) \rightarrow 0$ when $n \rightarrow \infty$ (see Theorem 1.2.1). Further, $\varphi(x_1) \neq 0$ for an $x_1 \in X, x_1 > x_0$, would imply, again by (2.3.1), that $\varphi(f^n(x_1)) \neq 0$ for $n \in \mathbb{N}_0$. But that is impossible, since $f^n(x_1) \rightarrow 0, n \rightarrow \infty$.

Thus either $\varphi = 0$ or $\varphi(x) \neq 0$ in X. Since $g > 0$, φ is of constant sign on every sequence $f^n(x), x \in X$. Being monotonic, φ preserves the sign in the whole X. ∎

Theorem 2.3.1. *Let hypotheses* (i)–(iii) *be fulfilled and let*

$$\inf_{x \in X} g(x) = 1. \tag{2.3.2}$$

Then any two monotonic solutions $\varphi_1, \varphi_2: X \rightarrow \mathbb{R}$ *of equation* (2.3.1) *differ by a constant factor.*

Proof. Let φ_1 and φ_2 be monotonic solutions of (2.3.1). Lemma 2.3.1 says that if φ_2 is not the zero function then it is either positive in X or negative in X. If $\varphi_2 = 0$, then $\varphi_2 = 0 \cdot \varphi_1$ and the theorem holds true. Since, together with a function φ, $-\varphi$ is also a solution of (2.3.1), we may assume that both φ_1 and φ_2 are positive in X. Moreover, (2.3.2) implies $g \geqslant 1$ in X, so that φ_1 and φ_2 are decreasing in X.

Write $\omega = \varphi_1/\varphi_2$. Then $\omega: X \rightarrow \mathbb{R}$ satisfies the equation

$$\omega(f(x)) = \omega(x), \tag{2.3.3}$$

whence it follows that for every $x_0 \in X$ (putting $X_0 := [f(x_0), x_0]$)

$$k := \inf_{X_0} \omega = \inf_{X} \omega \quad \text{and} \quad K := \sup_{X_0} \omega = \sup_{X} \omega.$$

By the monotonicity of φ_1 and φ_2 we have for $x \in X_0$

$$0 < \varphi_1(x_0)/\varphi_2(f(x_0)) \leqslant \omega(x) \leqslant \varphi_1(f(x_0))/\varphi_2(x_0) < \infty.$$

Hence $k > 0$ and $K < \infty$. For the proof of the theorem it is enough to show that $k = K$.

Suppose that $K > k$ and take a positive c such that $1 < c^2 < K/k$. In view of (2.3.2) we can choose an $x_0 \in X$ such that $g(x_0) < c$. Then we have for any $u, v \in X_0, u < v$,

$$\frac{\omega(v)}{\omega(u)} = \frac{\varphi_1(v)}{\varphi_1(u)} \cdot \frac{\varphi_2(u)}{\varphi_2(v)} \leqslant \frac{\varphi_2(u)}{\varphi_2(v)} \leqslant \frac{\varphi_2(f(x_0))}{\varphi_2(x_0)} = g(x_0) < c. \qquad (2.3.4)$$

Putting into (4) $v = f(x_0)$ we get by (2.3.3)

$$\frac{1}{k} \omega(x_0) = \sup_{u \in X_0} (\omega(x_0)/\omega(u)) \leqslant c.$$

Similarly, putting $u = x_0$ into (2.3.4), we get

$$K/\omega(x_0) = \sup_{v \in X_0} (\omega(v)/\omega(x_0)) \leqslant c.$$

Hence $K/k \leqslant c^2$, a contradiction. Thus necessarily $K = k$ and $\varphi_1 = k\varphi_2$. ∎

Remark 2.3.1. Clearly, condition (2.3.2) may be replaced by

$$\sup_{x \in X} g(x) = 1. \qquad (2.3.5)$$

Theorem 2.3.2. *Let hypotheses* (i)–(iii) *be fulfilled and let*

$$\lim_{n \to \infty} g(f^n(x)) = 1, \qquad x \in X. \qquad (2.3.6)$$

If $\varphi : X \to \mathbb{R}$ *is a monotonic solution of equation* (2.3.1), *then*

$$\varphi(x) = c \lim_{n \to \infty} \frac{G_n(x_0)}{G_n(x)} = c \prod_{n=0}^{\infty} \frac{g(f^n(x_0))}{g(f^n(x))} \qquad (2.3.7)$$

(see (2.3.2)), *where* $x_0 \in X$ *is arbitrarily fixed, and* $c = \varphi(x_0)$.

Proof. If $\varphi = 0$, then (2.3.7) holds with $c = 0$. If $\varphi \neq 0$, then it is of constant sign on X (Lemma 2.3.1). Assume that φ is positive and increasing (in the remaining cases the proof is analogous). Take an arbitrary $x \in X$. It follows from (2.3.1) by induction that

$$\varphi(f^n(x)) = G_n(x)\varphi(x), \qquad n \in \mathbb{N}. \qquad (2.3.8)$$

Put $y = \max(x_0, x)$. Since $f^n(y) \searrow 0$ (Theorem 1.2.1), there exists an $m \in \mathbb{N}$ such that $x_0, x \in [f^m(y), y]$. By the monotonicity of φ and (2.3.8) we have

$$G_m(f^n(y)) = \prod_{i=0}^{m-1} g(f^{n+i}(y)) = \frac{\varphi(f^{n+m}(y))}{\varphi(f^n(y))} \leqslant \frac{\varphi(f^n(x))}{\varphi(f^n(x_0))}$$

$$\leqslant \frac{\varphi(f^n(y))}{\varphi(f^{n+m}(y))} = [G_m(f^n(y))]^{-1},$$

whence by (2.3.6), on letting $n \to \infty$,

$$\lim_{n \to \infty} \frac{\varphi(f^n(x))}{\varphi(f^n(x_0))} = 1 \qquad \text{for } x \in X. \tag{2.3.9}$$

Using (2.3.8) for x and x_0, by (2.3.9) we obtain (2.3.7) with $c = \varphi(x_0)$. ∎

Theorem 2.3.3. *Let hypotheses* (i)–(iii) *be fulfilled. If the function g is monotonic and*

$$\lim_{x \to 0} g(x) = 1, \tag{2.3.10}$$

then equation (2.3.1) *has a unique one-parameter family of monotonic solutions $\varphi: X \to \mathbb{R}$. These solutions are given by formula* (2.3.7), *where $x_0 \in X$ is arbitrarily fixed, and $c \in \mathbb{R}$ is an arbitrary constant (a parameter).*

Proof. We may assume that g is increasing (in the other case the proof is similar). We have to prove that the sequence in (2.3.7) converges.

Fix an arbitrary $x_0 \in X$, take an arbitrary $x \in X$, and put $y = \max(x_0, x)$. There exists an $m \in \mathbb{N}$ such that $x_0, x \in [f^m(y), y]$. Hence

$$f^n(x_0), f^n(x) \in [f^{n+m}(y), f^n(y)] \qquad \text{for } n \in \mathbb{N}_0$$

and

$$\frac{g(f^{n+m}(y))}{g(f^n(y))} \leqslant \frac{g(f^n(x_0))}{g(f^n(x))} \leqslant \frac{g(f^n(y))}{g(f^{n+m}(y))} \qquad \text{for } n \in \mathbb{N}_0.$$

Therefore

$$\frac{g(f^{k+1}(y)) \cdots g(f^{k+m}(y))}{g(y) \cdots g(f^{m-1}(y))} \leqslant \prod_{n=0}^{k} \frac{g(f^n(x_0))}{g(f^n(x))} \leqslant \frac{g(y) \cdots g(f^{m-1}(y))}{g(f^{k+1}(y)) \cdots g(f^{k+m}(y))}$$

for $k \geqslant m$. Since (2.3.10) implies (2.3.6), the sequence whose terms are $G_n(x_0)/G_n(x)$ is bounded from below and from above by positive constants. Moreover, this sequence is monotonic, since $g(f^n(x_0))/g(f^n(x)) \geqslant 1$ or $\leqslant 1$ for all $n \in \mathbb{N}_0$, according as $x_0 \geqslant x$ or $x_0 \leqslant x$. Consequently the infinite product in (2.3.7) converges for every $x \in X$. For every $c \in \mathbb{R}$ formula (2.3.7) defines a function (nontrivial unless $c = 0$) $\varphi: X \to \mathbb{R}$, and φ is monotonic, since g is monotonic and f is increasing. It is easily checked that φ satisfies equation (2.3.1).

The uniqueness statement results from Theorem 2.3.2. ∎

2.3B. Special inhomogeneous equation

Modifying the proofs of Theorems 2.3.1–2.3.3 one can obtain the following results for the equation

$$\varphi(f(x)) = \varphi(x) + h(x), \tag{2.3.11}$$

(see Kuczma [6], [12], A. Smajdor [5]), where

(iv) $h: X \to \mathbb{R}$ is a function.

In Theorems 2.3.4–2.3.6 that follow we assume hypotheses (i), (ii) and (iv).

Theorem 2.3.4. *If* $\inf_X h = 0$ *or* $\sup_X h = 0$, *then any two monotonic solutions of equation (2.3.11) in X differ by a constant.*

Theorem 2.3.5. *Let* $\lim_{n \to \infty} h(f^n(x)) = 0$ *for all* $x \in X$. *If* $\varphi: X \to \mathbb{R}$ *is a monotonic solution of equation (2.3.11), then*

$$\varphi(x) = c - \sum_{n=0}^{\infty} (h(f^n(x)) - h(f^n(x_0))), \qquad (2.3.12)$$

where $x_0 \in X$ *is arbitrarily fixed, and* $c = \varphi(x_0)$ *is a constant.*

Theorem 2.3.6. *If the function* h *is monotonic and* $\lim_{x \to 0} h(x) = 0$, *then equation (2.3.11) has a unique one-parameter family of monotonic solutions* $\varphi: X \to \mathbb{R}$. *They are given by formula (2.3.12), where* $x_0 \in X$ *is arbitrarily fixed, and* $c \in \mathbb{R}$ *is an arbitrary constant (a parameter).*

2.3C. General inhomogeneous equation

For the equation

$$\varphi(f(x)) = g(x)\varphi(x) + h(x) \qquad (2.3.13)$$

the situation is much less satisfactory. We give here two results, chosen from among those due to A. Smajdor [5], concerning monotonic solutions of equation (2.3.13).

Theorem 2.3.7. *Let hypotheses (i)–(iv) be fulfilled and let the functions* g *and* h *be increasing. Further assume that the relations (2.3.10) and*

$$\lim_{n \to \infty} [h(f^n(x))/G_{n+1}(x)] = 0 \qquad \text{for all } x \in X \qquad (2.3.14)$$

hold. Then for every $x_0 \in X$ *there exists a one-parameter family of solutions* $\varphi: X \to \mathbb{R}$ *of equation (2.3.13) which are nonnegative and decreasing in* $(0, x_0]$. *These functions are given by*

$$\varphi(x) = c\varphi_0(x) + \sum_{n=0}^{\infty} \left[\frac{h(f^n(x_0))}{G_{n+1}(x_0)} \varphi_0(x) - \frac{h(f^n(x))}{G_{n+1}(x)} \right], \qquad (2.3.15)$$

where $\varphi_0(x) := \lim_{n \to \infty} [G_n(x_0)/G_n(x)]$, *and* $c \geq 0$ *is an arbitrary constant (a parameter).*

Proof. We shall look for solutions of (2.3.13) in the form of the product of functions satisfying equations (2.3.1) and (2.3.11) (with an appropriately defined function h).

The argument in the proof of Theorem 2.3.3 shows that the function φ_0 is a positive and decreasing solution of equation (2.3.1). Define $\hat{h}: X \to \mathbb{R}$ by

$\hat{h}(x) = h(x)/\varphi_0(f(x))$. Thus $\hat{h}$ is increasing in X, and we have by (2.3.14) and (2.3.8) (with φ replaced by φ_0)

$$\lim_{n \to \infty} \hat{h}(f^n(x)) = \lim_{n \to \infty} \frac{h(f^n(x))}{G_{n+1}(x)} \varphi_0(x) = 0$$

for all $x \in X$. The function

$$\hat{\phi}(x) = c - \sum_{n=0}^{\infty} [\hat{h}(f^n(x)) - \hat{h}(f^n(x_0))] \tag{2.3.16}$$

is a monotonic solution of the equation

$$\hat{\phi}(f(x)) = \hat{\phi}(x) + \hat{h}(x)$$

(Theorem 2.3.6), and it is readily seen that, for $c \geqslant 0$, $\hat{\phi}$ is nonnegative and decreasing in $(0, x_0]$. Hence the function $\varphi(x) := \varphi_0(x)\hat{\phi}(x)$ satisfies equation (2.3.13) and is nonnegative and decreasing in $(0, x_0]$. Formula (2.3.15) results from (2.3.16) and (2.3.8) (with φ replaced by φ_0). ■

It is not clear under what additional conditions functions (2.3.15) are monotonic in the whole of X, and whether they are the only monotonic (resp. monotonic in a vicinity of the origin) solutions of (2.3.13). However, the following theorem shows that in general monotonic solution of (2.3.13) is not unique.

Theorem 2.3.8. *Let hypotheses* (i)–(iv) *be fulfilled and let the functions g and h be increasing, $g \leqslant 1$ and $h \leqslant 0$. Then for every $x_0 \in X$ and for every nonnegative increasing function $\varphi_0: [f(x_0), x_0] \to \mathbb{R}$ such that*

$$\varphi_0(f(x_0)) = g(x_0)\varphi_0(x_0) + h(x_0) \tag{2.3.17}$$

there exists a unique extension φ of φ_0 onto X which satisfies equation (2.3.13); *this extension is increasing in* $(0, x_0]$. *Every solution $\varphi: X \to \mathbb{R}$ which is nonnegative and increasing in* $(0, x_0]$ *can be obtained in this manner.*

Remark 2.3.2. For functions f, g decreasing and $g \geqslant 1$, $h \geqslant 0$ we get the extension and the solution decreasing (if the φ_0 was decreasing).

The proof of the theorem is straightforward and is therefore omitted (see A. Smajdor [5]). We may observe that, if $g(x_0) = 1$ and $h(x_0) = 0$, then every monotonic function $\varphi_0: [f(x_0), x_0] \to \mathbb{R}$ fulfilling (2.3.17) must be constant and Theorem 2.3.8 describes only a one-parameter family of solutions. If $g(x_0) \neq 1$ or $h(x_0) \neq 0$, then the solution described in Theorem 2.3.8 depends on an arbitrary function.

2.3D. An example

Although for the general linear equation (2.3.13) the results are rather weak, in the particular case of equations (2.3.1) and (2.3.11) they are quite powerful.

Monotonic solutions of these equations may tend to infinity as $x \to 0$. So they could not be obtained with the aid of theorems contained in Chapter 3 which concern solutions regular at the origin.

As an example we consider the equation

$$\varphi\left(\frac{x}{x+1}\right) = (x+1)\varphi(x), \qquad x \in X := (0, a], \ a > 0. \qquad (2.3.18)$$

The reader may consult Chapter 3 to find that no regularity conditions imposed on solutions of (2.3.18) (either in X or in $X \cup \{0\}$) can give a one-parameter family of solutions. On the other hand, Theorem 2.3.3 says that equation (2.3.18) has a unique one-parameter family of monotonic solutions in X, viz. $\varphi(x) = c/x$, $c \in \mathbb{R}$.

Having a particular solution of (2.3.18) we may try to find a formula for its general solution (see Subsection 0.1D). To this end we need a particular invertible solution $\alpha: X \to \mathbb{R}$ of the Abel equation

$$\alpha\left(\frac{x}{x+1}\right) = \alpha(x) + 1. \qquad (2.3.19)$$

The function $\alpha(x) = 1/x$ serves as such a solution. Thus it furnishes a change of variables (see Subsection 0.2A) which transforms the function $f(x) = x/(x+1)$ into $\hat{f}(x) = x+1$. In other words, a function $\varphi: X \to \mathbb{R}$ satisfies equation (2.3.18) in X if and only if the function $\hat{\varphi} = \varphi \circ \alpha^{-1}$ satisfies in $\hat{X} := [1/a, \infty)$ the equation

$$\hat{\varphi}(x+1) = \frac{x+1}{x}\,\hat{\varphi}(x). \qquad (2.3.20)$$

Further, since $\varphi_0 = \alpha$ is also a solution of (2.3.18), the function $\hat{\varphi}_0 = \varphi_0 \circ \alpha^{-1} = \alpha \circ \alpha^{-1} = \mathrm{id}_{\hat{X}}$ fulfils (2.3.20). The quotient $\omega := \hat{\varphi}/\hat{\varphi}_0$ of two solutions of (2.3.20) is then a periodic function on $\hat{X}$, of period 1: $\omega(x+1) = \omega(x)$.

Conversely, every 1-periodic function $\omega: \hat{X} \to \mathbb{R}$ yields the solution $\hat{\varphi} = \omega\hat{\varphi}_0$ of equation (2.3.20). Since $\varphi = \hat{\varphi} \circ \alpha$, the general solution of equation (2.3.18) can be expressed as

$$\varphi = (\hat{\varphi}_0 \circ \alpha)(\omega \circ \alpha). \qquad (2.3.21)$$

Recalling that $\hat{\varphi}_0(x) = x$ and $\alpha(x) = 1/x$ we see that $\varphi(x) = (1/x)\omega(1/x)$, $x \in X$, where $\omega: \hat{X} \to \mathbb{R}$ is an arbitrary periodic function of period 1. Taking ω of a suitable regularity, we can obtain solutions of an arbitrary regularity on X (for another way of describing regular solutions of (2.3.18) see Section 3.5).

Remark 2.3.3. To get the general solution φ of equation (2.3.1) we need an invertible solution α of the Abel equation $\alpha(f(x)) = \alpha(x) + 1$ and a never vanishing φ_0 that satisfies (2.3.1). With $\hat{\varphi}_0 = \varphi_0 \circ \alpha^{-1}$ formula (2.3.21) becomes $\varphi = (\omega \circ \alpha)\varphi_0$ and gives all the solutions of (2.3.1).

2.3E. Homogeneous difference equation

The experience gained in the preceding subsection will be helpful in reducing the equation

$$\varphi(x+1) = g(x)\varphi(x) \tag{2.3.22}$$

to the case covered by Theorems 2.3.1–2.3.3.

Take $X = |b, \infty)$, $b \geqslant 0$, put $\alpha(x) = 1/x$ and write $\hat{\varphi} = \varphi \circ \alpha$, $\hat{g} = g \circ \alpha$. Further, let $\hat{X} = (0, a|$, where $a = 1/b$ if $b > 0$ and $a = \infty$ if $b = 0$. Then φ and/or g is defined and monotonic in X if and only if so is $\hat{\varphi}$ and/or $\hat{g}$ in $\hat{X}$, and φ satisfies equation (2.3.22) in X if and only if $\hat{\varphi}$ satisfies the equation

$$\hat{\varphi}\left(\frac{x}{x+1}\right) = \hat{g}(x)\hat{\varphi}(x)$$

in X. Thus we get from Theorems 2.3.1–2.3.3 the following results.

Theorem 2.3.9. *Let $X = |b, \infty)$, $b \geqslant 0$, and let $g: X \to \mathbb{R}$ be a positive function such that (2.3.2) or (2.3.5) holds. Then any two monotonic solutions of equation (2.3.22) in X differ by a constant factor.*

Theorem 2.3.10. *Let $X = |b, \infty)$, $b \geqslant 0$, and let $g: X \to \mathbb{R}$ be a positive function such that*

$$\lim_{n \to \infty} g(x+n) = 1 \qquad \text{for all } x \in X.$$

If $\varphi: X \to \mathbb{R}$ is a monotonic solution of equation (2.3.22), then

$$\varphi(x) = c \prod_{n=0}^{\infty} \frac{g(x_0+n)}{g(x+n)}, \tag{2.3.23}$$

where $x_0 \in X$ is arbitrarily fixed, and $c = \varphi(x_0)$.

Theorem 2.3.11. *Let $X = |b, \infty)$, $b \geqslant 0$, and let $g: X \to \mathbb{R}$ be a positive and monotonic function such that*

$$\lim_{x \to \infty} g(x) = 1.$$

Then equation (2.3.22) has a unique one-parameter family of monotonic solution $\varphi: X \to \mathbb{R}$. They are given by formula (2.3.23), where $x_0 \in X$ is arbitrarily fixed, and $c \in \mathbb{R}$ is an arbitrary constant (a parameter).

2.3F. Schröder's equation

Theorem 2.3.3 when applied to the Schröder equation

$$\sigma(f(x)) = s\sigma(x) \tag{2.3.24}$$

yields (see Kuczma [13]) the following.

Theorem 2.3.12. *Let $X = |0, a|$, $0 < a \leqslant \infty$, and assume that $f: X \to X$ is continuous and strictly increasing on X, $0 < f(x) < x$ in $X \setminus \{0\}$, the function*

$x \mapsto f(x)/x$ is monotonic in $X \setminus \{0\}$ and $\lim_{x \to 0}[f(x)/x] = s$, where $0 < s < 1$. Then equation (2.3.24) has a unique one-parameter family of solutions $\sigma: X \to \mathbb{R}$ such that the function $x \mapsto \sigma(x)/x$ is monotonic in $X \setminus \{0\}$. These solutions are given by the formula

$$\sigma(x) = c \lim_{n \to \infty} \frac{f^n(x)}{f^n(x_0)}, \tag{2.3.25}$$

where $c \in \mathbb{R}$ is any constant, and x_0 is an arbitrarily fixed point from $X \setminus \{0\}$.

Proof. If $0 \in X$, then $\sigma(0) = 0$ by (2.3.24), in accordance with (2.3.25). Thus $\sigma: X \to \mathbb{R}$ is a solution of (2.3.24) such that $\varphi(x) = \sigma(x)/x$, $x \in X \setminus \{0\}$ is monotonic if and only if the function φ is a monotonic solution of the equation

$$\varphi(f(x)) = \frac{sx}{f(x)} \varphi(x) \tag{2.3.26}$$

in $X \setminus \{0\}$. Now, monotonic solutions of (2.3.26) are determined by Theorem 2.3.3 (see (2.3.7) and note that $G_n(x) = s^n/f^n(x)$). ∎

This theorem is significant for applications in the theory of branching processes; see Sections 2.1 and 2.6.

Comments. For an exhaustive treatise of monotonic solutions of linear and nonlinear equations on abstract structures see A. Smajdor [5]. Further results concerning monotonic solutions of linear and general equations may be found in Bajraktarević [2], Brydak–Kordylewski [1], Burek–Kuczma [1], Czerwik [1], Hamilton [1], Hoppe [1], Kuczma [6], [8], [10], [26, Ch. V], Matkowski [1], A. Smajdor [2], [4], [7], among others.

2.4 Convex solutions

We shall confine our attention to the particular equation

$$\varphi(f(x)) = \varphi(x) + h(x). \tag{2.4.1}$$

After having proved two lemmas on the convergence of some sequences built out of iterates of f, we prove a basic theorem on the existence of a unique one-parameter family of convex solutions of (2.4.1) and its more refined version in the case where $f(x) = x + 1$. The section is concluded with theorems on convex solutions of the Abel and Schröder equations.

2.4A. Lemmas

For the lemmas here see Kuczma–A. Smajdor [1], Kuczma [19], Szekeres [2]. We assume that

(i) $X = (0, a|, \ 0 < a \leqslant \infty$,

(ii) $f: X \to X$ is concave and strictly increasing, $0 < f(x) < x$ in X and $\lim_{x \to 0} [f(x)/x] = 1$.

Lemma 2.4.1. *If hypotheses* (i) *and* (ii) *are fulfilled, then*

$$\lim_{n \to \infty} \frac{f^{n+1}(x) - f^n(x)}{f^{n+1}(y) - f^n(y)} = 1 \qquad \text{for all } x, y \in X. \qquad (2.4.2)$$

Proof. Let, e.g., $y \leqslant x$. To simplify the notation we put

$$x_n := f^n(x), \qquad y_n := f^n(y), \qquad n \in \mathbb{N}. \qquad (2.4.3)$$

We know that $x_n \to 0$ (Theorem 1.2.1). Thus there exists a $k \in \mathbb{N}$ such that $x_k \leqslant y \leqslant x$. Since f is concave, we have

$$f'_+(x_n) \leqslant \frac{f(x_{n+1}) - f(x_n)}{x_{n+1} - x_n} \leqslant f'_+(x_{n+1}), \qquad n \in \mathbb{N},$$

where f'_+ denotes the right-sided derivative. Condition (ii) implies that $\lim_{x \to 0} f'_+(x) = 1$, so we get from this

$$\lim_{n \to \infty} \frac{f(x_{n+k}) - x_{n+k}}{f(x_n) - x_n} = \lim_{n \to \infty} \prod_{i=0}^{k-1} \frac{f(x_{n+i+1}) - f(x_{n+i})}{x_{n+i+1} - x_{n+i}} = 1 \qquad (2.4.4)$$

(note that $f(x_p) = x_{p+1}$). Condition (ii) implies also that the function $x \mapsto f(x) - x$ is negative and decreasing in X. Since $x_n \leqslant y_n \leqslant x_{n+k}$, we have

$$1 \leqslant \frac{f(x_n) - x_n}{f(y_n) - y_n} \leqslant \frac{f(x_n) - x_n}{f(x_{n+k}) - x_{n+k}}, \qquad n \in \mathbb{N},$$

and (2.4.2) results from (2.4.4) and (2.4.3). ∎

Lemma 2.4.2. *If hypotheses* (i) *and* (i) *are fulfilled, then for every* $x, y \in X$ *there exists the limit*

$$\alpha(x, y) = \lim_{n \to \infty} \alpha_n(x, y), \qquad \text{where } \alpha_n(x, y) := \frac{f^n(x) - f^n(y)}{f^{n+1}(y) - f^n(y)} \qquad (2.4.5)$$

and, for every fixed $y \in X$, *the function* $\alpha(\cdot, y)$ *satisfies the Abel functional equation*

$$\alpha(f(x), y) = \alpha(x, y) + 1. \qquad (2.4.6)$$

Proof. We again use notation (2.4.3). First take $x \in [f(y), y)$. Thus $y_{n+1} \leqslant x_n < y_n$. Since f is concave, its difference quotient is decreasing:

$$\frac{f(y_{n+1}) - f(y_n)}{y_{n+1} - y_n} \geqslant \frac{f(x_n) - f(y_n)}{x_n - y_n}.$$

Hence $0 < \alpha_{n+1}(x, y) \leqslant \alpha_n(x, y)$ $(= (x_n - y_n)/(y_{n+1} - y_n))$ and the limit (2.4.5) exists in $[f(y), y)$. For other values of x in X the convergence results from the equality

$$\alpha_n(f(x), y) = \alpha_n(x, y) + \frac{f^{n+1}(x) - f^n(x)}{f^{n+1}(y) - f^n(y)}$$

and from Lemma 2.4.1. Letting $n \to \infty$ here we get (2.4.6). ∎

2.4B. Existence-and-uniqueness result

Now we add the assumption

(iii) $h: X \to \mathbb{R}$ is increasing and concave (resp. decreasing and convex) in X and there exists the limit $L := \lim_{x \to 0} h(x)$, $0 \leqslant L < \infty$ (resp. $-\infty < L \leqslant 0$).

The following result is due to M. Rozmus-Chmura [2].

Theorem 2.4.1. *If hypotheses* (i)–(iii) *are fulfilled, then equation* (2.4.1) *has a unique one-parameter family of convex (resp. concave) solutions* $\varphi: X \to \mathbb{R}$. *They are given by the formula*

$$\varphi(x) = c - \sum_{n=0}^{\infty} [h(f^n(x)) - h(f^n(x_0))] + L\alpha(x, x_0) \qquad (2.4.7)$$

(see (2.4.5))*, where* $x_0 \in X$ *is arbitrarily fixed, and* $c \in \mathbb{R}$ *is an arbitrary constant (a parameter).*

Proof. Now we shall write for short

$$x_n = f^n(x), \quad n \in \mathbb{N}, \qquad z_n := f^n(x_0), \quad n \in \mathbb{N}_0. \qquad (2.4.8)$$

By Theorem 2.3.6 the series (see (2.3.12) and (2.4.8))

$$\hat{\varphi}(x) = \sum_{n=0}^{\infty} [h(x_n) - h(z_n)] = \sum_{n=0}^{\infty} [(h(x_n) - L) - (h(z_n) - L)]$$

converges in X and its sum satisfies the equation

$$\hat{\varphi}(f(x)) = \hat{\varphi}(x) + (h(x) - L).$$

Thus, in virtue of Lemma 2.4.2, the functions φ given by (2.4.7) exist and satisfy equation (2.4.1) in X. It follows from (2.4.7) and from the properties of f and h that the functions φ are convex (resp. concave) in X.

To prove the uniqueness, suppose that a convex function $\varphi: X \to \mathbb{R}$ satisfies equation (2.4.1). We shall show that it is given by (2.4.7) with a certain c. (The argument below with the inequalities reversed applies to a concave φ.)

Fix an $x_0 \in X$ and take an $x \in [f(x_0), x_0)$. Then $z_{n+1} \leqslant x_n < z_n$, $n \in \mathbb{N}$, and, as the difference quotient of φ relative to the point $(z_n, \varphi(z_n))$ increases,

$$\frac{h(z_n)}{z_{n+1} - z_n} = \frac{\varphi(z_{n+1}) - \varphi(z_n)}{z_{n+1} - z_n} \leqslant \frac{\varphi(x_n) - \varphi(z_n)}{x_n - z_n}.$$

Similarly

$$\frac{h(z_{n-1})}{z_n - z_{n-1}} = \frac{\varphi(z_n) - \varphi(z_{n-1})}{z_n - z_{n-1}} \geq \frac{\varphi(x_n) - \varphi(z_n)}{x_n - z_n}.$$

Hence we obtain the estimation

$$\frac{x_n - z_n}{z_n - z_{n-1}} h(z_{n-1}) \leq \varphi(x_n) - \varphi(z_n) \leq \alpha_n(x, x_0) h(z_n). \tag{2.4.9}$$

Since $z_n \to 0$, by (2.4.5) and (iii) the expression on the right-hand side of (2.4.9) tends to $L\alpha(x, x_0)$ as $n \to \infty$. The same limit is attained by the left-hand side of (2.4.9). For, by (2.4.8) and (2.4.5), we have (with $y := f^{-1}(x_0)$)

$$\frac{x_n - z_n}{z_n - z_{n-1}} = \alpha_n(x, x_0) \frac{f^{n+1}(x_0) - f^n(x_0)}{f^{n+1}(y) - f^n(y)}$$

and Lemmas 2.4.1 and 2.4.2 apply. Therefore, (2.4.9) and (2.4.8) yield as $n \to \infty$ (recall (2.4.8))

$$\lim_{n \to \infty} [\varphi(f^n(x)) - \varphi(f^n(x_0))] = L\alpha(x, x_0). \tag{2.4.10}$$

But from (2.4.1) we get

$$\varphi(f^n(x)) - \varphi(f^n(x_0)) = \varphi(x) - \varphi(x_0) + \sum_{i=0}^{n-1} [h(f^i(x)) - h(f^i(x_0))]. \tag{2.4.11}$$

By (2.4.11), (2.4.10) amounts to (2.4.7) with $c = \varphi(x_0)$ and φ is uniquely determined (up to an additive constant) in $[f(x_0), x_0)$, and hence by (2.4.1) it is so in the whole X. ∎

2.4C. A difference equation

Theorem 2.4.1 can be slightly generalized in the case where $f(x) = x + 1$; see Krull [1], Hamilton [1], Kuczma [17], [26]. Equation (2.4.1) then becomes

$$\varphi(x + 1) = \varphi(x) + h(x). \tag{2.4.12}$$

Theorem 2.4.2. *Let* $X = [a, \infty)$, $-\infty \leq a < \infty$. *If the function* $h: X \to \mathbb{R}$ *is concave (resp. convex) and satisfies the condition*

$$\lim_{x \to 0} [h(x + 1) - h(x)] = 0, \tag{2.4.13}$$

then equation (2.4.12) *has a unique one-parameter family of convex (resp. concave) solutions* $\varphi: X \to \mathbb{R}$. *These solutions are given by the formula*

$$\varphi(x) = c + \lim_{n \to \infty} \left[(x - x_0)h(x_0 + n) - \sum_{i=0}^{n-1} (h(x + i) - h(x_0 + i)) \right]$$

$$= c + (x - x_0)h(x_0) + \sum_{n=0}^{\infty} [(x - x_0)(h(x_0 + n + 1) - h(x_0 + n))$$

$$- (h(x + n) - h(x_0 + n))], \tag{2.4.14}$$

where $x_0 \in X$ is arbitrarily fixed, and $c \in \mathbb{R}$ is an arbitrary constant.

Proof. Assume that h is convex (if h is concave, the argument is similar). Condition (2.4.13) then forces h to be decreasing. Fix an arbitrary $x_0 \in X$, and take an $x \in (x_0, x_0 + 1]$. Then the difference quotients of h, calculated consecutively for the intervals $[x_0 + n - 1, x_0 + n]$, $[x_0 + n, x + n]$, $[x_0 + n, x_0 + n + 1]$ become more and more large:

$$u_n := \frac{h(x_0 + n) - h(x_0 + n - 1)}{1} \leqslant \frac{h(x + n) - h(x_0 + n)}{x - x_0} \leqslant u_{n+1} \leqslant 0.$$

This implies that the terms $(x - x_0)u_{n+1} - [h(x + n) - h(x_0 + n)] \geqslant 0$ of the series in (2.4.14) are majorized by $(x - x_0)(u_{n+1} - u_n)$. The majorant is convergent, since by (2.4.13) $\lim_{n \to \infty} u_{n+1} = 0$. Now, with

$$\varphi_n(x) := \varphi(x_0) + (x - x_0)h(x_0 + n) - \sum_{i=0}^{n-1} (h(x + i) - h(x_0 + i)),$$

we have φ_n concave and

$$\varphi_n(x + 1) = \varphi_n(x) + h(x) - h(x + n) + h(x_0 + n).$$

Condition (2.4.13) implies that $\lim_{n \to \infty} [h(x + n) - h(x_0 + n)] = 0$, and consequently the sequence φ_n (being convergent in $(x_0, x_0 + 1]$) converges in X and its (concave) limit φ satisfies equation (2.4.12). The proof of the uniqueness follows the lines of that from Theorem 2.4.1. Note that in the present case we have $f^n(x) = x + n$, $f^{n+1}(x_0) - f^n(x_0) = 1$, $f^n(x) - f^n(x_0) = x - x_0$, $\alpha_n(x, x_0) = x - x_0$, $n \in \mathbb{N}$. ∎

2.4D. Abel's and Schröder's equations

For the Abel equation

$$\alpha(f(x)) = \alpha(x) + 1 \tag{2.4.15}$$

we obtain from Theorem 2.4.1 (see Kuczma [19]) the following.

Theorem 2.4.3. *Let hypotheses* (i) *and* (ii) *be fulfilled. Then equation* (2.4.15) *has a unique one-parameter family of convex solutions $\alpha: X \to \mathbb{R}$. These solutions are given by the formula*

$$\alpha(x) = c + \lim_{n \to \infty} \frac{f^n(x) - f^n(x_0)}{f^{n+1}(x_0) - f^n(x_0)}, \tag{2.4.16}$$

whose $x_0 \in X$ is arbitrarily fixed and $c \in \mathbb{R}$ is an arbitrary constant (a parameter). Moreover, they are strictly decreasing in X.

Proof. In view of Lemma 2.4.2 only the monotonicity requires proof. Since all f^n are increasing and $f^{n+1}(x_0) < f^n(x_0)$, α given by (2.4.16) is decreasing. Being convex, it is either constant (which is impossible in view of (2.4.15)) or strictly decreasing in a vicinity of the origin. Then (2.4.15) implies that α is strictly decreasing in X. ∎

We complete this section by a theorem (see Kuczma [13], [22]) on convex solutions of the Schröder equation

$$\sigma(f(x)) = s\sigma(x). \qquad (2.4.17)$$

Theorem 2.4.4. *Assume that*

(i') $X = |0, a|, 0 < a \leqslant \infty,$
(ii') $f: X \to X$ *is convex or concave and strictly increasing in* $X, 0 < f(x) < x$ *in* $X \setminus \{0\}$, $\lim_{x \to 0}[f(x)/x] = s, 0 < s < 1.$

Then equation (2.4.17) has a unique one-parameter family of solutions $\sigma: X \to \mathbb{R}$ *which are convex or concave in* X. *These solutions are given by the formula*

$$\sigma(x) = c \lim_{n \to \infty} \frac{f^n(x)}{f^n(x_0)}, \qquad (2.4.18)$$

where $c \in \mathbb{R}$ *is an arbitrary constant (a parameter) and* x_0 *is an arbitrarily chosen point from* $X \setminus \{0\}$.

Proof. Since the ratio $f(x)/x$ is monotonic for a convex or concave f, it follows from Theorem 2.3.12 that functions (2.4.18) are the only solutions of equation (2.4.17) such that the quotient $\sigma(x)/x$ is monotonic. All the functions f^n, $n \in \mathbb{N}$, are convex or concave, hence so is σ given by (2.4.18). Moreover if a solution $\sigma: X \to \mathbb{R}$ of (2.4.17) is convex or concave, then $\sigma(x)/x$ is monotonic and the uniqueness follows. ∎

Theorems 2.4.3 and 2.4.4 will be applied in Section 2.6.

Comments. There is a long list of papers devoted to convex solutions of iterative functional equations. Linear cases are studied in Anastassiadis [1]–[7], Czerwik [13], Dufresnoy–Pisot [1], John [1], [2], Krull [1], Kuczma [1], [2], [8], [9], [13], [14], [22], [26]. Regarding convex solutions of nonlinear functional equations see, in particular, Bajraktarević [1], Baron–Matkowski [1], Cooper [1], Czerwik [9], [10], Kominek–Matkowski [1], Kuczma [8], [26], [38], Kwapisz–Turo [3], Matkowski [1], Mayer [1], Pelczar [4], Rozmus-Chmura [1], Thielman [1], Vajzović [1].

2.5 Regularly varying solutions

In this section we shall deal with some particular cases of the equation

$$\varphi(f(x)) = g(x)\varphi(x) + h(x) \qquad (2.5.1)$$

under the following assumptions

(i) $X = (0, a|, 0 < a \leqslant \infty.$

(ii) $f: X \to X$ is continuous and increasing, $0 < f(x) < x$ in X, the function $x \mapsto f(x)/x$ is monotonic in X and $\lim_{x \to 0}[f(x)/x] = s$, $0 < s < 1$.

(iii) $g: X \to \mathbb{R}$ is positive, continuous and monotonic in X, $\lim_{x \to 0} g(x) = g_0$, $0 < g_0 < \infty$.

(iv) $h: X \to \mathbb{R}$ is continuous on X and has a constant sign in a vicinity of the origin.

2.5A. Regularly varying functions

The notion is due to J. Karamata [1].

Definition 2.5.1. A continuous and positive function $\varphi: X \to \mathbb{R}$ is said to be *regularly varying* (at the origin) iff for every $\lambda > 0$ there exists a positive limit of $\varphi(\lambda x)/\varphi(x)$ as $x \to 0$. If this is the case, the limit is necessarily of the form

$$\lim_{x \to 0} \frac{\varphi(\lambda x)}{\varphi(x)} = \lambda^\delta \tag{2.5.2}$$

with a real constant δ which is called the *index* of φ.

If $\delta = 0$, then φ is said to be *slowly varying*.

Lecture notes by E. Seneta [11] contain a comprehensive presentation of the theory of regularly and slowly varying functions. We shall need the following facts.

(a) The convergence in (2.5.2) is almost uniform with respect to λ in $(0, \infty)$.

(b) Every regularly varying function φ of the index δ has the representation

$$\varphi(x) = x^\delta L(x), \tag{2.5.3}$$

where L is a slowly varying function.

(c) If φ is regularly varying with the index δ, then for every $\varepsilon > 0$ the inequality

$$x^{\delta+\varepsilon} \leqslant \varphi(x) \leqslant x^{\delta-\varepsilon} \tag{2.5.4}$$

holds in a suitable vicinity of the origin.

If there exists a finite and positive limit of φ at the origin, then clearly φ is slowly varying. The converse is not true: a slowly varying function may approach zero or infinity, or even fail to approach any definite value (finite or infinite) as $x \to 0$. Some examples are exhibited in Note 2.8.11. Also, it follows from (2.5.4) that a regularly varying function cannot tend at the origin to zero or infinity too rapidly.

2.5B. Homogeneous equation

The main result for this equation, i.e., for

$$\varphi(f(x)) = g(x)\varphi(x), \tag{2.5.5}$$

which is due to E. Seneta [7], will be proved with the aid of two lemmas.

Lemma 2.5.1. *Under conditions* (i)–(iii) *with* $g_0 = 1$ *every positive monotonic solution* $\varphi\colon X \to \mathbb{R}$ *of equation* (2.5.5) *is slowly varying.*

Proof. Let $\varphi\colon X \to \mathbb{R}$ be a positive increasing solution of (2.5.5), and take a $\lambda \in (s, 1)$. Then $f(x) < \lambda x$ for sufficiently small positive x, and

$$g(x) = \frac{\varphi(f(x))}{\varphi(x)} \leqslant \frac{\varphi(\lambda x)}{\varphi(x)} \leqslant 1,$$

and limit (2.5.2) is equal to 1. If φ is decreasing, the inequalities are reversed, but the conclusion remains the same. If $0 < \lambda \leqslant s$, then there exists a $k \in \mathbb{N}$ such that $s < \lambda^{1/k} < 1$, and (with $\mu := \lambda^{1/k}$, $y_i := \mu^i x$)

$$\lim_{x \to 0} \frac{\varphi(\lambda x)}{\varphi(x)} = \lim_{x \to 0} \prod_{i=0}^{k-1} \frac{\varphi(\mu y_i)}{\varphi(y_i)} = 1.$$

If $\lambda > 1$, then $\lambda^{-1} < 1$, and

$$\lim_{x \to 0} \frac{\varphi(\lambda x)}{\varphi(x)} = \lim_{x \to 0} \frac{\varphi(\lambda^{-1}(\lambda x))}{\varphi(\lambda x)} = 1.$$

It remains to prove (see Definition 2.5.1) that φ is continuous. By Theorem 2.3.3 (formula (2.5.7)) positive monotonic solutions φ of equation (2.5.5) (exist and) are given by

$$\varphi(x) = \varphi(x_0) \prod_{n=0}^{\infty} \frac{g(f^n(x_0))}{g(f^n(x))},$$

where x_0 may be taken arbitrarily from X. Assume that g is increasing (leaving the other case to the reader); then we have

$$\left(\frac{g(f^n(x_0))}{g(f^n(x))} - 1 \right)(x - x_0) \leqslant 0$$

Hence φ, being the limit of a monotonic sequence of continuous functions, is lower semicontinuous for $x \leqslant x_0$ and upper semicontinuous for $x \geqslant x_0$. Since x_0 may be arbitrary, φ is continuous on X. ∎

Taking $g(x) = sx/f(x)$, we get as a corollary from Theorem 2.3.12 and Lemma 2.5.1 (see also Kuczma [13], Lundberg [1]) the following.

Lemma 2.5.2. *Under conditions* (i) *and* (ii), *for every* $x \in X$ *there exists a finite and positive limit*

$$\sigma(x) = \lim_{n \to \infty} \frac{f^n(x)}{f^n(x_0)} \tag{2.5.6}$$

(*where* x_0 *is arbitrarily fixed*). *The function* σ *is regularly varying with index* 1, *and satisfies the Schröder equation*

$$\sigma(f(x)) = s\sigma(x). \tag{2.5.7}$$

Theorem 2.5.1. *Let hypotheses* (i)–(iii) *be fulfilled. Then equation* (2.5.5) *has a unique one-parameter family of regularly varying solutions* $\varphi: X \to \mathbb{R}$. *These solutions are given by the formula*

$$\varphi(x) = c[\sigma(x)]^\delta \prod_{n=0}^{\infty} \frac{g(f^n(x_0))}{g(f^n(x))}, \tag{2.5.8}$$

where $x_0 \in X$ *is arbitrarily fixed,* $c > 0$ *is an arbitrary constant (a parameter),* σ *is given by* (2.5.6), *and*

$$\delta = \log g_0 / \log s \tag{2.5.9}$$

is the index of φ.

Proof. The product in (2.5.8) represents a positive monotonic solution $\hat{\varphi}$ of the equation

$$\hat{\varphi}(f(x)) = \frac{g(x)}{g_0} \hat{\varphi}(x) \tag{2.5.10}$$

(Theorem 2.3.3) and by Lemma 2.5.1 is a slowly varying function. By Lemma 2.5.2, $(\sigma)^\delta$ is regularly varying with index δ. Relation (2.5.5) results from (2.5.7), (2.5.9) and (2.5.10), since $\varphi(x) = [\sigma(x)]^\delta \hat{\varphi}(x)$.

To prove the uniqueness, suppose we are given a regularly varying solution $\varphi: X \to \mathbb{R}$ of (2.5.5), with an index δ. Then (see (a))

$$g_0 = \lim_{x \to 0} \frac{\varphi(f(x))}{\varphi(x)} = \lim_{x \to 0} \frac{\varphi\left(\frac{f(x)}{x} x\right)}{\varphi(x)} = s^\delta,$$

whence (2.5.9) follows. Thus we get by (2.5.5) and (2.5.3) (putting, for short, $x_n := f^n(x)$, $y_n := f^n(x_0)$)

$$\frac{\varphi(x_n)}{\varphi(y_n)} = \left[\frac{x_n}{y_n}\right]^\delta \frac{L[(x_n/y_n)y_n]}{L(y_n)},$$

where L is slowly varying. Hence, again by (a), according to (2.5.6)

$$\lim_{n \to \infty} \frac{\varphi(x_n)}{\varphi(y_n)} = [\sigma(x)]^\delta.$$

We obtain from (2.5.5) by induction

$$\frac{\varphi(x)}{\varphi(x_0)} = \frac{\varphi(x_k)}{\varphi(y_k)} \prod_{n=0}^{k-1} \frac{g(y_n)}{g(x_n)}, \qquad k \in \mathbb{N},$$

whence (2.5.8) follows (with $c = \varphi(x_0)$). ∎

Remark 2.5.1. If $\delta = 0$, then the solutions (2.5.8) are also monotonic.

2.5C. Special inhomogeneous equation

For the equation

$$\varphi(f(x)) = \varphi(x) + h(x) \tag{2.5.11}$$

we also have a uniqueness result (Theorem 2.5.2; see Kuczma [43]). We begin with the following.

Lemma 2.5.3. *Let hypotheses* (i), (ii) *and* (iv) *be fulfilled. If* $\varphi: X \to \mathbb{R}$ *is a regularly varying solution of equation* (2.5.11), *then there exists a (finite or not) limit* $\lim_{x \to 0} \varphi(x)$. *This limit is finite if and only if the series*

$$\Phi(x) = \sum_{n=0}^{\infty} h(f^n(x)) \tag{2.5.12}$$

converges.

Proof. (2.5.11) yields by induction

$$\varphi(f^k(x)) = \varphi(x) + \sum_{n=0}^{k-1} h(f^n(x)), \qquad k \in \mathbb{N}. \tag{2.5.13}$$

Since, for any fixed $x \in X$, $h(f^n(x))$ have a constant sign (at least for large n), consequently there exists a finite or infinite limit $\lim_{k \to \infty} \varphi(f^k(x))$, and this limit is finite if and only if series (2.5.12) converges. If the index of φ is $\neq 0$, the existence of a limit of φ at zero results from (2.5.4). Thus we need consider only the case where φ is slowly varying, and the lemma will be proved if we show that for every decreasing sequence $(t_n)_{n \in \mathbb{N}_0}$, $t_n \in X$, that converges to zero, we have

$$\lim_{n \to \infty} \varphi(t_n) = \lim_{n \to \infty} \varphi(f^n(t_0)). \tag{2.5.14}$$

Indeed, for every $n \in \mathbb{N}_0$ there exists an $m(n) \in \mathbb{N}_0$ such that $f(y_n) \leq t_n \leq y_n$, where $y_n := f^{m(n)}(t_0)$. The sequence $(y_n)_{n \in \mathbb{N}_0}$ is a subsequence of $(f^n(t_0))_{n \in \mathbb{N}_0}$, so that $y_n \to 0$ and $\lim_{n \to \infty} \varphi(y_n) = \lim_{n \to \infty} \varphi(f^n(t_0))$. Fix an s', $0 < s' < s$. Then $s' \leq f(y_n)/y_n \leq t_n/y_n \leq 1$ for sufficiently large n. Since φ is slowly varying, we have

$$\lim_{n \to \infty} \frac{\varphi(t_n)}{\varphi(y_n)} = \lim_{n \to \infty} \frac{\varphi\left(\dfrac{t_n}{y_n} y_n\right)}{\varphi(y_n)} = 1,$$

which implies (2.5.14). ∎

Remark 2.5.2. It follows from Lemma 2.5.3 and relation (2.5.13) that if $\varphi: X \to \mathbb{R}$ is a regularly varying solution of equation (2.5.11) and series (2.5.12) converges, then necessarily

$$\varphi(x) = C - \Phi(x), \tag{2.5.15}$$

where $C := \lim_{x \to 0} \varphi(x) \geq 0$ may be considered as a parameter. Thus in such a case equation (2.5.11) may have at most a one-parameter family of regularly varying solutions.

To get an existence theorem in the case just spoken about, we establish first the following.

Lemma 2.5.4. *Let hypotheses* (i), (ii) *and* (iv) *be fulfilled. If, moreover, h is regularly varying with the index* $\delta > 0$, *then so is* Φ *given by* (2.5.12), *with the same index* δ.

Proof. We shall estimate series (2.5.12) by two geometric ones with quotients s_1 and s_2 as close to s as one wishes and prove this way the existence of the limit wanted.

Fix an $\varepsilon > 0$, $\varepsilon < \min(s, 1 - s)$, so that $s_1 := (s + \varepsilon)^\delta (1 + \varepsilon) < 1$, and write $s_2 := (s - \varepsilon)^\delta (1 - \varepsilon)$. Fix a positive $\mu < \delta$, and write $L(x) = x^{-\delta} h(x)$. Then L is a slowly varying function. We can find an $A \in X$ such that the inequalities

$$s - \varepsilon < f(x)/x < s + \varepsilon, \tag{2.5.16}$$

$$1 - \varepsilon < L(f(x))/L(x) < 1 + \varepsilon, \tag{2.5.17}$$

$$0 < h(x) < x^\mu \tag{2.5.18}$$

hold for $x \in (0, A)$. It follows from (2.5.16) and (2.5.18) that series (2.5.12) converges a.u. in X, therefore Φ is continuous and positive in X.

Take a $\lambda > 0$. We get by (2.5.16) and (2.5.17) for $x \in (0, A/\lambda) \cap X$ and $n \in \mathbb{N}_0$

$$\lambda(s - \varepsilon)^n < f^n(\lambda x)/x < \lambda(s + \varepsilon)^n$$

and

$$(1 - \varepsilon)^n < L(f^n(\lambda x))/L(\lambda x) < (1 + \varepsilon)^n.$$

Hence

$$\frac{\Phi(\lambda x)}{h(x)} = \sum_{n=0}^\infty \frac{h(f^n(\lambda x))}{h(x)} = \sum_{n=0}^\infty \left[\frac{f^n(\lambda x)}{x} \right]^\delta \frac{L(f^n(\lambda x))}{L(x)}$$

$$< \sum_{n=0}^\infty \lambda^\delta (s + \varepsilon)^{n\delta} (1 + \varepsilon)^n \frac{L(\lambda x)}{L(x)} = \frac{\lambda^\delta}{1 - s_1} \frac{L(\lambda x)}{L(x)},$$

and similarly

$$\frac{\Phi(\lambda x)}{h(x)} > \frac{\lambda^\delta}{1 - s_2} \frac{L(\lambda x)}{L(x)},$$

$x \in (0, A/\lambda) \cap X$. Letting $x \to 0$ we obtain, since L is slowly varying

$$\frac{\lambda^\delta}{1 - s_2} \leqslant \liminf_{x \to 0} \frac{\Phi(\lambda x)}{h(x)} \leqslant \limsup_{x \to 0} \frac{\Phi(\lambda x)}{h(x)} \leqslant \frac{\lambda^\delta}{1 - s_1},$$

whence we get on letting $\varepsilon \to 0$

$$\lim_{x \to 0} \frac{\Phi(\lambda x)}{h(x)} = \frac{\lambda^\delta}{1 - s}. \tag{2.5.19}$$

For $\lambda = 1$ limit (2.5.19) becomes $1/(1-s)$, which together with (2.5.19) yields $\lim_{x\to 0}[\Phi(\lambda x)/\Phi(x)] = \lambda^{\delta}$. ∎

Theorem 2.5.2. *Let hypotheses* (i), (ii), (iv) *be fulfilled. If the function* $-h$ *is regularly varying with the index* $\delta > 0$, *then equation* (2.5.11) *has a unique one-parameter family of regularly varying solutions* $\varphi: X \to \mathbb{R}$. *They are given by formula* (2.5.15) *with* Φ *defined by* (2.5.12) *and an arbitrary constant* $C \geqslant 0$. *The index of* φ *is* δ *if* $C = 0$, *and the solutions are slowly varying if* $C > 0$.

Proof. This is a direct consequence of Lemmas 2.5.3 and 2.5.4. Note that, since $\delta > 0$, we have $\lim_{x\to 0} \Phi(x) = 0$, whence the limit of φ (at zero) is C and φ is slowly varying if $C > 0$. ∎

In the case where the series (2.5.12) diverges, we have the following result (Seneta [10]).

Theorem 2.5.3. *Let hypotheses* (i), (ii), (iv) *be fulfilled, let h be positive and monotonic in X and $h_0 := \lim_{x\to 0} h(x) < \infty$, and let series* (2.5.12) *diverge. Then there exists a slowly varying function L such that every continuous positive solution* $\varphi: X \to \mathbb{R}$ *of equation* (2.5.11) *has the property* $\lim_{x\to 0} \varphi(x)/L(x) = 1$ *and φ is therefore itself slowly varying.*

Proof. Write $g(x) = \exp h(x)$, $x \in X$. Then g fulfils hypothesis (iii), and by Theorem 2.5.1 equation (2.5.5) has a regularly varying solution $\varphi_0: X \to \mathbb{R}$ with the index $\delta = h_0/\log s$. If $h_0 = 0$, then $\delta = 0$ and φ_0 is monotonic in X; see Remark 2.5.1 (in this case necessarily decreasing). By (2.5.5)

$$\varphi_0(f^n(x)) = \left(\exp \sum_{i=0}^{n-1} h(f^i(x))\right)\varphi_0(x),$$

whence it follows by letting $n \to \infty$ that $\varphi_0(f^n(x)) \to \infty$, i.e.

$$\lim_{x\to 0} \varphi_0(x) = \infty. \tag{2.5.20}$$

If $h_0 > 0$, then $\delta < 0$ and (2.5.20) follows by (2.5.4). Thus the function $L(x) := \log \varphi_0(x)$ is continuous and positive in a vicinity of the origin, say in $(0, A) \subset X$, and satisfies equation (2.5.11) in $(0, A)$. Moreover, since φ_0 is regularly varying, L is slowly varying.

Now, if φ is an arbitrary continuous and positive solution of equation (2.5.11) on X, then the function $\omega(x) := \varphi(x) - L(x)$ is continuous on $(0, A)$ and satisfies the equation $\omega(f(x)) = \omega(x)$ in $(0, A)$. Hence ω is bounded on $(0, A)$ and

$$\lim_{x\to 0} \frac{\varphi(x)}{L(x)} = 1 + \lim_{x\to 0} \frac{\omega(x)}{L(x)} = 1,$$

since by (2.5.20) $\lim_{x\to 0} L(x) = \infty$. ∎

Remark 2.5.3. In the case where $h_0 > 0$ we have $\varphi_0(x) = x^\delta L_0(x)$ with $\delta = h_0/\log s$ and a slowly varying L_0. Then $\lim_{x \to 0} \log \varphi_0(x)/\log x = \delta$, and we may take as L in Theorem 2.5.3 the function $x \mapsto \delta \log x = h_0 \log x/\log s$.

Comments. Further results on regularly varying solutions of functional equations may be found in Coifman [1]–[3], Coifman–Kuczma [1], Hoppe [1], Kuczma [43], Seneta [7], [10].

The role of regularly varying functions in the theory of branching processes is explained in Seneta [9]. This class of functions is also useful in other branches of the probability theory; see Feller [1].

2.6 Application to branching processes

We return to the problems from the theory of Galton–Watson processes we have left without solutions in Section 2.1. For basic notations and facts on such processes the reader is referred to that section.

2.6A. Conditional limit probabilities

The probabilities a_j we have introduced in Subsection 2.1B can be determined as follows.

Theorem 2.6.1. *Let the RV Z_n represent the size of the nth generation in a simple Galton–Watson process (descending from a single ancestor). Assume that*

$$m \neq 1 \quad and \quad p_0 > 0.$$

Then there exist the limits of the conditional probabilities

$$a_j = \lim_{n \to \infty} P(Z_n = j \mid n < T < \infty), \quad j \in \mathbb{N}. \tag{2.6.1}$$

They satisfy the condition

$$\sum_{j=1}^{\infty} a_j = 1, \tag{2.6.2}$$

and their PGF $A: [0, 1] \to \mathbb{R}$, given by

$$A(t) = \sum_{j=1}^{\infty} a_j t^j \tag{2.6.3}$$

is the unique solution of the functional equation

$$1 - A(F(y)/q) = F'(q)(1 - A(y/q)), \quad y \in [0, q] \tag{2.6.4}$$

in the class of PGFs such that $A(0) = 0$.

Proof. First we find a suitable solution of equation (2.6.4). Substituting in (2.6.4)

$$y = q - x, \qquad f(x) = q - F(q - x), \qquad \sigma(x) = 1 - A\left(\frac{q - x}{q}\right),$$

all for $x \in [0, q]$, we see that it is equivalent to the Schröder equation

$$\sigma(f(x)) = s\sigma(x), \qquad s := f'(0), \qquad x \in [0, q]. \qquad (2.6.5)$$

We want to apply Theorem 2.3.12, therefore the function f is to be examined in more detail.

The properties of the PGF F (see Subsection 2.1A) imply the following: the function f maps the interval $[0, q]$ onto itself, $0 < f(x) < x$ in $(0, q]$ (since $t < F(t) < q$ in $[0, q)$); f is continuous and concave on $[0, q]$ (F is convex there), so that the function $x \mapsto f(x)/x$ is decreasing in $(0, q]$.

Moreover, f is analytic on $(0, q]$ (as F is on $[0, q)$) and $0 < s < 1$. Indeed, $s = f'(0) = F'(q)$, and $0 < F'(q) \leqslant 1$. Equality $F'(q) = 1$ would mean $m = F'(1) = 1$ if $q = 1$ or $F'(t) \geqslant 1$ for $t > q$ if $q < 1$ (since F is convex). Both cases are impossible, the first by the assumption $m \neq 1$, the other since it implies $F(t) \geqslant t$ for $t \geqslant q$, contrary to the actual property of F.

We see that Theorem 2.3.12 works in our case, yielding the existence of the limit

$$\sigma(x) = \lim_{n \to \infty} \frac{f^n(x)}{f^n(q)}, \qquad x \in [0, q],$$

which is a solution of equation (2.6.5) such that the quotient $\sigma(x)/x$ is monotonic in $(0, q]$. Consequently the sequence with the terms

$$A_n(t) = 1 - \frac{f^n(x)}{f^n(q)}, \qquad x = q(1 - t)$$

(see (2.1.5)), converges in $[0, 1]$ to $A(t) = 1 - \sigma(q(1 - t))$, being a solution of equation (2.6.4) fulfilling $A(0) = 1 - \sigma(q) = 0$.

It follows from the continuity theorem for the Laplace transform (see Note 2.8.13) that $A(t) = \lim_{n \to \infty} A_n(t)$ actually has expansion (2.6.3) with the coefficients which are the limits of the corresponding coefficients of the expansion of $A_n(t)$ (see Subsection 2.1B), i.e. (2.6.1) is satisfied. Moreover, we have $A(1) = \sum_{j=1}^{\infty} a_j = 1 - \sigma(0) = 1$, as stated in (2.6.2).

It remains to prove the uniqueness. Let A^* be another probability generating solution of (2.6.4) in $[0, 1]$, $A^*(0) = 0$. Then $\sigma^*(x) := 1 - A^*(1 - x/q)$ satisfies equation (2.6.5) in $[0, q]$ and the quotient $\sigma^*(x)/x$ is monotonic in $(0, q]$. In fact,

$$\frac{\sigma^*(x)}{x} = \frac{1 - A^*(t)}{q(1 - t)} = \frac{1}{q} \frac{A^*(1) - A^*(t)}{1 - t},$$

and the latter is the difference quotient of the convex A^*, thus it is necessarily monotonic. The uniqueness statement of Theorem 2.3.12 now yields the existence of a constant c such that $\sigma^*(x) = c\sigma(x)$, whence $1 = \sigma^*(q) = c\sigma(q) = c$. ∎

The following theorem concludes our discussion of the case $m = 1$; see the last part of Subsection 2.1B.

Theorem 2.6.2. *Let* $m = 1$. *Then for every* $i \in \mathbb{N}$ *and* $j \in \mathbb{N}$ *there exist the limits*

$$u_j(i) = \lim_{n \to \infty} P(Z_n = j \mid T = n + i),$$

determined by their PGFs

$$U_i(t) = \sum_{j=1}^{\infty} u_j(i)t^j, \qquad (2.6.6)$$

which are of the form

$$U_i(t) = \alpha(1 - tF^i(0)) - \alpha(1 - tF^{i-1}(0)), \qquad t \in [0, 1], \qquad (2.6.7)$$

where $\alpha \colon (0, 1] \to \mathbb{R}$ *is a convex solution of the Abel equation*

$$\alpha(f(x)) = \alpha(x) + 1, \qquad (2.6.8)$$

and $f(x) := 1 - F(1 - x)$. *Every function* U_i *fulfils the condition*

$$U_i(1) = \sum_{j=1}^{\infty} u_j(i) = 1. \qquad (2.6.9)$$

Proof. Since the PGF F is convex in $[0, 1]$, f is concave there. Moreover, $f'(0) = F'(1) = m = 1$. Theorem 2.4.3 yields the existence of limit (2.1.11), which is a convex solution of (2.6.8). Thus formula (2.6.7) results from (2.1.10) and (2.1.11). Convex solutions of (2.6.8) differ by a constant, whence U_i is uniquely determined. Expansion (2.6.6) again follows by the continuity theorem for the Laplace transform. Setting $t = 1$ in (2.6.7) and recalling that $x_0 = 1 - F^{i-1}(0)$ implies $f(x_0) = 1 - F^i(0)$ we get by (2.6.8) $U_i(1) = \alpha(f(x_0)) - \alpha(x_0) = 1$, i.e. (2.6.9) also holds true. ∎

2.6B. Stationary measures

We pass to the question of existence and uniqueness of stationary measures for a Galton–Watson process.

As we have seen in Subsection 2.1C, to determine a stationary measure for the Galton–Watson process $(\bar{Z}_n)$ with immigration it is enough to find a solution $V \colon [0, q) \to \mathbb{R}$ of the equation

$$V(t) = G(t)V(F(t)), \qquad t \in [0, q), \qquad (2.6.10)$$

having the expansion

$$V(t) = \sum_{j=0}^{\infty} v_j t^j, \qquad (2.6.11)$$

with nonnegative coefficients v_j, not all zero, convergent in a neighbourhood of the origin. Function (2.6.11) then generates the stationary measure $(v_j)_{j \in \mathbb{N}_0}$.

Let us put, for $x \in [0, q]$,

$$f(x) = q - F(q - x), \qquad g(x) = 1/G(q - x); \qquad \varphi(x) = V(q - x) \quad (x \neq 0).$$

Then equation (2.6.10) is equivalent to

$$\varphi(f(x)) = g(x)\varphi(x), \qquad x \in (0, q]. \tag{2.6.12}$$

Theorem 2.6.3. *Under the conditions*

$$0 < p_0 < 1, \qquad 0 < q_0 < 1 \quad and \quad m \leq 1, \qquad G(1) = 1$$

there exists a stationary measure for the Galton–Watson process $(\bar{Z}_n)$ with immigration. The measure is unique up to a constant factor and the generating functions $V: [0, 1] \to \mathbb{R}$ are given by the formula

$$V(t) = c \prod_{n=0}^{\infty} \frac{G(F^n(t))}{G(F^n(0))}, \qquad c > 0. \tag{2.6.13}$$

Proof. Since $m = F'(1) \leq 1$, we have $q = 1$ and $g(0) = 1/G(1) = 1$. By Theorem 2.3.3 equation (2.6.12) has a unique monotonic solution $\varphi: (0, 1] \to \mathbb{R}$ such that $\varphi(1) = 1$, given by (see (2.3.7))

$$\varphi(x) = \prod_{n=0}^{\infty} \frac{g(f^n(1))}{g(f^n(x))}.$$

The function $\tilde{V}: [0, 1) \to \mathbb{R}$, $\tilde{V}(t) = \varphi(1 - t)$ is given by (2.6.13) with $c = 1$ and satisfies equation (2.6.10). The partial products of the infinite product in (2.6.13) can be represented as series with nonnegative coefficients, convergent in $[0, 1)$ (since G and F are PGFs). By the continuity theorem for the Laplace transform the limit function $\tilde{V}$ has expansion (2.6.11) with nonnegative v_js, convergent in $[0, 1)$, i.e. it generates a stationary measure for $(\bar{Z}_n)$. Any monotonic solution of (2.6.12) is a constant multiple of φ. Thus (2.6.13) represents all the functions that generate stationary measures, i.e., any two such measures are proportional. ∎

If $m > 1$ and $q_0 < 1$, then $q < 1$, $G(q) < G(1) = 1$ (G is strictly increasing), and $g(0) = 1/G(q) > 1$. We have no uniqueness assertion for monotonic solutions of (2.6.12) in such a case. However, the uniqueness can be assured in the class of functions regularly varying at the origin. The same situation occurs if we have $m < 1$, but G generates a defective probability distribution: $G(1) < 1$. First we prove a theorem concerning the latter case.

Theorem 2.6.4. *Under the conditions*

$$0 < p_0 < 1, \qquad 0 < q_0 < 1 \quad and \quad m < 1, \qquad G(1) < 1$$

there exists a unique (up to a constant factor) stationary measure for $(\bar{Z}_n)$, *generated by a solution* $V: [0, 1) \to \mathbb{R}$ *of equation (2.6.10) such that the function* $x \mapsto V(1 - x)$ *is regularly varying at the origin.*

Proof. We may apply Theorem 2.5.1 to equation (2.6.12) (in particular, $s = f'(0) = F'(1) = m$). All regularly varying solutions of (2.6.12) are then given by the formula (see (2.5.8))

$$\varphi(x) = c[\sigma(x)]^\delta \prod_{n=0}^{\infty} \frac{g(f^n(1))}{g(f^n(x))} \tag{2.6.14}$$

with $c > 0$, $\delta = \log g(0)/\log m < 0$ as $g(0) = 1/G(1) > 1$, and

$$\sigma(x) = \lim_{n \to \infty} \frac{f^n(x)}{f^n(1)}. \tag{2.6.15}$$

The function $V: [0, 1) \to \mathbb{R}$, $V(t) = \varphi(1 - t)$, satisfies equation (2.6.10) and can be written in the form

$$V(t) = c(1 - A(t))^\delta V_0(t), \qquad t \in [0, 1). \tag{2.6.16}$$

Here $A(t) := 1 - \sigma(1 - t)$, with σ given by (2.6.15), is a solution of equation (2.6.4) in which $q = 1$, $F'(q) = m$. Inspecting the proof of Theorem 2.6.1 we find that A is just the PGF given by (2.6.3). We have to examine the power series expansion of function (2.6.16).

The function V_0 generates a stationary measure for the Galton–Watson process $(\tilde{Z}_n)$ with the PGF of immigration $G_0(t) := G(t)/G(1)$. To prove this use can be made of Theorem 2.6.3, which applies since $G_0(1) = 1$ and yields (see (2.6.13))

$$V_0(t) = \prod_{n=0}^{\infty} \frac{G_0(F^n(t))}{G_0(F^n(0))} = \prod_{n=0}^{\infty} \frac{G(F^n(t))}{G(F^n(0))}.$$

Putting here $t = 1 - x$ we get the infinite product from (2.6.14), and (2.6.16) follows. Therefore the V given by (2.6.16) as the product of the two series with nonnegative coefficients $V_0(t)$ and $(1 - A(t))^\delta$ (where $A(t)$ has expansion (2.6.3), $0 \le A(t) < 1$ for $t \in [0, 1)$, $\delta < 0$) itself represents such a series, convergent, as both factors are, in $[0, 1)$. By this, V actually generates a stationary measure for the process $(\bar{Z}_n)$. ∎

Theorem 2.6.5. *Under the conditions*

$$0 < p_0 < 1, \qquad 0 < q_0 < 1 \qquad and \qquad m > 1, \qquad G(1) = 1,$$

there is a unique stationary measure (up to a constant factor) of the process $(\bar{Z}_n)$ *and it is generated by the solution* $V: [0, q) \to \mathbb{R}$ *of equation (2.6.10) such that the function* $x \mapsto V(q(1 - x))$ *is regularly varying at the origin.*

Proof. The condition $m > 1$ implies $0 < q < 1$. With $W(t) := V(qt)$, $F_q(t) := F(qt)/q$, $G_q(t) := G(qt)$, all for $t \in [0, 1)$, equation (2.6.10) goes over into

$$W(t) = G_q(t) W(F_q(t)). \tag{2.6.17}$$

We have $m_q := F'_q(1) = F'(q) < 1$ (similarly as is shown in the proof of Theorem 2.6.1) and $G_q(1) = G(q) < 1$ (by the strict monotonicity of G). Thus for equation (2.6.17) Theorem 2.5.4 applies. ∎

Remark 2.6.1. In the case $m < 1$, $G(1) = 1$, the unique stationary measure (Theorem 2.6.3) always has a slowly varying generating function V ($\delta = 0$). This is a property of monotonic solutions of equation (2.6.12); see Lemma 2.5.1.

2.6C. Restricted stationary measures

Finally, we answer the questions of the existence of restricted stationary measures (RSMs) for the simple Galton–Watson process; see Subsection 2.1D.

An RSM $(u_j)_{j \in \mathbb{N}}$ is now defined on $\mathbb{N}$ and may be determined through the generating function

$$U(t) = \sum_{j=1}^{\infty} u_j t^j, \qquad u_j \geq 0, \ U(t) \neq 0. \tag{2.6.18}$$

The series (2.6.18) should converge in a neighbourhood of the origin. Let us recall (see Subsection 2.1D) that such Us are constant multiples of solutions of the Abel equation

$$U(F(t)) = U(t) + 1. \tag{2.6.19}$$

We start with the case $m = 1$, in which $q = 1$ and $f(x) = 1 - F(1 - x)$, $x \in [0, 1)$.

Theorem 2.6.6. Let $m = 1$. Then the numbers

$$u_j := \frac{u_j(i)}{[F^i(0)]^j - [F^{i-1}(0)]^j}, \qquad i \in \mathbb{N}, j \in \mathbb{N}, \tag{2.6.20}$$

where $u_j(i)$ are defined by (2.1.7), do not depend on i and constitute an RSM for the process (Z_n). The series of (2.6.18) converges in $[0, 1)$ and the function U generating the RSM is the unique convex solution of equation (2.6.19) satisfying the condition $U(0) = 0$. Any other RSM differs from that given by (2.6.20) by a constant factor.

Proof. The equality $(F^i(0))^j = p^i_{j0}$ (see (2.1.8)), when applied in the formula

$$u_j(i) = \lim_{n \to \infty} \frac{p^n_j(p^i_{j0} - p^{i-1}_{j0})}{\sum_{k=1}^{\infty} p^n_k(p^i_{k0} - p^{i-1}_{k0})}$$

(see (2.1.9)) yields, in view of (2.6.20),

$$u_j = \lim_{n \to \infty} \frac{p^n_j}{\sum_{k=1}^{\infty} p^n_k([F^i(0)]^k - [F^{i-1}(0)]^k)}$$

$$= \lim_{n \to \infty} \frac{p^n_j}{F^n(F^i(0)) - F^n(F^{i-1}(0))}.$$

The sequence with the terms

$$\frac{F^n(0) - F^{n-1}(0)}{F^n(F^i(0)) - F^{n-1}(F^i(0))} = \frac{f^n(1) - f^{n-1}(1)}{f^n(y) - f^{n-1}(y)}$$

(with $y := 1 - F^i(0)$, $f(x) = 1 - F(1-x)$) approaches, by Lemma 2.4.1, the limit 1. Consequently

$$u_j = \lim_{n \to \infty} \frac{p_j^n}{F^n(0) - F^{n-1}(0)} \qquad (2.6.21)$$

and the u_j are actually independent of i.

Next we prove the convergence of the series of (2.6.18) with u_j given by (2.6.21) (which is equivalent to (2.6.20)). Observe first the inequality (for any $z \in (0, 1)$)

$$p_j^n \leq z^{-j} \sum_{k=1}^{\infty} p_k^n z^k = z^{-j}(F^n(z) - F^n(0)). \qquad (2.6.22)$$

The limit

$$\lim_{n \to \infty} \frac{F^n(z) - F^n(0)}{F^n(0) - F^{n-1}(0)} = \lim_{n \to \infty} \frac{f^n(x) - f^n(1)}{f^n(1) - f^{n-1}(1)} \qquad (2.6.23)$$

(with $z = 1 - x$ and $f(x) = 1 - F(z)$) exists and represents a convex solution, say α, of the Abel equation

$$\alpha(f(x)) = \alpha(x) + 1, \qquad x \in (0, 1] \qquad (2.6.24)$$

(see Theorem 2.4.3). Hence the sequence from (2.6.23) is bounded by a constant, say $C(z)$. Take a $t \in [0, 1)$ and a $z \in (t, 1)$. Recalling (2.6.22), we get the estimate

$$u_j^n(t) := \frac{p_j^n t^j}{F^n(0) - F^{n-1}(0)} \leq \left(\frac{t}{z}\right)^j C(z)$$

and the series with the terms $u_j^n(t)$ converges, uniformly with respect to n. Thus (see (2.6.21) and (2.6.18))

$$\lim_{n \to \infty} \frac{F^n(t) - F^n(0)}{F^n(0) - F^{n-1}(0)} = \lim_{n \to \infty} \sum_{j=1}^{\infty} u_j^n(t) = \sum_{j=1}^{\infty} \lim_{n \to \infty} u_j^n(t)$$

$$= \sum_{j=1}^{\infty} u_j t^j = U(t).$$

In view of (2.6.23) the function $U(t) = \alpha(1 - t)$ for $t \in [0, 1)$ is a convex solution of equation (2.6.19) in $[0, 1)$ (under the substitutions introduced, equation (2.6.19) is equivalent to (2.6.24)), fulfilling $U(0) = 0$.

According to Theorem 2.4.3 convex solutions of (2.6.24) can differ only by a constant. The same is true for convex solutions of (2.6.19), and the condition $U(0) = 0$ determines the solution uniquely.

Finally, if U^* generates an RSM then $U(t) := U^*(t)/U^*(F(0))$ is a convex solution of (2.6.19) (see (2.1.20)) such that $U(0) = 0$. The only such function

U is given by (2.6.18) with coefficients (2.6.20), i.e. U^* differs from that U by the factor $U^*(F(0))$. ∎

In the case $m \neq 1$ RSMs exist, but need not be unique, as has been shown by example by J. F. C. Kingman [1]. However, similarly as in the case of stationary measures for the Galton–Watson process with immigration, the uniqueness can be assured by appealing to regularly varying functions.

Theorem 2.6.7. *If $m \neq 1$ and $p_0 > 0$, then there exists an RSM for the process* (Z_n).

Proof. We shall construct the RSM with the aid of the PGF A, determined in Theorem 2.6.1. Note that the function A is analytic in $[0, 1)$, $A(t) < 1$ for $t \in [0, 1)$. Let us put

$$U(t) = \log\left(1 - A\left(\frac{t}{q}\right)\right)\Big/ \log F'(q), \qquad t \in [0, q), \qquad (2.6.25)$$

where $0 < F'(q) < 1$. We learn from (2.6.25) that $U \; (\neq 0)$ has a power series expansion with nonnegative coefficients, convergent in a neighbourhood of zero. Moreover, it is a solution of equation (2.6.19), as A satisfies equation (2.6.4). Thus U generates an RSM for the process (Z_n). ∎

Let $U: [0, q) \to \mathbb{R}$ be the generating function of an RSM for the process (Z_n), $U(F(0)) = 1$. Thus U satisfies equation (2.6.19) which is equivalent to the equation (we put $F_q(y) = F(qy)/q$)

$$W(F_q(y)) = \exp W(y), \qquad y \in [0, 1), \qquad (2.6.26)$$

for the function

$$W(y) = \exp U(qy), \qquad y \in [0, 1). \qquad (2.6.27)$$

Equation (2.6.26) is of the form (2.6.10). RSMs for (Z_n) may then be found with the aid of stationary measures for the Galton–Watson process $(\bar{Z}_n)$ with immigration, determined by the offspring PGF F_q and that of immigration $G = 1/e$. For the process $(\bar{Z}_n)$ we have $\bar{p}_0 = F_q(0) = F(0)/q = p_0/q < 1$, $q_0 = 1/e$, the mean $\bar{m} = F_q'(1) = F'(q) < 1$ and $G(1) = 1/e < 1$. Thus Theorem 2.6.4 applies to the process $(\bar{Z}_n)$, yielding the following uniqueness result.

Theorem 2.6.8. *Let $m \neq 1$ and $0 < p_0 < 1$. Then there is a unique function $U: [0, q) \to \mathbb{R}$ that generates an RSM for the simple Galton–Watson process (Z_n) and has the property that the function $x \mapsto \exp U(q(1-x))$, $x \in (0, 1]$, is regularly varying at the origin. This function is given by formula (2.6.25), where A is the PGF (2.6.3).*

Proof. We learn from the foregoing analysis that we should examine equation (2.6.26). According to Theorem 2.6.4, there exists a unique solution W of (2.6.26) generating a stationary measure for the process $(\bar{Z}_n)$ related to $(Z_n,$

which has the property that the function $y \mapsto W(1-y)$, $y \in [0, 1)$ is regularly varying and $W(0) = 1$. Thus the function U we are looking for is uniquely determined by (2.6.27). Moreover, as follows from the proof of Theorem 2.6.4, the function W is given by the formula

$$W(y) = (1 - A(y))^{\delta}, \qquad y \in [0, 1), \qquad (2.6.28)$$

where $\delta = 1/\log F'(q)$. Formula (2.6.28) is obtained from (2.6.16) since now $c = 1$ (as $W(0) = 1$) and $V_0 = 1$ (as $G = 1/e$). Inserting (2.6.28) into (2.6.27) we get (2.6.25). ∎

Comments. We have focused our attention on showing that results on iterative functional equations yield solutions of problems from the theory of Galton–Watson processes without any additional assumptions on the process itself, contrary to the situation that occurs if we use probability theory to solve the same problems. As in Section 2.1 we followed Kuczma [36], where the greater part of the relevant material is derived from the original papers by Heathcote–Seneta–Vere-Jones [1] and Seneta [1] (results in Subsection 2.6A), Heathcote [1], [2] and Seneta [2] (in 2.6B) and Seneta [2], [7] (in 2.6C).

Other facts both on Galton–Watson processes and on related functional equations may be found in Athreya [1], Bingham–Doney [1], [2], Doney [1], Dubuc [1], Durham [1], Hoppe [1]–[3], Pakes [1]–[4], Seneta [3], [4], [6], [8], [9], Seneta–Vere-Jones [1], [2]. See also Barbour [1], Pakes–Kaplan [1] for other stochastic processes, and Lukacs [1], Lukacs–Laha [1] for other use of functional equations in probability theory.

2.7 Convex solutions of higher order

We conclude this chapter with a discussion of p-convex solutions of the equation (see (2.4.12))

$$\varphi(x + 1) - \varphi(x) = h(x). \qquad (2.7.1)$$

2.7A. Definitions and results

We start with the following.

Definition 2.7.1. A function $\varphi: X \to \mathbb{R}$, where X is an open interval, is called *convex* (resp. *concave*) *of order* p in X iff φ is measurable and $\Delta_y^{p+1} \varphi(x) \geq 0$ (resp. ≤ 0) for all $x \in X$ and $y \geq 0$ such that $x + (p + 1)y \in X$ (see Section 2.0).

If a continuous function φ is convex or concave of order $p \geq 1$ in X, then it is of class C^{p-1} in X (Popoviciu [1], see also Kuczma [51, p. 392]).

The result contained in Theorem 2.4.2 can be generalized to the case of convex functions of order p. As a consequence, we shall get a characterization of polynomials.

We need the notion of Bernoulli polynomials.

Definition 2.7.2. The *Bernoulli polynomials* are defined by the recurrence

$$B_1(x) = x, \qquad B_n(x) = x^n - \sum_{i=1}^{n-1} \frac{1}{i}\binom{n}{i-1} B_i(x) \qquad \text{for } n = 2, 3, \ldots$$

It is easily shown by induction that B_n is a polynomial of degree n, $B_n(0) = 0$, and

$$B_n(x+1) - B_n(x) = nx^{n-1}, \qquad n \in \mathbb{N}.$$

Theorem 2.7.1. *Let* $X = [a, \infty)$, $a \in \mathbb{R}$, *or* $X = \mathbb{R}$. *If the function* $h: X \to \mathbb{R}$ *is concave (resp. convex) of order* p *in* X, $p \geq 1$, *and fulfils the condition*

$$\lim_{x \to \infty} \Delta_1^p h(x) = 0,$$

then equation (2.7.1) has a unique one-parameter family of solutions $\varphi: X \to \mathbb{R}$ *which are convex (resp. concave) of order* p *in* X. *These solutions are given by the formula*

$$\varphi(x) = c + \sum_{i=0}^{p-1} \frac{h^{(i)}(x_0)}{(i+1)!} B_{i+1}(x - x_0) - \sum_{n=0}^{\infty} \left\{ H(x+n) - H(x_0 + n) \right.$$
$$\left. - \sum_{i=0}^{p-1} \frac{1}{(i+1)!} B_{i+1}(x - x_0)[H^{(i)}(x_0 + n + 1) - H^{(i)}(x_0 + n)] \right\},$$

where $x_0 \in X$ *is arbitrarily fixed,* $c \in \mathbb{R}$ *is an arbitrary constant (a parameter),* B_i *are the Bernoulli polynomials, and*

$$H(x) = h(x) - \sum_{i=0}^{p-1} \frac{h^{(i)}(x_0)}{i!} (x - x_0)^i.$$

Proof. Induction on p, starting from Theorem 2.4.2 which is the particular case of our theorem for $p = 1$. ∎

If P is a polynomial of degree p, then $\Delta_y^{p+1} P(x) = 0$, thus P is both convex and concave of order p in $(-\infty, \infty)$. The following is an easy consequence of Theorem 2.7.1.

Theorem 2.7.2. *If*

$$h(x) = \sum_{i=0}^{p-1} a_i x^i$$

is a polynomial of degree $p - 1$, $p \geq 1$, *then equation (2.7.1) has a unique one-parameter family of polynomial solutions* φ. *These solutions are given by*

$$\varphi(x) = c + \sum_{i=0}^{p-1} \frac{a_i}{i+1} B_{i+1}(x), \tag{2.7.2}$$

where B_i *are the Bernoulli polynomials, and* $c \in \mathbb{R}$ *is an arbitrary constant (a parameter).*

Actually, Theorem 2.7.2 can be established directly, by observing that functions (2.7.2) are polynomial solutions of equation (2.7.1) and that the difference of two polynomial solutions of (2.7.1) is a periodic polynomial with period 1, and hence a constant. This argument applies also to complex polynomials.

2.7B. A characterization of polynomials

The corresponding result is due to M. Kuczma [14].

Theorem 2.7.3. Let $X = |a, \infty)$, $a \in \mathbb{R}$, or $X = \mathbb{R}$, and let $\varphi: X \to \mathbb{R}$ be a measurable function such that $\Delta_y^{p+1}\varphi(x)$ has a constant sign for all $x \in X$, $y \geq 0$, and vanishes identically in x for $y = 1$, where $p \in \mathbb{N}_0$ is fixed. Then φ is a polynomial of degree less than or equal to p.

Proof. Put

$$\varphi_i(x) = \Delta_1^i \varphi(x), \qquad i = 0, \ldots, p.$$

Then $\varphi_i: X \to \mathbb{R}$ is convex or concave of order $p - i$ in X (see Popoviciu [1]) and satisfies the equation

$$\varphi_i(x + 1) - \varphi_i(x) = \varphi_{i+1}(x),$$

where $\varphi_{p+1}(x) = 0$. Thus φ_p is periodic with period 1 and convex or concave of order zero, i.e. monotonic in X. Clearly $\varphi_p = \text{const}$. In other words, it is a polynomial of degree ≤ 0. Using Theorem 2.7.1 we prove by induction that φ_{p-i} is a polynomial of degree $\leq i$, $i = 0, \ldots, p$, whence in particular $\varphi = \varphi_0$ is a polynomial of degree $\leq p$. ∎

Comments. Concerning generalizations of Theorem 2.4.2 (on convex solutions) to the case of p-convex functions see Krull [1], Kuczma [14], Dufresnoy–Pisot [1]. The details of the proof of Theorem 2.7.1 are found in Kuczma [14], [26, p. 119]. For characterizations of the Bernoulli polynomials via functional equations see also Dickey–Kairies–Shank [1], Kairies [1].

2.8 Notes

2.8.1. Let an RV X have the probability distribution

$$p_0 = a, \qquad p_j = bc^{j-1}, \qquad j \in \mathbb{N}, \, a, b, c \in (0, 1),$$

where $a + b + c - ac = 1$. Then the PGF of X is $F(t) = a + bt/(1 - ct)$. This model has been used by A. J. Lotka [1] in the statistical analysis of data concerning generations of male lines in some American families. It turned out that well-suited constants are $a = 0.4825$, $b = 0.2126$, $c = 0.5893$. For the Galton–Watson process originated by X, this yields the mean $m = 1.321$ and the probability of extinction $q = 0.819$.

2.8.2. Formula (2.1.8) may be obtained by interpreting the RV Z_{k+n} (under the condition $Z_k = i$) as the sum of i independent RVs, each of them describing the size of the nth generation in the Galton–Watson process starting from a single ancestor – an object of the kth generation of the original process. Any one of such RVs, which are independent of each other and together compose Z_{k+n}, has the same probability distribution as the original Z_n, i.e., its PGF is F^n. Therefore (similarly as for p_{ij} in (2.1.2)) the PGF for the $p_{ij}^n = P(Z_{k+n} = j \mid Z_k = i)$ is the ith power of F^n, which yields (2.1.8).

2.8.3. Conditions (2.2.3) and (2.2.5) of Theorem 2.2.1 are implied by $X = [0, a|, 0 < a \leqslant \infty$, $f: X \to X$ is upper semicontinuous from the right on X, $f(x) < x$ in $X \setminus \{0\}$, $g, h: X \to \mathbb{R}$ and h is increasing in X. Moreover, $g \leqslant -1$, $h \geqslant 0$, $\lim_{x \to 0} h(x) = 0$ or $g < \theta < -1$, $h \leqslant 0$, where θ is a constant.

2.8.4. Let $X = |0, a|$, $0 < a \leqslant \infty$, $f: X \to X$ be continuous and strictly increasing in X, $0 < f(x) < x$ in $X \setminus \{0\}$ and $g, h: X \to \mathbb{R}$ be such that $g > 1$, $h \leqslant 0$. If the series (2.2.4) converges in X, then the equation

$$\varphi(f(x)) = g(x)\varphi(x) + h(x) \tag{2.8.1}$$

has a nonnegative solution in X depending on an arbitrary function.

(This follows from Theorem 2.2.3 applied to the equation

$$\varphi(f(x)) = g(x)\varphi(x) \tag{2.8.2}$$

and from Theorem 2.2.2.)

2.8.5. Consider the generalized dynamic programming equation for a mapping $\Phi: D \to \mathbb{R}$, where D is a region in $\mathbb{R}^m$:

$$\Phi(p) = \sup_{t \in S} F(p, t, \Phi(T(p, t))), \qquad \Phi(0) = 0. \tag{2.8.3}$$

Here S is a set of reals, $T: D \times S \to D$ and $F: D \times S \times \mathbb{R} \to \mathbb{R}$ are some mappings subjected to the conditions

$$|F(p, t, x_1) - F(p, t, x_2)| \leqslant \bar{g}(|p|)|x_1 - x_2|$$

in $D \times S \times \mathbb{R}$, with a $\bar{g}: \mathbb{R}^+ \to \mathbb{R}^+$, and

$$|T(p, t)| \leqslant f(|p|),$$

where $f: \mathbb{R}^+ \to \mathbb{R}^+$ is a nonnegative and increasing function, $f(0) = 0$. Let us put for $x \in \mathbb{R}^+$

$$\bar{h}(x) = \sup_{|p| \leqslant x} \sup_{t \in S} |F(p, t, 0)|$$

and

$$\bar{G}_n(x) = \prod_{i=0}^{n-1} \bar{g}(f^i(x)), \qquad n \in \mathbb{N}_0.$$

With the aid of nonnegative solutions of the equation

$$\varphi(x) = \bar{g}(x)\varphi(f(x)) + \bar{h}(x), \qquad x \in \mathbb{R}^+, \tag{2.8.4}$$

M. Kwapisz [3] has obtained the following result.

Proposition 2.8.1. *Under the above mentioned assumptions, if the series* $\varphi(x) = \sum_{n=0}^{\infty} \bar{h}(f^n(x))\bar{G}_n(x)$ *converges in* $\mathbb{R}^+$, *then there exists a unique solution* $\Phi : D \to \mathbb{R}$ *of equation (2.8.3). The solution has the properties*

$$\sup_{|p| \leqslant x} |\Phi(p)| \leqslant \varphi(x), \qquad x \in \mathbb{R}^+,$$

$$\sup_{|p| \leqslant x} |\Phi(p) - \Phi_k(p)| \leqslant \sum_{n=k}^{\infty} \bar{h}(f^n(x))\bar{G}_n(x), \qquad x \in \mathbb{R}^+, n \in \mathbb{N}_0,$$

where Φ_k *are the terms of the sequence of successive approximations for (2.8.3),* $\Phi_0 = 0$.

(Theorem 2.2.2 can be applied, yielding the unique nonnegative solution φ of (2.8.4).)

Both Bellman's theorems (Bellman [1, pp. 145–8]) for dynamic programming equations of the first and second kind follow from Proposition 2.8.1. To see this, put $\bar{g}(x) = 1$, $f(x) = bx$, $0 \leqslant b < 1$, and, respectively, $\bar{g}(x) = c$, $0 \leqslant c < 1$, $f(x) = x$.

2.8.6. In a similar way as we have derived Theorems 2.3.9.–2.3.11 from Theorems 2.3.4–2.3.6 one can obtain results for monotonic solutions of the difference equation

$$\varphi(x + 1) = \varphi(x) + h(x). \tag{2.8.5}$$

For instance, we have (see Theorem 2.3.4) the following.

Proposition 2.8.2. *If* $X = |a, \infty)$, $-\infty \leqslant a < \infty$, *and for* $h : X \to \mathbb{R}$ *we have* $\inf_X h = 0$ *or* $\sup_X h = 0$, *then any two monotonic solutions of equation (2.8.5) in* X *differ by a constant.*

2.8.7. The theorems of Section 2.3 remain to a great extent true if the assumptions are made only locally, in a neighbourhood of zero, or of infinity. For instance, for the equation

$$\varphi(x + 1) = g(x)\varphi(x) \tag{2.8.6}$$

the following version of Theorem 2.3.11 can be proved.

Proposition 2.8.3. *Let* $X = |a, \infty)$, $-\infty \leqslant a < \infty$, *and let* $g : X \to \mathbb{R}$ *be positive and monotonic for sufficiently large* x, $\lim_{x \to \infty} g(x) = 1$. *Then equation (2.8.6) has a unique one-parameter family of solutions* $\varphi : X \to \mathbb{R}$ *which are monotonic for sufficiently large* x. *They are given by formula (2.3.23).*

2.8.8. Convex solutions of the Abel equation $\alpha(x/(x + 1)) = \alpha(x) + 1$ are given by $\alpha(x) = c + 1/x$, $x \in (0, \infty)$, $c \in \mathbb{R}$.

(It suffices to apply (2.4.16) with $f^n(x) = x/(1 + nx)$.)

2.8.9. Theorem 2.3.12 yields also the following result for the Schröder equation

$$\sigma(f(x)) = s\sigma(x). \tag{2.8.7}$$

Proposition 2.8.4. *Let* $X = (-\infty, a|$, $-\infty < a \leqslant \infty$, *and let* $f: X \to X$ *be continuous and strictly increasing in* X, $f(x) < x$ *in* X. *If* f *is convex, resp. concave in* X *and* $\lim_{x \to -\infty}[f(x)/x] = s > 1$, *then equation* (2.8.7) *has a unique one-parameter family of solutions* $\sigma: X \to \mathbb{R}$ *which are convex, resp. concave in* X. *These solutions are given by* (2.3.25).

(See Kuczma [22]. Check first that $f(x)/x$ is monotonic in an interval $(-\infty, b)$, $b < 0$. Put $\hat{X} := (0, -1/b)$. With $y: \hat{X} \to (-\infty, b)$, $y(t) = -1/t$ and $\hat{f} := y^{-1} \circ f^{-1} \circ y$, $\hat{\sigma} := \sigma \circ y$, equation (2.8.7) in $(-\infty, b)$ is equivalent to the equation $\hat{\sigma}(\hat{f}(t)) = (1/s)\hat{\sigma}(t)$ in X. And to this equation Theorem 2.3.12 applies.)

2.8.10. Formula (2.5.2) may be obtained as follows. Denote by $p(\lambda)$ the limit in (2.5.2) and observe that for every $\lambda, \mu > 0$ we have

$$p(\lambda\mu) = \lim_{x \to 0} \frac{\varphi(\lambda \cdot \mu x)}{\varphi(\mu x)} \frac{\varphi(\mu x)}{\varphi(x)} = p(\lambda)p(\mu).$$

The only solutions of this Cauchy equation (e.g. in the class of measurable functions on $\mathbb{R}^+$) are power functions $p(\lambda) = \lambda^\delta$, $\lambda > 0$, $\delta \in \mathbb{R}$ (see Aczél [2], Aczél–Dhombres [1], Kuczma [51, Section 13.1], Seneta [11]).

2.8.11. The functions $\varphi(x) = 1/\log(1/x)$, $\varphi(x) = \log(1/x)$, $\varphi(x) = \exp \sin \log \log(1/x)$, where $x \in (0, 1)$, are all slowly varying at the origin. This shows the variety of asymptotic behaviour at $x = 0$ of slowly varying functions.

2.8.12. Regularly varying solutions of the inhomogeneous equation (2.8.1) can also be found with the help of Theorems 2.5.1–2.5.3 (see Kuczma [43]). First one determines a regularly varying solution $\varphi_0: X \to \mathbb{R} \setminus \{0\}$ of the homogeneous equation (2.8.2). A function $\varphi: X \to \mathbb{R}$ is a regularly varying solution of (2.8.1) if and only if the function $\hat{\varphi} := \varphi/\varphi_0$ is a regularly varying solution of the equation $\hat{\varphi}(f(x)) = \hat{\varphi}(x) + h(x)/\varphi_0(f(x))$. But for equations of this form Theorems 2.5.2 and 2.5.3 work.

2.8.13. The continuity theorem for the Laplace transform (see Feller [1], Kuczma [36]) we have referred to in the proofs of Theorems 2.6.1–2.6.3, when adapted to PGFs, reads as follows.

Proposition 2.8.5. *Let* $p_{n,j} = P(X_n = j)$, $j \in \mathbb{N}_0$, $n \in \mathbb{N}$, *be the probability distribution (not necessarily proper) of an RV* X_n *and let* $F_n(t) = \sum_{j=0}^{\infty} p_{n,j} t^j$ *be the generating function of* X_n. *If there exist*

$$p_j = \lim_{n \to \infty} p_{n,j}, \qquad j \in \mathbb{N}_0, \tag{2.8.8}$$

then there exists the limit $F(t) = \lim_{n \to \infty} F_n(t)$ *and it is given by*

$$F(t) = \sum_{j=0}^{\infty} p_t t^j. \tag{2.8.9}$$

Conversely, if $F_n \to F$ as $n \to \infty$, then there exist the limits (2.8.8) and relation (2.8.9) holds true.

2.8.14. The problem of finding norming sequences for a simple Galton–Watson process in the so-called *supercritical* $(m > 1)$ and *explosive* $(m = \infty)$ cases can also be solved with the aid of results we have presented in Section 2.8.3. In the supercritical case B. P. Stigum [1] proved that the sequence $(Z_n m^{-n})_{n \in \mathbb{N}}$, where $m^n = E(Z_n)$ $(= (F^n)'(1))$, converges almost everywhere to a (proper) RV Y which, however, can be degenerate at zero: $P(Y = 0) = 1$. This deficiency can be removed by using other factors in place of those m^{-n}. A result to this effect is due to E. Seneta [3].

We need to introduce the cumulant generating function $k \colon \mathbb{R}^+ \to \mathbb{R}^+$, $k(t) = -\log F(e^{-t})$, related to the PGF F of the process (Z_n).

Proposition 2.8.6. *If $1 < m < \infty$ and $p_0 > 0$, then there exists a sequence $(c_n)_{n \in \mathbb{N}}$ of positive numbers, $\lim_{n \to \infty} c_n = \infty$, such that the sequence $(Z_n / c_n)_{n \in \mathbb{N}}$ of RVs converges (in the sense of weak convergence of distributions, i.e. in law) to a proper, nondegenerate RV, say W, such that $P(W = 0) = q$. Moreover, the cumulant generating function $K \colon \mathbb{R}^+ \to \mathbb{R}^+$ of W is determined as the strictly increasing and concave in $\mathbb{R}^+$ solution of the Poincaré functional equation*

$$K(mt) = k(K(t)), \qquad t \in \mathbb{R}^+, \tag{2.8.10}$$

satisfying the condition $K(0) = 0$. Every solution of (2.8.10) having the same properties as K is of the form $K(bt)$, $b > 0$.

(The inverse function to K, say L, satisfies the Schröder equation $L(k(t)) = mL(t)$, and the basic facts used in the proof of the proposition are those established in Theorem 2.3.12. For the explosive case no such normalization is possible; see Seneta [6].)

3

Regularity of solutions of
linear equations

3.0 Introduction

In this chapter we focus our attention on regular solutions of the linear
equation, in a real variable x,

$$\varphi(f(x)) = g(x)\varphi(x) + h(x). \tag{3.0.1}$$

'Regularity' stands here for such properties as, e.g. continuity, differentiability,
smoothness class and some others.

We start by presenting a complete description of continuous solutions of
the homogeneous equation

$$\varphi(f(x)) = g(x)\varphi(x) \tag{3.0.2}$$

and some results for the inhomogeneous equation (3.0.1). We are concerned
with solutions which are mappings from a real interval X into a Banach
space Y.

The problems which naturally arise are those of the existence, uniqueness
and continuous dependence on the data. Then we give a survey of results
on the asymptotic behaviour (at the fixed point of the function f) of
real-valued continuous solutions of equation (3.0.1) and (3.0.2). Further on,
we consider real-valued solutions of class C^r, $0 < r \leqslant \infty$, and we exhibit the
results concerning the Schröder and Abel equations and related ones.
Solutions of bounded variation are also discussed in this chapter.

We give applications to invariant measures, doubly stochastic measures
and ordinary differential equations. We end with a section containing
supplementary information and examples.

3.1 Continuous solutions

In the case where X is a real interval the family of continuous solutions of
the equation

$$\varphi(f(x)) = g(x)\varphi(x) + h(x) \tag{3.1.1}$$

depends, first of all, on whether the fixed point $(x = 0)$ of the function f belongs to X or not.

If $0 \notin X$ then the continuous solution of (3.1.1) depends on an arbitrary function. This is the point of our Theorem 3.1.1 below. The easy proof of the theorem will not be given here (see Kordylewski–Kuczma [4], Kuczma [26, p. 46] for the case $Y = \mathbb{R}$).

Theorem 3.1.1. *Let* $X = (0, a], 0 < a \leq \infty$, *and* Y *be a Banach space over* $\mathbb{K}$. *Assume that the functions* $f: X \to X$, $g: X \to \mathbb{K}$, $h: X \to Y$ *are continuous in* X, f *is strictly increasing,* $0 < f(x) < x$ *in* X *and* $g(x) \neq 0$ *in* X. *If* $x_0 \in X$ *is arbitrarily fixed,* $X_0 := [f(x_0), x_0]$, *then every function* $\varphi_0: X_0 \to Y$ *fulfilling the condition*

$$\varphi_0(f(x_0)) = g(x_0)\varphi_0(x_0) + h(x_0) \tag{3.1.1$_0$}$$

can be uniquely extended onto X *to a solution* $\varphi: X \to Y$ *of equation* (3.1.1). *Thus* φ *is continuous on* X *if the* φ_0 *is so on* X_0.

Remark 3.1.1. The unique extension statement of Theorem 3.1.1 remains true if we drop the assumption of continuity of g and h.

In Subsections 3.1A, 3.1B and 3.1C we consider the case where $0 \in X$.

3.1A. Homogeneous equation

We consider the equation

$$\varphi(f(x)) = g(x)\varphi(x) \tag{3.1.2}$$

and make the following assumptions.

(i) $X = [0, a], 0 < a \leq \infty$; Y is a Banach space over $\mathbb{K}$.
(ii) The functions $f: X \to X$ and $g: X \to \mathbb{K}$ are continuous on X. Moreover, $0 < f(x) < x$ and $g \neq 0$ in $X \setminus \{0\}$.
(iii) f is strictly increasing.

Simple induction shows that if $\varphi: X \to Y$ is a solution of equation (3.1.2) then also

$$\varphi(f^n(x)) = G_n(x)\varphi(x), \qquad x \in X, n \in \mathbb{N}, \tag{3.1.3}$$

where

$$G_n(x) := \prod_{i=0}^{n-1} g(f^i(x)), \qquad n \in \mathbb{N}, \quad x \in X. \tag{3.1.4}$$

It turns out that the number of continuous solutions of equation (3.1.2) in X depends on the behaviour of the sequence (3.1.4).

We distinguish three cases.

(A) The limit

$$G(x) := \lim_{n \to \infty} G_n(x)$$

exists in X, is continuous and different from zero in X.

(B) $\lim_{n \to \infty} G_n(x) = 0$ uniformly on a subinterval of X.

(C) Neither (A) nor (B) occurs.

Further considerations will be preceded by two lemmas which are due to R. Węgrzyk [1].

Lemma 3.1.1. *Let X be a regular topological space, Y be a metric space, and $G: X \to Y$ and $G_n: X \to Y$, $n \in \mathbb{N}$, be arbitrary functions. Then there exists the maximal open subset U of X such that $\lim_{n \to \infty} G_n(x) = G(x)$ almost uniformly on U.*

Proof. Suppose (G_n) converges to G a.u. on U_1 and on U_2, both being open subsets of X. Take an arbitrary compact $C \subset U_1 \cup U_2$. For every $x \in C$ there exists an open neighbourhood V_x of x such that cl $\bar{V}_x$ is contained either in U_1 or in U_2. Choose a finite covering $\{V_i, i = 1, \ldots, k\}$ of C from $\{V_x, x \in C\}$. Take the union of those cl V_i that are contained in U_1 and take its intersection with C to get a compact set C_1. Do the same with cl $V_i \subset U_2$ and denote the resulting compact set by C_2. We have $C = C_1 \cup C_2$ and (G_n) converges to G uniformly on both C_1 and C_2, and hence on C. This shows that (G_n) converges a.u. on $U_1 \cup U_2$. By induction this generalizes to arbitrary finite unions $U_1 \cup \cdots \cup U_m$.

Now, let $\mathcal{U}$ be the family of all open subsets V of X such that (G_n) converges to G a.u. on V. (Of course, $\mathcal{U}$ may be empty.) Let U be the union of all sets belonging to $\mathcal{U}$. Let $C \subset U$ be an arbitrary compact set. There exists a finite covering of C by some sets from $\mathcal{U}$. Thus (G_n) converges to G a.u. on the union of these sets, and hence uniformly on C. This shows that (G_n) converges to G a.u. on U, and clearly U is maximal. $\blacksquare$

Lemma 3.1.2. *Let hypotheses (i), (ii) be fulfilled and let $\varphi: X \to Y$ be a continuous solution of equation (3.1.2) such that $\varphi(0) = 0$. Let U be the maximal open (in the relative topology of X) subset of X such that the sequence (G_n) given by (3.1.4) converges to zero a.u. on U. Then $\varphi(x) = 0$ for $x \in X \setminus U$.*

Proof. Take an $x_0 \in X$ such that $\varphi(x_0) \neq 0$. Then there exist a $c > 0$ and a compact neighbourhood $V \subset X$ of x_0 such that $\|\varphi(x)\| \geq c$ on V. Given an $\varepsilon > 0$, we can find an $N \in \mathbb{N}$ such that $\|\varphi(f^n(x))\| < \varepsilon c$ for $x \in V$ and $n > N$, since $(f^n)_{n \in \mathbb{N}}$ converges to zero a.u. on X (see Theorem 1.2.1). From (3.1.3) we get $\|G_n(x)\| < \varepsilon$ for $x \in V$ and $n > N$. Consequently $x_0 \in U$. $\blacksquare$

3.1B. General continuous solution of the homogeneous equation

We now show that the family of continuous solutions of equation (3.1.2) either is a one-parameter family or depends on an arbitrary function or consists of a single function $\varphi = 0$, according to which of the cases (A) or (B) or (C) is dealt with.

Theorem 3.1.2. *Let hypotheses* (i), (ii) *be fulfilled and suppose that case* (A) *occurs. Then the general continuous solution* $\varphi: X \to Y$ *of equation* (3.1.2) *is given by the formula*

$$\varphi(x) = y/G(x), \tag{3.1.5}$$

where $y \in Y$ *is arbitrary (a parameter). Thus equation* (3.1.2) *has a unique one-parameter family of continuous solutions.*

Proof. If $\varphi: X \to Y$ is a continuous solution of (3.1.2), then (3.1.5) results from (3.1.3) on letting $n \to \infty$, with $y = \varphi(0)$. Conversely, every function of form (3.1.5) clearly is a continuous solution of equation (3.1.2) in X, as follows from the recurrence (see (3.1.4))

$$G_{n+1}(x) = g(x)G_n(f(x)), \qquad n \in \mathbb{N}_0. \quad \blacksquare \tag{3.1.6}$$

Note that in case (A) we must have $g(0) = 1$, whence also $G(0) = 1$ and $\varphi(0) = y$ for φ given by (3.1.5).

In order to deal with case (B) we introduce the function $m: U \to (0, \infty)$, where $U \neq \varnothing$ is the maximal open subset of X on which the sequence (G_n) tends a.u. to zero:

$$m(x) = \sup_{n \in \mathbb{N}} |G_n(x)|. \tag{3.1.7}$$

Theorem 3.1.3. *Let hypotheses* (i)–(iii) *be fulfilled, and suppose that case* (B) *occurs. Fix an* $x_0 \in X \setminus \{0\}$ *and write* $X_0 = [f(x_0), x_0]$. *Then every continuous function* $\varphi_0: X_0 \to Y$ *fulfilling the conditions*

$$\varphi_0(f(x_0)) = g(x_0)\varphi_0(x_0), \tag{3.1.8}$$

$$\varphi_0(x) = 0 \quad \text{for } x \in X_0 \setminus U, \tag{3.1.9}$$

$$\lim_{\substack{x \to u \\ x \in X_0 \cap U}} m(x)\varphi_0(x) = 0 \quad \text{for every } u \in X_0 \cap (\text{cl } U \setminus U), \tag{3.1.10}$$

can be uniquely extended to a continuous solutions $\varphi: X \to Y$ *of equation* (3.1.2) *in* X; *and all the continuous solutions* $\varphi: X \to Y$ *of* (3.1.2) *can be obtained in this manner. Thus the general continuous solution of* (3.1.2) *in* X *depends on an arbitrary function.*

Proof. Let $\varphi: X \to Y$ be a continuous solution of equation (3.1.2) in X.

Taking in (3.1.3) an $x \in U$ and letting $n \to \infty$ we obtain

$$\varphi(0) = 0. \tag{3.1.11}$$

Write $\varphi_0 = \varphi | X_0$. This function should fulfil conditions (3.1.8)–(3.1.10). Indeed, (3.1.8) results from (3.1.2) and (3.1.9) from (3.1.11) and Lemma 3.1.2. To prove (3.1.10), fix an $\varepsilon > 0$ and a $u \in X_0 \cap (\mathrm{cl}\, U \backslash U)$. There is an index N such that $\|\varphi(f^n(x))\| < \varepsilon$ for $x \in X_0$ and $n \geq N$. Write $M = \max_{1 \leq n \leq N} \sup_{x \in X_0} |G_n(x)| > 0$. By (3.1.9) $\varphi_0(u) = 0$, so we can find a neighbourhood V of u such that $\|\varphi_0(x)\| < \varepsilon/M$ for $x \in V_0 = V \cap X_0$. By (3.1.7) for every $x \in U$ there is an $n(x) \in \mathbb{N}_0$ such that $m(x) = |G_{n(x)}(x)|$. Hence we get by (3.1.3) for $x \in V_0 \cap U$

$$m(x) \|\varphi_0(x)\| = |G_{n(x)}(x)| \|\varphi_0(x)\| = \|\varphi(f^{n(x)}(x))\|,$$

whence $m(x) \|\varphi_0(x)\| < \varepsilon$. This proves (3.1.10).

Suppose now that a continuous function $\varphi_0 : X_0 \to Y$ satisfies conditions (3.1.8)–(3.1.10). By Theorem 3.1.1 φ_0 can be uniquely extended onto $X \backslash \{0\}$ to a solution $\varphi : X \backslash \{0\} \to Y$ of (3.1.2). We define φ at zero by (3.1.11). The function φ thus defined satisfies equation (3.1.2) in X and is continuous on $X \backslash \{0\}$. It remains to show that

$$\lim_{x \to 0} \varphi(x) = 0. \tag{3.1.12}$$

Take an arbitrary sequence $(x_p)_{p \in \mathbb{N}}$, $x_p \in X$, $x_p \to 0$. We may assume that all $x_p \in (0, x_0)$. For every $p \in \mathbb{N}$ there exist a point $z_p \in X_0$ and $k_p \in \mathbb{N}$ such that $x_p = f^{k_p}(z_p)$, and obviously $k_p \to \infty$. Since X_0 is compact, we may assume that the sequence $(z_p)_{p \in \mathbb{N}}$ converges to a $z_0 \in X_0$. Relation (3.1.3) yields

$$u_p := \varphi(x_p) = G_{k_p}(z_p) \varphi_0(z_p). \tag{3.1.13}$$

We shall distinguish three cases.

(1) $z_0 \in U$. Then $\lim_{p \to \infty} G_{k_p}(z_p) = 0$, whereas the sequence $(\varphi_0(z_p))$ is bounded, since φ_0 is continuous on X_0. Thus $u_p \to 0$ in virtue of (3.1.13).

(2) $z_0 \in X_0 \backslash \mathrm{cl}\, U$. Then φ_0 vanishes in a neighbourhood of z_0, and $u_p = 0$ for almost all p in virtue of (3.1.9) and (3.1.13).

(3) $z_0 \in X_0 \cap (\mathrm{cl}\, U \backslash U)$. Then $\lim_{p \to \infty} u_p = 0$. This follows from (3.1.10) and (3.1.13) whenever $z_p \in U$, and from (3.1.9) and (3.1.13) if $z_p \notin U$.

This proves relation (3.1.12).   ∎

Note that hypothesis (iii) (the monotonicity of f) was needed in this proof only when we used Theorem 3.1.1. In particular, relation (3.1.11) follows without appealing to (iii). Thus we have the following.

Corollary 3.1.1. *Under conditions* (i), (ii), *in case* **(B)** *we have* $\varphi(0) = 0$ *for every continuous solution* $\varphi : X \to Y$ *of equation* (3.1.2).

Finally we deal with case (C).

Theorem 3.1.4. *Let hypotheses* (i), (ii) *be fulfilled and suppose that case* (C) *occurs. Then the only continuous solution* $\varphi: X \to Y$ *of equation* (3.1.2) *is* $\varphi = 0$.

Proof. Suppose equation (3.1.2) has a nontrivial continuous solution $\varphi: X \to Y$. If $\varphi(0) = 0$, then by Lemma 3.1.2 $U \neq \varnothing$ and case (B) must occur. If $\varphi(0) = y \neq 0$, then $\varphi(x) \neq 0$ in the whole X. For, if $\varphi(x_0) = 0$ for an $x_0 \in X$, then by (3.1.3) also $\varphi(f^n(x_0)) = 0$ for $n \in \mathbb{N}$, and consequently $\varphi(0) = \lim_{n \to \infty} \varphi(f^n(x_0)) = 0$. Thus we get by (3.1.3)

$$G(x) = \lim_{n \to \infty} G_n(x) = \lim_{n \to \infty} \frac{\varphi(f^n(x))}{\varphi(x)} = \frac{y}{\varphi(x)},$$

and case (A) must occur. ∎

3.1C. Inhomogeneous equation

Let us turn to equation (3.1.1). We introduce the hypothesis

(iv) The function $h: X \to Y$ is continuous on X.

We have the following (obvious) result.

Theorem 3.1.5. *Let hypotheses* (i), (ii), (iv) *be fulfilled. Then the general continuous solution* $\varphi: X \to Y$ *of equation* (3.1.1) *is given by* $\varphi = \varphi_1 + \varphi_2$, *where* $\varphi_1: X \to Y$ *is a particular continuous solution of* (3.1.1), *and* $\varphi_2: X \to Y$ *is the general continuous solution of equation* (3.1.2).

Consequently, if equation (3.1.1) has a particular continuous solution in X, then in case (A) it has a unique one-parameter family of such solutions; in case (B), if (iii) holds, its continuous solution depends on an arbitrary function; and in case (C) that particular solution is the unique continuous solution of (3.1.1) in X.

It may well happen that equation (3.1.1) has no continuous solution $\varphi: X \to Y$. The next theorems give some conditions for the existence of such solutions in cases (A)–(C), consecutively.

Theorem 3.1.6. *Let hypotheses* (i), (ii), (iv) *be fulfilled and suppose that case* (A) *occurs. Equation* (3.1.1) *has a continuous solution* $\varphi: X \to Y$ *if and only if the series*

$$\varphi_0(x) = - \sum_{n=0}^{\infty} \frac{h(f^n(x))}{G_{n+1}(x)} \tag{3.1.14}$$

converges and φ_0 *is continuous on* X. *The general continuous solution* $\varphi: X \to Y$ *of equation* (3.1.1) *is then given by the formula*

$$\varphi(x) = \varphi_0(x) + y/G(x), \tag{3.1.15}$$

where y *is an arbitrary constant (a parameter).*

Proof. The 'if' part is easily checked, so suppose that equation (3.1.1) has a continuous solution $\varphi: X \to Y$. We get from (3.1.1) by induction

$$\varphi(x) = \frac{\varphi(f^n(x))}{G_n(x)} - \sum_{i=0}^{n-1} \frac{h(f^i(x))}{G_{i+1}(x)}, \qquad n \in \mathbb{N}, \qquad (3.1.16)$$

whence, on letting $n \to \infty$, we obtain (3.1.15) with $y = \varphi(0)$. Hence $\varphi_0(x) = \varphi(x) - y/G(x)$ is a continuous function. ∎

Passing to case (B), define the functions $h(\cdot\,;\cdot): X \times Y \to Y$ and $H_n(\cdot\,;\cdot): X \times Y \to Y, n \in \mathbb{N}$, by

$$h(x; y) = h(x) + y(g(x) - 1), \qquad (3.1.17)$$

and

$$H_n(x; y) = G_n(x) \sum_{i=0}^{n-1} \frac{h(f^i(x); y)}{G_{i+1}(x)} \qquad (3.1.18)$$

For an $x_0 \in X \setminus \{0\}$ write

$$X_0 = [f(x_0), x_0].$$

Theorem 3.1.7. *Let hypotheses* (i), (ii), (iv) *be fulfilled and suppose that case* (B) *occurs.*

(1) *If equation* (3.1.1) *has a continuous solution* $\varphi: X \to Y$, *then there exists a* $y \in Y$ *such that* $h(0; y) = 0$ *and the sequence* $(H_n(x; y))_{n \in \mathbb{N}}$ *converges to zero a.u. in* U.

(2) *If moreover hypothesis* (iii) *is fulfilled,* $h(0; y) = 0$ *for a* $y \in Y$ *and the sequences* $(G_n(x))_{n \in \mathbb{N}_0}$ *and* $(H_n(x; y))_{n \in \mathbb{N}}$ *converge to zero uniformly in* X_0, *then every continuous function* $\varphi_0: X_0 \to Y$ *fulfilling condition* (3.1.1_0) *can be uniquely extended onto* X *to a continuous solution* $\varphi: X \to Y$ *of equation* (3.1.1) *so that* $\varphi(0) = y$.

Proof. Suppose equation (3.1.1) has a continuous solution $\varphi: X \to Y$ and take $y = \varphi(0)$. Setting $x = 0$ in (3.1.1) we obtain $h(0; y) = 0$. Moreover, we get from (3.1.1) by induction (see (3.1.6) and (3.1.17) with (3.1.18)),

$$H_n(x; y) = \varphi(f^n(x)) - y - G_n(x)(\varphi(x) - y). \qquad (3.1.19)$$

Relation (3.1.19) implies that the sequence $(H_n(x; y))_{n \in \mathbb{N}}$ tends to zero a.u. in U, as does the sequence $(G_n(x))_{n \in \mathbb{N}_0}$ in case (B) (see Lemma 3.1.2).

Now accept the assumptions of (2). Let $\varphi_0: X_0 \to Y$ be a continuous function fulfilling condition (3.1.1_0). By Theorem 3.1.1 φ_0 can be uniquely extended onto $X \setminus \{0\}$ to a solution $\varphi: X \setminus \{0\} \to Y$ of equation (3.1.1). If we put $\varphi(0) = y$, then the function $\varphi: X \to Y$ thus defined satisfies equation (3.1.1) in X and is continuous on $X \setminus \{0\}$. Given an $\varepsilon > 0$, we can find an $N \in \mathbb{N}$ such that $|G_n(x)| \sup_{t \in X_0} \|\varphi_0(t) - y\| < \varepsilon/2$ and $\|H_n(x; y)\| < \varepsilon/2$ for

$x \in X_0$ and $n \geq N$. For every $x \in (0, f^N(x_0))$ there are a $z \in X_0$ and an $n \geq N$ such that $x = f^n(z)$. Thus we get from (3.1.19) for $x \in (0, f^N(x_0))$

$$\|\varphi(x) - y\| \leq |G_n(z)| \|\varphi_0(z) - y\| + \|H_n(z; y)\| < \varepsilon,$$

whence $\lim_{x \to 0} \varphi(x) = y$ and φ is continuous in the whole of X. ∎

Here is a consequence of Theorem 3.1.7.

Theorem 3.1.8. *Let hypotheses* (i)–(iv) *be fulfilled and suppose that* $|g(0)| < 1$. *Then every continuous function* $\varphi_0: X_0 \to Y$ *fulfilling condition* (3.1.1$_0$) *can be uniquely extended onto* X *to a continuous solution* $\varphi: X \to Y$ *of equation* (3.1.1).

Proof. We check the assumptions of (2) of Theorem 3.1.7. First, $h(0; y) = 0$ for $y = h(0)/(1 - g(0))$. Now fix a Θ, $|g(0)| < \Theta < 1$, and choose a $\delta > 0$ such that $|g(x)| < \Theta$ for $x \in [0, \delta)$. By Theorem 1.2.1 there is an $N \in \mathbb{N}$ such that $f^n(x) \in [0, \delta)$ for $x \in X_0$ and $n \in \mathbb{N}$. Thus, for these x and n we have

$$|G_n(x)| < \Theta^{n-N} \sup_{t \in X_0} |G_N(t)|$$

whence $\lim_{n \to \infty} G_n = 0$ uniformly on X_0 so that case (B) occurs. Further, for the same x and n (see (3.1.18))

$$\|H_n(x; y)\| \leq |G_n(x)| \sup_{t \in X_0} \left\| \sum_{i=0}^{N-1} \frac{h(f^i(t)); y)}{G_{i+1}(t)} \right\| + \frac{1 - \Theta^{n-N}}{1 - \Theta} \sup_{t \in [0,\delta)} \|h(t; y)\|.$$

Therefore $\lim_{n \to \infty} H_n(\cdot; y) = 0$ uniformly on X_0, since the second supremum above tends to zero as δ does. The theorem results from Theorem 3.1.7. ∎

In case (C) a necessary and sufficient condition for the existence of a continuous solution in X of equation (3.1.1) will be proved under the additional hypothesis

(v) The sequence $(1/G_n(x))_{n \in \mathbb{N}_0}$ is bounded at every point $x \in X$.

Theorem 3.1.9. *Let hypotheses* (i), (ii), (iv), (v) *be fulfilled and suppose that case* (C) *occurs. Equation* (3.1.1) *has a continuous solution* $\varphi: X \to Y$ *if and only if there exists a* $y \in Y$ *such that the series*

$$\varphi_0(x) = - \sum_{n=0}^{\infty} \frac{h(f^n(x); y)}{G_{n+1}(x)} \tag{3.1.20}$$

converges and φ_0 *is continuous on* X. *The only continuous solution* $\varphi: X \to Y$ *of equation* (3.1.1) *is then given by the formula*

$$\varphi(x) = y + \varphi_0(x). \tag{3.1.21}$$

Proof. The 'if' part is easily verified, so assume we are given a continuous solution $\varphi: X \to Y$ of equation (3.1.1). Write $y = \varphi(0)$. Relation (3.1.19) may be written as

$$\varphi(x) - y = \frac{\varphi(f''(x)) - y}{G_n(x)} - \sum_{i=0}^{n-1} \frac{h(f^i(x); y)}{G_{i+1}(x)},$$

whence, on letting $n \to \infty$, we obtain (3.1.21). Hence $\varphi_0 = \varphi - y$ is a continuous function. ∎

Hypothesis (v) is clearly fulfilled if $|g(0)| > 1$. The theorem that follows can then be proved by examining the convergence of the series (3.1.20). However, we supply an independent proof based on Banach's Theorem.

Theorem 3.1.10. *Let hypotheses* (i), (ii), (iv) *be fulfilled and assume that* $|g(0)| > 1$. *Then equation* (3.1.1) *has the unique continuous solution* $\varphi: X \to Y$ *given by formula* (3.1.21) *with* $y = h(0)/(1 - g(0))$.

Proof. Choose a number ϑ, $|g(0)| > \vartheta > 1$ and positive numbers $b < a$ and K so that $|g(x)| > \vartheta$ and $\|h(x)\| \leq K$ on $[0, b] \subset X$. Taking an $M > K/(\vartheta - 1)$ consider the space (endowed with the sup norm) $\mathscr{A} := \{\varphi: [0, b] \to Y, \varphi$ is continuous on $[0, b], \|\varphi(x)\| \leq M, x \in [0, b]\}$. A direct calculation shows that the formula

$$(T\varphi)(x) := (g(x))^{-1}(\varphi(f(x)) - h(x)), \qquad x \in [0, b], \qquad (3.1.22)$$

defines a contractive map of the space $\mathscr{A}$ into itself. By Banach's Theorem equation (3.1.1) has a unique continuous solution $\varphi_b \in \mathscr{A}$. By the extension Theorem 3.1.1 we get the unique continuous solution $\varphi: X \to Y$ of (3.1.1) such that $\varphi = \varphi_b$ on $[0, b]$. Since (v) holds, Theorem 3.1.9 gives formula (3.1.21) for this φ with $y = \varphi(0) = h(0)/(1 - g(0))$, as we get from (3.1.1) by setting $x = 0$. ∎

We shall also need solutions of (3.1.1) in the following class of functions:

$$\mathscr{B} := \{\varphi: X \to Y, \varphi \text{ is continuous on } X \setminus \{0\} \text{ and bounded on } X\}. \quad (3.1.23)$$

Modifying slightly the proof of Theorem 3.1.10 we obtain the following.

Theorem 3.1.11. *Let hypotheses* (i), (ii) *be fulfilled and* $h \in \mathscr{B}$. *If* $|g(0)| > 1$, *then equation* (3.1.1) *has a unique solution in the class* $\mathscr{B}$.

Remark 3.1.2. Notice that assumption (iii) on the monotonicity of f was used only in Theorems 3.1.3, 3.1.7 and 3.1.8 concerning case (B). This assumption is to some extent essential there; see Note 3.8.5. The following result, whose proof will not be given here, is due to W. Jarczyk [4].

Theorem 3.1.12. *Let hypotheses* (i), (ii) *be fulfilled and suppose that* $|g(0)| < 1$. *If for every* $h: X \to \mathbb{R}$ *satisfying* (iv) *there exists a continuous solution* $\varphi: X \to \mathbb{R}$ *of equation* (3.1.1), *then f is strictly increasing in a neighbourhood of the origin.*

Remark 3.1.3. We have seen that, under assumptions (i)–(iv), the condition $|g(0)| > 1$ implies case (C), whereas (B) holds if $|g(0)| < 1$. Moreover, if $|g(0)| \neq 1$, then equation (3.1.1) has continuous solutions in X; see Theorems 3.1.8 and 3.1.10. The case where $|g(0)| = 1$ is referred to as the *indeterminate* one. If it holds, then any of the possibilities (A)–(C) may occur, and equation (3.1.1) need not have continuous solutions in X unless further assumptions are imposed. The additional assumptions often concern the asymptotic behaviour of the given functions at the origin. This matter will be discussed in Section 3.3. Here we supply only one result of this kind.

Theorem 3.1.13. *Assume* (i)–(iii) *with* $Y = \mathbb{K} = \mathbb{R}$. *If, moreover, there exist numbers* $s \in (0, 1)$, $m > 0$, $k > 0$ *such that*

$$f(x) = sx + O(x^{1+m}), \qquad x \to 0, \\ g(x) = 1 + O(x^k), \qquad x \to 0, \Bigg\} \tag{3.1.24}$$

then case (A) *occurs, i.e., equation* (3.1.2) *has a unique one-parameter family of continuous solutions* $\varphi \colon X \to \mathbb{R}$. *The solutions are given by formula* (3.1.5).

Proof. We should examine the sequence (G_n) given by (3.1.4). Since $g(0) = 1$ (see (3.1.24)), assumption (ii) implies $g > 0$. In turn, from (3.1.24) for f and Theorem 1.3.1 we deduce that

$$\lim_{n \to \infty} s^{-n} f^n(x) \in (0, \infty). \tag{3.1.25}$$

Let $x_0 \in X \setminus \{0\}$ and $s_1 \in (s, 1)$ be arbitrarily fixed. Then we have by (3.1.24), (iii) and (3.1.25), with an $M > 0$,

$$|g(f^n(x)) - 1| \leqslant M(f^n(x))^k \leqslant M(f^n(x_0))^k \leqslant M_1,$$

$M_1 = M_1(x_0) > 0$. This estimation holds for every $x \in [0, x_0]$, provided n is large enough. We see that the series $\sum_{n=1}^{\infty} (g(f^n(x)) - 1)$ converges a.u. in X. Therefore there exists the limit $G(x) := \lim_{p \to \infty} G_p(x) = \prod_{n=1}^{\infty} g(f^n(x))$ (a.u. in X) and it is a continuous and positive function on X. Thus case (A) occurs and the theorem follows from Theorem 3.1.2. ∎

Remark 3.1.4. The case where f is strictly decreasing can be reduced to that of increasing f (Kordylewski [2]; see also Kuczma [26]).

Comments. In the determinate case $|g(0)| \neq 1$ the investigation of continuous solutions of linear iterative functional equations has been initiated by J. Kordylewski–M. Kuczma [4]. The indeterminate case $|g(0)| = 1$ has been first dealt with in Choczewski–Kuczma [1] (see also Kuczma [12]). R. Węgrzyk [1] contributed further to the theory and the general continuous solution of the homogeneous equation (3.1.2) has been found (Kuczma [46]). However, in the papers quoted $Y = \mathbb{K}$ was the case. Consult also Kravčenko [1].

Some sufficient conditions for the existence of continuous solutions of the inhomogeneous equation (3.1.1) in the indeterminate case may be found in Kuczma [26, Ch. II, §7], [47], Kordylewski–Kuczma [2], [5], Choczewski–Kuczma [1], Choczewski [10], Bajraktarević [4], Matkowski–Zdun [1], Zdun [5], [7], [8]. See also Section 3.6.

Equation (3.1.1) with $g = 1$ and an involutory f, $f^2 = \mathrm{id}$, has been dealt with by A. Smajdor [7].

Recently a qualitative theory of iterative functional equations is coming into being. Most results in this direction, involving continuity of solutions, also for nonlinear equations, are due to W. Jarczyk [1], [3], [6]; see Note 3.8.4. A measure-theoretical version of the result from Jarczyk [3] has been obtained by K. Baron [13]. See also Kuczma [53].

3.2 Continuous dependence of continuous solutions on given functions

In each case (A)–(C) one can prove that continuous solutions of equation (3.0.1) depend continuously on given functions f, g, h (Kordylewski–Kuczma [5], Czerwik [4], [6]). This will require strengthening the assumptions about these functions, especially in cases (A) and (C). Following S. Czerwik [6] we are going to deal with the equation

$$\varphi(f(x, t), t) = g(x, t)\varphi(x, t) + h(x, t). \qquad (3.2.1)$$

We make the following hypotheses.

(i) $X = [0, a|, 0 < a \leq \infty$; Y is a Banach space over $\mathbb{K}$; T is a metric space.
(ii) The functions $f: X \times T \to X$, $g: X \times T \to \mathbb{K}$ and $h: X \times T \to Y$ are continuous, $0 < f(x, t) < x$ and $g(x, t) \neq 0$ in $(X \setminus \{0\}) \times T$.

The iterates of f with respect to x will be denoted by f^n, $f^{n+1}(x, t) = f(f^n(x, t), t)$, and we shall define the functions $h(\cdot, t; y)$ and $G_n(\cdot, t)$ by formulae (3.1.17) and (3.1.4), respectively, with the functions $f^i(\cdot, t)$, $g(\cdot, t)$, $h(\cdot, t)$ in place of f^i, g, h.

Case (A) is the most delicate one, and we have to make the strongest assumptions in order to ensure the desired result.

(iii) For every compact $C \subset T$ there exist a $b \in X \setminus \{0\}$, positive constants M, μ and $\Theta \in (0, 1)$ and a bounded function $B: [0, b] \times C \to \mathbb{R}^+$ such that we have

$$f(x, t) \leq \Theta x,$$
$$|g(x, t) - 1| \leq M x^\mu,$$
$$\|h(x, t)\| \leq B(x, t)$$

and

$$B(f(x, t), t) \leq \Theta B(x, t) \tag{3.2.2}$$

in $[0, b] \times C$.

Remark 3.2.1. The function $B(\cdot, t)$ may be supposed increasing as one sees by replacing $B(x, t)$ by $\sup_{z \in [0,x]} B(z, t)$.

Theorem 3.2.1. *Let hypotheses* (i)–(iii) *be fulfilled. Then for every continuous function* $y: T \to Y$ *there exists a unique solution* $\varphi: X \times T \to Y$ *of equation* (3.2.1) *such that, for every* $t \in T$, *the function* $\varphi(\cdot, t)$ *is continuous in* X, *and* $\varphi(0, t) = y(t)$. *This solution is given by the formula*

$$\varphi(x, t) = \varphi_0(x, t) + y(t)/G(x, t),$$

where

$$\varphi_0(x, t) = -\sum_{n=0}^{\infty} \frac{h(f^n(x, t), t)}{G_{n+1}(x, t)}, \qquad G(x, t) = \lim_{n \to \infty} G_n(x, t), \tag{3.2.3}$$

and is continuous on $X \times T$.

Proof. Assumption (iii) guarantees that the series $\sum_{n=0}^{\infty} |g(f^n(x, t), t) - 1|$ converges a.u. in $X \times T$. Hence the function G exists, is continuous and different from zero in $X \times T$; in particular, for every fixed $t \in T$ case (A) occurs for equation (3.2.1). Similarly, by (iii) the series in (3.2.3) converges a.u. in $X \times T$. Thus the theorem results from Theorem 3.2.6. ■

To obtain a result for case (B) it is enough to reformulate part (2) of Theorem 3.2.7 (see Czerwik [6]).

For case (C) we make the following additional hypothesis.

(iv) There is a function $y: T \to Y$ such that for every compact $C \subset T$ there exist a $b \in X \setminus \{0\}$, a $\Theta \in (0, 1)$ and a bounded function $B: [0, b] \times C \to \mathbb{R}^+$ such that condition (3.2.2) and

$$\|h(x, t; y(t))\| \leq B(x, t)$$

hold in $[0, b] \times C$ and, for every fixed $x \in [0, b]$, the function $B(x, \cdot)$ is continuous on C.

Theorem 3.2.2. *Let hypotheses* (i), (ii), (iv) *be fulfilled, and either let* Y *be finite-dimensional or let the function* y *be continuous on* T. *Suppose that, for every fixed* $t \in T$, *case* (C) *occurs for equation* (3.2.1). *If the sequence* $(1/G_n)$ *is almost uniformly bounded in* $X \times T$, *then there exists a unique solution* $\varphi: X \times T \to Y$ *of equation* (3.2.1) *such that, for every fixed* $t \in T$, *the function* $\varphi(\cdot, t)$ *is continuous on* X. *This solution is given by the formula*

$$\varphi(x, t) = y(t) + \varphi_0(x, t) \tag{3.2.4}$$

where

$$\varphi_0(x, t) = -\sum_{n=0}^{\infty} \frac{h(f^n(x), t; y(t))}{G_{n+1}(x, t)}, \qquad (3.2.5)$$

and is continuous on $X \times T$.

Proof. It follows easily from (iv) that series (3.2.5) converges a.u. in $X \times T$. Thus, in virtue of Theorem 3.1.9, we need only prove that, if Y is finite dimensional, then the function y is continuous on T.

Fix a $t_0 \in T$ and a sequence $(t_n)_{n \in \mathbb{N}}$, $t_n \in T$, $t_n \to t_0$, and put $C = \{t_0, t_1, t_2, \ldots\}$. For this C choose b, Θ and B according to (iv). Since case (C) occurs for equation (3.2.1) with $t = t_0$, there exists an $x_0 \in [0, b]$ such that $g(x_0, t_0) \neq 1$. Hence also $g(x_0, t_n) \neq 1$ for n large enough, and it follows from the inequality

$$\|h(x_0, t_n) + y(t_n)(g(x_0, t_n) - 1)\| \leq B(x_0, t_n)$$

that the sequence $(y(t_n))_{n \in \mathbb{N}}$ is bounded. Suppose there is a sub-sequence $(y(t_{n_k}))_{k \in \mathbb{N}}$ convergent to a $y_0 \neq y(t_0)$. The inequality $\|h(x, t_{n_k}; y(t_{n_k}))\| \leq B(x, t_{n_k})$ yields on letting $k \to \infty$

$$\|h(x, t_0; y_0)\| \leq B(x, t_0) \qquad \text{for } x \in [0, b].$$

Hence we conclude that equation (3.2.1) with $t = t_0$ has a continuous solution $\varphi(\cdot, t_0): X \to Y$ such that $\varphi(0, t_0) = y_0 \neq y(t_0)$, which contradicts the uniqueness of solution (3.2.4). ∎

Comments. In the case where T is a real interval, the question of differentiability of solutions of equation (3.2.1) with respect to the parameter has also been studied; see Czerwik [3], [7], [8].

3.3 Asymptotic properties of solutions

3.3A. Solutions continuous at the origin

Let $\varphi: X \to \mathbb{R}$ be a continuous solution of the equation

$$\varphi(f(x)) = g(x)\varphi(x) + h(x) \qquad (3.3.1)$$

and let $0 \in X$, $f(0) = 0$. In this case one may be interested how fast $\varphi(x)$ converges to $\varphi(0)$ as $x \to 0$.

In this subsection we present some results to this effect. They are due to B. Choczewski [10], [12], M. Kuczma [47], and G. Szekeres [1]. Some proofs will be supplied in the next subsection.

We make the following suppositions.

(i) $X = [0, a|, 0 < a \leqslant \infty$.
(ii) The functions $f: X \to X$, $g: X \to \mathbb{R}$ and $h: X \to \mathbb{R}$ are continuous, f is

strictly increasing, $0 < f(x) < x$ in $X \setminus \{0\}$, $g(x) \neq 0$ in X. Moreover

$$f(x) = x - x^{m+1}u(x), \qquad u(x) = O(1), \qquad x \to 0,$$
$$g(x) = 1 + x^k v(x), \qquad v(x) = O(1), \qquad x \to 0,$$
$$h(x) = x^q w(x), \qquad w(x) = O(1), \qquad x \to 0,$$

where m, k, q are positive constants (not necessarily integers).

In the sequel all the asymptotic symbols will refer to $x \to 0$.

The three theorems below correspond to the three possible cases (A)–(C), which determine how large the set of continuous solutions $\varphi : X \to \mathbb{R}$ of equation (3.3.1) is. The theorems express the asymptotic properties of those solutions at the origin.

Let us fix a positive number p. The test-function will be the power function with the exponent p.

In Theorems 3.3.1 and 3.3.2 the functions v and w occurring in (ii) will be assumed to have further asymptotic properties:

$$v(x) = ru(x) + O(x^p), \qquad p > 0, \tag{3.3.2}$$
$$w(x) = su(x) + O(x^\sigma), \qquad \sigma > 0, \tag{3.3.3}$$

or

$$v(x) = r + O(x^p), \qquad r \neq 0, \, p > 0, \tag{3.3.4}$$
$$w(x) = s + O(x^\sigma), \qquad s \neq 0, \, \sigma > 0. \tag{3.3.5}$$

Theorem 3.3.1. *Let hypotheses* (i), (ii) *be fulfilled with*

$$m < \min(k, q) \qquad \text{and} \qquad \liminf_{x \to 0} u(x) > 0.$$

Then equation (3.3.1) *has a unique one-parameter family of continuous solutions* $\varphi_c : X \to \mathbb{R}$, $\varphi_c(0) = c$, $c \in \mathbb{R}$. *Asymptotic properties of these functions are put together in Table 3.1.*

Table 3.1

Additional assumptions on v and w	Conditions imposed upon k, m, q	Asymptotic properties of solutions	
None	$k > m + p, \, q > m + p$	$\varphi_c(x) = c + o(x^p)$	(3.3.6)
(3.3.2)	$k = m + p, \, q > m + p$	$\varphi_c(x) = c - \dfrac{cr}{p} x^p + o(x^p)$	
(3.3.3)	$k > m + p, \, q = m + p$	$\varphi_c(x) = c - \dfrac{s}{p} x^p + o(x^p)$	(3.3.7)
(3.3.2) and (3.3.3)	$k = m + p, \, q = m + p$	$\varphi_c(x) = c - \dfrac{cr + s}{p} x^p + o(x^p)$	
(3.3.4)	$k < p, \, q > m + p$	φ_0 satisfies (3.3.6) with $c = 0$	
(3.3.4) and (3.3.3)	$k < p, \, q = m + p$	φ_0 has property (3.3.7) with $c = 0$	

Moreover, in the two last cases none of φ_c for $c \neq 0$ satisfies the condition

$$\varphi_c(x) = c + O(x^p), \tag{3.3.8}$$

whereas if (3.3.4) and (3.3.5) are fulfilled and $k < p$ and $q < p$ then no φ_c satisfies condition (3.3.8).

Now denote by $\varphi_d \colon X \to \mathbb{R}$, $d \in \mathbb{R}$, the functions

$$\varphi_d(x) = \frac{s}{pt} - \sum_{n=0}^{\infty} \frac{h(f^n(x); s/pt)}{G_{n+1}(x)} + d \lim_{n \to \infty} (f^n(x))^p / G_n(x) \tag{3.3.9}$$

(see (3.1.17) and (3.1.4)) with some fixed s and $t \neq 0$.

Theorem 3.3.2. *Let hypotheses* (i) *and* (ii) *be fulfilled and assume that* $k \leqslant m = q$ *and*

$$u(x) = t + O(x^\tau), \qquad t > 0, \tau > 0, \tag{3.3.10}$$

v satisfies (3.3.4) *with* $r < 0$, *w satisfies* (3.3.5) *with* $\sigma > p$ *and with* $s = 0$ *if* $k < m$. *Then equation* (3.3.1) *has a continuous solution* $\varphi \colon X \to \mathbb{R}$ *depending on an arbitrary function,* $\varphi(0) = -s/r$. *Asymptotic behaviour of these solutions is described in Table 3.2.*

Table 3.2

Conditions for constants	Asymptotic properties of continuous solutions
(1) $k < m = q, s = 0$	Every φ has the property $$\varphi(x) = O(x^p)$$
$k = m = q, \rho > p, r < -pt$	Every φ has the property $$\varphi(x) = -\frac{s}{r} + O(x^p) \tag{3.3.11}$$
(2) $k = m = q, s = 0$ if $\rho \leqslant p, r = -pt$	(a) Only φ_d given by (3.3.9) have the properties $$\varphi_d(x) = \frac{s}{pt} + dx^p + o(x^p) \tag{3.3.12}$$ (b) Every φ fulfils (3.3.11) with $r = -pt$
(3) $k = m = q, r > -pt$	Only φ_0 given by (9) with $d = 0$ has property (3.3.12) (with $d = 0$)

Theorem 3.3.3. *Let hypotheses* (i), (ii) *be fulfilled with* $k = q \leqslant m$ *and* $\liminf_{x \to 0} v(x) > 0$. *Further, let*

$$w(x) = sv(x) + O(x^p)$$

with a positive $p \geqslant m - k$. *Then equation* (3.3.1) *has a unique continuous solution* $\varphi \colon X \to \mathbb{R}$ *and this solution has the property*

$$\varphi(x) = -s + O(x^p). \tag{3.3.13}$$

If, moreover, there are positive r *and* z *such that*

$$v(x) = r + o(1), \qquad w(x) = sv(x) + zx^p + o(x^p), \tag{3.3.14}$$

then φ has the property

$$\varphi(x) = -s - \frac{z}{r}x^p + o(x^p). \tag{3.3.15}$$

We conclude with a result on the homogeneous equation

$$\psi(f(x)) = g(x)\psi(x). \tag{3.3.16}$$

We follow the idea of G. Szekeres [1].

Theorem 3.3.4. *Let hypotheses* (i), (ii) *be fulfilled with* $h = 0$. *If* $k > m$ *and condition* (3.3.10) *is satisfied, then equation* (3.3.16) *has a unique one-parameter family of continuous solutions* $\psi: X \to \mathbb{R}$. *The solutions fulfil the condition*

$$\psi(x) = \psi(0) + O(x^{k-m}). \tag{3.3.17}$$

3.3B. Sample proofs

We shall prove Theorem 3.3.2 (Choczewski [10]), Theorem 3.3.3 (Kuczma [47]; see also Section 12.3) and Theorem 3.3.4 (adaptation of a method of G. Szekeres [1]). The proof of Theorem 3.3.1 is similar to that of Theorem 3.3.2 (Choczewski [8]).

In the proofs that follow we write $X_0 = [f(x_0), x_0]$, $x_0 \in X \setminus \{0\}$.

Proof of Theorem 3.3.2. To find continuous solutions of equation (3.3.1) we use Theorem 3.1.7. We are going to prove that the sequences (3.1.4) and (3.1.18) converge uniformly in an appropriate interval X_0.

Because of (ii), (3.3.4) and (3.3.10), for every positive $\bar{p}$ we have

$$g(x) - \left(\frac{f(x)}{x}\right)^{\bar{p}} = rx^k + \bar{p}tx^m + O(x^\kappa),$$

where $\kappa = \min(m + \tau, k + \rho, 2m) > k$. Thus an $x_0 \in X \setminus \{0\}$ may be found such that

$$0 < g(x) \leqslant \left(\frac{f(x)}{x}\right)^{\bar{p}} \qquad \text{for } x \in (0, x_0],$$

provided $r + \bar{p}t < 0$ if $k = m$. Hence we get the estimate

$$0 < G_n(x) = \prod_{i=0}^{n-1} g(f^i(x)) \leqslant \left(\frac{f^n(x)}{x}\right)^{\bar{p}}, \qquad x \in X_0. \tag{3.3.18}$$

This yields $0 < G_n(x) \leqslant [f^n(x_0)/f(x_0)]^{\bar{p}}$, $x \in X_0$, i.e. $\lim_{n \to \infty} G_n(x) = 0$ uniformly in X_0. Case (B) occurs.

Now, $G_n(x)/G_{i+1}(x) \leqslant [f^n(x)/f^{i+1}(x)]^{\bar{p}}$, $x \in X_0$ (see (3.3.18)), so that for the sequence (3.1.18) (with $y = -s/r$) we get for $x \in X_0$.

$$\left| H_n\left(x; -\frac{s}{r}\right) \right| = G_n(x) \left| \sum_{i=0}^{n-1} \frac{h(f^n(x); -s/r)}{G_{i+1}(x)} \right|$$

$$\leqslant [f^n(x)]^{\bar{p}} \sum_{i=0}^{n-1} [f^{i+1}(x)]^{-\bar{p}} |h(f^i(x); -s/r)|. \tag{3.3.19}$$

In view of (ii), (3.3.5) and the asymptotic equivalence $f(x) \sim x$ we get

$$h\left(x; -\frac{s}{r}\right) = h(x) - \frac{s}{r}(g(x) - 1) = O(x^\lambda) = O(f(x)^\lambda),$$

where $\lambda = \min(m + \sigma, m + \rho)$ if $k = m$ and $\lambda = m + \sigma$ if $k < m$ (since we have assumed that $s = 0$ whenever $k < m$). Consequently, there is a $C > 0$ such that

$$\left|h\left(x; -\frac{s}{r}\right)\right| \leqslant C(f(x))^\lambda, \qquad \lambda > m, \; x \in [0, x_0], \qquad (3.3.20)$$

provided x_0 has been chosen small enough. Now take in (3.3.18) a $\bar{p} > 0$ such that $\bar{p} < \lambda - m$ if $k < m$ and $\bar{p} < \min(\lambda - m, -r/t)$ if $k = m$. By Theorem 1.4.6 and by the monotonicity of f there exist positive constants c_1, c_2 and an $N \in \mathbb{N}$ such that

$$c_1 n^{-1/m} \leqslant f^n(x) \leqslant c_2 n^{-1/m}, \qquad x \in X_0, \; n \geqslant N. \qquad (3.3.21)$$

Using (3.3.20) and (3.3.21) in (3.3.19) we obtain

$$\left|H_n\left(x; -\frac{s}{r}\right)\right| \leqslant C c_2^{\lambda - \bar{p}}(f^n(x_0))^{\bar{p}} \sum_{i=0}^{n-1} (i+1)^{-(\lambda - \bar{p})/m}.$$

According to the choice of $\bar{p}$, we have $\lambda - \bar{p} > m$. Thus $\lim_{n \to \infty} H_n(x; -s/r) = 0$ uniformly in X_0. By Theorem 3.1.7 equation (3.3.1) has in X a continuous solution depending on an arbitrary function. All these solutions accept the value $-s/r$ at the origin.

Case (1). Let $\varphi: X \to \mathbb{R}$ be a continuous solution of (3.3.1). Write $M = \sup_{X_0} x^{-p}|\varphi(x) + s/r|$. To prove (3.3.11) we take an arbitrary $\bar{x} \in (0, x_0]$, find an $n \in \mathbb{N}_0$ and an $x \in X_0$ satisfying $\bar{x} = f^n(x)$ and use (3.1.19) with $y = -s/r$

$$\varphi(\bar{x}) + \frac{s}{r} = G_n(x)\left(\varphi(x) + \frac{s}{r}\right) + H_n\left(x, -\frac{s}{r}\right).$$

Thus

$$(\bar{x})^{-p}\left|\varphi(\bar{x}) + \frac{s}{r}\right| \leqslant G_n(x)x^p[f^n(x)]^{-p}M + [f^n(x)]^{-p}\left|H_n\left(x; -\frac{s}{r}\right)\right|.$$

By the assumptions on the constants ρ, σ and r we can take in (3.3.18) and (3.3.21) $\bar{p} = p$. With this choice of $\bar{p}$ (3.3.18), (3.3.19), (3.3.20) and (3.3.21) yield

$$(\bar{x})^{-p}\left|\varphi(\bar{x}) + \frac{s}{r}\right| \leqslant M + \sum_{i=1}^{\infty} (i+1)^{-(\lambda - p)/m} < \infty,$$

as $\lambda - p > m$. This proves the required asymptotic properties of φ.

The proofs of cases (2) and (3) are similar. We sketch here the proof of (2). Assertion (a). Note that now $r = -pt$ and write

$$\Phi(x) = x^{-p}(\varphi(x) - s/pt). \qquad (3.3.22)$$

A continuous function $\varphi: X \to \mathbb{R}$ satisfies (3.3.1) and (3.3.12) if and only if $\Phi: X \to \mathbb{R}$ given by (3.3.22) is a continuous solution of the equation

$$\Phi(f(x)) = (f(x)/x)^{-p}g(x)\Phi(x) + (f(x))^{-p}h(x; s/pt) \qquad (3.3.23)$$

such that $\Phi(0) = d$. The values at zero of the given functions occurring in (3.3.23) are defined to be equal to their limits at zero. In particular, we have

$$\tilde{g}(x) := (f(x)/x)^{-p} g(x), \qquad x \in X \setminus \{0\}, \; \tilde{g}(0) := 1.$$

To solve equation (3.3.23) we first examine the sequence (3.1.4) for $\tilde{g}$:

$$\tilde{G}_n(x) = \prod_{i=0}^{n-1} \tilde{g}(f^i(x)) = \left(\frac{f^n(x)}{x}\right)^{-p} \prod_{i=0}^{n-1} g(f^i(x)) = \frac{x^p G_n(x)}{(f^n(x))^p}. \qquad (3.3.24)$$

We have $\tilde{g}(x) = 1 + O(x^{m+\mu})$ with $\mu = \min(\rho, \tau, m) > 0$. This, together with (3.3.21) and the monotonicity of f, implies case (A) for equation (3.3.23) (see Note 3.8.8). In virtue of Theorem 3.1.6 the formula

$$\Phi_d(x) = - \sum_{n=0}^{\infty} (f^{n+1}(x))^{-p} h\left(f^n(x); \frac{s}{pt}\right) \bigg/ \tilde{G}_{n+1}(x) + d \bigg/ \lim_{n \to \infty} \tilde{G}_n(x) \qquad (3.3.25)$$

defines the unique one-parameter family of continuous solutions of equation (3.3.23), provided the series in (3.3.25) converges to a continuous function on X. This is actually so, as the series converges a.u. in X. Indeed (3.3.20) holds with $\lambda = m + \sigma > m$, and (3.3.21) can again be used. Both inequalities yield a convergent majorant $c(x_0) \sum (n+1)^{-(1+\sigma/m)}$ for our series in any interval $[0, x_0] \subset X$. Thus the a.u. convergence follows. Formula (3.3.9) results from (3.3.25) and (3.3.22).

Assertion (b). Each continuous solution $\varphi: X \to \mathbb{R}$ of (3.3.1) is of the form $\varphi = \varphi_0 + \psi$, where φ_0 is given by (3.3.9) with $d = 0$, and ψ is continuous and satisfies the homogeneous equation (3.3.16). We know that φ_0 obeys (3.3.11) with $r = -pt$ (put $d = 0$ in (3.3.12)). Thus it is enough to prove that $\psi(x) = O(x^p)$. To this end for a fixed $x \in (0, x_0]$ choose a $y \in X_0$ and an $n \in \mathbb{N}_0$ such that $x = f^n(y)$ and iterate (3.3.16) to get $\psi(x) = \psi(f^n(y)) = G_n(y)\psi(y)$. Further, $x^{-p}\psi(x) = (f^n(y))^{-p} G_n(y)\psi(y)$ and (3.3.24) yields

$$x^{-p}\psi(x) = y^{-p} \tilde{G}_n(y)\psi(y), \qquad y \in X_0.$$

Hence $x^{-p}\psi(x) = O(1)$, as the uniformly convergent sequence $(\tilde{G}_n(x))_{n \in \mathbb{N}}$ is bounded on X_0. ∎

Proof of Theorem 3.3.3. By Theorem 1.4.6 the first inequality in (3.3.21) is valid. Hence there is a $D > 0$ such that (see (ii))

$$g(f^n(x)) - 1 \geqslant Dn^{-k/m}, \qquad x \in X_0, \; n \geqslant N.$$

This proves that $\lim_{n \to \infty} G_n(x) = \infty$ in X_0. Since by (3.1.6) $\lim_{n \to \infty} G_n(x) = \infty$ also in $X \setminus \{0\}$, case (C) occurs for equation (3.3.1). According to Theorem 3.1.9 the formula

$$\varphi(x) = -s - \sum_{n=0}^{\infty} \frac{h(f^n(x); -s)}{G_{n+1}(x)} \qquad (3.3.26)$$

represents the unique continuous solution $\varphi: X \to \mathbb{R}$ of equation (3.3.1) provided the series in (3.3.26) uniformly converges in $[0, x_0]$.

Thanks to (ii) there exist positive constants C_1 and C_2 and an $x_0 \in X \setminus \{0\}$ such that

$$|h(x; -s)| \leqslant C_1 x^{k+p}, \qquad g(x) \geqslant 1 + C_2 x^k, \qquad x \in [0, x_0].$$

Consequently, the series in (3.3.26) is majorized by

$$\Phi(x) = \sum_{n=0}^{\infty} \left[\prod_{i=0}^{n} (1 + C_2(f^i(x))^k)^{-1} C_1(f^n(x))^{k+p} \right] \qquad (3.3.27)$$

whenever $x \in [0, x_0]$. The sum of the series (3.3.27) (if it exists) satisfies the equation

$$\Phi(x) = \hat{g}(x)\Phi(f(x)) + \hat{h}(x); \qquad \hat{g}(x) := (1 + C_2 x^k)^{-1}, \ \hat{h}(x) := C_1 \hat{g}(x) x^{k+p}. \tag{3.3.28}$$

In view of Theorem 12.3.1 equation (3.3.28) actually has a continuous solution in X, given by (3.3.27). Moreover (see (12.3.9)), we have

$$0 < \Phi(x) \leqslant C_1 C_2^{-1} x^p \tag{3.3.29}$$

for small x. Since the series (3.3.27) has positive terms, it converges uniformly in $[0, x_0]$. Thus the series in (3.3.26) uniformly converges in $[0, x_0]$, as $|\varphi(x) + s| \leqslant \Phi(x)$. Moreover, relation (3.3.29) yields (3.3.13).

Now assume that v and w satisfy (3.3.14). Then we may take in (3.3.27) $C_1 = z + \varepsilon$ and $C_2 = r - \varepsilon$, with $0 < \varepsilon < r$. This ε may be as small as we wish. We choose an x_0 such that

$$|\varphi(x) + s| \leqslant \Phi(x) \leqslant (z + \varepsilon)(r - \varepsilon)^{-1} x^p, \qquad x \in [0, x_0] \tag{3.3.30}$$

(see (3.3.29)) and, if $0 < \varepsilon < z$,

$$|h(x; -s)| \geqslant (z - \varepsilon) x^{k+p}, \qquad g(x) \leqslant 1 + (r + \varepsilon) x^k, \qquad x \in [0, x_0].$$

Of course, the series given by (3.3.27) with $C_1 = z - \varepsilon$, $C_2 = r + \varepsilon$ also converges to a continuous function $\Psi: X \to \mathbb{R}$ satisfying equation (3.3.28) (with the same C_1 and C_2). Since the function $\hat{h}$ occurring in (3.3.28) has the property

$$\hat{h}(f(x))/\hat{h}(x) = (f(x)/x)^{k+p}(1 + o(1)) = 1 + O(x^m),$$

therefore also $\inf_{(0,x]}[\hat{h}(f(t))/\hat{h}(t)] = 1 + O(x^m)$. By Theorem 12.3.3

$$\Psi(x) = (z - \varepsilon)(r + \varepsilon)^{-1} x^p + o(x^p),$$

and the series representing Ψ is a minorant for the series in (3.3.26), i.e. $\Psi(x) \leqslant -(\varphi(x) + s)$. Combining this with (3.3.30) we get

$$-\Phi(x) \leqslant \varphi(x) + s \leqslant -\Psi(x), \qquad x \in [0, x_0].$$

Recall the estimations for Ψ and Φ (see (3.3.30)) which yield (3.3.15), since ε can be arbitrarily small. ∎

Proof of Theorem 3.3.4. The assumptions imply (see Note 3.8.8) that for equation (3.3.16) case (A) occurs. Thus $\lim_{n \to \infty} G_n(x) = G(x)$ and $G: X \to \mathbb{R}$

is a continuous, never vanishing function on X. By Theorem 3.1.2 the formula

$$\psi(x) = y/G(x), \qquad y = \psi(0),$$

yields all the continuous solutions $\psi: X \to \mathbb{R}$ of equation (3.3.16). In order to prove (3.3.17) it is enough to show that

$$G(x) = 1 + O(x^{k-m}). \tag{3.3.31}$$

Let $x_0 \in X \setminus \{0\}$ be such that $|g(x) - 1| < \frac{1}{2}$ and $|G(x) - 1| < \frac{1}{2}$ for $x \in (0, x_0)$. Since $|u| < 2|\log(1 + u)| < 4|u|$ for $|u| < \frac{1}{2}$, we have for $x \in (0, x_0)$ (see (3.1.4))

$$|G(x) - 1| < 2|\log G(x)| \leqslant 2 \sum_{n=0}^{\infty} |\log g(f^n(x))|$$

$$\leqslant 4 \sum_{n=0}^{\infty} |g(f^n(x)) - 1| \leqslant 4D \sum_{n=0}^{\infty} (f^n(x))^k,$$

where D is a positive constant. Now, for every $x \in (0, f(x_0))$ there is an $N = N(x) \in \mathbb{N}$ such that $f^{N+1}(x_0) \leqslant x \leqslant f^N(x_0)$. Hence

$$x^{m-k}|G(x) - 1| < 4D(f^{N+1}(x_0))^{m-k} \sum_{n=0}^{\infty} (f^{n+N}(x_0))^k. \tag{3.3.32}$$

Inequalities (3.3.21) may be used again, and $\sum_{n=N}^{\infty} n^{-k/m} < [m/(k-m)]N^{1-k/m}$. Thus the right-hand side of (3.3.32) is bounded in $(0, f(x_0))$ by

$$\frac{4Dm}{k-m} c_2^k c_1^{k-m} \left(\frac{N(x)}{N(x)+1} \right)^{1-k/m} \leqslant L,$$

where L is independent on x, and (3.3.31) follows. ∎

3.3C. Asymptotic series expansions

Theorems 3.3.1–3.3.4 dealt with the case where $f'(0) = g(0) = 1$. Below we quote without proofs some results (Choczewski [10], [12]) concerning other cases.

Definition 3.3.1. We say that the formal power series $\sum_{n=0}^{\infty} A_n x^n$ is an *asymptotic series expansion at the origin of a function F defined in a vicinity of the origin* iff for every $k \in \mathbb{N}$

$$F(x) - \sum_{n=0}^{k} A_n x^n = O(x^{k+1})$$

(as $x \to 0$). We then write

$$F(x) \sim \sum_{n=0}^{\infty} A_n x^n.$$

Now, we strengthen hypothesis (ii) replacing it by

(iii) The functions $f: X \to X$, g: $X \to \mathbb{R}$ and $h: X \to \mathbb{R}$ are continuous on X, f is strictly increasing, $0 < f(x) < x$ and $g(x) \neq 0$ in $X \setminus \{0\}$. Moreover,

f, g and h have the asymptotic series expansions at the origin

$$f(x) \sim a_1 x + a_2 x^2 + \cdots, \qquad 0 < a_1 \leqslant 1,$$

$$g(x) \sim g_0 + g_1 x + \cdots, \qquad g_0 \neq 1 \text{ if } a_0 = 1,$$

$$h(x) \sim h_0 + h_1 x + \cdots.$$

The coefficients d_j of the formal power series solution

$$\varphi(x) = \sum_{k=0}^{\infty} d_j x^j$$

to (3.3.1) can be determined by inserting the expansions of f, g and h (as formal series) into equation (3.3.1). One obtains the recurrences

$$d_0 = (1 - g_0)^{-1} h_0,$$

$$d_n = (a_1^n - g_0)^{-1} \left\{ \sum_{i=1}^{n-1} (g_{n-i} - b_{n-i}(i)) d_i + g_n d_0 + h_n \right\}, \qquad n \in \mathbb{N},$$

where $b_0(n) = a_1^n$ and

$$b_k(n) = (ka_1)^{-1} \sum_{j=1}^{k} ((n+1)j - k) a_{j+1} b_{k-j}(n), \qquad k \in \mathbb{N}$$

(see Choczewski [10]). These formulae allow us to determine the coefficients d_n uniquely whenever $a_1^n \neq g_0$ for $n \in \mathbb{N}_0$. If $g_0 = 1$ and $h_0 = 0$, then $d_0 = t$ may be arbitrary and the remaining d_n depend on t (note that $g_0 = 1$ implies $0 < a_1 < 1$). Finally, if

$$a_1^p = g_0 \qquad \text{for a } p \in \mathbb{N} \tag{3.3.33}$$

and

$$\sum_{i=1}^{p-1} (g_{p-i} - b_{p-i}(i)) d_i + g_p d_0 + h_p = 0, \tag{3.3.34}$$

then $d_0, \ldots, d_{p-1}$ are determined uniquely, $d_p = t$ is arbitrary (a parameter), and the d_n for $n \geqslant p$ are functions of t.

In the theorems whose formulation follows the coefficients d_j in the expansion

$$\varphi(x) \sim d_0 + d_1 x + \cdots \tag{3.3.35}$$

are supposed to be determined in the above described way.

Theorem 3.3.5. *Let hypotheses* (i) *and* (iii) *be satisfied. Continuous solutions* $\varphi: X \to \mathbb{R}$ *of equation* (3.3.1) *and their asymptotic series expansions at the origin are given by Table 3.3.*

The case where $0 < a_1 < 1$ and $0 < |g_0| < 1$ will be dealt with in the last two theorems of this subsection. Let us put

$$q := \log_{a_1} |g_0|, \qquad r := [q]$$

(the integer part of q).

Table 3.3

Case	Continuous solutions	Asymptotic series expansions
$\lvert g_0\rvert > 1$ or $g_0 = -1$	Unique	The solution has expansion (3.3.35)
$g_0 = 1$ and $h_0 = 0$	One-parameter family φ_t	$\varphi_t(x) \sim t + \sum\limits_{j=1}^{\infty} d_j(t)x^j$
$\lvert g_0\rvert < 1$ and $a_1 = 1$	Depends on an arbitrary function	Every solution has expansion (3.3.35)

Theorem 3.3.6. *Let hypotheses* (i) *and* (iii) *be satisfied with*

$$0 < \lvert g_0\rvert < 1, \qquad 0 < a_1 < 1, \qquad r < q$$

(i.e. q is not an integer). Then equation (3.3.1) *has a continuous solution* $\varphi\colon X \to \mathbb{R}$ *depending on an arbitrary function. All these solutions have the property*

$$\varphi(x) = d_0 + d_1 x + \cdots + d_r x^r + O(x^\rho), \tag{3.3.36}$$

for every ρ, $r < \rho < q$; *only the members of a one-parameter family of them satisfy*

$$\varphi(x) = d_0 + d_1 x + \cdots + d_r x^r + O(x^q); \tag{3.3.37}$$

and only a single continuous solution has the property

$$\varphi(x) = d_0 + d_1 x + \cdots + d_r x^r + O(x^{r+1})$$

and only this solution has the asymptotic series expansion (3.3.35).

Theorem 3.3.7. *Let hypotheses* (i) *and* (iii) *be satisfied with*

$$0 < g_0 < 1, \qquad 0 < a_1 < 1,$$

and suppose that conditions (3.3.33) *and* (3.3.34) *hold. Then equation* (3.3.1) *has a continuous solution depending on an arbitrary function. All these solutions have property* (3.3.36) *with* $p - 1$ *and* p *in place of* r *and* q, *respectively. But only the members* φ_t, $t \in \mathbb{R}$, *of a one-parameter family of them have property* (3.3.7) *with the same replacement, and they have also the asymptotic series expansion*

$$\varphi_t(x) \sim d_0 + d_1 x + \cdots + d_{p-1}x^{p-1} + tx^p + d_{p+1}(t)x^{p+1} + \cdots.$$

3.3D. Solutions discontinuous at the origin

At this place we touch upon the question of asymptotic behaviour of those solutions of the linear functional equation which, being defined in an interval, say $X = [0, a)$, are continuous inside X, but no longer at $x = 0$. The family of such solutions can depend on an arbitrary function. In particular, for the homogeneous equation

$$\varphi(f(x)) = g(x)\varphi(x) \tag{3.3.38}$$

this is so if case (C) occurs. Then we know that among all solutions continuous on $X \setminus \{0\}$ (Theorem 3.1.1) only the function $\varphi = 0$ is continuous also at the origin (Theorem 3.1.4). Asymptotic behaviour of solutions of (3.3.38) will be dealt with just for this case (see Choczewski–Kuczma [3]).

We make the following suppositions.

(iv) The functions $f: X \to X$ and $g: X \to \mathbb{R}$ are continuous on X and $0 < f(x) < x, g(x) > 0$ in $X \setminus \{0\}$. Moreover, f and g have the asymptotic properties

$$f(x) = x - tx^{m+1} + O(x^{m+1+\tau}),$$
$$g(x) = 1 + rx^k + O(x^{k+\rho}),$$

where k, m, r, τ, ρ are positive constants.

Lemma 3.3.1. *Assume* (i) *and* (iv) *to hold except for the asymptotic relation for g which is replaced by*

$$g(x) = 1 + O(x^k), \qquad k > m.$$

Then case (A) occurs.

The lemma follows by a direct examination of the sequence (G_n) defined by (3.1.4) (see Note 3.8.8).

Without assuming anything on the asymptotic properties of f and g we have the following simple, but useful, result.

Theorem 3.3.8. *Make assumptions* (i) *and* (iv) *but with the asymptotic relations deleted. Let $\varphi_0: X \to \mathbb{R}$ be a solution of (3.3.38) that is continuous and positive on $X \setminus \{0\}$. Then every solution $\varphi: X \to \mathbb{R}$ of (3.3.38) that is continuous on $X \setminus \{0\}$ has the property*

$$\varphi(x) = \varphi_0(x)O(1).$$

Proof. Put $\omega(x) := \varphi(x)/\varphi_0(x), x \in X \setminus \{0\}$. Then ω is a continuous solution of the equation $\omega(f(x)) = \omega(x)$ in $X \setminus \{0\}$. This implies the boundedness of ω on $X \setminus \{0\}$. ∎

Theorem 3.3.9. *Assume* (i) *and* (iv). *If $k = m$, then every solution $\varphi: X \to \mathbb{R}$ of equation (3.3.38) continuous on $X \setminus \{0\}$ satisfies*

$$\varphi(x) = O(x^{-r/t}), \tag{3.3.39}$$

whereas if $k < m$ and $\min(k, \rho, \tau) > m - k$, then it has the property

$$\varphi(x) = O\left(\exp\left[\frac{r}{t(m-k)} x^{k-m}\right]\right). \tag{3.3.40}$$

Proof. By Theorem 3.3.8 it is enough to find a particular solution φ_0 of (3.3.38), continuous and positive in $X \setminus \{0\}$, with the property (3.3.39), resp.

(3.3.40). We shall look for solutions φ_0 such that $\varphi_0(x) = O(x^{-p})$ in the case $k = m$, resp. $\varphi_0(x) = O(\exp(sx^{-q}))$ in the other case, with suitable positive p, q and s. Putting

$$\psi(x) := x^p \varphi_0(x), \qquad \text{resp.} \quad \psi(x) := \exp(-sx^{-q})\varphi_0(x), \qquad (3.3.41)$$

and inserting (3.3.41) into (3.3.38) we see that it suffices to find a bounded solution ψ of the equation

$$\psi(f(x)) = g_1(x)\psi(x), \qquad (3.3.42)$$

where

$$g_1(x) := g(x)\left(\frac{f(x)}{x}\right)^p, \qquad x \in X \setminus \{0\}, \qquad (3.3.43)$$

resp.

$$g_1(x) := g(x) \exp(-s[(f(x))^{-q} - x^{-q}]), \qquad x \in X \setminus \{0\}, \qquad (3.3.44)$$

and $g_1(0) := 1$ in both cases.

By (3.3.43) and (ii) we have $(k = m)$

$$g_1(x) = g(x)(1 - tx^m + O(x^{m+v}))^p = 1 + O(x^{m+v}),$$

where $v = \min(\rho, \tau, m) > 0$ so that $m + v > m$; provided that $p = r/t$. Similarly, for the function (3.3.44) we obtain

$$g_1(x) = g(x)(1 - sqtx^m + O(x^{m+\delta})) = 1 + O(x^{k+\kappa}),$$

where $\delta = \min(\tau, m)$, $\kappa = \min(k, \rho, \tau) > m - k$; provided that $q = m - k$, $s = r/qt$. Thus in both cases Lemma 3.3.1 applies and for equation (3.3.42) case (A) occurs. By Theorem 3.1.2 equation (3.3.42) has a positive and continuous solution ψ in X. Obviously, this ψ is bounded on X and, when inserted into (3.3.41), yields the desired solution φ_0 of equation (3.3.38). ∎

Comments. Other results on the asymptotic behaviour of solutions of linear functional equations may be found in Choczewski [5], [8]–[10], Anczyk [2], Riekstiņš [1], Kuczma [47], [52], Rozmus–Chmura [3]. For a result on solutions of the inhomogeneous equation (3.3.1) discontinuous at the origin see Choczewski–Kuczma [4]. Regarding a generalization of Theorem 3.3.9 see Choczewski [15].

3.4 Differentiable solutions

Now it will be more convenient to write the equation in the form
$$\varphi(x) = g(x)\varphi(f(x)) + h(x). \qquad (3.4.1)$$
We make the following hypotheses.

(i) $X = [0, a]$, $0 < a \leq \infty$; Y is a Banach space over $\mathbb{R}$.
(ii) The functions $f: X \to X$, $g: X \to \mathbb{R}$ and $h: X \to Y$ are of class C^r, $1 \leq r \leq \infty$, on X; moreover, $0 < f(x) < x$, $f'(x) > 0$ and $g(x) \neq 0$ in $X \setminus \{0\}$.

Induction yields the following.

Lemma 3.4.1. *Let hypotheses* (i), (ii) *be fulfilled and let* $\varphi: X \to Y$ *be a function of class* C^r. *Then*

$$\frac{d^k}{dx^k}[g(x)\varphi(f(x)) + h(x)] = g(x)(f'(x))^k \varphi^{(k)}(f(x))$$

$$+ \sum_{i=0}^{k-1} P_{ki}(x)\varphi^{(i)}(f(x)) + h^{(k)}(x),$$

$k = 1, \ldots, r$, *where* $P_{ki}: X \to \mathbb{R}$ *are functions of class* C^{r-k}.

As in the case of continuous solutions (see Theorem 3.1.1), if $0 \notin X$ there is a lack of uniqueness of differentiable solutions of equation (3.4.1) in X. This is more precisely stated in the theorem below, the proof of which is straightforward and is therefore omitted.

Theorem 3.4.1. *Let hypotheses* (i), (ii) *be fulfilled with* $0 \notin X$. *Fix an* $x_0 \in X$ *and write* $X_0 = [f(x_0), x_0]$. *Then every function* $\varphi_0: X_0 \to Y$ *of class* C^r *in* X_0 *and fulfilling the conditions*

$$\varphi_0(x_0) = g(x_0)\varphi(f(x_0)) + h(x_0), \tag{3.4.2}$$

$$\varphi^{(k)}(x_0) = g(x_0)(f'(x_0))^k \varphi_0^{(k)}(f(x_0)) + h^{(k)}(x_0)$$

$$+ \sum_{i=0}^{k-1} P_{ki}(x_0)\varphi_0^{(i)}(f(x_0)), \qquad k = 1, \ldots, r, \tag{3.4.3}$$

can be uniquely extended onto X *to a solution* $\varphi: X \to Y$ *of equation* (3.4.1). *This solution is of class* C^r *in* X.

Theorem 3.4.1 says that equation (3.4.1) has in X a C^r solution depending on an arbitrary function. This may be true or not if $0 \in X$. We show two theorems to this effect.

If we want a C^r solution φ of equation (3.4.1) to be defined at the origin, then the values $c_k = \varphi^{(k)}(0)$ must fulfil the system of equations

$$c_k = g(0)(f'(0))^k c_k + \sum_{i=0}^{k-1} P_{ki}(0)c_i + h^{(k)}(0), \qquad k = 0, \ldots, r. \tag{3.4.4}$$

System (3.4.4) has a unique solution provided $g(0)(f'(0))^k \neq 1$ for $k = 0, \ldots, r$. If this is not the case for an integer p, $0 \leq p \leq r$, then the solution does exist if and only if

$$\sum_{i=0}^{p-1} P_{pi}(0)c_i + h^{(p)}(0) = 0$$

where $c_0, \ldots, c_{p-1}$ ($p \geq 1$) are determined from the first p equations (3.4.4) (the condition reduces to $h(0) = 0$ if $p = 0$). In this case c_p is arbitrary, and the c_k for $k > p$ depend on the parameter c_p.

Theorem 3.4.2. *Let hypotheses* (i), (ii) *be fulfilled with* $0 \in X$ *and suppose that*

$$|g(0)|(f'(0))^r < 1. \qquad (3.4.5)$$

Then for every system $c_0, \ldots, c_r$ *satisfying* (3.4.4) *there exists a unique* C^r
solution $\varphi \colon X \to Y$ *of equation* (3.4.1) *fulfilling the initial conditions*

$$\varphi^{(k)}(0) = c_k, \qquad k = 0, \ldots, r. \qquad (3.4.6)$$

Proof. Take a positive $c < a$. Let $\mathscr{C}$ denote the metric space of all functions
$\varphi \colon [0, c] \to Y$ of class C^r on $[0, c]$ and fulfilling conditions (3.4.6), endowed
with the metric

$$\sigma(\varphi_1, \varphi_2) = \sup_{x \in [0,c]} \left\| \varphi_1^{(r)}(x) - \varphi_2^{(r)}(x) \right\|.$$

Then equation (3.4.1) induces a map of $\mathscr{C}$ into itself, which is a contraction
provided that c has been chosen small enough. By Banach's Theorem
equation (3.4.1) has a unique solution $\varphi \colon [0, c] \to Y$, $\varphi \in \mathscr{C}$. This solution can
be uniquely extended onto X according to Theorem 3.4.1 (see Kuczma [7]
for details). ∎

Remark 3.4.1. The above proof does not apply if $r = \infty$. In such a case (3.4.5)
should hold for all $r \geq r_0$ and Theorem 3.4.2 yields the existence of a unique
C^r solution $\varphi_r \colon X \to Y$ of (3.4.1) for all $r \geq r_0$. All these solutions must be
identical because of the uniqueness, and consequently $\varphi = \varphi_r \in C^\infty(X, Y)$.

Theorem 3.4.3. *Let hypotheses* (i), (ii) *be fulfilled with* $0 \in X$, *and suppose that*

$$|g(0)|(f'(0))^r > 1. \qquad (3.4.7)$$

Then every function $\varphi_0 \colon X_0 \to Y$ $(X_0 = [f(x_0), x_0], x_0 \neq 0)$ *of class* C^r *in* X_0
and fulfilling conditions (3.4.2), (3.4.3) *can be uniquely extended onto* X *to a*
solution $\varphi \colon X \to Y$ *of equation* (3.4.1). *This solution is of class* C^r *in* X.

Proof. The function φ_0 can be uniquely extended onto $X \setminus \{0\}$ in virtue of
Theorem 3.4.1. Condition (3.4.7) implies that $|g(0)|(f'(0))^k > 1$ for
$k = 0, \ldots, r$. Thus system (3.4.4) has a unique solution $c_0, \ldots, c_r$. We define
an extension φ of φ_0 at $x = 0$ by $\varphi(0) = c_0$. By Theorem 3.1.8 φ is continuous
on X (note that the g in (3.1.1) is the reciprocal of the present g). Assuming
φ to be of class C^{k-1}, $0 < k \leq r$, we get by Lemma 3.4.1 that $\varphi^{(k)}$, when
extended onto X by putting $\varphi^{(k)}(0) = c_k$, satisfies in X the equation

$$\varphi^{(k)}(x) = g(x)[f'(x)]^k \varphi^{(k)}(f(x)) + \sum_{i=0}^{k-1} P_{ki}(x) \varphi^{(i)}(f(x)) + h^{(k)}(x).$$

By Theorem 3.1.8 $\varphi^{(k)}$ is continuous on X, and induction completes the
proof. ∎

Remark 3.4.2. Theorem 3.4.3 is true for class C^∞ if (3.4.7) is postulated for all $r \geq r_0$. On the other hand, it is not generally true that if, for every $r < \infty$, the C^r solution of equation (3.4.1) depends on an arbitrary function, then so does the C^∞ solution (see Note 3.8.10).

Comments. Theorem 3.4.2 is the first one in this theory on differentiable solutions. It has been proved with the aid of Banach's Theorem (Kuczma [7]). (For a use of topological methods in the theory of iterative functional equations see Choczewski [7].) Fixed-point theorems, in general, cannot be applied in the indeterminate case $(|g(0)|(f'(0))^r = 1)$ for C^r solutions. The results concerning this case are very scarce (see Choczewski [4], [6], Czerwik [5], [12]).

Linear equations have also been studied in the class of functions whose rth derivative is either Lipschitzian (Jelonek [1]) or absolutely continuous (Sieczko [1]).

3.5 Special equations

In this section we are going to apply some results from the preceding sections to the problem of uniqueness of solutions to the equations of Schröder, Abel and Julia. The first two of them have already been studied in Chapter 2, but in other function classes, and they are more thoroughly treated in Chapters 8 and 9. The reader is referred to the latter chapter for notes and further applications of the Schröder and Abel equations. The Julia equation

$$\lambda(f(x)) = f'(x)\lambda(x)$$

(Julia [1]) plays an important role in the theory of continuous iteration (see e.g. Ecalle [1], Dubuc [1]) but this aspect exceeds the scope of our book. In Chapter 8 Julia's equation is used to determine conjugate and permutable power series.

3.5A. Schröder's equation

We aim at proving a uniqueness theorem for differentiable solutions $\sigma \colon X \to \mathbb{R}$ of the equation

$$\sigma(f(x)) = s\sigma(x) \tag{3.5.1}$$

(Crum [1], Szekeres [1]). We assume that

(i) $X = [0, a|, 0 < a \leqslant \infty$,
(ii) $f \colon X \to X$ is of class C^1 in X, $0 < f(x) < x$, $f'(x) \neq 0$ in $X \setminus \{0\}$ and
$$f'(x) = s + O(x^\delta), \qquad x \to 0, \; \delta > 0, \; 0 < s < 1.$$

Remark 3.5.1. If $f \in C^2(X)$ and $f'(0) = s$, then this asymptotic relation is certainly fulfilled.

Theorem 3.5.1. *Let hypotheses* (i), (ii) *be fulfilled. Then equation* (3.5.1) *has a unique solution* $\sigma: X \to \mathbb{R}$ *of class* C^1 *on* X *and satisfying* $\sigma'(0) = 1$. *This solution is given by the formula*

$$\sigma(x) = \lim_{n \to \infty} s^{-n} f^n(x), \tag{3.5.2}$$

is strictly increasing in X, *and fulfils the condition*

$$\sigma'(x) = 1 + O(x^\delta), \qquad x \to 0. \tag{3.5.3}$$

Proof. First observe that $\sigma(0) = 0$ for any solution $\sigma: X \to \mathbb{R}$ of (1). Further, equation (3.5.1) has a C^1 solution σ in X such that $\sigma'(0) = 1$ if and only if the equation

$$\varphi(f(x)) = \frac{s}{f'(x)} \varphi(x) \tag{3.5.4}$$

has a continuous solution $\varphi: X \to \mathbb{R}$ such that $\varphi(0) = 1$. Clearly, $\sigma(x) = \int_0^x \varphi(t) dt$.

Since $f'(x) = s + O(x^\delta)$, we have $f(x) = sx + O(x^{1+\delta})$ and $s/f'(x) = 1 + O(x^\delta)$ as $x \to 0$. We see that Theorem 3.1.13 applies to equation (3.5.4). Consequently, a continuous solution $\varphi: X \to \mathbb{R}$ of (3.5.4) such that $\varphi(0) = 1$ does exist and is unique. Thus the same is true for the C^1 solution σ of (3.5.1) in X such that $\sigma'(0) = 1$.

To prove (3.5.3) for this σ we proceed as follows. The formula

$$\hat{\sigma}'(x) = 1 + x^\delta \hat{\phi}(x), \qquad x \in X \setminus \{0\}, \ \hat{\phi}(0) = 0,$$

links C^1 solutions $\hat{\sigma}$ of (1) satisfying (3.5.3) with solutions $\hat{\phi}$ belonging to the class $\mathscr{B}$ (with $Y = \mathbb{R}$; see (3.1.23)) of the equation

$$\hat{\phi}(f(x)) = \hat{g}(x)\hat{\phi}(x) + \hat{h}(x), \tag{3.5.3}$$

where

$$\hat{g}(x) := \begin{cases} (x/f(x))^\delta s/f'(x), & x \in X \setminus \{0\}, \\ s^{-\delta}, & x = 0, \end{cases}$$

$$\hat{h}(x) := \begin{cases} (s/f'(x) - 1)^\delta (f(x))^{-\delta}, & x \in X \setminus \{0\}, \\ 0, & x = 0. \end{cases}$$

Since $|\hat{g}(0)| > 1$, from Theorem 3.1.11 we infer that equation (3.5.5) has a unique solution in the class $\mathscr{B}$. The corresponding solution $\hat{\sigma}$ of (3.5.1) is also unique. But $\hat{\sigma}$ is of class C^1 in X and $\hat{\sigma}'(0) = 1$ so that $\hat{\sigma} = \sigma$. Thus the already determined σ has property (3.5.3).

Since $\sigma'(0) = 1$, σ is strictly increasing in a neighbourhood of the origin, and by (3.5.1) it is so in X.

The last thing to be proved is formula (3.5.2). Iterating (3.5.1) we obtain by induction

$$\sigma(x) = s^{-n} \sigma(f^n(x)) = (s^{-n} f^n(x))(\sigma(f^n(x))/f^n(x))$$

for $x \in X \setminus \{0\}$ and $n \in \mathbb{N}$. Hence (3.5.2) follows as $\lim_{n \to \infty} f^n(x) = 0$ and $\sigma'(0) = 1$. For $x = 0$ (3.5.2) is trivial. ∎

3.5B. Julia's equation

We start by presenting some results by M. C. Zdun [3] on solutions of the Julia equation

$$\lambda(f(x)) = f'(x)\lambda(x) \tag{3.5.6}$$

belonging to the following function class:

$$\mathscr{D} := \{\varphi: X \to \mathbb{R}, \; \varphi \text{ is continuous on } X \text{ and differentiable at } x = 0\}.$$

To this end assume that

(iii) $f: X \to X$ is convex or concave and of class C^1 on X, $0 < f(x) < x$ and $f'(x) \neq 0$ in $X \setminus \{0\}$.

In the sequel we write

$$s := f'(0).$$

By (iii) we have $0 \leqslant s \leqslant 1$.

Theorem 3.5.2. *Let hypotheses* (i), (iii) *be fulfilled. Then the solutions* $\lambda \in \mathscr{D}$ *of equation* (3.5.6) *are the following.*

(1) $0 < s < 1$. *All the solutions are given by*

$$\lambda(x) = c \lim_{n \to \infty} \frac{f^n(x)}{(f^n)'(x)}, \tag{3.5.7}$$

where $c \in \mathbb{R}$ *is an arbitrary constant (a parameter).*

(2) $s = 1$. *The solution depends on an arbitrary function and* $\lambda'(0) = 0$ *for every solution* λ.

(3) $s = 0$. *The only solution is* $\lambda = 0$.

Proof. If $s < 1$, then $\lambda(0) = 0$ for every solution λ of (3.5.6). If $s = 1$, then f must be concave, and so are all f^n, $n \in \mathbb{N}$. Hence $(f^n)'(x) \leqq f^n(x)/x$ and $\lim_{n \to \infty} (f^n)'(x) = 0$ in $X \setminus \{0\}$. From (3.5.6) we get

$$\lambda(f^n(x)) = \prod_{i=0}^{n-1} f'(f^i(x))\lambda(x) = (f^n)'(x)\lambda(x), \qquad n \in \mathbb{N},$$

whence $\lambda(0) = 0$ for every continuous solution λ of (3.5.6). Thus $\lambda \in \mathscr{D}$ is a solution of (3.5.6) if and only if the function $\varphi(x) = \lambda(x)/x$ $(\varphi(0) = \lambda'(0))$ is a continuous solution of the equation

$$\varphi(f(x)) = g(x)\varphi(x), \tag{3.5.8}$$

where $g(x) := xf'(x)/f(x)$ for $x \in X \setminus \{0\}$ and $g(0) := 1$. Then the function $g: X \to \mathbb{R}$ is continuous in X and $g \geq 1$ or $g \leq 1$ according as f is convex or concave. Therefore the sequence (G_n) defined by (3.1.4) is increasing or decreasing according as f is convex or concave.

If f is convex, then from the relation

$$\int_{f(x)}^{x} f'(t)\,dt = f(x) - f^2(x)$$

we get the estimation

$$(x - f(x))f'(f(x)) \leqq f(x) - f^2(x) \leqq (x - f(x))f'(x),$$

whence by induction

$$\frac{x - f(x)}{f'(x)} \prod_{i=0}^{n} f'(f^i(x)) \leqq f^n(x) - f^{n+1}(x) \leqq (x - f(x)) \prod_{i=0}^{n-1} f'(f^i(x))$$

for $x \in X \setminus \{0\}$, $n \in \mathbb{N}$. By the equality

$$G_n(x) = \frac{x}{f^n(x)} \prod_{i=0}^{n-1} f'(f^i(x))$$

this may be rewritten as

$$G_{n+1}(x) \frac{f^{n+1}(x)}{f^n(x)f'(x)} \leqq \frac{x(1 - f^{n+1}(x)/f^n(x))}{x - f(x)} \leqq G_n(x). \qquad (3.5.9)$$

If f is concave, then the inequalities in (3.5.9) are reversed.

Assume that $0 < s < 1$. Since the function $x \mapsto f(x)/x$, $x \in X \setminus \{0\}$ is monotonic and approaches s as $x \to 0$, we get from (3.5.9)

$$1 \leqq G_{n+1}(x) \leqq h(x) \qquad (3.5.10)$$

for the case where f is convex; and (3.5.10) with the inequalities reversed if f is concave. Here we have put

$$h(x) = \frac{(1 - s)xf'(x)}{s(x - f(x))}$$

so that $\lim_{x \to 0} h(x) = 1$. Thus we have by (3.5.10), in view of the monotonicity of the sequence (G_n),

$$|G_{n+p}(x) - G_n(x)| = G_n(x)|G_p(f^n(x)) - 1| \leqq \max(h(x), 1)|h(f^n(x)) - 1|$$

for $n, p \in \mathbb{N}$, $x \in X$. This implies (Theorem 1.2.4) that the sequence (G_n) converges a.u. in X to a continuous function vanishing nowhere on X. By Theorem 3.1.2 equation (3.5.8) has in X a unique one-parameter family of continuous solutions:

$$\varphi(x) = c \bigg/ \lim_{n \to \infty} G_n(x) = \frac{c}{x} \lim_{n \to \infty} (f^n(x)/(f^n)'(x)), \quad x \neq 0; \qquad \varphi(0) = c. \quad (3.5.11)$$

This proves part (1) of the theorem.

If $s = 1$, then f must be concave and it follows from (3.5.9) (inequalities reversed!) that $\lim_{n \to \infty} G_n(x) = 0$ a.u. in $X \setminus \{0\}$. By Theorem 3.1.3 equation (3.5.8) has in X a continuous solution depending on an arbitrary function and every such solution vanishes at the origin (Corollary 3.1.1). This gives the conclusion of part (2).

Finally, let $s = 0$ and suppose equation (3.5.1) has a nontrivial solution $\lambda \in \mathscr{D}$. The function f must be convex and so it follows by (3.5.9) that the sequence (G_n) is bounded away from zero in X. Thus necessarily case (A) occurs for equation (3.5.8). Through (3.5.11) we come to formula (3.5.7) for our λ, with a $c \neq 0$. Moreover, $\lambda(x) \neq 0$ in $X \setminus \{0\}$, by the argument we have used in the proof of Theorem 3.1.4. Since $f' \circ f^n \geqslant f^{n+1}/f^n$, the sequence $((f^n)'/f^n)$ is increasing and we get from (3.5.7) the inequality $0 < (f^n)'(x)/f^n(x) < c/\lambda(x)$ for $x \in X \setminus \{0\}$, whence

$$\int_{f(x)}^x \frac{c\,dt}{\lambda(t)} \geqslant \int_{f(x)}^x \frac{(f^n)'(t)}{f^n(t)}\,dt = \log \frac{f^n(x)}{f^{n+1}(x)} \to \infty$$

as $n \to \infty$, which is impossible. Thus $\lambda = 0$. ∎

The following theorem shows a connection between the equations of Julia and Schröder.

Theorem 3.5.3. *Let hypotheses* (i) *and* (iii) *with* $0 < s < 1$ *be fulfilled. Let* $\sigma: X \to \mathbb{R}$ *be a nontrivial convex or concave solution of equation* (3.5.1) *and let* $\lambda \in \mathscr{D}$ *be a solution of equation* (3.5.6). *Then* σ *is of class* C^1 *in* $X \setminus \{0\}$ *(and even in* X *if* $\lim_{x \to 0} \sigma'(x) < \infty$), *and the functions* λ *and* σ *are related by the formula*

$$\lambda(x) = \lambda'(0)\sigma(x)/\sigma'(x) \qquad \text{in } X \setminus \{0\}. \tag{3.5.12}$$

Proof. The solutions λ and σ exist in virtue of Theorems 3.5.2 and 2.4.4. The case of $\lambda = 0$ is obvious, so assume $\lambda'(0) \neq 0$. Then $\lambda(x) \neq 0$ in $X \setminus \{0\}$ (see the proof of part (1) of Theorem 3.5.2) and $\lambda'(0)/\lambda(x) = \lim_{n \to \infty}((f^n)'(x)/f^n(x))$ a.u. in $X \setminus \{0\}$. By integrating this relation we have for arbitrary x, x_0 from $X \setminus \{0\}$

$$\lambda'(0) \int_{x_0}^x \frac{dt}{\lambda(t)} = \lim_{n \to \infty} \log \frac{f^n(x)}{f^n(x_0)} = \log\left(\frac{1}{c}\sigma(x)\right)$$

(see formula (2.4.18)), whence

$$\sigma(x) = c \exp\left[\lambda'(0) \int_{x_0}^x (\lambda(t))^{-1}\,dt\right]. \tag{3.5.13}$$

It follows from (3.5.13) that σ is of class C^1 in $X \setminus \{0\}$ and that (3.5.12) holds. Since σ is convex or concave, there exists $\sigma'(0) = \lim_{x \to 0} \sigma'(x)$. If this limit is finite, then σ is of class C^1 in X. ∎

Finally, with the aid of Theorem 3.3.4 we shall prove the following theorem which is implicitly contained in a result by G. Szekeres [1].

Theorem 3.5.4. *Make the assumptions* (i) *and* (ii), *but with the asymptotic relation replaced by*

$$f'(x) = 1 - b(m+1)x^m + O(x^{m+\delta}), \qquad x \to 0, \tag{3.5.14}$$

where b, m, δ *are some positive constants. Then equation (3.5.6) has a unique one-parameter family of continuous solutions* $\lambda: X \to \mathbb{R}$ *such that*

$$\lambda(x) = x^{m+1}(c + O(x^{\tau})), \qquad x \to 0, c \in \mathbb{R}, \tag{3.5.15}$$

where $\tau = \min(m, \delta)$. *They are given by the formula*

$$\lambda(x) = c \lim_{n \to \infty} [(f^n(x))^{m+1}/(f^n)'(x)]. \tag{3.5.16}$$

Proof. Let us put

$$\lambda(x) = x^{m+1}\varphi(x). \tag{3.5.17}$$

To each solution $\lambda: X \to \mathbb{R}$ of equation (3.5.6) having the properties stated in the theorem there corresponds a continuous solution $\varphi: X \to \mathbb{R}$ such that $\varphi(x) = c + \sigma(x^{\tau})$, $x \to 0$, of equation (3.5.8) where

$$g(x) := x^{m+1} f'(x)(f(x))^{-m-1}, \qquad x \in X \setminus \{0\}, \qquad g(0) := 1,$$

and conversely. The two solutions are linked by (3.5.17).

To solve equation (3.5.8) with our g, note that relation (3.5.14) implies

$$f(x) = x - bx^{m+1} + O(x^{m+1+\delta}). \tag{3.5.18}$$

Consequently, by the definition of g,

$$g(x) = 1 + O(x^{m+\tau}), \qquad x \to 0.$$

By Theorem 3.3.4 (see also Theorem 3.1.2) equation (3.5.8) has a unique one-parameter family of continuous solutions $\varphi: X \to \mathbb{R}$. They are given by

$$\varphi(x) = c \lim_{n \to \infty} \prod_{i=0}^{n-1} (g(f^i(x)))^{-1} = c \lim_{n \to \infty} [x^{-m-1}(f^n(x))^{m+1}/(f^n)'(x)]$$

and they have the property $\varphi(x) = c + O(x^{\tau})$, $x \to 0$. The functions λ defined by (3.5.17) fulfil the conditions of the theorem. ■

3.5C. Abel's equation

The equation is of the form

$$\alpha(f(x)) = \alpha(x) + 1. \tag{3.5.19}$$

Clearly, equation (3.5.19) cannot have a solution defined at the fixed point of f. Therefore now $0 \notin X$. We then replace hypothesis (i) by

(i′) $X = (0, a|, 0 < a \leqq \infty$.

We aim at showing that if $0 < s < 1$ in (ii) then some solutions of equation (3.5.19) can be obtained via Theorem 3.5.1 with the aid of differentiable solutions of a Schröder equation.

Theorem 3.5.5. *Assume* (i′) *and* (ii). *Then equation* (3.5.19) *has a unique one-parameter family (with an additive parameter) of solutions* $\alpha: X \to \mathbb{R}$ *which* *fulfil the condition*

$$\alpha(x) = \log x/\log s + \varphi(x)$$

where $\lim_{x\to 0} \varphi(x)$ *exists and is finite. These solutions are given by the formula*

$$\alpha(x) = \log \sigma(x)/\log s + c, \qquad c \in \mathbb{R}, \tag{3.5.20}$$

where $\sigma: X \cup \{0\} \to \mathbb{R}$ *is a* C^1 *solution of the Schröder equation* (3.5.1) *in* $X \cup \{0\}$ *such that* $\sigma'(0) = 1$.

Proof. We can extend f onto $X \cup \{0\}$ by putting $f(0) = 0$. By Theorem 3.5.1 equation (3.5.1) has a unique C^1 solution $\sigma: X \cup \{0\} \to \mathbb{R}$ fulfilling the condition $\sigma'(0) = 1$. It is easily seen that α given by (3.5.20) with $c = 0$ has all the required properties.

Now let $\tilde{\alpha}: X \to \mathbb{R}$ be another solution of equation (3.5.19) satisfying $\tilde{\alpha}(x) = \log x/\log s + \tilde{\varphi}(x)$, where $\tilde{\varphi}$ approaches a finite limit as $x \to 0$. Then the function $\omega(x) := \tilde{\alpha}(x) - \alpha(x) = \tilde{\varphi}(x) - \varphi(x)$ satisfies the equation $\omega(f(x)) = \omega(x)$ and has a finite limit at zero, whence $\omega = \text{const}$. Thus the solution α is unique up to an additive constant. ■

Differentiating both sides of (3.5.19) we get

$$\alpha'(f(x))f'(x) = \alpha'(x).$$

Thus the Julia equation is related also to that of Abel. In the case where $s = 1$ we have the following theorem (Szekeres [1]).

Theorem 3.5.6. *Assume* (i′) *and* (ii) *but with the asymptotic relation replaced by* (3.5.14). *Then equation* (3.5.19) *has a unique one-parameter family of solutions* $\alpha: X \to \mathbb{R}$ *which are of class* C^1 *in* X *and fulfil the condition*

$$\alpha'(x) = x^{-m-1}\varphi(x), \tag{3.5.21}$$

where $\lim_{x\to 0} \varphi(x)$ *exists and is finite. These solutions are strictly decreasing in* X, *and are given by the formula (of Lévy; see Remark 9.1.2)*

$$\alpha(x) = c + \lim_{n\to\infty} \frac{f^n(x) - f^n(x_0)}{f^{n+1}(x_0) - f^n(x_0)}, \tag{3.5.22}$$

where $x_0 \in X$ *is arbitrarily fixed and* c *is an arbitrary constant (a parameter). Moreover*

$$\alpha'(x) = -b^{-1}x^{-m-1} + O(x^{-m-1+\tau}), \qquad x \to 0, \tag{3.5.23}$$

where $\tau = \min(m, \delta)$.

Proof. This will be sketched only. Fix an $x_0 \in X$ and put

$$\alpha(x) = r \int_{x_0}^x \frac{dt}{\lambda(t)} \tag{3.5.24}$$

where $\lambda: X \to \mathbb{R}$ is given by (3.5.16) with $c = 1$ and has the properties stated in Theorem 3.5.4. The constant r is to be determined.

The following facts can be verified:

- the convergence in (3.5.16) is almost uniform in X,
- by Theorem 1.3.6, (3.5.14) and the monotonicity of f^n

$$\lim_{n \to \infty} (nbm)^{1/m} f^n(x) = 1 \qquad (3.5.25)$$

- by (3.5.25), formula (3.5.16) goes over into

$$\lambda(x) = \lim_{n \to \infty} (nbm)^{-1-1/m}/(f^n)'(x) \qquad \text{a.u. in } X,$$

so that interchanging 'lim' with '∫' signs in (3.5.24) yields

$$\alpha(x) = r \lim_{n \to \infty} (nbm)^{1+1/m}(f^n(x) - f^n(x_0)), \qquad (3.5.26)$$

- since λ satisfies (3.5.6), we have $\alpha(f(x)) - \alpha(x) = \alpha(f(x_0))$,
- by (3.5.18) (see (3.5.14)), (3.5.26) and (3.5.25) yield $\alpha(f(x_0)) = -rb$,
- taking in (3.5.26) $r = -1/b$ we get (3.5.19),
- formula (3.5.22) with $c = 0$ is obtained by using (3.5.26) $(r = -1/b)$, for both $\alpha(x)$ and $\alpha(f(x_0)) = 1$, and then forming their quotient,
- the regularity and monotonicity of α follow from (3.5.24) $(r = -1/b)$,
- relation (3.5.23) is a consequence of (3.5.21) that results from (3.5.24) $(r = -1/b)$ and (3.5.15) $(c = 1)$.

To prove the uniqueness, observe that if $\dot\alpha$ is a C^1 solution of (3.5.19) in X satisfying (3.5.21), then $\hat\lambda = 1/\dot\alpha'$ (defined as 0 at the origin) must be a continuous solution of equation (3.5.6) in $X \cup \{0\}$, fulfilling condition (3.5.15). By Theorem 3.5.4, $\lambda = q\hat\lambda$. Thus $\hat\alpha(x) = (q/r)(c + \alpha(x))$ and $q = r$ as $\dot\alpha$ satisfies (3.5.19). ∎

3.5D. A characterization of the cross ratio

Let us put

$$D := \{(x_1, x_2, x_3, x_4) \in \mathbb{R}^4 : x_i \neq x_j \text{ for } i \neq j\}.$$

The cross ratio $s: D \to \mathbb{R}$ of four points of the projective line is given by the formula

$$s(x_1, x_2, x_3, x_4) = (x_1 - x_3)(x_2 - x_4)/(x_2 - x_3)(x_1 - x_4).$$

S. Gołąb [2] considered the following system of functional equations:

$$S(x_3, x_4, x_1, x_2) = S(x_1, x_2, x_3, x_4), \qquad (3.5.27)$$

$$S(x_1, x_3, x_2, x_4) + S(x_1, x_2, x_3, x_4) = 1, \qquad (3.5.28)$$

$$S(x_1, x_2, x_3, x_4) \cdot S(x_1, x_2, x_4, x_5) = S(x_1, x_2, x_3, x_5). \qquad (3.5.29)$$

Of course, $S = s$ satisfies (3.5.27)–(3.5.29), but the general solution of this system, without any regularity assumptions, is given by (see Gołąb [2])

$$S(x_1, x_2, x_3, x_4) = s(\gamma(x_1), \gamma(x_2), \gamma(x_3), \gamma(x_4)), \qquad (3.5.30)$$

where $\gamma: \mathbb{R} \to \mathbb{R}$ is an arbitrary injection on $\mathbb{R}$.

To get a characterization of the cross ratio use can be made of the condition

$S = s$ imposed on a one-parameter family of harmonic points of the projective line, i.e.,

$$S\left(x, 1, \frac{2x}{x+1}, 0\right) = -1, \qquad x \in \mathbb{R}' := \mathbb{R} \setminus \{-1, 0, 1\}. \qquad (3.5.31)$$

Condition (3.5.31) leads to a functional equation for γ which in turn can be transformed to the Schröder equation

$$\sigma\left(\frac{x}{x+2}\right) = \tfrac{1}{2}\sigma(x) \qquad x \in \mathbb{R}'' := \mathbb{R} \setminus \{-2, -1, 0\}. \qquad (3.5.32)$$

If $\sigma: \mathbb{R} \to \mathbb{R}$ is an injection satisfying (3.5.32), then the function S given by (3.5.30) with γ defined by

$$\gamma(x) := b + (\sigma(x+1) - \sigma(1))^{-1}, \qquad x \neq 0, \gamma(0) := b \qquad (3.5.33)$$

(b may be arbitrary) has property (3.5.31). The converse is also true.

The injections $\sigma: \mathbb{R} \to \mathbb{R}$ given by

$$\sigma(x) = c\frac{x}{x+1}, \quad x \neq -1, \sigma(-1) = c, c \in \mathbb{R}, \qquad (3.5.34)$$

satisfy equation (3.5.32). Using them in (3.5.33) we get homographies, too, and $S = s$ with such γs. Since the function $f: X \to X$, $X := [0, a)$, $a > 0$, $f(x) = x/(x+2)$ satisfies the assumption of Theorem 3.5.1 (in particular, $f'(x) = \tfrac{1}{2} + O(x)$, $x \to 0$), formula (3.5.34) yields all solutions $\sigma \in C^1(X)$ of (3.5.32). Thus we have the following (Choczewski [13]).

Theorem 3.5.7. *Let $S: D \to \mathbb{R}$ be a function given by (3.5.30) with a $\gamma: \mathbb{R} \to \mathbb{R}$ injective and of class C^1 in a neighbourhood of $x = 1$. If S satisfies (3.5.31), then $S = s$. In other words: equations (3.5.27), (3.5.28), (3.5.29) and (3.5.31) then characterize the cross ratio s.*

3.6 Solutions of bounded variation

The theory of linear equations in this function class is also rather widely developed. An extensive study of solutions of bounded variation has been carried out by M. C. Zdun [4], [5], [7], [8], [15] (see also Matkowski–Zdun [1], Lasota [1]). Here we present a few of Zdun's results.

3.6A. Preliminaries

We denote by Var $\varphi | X$ the *variation* of the function $\varphi: X \to \mathbb{R}$ on the interval X. If X is not compact, then Var $\varphi | X$ is meant as the supremum of Var $\varphi | I$ taken over all compact subintervals I of X.

We shall look for solutions in the class of functions

$$\mathrm{BV}X := \{\varphi: X \to \mathbb{R}, \text{ Var } \varphi | X < \infty\}$$

that are of *bounded variation* on X. For the homogeneous equation

$$\varphi(f(x)) = g(x)\varphi(x) \tag{3.6.1}$$

we make the following hypotheses.

(i) $X = (0, a]$, $0 < a \leqslant \infty$.

(ii) $f : X \to \mathbb{R}$ is continuous and strictly increasing, $0 < f(x) < x$ in X.

(iii) $g \in \mathrm{BV}X$ and $m := \inf_X g > 0$.

As usual, the sequence $(G_n)_{n \in \mathbb{N}_0}$ is defined by (3.1.4). Moreover, throughout this section we use the following notations ($x_0 \in X$, $i \in \mathbb{N}_0$):

$$X_i = [f^{i+1}(x_0), f^i(x_0)];$$

$$a_i = \sup_{X_i} g = \sup_{X_0} g \circ f^i;$$

$$b_i = \sup_{X_0} G_i;$$

$$v_i = \mathrm{Var}\, g \,|\, X_i = \mathrm{Var}\, g \circ f^i \,|\, X_0;$$

$$A_i = \prod_{j=0}^{i-1} a_j.$$

The lemma that follows shows the role of the sequence (G_n).

Lemma 3.6.1. *Under hypotheses* (i)–(iii) *we have*

$$0 < L^{-1} G_n(x_1) \leqq A_n \leqq L G_n(x_2) \qquad \text{for } x_1, x_2 \in X_0, \, n \in \mathbb{N}_0, \tag{3.6.2}$$

where

$$L := \exp \mathrm{Var} \log g \,|\, X \tag{3.6.3}$$

does not depend on x_0; and

$$\mathrm{Var}\, G_n \,|\, X_0 \leqq K G_n(x_0), \qquad n \in \mathbb{N}_0, \tag{3.6.4}$$

where K is a positive constant independent of x_0.

Proof. We have for $x \in X_0$, $n \in \mathbb{N}_0$,

$$\left| \log \frac{A_n}{G_n(x)} \right| \leqq \sum_{i=0}^{n-1} |\log a_i - \log g(f^i(x))| \leqq \mathrm{Var} \log g \,|\, (0, x_0],$$

whence, according to (3.6.3), $L^{-1} \leqq A_n / G_n(x) \leqq L$ and (3.6.2) follows.

Further, $G_{n+1}(x) = g(f^n(x)) G_n(x)$, by (3.1.4). Therefore

$$\mathrm{Var}\, G_{n+1} \,|\, X_0 \leqq a_n \,\mathrm{Var}\, G_n \,|\, X_0 + v_n b_n, \qquad n \in \mathbb{N}_0,$$

whence, first by induction and next by the inequality $A_{i+1} \geqq b_{i+1}$,

$$\mathrm{Var}\, G_n \,|\, X_0 \leqq A_n \sum_{i=0}^{n-1} \frac{v_i b_i}{A_{i+1}} \leqq A_n \sum_{i=1}^{n-1} \frac{b_i}{b_{i+1}} v_i, \qquad n \in \mathbb{N}.$$

Take in the inequalities of (3.6.2) first $x_1 = x$, $x_2 = x_0$ and then $x_1 = x_0$, $x_2 = x$ (where $x \in X_0$) to get the inequalities

$$L^{-2} G_n(x_0) \leqq G_n(x)) \leqq L^2 G_n(x_0), \qquad x \in X_0, \, n \in \mathbb{N}_0. \tag{3.6.5}$$

Using (3.6.5) to estimate the suprema b_i we obtain $b_i/b_{i+1} \leqq L^4 m^{-1}$ (see (iii)). Therefore

$$\text{Var } G_n | X_0 \leqq M A_n, \qquad M := L^4 m^{-1} \text{ Var } g | X, \qquad 0 < M < \infty,$$

and (3.6.4) follows in view of the second inequality in (3.6.2) with $x_2 = x_0$. ∎

Lemma 3.6.2. *Under conditions* (i)–(iii), *if*

$$\lim_{n \to \infty} G_n(t) = 0 \tag{3.6.6}$$

for an $x \in X$, then (3.6.6) *holds a.u. in X. Moreover, if*

$$\sum_{n=0}^{\infty} G_n(x) < \infty \tag{3.6.7}$$

for an $x \in X$, then (3.6.7) *holds for all $x \in X$.*

Proof. Indeed, (3.6.5) implies that $\lim_{n \to \infty} G_n(t) = 0$ uniformly in the interval $[f(x), x]$. The rest is done by the relation

$$G_n(f^k(x)) = G_{n+k}(x)/G_k(x), \qquad n, k \in \mathbb{N}_0. \tag{3.6.8}$$

Similarly, (3.6.2) and (3.6.8) together imply our claim for the series in (3.6.7). ∎

3.6B. Homogeneous equation

It turns out that a necessary and sufficient condition for the uniqueness of bounded variation solutions of equation (3.6.1) is the divergence of the series $\sum G_n(x)$. Moreover, we may consider only the case where condition (3.6.6) holds. For, if (3.6.6) is not fulfilled, then the uniqueness follows from the following two facts: (1) every function $\varphi \in \text{BV}X$ tends to a finite limit as $x \to 0$; (2) this limit condition implies the uniqueness (possibly up to a constant factor) of solutions of equation (3.6.1) (see Section 3.1). To have the motivation completed, notice that the argument employed in the proofs of Theorems 3.1.2 (case (A)) and 3.1.4 (case (C)) can be modified so as to yield continuous solutions $\varphi: X \to \mathbb{R}$ with $\lim_{x \to 0} \varphi(x) < \infty$.

If series (3.6.7) converges, then we have the lack of uniqueness.

Theorem 3.6.1. *Let hypotheses* (i)–(iii) *and condition* (3.6.7) *be fulfilled. Then every function $\varphi_0 \in \text{BV}X_0$ such that*

$$\varphi_0(f(x_0)) = g(x_0)\varphi_0(x_0)$$

can be uniquely extended onto X to a solution $\varphi: X \to \mathbb{R}$ of equation (3.6.1). *This extension belongs to* $\text{BV}(0, x_0]$.

Proof. A standard construction yields the unique extension $\varphi: X \to \mathbb{R}$ of φ_0, satisfying (3.6.1) in X. To check that $\text{Var } \varphi | (0, x_0] < \infty$ observe that

$(0, x_0] = \bigcup_{n=0}^{\infty} X_n$. Thus we need to estimate $\text{Var } \varphi | X_n$. Since φ satisfies (3.6.1), we have equalities (3.1.3), whence

$$\text{Var } \varphi | X_n \leq (\text{Var } G_n | X_0) \sup_{X_0} |\varphi_0| + b_n \text{ Var } \varphi_0 | X_0, \qquad n \in \mathbb{N}_0. \quad (3.6.9)$$

The conditions $b_n \leq A_n$ for $n \in \mathbb{N}_0$ and (3.6.9) show that $\varphi \in \text{BV}(0, x_0]$ since

$$\text{Var } \varphi | (0, x_0] = \sum_{n=1}^{\infty} \text{Var } \varphi | X_n + \text{Var } \varphi_0 | X_0 < \infty$$

on account of Lemma 3.6.1 and (3.6.7). ∎

Remark 3.6.1. The question arises how to determine the solutions which are of bounded variation in the whole of X. This is easy to do if $a \notin X$ and $\lim_{x \to a} f(x) < a$. We simply extend f and g to $X \cup \{a\}$ by assigning them their limit values at $x = a$ and we apply Theorem 3.6.1 to the interval $X \cup \{a\}$ with $x_0 = a$. In the case where $\lim_{x \to a} f(x) = a$ condition (3.6.7) alone does not guarantee that $\text{Var } \varphi | X < \infty$. This case is discussed in Note 3.8.14.

Now we turn to the case

$$\sum_{n=0}^{\infty} G_n(x) = \infty \quad (3.6.10)$$

in which an explicit formula for BV-solutions of (3.6.1) may be found.

Lemma 3.6.3. *Let hypotheses* (i)–(iii) *be fulfilled and assume* (3.6.6) *and* (3.6.10) *to hold. If a function* $\varphi \in \text{BV} X$ *satisfies equation* (3.6.1) *and* $\varphi(t) = 0$ *for a* $t \in X$, *then* $\varphi = 0$.

Proof. It follows by (3.6.1) that if $f^k(t) \in X$, then $\varphi(f^k(t)) = 0$, $k \in \mathbb{Z}$, so we may assume that $t \in X_0$. We have by (3.1.3)

$$\text{Var } \varphi | X_i = \text{Var } \varphi \circ f^i | X_0 \geq |\varphi(f^i(t)) - \varphi(f^i(x_0))|$$
$$= |\varphi(f^i(x_0))| = G_i(x_0) |\varphi(x_0)|,$$

whence

$$\text{Var } \varphi | X \geq \sum_{i=0}^{\infty} \text{Var } \varphi | X_i \geq |\varphi(x_0)| \sum_{i=0}^{\infty} G_i(x_0), \qquad n \in \mathbb{N},$$

which implies in view of (3.6.10) that $\varphi(x_0) = 0$. Since $x_0 \in X$ could have been chosen arbitrarily, we have $\varphi = 0$. ∎

Lemma 3.6.4. *Let hypotheses* (i)–(iii) *be fulfilled. Then, for every* $x \in X$, *there exists a finite and positive limit of the sequence* $(G_n(x_0)/G_n(x))$ *as* $n \to \infty$.

Proof. At first assume that $x \in X_0$. Write $G_n(x_0)/G_n(x) = \exp \sum_{k=0}^{n-1} d_k(x)$, where $d_k(x) = \log g(f^k(x_0)) - \log g(f^k(x))$, $k \in \mathbb{N}_0$. Then

$$\sum_{n=0}^{\infty} |d_n(x)| \leq \sum_{n=0}^{\infty} \text{Var } \log g | X_n = \text{Var } \log g | (0, x_0] < \infty,$$

and the lemma follows. If $x \notin X_0$, we appeal to relation (3.6.8). ∎

Theorem 3.6.2. *Let hypotheses* (i)–(iii) *be fulfilled and assume* (3.6.6) *and* (3.6.10) *to hold. If a function* $\varphi: X \to \mathbb{R}$ *satisfies equation* (3.6.1) *and* $\operatorname{Var} \varphi |(0, x_0] < \infty$, *then*

$$\varphi(x) = d \lim_{n \to \infty} \frac{G_n(x_0)}{G_n(x)}, \qquad (3.6.11)$$

where $d \in \mathbb{R}$ *is a constant.*

Proof. If φ vanishes anywhere on X, then it has to vanish at a point of $(0, x_0]$. By Lemma 3.6.3 $\varphi = 0$ in $(0, x_0]$, and by (3.6.1) also in the whole X. Thus (3.6.11) holds with $d = 0$.

Now assume φ to be nowhere zero in X. We have by (3.1.3)

$$\frac{\varphi(f^n(x))}{\varphi(f^n(x_0))} = \frac{G_n(x)}{G_n(x_0)} \frac{\varphi(x)}{\varphi(x_0)}, \qquad n \in \mathbb{N}, \qquad (3.6.12)$$

whence it follows in view of Lemma 3.6.4 that there exists a finite limit of the quotient on the left-hand side of (3.6.12). The theorem will be proved if we show that this limit equals unity.

Let $x \in X_0$ and put

$$c_n(x) = \varphi(f^n(x)) - \varphi(f^n(x_0)), \qquad n \in \mathbb{N}_0.$$

We have $|c_n(x)| \leq \operatorname{Var} \varphi | X_n$, so that the series $\sum_{n=0}^{\infty} |c_n(x)|$ converges. By (3.6.10) and (3.1.3) $\liminf_{n \to \infty} |c_n(x)/\varphi(f^n(x_0))| = 0$. Hence

$$\lim_{n \to \infty} \frac{\varphi(f^n(x))}{\varphi(f^n(x_0))} = \liminf_{n \to \infty} \frac{\varphi(f^n(x))}{\varphi(f^n(x_0))} = 1 + \liminf_{n \to \infty} \frac{c_n(x)}{\varphi(f^n(x_0))} = 1.$$

By (iii) there exists a finite limit $\lim_{x \to 0} g(x)$, and (3.6.6) and (3.6.10) imply that it equals 1, whence also $\lim_{x \to 0} G_k(x) = 1$ for every $k \in \mathbb{N}_0$. Once again we use (3.1.3); now with $f^n(x)$ in place of x ($t = f^k(x)$)

$$\frac{\varphi(f^n(t))}{\varphi(f^n(x_0))} = \frac{\varphi(f^k(f^n(x)))}{\varphi(f^n(x_0))} = G_k(f^n(x)) \frac{\varphi(f^n(x))}{\varphi(f^n(x_0))}.$$

This yields $\lim_{n \to \infty} [\varphi(f^n(x))/\varphi(f^n(x_0))] = 1$ for all $x \in X$. ∎

This is a uniqueness result only. Under the hypotheses of Theorem 3.6.2 nontrivial solutions $\varphi: X \to \mathbb{R}$, $\varphi \in \mathrm{BV}(0, x_0]$, of equation (3.6.1) still need not exist; see Note 3.8.14.

3.6C. Inhomogeneous equation

Since the difference of two solutions of bounded variation of the inhomogeneous equation

$$\varphi(f(x)) = g(x)\varphi(x) + h(x) \qquad (3.6.13)$$

is a solution of bounded variation of the homogeneous equation (3.6.1),

Theorems 3.6.1 and 3.6.2 give conditions for the uniqueness or nonuniqueness of such solutions of equation (3.6.13).

Concerning the existence of solutions of bounded variation for equation (3.6.13) we have the following result, in which we shall use (3.1.18) for $y = 0$, i.e.

$$H_n(x) := G_n(x) \sum_{i=0}^{n-1} \frac{h(f^i(x))}{G_{i+1}(x)}, \qquad n \in \mathbb{N}. \tag{3.6.14}$$

Theorem 3.6.3. *Let hypotheses* (i)–(iii) *be fulfilled and assume that condition* (3.6.7) *holds and* $h \in BVX$. *Equation* (3.6.13) *has a solution* $\varphi: X \to \mathbb{R}$ *such that* Var $\varphi|(0, x_0] < \infty$ *if and only if*

$$\sum_{n=1}^{\infty} \text{Var } H_n|X_0 < \infty. \tag{3.6.15}$$

Proof. Let $\varphi: X \to \mathbb{R}$ be a solution of (3.6.13) such that Var $\varphi|(0, x_0] < \infty$. By induction

$$H_n(x) = \varphi(f^n(x)) - G_n(x)\varphi(x), \tag{3.6.16}$$

whence

$$\text{Var } H_n|X_0 \le \text{Var } \varphi|X_n + b_n \text{ Var } \varphi|X_0 + (\text{Var } G_n|X_0) \sup_{x_0} |\varphi|.$$

We get (3.6.15) as a result of Lemma 3.6.1, the inequality $b_n \le A_n$ and the condition $\sum_{n=0}^{\infty} \text{Var } \varphi|X_n = \text{Var } \phi|(0, x_0] < \infty$.

Conversely, let $\varphi_0 \in BVX_0$ be an arbitrary function satisfying (3.1.1$_0$). By Theorem 3.1.1 (see Remark 3.1.1), φ_0 can be extended onto X to a solution $\varphi: X \to \mathbb{R}$ of equation (3.6.13). Thus (3.6.16) holds, and consequently

$$\text{Var } \varphi|X_n = \text{Var } \varphi \circ f^n|X_0 \le \text{Var } H_n|X_0 + b_n \text{ Var } \varphi_0|X_0$$
$$+ (\text{Var } G_n|X_0) \sup_{x_0} |\varphi_0|,$$

and (3.6.15) implies Var $\varphi|(0, x_0] = \sum_{n=0}^{\infty} \text{Var } \varphi|X_n < \infty$. ∎

3.6D. Solutions of almost bounded variation

Consider the case $0 \in X$ and introduce the class of functions

$$ABVX := \{\varphi: X \to \mathbb{R}, \text{ Var } \varphi|I < \infty \text{ for every compact interval } I \subset X \setminus \{0\}\}.$$

We say that the functions in ABVX are of *almost bounded variation* on the interval X.

The results we state here without proofs can be proved similarly to the corresponding ones of M. C. Zdun [6]. An interesting aspect of these results is that they give conditions for the existence of continuous solutions of (3.6.13) in X.

Assume the following.

(iv) $X = [0, a|, 0 < a \leqq \infty$.
(v) The function $f : X \to X$ is continuous and strictly increasing, $0 < f(x) < x$ in $X \setminus \{0\}$.
(vi) The functions g and h belong to the class $\mathrm{BV}X$ and are continuous, $g(0) = 1$, $h(0) = 0$ and $\inf_X g > 0$.

For $y \in \mathbb{R}$ we shall use notation (3.1.17), i.e. we put

$$h(x; y) = h(x) + y(g(x) - 1). \tag{3.6.17}$$

Under conditions (iv)–(vi) the behaviour of the sequence $(G_n(x))$ is determined by its behaviour at a single point $x_0 \in X \setminus \{0\}$ (see Lemma 3.6.1). So we need only consider modified cases (A'), (B'), (C') (see Section 3.1), as follows.

(A') There exists an $x_0 \in X \setminus \{0\}$ such that $\lim_{n \to \infty} G_n(x_0)$ exists, is finite and positive.
(B') There exists an $x_0 \in X \setminus \{0\}$ such that $\lim_{n \to \infty} G_n(x_0) = 0$.
(C') Neither (A') nor (B') occurs.

It can be shown that cases (A'), (B'), (C') are, under our assumptions, equivalent to (A), (B), (C) of Section 3.1, respectively.

Theorem 3.6.4. *Let hypotheses* (iv)–(vi) *be fulfilled. Then the set* $\mathbf{\Phi}$ *of the solutions* $\varphi : X \to \mathbb{R}$ *of equation* (3.6.13) *that are continuous on X and belong to the class* $\mathrm{ABV}X$ *is determined as follows.*

(a) *If* (A') *occurs and* $\sum_{n=0}^{\infty} [h(f^n(x_0))/G_{n+1}(x_0)] < \infty$, *then* $\mathbf{\Phi} = \{\varphi_y; y \in \mathbb{R}\}$, *where*

$$\varphi_y(x) = y \lim_{n \to \infty} (G_n(x))^{-1} - \sum_{n=0}^{\infty} \frac{h(f^n(x))}{G_{n+1}(x)}.$$

(b) *If* $g(f^n(x_0)) < 1$, $n \in \mathbb{N}_0$, *and* $\lim_{n \to \infty} H_n(x_0) = y$ (H_n *is given by* (3.6.14)), *then every* $\varphi_0 \in \mathrm{BV}X_0$ *satisfying* (3.1.1$_0$) *can be uniquely extended to a solution* $\varphi \in \mathbf{\Phi}$ *and* $\varphi(0) = y$ *for every* $\varphi \in \mathbf{\Phi}$.
(c) *If* $g(f^n(x_0)) > 1$, $n \in \mathbb{N}_0$, *and* $\lim_{x \to 0} \{h(x)/[1 - g(x)]\} = y$, *then* $\mathbf{\Phi} = \{\varphi\}$, *where*

$$\varphi(x) = y - \sum_{n=0}^{\infty} \frac{h(f^n(x); y)}{G_{n+1}(x)},$$

and $h(x; y)$ *is given by* (3.6.17).

Comments. For further results on BV solutions the reader is referred to Zdun [5], Dyjak [3], and to Dyjak–Matkowski [1] for nonlinear equations.

3.7 Applications

Here we shall present three problems which lead to linear equations and function classes discussed in this chapter. We consider invariant measures, doubly stochastic measures and a differential equation of second order. For further applications in ergodic theory see Lasota [1], [2], Livšic [1], Ruelle [1].

3.7A. An Anosov diffeomorphism without invariant measure

Let $T^2 = \mathbb{R}^2$ (mod 1) be the torus and let a mapping $A: T^2 \to T^2$ be given by $A(x, y) = (x', y')$ with

$$x' = x + y \quad (\text{mod } 1), \qquad y' = x + 2y \quad (\text{mod } 1).$$

The matrix $\begin{pmatrix} 1 & 1 \\ 1 & 2 \end{pmatrix}$ has the eigenvalue $\lambda = \frac{1}{2}(3 - \sqrt{5})$ and the corresponding eigenvector $(1, b)$, $b = \frac{1}{2}(1 - \sqrt{5})$. Let $S_\varepsilon: T^2 \to T^2$ be given by $S_\varepsilon(x, y) = (x', y')$ with

$$x' = x - \varepsilon \sin 2\pi x \quad (\text{mod } 1), \qquad y' = y - \varepsilon b \sin 2\pi x \quad (\text{mod } 1).$$

Finally, put

$$Q := A \circ S_\varepsilon.$$

It follows from the general theorems of ergodic theory that, for small $\varepsilon > 0$, Q is an Anosov diffeomorphism (see Arnold–Avéz [1, p. 62]). Following Lasota–Yorke [1] we shall show that for $0 < \varepsilon < 1/(2\pi)$ the diffeomorphism Q admits no nontrivial invariant measure with a continuous density.

Suppose that $D: T^2 \to \mathbb{R}^+$ is a continuous function such that the measure

$$\mu(E) = \iint_E D(x, y)\,dx\,dy \tag{3.7.1}$$

is invariant under Q, i.e.,

$$\mu(E) = \mu(Q(E)) \tag{3.7.2}$$

for every measurable set $E \subset T^2$. The function D may be regarded as a doubly periodic function from $\mathbb{R}^2$ into $\mathbb{R}^+$ with period 1 in each variable, continuous on $\mathbb{R}^2$.

We have to show that $D = 0$. Put

$$\varphi(x) = D(x, bx), \qquad x \in \mathbb{R}^+. \tag{3.7.3}$$

The function φ is continuous on $\mathbb{R}^+$ and, since the set

$$\{(u, v) \in T^2: u = x \ (\text{mod } 1), v = bx \ (\text{mod } 1), x \in \mathbb{R}^+\}$$

is dense in T^2, the closure of the range of φ is the same as that of the range of D. Thus it is enough to show that $\varphi = 0$.

It follows from (3.7.1), (3.7.2) and the continuity of D that

$$D(x, y) = D(Q(x, y)) \det Q'(x, y).$$

Putting here $y = bx$ and using (3.7.3) we obtain

$$\varphi(f(x)) = g(x)\varphi(x), \tag{3.7.4}$$

where $f(x) = \lambda(x - \varepsilon \sin 2\pi x)$, $g(x) = (1 - 2\pi\varepsilon \cos 2\pi x)^{-1}$. Thus $0 < f(x) < x$, $f'(x) > 0$ and $g(x) > 0$ in $(0, \infty)$. Moreover, since $g(0) = (1 - 2\pi\varepsilon)^{-1} > 1$, the sequence (G_n) defined by (3.1.4) diverges to infinity. By Theorem 3.1.4 the function $\varphi = 0$ is the unique solution of (3.7.4) continuous on $\mathbb{R}^+$. ∎

3.7B. Doubly stochastic measures supported on a hairpin

A measure μ defined on Borel subsets of the unit square is said to be *doubly stochastic* (d.s.) iff

$$\mu(A \times [0, 1]) = \mu([0, 1] \times A) = l_1(A) \tag{3.7.5}$$

for every Borel set $A \subset [0, 1]$, where l_1 stands for the one-dimensional Lebesgue measure. Thus a d.s. measure may be considered as a continuous analogue of a doubly stochastic matrix.

Finding d.s. measures is of importance in probability theory (see, e.g., Schweizer–Sklar [5], Seethoff–Shiflett [1]). Here we present the approach and some of the results by H. Sherwood–M. D. Taylor [1].

We say that a measure μ is *supported* on a set S (or S is the *support* of μ) iff

$$S = \bigcap \{B \subset [0, 1]^2 : B \text{ is closed and } \mu([0, 1]^2 \setminus B) = 0\}.$$

T. L. Seethoff–R. C. Shiflett [1] considered d.s. measures supported on the union of graphs of two functions. They obtained strong uniqueness results but the existence theorems were less satisfactory.

Assume that

(i) $X = [0, 1]$ and f is an increasing homeomorphism from X onto itself such that $f(x) < x$ whenever $x \in (0, 1)$.

Fig. 3.1

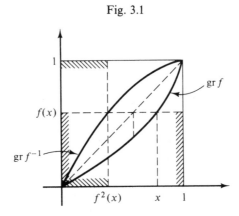

With the notation $\operatorname{gr} f := \{(t, f(t)) \in \mathbb{R}^2 : t \in X\}$, the union

$$H := \operatorname{gr} f \cup \operatorname{gr} f^{-1}$$

is called a *hairpin*; see Fig. 3.1. The symbol $\operatorname{gr} f | [0, x]$ will denote the graph of the restriction of f to the interval $[0, x] \subset X$.

The existence and uniqueness of d.s. measures supported on a hairpin are just dealt with in Sherwood–Taylor [1]. Finding such a measure turns out to be equivalent to finding a function φ such that $\varphi(x)$ is the amount of mass spread on the graph of f restricted to the interval $[0, x]$:

$$\varphi(x) = \mu(\operatorname{gr} f | [0, x]), \qquad x \in X. \tag{3.7.6}$$

The subsequent theorem says that φ can be determined with the aid of the functional equation

$$\varphi(f(x)) = -\varphi(x) + f(x), \qquad x \in X. \tag{3.7.7}$$

Remark 3.7.1. If (i) is fulfilled, then by Theorem 2.2.1 equation (3.7.7) has the unique nonnegative solution $\tilde{\varphi} \colon X \setminus \{1\} \to \mathbb{R}$ given by

$$\tilde{\varphi}(x) = \sum_{n=1}^{\infty} (-1)^{n+1} f^n(x), \qquad x \in X \setminus \{1\}. \tag{3.7.8}$$

Since the nth term of the sequence of remainders of the series in (3.7.8) is majorized by $f^{n+1}(x_0)$ in every interval $[0, x_0] \subset X \setminus \{1\}$ and $\lim_{n \to \infty} f^n(x_0) = 0$ (Theorem 1.1.1), the series converges a.u. in $X \setminus \{1\}$. Thus the function $\tilde{\varphi}$ is also the unique continuous solution of (3.7.7) in $X \setminus \{1\}$; see Theorem 3.1.9 and Note 3.8.3.

Theorem 3.7.1. *Let hypothesis* (i) *be satisfied and suppose that μ is a doubly stochastic measure supported on the hairpin $H := \operatorname{gr} f \cup \operatorname{gr} f^{-1}$. Then the function $\varphi \colon X \to \mathbb{R}$ given by formula* (3.7.6) *yields the nonnegative continuous increasing solution of equation* (3.7.7). *Conversely, if* (3.7.7) *has a nonnegative continuous and increasing solution $\varphi \colon X \to \mathbb{R}$, then there exists exactly one d.s. measure μ supported on H and such that* (3.7.6) *holds.*

Proof. By its definition (3.7.6), φ is increasing and nonnegative. We shall show that it is Lipschitzian (hence continuous). Indeed, for fixed $x, \bar{x} \in X$, $x \leqslant \bar{x}$, we have

$$|\varphi(x) - \varphi(\bar{x})| = |\mu(\operatorname{gr} f | [0, x]) - \mu(\operatorname{gr} f | [0, \bar{x}])| = \mu(\operatorname{gr} f | (x, \bar{x}])$$
$$\leqslant \mu((x, \bar{x}] \times X) = l_1((x, \bar{x}]) = |x - \bar{x}|.$$

Further, we obtain (see Fig. 3.1)

$$\mu(X \times [0, f(x)]) = \mu(\operatorname{gr} f | [0, x]) + \mu(\operatorname{gr} f^{-1} | [0, f^2(x)])$$

and

$$\mu([0, f^2(x)] \times X) = \mu(\operatorname{gr} f | [0, f^2(x)]) + \mu(\operatorname{gr} f^{-1} | [0, f^2(x)]).$$

Thus, according to (3.7.5) and (3.7.6),

$$f(x) - \varphi(x) = f^2(x) - \varphi(f^2(x)). \qquad (3.7.9)$$

Let us put

$$\omega(x) = f(x) - \varphi(f(x)) - \varphi(x), \qquad x \in X.$$

Subtracting $\varphi(f(x))$ from both sides of (3.7.9) we get $\omega(x) = \omega(f(x))$. Since f and φ are continuous in X and $f(0) = \varphi(0) = 0$, the automorphic function $\omega: X \to \mathbb{R}$ is constant, $\omega(x) = \lim_{x \to 0} \omega(f^n(x)) = 0$ (see Theorem 1.1.1). Thus (3.7.7) holds true.

We omit the standard but rather tedious extending procedure showing the other assertion of the theorem. ∎

In view of Theorem 3.7.1 and Remark 3.7.1, function (3.7.6), if it exists, coincides with function (3.7.8) in $X \setminus \{1\}$. On the other hand, if the function φ given by the formula

$$\varphi(x) = \sum_{n=1}^{\infty} (-1)^{n+1} f^n(x), \qquad x \in X \setminus \{1\}, \; \varphi(1) = \tfrac{1}{2}, \qquad (3.7.10)$$

is increasing in $(0, 1)$, then it generates a d.s. measure μ via relation (3.7.6). However, the function (3.7.10) need not generate any d.s. measure supported on H as it may fail to be increasing; see e.g. the case of $f(x) = x^2$, $x \in X$ (Seethoff–Shiflett [1]). Conditions on the homeomorphism f for φ given by (3.7.10) to be increasing are also given in Sherwood–Taylor [1]. The following result gives a necessary condition of that sort. The short and elegant proof we supply is due to J. H. B. Kemperman. It has been included in Sherwood–Taylor [1] as communicated to the authors privately.

Theorem 3.7.1. *If hypothesis* (i) *is fulfilled and the function given by* (3.7.10) *is increasing, then*

$$m(x) := \sum_{n=-\infty}^{\infty} [f^{2n}(x) - f^{2n+1}(x)] = \tfrac{1}{2} \qquad \text{for all } x \in (0, 1). \quad (3.7.11)$$

Proof. Let μ be the doubly stochastic measure generated by φ and let

$$A(x) := \bigcup_{n=-\infty}^{\infty} [f^{2n+1}(x), f^{2n}(x)), \qquad x \in (0, 1),$$

so that $m(x) = l_1(A(x))$. Because of the inclusion, valid for $x \in (0, 1)$,

$$H = \mathrm{gr}\, f \cup \mathrm{gr}\, f^{-1} \subset \{(0, 0)\} \cup (A(x) \times X) \cup (X \times A(x)) \cup \{(1, 1)\}$$

and of (3.7.5), we have

$$1 = \mu(H) \leqslant \mu(\{0\} \times X) + \mu(A(x) \times X) + \mu(X \times A(x)) + \mu(X \times \{1\})$$
$$= 2l_1(A(x)).$$

Consequently,

$$m(x) = l_1(A(x)) \geq \tfrac{1}{2} \qquad \text{for all } x \in (0, 1).$$

A similar argument with $B(x) = X \setminus A(x) = \bigcup_{n=-\infty}^{\infty} [f^{2n}(x), f^{2n-1}(x))$ in place of $A(x)$ shows that also $l_1(B(x)) \geq \tfrac{1}{2}$. But $l_1(A(x)) + l_1(B(x)) = l_1(X) = 1$, $x \in (0, 1)$. Thus $l_1(A(x)) = \tfrac{1}{2}$ for $x \in (0, 1)$ and (3.7.11) follows. ■

We conclude by formulating without proofs two theorems from Sherwood–Taylor [1]. The first (see also Note 3.8.16) is crucial for the problem of monotonicity of φ and the other shows possible forms of hairpins that may support d.s. measures.

Theorem 3.7.3. *Let hypothesis* (i) *be fulfilled. If* f *is convex, then function* (3.7.10) *is increasing if and only if condition* (3.7.11) *holds true.*

Theorem 3.7.4. *A hairpin* $H = \operatorname{gr} f \cup \operatorname{gr} f^{-1}$ *supports a d.s. measure* μ *if and only if* $f = (\operatorname{id}_X - \varphi)^{-1} \circ \varphi$ *where* φ *and* $\operatorname{id}_X - \varphi$ *are strictly increasing homeomorphisms from* X *onto* $[0, \tfrac{1}{2}]$ *and* $\varphi(x) < x/2$ *whenever* $x \in (0, 1)$. *The pair* (f, φ) *then satisfies* (3.7.7) *and* f, φ *and* μ *are related by formula* (3.7.6).

Remark 3.7.2. The condition $\varphi(1) = \tfrac{1}{2}$ means that the d.s. measure μ generated by φ has to satisfy $\tfrac{1}{2} = \mu(\operatorname{gr} f) = \mu(\operatorname{gr} f^{-1})$. An inspection of the proof of Theorem 3.7.1 shows that μ must be distributed symmetrically with respect to the diagonal of X, i.e., $\mu(\operatorname{gr} f \,|\, [0, x]) = \mu(\operatorname{gr} f^{-1} \,|\, [0, f(x)])$ for $x \in X$.

3.7C. Phase and dispersion for second-order differential equations

Consider the differential equation

$$u'' = q(x)u, \qquad x \in X := (0, a),\ 0 < a \leq \infty, \tag{3.7.12}$$

where $q: X \to \mathbb{R}$ is a continuous function. Following O. Borůvka [1] (see also Neuman [3]) we shall introduce the notion of a phase and a dispersion for equation (3.7.12) and find an Abel equation stemming from these notions.

Definition 3.7.1. Let $u_1: X \to \mathbb{R}$ and $u_2: X \to \mathbb{R}$ be two linearly independent solutions of equation (3.7.12).

(1) A function $\alpha: X \to \mathbb{R}$ which is continuous on X and such that the relation

$$\tan \alpha(x) = u_1(x)/u_2(x) \tag{3.7.13}$$

holds whenever $u_2(x) \neq 0$ is called the *phase* of equation (3.7.12).

(2) Let $x_0 \in X$. If a nontrivial solution of (3.7.12) vanishes at x_0, then let $f(x_0)$ be the first zero of this solution that lies to the right of x_0. The function f is called the *dispersion* of equation (3.7.12).

The dispersion is well defined. For if we have two nontrivial solutions of (3.7.12), both vanishing at x_0, then they are linearly dependent and thus have the same set of zeros. However, the domain of f is, in general, a subinterval $X' = (0, a')$ of X and X' may also be empty. We have $X' = X$ if and only if (3.7.12) is oscillatory for $x \to a$, i.e., every solution of (3.7.12) has infinitely many zeros greater than $x_0 \in X$. As for the phase, if $u_2(x_0) = 0$ for an $x_0 \in X$, then both one-sided limits of the quotient in (3.7.13) at x_0 are infinite and differ in the signs. Thus a function $k \colon X \to \mathbb{Z}$ can be found which is constant on each interval whose endpoints are consecutive zeros of u_2 and such that the function $\alpha(x) = \tan^{-1}(u_1(x)/u_2(x)) + k(x)\pi$ if $u_2(x) \neq 0$, $\alpha(x) = (2k(x) + 1)\pi/2$ if $u_2(x) = 0$ is continuous on X.

The phase α is even of class C^3 on X, $\alpha'(x) \neq 0$ in X. Any solution of (3.7.12) may be expressed by the formula (with some $c_1, c_2 \in \mathbb{R}$)

$$u(x) = c_1 |\alpha'(x)|^{-1/2} \sin(\alpha(x) + c_2).$$

Therefore, if $u(x_0) = 0$, $c_1 \neq 0$, then $\alpha(x_0) + c_2 = m\pi$ for an $m \in \mathbb{Z}$. Let $x_0 \in X'$, so that $u(f(x_0)) = 0$, too. Thus $\alpha(f(x_0)) + c_2 = n\pi$ with an $n \in \mathbb{Z}$. But, by Definition 3.7.1, $f(x_0)$ being the first zero lying to the right of x_0, we have either $n = m + 1$ or $n = m - 1$ according as α is increasing or decreasing. In this way we arrive at the Abel equation

$$\alpha(f(x)) = \alpha(x) + \pi \operatorname{sgn} \alpha', \qquad x \in X'. \tag{3.7.14}$$

The dispersion f has the properties

$$f \colon X' \to X, \quad f(x) > x, \quad f'(x) > 0, \quad f \text{ is of class } C^3 \text{ on } X', \tag{3.7.15}$$

which follow from Definition 3.7.1 and the properties of the phase α.

Theorem 3.7.5. *Assume that $X' = X$ and that f satisfies (3.7.15). Then the C^3 solution $\alpha \colon X \to \mathbb{R}$ of equation (3.7.14), $\alpha'(x) \neq 0$ in X, depends on an arbitrary function.*

This follows from Theorem 3.4.1 (we need to take f^{-1} in place of f). The result remains true in the case where $X' \neq X$, but then we get solutions of (3.7.14) which are defined on $X' \cup f(X')$. The latter fact has been first proved by E. Barvínek [1].

Remark 3.7.3. Answering in negative the question by F. Neuman [11], K. Baron–J. Walorski [1] showed that it is impossible to find conditions on f under which there exists a unique solution $\alpha \colon X \to \mathbb{R}$ of the Abel equation $\alpha(f(x)) = \alpha(x) + 1$ being of class C^3 on X, $\alpha' > 0$. Namely, if α is such a solution, then so is the function $\tilde{\alpha} = h \circ \alpha$, where $h(x) := x + (1/4\pi) \sin 2\pi x$.

Remark 3.7.4. Using the Abel equation, F. Neuman [3] found necessary and sufficient conditions for a plane curve to be closed or simply closed or centrosymmetric. Moreover, an explicit form of equation (3.7.12), all solutions of which are periodic or half-periodic can be given (Neuman [1]). Several other problems that lead to Schröder and Abel equations are discussed in Neuman [2], [4]–[8] and in the book of F. Neuman [13].

3.8 Notes

3.8.1. It can well happen that the sequence (G_n) defined by (3.1.4) converges uniformly to zero on an interval $X_0 = [f(x_0), x_0]$, $x_0 \in X \setminus \{0\}$. In this case all the continuous solutions $\varphi : X \to Y$ of the equation

$$\varphi(f(x)) = g(x)\varphi(x) \tag{3.8.1}$$

can be obtained by taking any continuous function $\varphi_0 : X_0 \to Y$ subjected only to condition (3.1.8) and then extending it onto X.

(See Theorem 3.1.3. By (3.1.7) we have $U = \text{int } X$.)

3.8.2. The following is a corollary from Theorem 3.1.2 and 3.1.3.

Proposition 3.8.1. *Assume conditions* (i)–(iii) *in Section 3.1 to hold. Let equation* (3.8.1) *have a nontrivial continuous solution* $\varphi : X \to Y$. *If* $\varphi(0) \neq 0$, *then case* (A) *holds; whereas if* $\varphi(0) = 0$, *then case* (B) *holds.*

3.8.3. For the particular equations

$$\varphi(f(x)) - \varphi(x) = h(x), \tag{3.8.2}$$

$$\varphi(f(x)) + \varphi(x) = h(x) \tag{3.8.3}$$

Theorems 3.1.6 and 3.1.9 yield the following.

Proposition 3.8.2. *Let hypotheses* (i) *and* (iv) *in Section 3.1 be satisfied and let* f *satisfy* (ii). *If* $\varphi : X \to Y$ *is a continuous solution of equation* (3.8.1), *then* $h(0) = 0$ *and*

$$\varphi(x) = y - \sum_{n=0}^{\infty} h(f^n(x)), \tag{3.8.4}$$

where $y = \varphi(0)$ *is a constant (a parameter). Similarly, if* $\varphi : X \to Y$ *is a continuous solution of equation* (3.8.2), *then*

$$\varphi(x) = \tfrac{1}{2}h(0) + \sum_{n=0}^{\infty} (-1)^n [h(f^n(x)) - h(0)]. \tag{3.8.5}$$

Conversely, if formula (3.8.4) *(resp.* (3.8.5)) *defines a continuous function* $\varphi : X \to Y$, *then this function satisfies equation* (3.8.2) *(resp.* (3.8.3)).

3.8.4. Under assumptions (i), (ii), (iv) (Section 3.1) in the indeterminate case $|g(0)| = 1$, for most cases equation (3.8.1) does not have any continuous solution in X, 'most' being meant in the sense of Baire category. More exactly, in the space $C(X, \mathbb{R})$ endowed with the usual metric, the set of those triples (f, g, h) of functions for which the corresponding equation

$$\varphi(f(x)) = g(x)\varphi(x) + h(x) \tag{3.8.6}$$

has a continuous solution $\varphi: X \to \mathbb{R}$ is of the first Baire category. In other words, nonexistence of continuous solutions on X is a generic property of inhomogeneous equation (3.8.6).

(This, and some related results, are due to W. Jarczyk [1], [3].)

3.8.5. Further arguments in favour of the fact that the assumption 'f is strictly increasing' (see Remark 3.1.2) is essential in Theorems 3.1.1 and 3.1.3 are contained in Kalinowski–Kuczma [1]. The authors exhibit examples showing that, in case (B), when the monotonicity assumption on f is dropped it may happen that the equation in question has only one continuous solution.

3.8.6. The following equation appears in statics:

$$\varphi(x/2) = 2\varphi(x) - x/2, \qquad x \in \mathbb{R}^+ \tag{3.8.7}$$

(see Kuczma [26, p. 65]). It is seen that the function $\varphi(x) = \frac{1}{3}x$ satisfies (3.8.7). Theorem 3.1.10, when applied to equation (3.8.7), says that this function is the only continuous solution of (3.8.7) in $\mathbb{R}^+$.

3.8.7. If in Theorems 3.2.1 and 3.2.2 the set T consists of a single point, then the theorems give some criteria for the existence of continuous solutions of equation (3.8.6) in cases (C) and (A).

3.8.8. Asymptotic properties imposed on the function f and g in Section 3.3 influence the behaviour of the sequence (G_n). For instance, take the following version of (ii) in Section 3.3:

$$f(x) = x - tx^{m+1} + O(x^{m+1+\tau}), \qquad m, t, \tau > 0, x \to 0,$$

$$g(x) = 1 + rx^k + O(x^{k+\rho}), \qquad k, \rho > 0, r \neq 0, x \to 0.$$

The question of which one of cases (A)–(C) occurs can be settled by considering the series

$$\sum_{n=0}^{\infty} a_n(x), \qquad a_n(x) := g(f^n(x)) - 1.$$

In view of Theorem 1.4.5 $f^n(x) \sim n^{-1/m}, n \to \infty$, so that $a_n(x) \sim n^{-k/m}, n \to \infty$. This shows that the infinite product

$$\prod_{n=0}^{\infty} g(f^n(x)) = \prod_{n=0}^{\infty} (1 + a_n(x))$$

converges a.u. in X if $k > m$; it diverges to zero (uniformly on an interval $X_0 \subset X$) if $k \leq m$ and $r < 0$; and its absolute value diverges to infinity if $k \leq m$ and $r > 0$. Consequently, $(k > m) \Rightarrow (A), (k \leq m$ and $r < 0) \Rightarrow (B)$ and $(k \leq m$ and $r > 0) \Rightarrow (C)$.

3.8.9. Theorems 3.4.1 and 3.4.2 clearly show that for every positive integer r one can construct a linear functional equation whose solution of class C^r

is unique, whereas the solution of class C^{r-1} depends on an arbitrary function. However, it is not so easy to find an equation which for every $r \in \mathbb{N}$ has a C^r solution depending on an arbitrary function while its C^∞ solution is unique. An example of such an equation (of form (3.8.1)) has been exhibited in Kuczma–W. Smajdor [1]. In that example the interval on which the C^r solution can be arbitrarily prescribed depends on r. Moreover, the function g is discontinuous at the origin. It is an open problem whether such an example could be constructed with a g of class C^∞ in X.

3.8.10. The unique C^r solution of equation (3.4.1) spoken of in Theorem 3.4.2 depends continuously on the data. The key assumption is the inequality $|g_m(0)|(f'_m(0))^r < 1$ imposed on the given functions f_m, g_m occurring in the sequence of equations of form (3.4.1).

3.8.11. It is easily checked that if the series in (3.8.5) converges, then its sum φ is a solution of equation (3.8.3) (see Kuczma [26, p. 58]).

The question of the existence of solutions to this equation in the case where the series in (3.8.5) diverges is investigated by M. Malenica [1]–[5]. Following the idea of M. Bajraktarević [4] she deals with cases where this series is summable by a regular summability method.

3.8.12. The following complements Theorem 3.5.1.

Proposition 3.8.3. *Under conditions of Theorem 3.5.1, the general C^1 solution $\sigma: X \to \mathbb{R}$ of the Schröder equation*

$$\sigma(f(x)) = s\sigma(x) \tag{3.8.8}$$

is given by $\sigma(x) = c\sigma_0(x)$, where σ_0 is the solution defined by (3.5.2), and c is an arbitrary constant (a parameter).

However, if the asymptotic condition in (ii) of Section 3.5 is not fulfilled, the proposition is no longer valid. For (see Kuczma [11]), if

$$f(x) = s \int_0^x (1 - \varepsilon/\log t)dt \qquad \text{for } x \in (0, a), \ f(0) = 0,$$

where $a, s \in (0, 1)$, and either $\varepsilon = 1$ or $\varepsilon = -1$, then equation (3.8.8) has in $X = [0, a)$ a C^1 solution depending on an arbitrary function if $\varepsilon = 1$, while $\sigma = 0$ is the only C^1 solution of (3.8.8) if $\varepsilon = -1$. This follows from the fact that $\sigma: X \to \mathbb{R}$ is a C^1 solution of (3.8.8) if and only if $\varphi = \sigma'$ is a continuous solution of the equation

$$\varphi(f(x)) = \frac{s}{f'(x)} \varphi(x)$$

in X, and from Theorems 3.1.3 and 3.1.4.

3.8.13. The following result has been obtained with the aid of Theorem 3.4.2.

Proposition 3.8.4. *Let* $X = [0, 1]$. *For every positive integer r there exists a continuous strictly increasing bijective mapping* $p: X \to X$ *and infinitely many nontrivial functions* $\varphi: X \to \mathbb{R}$ *of class* C^r *on* X *and such that the equation*

$$\int_X \varphi(tp(u))du = 0$$

is satisfied for any $t \in X$.

This problem, stemming from the theory of convex bodies, has ben raised by S. Gołąb [1], and then solved by J. S. Lipiński [1] for continuous φ. The proposition is taken from Choczewski–Kuczma [2]. At present the proposition has been proved even for real analytic φ (Gulgowska–Gulgowski [1]).

(The functional equation used in the proof of Proposition 3.8.4 is $\varphi(x/2) = -k^{-1}\varphi(x)$, $x \in X$, where k is a positive number. Given an $r \in \mathbb{N}$ we take a $k > 2^r$ to ensure that condition (3.4.5) of Theorem 3.4.2 is fulfilled.)

3.8.14. According to Lemma 3.6.4, formula (3.6.11) defines a nontrivial one-parameter family of functions. They all satisfy equation (3.8.1), but it may happen that Var $\varphi|(0, x_0] = \infty$ for φ defined by (3.6.11) with $c \neq 0$ (see an example in Zdun [5]). It may be shown that if we add to those of Theorem 3.6.2 the condition

$$\sum_{n=0}^{\infty} G_n(x_0) \text{ Var } g|(0, f^n(x_0))] < \infty,$$

then φ given by (3.6.11) belongs to $BV(0, x_0]$. The last condition is certainly fulfilled if $\sum_{n=0}^{\infty} n \text{ Var } g|X_n < \infty$.

(All these facts are reproduced from Zdun [5], where also further conditions for the existence of BV solutions of equation (3.8.1) may be found.)

3.8.15. The assertions (a)–(c) of Theorem 3.6.4 corresponding to cases (A′)–(C′) remain valid if we replace some of its assumptions by the following (see Zdun [6]):

In (a): '(A′) occurs' by 'the series $\sum_{n=0}^{\infty} |g(f^n(x_0) - 1)|$ converges',

In (b): '$\lim_{n \to \infty} H_n(x) = y$' by '$\lim_{x \to 0}[h(x)/(1 - g(x))] = y$',

In (c): '$|g(f^n(x_0))| > 1$, $n \in \mathbb{N}_0$' by '$G_n(x) \geqslant M > 0$ for $x \in X$, $n \in \mathbb{N}$, and the series $\sum_{n=0}^{\infty} |h(f^n(x); y)|$ converges'.

3.8.16. Under assumption (i) of Section 3.7 equation (3.7.7),

$$\varphi(f(x)) = -\varphi(x) + f(x), \tag{3.8.9}$$

is equivalent to

$$\varphi(f^{-1}(x)) = -\varphi(x) + x, \qquad x \in X := [0, 1]. \tag{3.8.10}$$

On putting $\hat{f}(x) := 1 - f^{-1}(1 - x)$, $\hat{\varphi}(x) := \varphi(1 - x)$, $x \in X$, (3.8.10) is rewritten as $\hat{\varphi}(\hat{f}(x)) = -\hat{\varphi}(x) + 1 - x$, and to this equation Proposition 3.8.2 applies.

In this way we get the unique solution φ^* of equation (3.8.9) in X, continuous on $X \setminus \{0\}$, given by the formula (see (3.8.5))

$$\varphi^*(x) = \tfrac{1}{2} + \sum_{n=0}^{\infty} (-1)^{n+1}(1 - f^{-n}(x)), \qquad x \in X \setminus \{0\}; \ \varphi^*(0) = 0. \quad (3.8.11)$$

If the condition (see (3.7.11))

$$\sum_{n=-\infty}^{+\infty} (f^{2n}(x) - f^{2n+1}(x)) = \tfrac{1}{2}, \qquad x \in (0, 1), \quad (3.8.12)$$

holds, then the functions φ^* given by (3.8.11) and φ defined by (see (3.7.10))

$$\varphi(x) = \sum_{n=1}^{\infty} (-1)^{n+1} f^n(x), \qquad x \in [0, 1), \ \varphi(1) = \tfrac{1}{2}, \quad (3.8.13)$$

coincide on X and thus yield the unique continuous solution of (3.8.9) (or (3.8.10)) on X. The converse is also true. This results from the following.

Proposition 3.8.5. *Let* (i) *of Section 3.7 be satisfied and assume that*

(ξ)*There exists a* $\xi \in (0, 1)$ *such that the function* $\mathrm{id}_X - f$ *is increasing in* $(0, \xi)$ *and decreasing in* $(\xi, 1)$.

Let $\varphi: X \to \mathbb{R}$ *be a solution of equation* (3.8.9). *Then the conditions*

(1) φ *is monotonic in* X,
(2) φ *is continuous on* X,
(3) $0 \leq \varphi(x) \leq \tfrac{1}{2}, x \in X$,

are equivalent to each other.

A solution $\varphi: X \to R$ *satisfying* (1)–(3) *exists if and only if condition* (3.7.12) *holds true. When it exists, the* φ *is given by formula* (3.8.13) *or* (3.8.11).

If the function f is convex, then the function $\mathrm{id}_X - f$ satisfies condition (ξ), but (ξ) can also be satisfied in other situations. Thus Proposition 3.8.5 improves on Theorem 3.7.3.

(Propostion 3.8.5 is a consequence of a result from Choczewski–Kuczma [6] concerning solutions of equation (3.8.3) having properties (1)–(3).)

3.8.17. The functional equation stemming from gas dynamics

$$\varphi(x) = A(x + c)^2 \left[\varphi\left(a + \frac{b}{x + c}\right) + d \right],$$

where all the occurring constants are some reals, has been solved by J. A. Baker [1] in the class of real analytic functions on $(0, \infty)$.

4

Analytic and integrable solutions of linear equations

4.0 Introduction

Our main goal here is to introduce a more advanced theory of linear equations. A major part of this chapter is devoted to complex-valued solutions (Sections 4.2–4.6). Analytic solutions of the equation

$$\varphi(f(x)) = g(x)\varphi(x) + h(x) \tag{4.0.1}$$

are found in Sections 4.2, 4.3 and 4.4, according to the value of $|f'(0)|$, the multiplier of f. Further we discuss meromorphic solutions (Section 4.5). In Section 4.6 general results are applied to the Schröder and Abel equations.

Another function class we deal with is that of functions that are integrable with respect to an arbitrary measure (Section 4.7). Next absolutely continuous solutions of a special equation are considered and their applications to partial differential equations (the Goursat problem) are shown. A selection of comments, examples and notes is given in the last section.

We start with general theorems on solutions of (4.0.1) which are mappings of a topological space into a Banach space.

4.1 Linear equation in a topological space

We shall deal with the equation

$$\varphi(x) = g(x)\varphi(f(x)) + h(x). \tag{4.1.1}$$

We assume the following hypotheses.

(i) X is a topological space, Y is a Banach space over $\mathbb{K}$.
(ii) $f: X \to X$, $g: X \to \mathbb{K}$, $h: X \to Y$ and there exist constants $\theta \in (0, 1)$ and $L > 0$ such that

$$|g(x)| \leqslant \theta \qquad \text{and} \qquad \|h(x)\| \leqslant L \qquad \text{on } X.$$

We provide a theorem on bounded solutions of equation (4.1.1).

Theorem 4.1.1. *Let hypotheses* (i), (ii) *be fulfilled. Then equation* (4.1.1) *has a unique solution* $\varphi: X \to Y$ *that is bounded on* X. *This solution is given by the formula*

$$\varphi(x) = \sum_{n=0}^{\infty} \left(\prod_{i=0}^{n-1} g(f^i(x)) \right) h(f^n(x)). \tag{4.1.2}$$

Moreover, if f, g, h *are continuous at a point* $\xi \in X$ *such that* $f(\xi) = \xi$, *then so is* φ. *If* f, g, h *are continuous on* X, *then so is* φ.

Proof. Let $M > L(1-\theta)^{-1}$ be a constant. Let Φ_M denote the class of those functions $\varphi: X \to Y$ that are bounded by M on X. We endow Φ_M with the sup norm so that it becomes a complete metric space.

Applying Banach's Theorem to the self-mapping T of Φ_M given by $T[\varphi] := g \cdot (\varphi \circ f) + h$ we obtain the existence of a unique solution $\varphi \in \Phi_M$ of (4.1.1). The solution is given as the limit of the sequence of successive approximations, i.e., by (4.1.2). Further, $M < M'$ implies $\Phi_M \subset \Phi_{M'}$ so that φ is unique in the union of all Φ_Ms with $M > L/(1-\theta)$.

If f, g, h are continuous at a $\xi \in X$ (resp. on X), then we take instead of Φ_M its subspace Φ_M^1 (resp. Φ_M^2) consisting of those $\varphi \in \Phi_M$ which are continuous at ξ (resp. on X) and again use Banach's Theorem. ∎

Remark 4.1.1. In most cases Theorem 4.1.1 is first applied locally in a neighbourhood U of a fixed point ξ of the function f, $f(U) \subset U$. Next, by Baron's Extension Theorem 7.1.1, the solution obtained is uniquely extended to a solution of (4.1.1) in X.

Under somewhat stronger conditions we can prove the continuous dependence of the continuous solution of equation (4.1.1) on the given functions. Consider a sequence of equations

$$\varphi_m(x) = g_m(x)\varphi_m(f_m(x)) + h_m(x), \qquad m \in \mathbb{N}_0, \tag{4.1.3$_m$}$$

and replace hypotheses (i), (ii) by the following ones.

(i') X is a metric space, Y is a Banach space over $\mathbb{K}$, $\xi \in X$.

(ii') The functions $f_m: X \to X$, $g_m: X \to \mathbb{K}$ and $h_m: X \to Y$, $m \in \mathbb{N}_0$, are continuous in X.

(iii') For every $x \in X$ and every $m \in \mathbb{N}_0$ we have $\lim_{n \to \infty} f_m^n(x) = \xi$. For $m = 0$ the convergence is almost uniform.

(iv') For every neighbourhood U of ξ there exists a compact neighbourhood C of ξ such that for every $m \in \mathbb{N}_0$ we have $f_m(C) \subset C \subset U$.

(v') $\lim_{m \to \infty} f_m = f_0$, $\lim_{m \to \infty} g_m = g_0$, $\lim_{m \to \infty} h_m = h_0$ a.u. in X.

(vi') There is an increasing sequence of sets $(V_p)_{p \in \mathbb{N}}$ with compact closures cl V_p such that $X = \bigcup_{p=1}^{\infty} V_p$ and $f_m(V_p) \subset V_p$, $p \in \mathbb{N}$, $m \in \mathbb{N}_0$.

Theorem 4.1.2. *Make assumptions* (i')–(vi'). *If, moreover,*

$$|g_0(\xi)| < 1,$$

then, for $m = 0$ *and for* m *sufficiently large, equation* (4.1.3$_m$) *has a unique continuous solution* $\varphi_m : X \to Y$, *and* $\lim_{m \to \infty} \varphi_m = \varphi_0$ *a.u. in* X.

Proof. By (ii') and (v') there are an $m' \in \mathbb{N}$, constants $L > 0$, $\Theta \in (0, 1)$ and a compact neighbourhood C' of ξ (independent of m) such that

$$|g_m(x)| \le \Theta, \qquad \|h_m(x)\| \le L$$

for $m \ge m'$ and $x \in C'$. In virtue of (iv') we now choose a C such that $f_m(C) \subset C \subset C'$, $m \in \mathbb{N}_0$. The existence and uniqueness of a continuous solution $\varphi_m : X \to Y$ to equation (4.1.3$_m$) for $m = 0$ and $m \ge m'$ result from Theorem 4.1.1 (with $X = C$) and Remark 4.1.1. By Theorem 1.6.4 we have $\lim_{m \to \infty} \varphi_m = \varphi_0$ uniformly on C.

Fix a $p \in \mathbb{N}$. It follows from (iii') that there exists an $n \in \mathbb{N}$ such that $f_0^n(\text{cl } V_p) \subset \text{int } C$, and hence, by (v'), $f_m^n(\text{cl } V_p) \subset C$ for $m = 0$ and m large enough. Since the φ_m satisfy (4.1.3$_m$), we have (induction)

$$\varphi_m(x) = \varphi_m(f_m^n(x)) \prod_{i=0}^{n-1} g_m(f_m^i(x)) + \sum_{i=0}^{n-1} \left[\prod_{j=0}^{i-1} g_m(f_m^j(x)) \right] h_m(f_m^i(x)). \quad (4.1.4)$$

Using formula (4.1.4) we can easily check that $\lim_{m \to \infty} \varphi_m = \varphi_0$ uniformly on cl V_p. Hence the theorem follows in view of (vi'). ∎

4.2 Analytic solutions; the case $|f'(0)| < 1$

Now we consider complex-valued functions defined on subsets of $\mathbb{C}$ – the set of complex numbers. In this and the next two sections we use abbreviation 'LAS' for 'local analytic solution (of the underlying equation) in a neighbourhood of the origin'.

4.2A. Extension theorems

The question arises whether a solution analytic in a neighbourhood of the origin can be extended to a solution on a larger set with the analyticity preserved. We present two theorems to this effect which are due to J. Matkowski [2]. The equation in question is

$$p(x)\varphi(x) = g(x)\varphi(f(x)) + h(x). \quad (4.2.1)$$

Assume the following.

(i) $X \subset \mathbb{C}$ is an open set and $0 \in X$. The boundary of X contains at least two finite points.

(ii) $f : X \to X$ is an analytic function, $f(0) = 0$, $|f'(0)| < 1$, and $f(x) \ne x$ for $x \in X \setminus \{0\}$.

(iii) The functions $p : X \to \mathbb{C}$, $g : X \to \mathbb{C}$ and $h : X \to \mathbb{C}$ are analytic in X, $p(x) \ne 0$ in X.

Theorem 4.2.1. *Let hypotheses* (i)–(iii) *be fulfilled and let* $U \subset X$ *be a neighbourhood of the origin such that* $f(U) \subset U$. *If an analytic function* $\varphi_0: U \to \mathbb{C}$ *satisfies equation* (4.2.1) *in* U, *then there exists a unique solution* $\varphi: X \to \mathbb{C}$ *of equation* (4.2.1) *in* X *such that*

$$\varphi(x) = \varphi_0(x) \qquad \text{for } x \in U. \tag{4.2.2}$$

This function φ *is analytic in* X.

Proof. Take an arbitrary compact set $C_0 \subset X$ and write $C_n = f^n(C_0)$, $n \in \mathbb{N}$. By Theorem 1.2.6 we have $C_N \subset U$ for an index $N \in \mathbb{N}$. Define the functions $\varphi_i: C_{N-i} \to \mathbb{C}$ by the recurrence

$$\varphi_{i+1}(x) = (g(x)\varphi_i(f(x)) + h(x))/p(x), \qquad x \in C_{N-i-1}, \tag{4.2.3}$$

where φ_0 is the given solution of (4.2.1). By (4.2.3), φ_i is analytic in C_{N-i}, in particular so is φ_N in C_0. Denote this φ_N by $\varphi(C_0; \cdot)$. Since φ_0 satisfies equation (4.2.1) in U, the function $\varphi(C_0; \cdot)$ is independent of the choice of N.

Let K_1, K_2 be compact sets, $K_1 \subset K_2 \subset X$. We claim that

$$\varphi(K_1; x) = \varphi(K_2; x) \qquad \text{for } x \in K_1. \tag{4.2.4}$$

Indeed, if $N \in \mathbb{N}$ is such that $f^N(K_2) \subset U$, then also $f^N(K_1) \subset f^N(K_2) \subset U$. Thus $\varphi_0(K_1; x) = \varphi_0(K_2; x) = \varphi_0(x)$ for $x \in f^N(K_1)$. Making use of (4.2.3) we check that $\varphi_i(K_1; x) = \varphi_i(K_2; x)$ for $x \in f^{N-i}(K_1)$, $i = 1, \ldots, N$, and, in particular $(i = N)$, $\varphi(K_1; x) = \varphi_N(K_1; x) = \varphi_N(K_2; x) = \varphi(K_2; x)$ for $x \in K_1$.

Let $(K_j)_{j \in \mathbb{N}}$ be an increasing sequence of compact sets, the union of which is X. In virtue of (4.2.4) the function $\varphi: X \to \mathbb{C}$

$$\varphi(x) = \varphi(K_j; x) \qquad \text{for } x \in K_j, j \in \mathbb{N},$$

is well defined. We shall prove that it is a solution of (4.2.1) in X. Take an $x \in X$. Since $X = \bigcup_{j=1}^\infty K_j$, we have $x \in K_j$ for a certain j. Further $\varphi(x) = \varphi(K_j; x) = \varphi_N(K_j; x)$ with N such that $f^N(K_j) \subset U$. Of course $f(x) \in f(K_j)$ and $f^{N-1}(f(K_j)) = f^N(K_j) \subset U$. Thus

$$\varphi(f(K_j); f(x)) = \varphi_{N-1}(f(K_j); f(x)) = \varphi_{N-1}(K_j; f(x)).$$

By (4.2.3) one gets

$$g(x)\varphi_{N-1}(K_j; f(x)) = p(x)\varphi_N(K_j; x) - h(x) = p(x)\varphi(x) - h(x). \tag{4.2.5}$$

There is an index $k \in \mathbb{N}$ such that $f(x) \in K_k$. Relation (4.2.4) yields

$$\varphi_{N-1}(K_j; f(x)) = \varphi(f(K_j); f(x)) = \varphi(f(K_j) \cup K_k; f(x)) = \varphi(K_k; f(x))$$
$$= \varphi(f(x)),$$

which together with (4.2.5) shows that (4.2.1) holds. Our claim is proved. In order to verify the relation (4.2.2) take an $x \in U$ and a compact set K such that $x \in K \subset U$. Since φ_0 satisfies (4.2.1) in U, we get by (4.2.3) $\varphi(K, x) = \varphi_0(x)$. There is a $j \in \mathbb{N}$ such that $x \in K_j$. Hence by (4.2.4)

$$\varphi(x) = \varphi(K_j; x) = \varphi(K_j \cup K; x) = \varphi(K; x) = \varphi_0(x),$$

i.e. φ is an extension of φ_0.

To prove the uniqueness, suppose $\varphi_1 : X \to \mathbb{C}$ and $\varphi_2 : X \to \mathbb{C}$ are two solutions of (4.2.1) with the property $\varphi_1(x) = \varphi_2(x)$ for $x \in U$. Take an arbitrary $x \in X$. There is an $N \in \mathbb{N}$ such that $f^N(x) \in U$, whence $\varphi_1(f^N(x)) = \varphi_2(f^N(x))$. Using (4.2.1) we prove that $\varphi_1(f^{N-i}(x)) = \varphi_2(f^{N-i}(x))$ for $i = 1, \ldots, N$, whence, for $i = N$, we obtain $\varphi_1(x) = \varphi_2(x)$. Thus $\varphi_1 = \varphi_2$. ∎

For meromorphic solutions we have an analogous result.

Theorem 4.2.2. *Let hypotheses* (i), (ii) *be fulfilled and let* $U \subset X$, $f(U) \subset U$ *be a neighbourhood of the origin. If the functions* p, g *and* h *are meromorphic in* X, $g(x) \neq 0$ *in* X, *and* φ_0 *is a meromorphic solution of equation* (4.2.1) *in* U, *then there exists a unique meromorphic function* $\varphi : X \to \mathbb{C}$ *satisfying equation* (4.2.1) *in* X *and condition* (4.2.2) *in* U.

Remark 4.2.1. Note that a part of the conditions in (i) and (ii) were necessary only to enable us to apply Theorem 1.2.6. Thus in Theorems 4.2.1 and 4.2.2 hypotheses (i) and (ii) may be replaced by the following one:

$X \subset \mathbb{C}$ is an open set, $0 \in X$; $f : X \to X$ is analytic on X, $f(0) = 0$, and the sequence (f^n) converges to 0 a.u. in X.

4.2B. Existence and uniqueness results

The methods we have used to investigate real-valued C^r solutions (see Section 3.4) can be adapted to yield the existence and uniqueness of the LAS of the equation

$$\varphi(x) = g(x)\varphi(f(x)) + h(x). \tag{4.2.6}$$

Assume the following.

(iv) $X \subset \mathbb{C}$ is an open set containing the origin.

(v) $f : X \to X$, $g : X \to \mathbb{C}$ and $h : X \to \mathbb{C}$ are analytic in X, $f(0) = 0$.

Under conditions (iv), (v) Lemma 3.4.1 holds true for analytic $\varphi : X \to \mathbb{C}$. The functions P_{ki} that appear there now are analytic in X. Also the remarks from Section 3.4 concerning the (now infinite) system of equations (3.4.4)

$$c_k = g(0)(f'(0))^k c_k + \sum_{i=0}^{k-1} P_{ki}(0)c_i + h^{(k)}(0), \qquad k \in \mathbb{N}_0, \tag{4.2.7}$$

remain valid.

The following corresponds to Theorem 3.4.1, and is a particular case of a result by W. Smajdor [1] (see also Matkowski [7]).

Theorem 4.2.3. *Let hypotheses* (iv), (v) *be fulfilled and*

$$\left| f'(0) \right| < 1.$$

Then for every sequence $(c_k)_{k \in \mathbb{N}_0}$ *of numbers satisfying* (4.2.7) *there exists a unique LAS* $\varphi: U \to \mathbb{C}$ *of equation* (4.2.6), *fulfilling the conditions*

$$\varphi^{(k)}(0) = c_k, \qquad k \in \mathbb{N}_0. \tag{4.2.7'}$$

Proof. We can find an $r \in \mathbb{N}$, a $\Theta \in (0, 1)$, and a disc $U = \{x \in \mathbb{C}: |x| < \delta\}$, $\delta > 0$, such that cl $U \subset X$ and

$$|f'(x)|^r |g(x)| < \Theta \qquad \text{for } x \in U. \tag{4.2.8}$$

Write $\varphi(x) = Q(x) + \psi(x)$, where

$$Q(x) = \sum_{i=0}^{r} c_i \frac{x^i}{i!} \tag{4.2.9}$$

and $\psi: U \to \mathbb{C}$ is an analytic function that satisfies

$$\psi^{(k)}(0) = 0 \qquad \text{for } k = 0, \ldots, r. \tag{4.2.10}$$

Then ψ should satisfy the equation

$$\psi(x) = g(x)\psi(f(x)) + \hat{h}(x), \tag{4.2.11}$$

where

$$\hat{h}(x) = h(x) + Q(x) - g(x)Q(f(x)) \tag{4.2.12}$$

is an analytic function in X such that $\hat{h}^{(k)}(0) = 0$ for $k = 0, \ldots, r$.

Let Ψ be the set of those functions $\psi: \text{cl } U \to \mathbb{C}$ that are analytic in U, that fulfil condition (4.2.10), and whose derivatives $\psi^{(r)}$ are continuous in cl U. This Ψ endowed with the norm $\|\psi\| = \sup_{\text{cl } U} |\psi^{(r)}(x)|$ is a Banach space. Consider the map $T: \Psi \to \Psi$ defined by

$$T(\psi)(x) = g(x)\psi(f(x)) + \hat{h}(x), \qquad x \in \text{cl } U. \tag{4.2.13}$$

Choosing an $M > 0$ so that

$$|P_{ri}(x)| < M \qquad \text{in cl } U, \ i = 1, \ldots, r-1, \tag{4.2.14}$$

we have by Lemma 3.4.1 for $\psi_1, \psi_2 \in \Psi$

$$\|T(\psi_1) - T(\psi_2)\| \leq \left(\Theta + \sum_{i=0}^{r-1} \delta^{r-i} \right) \|\psi_1 - \psi_2\|.$$

By taking δ small enough we can force T to be a contraction. (Notice that for a smaller δ inequalities (4.2.8) and (4.2.14) are satisfied as well.)

By Banach's Theorem there exists a unique solution $\psi \in \Psi$ of equation (4.2.11), and hence also a unique analytic solution $\varphi: U \to \mathbb{C}$ of equation (4.2.6) satisfying (4.2.7'). $\blacksquare$

Remark 4.2.2. As also follows from Banach's Theorem, the assertion of Theorem 4.2.3 could be complemented by the formula for the solution φ of (4.2.6):

$$\varphi(x) = Q(x) + \sum_{n=0}^{\infty} \left[\prod_{j=0}^{n-1} g(f^j(x)) \right] \hat{h}(f^n(x)), \tag{4.2.15}$$

where Q and $\hat{h}$ are given by (4.2.9) and (4.2.12), respectively (see also (4.1.2)).

Formula (4.2.15) represents the limit of the sequence of iterates of the map T defined by (4.2.13).

Making use of Theorem 4.2.1 we deduce from Theorem 4.2.3 the following.

Theorem 4.2.4. *Let hypotheses* (i), (ii), (v) *be fulfilled. Then for every sequence of numbers* (c_k) *satisfying* (4.2.7) *there exists a unique analytic solution* $\varphi\colon X \to \mathbb{C}$ *of equation* (4.2.6) *fulfilling conditions* (4.2.7'). *This solution is given by formula* (4.2.15).

The last sentence here results from Theorem 1.2.5 and from the fact that if (4.2.15) holds for $x = f(z)$ and φ satisfies (4.2.6), then (4.2.15) holds also for $x = z$.

Remark 4.2.3. In the case where
$$|f'(0)| > 1,$$
the function f is locally invertible. If $g(x) \neq 0$ in X, we can replace in (4.2.6) x by $f^{-1}(x)$ to get the equation
$$\varphi(x) = (g(f^{-1}(x)))^{-1}\varphi(f^{-1}(x)) - h(f^{-1}(x))/g(f^{-1}(x)),$$
to which Theorem 4.2.4 applies.

4.2C. Continuous dependence on the data

The problem considered here is to find conditions under which the sequence $(\varphi_m)_{m\in\mathbb{N}}$ of analytic solutions of the equations
$$\varphi_m(x) = g_m(x)\varphi_m(f_m(x)) + h_m(x) \tag{4.2.16}$$
converges, if those of the given functions (f_m), (g_m), (h_m) do. We have (see Matkowski [3]) the following.

Theorem 4.2.5. *Let hypothesis* (i) *be fulfilled and assume that the functions* $f_m\colon X \to X, g_m\colon X \to \mathbb{C}$ *and* $h_m\colon X \to \mathbb{C}$ *are analytic in* $X, f_m(0) = 0, |f'_m(0)| < 1,$ $m \in \mathbb{N}_0$. *Let the sequences* (f_m), (g_m), (h_m), *all with* $m \in \mathbb{N}$ *converge a.u. in* X *to the functions* f_0, g_0, h_0, *respectively. Suppose that the system* (see (4.2.7))
$$c_{km} = g_m(0)\left((f'_m(0))^k c_{km} + \sum_{i=0}^{k-1} P_{ki}^m(0)c_{im} + h_m^{(k)}(0)\right), \qquad k \in \mathbb{N}_0, \tag{4.2.17}$$
has a solution (c_{km}), *and* $\lim_{m\to\infty} c_{km} = c_{k0}, k \in \mathbb{N}_0$. *Then, for every* $m \in \mathbb{N}_0$, *equation* (4.2.16) *has a unique analytic solution* $\varphi_m\colon U \to \mathbb{C}$ *in a neighbourhood* U *(independent of* m*) of the origin, fulfilling the conditions*
$$\varphi_m^{(k)}(0) = c_{km}, \qquad k \in \mathbb{N}_0,$$
and $\lim_{m\to\infty} \varphi_m = \varphi_0$ *uniformly in* U.

Proof. By Theorem 4.2.3 the desired solutions φ_m, $m \in \mathbb{N}_0$, exist in a neighbourhood U which can be chosen so that inequalities (4.2.8) and (4.2.14) hold true (with obvious changes) for every $m \in \mathbb{N}_0$. Since (f_m), (g_m) and hence also $(P_{ki}^m)_{m \in \mathbb{N}}$ converge a.u. in X to f_0, g_0 and P_{ki}^0, respectively, U may be chosen to be independent of m. Now, φ_m can be written as (see the proof of Theorem 4.2.3) $\varphi_m = Q_m + \psi_m$ with

$$Q_m(x) = \sum_{i=0}^{r} c_{im} \frac{x^i}{i!}.$$

By Theorem 1.6.4 we have $\lim_{m \to \infty} \psi_m = \psi_0$ uniformly in U, and $\lim_{m \to \infty} Q_m = Q_0$ because of the convergence of $(c_{im})_{m \in \mathbb{N}}$. ∎

An analogous theorem for global analytic solutions, patterned upon Theorem 4.2.5, can also be formulated; see Note 4.9.1.

4.3 Analytic solutions; the case $\left|f'(0)\right| = 1$

Now we shall deal with the linear equation written in the form

$$\varphi(f(x)) = g(x)\varphi(x) + h(x) \tag{4.3.1}$$

instead of (4.2.6). We preserve the abbreviation 'LAS' from the preceding section. If the condition

$$|f'(0)| = 1 \tag{4.3.2}$$

holds, the problem of determining local analytic solutions of (4.3.1) becomes much more involved. For instance, for the Schröder equation

$$\sigma(f(x)) = s\sigma(x), \tag{4.3.3}$$

where

$$f(0) = 0, \qquad f'(0) = s, \tag{4.3.4}$$

if s is not a root of unity, system (4.2.7) has always a solution $(c_k)_{k \in \mathbb{N}_0}$ (c_1 is arbitrary), since $s^{-1}(f'(0))^k = s^{k-1} \neq 1$ for $k \neq 1$. In other words, equation (4.3.3) has always a unique one-parameter family of formal solutions (with c_1 as a parameter):

$$\sigma(x) = \sum_{n=1}^{\infty} \frac{c_n}{n!} x^n. \tag{4.3.5}$$

However, it may well happen that series (4.3.5) diverges for all $x \neq 0$, so that (4.3.3) may have no LAS (see Cremer [1], Siegel [2], Kuczma [26, p. 149]).

4.3A. The Siegel set

It turns out that the lack of LAS of (4.3.3) is rather an exceptional case. We quote here a result of C. L. Siegel [1] to this effect. There exist several proofs of this theorem: Siegel [2] (see also Kuczma [26, pp. 149–56]), Rüssmann [1], Moser [1], but they all are too long and involved to be presented here.

Theorem 4.3.1. *There exists a subset S of the unit circle, of measure 2π, such that if f is analytic in a neighbourhood of the origin and satisfies (4.3.4) with an $s \in S$ then the formal solutions (4.3.5) of equation (4.3.3) have a positive radius of convergence.*

In other words, if $s \in S$, then equation (4.3.3) has a unique one-parameter family of LASs: for every $c_1 \in \mathbb{C}$ there exists a unique LAS σ of (4.3.3) such that $\sigma'(0) = c_1$.

Definition 4.3.1. The set of all s in the unit circle for which the assertion of Theorem 4.3.1 is true and such that $s^n \neq 1, n \in \mathbb{N}$, is called the *Siegel set*.

The following properties of the Siegel set are known.

Siegel [1]: $\log|s^n - 1| = O(\log n)$, $n \to \infty$, implies $s \in S$.

Rüssmann [1]: if $\sum_{n=0}^{\infty} q_n^{-1} \log q_n < \infty$, where q_n is the denominator of the nth convergent of the continued fraction of $w = (2\pi)^{-1} \arg s$, then $s \in S$.

Moser [1]: $|s^n - 1|^{-1} = O(n^2)$, $n \to \infty$, implies $s \in S$.

Rüssmann's condition is fulfilled by some Liouville numbers (see Note 4.9.2) and hence is more general than Siegel's one. On the other hand, Moser's condition is slightly weaker than that of Siegel.

4.3B. Special inhomogeneous equation

Now, following D. Vaida [1], we are going to deal with the special case of equation (4.3.1):

$$\varphi(sx) = \varphi(x) + h(x), \tag{4.3.6}$$

where $|s| = 1$, $s^n \neq 1$ for $n \in \mathbb{N}$. We aim at determining a set V on the unit circle, similar to the Siegel set S, but corresponding to equation (4.3.6).

We need several auxiliary notions. Put

$$\arg s = 2\pi w, \qquad 0 < w < 1,$$

so that ($[x]$ denoting the integer part of x)

$$w_n = \arg(s^n - 1) = \pi(nw - [nw]), \qquad x_n = |s^n - 1| = 2 \sin w_n, \qquad n \in \mathbb{N}.$$

Take an ε subject to the condition

$$0 < \varepsilon < \min(\pi/3, \pi w, \pi(1 - w)), \tag{4.3.7}$$

and distinguish those indices n for which w_n is close either to 0 or to π:

$$A_\varepsilon^1 = \{n \in \mathbb{N} : 0 < w_n < \varepsilon\}, \qquad A_\varepsilon^2 = \{n \in \mathbb{N} : \pi - \varepsilon < w_n < \pi\}, \qquad A_\varepsilon = A_\varepsilon^1 \cup A_\varepsilon^2.$$

Observe that, since the sequence $(w_n)_{n \in \mathbb{N}}$ is dense in $(0, \pi)$, the set A_ε is infinite.

Finally we put

$$k_n = \begin{cases} [nw] & \text{if } n \in A_\varepsilon^1, \\ [nw] + 1 & \text{if } n \in A_\varepsilon^2. \end{cases}$$

Then we have $0 < k_n/n < 1$ for all $n \in A_\varepsilon$ (see (4.3.7)). Now we will prove

Lemma 4.3.1. *For every ε satisfying (4.3.7) there exist positive numbers a, b, c such that we have*

$$x_n^{1/n} \geq a \tag{4.3.8}$$

for sufficiently large n outside A_ε, whereas

$$b \left| w - \frac{k_n}{n} \right|^{1/n} \leq x_n^{1/n} \leq c \left| w - \frac{k_n}{n} \right|^{1/n} \tag{4.3.9}$$

for sufficiently large n in A_ε.

Proof. For $n \in A_\varepsilon$, the inequality $x_n = 2 \sin w_n \geq 2 \sin \varepsilon$ implies (4.3.8). Further, for $n \in A_\varepsilon^1$ we have $\frac{1}{2} w_n \leq \sin w_n \leq w_n$, i.e. $\pi n(w - k_n/n) \leq x_n \leq 2\pi n(w - k_n/n)$, which implies (4.3.9). Similarly, if $n \in A_\varepsilon^2$, then $\frac{1}{2}(\pi - w_n) \leq \sin w_n \leq \pi - w_n$, i.e. $\pi n(-w + k_n/n) \leq x_n \leq 2\pi n(-w + k_n/n)$, which again implies (4.3.9). ∎

Put

$$B_\varepsilon = \left\{ n \in A_\varepsilon : \left| w - \frac{k_n}{n} \right| < \tfrac{1}{2} n^{-2} \right\}.$$

Let us now recollect some facts from the theory of continued fractions. We quote them according to Hinčin [1].

(1) If $n \in B_\varepsilon$, then $k_n/n = p_m/q_m$ is the mth convergent of the continued fraction of w, where m depends on n.

(2) The inequalities

$$q_m q_{m+1} \leq |w - p_m/q_m|^{-1} \leq q_m(q_m + q_{m+1}) \leq 2q_m q_{m+1} \tag{4.3.10}$$

hold for $m \in \mathbb{N}$.

(3) There is a universal constant B such that for almost all s on the unit circle the inequality

$$q_m < e^{Bm} \tag{4.3.11}$$

holds for sufficiently large m, say $m \geq m_0$.

The following will help in determining the set V.

Theorem 4.3.2. *Let*

$$h(x) = \sum_{n=1}^\infty h_n x^n \tag{4.3.12}$$

be an analytic function in a neighbourhood of the origin, and let $|s| = 1$, $s^n \neq 1$

for $n \in \mathbb{N}$. *Equation* (4.3.6) *has an LAS* φ *if and only if the sequence* $((|h_n| q_m q_{m+1})^{1/n})$ *is bounded for* $n \in B_\varepsilon$.

Proof. Equation (4.3.6) has a unique one-parameter family of formal solutions

$$\varphi(x) = c_0 + \sum_{n=1}^{\infty} h_n (s^n - 1)^{-1} x^n \qquad (4.3.13)$$

(c_0 is the parameter). These solutions represent analytic functions in a neighbourhood of the origin if and only if the sequence $(|h_n(s^n - 1)^{-1}|^{1/n}) = (|h_n|^{1/n} x_n^{-1/n})$ is bounded. By Lemma 4.3.1 it is so if and only if the sequence with terms

$$|h_n|^{1/n} |w - k_n/n|^{-1/n} \qquad (4.3.14)$$

is bounded for $n \in B_\varepsilon$. Indeed, for $n \notin A_\varepsilon$ we have by (4.3.8) $|h_n|^{1/n} x_n^{-1/n} \leqq a^{-1} |h_n|^{1/n}$ and the boundedness of (4.3.14) follows from the analyticity of (4.3.12). Further, for $n \in A_\varepsilon \setminus B_\varepsilon$ this sequence is majorized by $|h_n|^{1/n}(2n^2)^{1/n}$, thus it is again bounded. By (4.3.9), it remains to establish the equivalence of the boundedness of (4.3.14) for $n \in B_\varepsilon$ with the boundedness of the sequence $((|h_n| q_m q_{m+1})^{1/n})$ for $n \in B_\varepsilon$. But this fact follows from inequalities (4.3.10). ■

Now we can prove the following.

Theorem 4.3.3. *There exists a set* V *on the unit circle, of measure* 2π, *such that if the function* h *is analytic in a neighbourhood of the origin,* $h(0) = 0$, *and if* $s \in V$, *then the formal solutions* (4.3.13) *of* (4.3.6) *have a positive radius of convergence.*

Proof. We take as V the set of those s on the unit circle for which inequality (4.3.11) holds. By property (3) the measure of V equals 2π. Let $s \in V$. If $m \geqslant 7$, then the denominator q_m of the mth convergent of the continued fraction of w satisfies $q_m \geqslant 2^{(m-1)/2} > m$, whence, since p_m/q_m is in its lower terms, $n \geqslant q_m > m$. Thus by (4.3.11)

$$(|h_n| q_m q_{m+1})^{1/n} \leqslant |h_n|^{1/n} e^{2Bm/n} e^{B/n} \leqslant |h_n|^{1/n} e^{3B}$$

whenever $m \geqslant m_0$, and the theorem follows from Theorem 4.3.2. ■

Remark 4.3.1. By Roth's Theorem (see Note 4.9.3), if w is an irrational algebraic number, then $|w - k_n/n| \geqslant n^{-3}$ for sufficiently large $n \in \mathbb{N}$. This implies that sequence (4.3.14) is bounded and consequently $s \in V$. Thus the set V contains in particular all s on the unit circle such that $w = (2\pi)^{-1} \arg s$ is irrational algebraic.

4.3C. General homogeneous and inhomogeneous equations

The material of this subsection is taken from Smajdor–Smajdor [1]. First we consider the homogeneous equation

$$\varphi(f(x)) = g(x)\varphi(x) \qquad (4.3.15)$$

with f fulfilling (4.3.2). We have the following.

Lemma 4.3.2. *Let $f, g; \varphi \neq 0$ be analytic functions in a neighbourhood U of the origin satisfying (4.3.15) in U. If $f(0) = 0$ and $\varphi(x) = x^k \Phi(x)$ with $\Phi(0) \neq 0$, $k \geq 0$, then*

$$g(0) = (f'(0))^k. \qquad (4.3.16)$$

Proof. In view of (4.3.15), the function Φ satisfies $g(x)\Phi(x) = (f(x)/x)^k \Phi(f(x))$, $x \neq 0$. Hence (4.3.16) follows by (4.3.4) on letting $x \to 0$. ∎

By combining results for equations (4.3.3) and (4.3.6) we obtain the following.

Theorem 4.3.4. *Let f and g be analytic functions in a neighbourhood of the origin, $f(0) = 0$, $s := f'(0) \in S \cap V$, $g(0) = s^k$ with a $k \in \mathbb{N}_0$. Then equation (4.3.15) has a nontrivial LAS φ. This solution is unique up to a multiplicative constant.*

Proof. By Theorem 4.3.1 equation (4.3.3) has an LAS σ such that $\sigma(0) = 0$, $\sigma'(0) = 1$. The function

$$h(x) := \log[g(\sigma^{-1}(x))(\sigma^{-1}(x)/f(\sigma^{-1}(x)))^k], \qquad x \neq 0, \; h(0) := 0, \quad (4.3.17)$$

is analytic in a neighbourhood of the origin (we take $\log 1 = 0$). By Theorem 4.3.3 equation (4.3.6) (with h given by (4.3.17)) has an LAS $\tilde{\varphi}$. The function $x \mapsto \varphi(x) = x^k \exp \tilde{\varphi}(\sigma(x))$ is a nontrivial LAS of (4.3.15).

As for the uniqueness, let φ_i, $i = 1, 2$, be nontrivial LASs of (4.3.15). In virtue of Lemma 4.3.2 we have $\varphi_i(x) = x^k \Phi_i(x)$, where $\Phi_i(0) \neq 0$. Thus $\omega := \Phi_1/\Phi_2$ is an LAS of the equation $\omega(f(x)) = \omega(x)$. Hence $\omega = \omega(0) = \text{const}$ (see Note 4.9.17). This means that φ_1 is a constant multiple of φ_2. ∎

Finally for the inhomogeneous equation (4.3.1) we consider the case where the functions f, g, h are analytic in a neighbourhood of the origin and fulfil (4.3.4) and (4.3.16) with an $s \in S \cap V$ and a $k \in \mathbb{N}_0$. Write

$$h(x) = x^m H(x), \qquad H(0) \neq 0, \; m \in \mathbb{N}_0. \qquad (4.3.18)$$

We can always assume that $m \geq k$. For if $m < k$, we introduce the functions

$$\left. \begin{array}{l} h_m(x) = h(x), \\[2mm] h_{j+1}(x) = h_j(x) + \dfrac{h_j^{(j)}(0)}{j!(s^k - s^j)} [(f(x))^j - g(x)x^j], \qquad j = m, \ldots, k-1. \end{array} \right\} \quad (4.3.19)$$

Induction shows that the last of them, h_k, can be written as $h_k(x) = x^k H_k(x)$, where H_k is a function analytic at the origin. Thus instead of equation (4.3.1) we may consider the equation

$$\psi(f(x)) = g(x)\psi(x) + h_k(x), \qquad (4.3.20)$$

where h_k has at the origin a zero of order not smaller than k, as needed.

There is a one-to-one correspondence between LASs φ of equation (4.3.1) and those ψ of (4.3.20), viz. the solutions are linked by the formula

$$\psi(x) = \varphi(x) + \sum_{j=m}^{k-1} \frac{h_j^{(j)}(0)}{j!(s^k - s^j)} x^j.$$

By combining Theorems 4.3.4, 4.3.2 and 4.3.1 we arrive at the following.

Theorem 4.3.5. *Let f, g, h be analytic functions in a neighbourhood of the origin, $f(0) = 0$, $s := f'(0)$, $g(0) = s^k$ with $s \in S \cap V$ and $k \in \mathbb{N}_0$. Equation (4.3.1) has an LAS φ if and only if the function h fulfils (4.3.18) with $m \neq k$ and when $m < k$ the function h_k defined by (4.3.19) has at the origin a zero of order $M > k$.*

Proof. (1) Let $m > k$. By Theorem 4.3.4 and Lemma 4.3.2 equation (4.3.15) has an LAS ϕ_0 such that $\phi_0(x) = x^k \Phi_0(x)$, $\Phi_0(0) \neq 0$. By Theorem 4.3.1 equation (4.3.3) has an LAS σ such that $\sigma(0) = 0$, $\sigma'(0) = 1$. By Theorem 4.3.3 the equation

$$\phi(sx) = \phi(x) + h(\sigma^{-1}(x))/\phi_0(\sigma^{-1}(sx))$$

has an LAS ϕ. The function $\varphi := \phi_0 \cdot (\phi \circ \sigma)$ is an LAS of (4.3.1).

(2) Let $m = k$. Suppose that equation (4.3.1) has an LAS φ, $\varphi(x) = x^p \Phi(x)$ with a $p \in \mathbb{N}$ and $\Phi(0) \neq 0$. We get by (4.3.1) and (4.3.18)

$$x^p[\Phi(f(x))(f(x)/x)^p - g(x)\Phi(x)] = x^m H(x). \qquad (4.3.21)$$

Now, $p \neq m = k$ implies that the left-hand side of (4.3.21) has at the origin a zero of order p, whereas if $p = m = k$ it has a zero of order greater than m. Thus (4.3.21) is impossible.

(3) Let $m < k$. We consider equation (4.3.20) instead of (4.3.1) (see the discussion before the formulation of the theorem). By the first two parts of the proof, equation (4.3.20) with $h_k(x) = x^M H_M(x)$, $H_M(0) \neq 0$ given by (4.3.19) has an LAS ψ if and only if $M > k$. ■

Remark 4.3.2. It follows from Theorem 4.3.4 that, under the conditions of Theorem 4.3.5, if equation (4.3.1) has LASs, then they form a one-parameter family of functions.

4.4 Analytic solutions; the case $f'(0) = 0$

In our study of local analytic solutions (LASs, see Section 4.2) of the equation

$$\varphi(f(x)) = g(x)\varphi(x) + h(x) \qquad (4.4.1)$$

we are left with the case where $|f'(0)| < 1$ and $g(0) = 0$. First we consider the situation

$$f(0) = f'(0) = g(0) = 0. \qquad (4.4.2)$$

Of course, we disregard the trivial cases where either $f = 0$ or $g = 0$. There exist integers p, q, r fulfilling

$$f(x) = x^p F(x), \qquad g(x) = x^q G(x), \qquad h(x) = x^r H(x),$$

where

$$F(0) \neq 0, \qquad G(0) \neq 0 \quad \text{and} \quad H(0) \neq 0 \quad \text{unless } h = 0.$$

By (4.4.2) we have $p \geq 2$, $q \geq 1$, whereas $r \geq 0$ (we take $r = \infty$ if $h = 0$).

We exclude from further considerations the trivial LAS $\varphi = 0$ of equation (4.4.1) (occurring iff $h = 0$). Thus we seek the LAS of (4.4.1) in the form

$$\varphi(x) = x^s \Phi(x), \qquad \Phi(0) \neq 0, \qquad s \in \mathbb{N}_0,$$

so that (4.4.1) can be written as

$$x^{ps}(F(x))^s \Phi(f(x)) = x^{q+s} G(x) \Phi(x) + x^r H(x). \qquad (4.4.3)$$

Comparing here the orders of the zeros at $x = 0$ we arrive at the four possibilities

(1) $ps = q + s = r$,
(2) $ps = q + s < r$,
(3) $ps > q + s = r$,
(4) $ps = r < q + s$.

Note that the homogeneous equation ($h = 0$) comes under (2).

We need to find nonnegative integers s fulfilling one of the relations (1)–(4). This leads to the following conditions for p, q, r:

$$p - 1 \text{ divides } q \qquad \text{and} \qquad pq = (p-1)r \qquad (s = q/(p-1)); \quad (4.4.4)$$

$$p - 1 \text{ divides } q \qquad \text{and} \qquad pq < (p-1)r \qquad (s = q/(p-1)); \quad (4.4.5)$$

$$p(r - q) > r \qquad\qquad\qquad\qquad\qquad\qquad (s = r - q); \qquad (4.4.6)$$

$$p \text{ divides } r \qquad \text{and} \qquad pq > (p-1)r \qquad (s = r/p). \qquad (4.4.7)$$

In each of instances (4.4.4)–(4.4.7) there is a unique $s \in \mathbb{N}_0$ (given in the parentheses) satisfying the corresponding one of relations (1)–(4). Conditions (4.4.4)–(4.4.7) exclude each other, with the only exception that (4.4.5) is a subcase of (4.4.6) (the inequalities in (4.4.5) and (4.4.6) are equivalent). Of course, the s from (4.4.5) is smaller than that from (4.4.6), when calculated for the same p, q, r. Moreover, each of conditions (4.4.4)–(4.4.6) implies $q < r$.

The first thing to be determined now is the value $c_0 = \Phi(0)$ of an LAS Φ of (4.4.3) in each of cases (4.4.4)–(4.4.7). Whenever it exists, all the remaining $c_k = \Phi^{(k)}(0)$, $k \geq 1$, are uniquely determined from equations (4.2.7), as we have $f'(0) = 0$. They will depend on $c_0 = t$ if it happens to be arbitrary. The rest is obtained by applying Theorem 4.2.3 to equation (4.4.3).

Let us examine cases (4.4.4)–(4.4.6).

Case (4.4.4). Equation (4.4.3) becomes

$$\Phi(x) = \frac{(F(x))^s}{G(x)} \Phi(f(x)) - \frac{H(x)}{G(x)}, \qquad s = \frac{q}{p-1}.$$

Since $H(0) \neq 0$, the (unique) c_0 exists if and only if $(F(0))^s \neq G(0)$.

Case (4.4.5). We get by (4.4.3)

$$\Phi(x) = \frac{(F(x))^s}{G(x)} \Phi(f(x)) - \frac{x^{r-ps}H(x)}{G(x)}, \qquad s = \frac{q}{p-1}. \qquad (4.4.8)$$

Now, if $(F(0))^s = G(0)$, then $c_0 = t$ may be arbitrary, otherwise there is no $c_0 = \Phi(0)$ satisfying (4.4.8) for $x = 0$. Let $(F(0))^s = G(0)$. Then equation (4.4.1) has a unique one-parameter family of LASs $\varphi_t(x) = x^s \Phi_t(x)$, where $s = q/(p-1)$. For $t = 0$ we obtain the solution φ_0 which at the origin has a zero of an order higher than $q/(p-1)$. If $h \neq 0$, this order must be $r - q$ (case (4.4.6)); if $h = 0$, then $\varphi_0 = 0$.

Case (4.4.6). Let us exclude (4.4.5). Now (4.4.3) may be written as

$$\Phi(x) = \frac{(F(x))^s}{G(x)} x^{ps-r} \Phi(f(x)) - \frac{H(x)}{G(x)}, \qquad s = r - q.$$

Then $c_0 = \Phi(0) = -H(0)/G(0)$ always exists.

The results obtained so far are summarized in the following.

Theorem 4.4.1. *Let the functions* $f(x) = x^p F(x)$, $g(x) = x^q G(x)$, $h(x) = x^r H(x)$ *be analytic in a neighbourhood of the origin,* $F(0) \neq 0$, $G(0) \neq 0$, *and* $H(0) \neq 0$ *except when* $h = 0$. *Further assume that* $p \geq 2$, $q \geq 1$ *and* $r \geq pq/(p-1)$. *Table 4.1 describes the family of LASs of equation* (4.4.1).

Table 4.1

Conditions	LAS	Order of its zero at $x = 0$
(4.4.4) and $(F(0))^{q/(p-1)} \neq G(0)$	Unique	$s = q/(p-1)$
(4.4.5) and $(F(0))^{q/(p-1)} = G(0)$	Unique one-parameter family φ_t	$s = q/(p-1)$ if $t \neq 0$; $s = r - q$ if $t = 0$, provided $h \neq 0$, otherwise $\varphi_0 = 0$
(4.4.6) and $q/(p-1) \notin \mathbb{N}$	Unique	$s = r - q$
Other cases covered by (4.4.4), (4.4.5), (4.4.6)	Does not exist	—

Now we pass to case (4.4.7). Then we cannot reduce equation (4.4.3) directly to a form to which we could apply Theorem 4.2.3. Put $s = r/p$ and let m be the smallest integer fulfilling

$$m \geq q/(p-1). \qquad (4.4.9)$$

Of course, $m > s$ by (4.4.7). Suppose that equation (4.4.1) has an LAS φ. Since the order of zero of φ at the origin must be s, we can write

$$\varphi(x) = P(x) + \varphi^*(x), \tag{4.4.10}$$

where

$$P(x) = d_s x^s + \cdots + d_{m-1} x^{m-1}, \tag{4.4.11}$$

and $\varphi^*(x) = x^S \Phi^*(x)$, $S \geq m$, $\Phi^*(0) \neq 0$. It follows from (4.4.1) that

$$\varphi^*(f(x)) = g(x)\varphi^*(x) + h^*(x), \tag{4.4.12}$$

where

$$h^*(x) = h(x) - P(f(x)) + g(x)P(x). \tag{4.4.13}$$

Write $h^*(x) = x^R H^*(x)$, where $H^*(0) \neq 0$ unless $h^* = 0$. By (4.4.9) we have $Sp \geq q + S$, whence $R \geq q + S \geq q + m$.

We conclude from the above remarks that equation (4.4.1) cannot have LASs unless there exists a polynomial (4.4.11) such that function (4.4.13) has at the origin a zero of an order at least $m + q$. If such a polynomial does exist, it is uniquely determined by (4.4.13). Indeed, we have by (4.4.13) and (4.4.11)

$$h^*(x) = h(x) - \sum_{i=s}^{m-1} d_i x^{ip}(F(x))^i + \sum_{i=s}^{m-1} d_i x^{i+q} G(x). \tag{4.4.14}$$

The integer m being the smallest one to fulfil (4.4.9), we have $ip < i + q$ for $i < m$. Thus the coefficient of x^{ip} on the right-hand side of (4.4.14) has the form $-d_i(F(0))^i + A_i$, where A_i depends only on f, g, h and on $d_j, j < i$. All the coefficients of x^k in (4.4.14) must vanish whenever $k < m + q$. If $k = ip$, $i = s, \ldots, m - 1$; this means that $d_i = A_i(F(0))^{-i}$ and the $d_s, \ldots, d_{m-1}$ in (4.4.11) are uniquely determined (if they exist). The existence of P depends on whether the coefficients of the remaining x^k in (4.4.14), i.e. of those where k is not an integral multiple of p and $k < m + q$, vanish for the d_i just determined.

Suppose the polynomial P exists. Then equation (4.4.1) has LASs if and only if equation (4.4.12) has, and these solutions are linked by formula (4.4.10). As we have $R \geq m + q > q$ for h^* given by (4.4.13), Theorem 4.4.1 applies to equation (4.4.12), yielding the following result.

Theorem 4.4.2. *Let the hypotheses of Theorem 4.4.1 be satisfied, except that now $r < pq/(p - 1)$. Assume that there exists a polynomial (4.4.11) such that function (4.4.13) has at the origin a zero of an order $R \geq q + m$, where m is the smallest integer fulfilling (4.4.9). The LASs of equation (4.4.1) are described in Table 4.2.*

Moreover, the orders of zeros of these solutions at the origin are equal to $s = r/p$.

Table 4.2

Conditions			LAS
$q/(p-1) \in \mathbb{N}$,	$pq = R(p-1)$,	$(F(0))^{q/(p-1)} \neq G(0)$	Unique
$q/(p-1) \in \mathbb{N}$,	$pq < R(p-1)$,	$(F(0))^{q/(p-1)} = G(0)$	Unique one-parameter family
$q/(p-1) \notin \mathbb{N}$,	$pq < R(p-1)$		Unique
Remaining cases			Does not exist

It is clear from our considerations that equation (4.4.1) which has no LASs has not even a formal solution. Hence we get the following.

Theorem 4.4.3. *Let the functions f, g, h be analytic in a neighbourhood of the origin and fulfill (4.4.2). Then every formal solution*

$$\varphi(x) = \sum_{n=0}^{\infty} c_n x^n \qquad (4.4.15)$$

of equation (4.4.1) has a positive radius of convergence.

This is no longer true if $f'(0) \neq 0$. We have the following.

Theorem 4.4.4. *Let b be a real number, $0 < b < 1$, let $q \in \mathbb{N}$ and let $C \neq 0$ be a complex constant. If*

$$h(x) = C \sum_{n=0}^{\infty} a_n x^n, \qquad a_n \geq 0, \quad h \neq 0,$$

is an analytic function in a neighbourhood of the origin, then the functional equation

$$\varphi(bx) = x^q \varphi(x) + h(x)$$

has a unique formal solution (4.4.15) which, however, diverges for all $x \neq 0$.

The proof of this theorem (Kuczma [30]) will not be given here. It turns out that the theorem fails to hold if we assume $a_n \geq 0$ for sufficiently large n only.

Comments. Results for the case (4.4.2) are due to M. Kuczma [29]. For analytic solutions of the equation

$$\varphi(x^2) = (\varphi(x))^2 + h(x)$$

see I. N. Baker [6]. Regarding the general nonlinear equation in the case of multiplier zero see J. Ger [2]–[5].

Analytic, in particular entire, solutions of iterative functional equations have been dealt with, among others, in Fatou [1]–[5], Gross [1], Kuczma [28], Kuczma–W. Smajdor [2], [3], Myrberg [2]. See also the bibliography in Kuczma [26, p. 180].

4.5 Meromorphic solutions

The results of the preceding sections help to find meromorphic solutions φ of the equation

$$\varphi(f(x)) = g(x)\varphi(x) + h(x). \tag{4.5.1}$$

The material of this section is taken from Matkowski [2].

We assume the following.

(i) $X \subset \mathbb{C}$ is an open set and $0 \in X$. The boundary of X contains at least two finite points.

(ii) $f: X \to X$ is an analytic function, $f(0) = 0$, $0 < |f'(0)| < 1$.

(iii) The functions g and h are meromorphic in X.

By (ii) and (iii) there are functions F, G, H analytic at the origin and integers q and r fulfilling

$$f(x) = xF(x), \qquad g(x) = x^q G(x), \qquad h(x) = x^r H(x), \tag{4.5.2}$$

where $F(0) = f'(0) \neq 0$, $G(0) \neq 0$ and $H(0) \neq 0$.

We shall seek solutions φ of (4.5.1) in the form

$$\varphi(x) = x^k \Phi(x), \qquad \Phi(0) \neq 0, \tag{4.5.3}$$

where $k \in \mathbb{Z}$ and Φ is meromorphic in X and analytic in a neighbourhood of the origin (in short: Φ is *regular at the origin*).

To begin with we take the homogeneous equation

$$\varphi(f(x)) = g(x)\varphi(x). \tag{4.5.4}$$

We insert (4.5.2) and (4.5.3) into (4.5.4) to get

$$x^k (F(x))^k \Phi(f(x)) = x^{k+q} G(x)\Phi(x), \tag{4.5.5}$$

whence $q = 0$ and $(F(0))^k = G(0)$, i.e. $g(0) = (f'(0))^k$. If this equality holds for a $k \in \mathbb{Z}$, then it does not hold for any other integer and, by Theorem 4.2.3, equation (4.5.5) has a unique one-parameter family of local analytic solutions in a neighbourhood of the origin. To extend these solutions to all of X use can be made of Theorem 4.2.2. In this way we arrive at the following.

Theorem 4.5.1. *Let hypotheses* (i), (ii) *be fulfilled and let g be a meromorphic function in X. If*

$$g(0) = [f'(0)]^m \qquad \text{for an } m \in \mathbb{Z} \tag{4.5.6}$$

then equation (4.5.4) *has a unique one-parameter family of meromorphic solutions φ in X. Except for $\varphi = 0$, they can be written in the form* (4.5.3), *where $k = m$ and Φ is regular at the origin. Moreover, if*

$$g(0) \neq [f'(0)]^p \qquad \text{for every } p \in \mathbb{Z} \tag{4.5.7}$$

(in particular, if g has a pole at zero), then $\varphi = 0$ is the only meromorphic solution of (4.5.4) *in X.*

Now we pass to the general linear equation (4.5.1), assuming thus $h \neq 0$.

Theorem 4.5.2. *Let hypotheses* (i)–(iii) *be fulfilled with* f, g, h *of form* (4.5.2) *and* $q \leq 0$. *The set* Φ *of meromorphic solutions of equation* (4.5.1) *is determined as follows.*

(1) *If either* $q < 0$ *or* (4.5.7) *holds, then* Φ *consists of the single* φ *which is of form* (4.5.3) *with* $k = r - q$ *and* Φ *regular at the origin.*

(2) *If* (4.5.6) *holds with an* $m < 0$, *then* $\Phi = \varnothing$ *when* $r = m$, *and when* $r > m$ Φ *forms a one-parameter family whose members* φ *satisfy* (4.5.3) *with* $k = m$ *and* Φs *regular at the origin except for a unique* φ_0 *of the form*

$$\varphi_0(x) = x^r \Phi_0(x), \qquad \Phi_0(0) \neq 0,$$

with Φ_0 *regular at the origin.*

Proof. Inserting (4.5.2) and (4.5.3) into (4.5.1) we get

$$x^k (F(x))^k \Phi(f(x)) = x^{k+q} G(x)\Phi(x) + x^r H(x). \tag{4.5.8}$$

(1) If $q = 0$ and (4.5.7) holds then we must have $k = r = r - q$, and Φ can be uniquely determined from (4.5.8) in virtue of Theorem 4.2.3. If $q < 0$, then necessarily $k = r - q$ and Theorem 4.2.3 again works for equation (4.5.8), but now in a neighbourhood of the origin. The unique Φ resulting from Theorem 4.2.3 can be then uniquely extended onto X with the aid of Theorem 4.2.2.

(2) If (4.5.6) holds with an $m < 0$, then obviously $q = 0$, and (4.5.8) can be written as

$$x^k [(F(x))^k \Phi(f(x)) - G(x)\Phi(x)] = x^r H(x). \tag{4.5.9}$$

If $r = m$ and $k \neq m$, then the left-hand side of (4.5.9) has at the origin a pole of order different from r, a contradiction. For $k = m = r$ (4.5.9) yields $0 \neq H(0)/\Phi(0) = F(0)^m - G(0) = [f'(0)]^m - g(0) = 0$ which is also impossible. Thus if $r = m$ equation (4.5.1) has no solutions of the form (4.5.3).

If $r < m$, we take $k = r$ in (4.5.9) and determine a unique $\Phi = \Phi_0$ on account of Theorems 4.2.3 and 4.2.2. The theorem results now from Theorem 4.5.1. ∎

Remark 4.5.1. If in (4.5.6) we have $m \geq 0$ or $r < m < 0$, Theorem 4.5.2 does not work. Now $q = 0$ in (4.5.2) and $k = r$ in (4.5.3), so that a meromorphic φ satisfies (4.5.1) if and only if Φ is a local analytic solution of the equation

$$\Phi(x) = \frac{(F(x))^r}{G(x)} \Phi(f(x)) - \frac{H(x)}{G(x)}, \tag{4.5.10}$$

resulting from (4.5.8) on taking $k = r$. Equation (4.5.10), in turn, has such solutions if and only if it has formal ones, i.e. if system (4.2.6) associated with equation (4.5.10) has a solution (c_i).

We learn from Theorem 4.4.4 that we cannot handle the case of $q > 0$. Equation (4.5.1) may then have divergent formal solutions.

Comments. Meromorphic solutions of linear and nonlinear functional equations were investigated also by R. Raclis [1], W. Pranger [1], R. Goldstein [1]–[6].

4.6 Special equations

Some of the results of Sections 4.2–4.4 will be used here for finding local analytic solutions of the Schröder and Abel equations. We preserve the abbreviation 'LAS' from Section 4.2.

4.6A. The Schröder equation

Theorem 4.2.3 when applied to the equation

$$\sigma(f(x)) = s\sigma(x) \tag{4.6.1}$$

yields the famous theorem of G. Koenigs [1].

Theorem 4.6.1. *Let $X \subset \mathbb{C}$ be a neighbourhood of the origin, and let $f : X \to \mathbb{C}$ be an analytic function,*

$$f(0) = 0, \qquad f'(0) = s, \qquad 0 < |s| < 1.$$

Then equation (4.6.1) has a unique LAS σ fulfilling the condition $\sigma'(0) = 1$. This solution is given by the formula

$$\sigma(x) = \lim_{n \to \infty} s^{-n} f^n(x). \tag{4.6.2}$$

Proof. Because $s(f'(0))^k = s^{k+1} \neq 1$ for $k \in \mathbb{N}$, system (4.2.7), when written for equation (4.6.1), is uniquely solvable. Thus the existence of a unique LAS σ of (4.6.1) actually follows from Theorem 4.2.3. Formula (4.6.2) may be obtained by the argument we have used in the proof of Theorem 3.5.1 (formula (3.5.2)). ∎

Remark 4.6.1. If an invertible function σ satisfies equation (4.6.1), then its inverse $\varphi = \sigma^{-1}$ satisfies the *Poincaré equation* (Poincaré [1], [2])

$$\varphi(sx) = f(\varphi(x)).$$

The case where $|s| = 1$, but s is not a root of unity, is covered by Theorem 4.3.1. For the case where s is a root of unity we have the following result (see Rausenberger [1], Muckenhoupt [1]).

Theorem 4.6.2. *Let the conditions of Theorem 4.6.1. be fulfilled except that now s is a pth root of unity. Then equation (4.6.1) has an LAS σ such that $\sigma(0) = 0$, $\sigma'(0) = 1$ if and only if $f^p = \mathrm{id}$. The solution is not unique.*

Proof. It follows from (4.6.1) that $\sigma(f^p(x)) = \sigma(x)$, whence $f^p = \mathrm{id}$ if σ is an invertible LAS of (4.6.1). On the other hand, if $f^p = \mathrm{id}$ and if $g: X \to \mathbb{C}$ is an arbitrary analytic function, then

$$\sigma(x) = \sum_{i=0}^{p-1} s^{-i} g(f^i(x)) \tag{4.6.3}$$

is an LAS of (4.6.1). If $g(0) = 0$ and $g'(0) = p^{-1}$, then $\sigma(0) = 0$ and $\sigma'(0) = 1$. ∎

Actually formula (4.6.3), with g ranging over the family of all analytic functions at the origin, gives the general LAS of equation (4.6.1) in the case where $f^p = \mathrm{id}$. In fact, every such solution σ can be represented in form (4.6.3) with $g(x) = p^{-1}\sigma(x)$.

Returning to the case where s is not a root of unity, let us note the following facts.

Theorem 4.6.3. *Let $X \subset \mathbb{C}$ be a neighbourhood of the origin, and let $f: X \to \mathbb{C}$ be an analytic function, $f(0) = 0$, $f'(0) = s \neq 0$. If s is not a root of unity and σ_0 is a nontrivial LAS of equation (4.6.1), then $\sigma_0'(0) \neq 0$ and the general LAS of (4.6.1) is given by $\sigma(x) = c\sigma_0(x)$, where $c \in \mathbb{C}$ is arbitrary.*

Proof. Suppose that $\sigma_0'(0) = 0$. Since $s \neq 1$, we have $\sigma_0(0) = 0$ and thus $\sigma_0(x) = x^p \varphi(x)$, with $p \geq 1$, $\varphi(0) \neq 0$. Hence $\phi(f(x)) = s(x/f(x))^p \varphi(x)$ and on setting $x = 0$ we obtain $\varphi(0) = \varphi(0)s^{1-p}$, a contradiction. Now let σ be an arbitrary LAS of equation (4.6.1). Again $\sigma(0) = 0$ and thus the function $\omega = \sigma/\sigma_0 - c$, where $c = \sigma'(0)/\sigma_0'(0)$, is analytic in a neighbourhood of the origin and $\omega(0) = 0$. Moreover, by (4.6.1) for σ and σ_0 we have $\omega(f(x)) = \omega(x)$ in X. Consult Note 4.9.17 to find that $\omega = 0$, whence $\sigma = c\sigma_0$. ∎

4.6B. The Abel equation

Since the Schröder equation (4.6.1) cannot have interesting solutions in the case where $s = f'(0) = 1$, the Abel equation

$$\alpha(f(x)) = \alpha(x) + 1 \tag{4.6.4}$$

is particularly useful then, see Subsection 8.5D. Here we discuss other cases of $|s| \leqslant 1$. We start with the case where s is not a root of unity. We assume the following hypotheses.

(i) $X \subset \mathbb{C}$ is a neighbourhood of the origin.
(ii) $f: X \to \mathbb{C}$ is an analytic function, $f(0) = 0$, $s := f'(0)$.

Anticipating the formulations of Theorems 4.6.4, 4.6.5 let us point out that the solutions α of (4.6.4) are, of course, multivalued. Equation (4.6.4) is to be understood as satisfied in the following sense: for every x, such that

x and $f(x)$ lie in the domain of definition of α, and for every value of α at $f(x)$, a branch of α at x can be chosen so that (4.6.4) holds.

Theorem 4.6.4. *Let hypotheses* (i), (ii) *be fulfilled, and assume that either* $0 < |s| < 1$, *or* $s \in S$, *where S is the Siegel set (see Subsection 4.3A). Then equation (4.6.4) has a unique one-parameter family (with an additive parameter) of solutions α defined in a vicinity of the origin and such that*

$$\alpha(x) = \log x / \log s + \varphi(x), \qquad (4.6.5)$$

where φ is an analytic function in a neighbourhood of the origin. These solutions are given by (log s *being an arbitrarily fixed value of the logarithm at s):*

$$\alpha(x) = \log \sigma(x) / \log s, \qquad (4.6.6)$$

where σ is a nontrivial LAS of the Schröder equation (4.6.1).

Proof. Write $\psi(x) = \varphi(x) \log s$. If α has form (4.6.5) and satisfies equation (4.6.4) in a vicinity of $x = 0$, then the function σ given by

$$\sigma(x) = \exp[\alpha(x) \log s] = x \exp \psi(x)$$

is an LAS of equation (4.6.1). Conversely, if σ is a solution of (4.6.1) as required, then by Theorem 4.6.3 $\sigma'(0) \neq 0$ and thus α given by (4.6.6) has property (4.6.5) and satisfies (4.6.4). Since by Theorems 4.6.3 and 4.3.1 the LAS σ of (4.6.1) is determined uniquely up to a multiplicative constant, so is α up to an additive constant. ∎

The same argument, together with Theorem 4.6.2, yields the following.

Theorem 4.6.5. *Let hypotheses* (i), (ii) *be fulfilled, and assume that $s \neq 1$ is a pth root of unity. Then equation (4.6.4) has solutions α of form (4.6.5) with φ analytic in a neighbourhood of the origin if and only if $f^p = \text{id}$. The solution is not unique.*

Remark 4.6.2. Let us recall that no single-valued solution of (4.6.4) exists in a domain containing fixed points of f of any order (see Subsection 3.5C). This is no longer the case for multivalued solutions. For example, the equation $\alpha(-x) = \alpha(x) + 1$ cannot have a single-valued solution, since applying it twice we get $\alpha(x) = \alpha(x) + 2$. Nevertheless, both equations are satisfied by the multivalued function $\alpha(x) = (\pi i)^{-1} \log x$.

4.7 Integrable solutions

Following the exhaustive work of J. Matkowski [13] we shall investigate here integrable solutions of the equation

$$\varphi(x) = g(x)\varphi(f(x)) + h(x) \qquad (4.7.1)$$

(see also Kuczma [35]).

4.7A. Preliminaries

Since in the problems of integration the values of functions on sets of measure zero are irrelevant, it appears natural that equations should be postulated only almost everywhere. Thus in this section by a *solution* we understand a function which satisfies the equation in question a.e. in the set considered. A solution is *essentially unique* iff it is determined uniquely up to a set of measure zero.

We assume the following.

(i) $(X, \mathbf{S}, \mu)$ is a σ-finite measure space, $\mu(X) > 0$.

(ii) f is a bijective mapping from X onto itself, and both f and f^{-1} are $\mathbf{S}$-measurable. The measures μf and μf^{-1} –

$$\mu f(A) = \mu(f(A)), \qquad \mu f^{-1}(A) = \mu(f^{-1}(A)), \qquad A \in \mathbf{S}$$

– are absolutely continuous with respect to μ, and

$$\mu\left(X \setminus \bigcup_{n=0}^{\infty} A_n\right) = 0, \tag{4.7.2}$$

where

$$A_n := f^n(X) \setminus f^{n+1}(X), \qquad n \in \mathbb{N}_0. \tag{4.7.3}$$

We shall freely use the standard results from the theory of measure and integral (see e.g. Halmos [1]). In particular, we shall need the following notions and facts.

For any two measures μ, ν on $\mathbf{S}$, where ν is absolutely continuous with respect to μ, the symbol $d\nu/d\mu$ stands for the Radon–Nikodym derivative of ν with respect to μ; see Note 4.9.8.

The formulae

$$\int_A \psi \, d\mu = \int_{f^{-1}(A)} (\psi \circ f) d(\mu f),$$

and

$$\int_A \psi \, d(\mu f^k) = \int_A \psi \, \frac{d(\mu f^k)}{d(\mu f^m)} d(\mu f^m)$$

are valid for $A \in \mathbf{S}$, measurable $\psi : A \to \mathbb{R}^+$ and $k, m \in \mathbb{Z}$, all arbitrary (see Halmos [1, §§ 32 and 39]).

We have the relations

$$\frac{d(\mu f)}{d\mu} \circ f^i = \frac{d(\mu f^{i+1})}{d(\mu f^i)} \qquad \text{a.e. in } X, \ i \in \mathbb{N}_0, \tag{4.7.4}$$

and

$$\prod_{i=0}^{n-1} \left(\frac{d(\mu f)}{d\mu} \circ f^i\right) = \frac{d(\mu f^n)}{d\mu} \qquad \text{a.e. in } X, \ n \in \mathbb{N}. \tag{4.7.5}$$

The symbol L^p, $p > 0$, stands for $L^p(X, \mathbf{S}, \mu)$, i.e. the space of all functions $\varphi : X \to \mathbb{R}$ such that $\int_X |\varphi|^p d\mu < \infty$.

With

$$\alpha(p) = \begin{cases} 1 & \text{for } 0 < p \leqslant 1, \\ 1/p & \text{for } 1 < p < \infty, \end{cases} \tag{4.7.6}$$

the formula

$$\|\varphi\| = \left(\int_X |\varphi|^p \, d\mu \right)^{\alpha(p)}, \qquad \varphi \in L^p, \tag{4.7.7}$$

defines a paranorm, the convergence in which is equivalent to the convergence in measure.

In addition to the latter fact let us note the following.

Lemma 4.7.1. *Let hypothesis* (i) *be fulfilled and let* $\varphi_n \in L^p$, $n \in \mathbb{N}$. *If* $\sum_{n=1}^{\infty} \|\varphi_n\| < \infty$, *then the series* $\sum_{n=1}^{\infty} \varphi_n = \varphi$ *converges a.e. in* X *and* $\varphi \in L^p$.

Proof. Put $\psi_n = |\varphi_1| + \cdots + |\varphi_n| \in L^p$. Then

$$\int_X (\psi_n)^p \, d\mu = \|\psi_n\|^{1/\alpha(p)} \leqslant \left(\sum_{i=1}^{n} \|\varphi_i\| \right)^{1/\alpha(p)},$$

which implies the convergence of the increasing sequence $(\int_X (\psi_n)^p \, d\mu)$. It follows that (ψ_n) converges a.e. in X, and hence also the series $\sum_{n=1}^{\infty} \varphi_n$ does. Now we have by Fatou's Lemma

$$\|\varphi\| = \left(\int_X \lim_{n \to \infty} \left| \sum_{i=1}^{n} \varphi_i \right|^p \, d\mu \right)^{\alpha(p)}$$

$$\leqslant \left(\liminf_{n \to \infty} \int_X \left| \sum_{i=1}^{n} \varphi_i \right|^p \, d\mu \right)^{\alpha(p)}$$

$$\leqslant \liminf_{n \to \infty} \sum_{i=1}^{n} \|\varphi_i\| = \sum_{i=1}^{\infty} \|\varphi_i\| < \infty. \quad \blacksquare$$

The members of the family of measures $\{\mu f^k : k \in \mathbb{Z}\}$, are mutually absolutely continuous.

Lemma 4.7.2. *Under conditions* (i), (ii), *any two measures* μf^m, μf^k, $m, k \in \mathbb{Z}$, *are absolutely continuous with respect to each other and* $d(\mu f^m)/d(\mu f^k) > 0$ *a.e. in* X.

Proof. If $\mu f^n(A) = 0$ for an $n \in \mathbb{Z}$, then $\mu f^{n-1}(A) = \mu f^{-1}(f^n(A)) = 0$ and $\mu f^{n+1}(A) = \mu f(f^n(A)) = 0$, which implies the first part of the lemma. Now put

$$A = \left\{ x \in X : \frac{d(\mu f^m)}{d(\mu f^k)} (x) = 0 \right\}$$

and calculate

$$\mu f^m(A) = \int_A \frac{d(\mu f^m)}{d(\mu f^k)} \, d(\mu f^k) = 0.$$

Hence $\mu(A) = 0$ by the absolute continuity of $\mu = \mu f^0$ with respect to μf^m. $\quad \blacksquare$

The sets A_n defined by (4.7.3) have positive measures.

Lemma 4.7.3. *Under conditions* (i), (ii), *we have* $\mu(A_n) > 0$ *for* $n \in \mathbb{N}_0$.

Proof. If we had $\mu(A_0) = 0$, then we would get by (4.7.3) and Lemma 4.7.2 $\mu(A_n) = \mu(f^n(A_0)) = \mu f^n(A_0) = 0$, $n \in \mathbb{N}_0$, and by (4.7.2) that $\mu(X) = 0$, a contradiction. Thus $\mu(A_0) > 0$ and the lemma results from Lemma 4.7.2. ∎

4.7B. A functional inequality

Let $g: X \to \mathbb{R}$ be a measurable function and let

$$N_g := \{x \in X : g(x) = 0\}.$$

For A_n defined by (4.7.3), we put

$$\left.\begin{aligned}
u_n &= \operatorname*{essinf}_{A_n \setminus N_g} |g|^{-p} \frac{d(\mu f)}{d\mu}, \qquad n \in \mathbb{N}_0, \\
v_n &= \operatorname*{esssup}_{A_n \setminus N_g} |g|^{-p} \frac{d(\mu f)}{d\mu}, \qquad n \in \mathbb{N}_0,
\end{aligned}\right\} \tag{4.7.8}$$

and

$$w_n = \operatorname*{esssup}_{f^n(X)} |g|^p \left(\frac{d(\mu f)}{d\mu}\right)^{-1}, \qquad n \in \mathbb{N}_0. \tag{4.7.9}$$

The two theorems that follow deal with the functional inequality

$$|\varphi(x)| \leqslant |g(x)| |\varphi(f(x))| \tag{4.7.10}$$

and will be useful in the sequel.

Theorem 4.7.1. *Let hypotheses* (i), (ii) *be fulfilled and let* $g: X \to \mathbb{R}$ *be a measurable function. If the series*

$$\sum_{n=0}^{\infty} \prod_{i=0}^{n-1} u_i \tag{4.7.11}$$

diverges and if $\varphi \in L^p$ *satisfies inequality* (4.7.10) *a.e. in* X, *then* $\varphi = 0$ *a.e. in* X.

Proof. By (4.7.10) we have $\varphi = 0$ a.e. in N_g, whence

$$\int_{A_n} |\varphi|^p d\mu = \int_{f^{-1}(A_n)} |\varphi \circ f|^p d(\mu f) = \int_{A_{n-1}} |\varphi \circ f|^p \frac{d(\mu f)}{d\mu} d\mu$$

$$\geqslant \int_{A_{n-1} \setminus N_g} |g|^{-p} |\varphi|^p \frac{d(\mu f)}{d\mu} d\mu \geqslant u_{n-1} \int_{A_{n-1}} |\varphi|^p d\mu,$$

and

$$\int_{A_n} |\varphi|^p d\mu \geqslant \left(\prod_{i=0}^{n-1} u_i\right) \int_{A_0} |\varphi|^p d\mu, \qquad n \in \mathbb{N}_0.$$

Hence we get by (4.7.2)

$$\left(\sum_{n=0}^{\infty}\prod_{i=0}^{n-1}u_i\right)\int_{A_0}|\varphi|^p\,d\mu\leqslant\sum_{n=0}^{\infty}\int_{A_n}|\varphi|^p\,d\mu=\int_X|\varphi|^p\,d\mu<\infty,$$

and the divergence of series (4.7.11) implies $\varphi=0$ a.e. in A_0. Since

$$\sum_{n=0}^{\infty}\prod_{i=0}^{n-1}u_i=\sum_{n=0}^{k}\prod_{i=0}^{n-1}u_i+\left(\prod_{i=0}^{k}u_i\right)\sum_{n=0}^{\infty}\prod_{i=0}^{n-1}u_{k+i+1},\qquad(4.7.12)$$

the series on the right-hand side of (4.7.12) diverges. Replacing in the above argument X by $f^k(X)$ and A_0 by A_k, we obtain $\varphi=0$ a.e. in A_k, $k\in\mathbb{N}_0$. Relation (4.7.2) completes the proof. ∎

Remark 4.7.1. Series (4.7.11) diverges, in particular, if $|g|^p\leqslant d(\mu f)/d\mu$ a.e. in X; see (4.7.8).

Theorem 4.7.2. *Let hypotheses* (i), (ii) *be fulfilled and let* $g:X\to\bar{\mathbb{R}}$ *be measurable. If there exists an* $M>0$ *such that*

$$\prod_{i=0}^{n-1}w_i\leqslant M\qquad\text{for }n\in\mathbb{N},\qquad(4.7.13)$$

and if $\varphi\in L^p$ *satisfies inequality* (4.7.10) *a.e. in* X, *then* $\varphi=0$ *a.e. in* X.

Proof. We have from (4.7.10) by induction

$$|\varphi(x)|\leqslant\prod_{i=0}^{n-1}|g(f^i(x))|\,|\varphi(f^n(x))|\qquad\text{a.e. in }X,\,n\in\mathbb{N},$$

whence by (4.7.4) and (4.7.5)

$$\int_X|\varphi|^p\,d\mu\leqslant\int_X\left|\prod_{i=0}^{n-1}\frac{(g\circ f^i)^p}{d(\mu f^{i+1})/d(\mu f^i)}\right||\varphi\circ f^n|^p\frac{d(\mu f^n)}{d\mu}\,d\mu$$

$$=\int_X\left|\prod_{i=0}^{n-1}\frac{(g)^p}{d(\mu f)/d\mu}\circ f^i\right||\varphi\circ f^n|^p\frac{d(\mu f^n)}{d\mu}\,d\mu$$

$$\leqslant\int_X M|\varphi\circ f^n|^p\frac{d(\mu f^n)}{d\mu}\,d\mu=M\int_{f^n(X)}|\varphi|^p\,d\mu.$$

We have by (4.7.2)

$$\sum_{k=0}^{\infty}\int_{A_k}|\varphi|^p\,d\mu=\int_X|\varphi|^p\,d\mu<\infty,$$

whence

$$\lim_{n\to\infty}\int_{f^n(X)}|\varphi|^p\,d\mu=\lim_{n\to\infty}\sum_{k=n}^{\infty}\int_{A_k}|\varphi|^p\,d\mu=0,$$

and $\varphi=0$ a.e. in X. ∎

4.7C. Homogeneous equation

Now we are able to prove an existence theorem for integrable solutions of the equation

$$\varphi(x) = g(x)\varphi(f(x)). \qquad (4.7.14)$$

Theorem 4.7.3. *Let hypotheses* (i), (ii) *be fulfilled and let* $g: X \to \mathbb{R}$ *be measurable,* $g(x) \neq 0$ *a.e. in* X. *If the series*

$$\sum_{n=0}^{\infty} \prod_{i=0}^{n-1} v_i \qquad (4.7.15)$$

converges, then every measurable function $\varphi_0: A_0 \to \mathbb{R}$ *such that*

$$\int_{A_0} |\varphi_0|^p \mathrm{d}\mu < \infty \qquad (4.7.16)$$

can be extended onto X *to a function* $\varphi \in L^p$ *satisfying equation* (4.7.14) *a.e. in* X.

Proof. Define the functions $\varphi_k: A_k \to \mathbb{R}$ by the recurrence

$$\varphi_{k+1}(x) = \varphi_k(f^{-1}(x))/g(f^{-1}(x)) \qquad \text{for } x \in A_{k+1}, \; k \in \mathbb{N}_0,$$

and put

$$\varphi(x) = \varphi_k(x) \qquad \text{for } x \in A_k, \; k \in \mathbb{N}_0.$$

According to (4.7.2) the function φ is defined a.e. in X and satisfies equation (4.7.14) a.e. in X. We shall prove that it is an L^p function. We have

$$\int_{A_n} |\varphi|^p \mathrm{d}\mu = \int_{f^{-1}(A_n)} |\varphi \circ f|^p \mathrm{d}(\mu f) = \int_{A_{n-1} \setminus N_g} |\varphi|^p |g|^{-p} \frac{\mathrm{d}(\mu f)}{\mathrm{d}\mu} \mathrm{d}\mu$$

$$\leqslant v_{n-1} \int_{A_{n-1}} |\varphi|^p \mathrm{d}\mu,$$

whence

$$\int_{A_n} |\varphi|^p \mathrm{d}\mu \leqslant \left(\prod_{i=0}^{n-1} v_i \right) \int_{A_0} |\varphi_0|^p \mathrm{d}\mu, \qquad n \in \mathbb{N}_0,$$

and by (4.7.16)

$$\int_X |\varphi|^p \mathrm{d}\mu \leqslant \left(\sum_{n=0}^{\infty} \prod_{i=0}^{n-1} v_i \right) \int_{A_0} |\varphi_0|^p \mathrm{d}\mu < \infty. \qquad \blacksquare$$

Remark 4.7.2. The existence of an $s > 1$ such that $|g|^p \geqslant s\mathrm{d}(\mu f)/\mathrm{d}\mu$ a.e. in X suffices for the convergence of series (4.7.15); see (4.7.8).

4.7D. Inhomogeneous equation

Passing to the general linear equation (4.7.1), we assume additionally that

(iii) The functions $g, h: X \to \mathbb{R}$ are measurable and finite a.e. in X. Moreover, $h \in L^p$.

We present two existence-and-uniqueness results for integrable solutions of equation (4.7.1).

Theorem 4.7.4. *Let hypotheses* (i)–(iii) *be fulfilled. If there exists an* s, $0 < s < 1$, *such that*

$$|g|^p \leqslant s\,\mathrm{d}(\mu f)/\mathrm{d}\mu \qquad a.e. \text{ in } X, \tag{4.7.17}$$

then equation (4.7.1) *has an essentially unique solution* $\varphi \in L^p$. *This solution is given by the formula*

$$\varphi(x) = \sum_{n=0}^{\infty} \left[\prod_{i=0}^{n-1} g(f^i(x)) \right] h(f^n(x)) \qquad a.e. \text{ in } X. \tag{4.7.18}$$

Proof. We have by (4.7.17), (4.7.4) and (4.7.5)

$$\left\| \left(\prod_{i=0}^{n-1} g \circ f^i \right) \cdot (h \circ f^n) \right\| = \left(\int_X \left| \left(\prod_{i=0}^{n-1} g \circ f^i \right) h \circ f^n \right|^p \mathrm{d}\mu \right)^{\alpha(p)}$$

$$\leqslant (s^{\alpha(p)})^n \left(\int_X |h \circ f^n|^p \frac{\mathrm{d}(\mu f^n)}{\mathrm{d}\mu} \,\mathrm{d}\mu \right)^{\alpha(p)}$$

$$= (s^{\alpha(p)})^n \left(\int_{f^n(X)} |h|^p \mathrm{d}\mu \right)^{\alpha(p)}$$

$$\leqslant (s^{\alpha(p)})^n \|h\|.$$

Hence it follows in virtue of Lemma 4.7.1 that series (4.7.18) converges in a set $X \setminus A$, where $\mu(A) = 0$. By Lemma 4.7.2, also the set $B = \bigcup_{0}^{\infty} f^k(A)$ has measure zero. It is easily seen that the function φ is well defined by (4.7.18) on $X \setminus B$ and satisfies equation (4.7.1) on $X \setminus B$. By Lemma 4.7.1 $\varphi \in L^p$. The uniqueness statement results from Theorem 4.7.1. ∎

Theorem 4.7.5. *Let hypotheses* (i)–(iii) *be fulfilled. If the series*

$$\sum_{n=0}^{\infty} \left[\left(\prod_{i=0}^{n-1} w_i \right) \int_{f^n(X)} |h|^p \mathrm{d}\mu \right]^{\alpha(p)} \tag{4.7.19}$$

converges, then equation (4.7.1) *has a solution* $\varphi \in L^p$ *given by formula* (4.7.18). *This solution is essentially unique if series* (4.7.11) *diverges or if condition* (4.7.13) *holds, whereas the solution* $\varphi \in L^p$ *of* (4.7.1) *can be prescribed arbitrarily on* A_0 *(with condition* (4.7.16) *for* $\varphi_0 = \varphi | A_0$*) if series* (4.7.15) *converges.*

The proof of this theorem is similar to that of Theorem 4.7.4. The uniqueness part results from Theorems 4.7.1–4.7.3.

4.7E. Lebesgue measure

Now let X be a real interval, $X = (0, a]$, $0 < a < \infty$, let S be the σ-algebra of Lebesgue measurable subsets of X, and let μ be the Lebesgue measure. If the function $f: X \to X$ is strictly increasing, $0 < f(x) < x$ in X, and both

f and f^{-1} are absolutely continuous, then f fulfils hypothesis (ii) and we have $d(\mu f)/d\mu = f'$.

Let $u: X \to \mathbb{R}$ be a positive decreasing function. If $|g|^{-p}f' \geq u$ a.e. in a (right) vicinity of zero, then the series (4.7.11) diverges whenever the series

$$\sum_{n=0}^{\infty} \prod_{i=0}^{n-1} u(f^i(x)). \tag{4.7.20}$$

does. In the case where $|g|^{-p}f' \leq u$ a.e. in a (right) vicinity of zero, series (4.7.15) converges whenever (4.7.20) does. Note that the monotonicity of f and u implies that whatever is the behaviour of series (4.7.20) at an $x \in X$, it is the same in all of X.

Clearly (4.7.20) converges if $\lim_{x \to 0} u(x) < 1$, and diverges if the limit is larger than unity. In the more delicate case where this limit is just unity we can deduce the convergence or divergence of series (4.7.20) from the asymptotic behaviour of the functions f, g at the origin and from the results of Section 1.4.

We quote here without proof some results (Kuczma [37]) for some typical test-functions u. Let us assume the following.

(iv) $X = (0, a]$, $0 < a < \infty$, and $f: X \to X$ is continuous and strictly increasing, $0 < f(x) < x$ in X. Moreover,

$$f(x) = sx + x^{m+1}F(x),$$

where $0 \leq s \leq 1$, $m > 0$, and

$$0 \leq l = \liminf_{x \to 0} F(x) \leq \limsup_{x \to 0} F(x) = L < \infty.$$

Let c and q be positive constants.

Theorem 4.7.6. *Let hypothesis* (iv) *be fulfilled. Conditions for the convergence or divergence of series* (4.7.20) *are collected in Table 4.3.*

Table 4.3

		Conditions for constants		
Case	Function u	either	or	Series (4.7.20)
$s = 1$	$u(x) = 1 - Cx^q$	$m > q$	$m = q$ and $C > Lm$	converges
		$m < q$ and $l > 0$	$m = q$ and $C < lm$	diverges
$0 < s < 1$	$u(x) = 1 - C(\log x^{-1})^{-q}$	$q < 1$	$q = 1$ and $C > \log s^{-1}$	converges
		$q > 1$	$q = 1$ and $C < \log s^{-1}$	diverges
$s = 0$	$u(x) = 1 - C(\log \log x^{-1})^{-q}$	$q < 1$ and $l > 0$	$q = 1$, $l > 0$ and $C > \log(m + 1)$	converges
		$q > 1$	$q = 1$ and $C < \log(m + 1)$	diverges

Comments. The functional-integral equation

$$\varphi(x) = \int_0^x g(t)d\varphi(f(t)) + h(x)$$

is considered by W. Jarczyk [7]. Results he obtained yield a uniqueness theorem for Lebesgue integrable solutions of equation (4.7.1) (different from Theorem 4.7.4) and a generalization of Theorem 4.7.5 (the nonuniqueness case for L^1 solutions). For L^1 solutions of Abel's equation see Zdun [11].

4.8 Absolutely continuous solutions

We shall deal with the particular linear equation

$$\varphi(x) = s\varphi(f(x)) + h(x), \tag{4.8.1}$$

where $s \in \mathbb{R}$ is a constant. If $\varphi: X \to \mathbb{R}$ is *absolutely continuous* (for short a.c.), then it is continuous on X and (if X is compact) of bounded variation on X. Thus the main interest of the only theorem in this section lies in the existence rather than in the uniqueness. The result is due to J. Matkowski [13]. It will be used for solving a Goursat problem.

4.8A. Existence-and-uniqueness result

Assume the following.

(i) $X = [0, a], 0 < a < \infty$.
(ii) The function $f: X \to X$ is strictly increasing, both f and f^{-1} are a.c. in X, $0 < f(x) < x$ in $X \setminus \{0\}$.
(iii) $h: X \to \mathbb{R}$ is a.c. in X.

Theorem 4.8.1. *Let hypotheses* (i)–(iii) *be fulfilled and let* $|s| \leq 1$. *The family* Φ *of a.c. solutions* $\varphi: X \to \mathbb{R}$ *of equation* (4.8.1) *is determined as follows.*

(1) $|s| < 1$. *Then* $\Phi = \{\varphi\}$, *where*

$$\varphi(x) = \sum_{n=0}^{\infty} s^n h(f^n(x)). \tag{4.8.2}$$

(2) $s = -1$. *If*

$$\sum_{n=0}^{\infty} \operatorname{Var} h|[0, f^n(x_0)] < \infty \qquad \text{for an } x_0 \in X \setminus \{0\}, \tag{4.8.3}$$

then $\Phi = \{\varphi\}$, *where*

$$\varphi(x) = \frac{h(0)}{2} + \sum_{n=0}^{\infty} (-1)^n [h(f^n(x)) - h(0)]. \tag{4.8.4}$$

(3) $s = 1$. *If* $h(0) = 0$ *and* (4.8.3) *holds, then* $\Phi = \{\varphi_c: c \in \mathbb{R}\}$, *where*

$$\varphi_c(x) = c + \sum_{n=0}^{\infty} h(f^n(x)). \tag{4.8.5}$$

Proof. The uniqueness and formulae (4.8.2), (4.8.4), (4.8.5) result from Theorems 3.1.10, 3.1.9 and 3.1.6 (see also Note 3.9.3). To prove the existence consider the equation

$$\psi(x) = sf'(x)\psi(f(x)) + h'(x), \qquad (4.8.6)$$

where the derivatives f' and h' are defined almost everywhere in X. If ψ is defined and satisfies equation (4.8.6) a.e. in X, and is integrable in $[0, x]$ for every $x \in X \setminus \{0\}$, then the function

$$\hat{\phi}(x) = \int_0^x \psi(t)\,dt \qquad (4.8.7)$$

is defined and a.c. in X, and satisfies the equation

$$\hat{\phi}(x) = s\hat{\phi}(f(x)) + h(x) - h(0).$$

If $s = 1$, then we have $h(0) = 0$ and $\varphi = \hat{\phi}$ is an a.c. solution of (4.8.1) in X, whereas if $s \neq 1$ then the function $\varphi = \hat{\phi} + (1 - s)^{-1}h(0)$ serves as such a solution. Thus we need only to show that equation (4.8.6) has an integrable solution in $[0, x]$ for every $x \in X \setminus \{0\}$. If $|s| < 1$, this results directly from Theorem 4.7.4. If $|s| = 1$, we apply Theorem 4.7.5 (see Note 4.9.18). ∎

Remark 4.8.1. If (i)–(iii) hold and $|s| < 1$, then the a.c. solution of equation (4.8.1) depends on an arbitrary function (Matkowski [13]). For a different approach to study a.c. solutions of (4.8.1) see Jarczyk [7].

4.8B. A Goursat problem

We consider the partial differential equation

$$u''_{xy} = G(x, y) \qquad \text{a.e. in } X \times Y, \qquad (4.8.8)$$

where $X = [0, a]$, $a > 0$, $Y = [0, b]$, $b > 0$, $G: X \times Y \to \mathbb{R}$ is integrable with respect to the two-dimensional Lebesgue measure, and the unknown function u has in $X \times Y$ Lebesgue integrable derivatives u''_{xy} and u''_{yx} such that $u''_{xy} = u''_{yx}$ a.e. in $X \times Y$.

Let $f_1: X \to Y$ and $f_2: Y \to X$ be a.c. strictly increasing injections whose inverse functions are also a.c.

Further, we assume that the curves $y = f_1(x)$ and $x = f_2(y)$ do not intersect except at the origin. In terms of the function $f = f_2 \circ f_1$ this means that $f(x) \neq x$. Since for $x = a$ there is $f(a) = f_2(f_1(a)) \leq f_2(b) \leq a$, we must have $f(x) < x$ in $X \setminus \{0\}$. On the other hand, since f_1 and f_2 are strictly increasing, we get $f(x) > 0$ in $X \setminus \{0\}$.

The Goursat problem for equation (4.8.8) consists in finding a solution u subjected to the additional conditions

$$u(x, f_1(x)) = P(x), \qquad u(f_2(y), y) = Q(y), \qquad x \in X, \quad y \in Y, \quad (4.8.9)$$

where $P: X \to \mathbb{R}$ and $Q: Y \to \mathbb{R}$ are given a.c. functions and $P(0) = Q(0) = 0$.

Write

$$U(x, y) = \int_0^x \int_0^y G(s, t)\,ds\,dt.$$

Then (4.8.8) implies $u(x, y) = U(x, y) + \varphi(x) + \psi(y)$, where $\varphi: X \to \mathbb{R}$ and $\psi: Y \to \mathbb{R}$ are a.c. From (4.8.9) we get

$$\left.\begin{array}{l} \varphi(x) + \psi(f_1(x)) = P(x) - U(x, f_1(x)), \\ \varphi(f_2(y)) + \psi(y) = Q(y) - U(f_2(y), y), \end{array}\right\} \tag{4.8.10}$$

or

$$\varphi(x) = \varphi(f(x)) + h(x), \tag{4.8.11}$$

where $h(x) = P(x) - Q(f_1(x)) - U(x, f_1(x)) + U(f(x), f_1(x))$, and $f = f_2 \circ f_1$. If h is a.c. and condition (4.8.3) holds, then by Theorem 4.8.1 equation (4.8.11) has a unique one-parameter family (with an additive parameter) of a.c. solutions φ. The function ψ can be determined from the second equation in (4.8.10). Moreover, equations (4.8.10) imply that the sum $\varphi + \psi$ is independent of the additive parameter involved in φ. Thus u is uniquely determined. Consequently, under the conditions specified, the Goursat problem (4.8.9) for equation (4.8.8) has a unique solution $u: X \times Y \to \mathbb{R}$ (Matkowski [13]).

The function h is certainly a.c. whenever the function U fulfils a Lipschitz condition

$$|U(x, y) - U(\bar{x}, \bar{y})| \leqslant K|x - \bar{x}| + M|y - \bar{y}|,$$

which, in turn, certainly is the case if

$$\int_0^b |G(x, y)|\,dy \leqslant K \qquad \text{and} \qquad \int_0^a |G(x, y)|\,dx \leqslant M$$

a.e. in X and in Y, respectively.

Comments. Similar problems have also been studied by A. Bielecki–J. Kisyński [1], K. Deimling [1], G. Majcher [2], [4], [5] and others (see Majcher [5] and Kuczma [26, p. 102] for further references).

Some results concerning Goursat-like problems for partial differential equations of order greater than 2 are found in a series of papers by A. Borzymowski. In particular, for the poly-vibrating equation

$$L^p u(x, y) = c(x, y), \qquad (x, y) \in X \times Y, \tag{4.8.12}$$

where $L := \partial^2/\partial x \partial y$, a Goursat problem consists in finding a solution u subjected to $2p$ boundary conditions of form (4.8.9) set on $2p$ curves f_i emanating from the origin. In Borzymowski [1], [2] this problem is reduced to that of solving a system of linear iterative functional equations with 'inner' functions f_i, and then the solution of (4.8.12) is found in the form of a series.

4.9 Notes

4.9.1. A global version of Theorem 4.2.5 on the continuous dependence on the data of local analytic solutions of linear equations is the following (see Matkowski [3]).

Proposition 4.9.1. *Let the hypotheses of Theorem 4.2.5 be fulfilled and let the boundary of X contain at least two finite points, $f_m(x) \neq x$ in $X \setminus \{0\}$, and $g_m(x) \neq 0$ in X, $m \in \mathbb{N}_0$. Then for every $m \in \mathbb{N}_0$ equation (4.2.16) has a unique analytic solution $\varphi_m \colon X \to \mathbb{C}$ fulfilling the conditions $\varphi_m^{(k)}(0) = c_{km}$ (with c_{km} determined from (4.2.17)), and $\lim_{n \to \infty} \varphi_m = \varphi_0$ a.u. in X.*

4.9.2. By the Liouville transcendental numbers (see Subsection 4.3A) we mean the numbers α that are characterized by the following property.

> For every positive c and every $n \in \mathbb{N}$ there exist integers p and q, $q > 0$, such that $|\alpha - p/q| \leqslant cq^{-n}$.

(Liouville's theorem on the rational approximation to algebraic numbers shows that the numbers α determined by the property actually are not algebraic. Such numbers are easily constructed with a use of continued fractions. See Hinčin [1].)

4.9.3. The following is the famous theorem of K. F. Roth we have referred to in Subsection 4.3B (see e.g. Cassels [1]).

Proposition 4.9.2. *Let β be an irrational algebraic number and let $\delta > 0$. Then there exist only a finite number of pairs of integers $q > 0$ and p such that $|\beta - p/q| < q^{-2-\delta}$.*

4.9.4. The inhomogeneous equation

$$\varphi(f(x)) = g(x)\varphi(x) + h(x) \qquad (4.9.1)$$

may have local analytic solutions also if $g(0) \neq s^k$ for every $k \in \mathbb{N}_0$, $s := f'(0)$. A suitable equation can be constructed by taking appropriate f, g and φ and then defining h by (4.9.1).

In the case where $|g(0)| = 1 \neq s^k$ for every $k \in \mathbb{N}$, $|s| = 1$, no criteria are known for the existence of local analytic solutions of equation (4.9.1). Note that the uniqueness in this case is assured by Lemma 4.3.2.

4.9.5. If $s = f'(0)$ is an element of the Siegel set S, then the following result can be added to those contained in Subsection 4.3D.

Proposition 4.9.3. *Let f, g, h be analytic functions in a neighbourhood of the origin, $f(0) = 0$, $f'(0) = s$ with $s \in S$, and $0 \neq |g(0)| \neq 1$. Then equation (4.9.1) has a unique local analytic solution φ in a neighbourhood of the origin.*

(Since by Theorem 4.3.1 there exists a local analytic solution σ of the

Schröder equation (4.3.3) such that $\sigma(0) = 0$ and $\sigma'(0) = 1$, the substitution $\tilde{\varphi} = \varphi \circ \sigma^{-1}$ reduces the problem to that of finding local analytic solutions of the equation

$$\tilde{\varphi}(sx) = g_1(x)\tilde{\varphi}(x) + h_1(x), \qquad (4.9.2)$$

where $g_1 = g \circ \sigma^{-1}$, $h_1 = h \circ \sigma^{-1}$. Now it suffices to apply Banach's Theorem to equation (4.9.2) (solved with respect to $\tilde{\varphi}(x)$ if $|g(0)| > 1$, and written as $\tilde{\varphi}(x) = g_1(x/s)\tilde{\varphi}(x/s) + h_1(x/s)$ if $0 < |g(0)| < 1$). See W. Smajdor [3].)

4.9.6. Analytic solutions of equation (4.9.1) in a neighbourhood of the origin depend mainly on the values $f'(0)$ and $g(0)$. Note that we know how to proceed in any of possible situations. Indeed, the case where $|f'(0)| = 1$ (resp. $|f'(0)| < 1$ and $g(0) = 0$) is dealt with in Section 4.3 (resp. 4.4). On the other hand, if $|f'(0)| > 1$ (resp. $|f'(0)| < 1$ and $g(0) \neq 0$), then on replacing in (4.9.1) f by f^{-1} (resp. on dividing (4.9.1) by $g(x)$) we obtain an equation of the form (4.2.6), to which the results of Section 4.2 apply.

4.9.7. Meromorphic solutions of (4.9.1) are studied in Section 4.5 under the assumption $0 < |f'(0)| < 1$. The case where $f'(0) = 0$ can be reduced to that we discussed in Section 4.4. The argument is similar to that employed in the proofs of Theorems 4.5.1 and 4.5.2. The formulation of results, although lengthy, presents no essential difficulties; see Matkowski [2].

4.9.8. Let λ and μ be σ-finite measures on $\mathbf{S}$. The function $f: X \to \mathbb{R}$, measurable and integrable with respect to μ, such that

$$\lambda(A) = \int_A f \, d\mu \qquad \text{for } A \in \mathbf{S} \qquad (4.9.3)$$

is said to be the *Radon–Nikodym derivative*, denoted by $d\lambda/d\mu$, of the measure λ with respect to the measure μ. It actually exists if λ is absolutely continuous with respect to μ.

4.9.9. We supply a proof of the relation

$$\frac{d(\mu f)}{d\mu} \circ f^i = \frac{d(\mu f^{i+1})}{d(\mu f^i)} \qquad \text{a.e. in } X, \ i \in \mathbb{N}_0 \qquad (4.9.4)$$

(see (4.7.4)), where the measure μ and the function f are as in Subsection 4.7A.

The following equalities are valid for arbitrary $A \in \mathbf{S}$ (see (4.9.3)):

$$\int_A \left(\frac{d(\mu f)}{d\mu} \circ f^i \right) d(\mu f^i) = \int_{f^i(A)} \frac{d(\mu f)}{d\mu} \, d\mu$$

$$= \mu(f^{i+1}(A)) = (\mu f^{i+1})(A)$$

$$= \int_A \frac{d(\mu f^{i+1})}{d(\mu f^i)} \, d(\mu f^i).$$

Now (4.9.4) follows in view of arbitrariness of the set A.

4.9.10. The following is a version of Theorem 4.7.5.

Proposition 4.9.4. *Let hypotheses* (i)–(iii) *in Section 4.7 be fulfilled. If there exists an* $s \in (0, 1)$ *such that*

$$\left| g(x) \frac{h(f(x))}{h(x)} \right| \leqslant s \qquad \text{a.e. in } X, \tag{4.9.5}$$

then the equation

$$\varphi(x) = g(x)\varphi(f(x)) + h(x) \tag{4.9.6}$$

has a solution $\varphi \in L^p$ *given by formula* (4.7.18). *This solution is essentially unique if the series* (4.7.11) *diverges, whereas the solution* $\varphi \in L^p$ *of* (4.9.6) *can be prescribed arbitrarily on* A_0 *(with condition* (4.7.16) *for* $\varphi_0 = \varphi | A_0$*) if the series* (4.7.15) *converges.*

(Relation (4.9.5) yields the estimate

$$\left\| \left(\prod_{i=0}^{n-1} g \circ f^i \right) \cdot (h \circ f^n) \right\| \leqslant s^n \| h \|$$

and the proof follows the lines of that of Theorem 4.7.4. See Matkowski [13].)

4.9.11. As a particular case of Theorem 5.8.8 (for the nonlinear equation) we obtain the following.

Proposition 4.9.5. *Let hypotheses* (i)–(iii) *in Section 4.7 be fulfilled. If there exists an* $s > 1$ *such that*

$$|g|^p \geqslant s \frac{d(\mu f)}{d\mu} \qquad \text{a.e. in } X,$$

and the series

$$\sum_{n=0}^{\infty} \left(\int_{A_n} \left| \frac{h \circ f^{-1}}{g \circ f^{-1}} \right|^p d\mu \right)^{\alpha(p)}$$

converges, then every measurable function $\varphi_0 \in L^p(A_0)$ *can be uniquely extended to an* L^p *solution* $\varphi \colon X \to \mathbb{R}$ *of equation* (4.9.6).

4.9.12. It is a peculiarity of the theory of integrable solutions of iterative functional equations that it never yields a finite parameter family of solutions. The integrable solution either is unique, or depends on an arbitrary function. We illustrate this feature on the following.

Example 4.9.1. Let $X = [0, 1]$ and let μ be the Lebesgue measure. Consider the equation

$$\varphi(x) = \frac{1}{1+x} \varphi \left(\frac{x}{1+x} \right). \tag{4.9.7}$$

Write for short $a_n := (1+n)^{-1}$ and $b_n = a_{n+1}/a_n$, $n \in \mathbb{N}_0$. We may see that for $f(x) = x/(1+x)$ (see (4.7.3)) one gets

$$A_n = f^n(X) \setminus f^{n+1}(X) = (a_{n+1}, a_n], \qquad n \in \mathbb{N}_0.$$

The extreme values on the interval A_n of the function $|g(x)|^{-p}f'(x) = (x+1)^{p-2}$, i.e. u_n and v_n from (4.7.8), are also easily found:

$$u_n = b_n^{p-2}, \qquad v_n = b_{n+1}^{p-2} \qquad \text{for } p < 2,$$

$$u_n = b_{n+1}^{p-2}, \qquad v_n = b_n^{p-2} \qquad \text{for } p \geq 2.$$

Thus the series (4.7.11) diverges if $p \geq 1$, whereas the series (4.7.15) converges if $p < 1$. Consequently equation (4.9.7) has in the class $L^p(X)$ a solution depending on an arbitrary function if $p < 1$, and the unique solution $\varphi = 0$ if $p \geq 1$ (Theorems 4.7.3 and 4.7.1).

4.9.13. Consider, for $x \in \mathbb{R}$, the functional equation

$$\varphi(x) = a\varphi(bx) + h(x), \tag{4.9.8}$$

where $0 < a < 1$, $0 < b \neq 1$, and h is a bounded function on $\mathbb{R}$. By Theorem 4.1.1 equation (4.9.8) has the unique bounded solution $\varphi : \mathbb{R} \to \mathbb{R}$, $\varphi(x) = \sum_{n=0}^{\infty} a^n h(b^n x)$ (see 4.1.2). Clearly, φ is continuous. Take here $h(x) = \cos x$, b an odd integer and $ab > 1 + \frac{3}{2}\pi$, $0 < a < 1$, to get the Weierstrass c.n.d. functions (see Kuczma [26, pp. 81–2]).

In fact, the unique continuous solution of most (in the Baire category sense) equations (4.9.8) (with bx replaced by $f(x)$) is a c.n.d. function (see Kuczma [53]). C.n.d. functions are dealt with in Section 10.5.

4.9.14. The functional equation

$$\zeta(1 - x) = 2(2\pi)^{-x} \cos(\tfrac{1}{2}\pi x)\Gamma(x)\zeta(x) \tag{4.9.9}$$

(Γ is the Euler gamma function, see Section 10.4) is known since 1768 (Euler [1]). The proof that the Riemann zeta function, defined for Re $x > 1$ by

$$\zeta(x) = \sum_{n=1}^{\infty} n^{-x}$$

and then continued as a meromorphic function to the whole $\mathbb{C} \setminus \{1\}$, satisfies (4.9.9) was first given by Riemann in 1859.

(For a characterization of ζ with the aid of equation (4.9.9) see Kuczma [26, § 8.6]. A number of other functional equations also arise in the analytic number theory; see the references in Kuczma [26, p. 194].)

4.9.15. The generalized Fox integral equation

$$ap(t) + \int_{\mathbb{R}} F(t - s)p(s)\mathrm{d}s + \int_{\mathbb{R}} G(t - ks)p(s)\mathrm{d}s = q(t), \qquad t \in \mathbb{R}, \tag{4.9.10}$$

where a and k are nonzero constants, is considered by D. Przeworska-Rolewicz [1]. By the Fourier transform

$$\hat{p}(s) := (2\pi)^{-1/2} \int_{\mathbb{R}} \exp(-its)p(t)\mathrm{d}t,$$

equation (4.9.10) is reduced to the functional equation for $\hat{p} \in L^2(\mathbb{R})$:

$$(a + \sqrt{(2\pi)} \cdot \hat{F}(s))\hat{p}(s) + \sqrt{(2\pi)} \cdot \hat{G}(s)\hat{p}(ks) = \hat{q}(s), \qquad |k| \neq 1. \tag{4.9.11}$$

L^2 solutions of the homogeneous equation related to (4.9.11) are found in Przeworska-Rolewicz [1] with the aid of the functional equation

$$\varphi(s) = \varphi(ks) + h(s), \qquad s \in \mathbb{R}, \qquad (4.9.12)$$

where $\varphi := \log \hat{p}$ and $h := \log(-\hat{G}/(\hat{F} + a\sqrt{(2\pi)}))$, provided that $\exp \varphi \in L^2(\mathbb{R})$.

(Equations (4.9.11) and (4.9.12) are special cases of equation (4.9.6), therefore see Subsections 4.7D and 4.7E.)

4.9.16. The integro-differential equation ($n, k \in \mathbb{N}$, $\lambda \in \mathbb{R}$, are fixed)

$$\varphi(t) + \lambda \int_0^\infty J_n(2\sqrt{(tx)})(t/u)^{n/2} \varphi^{(k)}(u)\, du = h(t), \qquad (4.9.13)$$

where J_n is the Bessel function of the first order, was studied by I. Fenyő [1].

Equation (4.9.13) is reduced to the cyclic equation for the Laplace transform Φ of φ

$$\Phi(s) + \lambda s^{-p} \Phi(1/s) = H(s) \qquad (4.9.14)$$

with $p = n - k + 1$. The general solution of equation (4.9.14) was found on a set $D \subset \mathbb{K}$ which is closed under taking reciprocals. When $D = (0, \infty)$ this yields the general solution of the equation (4.9.13).

4.9.17. Under the assumptions of Theorem 4.3.4 the equation $\omega(f(x)) = \omega(x)$ has only constant LASs, $\omega = \omega(0)$. For if $\omega(x) - \omega(0) = x^k \Omega(x)$, $k \in \mathbb{N}$, $\Omega(0) \neq 0$, then $(f(x)/x)^k \Omega(f(x)) = \Omega(x)$, whence $s^k \Omega(0) = \Omega(0)$ as $x \to 0$. Since $s^k \neq 1$, we have $\Omega(0) = 0$, a contradiction.

4.9.18. We supply calculations to complete the proof of Theorem 4.8.1 in the case $|s| = 1$, i.e. to show that Theorem 4.7.5 is applicable there.

Since μ is the Lebesgue measure, we have $d(\mu f)/d\mu = f'$. We look for L^1 solutions of equation (4.8.6) in the interval $[0, x] \subset X$ with a fixed x. Thus the thing to check is the convergence of the series (4.7.19) for $\alpha(p) = p = 1$ (see (4.7.6)) and $f^n(X)$ replaced by $f^n([0, x]) = [0, f^n(x)]$. Equation (4.8.6) is of form (4.7.1) ($= (4.9.6)$) with $g = sf'$. Thus $w_n = |s| = 1$, $n \in \mathbb{N}$ (see (4.7.6)), and the series (4.7.19) reduces to

$$\sum_{n=0}^\infty \int_{[0, f^n(x)]} |h'(t)|\, dt = \sum_{n=0}^\infty \text{Var}\, h|[0, f^n(x)],$$

as h is assumed to be absolutely continuous on X. By the assumption (4.8.3) the latter series converges at an $x_0 \in X \setminus \{0\}$. In view of (ii) in Section 4.8, it converges for every $x \in X$. Thus Theorem 4.7.5 works, as claimed on p. 178.

5

Theory of nonlinear equations

5.0 Introduction

Most results obtained in Chapters 3 and 4 for linear equations can be extended to the case of the equations

$$\varphi(x) = h(x, \varphi(f(x))) \tag{5.0.1}$$

or

$$\varphi(f(x)) = g(x, \varphi(x)) \tag{5.0.2}$$

with a nonlinear function h, resp. g.

In this chapter we shall be concerned mainly with continuous, differentiable, analytic and integrable solutions φ of equations (5.0.1) or (5.0.2). The case of φ mapping real intervals into reals is considered in the first six sections. Next come local analytic solutions in a complex domain. Finally, in Section 5.8 the domain is a measure space.

The case of more general spaces, both in the domain and in the range of φ, and of equations of an arbitrary order is discussed in Chapter 7.

For some applications of nonlinear equations the reader is referred to Chapters 8 and 10.

5.1 An extension theorem

In the linear case of the equation

$$\varphi(x) = h(x, \varphi(f(x))) \tag{5.1.1}$$

we had $h(x, y) = \tilde{g}(x)y + \tilde{h}(x)$, and the function h was defined on the strip $X \times \mathbb{R}$, where X was the interval of definition of the functions $\tilde{g}$ and $\tilde{h}$. In the nonlinear case the domain of h need not be so simple. Let $\Omega \subset \mathbb{K}^2$, $X, Y \subset \mathbb{K}$ be some sets. Let h map Ω into Y and let, for a fixed x, Ω_x denote the section

$$\Omega_x := \{y \colon (x, y) \in \Omega\}.$$

Definition 5.1.1. By a *solution* of equation (5.1.1) in X we mean a function $\varphi: X \to Y$ satisfying (5.1.1) for all $x \in X$ and such that for every $x \in X$ we have $\varphi(f(x)) \in \Omega_x$.

If φ is a solution of (5.1.1), then $\varphi(f(x)) \in h(f(x), \Omega_{f(x)})$. We will always assume that

$$h(f(x), \Omega_{f(x)}) \subset \Omega_x \qquad \text{for every } x \in X. \tag{5.1.2}$$

Condition (5.1.2) guarantees that we can insert $\varphi(f(x))$ in place of y in $h(x, y)$ and in this way to construct the solution φ by a repeated use of (5.1.1).

Assuming that 0 is a fixed point of f, i.e. $f(0) = 0$, and putting $x = 0$ in (5.1.1), we obtain for $d = \varphi(0)$ the condition

$$d = h(0, d). \tag{5.1.3}$$

If $h(x, y) = \tilde{g}(x)y + \tilde{h}(x)$, then (5.1.3) either has a unique solution or none or it is fulfilled by all $d \in \mathbb{K}$. In the nonlinear case this is no longer true. Therefore every theorem on solutions φ of (5.1.1) fulfilling the condition $\varphi(0) = d$ will refer to a specified solution d of (5.1.3). For another d and the corresponding solution of (5.1.1) usually one has to consult another theorem. This will be explained in Note 5.9.2.

What follows is a useful extension theorem which is essentially due to J. Kordylewski [3]. We make the following hypotheses.

(i) $X = |0, a|, 0 < a \leq \infty$.
(ii) $f: X \to X$ is continuous, $0 < f(x) < x$ in $X \setminus \{0\}$.
(iii) $h: \Omega \to \mathbb{R}$, where $\Omega \subset \mathbb{R}^2$, satisfies (5.1.2). Moreover, $\Omega_x \neq \varnothing$ for every $x \in X$.

Let $X_b := |0, b), 0 < b < a$.

Theorem 5.1.1. *Let hypotheses* (i)–(iii) *be fulfilled and let* $\varphi_b: X_b \to \mathbb{R}$ *be a solution of equation* (5.1.1) *in* X_b. *Then there exists a unique extension* $\varphi: X \to \mathbb{R}$ *of* φ_b *onto* X *satisfying equation* (5.1.1) *in* X.

(a) *If* h *and* φ_b *are continuous in their domains, then* φ *is continuous in* X.

(b) *If* h *is of class* C^r, $1 \leq r \leq \infty$, *in* Ω, φ_b *is of class* C^r *in* X_b, *and for every* $x \in X \setminus \{0\}$ *the set* Ω_x *is a nondegenerate interval, then* φ *is of class* C^r *in* X.

(c) *If* f *and* φ_b *are increasing in* X *and* X_b, *resp., and* h *is increasing with respect to either variable in* Ω, *then* φ *is increasing in* X.

(d) *If* Ω *is convex,* f *and* φ_b *are increasing and convex in* X *and* X_b, *resp., and* h *is increasing with respect to either variable and convex in* Ω, *then* φ *is increasing and convex in* X.

Proof. First we need to replace f by a function F such that

$$f(x) \leq F(x) \qquad \text{for } x \in X \tag{5.1.4}$$

which is increasing in X and also satisfies hypothesis (ii). It can be seen that the following function $F: X \to \mathbb{R}$ will do:

$$F(x) := \sup_{(0,x]} f(t), \qquad x \in X \setminus \{0\}, \ F(0) = 0 \text{ if } 0 \in X.$$

Choose an $x_0 \in X_b \setminus \{0\}$ and write $x_{n+1} := \sup\{x \in X: F(x) < x_n\}$, $n \in \mathbb{N}_0$. It follows that

$$F(x) < x_n \qquad \text{for } x < x_{n+1}, \ x \in X, \ n \in \mathbb{N}_0, \tag{5.1.5}$$

since F is continuous and increasing. The sequence $(x_n)_{n \in \mathbb{N}_0}$ is increasing and converges to a. (It may happen that $x_n = a$ from some n on.) Introducing $X_0 := X \cap [0, x_0] \subset X_b$, and $X_n := (x_{n-1}, x_n]$, or (x_{n-1}, x_n) if $x_n = a$ and $a \notin X$, $n \in \mathbb{N}$, we have

$$X = \bigcup_{n=0}^{\infty} X_n. \tag{5.1.6}$$

We put $\varphi(x) = \varphi_b(x)$ for $x \in X_0$ and define φ inductively on X_n, $n \in \mathbb{N}$, by formula (5.1.1). The definition is correct, as (5.1.4) and (5.1.5) imply $f(x) \in X_0 \cup \cdots \cup X_{n-1}$ for $x \in X_n$. It follows by (5.1.6) that φ is well defined on X. This construction shows that φ satisfies (5.1.1) in X, as φ_b satisfies (5.1.1) in X_b, and φ coincides with φ_b on X_b.

The uniqueness of the extension is obvious.

The assertions (a)–(d) are also easily verified. ∎

Comments. Some results concerning the case where condition (5.1.2) does not hold may be found in Kuczma–Vopěnka [1] and Krzeszowiak-Dybiec [1]–[4].

5.2 Existence and uniqueness of continuous solutions

We continue to deal with the equation

$$\varphi(x) = h(x, \varphi(f(x))). \tag{5.2.1}$$

First we consider the case where h fulfils a Lipschitz condition. Other conditions implying existence and/or uniqueness of continuous solutions of (5.2.1) are shown in the next two sections.

5.2A. Lipschitzian h

We start with a theorem on bounded solutions of (5.2.1), where the function h satisfies the condition

$$|h(x, y) - h(x, \bar{y})| \le \Theta |y - \bar{y}|, \qquad 0 < \Theta < 1, \tag{5.2.2}$$

in a strip.

Theorem 5.2.1. *Let* $X = |0, a|$, $0 < a \le \infty$, *and let* f *be a self-mapping of* X.

Moreover, assume that $h: X \times \mathbb{R} \to \mathbb{R}$ *is bounded on the rectangles* $X \times [-\delta, \delta]$, $\delta > 0$ *(the bound may depend on* δ*) and it fulfils the Lipschitz condition* (5.2.2) *for* $x \in X$, y, $\bar{y} \in \mathbb{R}$. *Then equation* (5.2.1) *has a unique bounded solution* $\varphi: X \to \mathbb{R}$.

Proof. Let Φ be the Banach space of all bounded functions $\varphi: X \to \mathbb{R}$ with the usual sup norm. Equation (5.2.1) induces a self-mapping of Φ which, due to (5.2.2), is a contraction. By Banach's Theorem equation (5.2.1) has a unique solution $\varphi \in \Phi$. ∎

In the sequel we shall be interested in those solutions of (5.2.1) which are continuous at least at the origin. Therefore we make the following assumptions.

(i) $X = [0, a|$, $0 < a \leq \infty$.
(ii) $f: X \to X$ is continuous, $0 < f(x) < x$ in $X \setminus \{0\}$.
(iii) $h: X \times Y \to Y$ fulfils condition (5.2.2) for $x \in X$ and y, $\bar{y} \in Y$, where $Y \subset \mathbb{R}$ is an interval containing a d such that $d = h(0, d)$. Moreover, the function $h(\cdot, d)$ is continuous at $x = 0$.

The following theorem has been proved by J. Matkowski [1].

Theorem 5.2.2. *Let hypotheses* (i)–(iii) *be fulfilled. Then equation* (5.2.1) *has a unique solution* $\varphi: X \to \mathbb{R}$ *fulfilling the condition* $\varphi(0) = d$ *and continuous at the origin.*

(a) *If h is continuous in* $X \times Y$, *then* φ *is continuous in* X.

(b) *If f is increasing in* X, *and h is increasing (with respect to either variable) in* $X \times Y$, *then* φ *is increasing in* X.

(c) *If, in addition to the hypotheses of* (b), *for every* $y \in Y$ *the function* $h(\cdot, y)$ *is strictly increasing in* X, *then* φ *is strictly increasing in* X.

(d) *If f is increasing and convex in* X *and h is increasing (with respect to either variable) and convex in* $X \times Y$, *then* φ *is increasing and convex in* X.

Proof. Choose a $c \in (0, a)$ and a $\delta > 0$ such that
$$|h(x, d) - d| \leq (1 - \Theta)\delta \qquad \text{for } x \in [0, c].$$
Let Φ be the metric space of all functions $\varphi: [0, c] \to Y$ which are continuous at $x = 0$, and such that $\varphi(0) = d$ and $|\varphi(x) - d| \leq \delta$ in $[0, c]$. The metric in Φ is the usual one:
$$\rho(\varphi_1, \varphi_2) = \sup_{[0,c]} |\varphi_1(x) - \varphi_2(x)|.$$
Let T be the map defined for $\varphi \in \Phi$ by
$$T(\varphi)(x) = h(x, \varphi(f(x))), \qquad x \in [0, c].$$
If $\varphi \in \Phi$ and $\psi := T(\varphi)$, then, first of all, $\psi(x) \in Y$ for $x \in [0, c]$, by (iii). Further,

$\psi(0) = d$ by (5.1.3), ψ is continuous at zero, and

$$|\psi(x) - d| \leq |h(x, \varphi(f(x))) - h(x, d)| + |h(x, d) - d|$$
$$\leq \Theta|\varphi(f(x)) - d| + (1 - \Theta)\delta \leq \delta.$$

Thus T is a self-mapping of Φ, and in virtue of (5.2.2) T is a contraction. The existence and uniqueness of the desired solution φ of (5.2.1) result from Banach's Theorem and from Theorem 5.1.1.

Constructing suitable subspaces of Φ for the cases (b) and (d) we similarly get the assertions. To prove (c), observe that, by (b), the solution φ is increasing in X. Hence, for any $x_1 < x_2$ taken from X, we have $\varphi(f(x_1)) \leq \varphi(f(x_2))$. Now, the strict monotonicity of $h(\cdot, y)$ implies

$$\varphi(x_1) = h(x_1, \varphi(f(x_1))) < h(x_2, \varphi(f(x_1))) \leq h(x_2, \varphi(f(x_2))) = \varphi(x_2).$$

We see that φ is strictly increasing. ∎

Now replace condition (iii) by the following one.

(iv) $h: \Omega \to \mathbb{R}$ is continuous, $\Omega_x \neq \varnothing$ for $x \in X$, and h satisfies (5.1.2). Moreover, there exist d satisfying (5.1.3), $c \in (0, a)$ and $\delta > 0$ such that $R := [0, c] \times [d - \delta, d + \delta]$ is contained in Ω, and h fulfils condition (5.2.2) in R.

Remark 5.2.1. The rectangle R may also be of the form $R = [0, c] \times [d - \delta, d]$ (or $[0, c] \times [d, d + \delta]$) if $h(x, y) \leq d$ (resp. $h(x, y) \geq d$) for $(x, y) \in \Omega$.

Theorem 5.2.3. *Let hypotheses* (i), (ii) *and* (iv) *be fulfilled. Then equation* (5.2.1) *has a unique solution* $\varphi: X \to \mathbb{R}$ *continuous on* X *and such that* $\varphi(0) = d$.

Proof. Taking a smaller c, if necessary, we may have $|h(x, d) - d| \leq (1 - \Theta)\delta$ for $x \in [0, c]$. This, together with (5.2.2), yields $|h(x, y) - d| \leq \delta$ for $(x, y) \in R$. Now, the theorem follows from Theorem 5.2.2. applied to $[0, c] \times Y$, where $Y := [d - \delta, d + \delta]$, and from Theorem 5.1.1(a). ∎

Remark 5.2.2. The solution from Theorem 5.2.3 can be obtained as the limit of the sequence of successive approximations

$$\varphi(x) = \lim_{n \to \infty} \varphi_n(x), \tag{5.2.3}$$

where $\varphi_0: X \to \mathbb{R}$ is an arbitrary continuous function such that $\varphi_0(f(x)) \in \Omega_x$ for $x \in X$ and $\varphi_0(0) = d$, and

$$\varphi_{n+1}(x) = h(x, \varphi_n(f(x))), \qquad n \in \mathbb{N}_0. \tag{5.2.4}$$

Formulae (5.2.3) and (5.2.4) for x sufficiently close to zero result from Banach's Theorem, and then it is easily checked that they hold in the whole of X.

5.2B. Continuous dependence on the data

The solution obtained in Theorem 5.2.3 depends continuously on the given functions. Namely, for the sequence of equations

$$\varphi_m(x) = h_m(x, \varphi_m(f_m(x))), \qquad m \in \mathbb{N}_0, \tag{5.2.5}$$

we have the following result, the first version of which is found in Kordylewski–Kuczma [5].

Theorem 5.2.4. *Let hypothesis* (i) *be fulfilled, and assume that*

$$f_m: X \to X, \; m \in \mathbb{N}_0, \; \text{are continuous in } X, \; 0 < f_m(x) < x \; \text{in}$$
$$X \backslash \{0\}, \; m \in \mathbb{N}_0,$$

$$h_m: \Omega \to \mathbb{R}, \; m \in \mathbb{N}_0, \; \text{are continuous in } \Omega, \; \Omega_x \neq \varnothing \; \text{for } x \in X,$$

and

$$h_m(f_m(x), \Omega_{f_m(x)}) \subset \Omega_x, \; x \in X.$$

Moreover, there exist d_m *such that* $d_m = h_m(0, d_m)$, $m \in \mathbb{N}_0$, *and there exist* $c \in (0, a)$ *and* $\delta > 0$ *for the rectangles* $R_m := [0, c] \times [d_m - \delta, d_m + \delta]$ *to be contained in* Ω *and all* h_m *fulfil condition* (5.2.2) *in the respective* R_m *with the same* $\Theta < 1$,

$$\lim_{m \to \infty} f_m = f_0 \;\; \text{a.u. in } X, \qquad \lim_{m \to \infty} h_m = h_0 \;\; \text{a.u. in } \Omega \qquad \text{and} \qquad \lim_{m \to \infty} d_m = d_0.$$

Then for every $m \in \mathbb{N}_0$ *equation* (5.2.5) *has a unique continuous solution* $\varphi_m: X \to \mathbb{R}$ *fulfilling the condition* $\varphi_m(0) = d_m$, *and* $\lim_{m \to \infty} \varphi_m = \varphi_0$ *a.u. in* X.

Proof. The existence and uniqueness of φ_m result from Theorem 5.2.3. By introducing $\bar{h}_m(x, y) = h_m(x, y + d_m - d_0) - d_m + d_0$, $\psi_m(x) = \varphi_m(x) - d_m + d_0$, $m \in \mathbb{N}_0$, and replacing equations (5.2.5) by $\psi_m(x) = \bar{h}_m(x, \psi_m(f_m(x)))$, we can make all d_m equal to d_0. The uniform convergence of (ψ_m) to ψ_0 (implying that of (φ_m) to φ_0) in a right neighbourhood of zero results from Theorem 1.6.3. Now arguing as in the proof of Theorem 5.1.1 we check that the convergence holds a.u. in X. ∎

5.2C. Non-Lipschitzian h

Following Czaja-Pośpiech–Kuczma [1] we shall examine continuous solutions of equation (5.2.1) under conditions weaker than (5.2.2). Namely, we assume the following.

(v) The function h fulfils hypothesis (iv) but with condition (5.2.2) replaced by

$$|h(x, y) - h(x, \bar{y})| \leq \gamma(x)|y - \bar{y}| \tag{5.2.6}$$

in $\mathbb{R}$, with a $\gamma: [0, c] \to \mathbb{R}^+$.

Write

$$G_n(x) = \prod_{i=0}^{n-1} \gamma(f^i(x)) \qquad \text{for } n \in \mathbb{N}_0, \; x \in [0, c]. \tag{5.2.7}$$

Sequences of this type play an essential role in the theory of linear equations; see Chapter 3. Here we have the following.

Theorem 5.2.5. *Let hypotheses* (i), (ii) *and* (v) *be fulfilled. If for every* $x \in [0, c]$ *sequence* (5.2.7) *is bounded, then there may exist at most one solution* $\varphi: X \to \mathbb{R}$ *of equation* (5.2.1) *which is continuous at* $x = 0$ *and fulfils* $\varphi(0) = d$.

Proof. Let φ_1, φ_2, both from X into $\mathbb{R}$, be solutions of equation (5.2.1) continuous at $x = 0$ and such that $\varphi_1(0) = \varphi_2(0) = d$. By the continuity we may choose c to have $|\varphi_1(x) - d| \leq \delta$ and $|\varphi_2(x) - d| \leq \delta$ in $[0, c]$. Thus we obtain from (5.2.1), (5.2.6) and (5.2.7) by induction (for $n \in \mathbb{N}$ and $x \in [0, c]$)

$$|\varphi_1(x) - \varphi_2(x)| \leq G_n(x)|\varphi_1(f^n(x)) - \varphi_2(f^n(x))|.$$

On letting $n \to \infty$ we get from this $\varphi_1(x) = \varphi_2(x)$ in $[0, c]$. It follows from Theorem 5.1.1 that $\varphi_1 = \varphi_2$. ∎

Remark 5.2.3. If the sequence (G_n) is uniformly bounded in $[0, c]$ –

$$G_n(x) \leq M \qquad \text{for } x \in [0, c], \; n \in \mathbb{N}_0$$

– and there exists the solution φ spoken of in Theorem 5.2.5, then it can be obtained as limit (5.2.3) of the sequence of successive approximations (5.2.4), where φ_0 is such as described in Remark 5.2.2.

In fact, assume that c has been chosen so small that for $x \in [0, c]$ we have $|\varphi(x) - d| < \delta/(M + 1)$ and $|\varphi(x) - \varphi_0(x)| < \delta/(M + 1)$. Then it follows easily by induction that $|\varphi_n(x) - d| < \delta$ and $|\varphi(x) - \varphi_n(x)| \leq G_n(x)|\varphi(f^n(x)) - \varphi_0(f^n(x))|$ for $x \in [0, c]$ and $n \in \mathbb{N}$. Hence we see that (5.2.3) holds in $[0, c]$, and by (5.2.1) and (5.2.4) it holds also in the whole of X.

Now we shall give some conditions for the existence of a continuous solution φ of (5.2.1).

Let us write for short

$$H(x) := |h(x, d) - d|.$$

Theorem 5.2.6. *Let hypotheses* (i), (ii) *and* (v) *be fulfilled. If for every* $x \in [0, c]$ *sequence* (5.2.7) *is bounded, and if the series*

$$K(x) := \sum_{n=0}^{\infty} G_n(x) H(f^n(x)) \tag{5.2.8}$$

uniformly converges in $[0, c]$, *then equation* (5.2.1) *has a unique continuous solution* $\varphi: X \to \mathbb{R}$ *such that* $\varphi(0) = d$.

Proof. The sum $K: [0, c] \to \mathbb{R}^+$ of series (5.2.8) is a continuous function which, by (5.1.3), vanishes at $x = 0$. Thus we may assume, diminishing c, if necessary, that $K(x) < \delta$ for $x \in [0, c]$.

Write $\varphi_0(x) = d$ and define $\varphi_n \colon [0, c] \to \mathbb{R}$, $n \in \mathbb{N}$, by (5.2.4). We will show that

$$|\varphi_n(x) - d| < \delta \qquad \text{for } x \in [0, c],\ n \in \mathbb{N}_0, \tag{5.2.9}$$

$$|\varphi_{n+1}(x) - \varphi_n(x)| \leq G_n(x) H(f^n(x)) \qquad \text{for } x \in [0, c],\ n \in \mathbb{N}_0. \tag{5.2.10}$$

Both inequalities are obvious for $n = 0$. Assuming them to hold for $n = 0, \ldots, m$, we have for $x \in [0, c]$

$$|\varphi_{m+1}(x) - d| \leq \sum_{n=0}^{m} |\varphi_{n+1}(x) - \varphi_n(x)| \leq K(x) < \delta$$

whence, by (5.2.7),

$$
\begin{aligned}
|\varphi_{m+2}(x) - \varphi_{m+1}(x)| &= |h(x, \varphi_{m+1}(f(x))) - h(x, \varphi_m(f(x)))| \\
&\leq \gamma(x)|\varphi_{m+1}(f(x)) - \varphi_m(f(x))| \leq \gamma(x) G_m(f(x)) H(f^{m+1}(x)) \\
&= G_{m+1}(x) H(f^{m+1}(x)).
\end{aligned}
$$

Relation (5.2.10) implies the uniform convergence of the sequence (φ_m) to a continuous function $\varphi \colon [0, c] \to \mathbb{R}$. By (5.2.9) we have

$$\varphi(f(x)) \in [d - \delta, d + \delta] \subset \Omega_x \qquad \text{for } x \in [0, c].$$

It follows from (5.2.4) that φ satisfies equation (5.2.1) in $[0, c]$. The theorem results now from Theorem 5.2.5 and Theorem 5.1.1. ∎

Remark 5.2.4. Also in this case the solution φ may be obtained as the limit of the sequence of successive approximations (5.2.4). As the initial term of the sequence we can take any continuous function $\varphi_0 \colon X \to \mathbb{R}$ satisfying $\varphi_0(f(x)) \in \Omega_x$ for $x \in X$ and $\varphi_0(x) = d$ in a right neighbourhood of zero.

5.2D. Existence via solutions of inequalities

The method and the results we present here are due to K. Baron [1]. We introduce the functional inequalities

$$\varphi_1(x) \leq h(x, \varphi_1(f(x))) \tag{5.2.11}$$

and

$$\varphi_2(x) \geq h(x, \varphi_2(f(x))) \tag{5.2.12}$$

and we assume the following.

(vi) Inequality (5.2.11) has a lower semicontinuous solution $\varphi_1^* \colon [0, c] \to \mathbb{R}$, and inequality (5.2.12) has an upper semicontinuous solution $\varphi_2^* \colon [0, c] \to \mathbb{R}$. Moreover, $\varphi_1^*(x) \leq \varphi_2^*(x)$ in $[0, c]$.

(vii) For every $x \in [0, c]$ we have

$$I(x) := [\varphi_1^*(f(x)), \varphi_2^*(f(x))] \subset \Omega_x,$$

and d (occurring in (v)) belongs to $I(0)$ and is the unique solution of equation (5.1.3) in this interval. Moreover, for every $x \in [0, c]$ the function $h(x, \cdot)$ is increasing in $I(x)$.

Theorem 5.2.7. *Let hypotheses* (i), (ii), (v)–(vii) *be fulfilled. If for every* $x \in [0, c]$ *sequence* (5.2.7) *is bounded, then equation* (5.2.1) *has a unique continuous solution* $\varphi: X \to \mathbb{R}$ *such that* $\varphi(0) = d$.

Proof. Define two sequences of functions $\varphi_{jn}: [0, c] \to \mathbb{R}, j = 1, 2, n \in \mathbb{N}_0$, by the formulae

$$\varphi_{j0}(x) = \varphi_j^*(x), \qquad \varphi_{j,n+1}(x) = h(x, \varphi_{jn}(f(x))), \qquad j = 1, 2, n \in \mathbb{N}_0.$$

We get an increasing sequence (φ_{1n}) of lower semicontinuous functions, and a decreasing sequence (φ_{2n}) of upper semicontinuous functions. Moreover, $\varphi_1^*(x) \leq \varphi_{jn}(x) \leq \varphi_2^*(x)$ for $n \in \mathbb{N}_0$, $x \in [0, c]$, $j = 1, 2$. Hence there exist the limits

$$\varphi_j(x) = \lim_{n \to \infty} \varphi_{jn}(x), \qquad j = 1, 2,$$

the first of which is lower, and the second upper, semicontinuous, and $\varphi_1^*(x) \leq \varphi_j(x) \leq \varphi_2^*(x)$ for $x \in [0, c]$, $j = 1, 2$. Further, φ_j satisfy (5.2.1) in $[0, c]$, and

$$\varphi_1(0) = \varphi_2(0) = d, \tag{5.2.13}$$

since d is unique. We shall show that

$$a_0 := \limsup_{x \to 0} \varphi_1(x) \leq d \leq \liminf_{x \to 0} \varphi_2(x). \tag{5.2.14}$$

In fact, there exist numbers $x_n \in [0, c]$ such that $\lim_{n \to \infty} x_n = 0$ and $\lim_{n \to \infty} \varphi_1(x_n) = a_0 \in I(0)$. From the sequence $(\varphi_1(f(x_n)))$ we can choose a convergent subsequence (y_n) to fulfil $\lim_{n \to \infty} y_n = b \leq a_0$. Since φ_1 satisfies (5.2.1), we have $\varphi_1(x_n) = h(x_n, y_n)$, whence $a_0 = h(0, b) \leq h(0, a_0)$. Putting $a_{n+1} := h(0, a_n), n \in \mathbb{N}_0$, we obtain by induction $a_n \leq a_{n+1} \in I(0), n \in \mathbb{N}_0$, and hence the sequence (a_n) converges to a solution of equation (5.1.3) belonging to $I(0)$. Thus $\lim_{n \to \infty} a_n = d$, which implies $a_0 \leq d$. The second inequality in (5.2.14) follows in a similar way.

Relations (5.2.13), (5.2.14) and the semicontinuity of φ_1 and φ_2 imply that they are both continuous at zero. By Theorem 5.2.5 we have $\varphi_1 = \varphi_2$ in $[0, c]$. Thus $\varphi = \varphi_1 = \varphi_2$ is a continuous (as both lower and upper semicontinuous) solution of equation (5.2.1) in $[0, c]$. The existence of the desired solution $\varphi: X \to \mathbb{R}$ results from Theorem 5.1.1, and the uniqueness from Theorem 5.2.5. ∎

As a corollary we obtain the following result.

Theorem 5.2.8. *Let hypotheses* (i), (ii) *and* (v) *be fulfilled. If for every* $x \in [0, c]$ *sequence* (5.2.7) *is bounded, the function* $h(x, \cdot)$ *is increasing in* $[d - \delta, d + \delta]$, *and if*

$$|h(0, y) - d| < |y - d| \qquad \text{for } y \in [d - \delta, d + \delta], y \neq d, \tag{5.2.15}$$

then equation (5.2.1) has a unique continuous solution $\varphi \colon X \to \mathbb{R}$ *such that* $\varphi(0) = d$.

Proof. By (5.2.15) the functions $\varphi_1^*(x) = d - \delta$ and $\varphi_2^*(x) = d + \delta$ are continuous solutions of inequalities (5.2.11) and (5.2.12), respectively, and d is the unique root of equation (5.1.3) in the interval $[d - \delta, d + \delta]$. In the case where $R = [0, c] \times [d - \delta, d]$, resp. $[0, c] \times [d, d + \delta]$ (see Remark 5.2.1) we take $\varphi_2^*(x) = d$, resp. $\varphi_1^*(x) = d$. Thus the theorem actually results from Theorem 5.2.7. ∎

Remark 5.2.5. We may observe that the solution obtained in Theorems 5.2.6–5.2.8 is increasing if we assume the functions f, h and (in Theorem 5.2.7) φ_1^*, φ_2^* to be increasing in their domains, the h with respect to either variable. This results from the fact that in $[0, c]$ this solution is given as the limit of sequence (5.2.4) of increasing functions, and from Theorem 5.1.1 on extending solutions. The same is true if we replace the word 'increasing' above by 'increasing and convex'.

Remark 5.2.6. Hypothesis (ii) implies that X is the domain of attraction of the unique fixed point 0 of f. Is this property of f essential for the existence of continuous solutions to equation (5.2.1)? This problem is considered by W. Jarczyk [8]. In particular, Corollary 2 from this paper implies the following.

Let $X = [0, a]$, $f \colon X \to X$ be continuous, $f(0) = 0$, $f(x) \neq x$ for $x \in X \setminus \{0\}$. Fix a $d \in \mathbb{R}$ and consider the set $\mathcal{H}$ of all continuous functions $h \colon X \times \mathbb{R} \to \mathbb{R}$ satisfying $h(0, d) = d$. Further, denote by $\mathcal{R}$ the set of those $h \in \mathcal{H}$ such that there exists a continuous solution $\varphi \colon X \to \mathbb{R}$ of equation (5.2.1) such that $\varphi(0) = d$. If $\mathcal{R}$ is of second Baire category in $\mathcal{H}$, then $0 < f(x) < x$ in $X \setminus \{0\}$, i.e. hypothesis (ii) holds true.

Comments. The question whether such properties of continuous solutions of equation (5.2.1) as existence, uniqueness, continuous dependence on the data, convergence of sequences of successive approximations are generic ones for this equation in a suitable space of functions h is discussed in Jarczyk [5]. (By 'generic' we mean a property that holds in the space considered except of a subset of the first Baire category in the space; see Note 3.8.4).

5.3 Continuous solution depending on an arbitrary function

We rewrite equation (5.2.1) in the form

$$\varphi(f(x)) = g(x, \varphi(x)) \tag{5.3.1}$$

which is more convenient to handle in the case of the lack of uniqueness of solutions.

Assuming that the function g is defined in a domain $\Omega \subset \mathbb{R}^2$, we will mean by a solution of equation (5.3.1) a function $\varphi: X \to \mathbb{R}$ satisfying (5.3.1) in X and such that $\varphi(x) \in \Omega_x$ for $x \in X$.

Condition (5.1.2) is now to be replaced by

$$g(x, \Omega_x) \subset \Omega_{f(x)} \tag{5.3.2}$$

which, however, turns out to be insufficient in global theorems. Thus we shall rather use the stronger condition

$$g(x, \Omega_x) = \Omega_{f(x)}. \tag{5.3.3}$$

5.3A. Extension of solutions

Let us fix an $x_0 \in X$ and write $X_0 := [f(x_0), x_0]$. It turns out that, given a suitable function $\varphi_0: X_0 \to \mathbb{R}$ we can construct its extension, using (5.3.1), successively in the intervals $f^k(X_0)$ for $k \in \mathbb{Z}$. This extension is unique and it represents a solution of (5.3.1) in X. In this way one can prove the following extension theorem (see Kordylewski–Kuczma [1]).

Theorem 5.3.1. *Let us assume that $X = (0, a|, 0 < a \leqslant \infty$, and that*

> $f: X \to X$ *is continuous and strictly increasing, $0 < f(x) < x$ in X,*

> $g: \Omega \to \mathbb{R}$ *is continuous, $\Omega_x \neq \varnothing$ for $x \in X$, condition (5.3.3) is fulfilled and every $x \in X$ the function $g(x, \cdot)$ is invertible in Ω_x.*

Then every function $\varphi_0: X_0 \to \mathbb{R}$ such that $\varphi_0(x) \in \Omega_x$ for $x \in X_0$ and

$$\varphi(f(x_0)) = g(x_0, \varphi_0(x_0)) \tag{5.3.4}$$

can be uniquely extended onto X to a solution $\varphi: X \to \mathbb{R}$ of equation (5.3.1). If φ_0 is continuous in X_0, then φ is so in X. Thus equation (5.3.1) has in X a continuous solution depending on an arbitrary function.

Remark 5.3.1. If in Theorem 5.3.1 we replace equality (5.3.3) by inclusion (5.3.2), we shall be able to extend φ_0 only to the left, i.e. onto the interval $(0, x_0]$. The condition of the invertibility of $g(x, \cdot)$ may then be dropped. However, if $a \in X$, then we may choose $x_0 = a$, so that $(0, x_0] = X$.

5.3B. Nonuniqueness theorem

Now we shall consider the case where $0 \in X$. We make the following hypotheses.

(i) $X = [0, a|, 0 < a \leqslant \infty$.

(ii) $f: X \to X$ is continuous and strictly increasing, $0 < f(x) < x$ in $X \setminus \{0\}$.

(iii) $g: \Omega \to \mathbb{R}$ is continuous, $\Omega_x \neq \emptyset$ for $x \in X$, and g fulfils condition (5.3.3). For every $x \in X$ the function $g(x, \cdot)$ is invertible in Ω_x. Moreover, there exist $d \in \Omega_0$ fulfilling

$$d = g(0, d), \qquad\qquad (5.3.4)$$

$c \in (0, a)$ and $\delta > 0$ such that g fulfils the condition

$$|g(x, y) - g(x, \bar{y})| \leqslant \gamma(x)|y - \bar{y}| \qquad\qquad (5.3.6)$$

in $\Omega \cap R$, where $R := [0, c] \times [d - \delta, d + \delta]$, with a $\gamma: [0, c] \to [0, \infty)$ which has a positive lower bound in $[0, c]$.

Remark 5.3.2. The rectangle R may also be of the form $R = [0, c] \times [d - \delta, d]$ (resp. $[0, c] \times [d, d + \delta]$) if $g(x, y) \leqslant d$ (resp. $g(x, y) \geqslant d$) for $(x, y) \in \Omega$.

Define the sequence (G_n) by (5.2.7). The theorem which follows says that there is no uniqueness of continuous solutions of (5.3.1).

Theorem 5.3.2. *Let hypotheses* (i)–(iii) *be fulfilled. If the sequence* (G_n) *tends to zero uniformly on a subinterval* I *of* X, *and equation* (5.3.1) *has a continuous solution* $\varphi: X \to \mathbb{R}$ *such that*

$$\varphi(0) = d, \qquad\qquad (5.3.7)$$

then all such solutions form a family depending on an arbitrary function.

Proof. We are going to show how to construct the extension of an almost arbitrary continuous function to a continuous solution of (5.3.1) on X.

Assume that equation (5.3.1) has a continuous solution $\varphi^*: X \to \mathbb{R}$ fulfilling $\varphi^*(0) = d$. It follows from the relation (see (5.2.7))

$$G_{n+1}(x) = \gamma(x)G_n(f(x))$$

that for every $k \in \mathbb{N}_0$ the sequence (G_n) tends to zero uniformly on $f^k(I)$. Therefore we may find an $x_0 \in (0, c)$, as close to zero as we wish, and an interval $[u, v] \subset X_0 = [f(x_0), x_0]$ such that $\lim_{n \to \infty} G_n(x) = 0$ uniformly on $[u, v]$. Of course, the sequence (G_n) is bounded on $[u, v]$:

$$G_n(x) \leqslant M \qquad \text{for } x \in [u, v],\, n \in \mathbb{N}_0.$$

We may also have $|\varphi^*(x) - d| < \tfrac{1}{2}\delta$ in $[0, x_0]$.

Let $\varphi_0: X_0 \to \mathbb{R}$ be an arbitrary continuous function satisfying (5.3.4) and such that $\varphi_0(x) \in \Omega_x$ for $x \in X_0$ and

$$\varphi_0(x) = \varphi^*(x) \quad \text{in } X_0 \setminus [u, v], \qquad |\varphi_0(x) - \varphi^*(x)| < \delta/2M \quad \text{in } [u, v].$$

By Theorem 5.3.1 this φ_0 can be extended to a continuous solution $\varphi: X \setminus \{0\} \to \mathbb{R}$ of equation (5.3.1). We put $\varphi(0) := d$. Then φ fulfils (5.3.1) in X and it remains to prove that

$$\lim_{x \to 0} \varphi(x) = d. \qquad\qquad (5.3.8)$$

To this end we show at first that for $x \in X_0$

$$\left| \varphi(f^n(x)) - \varphi*(f^n(x)) \right| \leqslant G_n(x) \left| \varphi_0(x) - \varphi*(x) \right|, \qquad n \in \mathbb{N}_0. \quad (5.3.9)$$

This is trivial for $n = 0$. Assuming (5.3.9) to hold for an $n \in \mathbb{N}_0$, we have

$$\left| \varphi(f^n(x)) - d \right| \leqslant \left| \varphi(f^n(x)) - \varphi*(f^n(x)) \right| + \left| \varphi*(f^n(x)) - d \right|$$

$$\leqslant G_n(x) \left| \varphi_0(x) - \varphi*(x) \right| + \frac{\delta}{2} \leqslant M \frac{\delta}{2M} + \frac{\delta}{2} = \delta.$$

Hence we may use (5.3.6) in the estimate

$$\left| \varphi(f^{n+1}(x)) - \varphi*(f^{n+1}(x)) \right| = \left| g(f^n(x)) \varphi(f^n(x))) - g(f^n(x), \varphi*(f^n(x))) \right|$$

$$\leqslant \gamma(f^n(x)) \left| \varphi(f^n(x)) - \varphi*(f^n(x)) \right|$$

$$\leqslant G_{n+1}(x) \left| \varphi_0(x) - \varphi*(x) \right|,$$

and induction completes the proof of (5.3.9).

Given an $\varepsilon > 0$, we can find an $N \in \mathbb{N}$ such that

$$G_n(x) < 2M\varepsilon/\delta \qquad \text{for } x \in [u, v], \, n \geqslant N. \quad (5.3.10)$$

For every $x \in (0, f^N(x_0))$ there is an $\bar{x} \in X_0$ such that $x = f^n(\bar{x})$ with an $n \geqslant N$. Thus by (5.3.9)

$$\left| \varphi(x) - \varphi*(x) \right| \leqslant G_n(\bar{x}) \left| \varphi_0(\bar{x}) - \varphi*(\bar{x}) \right|. \quad (5.3.11)$$

The right-hand side of (5.3.11) is zero if $\bar{x} \in X_0 \setminus [u, v]$, whilst it is less than ε if $\bar{x} \in [u, v]$, by (5.3.10). This means that $\lim_{x \to 0}(\varphi(x) - \varphi*(x)) = 0$, which implies (5.3.8). ∎

5.3C. Existence theorem

We look for conditions ensuring the existence of a continuous solution $\varphi: X \to \mathbb{R}$ of equation (5.3.1). Write

$$F(x) := \left| g(x, d) - d \right|$$

and

$$H_n(x) := \sum_{i=0}^{n-1} \frac{G_n(x)}{G_{i+1}(x)} F(f^i(x)), \qquad n \in \mathbb{N}.$$

As previously, we put $X_0 = [f(x_0), x_0]$.

Theorem 5.3.3. *Let hypotheses* (i)–(iii) *be fulfilled and assume that $R \subset \Omega$. If there exists an $x_0 \in X \setminus \{0\}$ such that*

$$\lim_{n \to \infty} G_n(x) = \lim_{n \to \infty} H_n(x) = 0 \quad (5.3.12)$$

uniformly on X_0, then equation (5.3.1) *has a continuous solution $\varphi: X \to \mathbb{R}$ fulfilling* (5.3.7) *depending on an arbitrary function.*

Proof. This will run as follows. Given a suitable φ_0 defined on X_0 we uniquely extend it, by Theorem 5.3.1, to a solution φ of (5.3.1) on $X \setminus \{0\}$. Then we define $\varphi(0) = d$ and prove the continuity of φ at zero, using the inequality

$$|\varphi(f^n(x)) - d| \leqslant H_n(x) + G_n(x)|\varphi_0(x) - d|, \qquad n \in \mathbb{N}, \qquad (5.3.13)$$

the inductive proof of which will not be given here.

What we need is the possibility of replacing the interval X_0 by another one, arbitrarily close to the origin. Actually, if $\lim_{n \to \infty} G_n(x) = 0$ on X_0, then it is so also on $f^k(X_0)$, $k \in \mathbb{N}$; see the proof of Theorem 5.3.2. To get the same conclusion for (H_n) consider the estimation

$$H_n(f(x)) = H_{n+1}(x) - F(x)G_{n+1}(x)/\gamma(x) \leqslant H_{n+1}(x),$$

whence

$$\sup_{f^k(X_0)} H_n(x) \leqslant \sup_{X_0} H_{n+k}(x), \qquad n, k \in \mathbb{N}.$$

Consequently, the convergence in (5.3.12) is uniform in every interval $f^k(X_0)$, $k \in \mathbb{N}$, too.

Replacing, if necessary, x_0 by $f^k(x_0)$ with $k \in \mathbb{N}$ large enough, we may have $x_0 \in (0, c)$, $H_n(x) < \delta/3$ for $x \in X_0$, $n \in \mathbb{N}$, $G_n(x) \leqslant M$ for $x \in X_0$, $n \in \mathbb{N}_0$, where $M \geqslant 1$, and c and δ are the numbers from (iii).

Let $\varphi_0 \colon X_0 \to \mathbb{R}$ be an arbitrary continuous function fulfilling condition (5.3.4) and such that

$$|\varphi_0(x) - d| \leqslant \delta/2M \qquad \text{for } x \in X_0. \qquad (5.3.14)$$

Just for this φ_0 and its extension φ inequality (5.3.13) holds.

Given an $\varepsilon > 0$, we can find an $N \in \mathbb{N}$ such that for $x \in X_0$ and $n \geqslant N$ we have $H_n(x) < \varepsilon/2$ and $G_n(x) < M\varepsilon/\delta$. For every $x \in I_N := (0, f^N(x_0))$ there is an $\bar{x} \in X_0$ and an $n \geqslant N$ satisfying $x = f^n(\bar{x})$. Hence for $x \in I_N$ inequalities (5.3.13) and (5.3.14) imply

$$|\varphi(x) - d| = |\varphi(f^n(\bar{x})) - d| \leqslant \frac{\varepsilon}{2} + \frac{M\varepsilon}{\delta}\frac{\delta}{2M} = \varepsilon$$

which proves (5.3.8), i.e. the continuity of φ at $x = 0$. ∎

Remark 5.3.3. The proof of Theorem 5.3.3 shows that under the assumptions of that theorem there exist positive constants, c_0, δ_0 such that every continuous function $\varphi_0 \colon X_0 \to \mathbb{R}$, where $x_0 \in (0, c_0)$, fulfilling condition (5.3.4) and such that $|\varphi_0 - d| < \delta_0$ for $x \in X_0$ can be uniquely extended to a continuous solution $\varphi \colon X \to \mathbb{R}$ of equation (5.3.1) satisfying (5.3.7).

Finally, let us note that if in (iii) we have $\gamma = \Theta = \text{const}$ and $\Theta < 1$, then conditions (5.3.12) hold uniformly in a neighbourhood of the origin. Thus Theorem 5.3.3 yields (see Kuczma [26, p. 75]) the following.

Theorem 5.3.4. *Let hypotheses* (i)–(iii) *be satisfied with* $\gamma = \Theta < 1$ *and* $R \subset \Omega$. *Then the continuous solution* $\varphi \colon X \to \mathbb{R}$ *of equation* (5.3.1) *satisfying* (5.3.7) *depends on an arbitrary function.*

5.3D. Comparison with the linear case

If the function g is linear with respect to the second variable, then for a given d fulfilling (5.3.5) there may exist either no continuous solution φ of (5.3.1) in X satisfying (5.3.7), or a unique such solution, or a solution depending on an arbitrary function (see Theorem 2.1.5). The same is true in the nonlinear case under conditions (i)–(iii) (Kuczma [45]).

The example below shows that if we drop the assumption of the invertibility of the function $g(x, \cdot)$, then other situations are also possible.

Example 5.3.1. Take $X = \mathbb{R}^+$, $\Omega = X \times \mathbb{R}$, and define $g: \Omega \to \mathbb{R}$ by the formula

$$g(x, y) = \begin{cases} 2y & \text{for } y \in (-\infty, \tfrac{1}{2}x] \cup [2x, \infty), \\ \tfrac{3}{2}x - y & \text{for } y \in (\tfrac{1}{2}x, x], \\ -3x + \tfrac{7}{2}y & \text{for } y \in (x, 2x). \end{cases} \tag{5.3.15}$$

The function g is continuous in Ω and

$$g(x, y) - g(x, \bar{y}) \geq 2(y - \bar{y}) \qquad \text{for } \bar{y} \leq y \leq \tfrac{1}{2}x \text{ or } x \leq \bar{y} \leq y. \tag{5.3.16}$$

Consider the equation

$$\varphi(\tfrac{1}{2}x) = g(x, \varphi(x)), \tag{5.3.17}$$

with g defined by (5.3.15). Thus $f(x) = \tfrac{1}{2}x$, and conditions (i)–(iii) are fulfilled, except that $g(x, \cdot)$ is not invertible. Equation (5.3.5) becomes $d = 2d$, i.e. $d = 0$, and $\varphi = 0$ and $\varphi = \mathrm{id}$ clearly are continuous solutions of (5.3.17) in X. We will show that these are the only continuous solutions of (5.3.17) in X.

Let $\varphi: X \to \mathbb{R}$ be a continuous solution of equation (5.3.17).

If we had $\varphi(t) < 0$ for a $t \in X$, then $\varphi(\tfrac{1}{2}t) = g(t, \varphi(t)) = 2\varphi(t) < 0$, and by induction $\varphi(2^{-n}t) = 2^n\varphi(t) \to -\infty$ as $n \to \infty$. Thus φ could not be continuous at zero.

If we had $\varphi(t) > t$, then by (5.3.16) with $y := \varphi(t)$, $\bar{y} = t$ we would get

$$\varphi(\tfrac{1}{2}t) = g(t, y) = \tfrac{1}{2}t + (g(t, y) - g(t, t)) \geq \tfrac{1}{2}t + 2(y - t).$$

This yields

$$\varphi(\tfrac{1}{2}t) - \tfrac{1}{2}t \geq 2(\varphi(t) - t) > 0$$

and $\varphi(\tfrac{1}{2}t) > \tfrac{1}{2}t$. Thus (5.3.16) again works and induction shows that

$$\varphi(2^{-n}t) - 2^{-n}t \geq 2^n(\varphi(t) - t) \to \infty$$

as n does, and φ cannot be continuous at zero, too.

If we had $\tfrac{1}{4}t < \varphi(t) < t$, then $\varphi(\tfrac{1}{2}t) = \tfrac{3}{2}t - \varphi(t) > \tfrac{1}{2}t$, which is impossible as we have already shown.

Finally, if we had $0 < \varphi(t) \leq \tfrac{1}{4}t$, then $\varphi(\tfrac{1}{2}t) = g(t, \varphi(t)) = 2\varphi(t)$. If it happens that $2\varphi(t) \leq \tfrac{1}{4}t$, too, i.e. $\varphi(t) \leq \tfrac{1}{4}(t/2)$, then we get $\varphi(2^{-2}t) = 2\varphi(2^{-1}t) = 2^2\varphi(t)$; and so on. By this argument we see that $\varphi(2^{-n}t) = 2^n\varphi(t)$ as long as $\varphi(2^{-n+1}t) \leq \tfrac{1}{4} \cdot 2^{-n+1}t$. So after a finite number of steps the graph of φ must get over the line $y = \frac{1}{4}x$. But then it must jump to the line $y = x$, for as we have just seen any intermediate values are impossible. This again contradicts the continuity of φ.

So the only possible values of φ, at any point $x \in X$, are $\varphi(x) = 0$ or $\varphi(x) = x$. Since φ is continuous, we must have either $\varphi = 0$ or $\varphi = \mathrm{id}$. Consequently equation (5.3.17) has exactly two continuous solutions $\varphi \colon X \to \mathbb{R}$, and they both fulfil the same condition $\varphi(0) = 0$.

Comments. The results we presented in Subsections 5.3B and 5.3C are due to Czaja-Pośpiech–Kuczma [1]. Example 5.3.1 is taken from Kuczma [45].

5.4 Asymptotic properties of solutions

In the present section we consider solutions of the equation

$$\varphi(x) = h(x, \varphi(f(x))) \qquad (5.4.1)$$

which differ from a polynomial by terms of higher order than the degree of the polynomial.

5.4A. Coincidence and existence theorems

Both theorems are due to J. Matkowski [10]. We assume the following.

(i) $X = (0, a|, \; 0 < a \leqslant \infty$,
(ii) $f \colon X \to X$ satisfies $0 < f(x) < x$ in X and there exist positive c and s such that $f(x) \leqslant sx$ in $(0, c] \subset X$.
(iii) $h \colon \Omega \to \mathbb{R}$ satisfies $h(f(x), \Omega_{f(x)}) \subset \Omega_x \neq \varnothing$ for $x \in X$ and there exist positive numbers Θ and δ and a $d \in \mathbb{R}$ such that

$$|h(x, y) - h(x, \bar{y})| \leqslant \Theta |y - \bar{y}| \qquad \text{for } (x, y), (x, \bar{y}) \in \Omega \times R,$$

where $R := (0, c] \times [d - \delta, d + \delta]$ and $s^r \Theta < 1$ with an $r \in \mathbb{N}$. Moreover, $h(x, d) = d + o(x), \; x \to 0$.

Remark 5.4.1. The rectangle R may also be of other forms, as described in Remark 5.2.1.

Let us fix numbers $d_1, \ldots, d_{r-1}$ and write

$$P(x) := d + \sum_{i=1}^{r-1} d_i x^i.$$

We are interested in solutions φ on (5.4.1) that satisfy

$$\varphi(x) = P(x) + O(x^r), \qquad x \to 0. \qquad (5.4.2)$$

We start with the following coincidence theorem.

Theorem 5.4.1. *Let hypotheses* (i)–(iii) *be satisfied. If solutions* $\varphi_j \colon X \to \mathbb{R}$,

$j = 1, 2$, of equation (5.4.1) *both have property* (5.4.2), *then the* φ_j *coincide in a right vicinity of* $x = 0$. *If, moreover, f is continuous on X, then the φ_j coincide in the whole of X.*

Proof. According to (5.4.2) we may write

$$\varphi_j(x) = P(x) + x^r Q_j(x), \qquad |Q_j(x)| \leq M, \qquad j = 1, 2, \qquad (5.4.3)$$

in a right vicinity of zero. We shall show that the Q_j coincide near the origin.

We may assume that $c < 1$ is so small that the inequality in (5.4.3) holds in $(0, c]$, and

$$|P(f(x)) + w(f(x))^r - d| \leq \delta \qquad (5.4.4)$$

for $x \in (0, c]$ and $|w| \leq M$. The functions Q_j satisfy the relation

$$Q_j(x) = H(x, Q_j(f(x))), \qquad j = 1, 2, \qquad (5.4.5)$$

in $(0, c]$. Here we put

$$H(x, w) := x^{-r}[h(x, P(f(x)) + w(f(x))^r) - P(x)].$$

The function H fulfils the Lipschitz condition (see (iii) and (ii))

$$|H(x, w) - H(x, \bar{w})| \leq s^r \Theta |w - \bar{w}| \qquad (5.4.6)$$

for $x \in (0, c]$ and $w, \bar{w} \in [-M, M]$. It follows from (5.4.3), (5.4.5) and (5.4.6) by induction that for $x \in (0, c]$ we have

$$|Q_1(x) - Q_2(x)| \leq (s^r \Theta)^n |Q_1(f^n(x)) - Q_2(f^n(x))|, \qquad n \in \mathbb{N}_0,$$

whence by (5.4.3)

$$|Q_1(x) - Q_2(x)| \leq 2M(s^r \Theta)^n, \qquad n \in \mathbb{N}_0.$$

Letting $n \to \infty$ we obtain $Q_1 = Q_2$ in $(0, c]$, since $s^r \Theta < 1$. Hence also $\varphi_1 = \varphi_2$ in $(0, c]$, in view of (5.4.3). If f is continuous, then by Theorem 5.1.1 these solutions coincide in X. ∎

Without the assumption of the continuity of f, the solutions φ_1 and φ_2 have been proved to coincide in $(0, c]$, but in the course of the proof above this c was modified so that it depended on φ_1 and φ_2. In general, there need not exist a common interval on which all solutions $\varphi: X \to \mathbb{R}$ of (5.4.1) with property (5.4.2) coincide, as may be seen from the following example.

Example 5.4.1. Take $X = (0, 1]$, $\Omega = X \times \mathbb{R}$, and put $x_n = 2^{-n}$, $n \in \mathbb{N}_0$. Define the function $f: X \to X$ by

$$f(x) = x_n + \tfrac{1}{2}(x - x_n) \qquad \text{for } x_n < x \leq x_{n-1}, n \in \mathbb{N},$$

and consider the equation

$$\varphi(x) = \tfrac{1}{2}\varphi(f(x)). \qquad (5.4.7)$$

Here $h(x, y) = \tfrac{1}{2}y$, $\Theta = \tfrac{1}{2}$, $s = 1$, and $s^r \Theta < 1$ for every $r \in \mathbb{N}$. Further, we have

$d = 0$. Let us take $d_1 = \cdots = d_{r-1} = 0$ and define for $N \in \mathbb{N}$ the functions

$$\varphi_N(x) = \begin{cases} N/(x - x_n) & \text{for } x_n < x \le x_{n-1}, n = 1, \ldots, N, \\ 0 & \text{for } 0 < x \le x_N. \end{cases}$$

They all satisfy equation (5.4.7) and condition (5.4.2), but there is no subset of X on which all the φ_N would coincide.

In order to prove the existence of solutions spoken of in Theorem 5.4.1 we have to assume something more.

Theorem 5.4.2. *Let hypotheses* (i)–(iii) *be satisfied and let* $R \subset \Omega$. *Further suppose that* $P(f(x)) \in \Omega_x$ *and*

$$h(x, P(f(x))) = P(x) + O(x^r), \qquad x \to 0. \tag{5.4.8}$$

Then equation (5.4.1) *has in a right vicinity of zero a solution* φ *fulfilling condition* (5.4.2). *If, moreover, f is continuous in X, then equation* (5.4.1) *has a unique solution* $\varphi: X \to \mathbb{R}$ *satisfying* (5.4.2). *If h is continuous in Ω, then φ is continuous in X.*

Proof. First we shall find the desired solution in a vicinity $(0, c]$ of zero. It will be the limit of the sequence of functions $\varphi_n: (0, c] \to \mathbb{R}$ defined by the recurrence

$$\varphi_0(x) = P(x), \qquad \varphi_{n+1}(x) = h(x, \varphi_n(f(x))), \qquad n \in \mathbb{N}_0, x \in (0, c]. \tag{5.4.9}$$

We may assume that c is so small that (5.4.4) and

$$x^{-r}|h(x, P(f(x))) - P(x)| \le k \tag{5.4.10}$$

hold for $x \in (0, c]$ and $|w| \le M$, where $k > 0$ and $M := k(1 - s^r\Theta)^{-1}$. By (5.4.8) such a vicinity of zero and a constant k do exist.

We want to estimate the distance between two consecutive terms of sequence (5.4.9) with the aid of the Lipschitz condition in (iii). Thus we must check whether $\varphi_n(f(x))$ lie in the interval $[d - \delta, d + \delta]$. To see they really do, we will prove that for $x \in (0, c]$

$$x^{-r}|\varphi_n(x) - P(x)| \le M, \qquad n \in \mathbb{N}_0. \tag{5.4.11}$$

If (5.4.11) holds for an $n \in \mathbb{N}_0$, then

$$|\varphi_n(f(x)) - d| \le |\varphi_n(f(x)) - P(f(x))| + |P(f(x)) - d|$$

$$\le M(f(x))^r + |P(f(x)) - d| \le \delta,$$

because of $P(f(x)) = d + o(x)$, $x \to 0$, and (ii), taking possibly a yet smaller c. And this is what we need.

Returning to (5.4.11) we find it trivial for $n = 0$. Assuming (5.4.11) true for an $n \in \mathbb{N}_0$, we have, by (5.4.4), (iii), (ii) and (5.4.10),

$$x^{-r}|\varphi_{n+1}(x) - P(x)|$$
$$\leq x^{-r}|h(x, \varphi_n(f(x))) - h(x, P(f(x)))| + x^{-r}|h(x, P(f(x))) - P(x)|$$
$$\leq x^{-r}\Theta|\varphi_n(f(x)) - P(f(x))| + k \leq s^r\Theta M + k = M.$$

Induction yields (5.4.11).

Applying the inequality of (iii) in (5.4.9) we obtain

$$|\varphi_{n+1}(x) - \varphi_n(x)| \leq \Theta|\varphi_n(f(x)) - \varphi_{n-1}(f(x))|, \qquad n \in \mathbb{N},$$

whence by (5.4.11), since $f^n(x) \leq s^n x$, we have for $x \in (0, c]$

$$|\varphi_{n+1}(x) - \varphi_n(x)| \leq \Theta^n|\varphi_1(f^n(x)) - P(f^n(x))| \leq (s^r\Theta)^n Mc^r, \qquad n \in \mathbb{N}.$$

Thus the sequence (φ_n) uniformly converges in $(0, c]$. Since by the Lipschitz condition in (iii) the function $h(x, \cdot)$ is continuous, the limit $\varphi(x) = \lim_{n \to \infty} \varphi_n(x)$ satisfies equation (5.4.1) in $(0, c]$. By (5.4.11) this φ has property (5.4.2). If h is continuous in Ω and f is so in X, then all φ_n are continuous in $(0, c]$ and so is also φ. Theorems 5.4.1 and 5.1.1 complete the proof. ∎

5.4B. Solutions differentiable at the origin

Now we are going to deal with solutions φ of (5.4.1) satisfying

$$\varphi(x) = d + d_1 x + o(x), \qquad x \to 0,$$

i.e. with those that are differentiable at $x = 0$.

Theorem 5.4.3. *Let hypotheses* (i)–(iii) *be satisfied with* $r = 1$ *and* $R \subset \Omega$. *Suppose that there exist the limit* $\lim_{x \to 0}(f(x)/x) = \bar{s}$ *and* $A, B \in \mathbb{R}$ *such that*

$$h(x, y) = d + Ax + B(y - d) + o(x + |y - d|), \qquad x \to 0, \, y \to d. \quad (5.4.12)$$

Then equation (5.4.1) *has in a right vicinity of zero a solution with the property*

$$\varphi(x) = d + \frac{A}{1 - B\bar{s}} x + o(x), \qquad x \to 0. \quad (5.4.13)$$

If, moreover, f is continuous in X, then equation (5.4.1) *has a unique solution* $\varphi: X \to \mathbb{R}$ *satisfying* (5.4.13). *If h is continuous in Ω, then φ is continuous in X.*

Proof. The proof follows the lines of that of Theorem 5.4.2. First note that $B\bar{s} \neq 1$. For, by (5.4.12), (iii) and (ii) we have $B\bar{s} \leq |B|\bar{s} \leq \Theta s < 1$. Let us put

$$p(x) := d + d_1 x, \qquad d_1 := A/(1 - B\bar{s}).$$

It is seen that d_1 has been chosen to ensure the relation

$$h(x, p(f(x))) - p(x) = o(x), \qquad x \to 0.$$

Fix a $K > 0$. We may assume that c is such that for $x \in (0, c]$ we have

$$|h(x, p(f(x))) - p(x)| \leq (1 - \Theta s)K.$$

This inequality yields

$$|\varphi_n(x) - p(x)| \leq Kx, \qquad x \in (0, c], \, n \in \mathbb{N}_0, \qquad (5.4.14)$$

similarly as (5.4.10) implied (5.4.11) in the proof of Theorem 5.4.2. Here $\varphi_0 = p$ and for $n \in \mathbb{N}$ the functions $\varphi_n: (0, c] \to \mathbb{R}$ are defined by (5.4.9).

In particular, relation (5.4.14) implies

$$|\varphi_n(x) - d| \leq |\varphi_n(x) - p(x)| + |p(x) - d| \leq c(K + |d_1|).$$

Therefore, taking a yet smaller c, if need be, we obtain $|\varphi_n(x) - d| < \delta$ and estimation from (iii) works for $y = \varphi_n$ as in the proof of Theorem 5.4.2. We conclude that the sequence $(\varphi_n)_{n \in \mathbb{N}_0}$ uniformly converges on $(0, c]$ to a function φ. Of course, φ is a solution of (5.4.1) in $(0, c]$, and (5.4.14) becomes

$$|\varphi(x) - p(x)| \leq Kx. \qquad (5.4.15)$$

Now take a positive $K^* < K$. The same argument yields the existence of a $c^* > 0$ and of a function $\varphi^*: (0, c^*] \to \mathbb{R}$ satisfying equation (5.4.1) and condition (5.4.15) (with K replaced by K^*) in $(0, c^*]$. By Theorem 5.4.1 ($r = 1$) the functions φ and φ^* coincide in a right vicinity of zero. This means that φ satisfies (5.4.15) with arbitrarily small $K > 0$ in a right vicinity (depending on K) of zero. Consequently, φ has property (5.4.13). The remaining statements result from Theorems 5.4.1 and 5.1.1. (Note that φ is continuous in $(0, c]$ whenever f and h are continuous.) ∎

Comments. Theorem 5.4.2 has been proved in Kuczma [38] with the aid of Banach's Theorem. Another existence result (for $r = 1$) related to Theorem 5.4.2, is found in Kuczma–Matkowski [1]. Consult also the papers by M. Kwapisz–J. Turo [1], [3].

5.5 Lipschitzian solutions

We leave outside considerations the problem of absolutely continuous (a.c.) solutions of the equation

$$\varphi(x) = h(x, \varphi(f(x))). \qquad (5.5.1)$$

Conditions for the uniqueness of such solutions are essentially the same as those ensuring the uniqueness of continuous solutions.

If, however, we restrict the class of a.c. functions to those fulfilling a Lipschitz condition, then we may obtain uniqueness under slightly weaker assumptions. Indeed, a Lipschitzian function $\varphi: X \to \mathbb{R}$ also satisfies (if $0 \in X$) condition (5.4.2) for $r = 1$ and Theorem 5.4.1 applies.

In this section we prove a theorem on the existence and uniqueness of Lipschitzian solutions of (5.5.1) (Matkowski [11]; see also Pelczar [1]). We also show that Lipschitzian Nemitskiĭ operators acting on the space of Lipschitzian functions are all affine (Matkowski [14]).

5.5A. Existence and uniqueness

Let us suppose the following.

(i) $X = [0, a|, 0 < a \leq \infty$.

(ii) $f: X \rightarrow X$ satisfies $0 < f(x) < x$ in $X \setminus \{0\}$ and the Lipschitz condition
$$|f(x) - f(\bar{x})| \leq s|x - \bar{x}|, \qquad x, \bar{x} \in X.$$

(iii) $h: \Omega \rightarrow \mathbb{R}$ satisfies (5.1.2), $\Omega_x \neq \emptyset$ for $x \in X$. Further, there exist d satisfying (5.1.3), $c \in X \setminus \{0\}$, and $\delta > 0$ such that $R := [0, c] \times [d - \delta, d + \delta] \subset \Omega^1$. Moreover, h fulfils in R the Lipschitz condition
$$|h(x, y) - h(\bar{x}, \bar{y})| \leq p|x - \bar{x}| + q|y - \bar{y}|$$
and $sq < 1$.

Theorem 5.5.1. *Let hypotheses* (i)–(iii) *be satisfied. Then equation* (5.5.1) *has a unique solution* $\varphi: X \rightarrow \mathbb{R}$ *fulfilling in* X *a Lipschitz condition and such that* $\varphi(0) = d$.

Proof. Since Lipschitzian φ have the property $\varphi(x) = \varphi(0) + O(x), x \rightarrow 0$, the uniqueness results from Theorem 5.4.1 $(r = 1)$.

To prove the existence we shall use Schauder's Theorem. First put $k := p/(1 - sq)$ and force c to satisfy $kc < \delta$.

Let Φ be the Banach space of all continuous functions $\varphi: [0, c] \rightarrow \mathbb{R}$, endowed with the standard sup norm. Let Φ_0 be the set of those $\varphi \in \Phi$ that fulfil $\varphi(0) = d$ and the Lipschitz condition
$$|\varphi(x) - \varphi(\bar{x})| \leq k|x - \bar{x}| \tag{5.5.2}$$
in $[0, c]$. Clearly Φ_0 is a convex and, by Arzelà's Theorem, compact subset of Φ.

Now define the transform T on Φ_0 by
$$T(\varphi)(x) = h(x, \varphi(f(x))).$$
Note that if $\varphi \in \Phi_0$ then $|\varphi(x) - d| \leq \delta$ for $x \in [0, c]$ (put $\bar{x} = 0$ in (5.5.2) and recall $kc < \delta$). Thus the Lipschitz condition on h from (iii) may be used and it is easy to check that T is a continuous self-mapping of Φ_0. Schauder's Theorem says that there is a fixed point $\varphi \in \Phi_0$ of T. This φ represents a Lipschitzian solution of equation (5.5.1) in $[0, c]$ satisfying $\varphi(0) = d$. By Theorem 5.1.1 this solution can be extended to a solution $\varphi: X \rightarrow \mathbb{R}$ of (5.5.1) in the whole of X.

What remains to be proved is that φ is Lipschitzian in X. Put
$$b_0 := \sup\{b \in X: \varphi \text{ satisfies } (5.5.2) \text{ in } [0, b]\}.$$

Thus $c \leq b_0 \leq a$. Suppose $b_0 < a$. The Lipschitz condition in (ii) implies that there is a $b', b_0 < b' < a$, such that $f(x) \leq f(b_0) + s(b' - b_0) \leq b_0$ for $x \in [0, b']$.

---

[1] Remark 5.2.1 on other possible forms of R applies.

Thus we have for $x, \bar{x} \in [0, b']$, with $y := \varphi(f(x))$, $\bar{y} := \varphi(f(\bar{x}))$,

$$|\varphi(x) - \varphi(\bar{x})| = |h(x, y) - h(\bar{x}, \bar{y})| \leq p|x - \bar{x}| + q|y - \bar{y}|$$

$$\leq p|x - \bar{x}| + qk|f(x) - f(\bar{x})| \leq (p + qks)|x - \bar{x}| = k|x - \bar{x}|$$

and φ satisfies (5.5.2) in $[0, b']$, which contradicts the definition of b_0. Hence $b_0 = a$, and φ satisfies (5.5.2) in X. ∎

5.5B. Lipschitzian Nemitskiĭ operators

Theorem 5.5.1 does not say how to construct the unique Lipschitzian solution of (5.5.1). This is an inconvenience of proofs using Schauder's Theorem. Because of the uniqueness one can expect that Banach's Theorem should also apply in our case.

With the aid of a Nemitskiĭ operator associated to equation (5.5.1) we are going to prove that the latter method of proof is applicable in the linear case only.

We start by introducing the Nemitskiĭ operator.

Definition 5.5.1. Let $X = [0, a]$, $a > 0$, and let $\mathscr{B}$ be a Banach space of functions $\varphi: X \to \mathbb{R}$. A function $h: X \times \mathbb{R} \to \mathbb{R}$ generates the *Nemitskiĭ operator* of substitution N on $\mathscr{B}$ which is defined by the formula

$$(N\varphi)(x) = h(x, \varphi(x)), \qquad \varphi \in \mathscr{B}.$$

Using the operator N we can rewrite equation (5.5.1) in the form

$$\varphi = (N \circ S)\varphi, \tag{5.5.3}$$

where $S\varphi = \varphi \circ f$ is a linear and, in general, continuous mapping of $\mathscr{B}$ into itself.

Now, if $N \circ S$ is a contraction map of $\mathscr{B}$ into itself, then N 'practically' has to be Lipschitzian, and certainly is if $f: X \to X$ is a bijection so that $S: \mathscr{B} \to \mathscr{B}$ also is. Then by the Open Mapping Theorem S^{-1} is continuous. Take $\varphi_1, \varphi_2 \in \mathscr{B}$ and evaluate

$$\|N\varphi_1 - N\varphi_2\| = \|(N \circ S)(S^{-1}\varphi_1) - (N \circ S)(S^{-1}\varphi_2)\|$$

$$\leq k\|S^{-1}\varphi_1 - S^{-1}\varphi_2\| = k\|S^{-1}(\varphi_1 - \varphi_2)\|$$

$$\leq k\|S^{-1}\|\|\varphi_1 - \varphi_2\|,$$

where $k < 1$, since $N \circ S$ was a contraction.

Now we shall work in the space $\mathscr{B} = \text{Lip } X$ of the Lipschitzian functions $\varphi: X \to \mathbb{R}$ endowed with the norm

$$\|\varphi\| := |\varphi(0)| + \sup\left\{\frac{|\varphi(x) - \varphi(\bar{x})|}{|x - \bar{x}|} : x, \bar{x} \in X, x \neq \bar{x}\right\}. \tag{5.5.4}$$

We are going to prove the following.

Theorem 5.5.2. *Let* $X = [0, a]$, $h: X \times \mathbb{R} \to \mathbb{R}$ *and let* $N: \operatorname{Lip} X \to \operatorname{Lip} X$ *be the Nemitskiĭ operator generated by* h. *The operator* N *is Lipschitzian, i.e.*

$$\|N\varphi_1 - N\varphi_2\| \leq L\|\varphi_1 - \varphi_2\|, \qquad L > 0, \ \varphi_1, \varphi_2 \in \operatorname{Lip} X, \qquad (5.5.5)$$

if and only if there exist functions $G, H \in \operatorname{Lip} X$ *such that*

$$h(x, y) = G(x)y + H(x), \qquad x \in X, \ y \in \mathbb{R}. \qquad (5.5.6)$$

Proof. Suppose N satisfies inequality (5.5.5). Thus the following is also fulfilled, for each $t, \bar{t} \in X$, $\varphi_1, \varphi_2 \in \operatorname{Lip} X$:

$$|h(0, \varphi_1(0)) - h(0, \varphi_2(0))|$$
$$+ |h(t, \varphi_1(t)) - h(t, \varphi_2(t)) - h(\bar{t}, \varphi_1(\bar{t})) + h(\bar{t}, \varphi_2(\bar{t}))||t - \bar{t}|^{-1}$$
$$\leqq L\|\varphi_1 - \varphi_2\|.$$

Let us fix $x, \bar{x} \in X$, $x < \bar{x}$, and $y_1, y_2, \bar{y}_1, \bar{y}_2 \in \mathbb{R}$. We shall use the inequality above for $t = x$, $\bar{t} = \bar{x}$ and for the following two broken linear functions φ_1 and φ_2:

$$\varphi_j(t) = \begin{cases} y_j & \text{for } 0 \leqq t < x, \\ \bar{y}_j + (y_j - \bar{y}_j)(t - \bar{x})/(x - \bar{x}) & \text{for } x \leqq t \leqq \bar{x}, \\ \bar{y}_j & \text{for } \bar{x} < t \leqq a, \end{cases}$$

$j = 1, 2$. Of course, $\varphi_1, \varphi_2 \in \operatorname{Lip} X$. The inequality becomes

$$|h(0, y_1) - h(0, y_2)| + |x - \bar{x}|^{-1}|h(x, y_1) - h(x, y_2) - h(\bar{x}, \bar{y}_1) + h(\bar{x}, \bar{y}_2)|$$
$$\leqq L(|y_1 - y_2| + |x - \bar{x}|^{-1}|y_1 - y_2 - \bar{y}_1 + \bar{y}_2|).$$

Now, since $N(\operatorname{Lip} X) \subset \operatorname{Lip} X$ and constant functions are Lipschitzian, the function $h(\cdot, y) \in \operatorname{Lip} X$ for every $y \in \mathbb{R}$ and is therefore continuous. Multiply the last inequality by $|x - \bar{x}|$ and let $\bar{x} \to x$ to obtain, by the continuity of $h(\cdot, y)$,

$$|h(x, y_1) - h(x, y_2) - h(x, \bar{y}_1) + h(x, \bar{y}_2)| \leqq L|y_1 - y_2 - \bar{y}_1 + \bar{y}_2| \quad (5.5.7)$$

for all $x \in X$, $y_1, y_2, \bar{y}_1, \bar{y}_2 \in \mathbb{R}$. Putting here $y_1 = u + v$, $y_2 = u$, $\bar{y}_1 = v$, $\bar{y}_2 = 0$ we make the right-hand side of (5.5.7) vanish. Thus

$$h(x, u + v) = h(x, u) + h(x, v) - h(x, 0).$$

This means that the function $g_x: \mathbb{R} \to \mathbb{R}$ defined by

$$g_z(y) = h(x, y) - h(x, 0) \qquad (5.5.8)$$

is additive for every $x \in X$:

$$g_x(u + v) = g_x(u) + g_x(v).$$

Setting $\bar{y}_1 = \bar{y}_2 = 0$ in (5.5.7) we find that g_x is Lipschitzian, and therefore continuous. Thus g_x is linear (see Aczél [2]), $g_x(y) = G(x)y$ with a $G: X \to \mathbb{R}$. Formula (5.5.6) results from (5.5.8) by putting $H(x) := h(x, 0)$. Finally, $G, H \in \operatorname{Lip} X$ (G because of $G(x) = h(x, 1) - h(x, 0)$).

Conversely, given $G, H \in \operatorname{Lip} X$, the Nemitskiĭ operator generated by function (5.5.6) is Lipschitzian:

$$\|N\varphi_1 - N\varphi_2\| \leq (L + bM)\|\varphi_1 - \varphi_2\|,$$

where $L := \sup_X G$, $b := \max(1, a)$ and $M = \|G\|$. ∎

Remark 5.5.1. From Theorem 5.5.2 we infer that Banach's Theorem can be applied to equation (5.5.1) (see (5.5.3)) in the space Lip X if and only if the function h is affine with respect to y.

Remark 5.5.2. In Theorem 5.5.2 the space Lip X cannot be replaced by $C(X)$ or $L^p(X)$. For, observe that the function $h(x, y) = \sin y$ generates a Lipschitzian Nemitskiĭ operator in both spaces. On the other hand, Theorem 5.5.2 works in the Banach space $C^n(X)$ of n times continuously differentiable functions (Matkowski [15]) as well as in the space $\mathrm{Lip}^\alpha(X)$, $\alpha \in (0, 1]$, of Hölderian functions (Matkowska [1]).

Comments. Lipschitzian solutions of iterative functional equations have been dealt with for the first time by A. Pelczar [1]–[3]. See also Jakowska-Suwalska [1], Miś [1], Matkowski–Miś [1].

Regarding properties of nonlinear Nemitskiĭ operators acting in spaces of integrable functions consult Martin [1].

5.6 Smooth solutions

The problem of differentiable solutions is more difficult to handle for the nonlinear equation

$$\varphi(x) = h(x, \varphi(f(x))) \tag{5.6.1}$$

than for the linear one we considered in Section 3.4. We begin with the basic existence-and-uniqueness theorem on C^r solutions of (5.6.1) and then discuss the case of the lack of uniqueness of C^r solutions.

5.6A. Preliminaries

We make the following assumptions.

(i) $X = [0, a|, 0 < a \leq \infty$.
(ii) $f: X \to X$ is of class C^r in X, $1 \leq r \leq \infty$, and $0 < f(x) < x$ in $X \setminus \{0\}$.
(iii) $h: \Omega \to \mathbb{R}$ is of class C^r in Ω. For every $x \in X$ the set Ω_x is a nondegenerate interval and $h(f(x), \Omega_{f(x)}) \subset \Omega_x$. There are numbers d satisfying $d = h(0, d)$ and $c \in (0, a)$, $\delta > 0$ such that $R := [0, c] \times [d - \delta, d + \delta] \subset \Omega$ (Remark 5.2.1 applies).

We put $\hat{\Omega} := \Omega \cap (X \times \mathbb{R})$ and define the functions $h_k : \hat{\Omega} \times \mathbb{R}^k \to \mathbb{R}$,

$k = 1, \ldots, r$, by the recurrence

$$
\left.
\begin{aligned}
h_1(x, y, y_1) &= \frac{\partial h}{\partial x}(x, y) + f'(x)\frac{\partial h}{\partial y}(x, y)y_1, \\[2mm]
h_{k+1}&(x, y, y_1, \ldots, y_{k+1}) \\[1mm]
&= \frac{\partial h_k}{\partial x} + f'(x)\left(\frac{\partial h_k}{\partial y}y_1 + \frac{\partial h_k}{\partial y_1}y_2 + \cdots + \frac{\partial h_k}{\partial y_k}y_{k+1}\right), \\[2mm]
&\qquad\qquad\qquad\qquad\qquad\qquad\qquad k = 1, \ldots, r-1.
\end{aligned}
\right\} \quad (5.6.2)
$$

Putting here $y = y(x)$ and $y_j = y^{(j)}(x)$ we see that h_k represents the kth derivative of the composite function $h(\cdot, y \circ f)$. This explains the origin of formulae (5.6.2).

The following two lemmas are easily established by induction.

Lemma 5.6.1. *Let hypotheses* (i)–(iii) *be satisfied. Then, for every* $k = 1, \ldots, r$, *the function* h_k *is of class* C^{r-k} *on* $\hat{\Omega} \times \mathbb{R}^k$, *and*

$$
h_k(x, y, y_1, \ldots, y_k) = u_k(x, y, y_1, \ldots, y_{k-1}) + (f'(x))^k\frac{\partial h}{\partial y}(x, y)y_k, \quad (5.6.3)
$$

where $u_k: \hat{\Omega} \times \mathbb{R}^{k-1} \to \mathbb{R}$ *belongs to the class* C^{r-k}.

Lemma 5.6.2. *Let hypotheses* (i)–(iii) *be satisfied, and let* $\varphi: X \to \mathbb{R}$ *be a function of class* C^r *in* X. *If*

$$
\psi(x) := h(x, \varphi(f(x)))
$$

on X, *then the derivatives of* φ *and* ψ *satisfy the relations*

$$
\psi^{(k)}(x) = h_k(x, \varphi(f(x)), \varphi'(f(x)), \ldots, \varphi^{(k)}(f(x))), \qquad k = 1, \ldots, r.
$$

It follows from Lemma 5.6.2 that if $\varphi: X \to \mathbb{R}$ is a C^r solution of equation (5.6.1) such that

$$
\varphi(0) = d \tag{5.6.4}
$$

then the values

$$
d_k = \varphi^{(k)}(0), \qquad k = 1, \ldots, r, \tag{5.6.5}
$$

satisfy the system of equations

$$
d_k = h_k(0, d, d_1, \ldots, d_k), \qquad k = 1, \ldots, r. \tag{5.6.6}
$$

According to (5.6.3), if $(f'(0))^k h'_y(0, d) \neq 1$ for $k = 1, \ldots, r$, then system (5.6.6) has a unique solution $(d_1, \ldots, d_r)$. If this expression equals 1 for some k, then system (5.6.6) has either no solution, or a family of solutions depending on the parameter d_k.

The following lemma makes it evident that we may always assume, without loss of generality, that $d = 0$ and system (5.6.6) has the solution $(0, \ldots, 0) \in \mathbb{R}^r$.

Lemma 5.6.3. *Let hypotheses* (i)–(iii) *be satisfied. Let us fix a solution* $(d_1, \ldots, d_r)$ *of* (5.6.6). *Equation* (5.6.1) *has a* C^r *solution* $\varphi \colon X \to \mathbb{R}$ *satisfying* (5.6.4) *and* (5.6.5) *if and only if* $\Phi \colon X \to \mathbb{R}$ *is a* C^r *solution of the equation*

$$\Phi(x) = H(x, \Phi(f(x)))$$

satisfying the condition

$$\Phi(0) = \Phi'(0) = \cdots = \Phi^{(r)}(0) = 0, \tag{5.6.7}$$

where

$$H(x, y) = h(x, y + P(f(x))) - P(x), \qquad P(x) = d + \sum_{k=1}^{r} \frac{d_k}{k!} x^k.$$

The two solutions are linked by the formula

$$\varphi(x) = P(x) + \Phi(x), \qquad x \in X.$$

Proof. Straightforward verification. Note that $H(0, 0) = 0$ and $(0, \ldots, 0) \in \mathbb{R}^r$ is a solution of system (5.6.6) with h_k calculated according to (5.6.2) for the function H. ∎

5.6B. Existence and uniqueness of C^r solutions

Now comes

Theorem 5.6.1. *Let hypotheses* (i)–(iii) *be satisfied with* $d = 0$, *and let*

$$h_k(0, \ldots, 0) = 0, \qquad k = 1, \ldots, r. \tag{5.6.8}$$

If

$$\left| (f'(0))^r \frac{\partial h}{\partial y} (0, 0) \right| < 1, \tag{5.6.9}$$

then equation (5.6.1) *has a unique* C^r *solution* $\varphi \colon X \to \mathbb{R}$ *fulfilling condition* (5.6.7) ($\Phi = \varphi$).

Proof. By (5.6.8) every C^r solution $\varphi \colon X \to \mathbb{R}$ of (5.6.1) has the property $\varphi(x) = O(x^r)$, $x \to 0$. Thus the uniqueness results from Theorem 5.4.1.

To prove the existence we first consider the case where

$$f'(x) \leqslant 1 \tag{5.6.10}$$

in a right neighbourhood of zero.

We want to apply Schauder's Theorem.

To begin with, note that relations (5.6.8) imply $u_k(0, \ldots, 0) = 0$ for $k = 1, \ldots, r$. We may assume that c and δ have been chosen in such a manner that $c < 1$, $c < \delta$, inequality (5.6.10) holds in $[0, c]$, and

$$\left| (f'(x))^r \frac{\partial h}{\partial y} (x, y) \right| \leqslant \Theta < 1 \qquad \text{in } R, \tag{5.6.11}$$

$$\left| u_r(x, y, y_1, \ldots, y_{r-1}) \right| \leqslant 1 - \Theta \qquad \text{in } [0, c] \times [-\delta, \delta]^r, \tag{5.6.12}$$

$$\left| (f'(x))^r \frac{\partial h}{\partial y} (x, y) - (f'(x))^r \frac{\partial h}{\partial y} (0, 0) \right| \leqslant 1 - \Theta \qquad \text{in } R. \tag{5.6.13}$$

The functions $u_r, f', \partial h/\partial y$ are uniformly continuous in the sets $[0, c] \times [-\delta, \delta]^r$, $[0, c]$, R, respectively. Thus there exists a function $\eta: (0, \infty) \to (0, \infty)$ such that for every $\varepsilon > 0$

$$|u_r(x, y, y_1, \ldots, y_{r-1}) - u_r(\bar{x}, \bar{y}, \bar{y}_1, \ldots, \bar{y}_{r-1})| \le (1 - \Theta)\varepsilon/2, \quad (5.6.14)$$

$$\left|(f'(x))^r \frac{\partial h}{\partial y}(x, y) - (f'(\bar{x}))^r \frac{\partial h}{\partial y}(\bar{x}, \bar{y})\right| \le (1 - \Theta)\varepsilon/2 \quad (5.6.15)$$

for $|x - \bar{x}| \le \eta(\varepsilon), |y - \bar{y}| \le \eta(\varepsilon), |y_k - \bar{y}_k| \le \eta(\varepsilon), x, \bar{x} \in [0, c], y, \bar{y}, y_k, \bar{y}_k \in [-\delta, \delta]$, $k = 1, \ldots, r - 1$.

Let Φ be the Banach space of all functions $\varphi: [0, c] \to \mathbb{R}$ of class C^r in $[0, c]$, with the norm

$$\|\varphi\| = \max\left(\sup_{[0,c]} |\varphi(x)|, \sup_{[0,c]} |\varphi'(x)|, \ldots, \sup_{[0,c]} |\varphi^{(r)}(x)|\right).$$

Let Φ_0 be the set of those $\varphi \in \Phi$ which satisfy (5.6.7) ($\Phi = \varphi$), and also

$$|\varphi^{(r)}(x)| \le 1 \qquad \text{for } x \in [0, c], \quad (5.6.16)$$

$$|\varphi^{(r)}(x) - \varphi^{(r)}(\bar{x})| \le \varepsilon \qquad \text{for } |x - \bar{x}| \le \eta(\varepsilon), x, \bar{x} \in [0, c]. \quad (5.6.17)$$

We have

$$\|\varphi_1 - \varphi_2\| = \sup_{[0,c]} |\varphi_1^{(r)}(x) - \varphi_2^{(r)}(x)| \qquad \text{for } \varphi_1, \varphi_2 \in \Phi_0. \quad (5.6.18)$$

Indeed, by the Mean Value Theorem we have for $\varphi_1, \varphi_2 \in \Phi_0$, with $\Phi := \varphi_1 - \varphi_2$,

$$|\Phi^{(k-1)}(x)| = x|\Phi^{(k)}(x')|, \qquad k = 1, \ldots, r, \quad (5.6.19)$$

with an $x' \in (0, x) \subset [0, c]$. Hence, since $x \le c < 1$, we get

$$\sup_{[0,c]} |\Phi^{(k-1)}(x)| \le \sup_{[0,c]} |\Phi^{(k)}(x)|, \qquad k = 1, \ldots, r,$$

and (5.6.18) follows by the definition of the norm in Φ.

Condition (5.6.17) shows that the derivatives $\varphi^{(r)}$ of the functions φ from Φ_0 are equicontinuous, and by (5.6.16) they are equibonded in $[0, c]$. Thus in view of (5.6.18) Φ_0 is a compact subset of Φ, and evidently Φ_0 is convex.

It remains to prove that the transform $T: \Phi \to \Phi$ defined by

$$T(\varphi)(x) = h(x, \varphi(f(x))) \quad (5.6.20)$$

maps the set Φ_0 into itself and is continuous on Φ_0.

Take a $\varphi \in \Phi_0$ and write $\psi = T(\varphi)$. Clearly ψ satisfies condition (5.6.7) ($\Phi = \psi$); see (5.6.8) and Lemma 5.6.2. We check condition (5.6.16) for ψ. First observe that (5.6.19) implies (take $\Phi = \varphi - 0$)

$$|\varphi^{(k-1)}(x)| \le c^{r-k+1}|\varphi^{(r)}(\bar{x})|, \qquad k = 1, \ldots, r, \bar{x} \in [0, c].$$

From this, by $c < \min(1, \delta)$ and (5.6.16), we get

$$|\varphi(f(x))| \le \delta, \qquad |\varphi^{(k)}(x)| \le \delta, \qquad k = 1, \ldots, r - 1, x \in [0, c].$$

Now we are in position to apply Lemmas 5.6.1 and 5.6.2. By (5.6.12), (5.6.11), (5.6.16),

$$\left|\psi^{(r)}(x)\right| = \left|h_r(x, \varphi(f(x)), \varphi'(f(x)), \ldots, \varphi^{(r)}(f(x)))\right|$$

$$\leqslant \left|u_r(x, \varphi(f(x)), \varphi'(f(x)), \ldots, \varphi^{(r-1)}(f(x)))\right|$$

$$+ \left|(f'(x))^r \frac{\partial h}{\partial y}(x, \varphi(f(x)))\varphi^{(r)}(f(x))\right| \leqslant (1 - \Theta) + \Theta = 1.$$

We have obtained (5.6.16) for ψ.

To prove (5.6.17) for ψ we first use Lemmas 5.6.2 and 5.6.1. Take $x, \bar{x} \in [0, c], |x - \bar{x}| \leqslant \eta(\varepsilon)$, and write $y = \varphi(f(x)), \bar{y} = \varphi(f(\bar{x})), y_k = \varphi^{(k)}(f(x))$, $\bar{y}_k = \psi^{(k)}(f(\bar{x}))$, for short. We arrive at

$$\left|\psi^{(r)}(x) - \psi^{(r)}(\bar{x})\right| = \left|h_r(x, y, \ldots, y_r) - h_r(\bar{x}, \bar{y}, \ldots, \bar{y}_r)\right|$$

$$\leqslant \left|u_r(x, y, \ldots, y_{r-1}) - u_r(\bar{x}, \bar{y}, \ldots, \bar{y}_{r-1})\right|$$

$$+ \left|y_r\right| \left|(f'(x))^r \frac{\partial h}{\partial y}(x, y) - (f'(\bar{x}))^r \frac{\partial h}{\partial y}(\bar{x}, \bar{y})\right|$$

$$+ \left|(f'(\bar{x}))^r \frac{\partial h}{\partial y}(\bar{x}, \bar{y})\right| \left|y_r - \bar{y}_r\right|. \qquad (5.6.21)$$

We should like to use here (5.6.14) and (5.6.15). To this end we must have $|y - \bar{y}| \leqslant \eta(\varepsilon), |y_k - \bar{y}_k| \leqslant \eta(\varepsilon), k = 1, \ldots, r - 1$. These inequalities follow from the Mean Value Theorem, (5.6.16) and (5.6.18) (since by (5.6.10) we have $|f(x) - f(\bar{x})| \leqslant |x - \bar{x}| \leqslant \eta(\varepsilon)$):

$$\left|\varphi^{(k)}(f(x)) - \varphi^{(k)}(f(\bar{x}))\right| = \left|\varphi^{(k+1)}(u)\right| \left|f(x) - f(\bar{x})\right| \leqslant \|\varphi\| \eta(\varepsilon) \leqslant \eta(\varepsilon)$$

where u lies between $f(x)$ and $f(\bar{x})$ and $k = 0, 1, \ldots, r - 1$. The left-hand side here equals $|y - \bar{y}|$ or $|y_k - \bar{y}_k|$. Our goal is achieved.

Return then to (5.6.21) and use consecutively (5.6.14), (5.6.16), (5.6.15), (5.6.11), (5.6.17) and (5.6.10) to obtain

$$\left|\psi^{(r)}(x) - \psi^{(r)}(\bar{x})\right| \leqslant (1 - \Theta)\varepsilon/2 + (1 - \Theta)\varepsilon/2 + \Theta\varepsilon = \varepsilon.$$

Thus $\psi \in \Phi_0$, i.e. $T(\Phi_0) \subset \Phi_0$.

Checking the continuity of T in Φ_0 makes no difficulties. If $\varphi_n \in \Phi_0, n \in \mathbb{N}_0$, and $\|\varphi_n - \varphi_0\| \to 0$ as $n \to \infty$, then, by (5.6.18), $\lim_{n \to \infty} \varphi_n^{(k)}(x) = \varphi_0^{(k)}(x)$ uniformly in $[0, c]$ for $k = 0, 1, \ldots, r$. This uniform convergence and the continuity of h_r (see Lemma 5.6.1) yield (we put $\psi_n = T(\varphi_n)$ and $\psi_0 = T(\varphi_0)$)

$$\lim_{n \to \infty} \psi_n^{(r)}(x) = \lim_{n \to \infty} h_r(x, \varphi_n(f(x)), \ldots, \varphi_n^{(r)}(f(x)))$$

$$= h_r(x, \varphi_0(f(x)), \ldots, \varphi_0^{(r)}(f(x))) = \psi_0^{(r)}(x)$$

uniformly in $[0, c]$. Recalling again (5.6.18) we see that $\|\psi_n - \psi\| \to 0$.

By Schauder's Theorem transform (5.6.20) has a fixed point $\varphi \in \Phi_0$. In other words, equation (5.6.1) has in $[0, c]$ a C^r solution φ fulfilling (5.6.7)

$(\Phi = \varphi)$. By Theorem 5.1.1(b) this solution can be extended to a C^r solution on X.

Now assume that (5.6.10) does not hold. Then necessarily $f'(0) = 1$ (see (ii)) and by (5.6.9)

$$\left| \frac{\partial h}{\partial y}(0, 0) \right| < 1. \qquad (5.6.22)$$

By Theorem 5.2.3 equation (5.6.1) has a unique continuous solution $\varphi: X \to \mathbb{R}$ such that $\varphi(0) = 0$. We shall prove that this solution is in fact of class C^r. Our tool will be Theorem 5.2.4 on the continuous dependence of continuous solutions on the data.

Take a sequence (t_n), $0 < t_n < 1$, $\lim_{n \to \infty} t_n = t_0 = 1$ and write $f_n(x) = t_n f(x)$ for $n \in \mathbb{N}_0$. Consider the sequence of equations

$$\varphi_n(x) = h(x, \varphi_n(f_n(x))), \qquad n \in \mathbb{N}_0. \qquad (5.6.23)$$

For $n = 0$ equation (5.6.23) is identical with (5.6.1). For $n \geq 1$ we have

$$\left| (f_n'(0))^k \frac{\partial h}{\partial y}(0, 0) \right| = t_n^k \left| \frac{\partial h}{\partial y}(0, 0) \right| < 1, \qquad k = 1, \dots, r,$$

and $f_n'(0) = t_n < 1$, whence $f_n'(x) \leq 1$ in a right neighbourhood of zero. Moreover, $h_{nk}(0, \dots, 0) = 0$, $k = 1, \dots, r$, $n \in \mathbb{N}_0$, where the h_{nk} are defined by (5.6.2) with the function f replaced by f_n, i.e.

$$h_{nk}(x, y, y_1, \dots, y_k) = h_k(x, y, t_n^k y_1, \dots, t_n^k y_k).$$

All this shows that for $n \geq 1$ equations (5.6.23) fall under the case considered in the first part of the proof.

Thus we know that for every $n \in \mathbb{N}$ equation (5.6.23) has a unique C^r solution $\varphi_n: X \to \mathbb{R}$ fulfilling condition (5.6.7) $(\Phi = \varphi_n)$. By Theorem 5.2.4 we have $\lim_{n \to \infty} \varphi_n(x) = \varphi(x)$ a.u. in X. Moreover, by Lemmas 5.6.1 and 5.6.2,

$$\varphi_n^{(k)}(x) = t_n^k (f'(x))^k \frac{\partial h}{\partial y}(x, \varphi_n(f_n(x))) \varphi_n^{(k)}(f_n(x))$$

$$+ u_{nk}(x, \varphi_n(f_n(x)), \dots, \varphi_n^{(k-1)}(f_n(x))),$$

$k = 1, \dots, r$. The limit case of this equation is the linear one for the unknown φ_{0k}:

$$\varphi_{0k}(x) = (f'(x))^k \frac{\partial h}{\partial y}(x, \varphi(f(x))) \varphi_{0k}(f(x))$$

$$+ u_k(x, \varphi(f(x)), \varphi_{01}(f(x)), \dots, \varphi_{0,k-1}(f(x))),$$

$k = 1, \dots, r$. Making use of Theorems 4.1.1 and 4.1.2 we easily prove by induction that, for every $k = 1, \dots, r$, equation (5.6.24) has a unique continuous solution $\varphi_{0k}: X \to \mathbb{R}$ and $\lim_{n \to \infty} \varphi_n^{(k)}(x) = \varphi_{0k}(x)$ a.u. in X. This implies that the $\varphi^{(k)}$ exist and are continuous in X, $k = 1, \dots, r$, viz. $\varphi^{(k)}(x) = \varphi_{0k}(x)$. Thus φ is of class C^r in X. ∎

Remark 5.6.1. In the case where $r = \infty$ the above proof does not apply directly, but we can argue as in the case of the linear equation; see Remark 3.4.1.

Remark 5.6.2. Lemma 5.6.3 shows that, under assumptions (i)–(iii), if $|(f'(0))^r(\partial h/\partial y)(0, d)| < 1$, then there exists a unique solution of (5.6.1) for any choice of $(d_1, \ldots, d_r)$ satisfying (5.6.6).

Remark 5.6.3. Under additional assumption (5.6.10) one can prove that the solution spoken of in Theorem 5.6.1 depends on given functions in a continuous manner; see Note 5.9.6.

5.6C. Lack of uniqueness of C^r solutions

In order to deal with the case where the C^r solution of (5.6.1) depends on an arbitrary function it will be more convenient to write this equation in the form

$$\varphi(f(x)) = g(x, \varphi(x)). \tag{5.6.25}$$

We start with the formulation of the assumptions.

(I) $X = [0, a], \ 0 < a \leqslant \infty$.

(II) $f: X \to X$ is of class C^r in X, $1 \leqslant r \leqslant \infty$. Moreover, $0 < f(x) < x$ in $X \setminus \{0\}$, and $f'(x) \neq 0$ in X.

(III) $g: \Omega \to \mathbb{R}$ is of class C^r in Ω. For every $x \in X$ the set Ω_x is a nondegenerate interval, $g(x, \Omega_x) = \Omega_{f(x)}$, and $(\partial g/\partial y)(x, y) \neq 0$ in Ω.

We put $\hat{\Omega} = (X \times \mathbb{R}) \cap \Omega$ and define the functions $g_k: \hat{\Omega} \times \mathbb{R}^k \to \mathbb{R}$ by the recurrence

$$g_1(x, y, y_1) = (f'(x))^{-1}\left(\frac{\partial g}{\partial x}(x, y) + \frac{\partial g}{\partial y}(x, y)y_1\right),$$

$$g_{k+1}(x, y, y_1, \ldots, y_{k+1}) = (f'(x))^{-1}\left(\frac{\partial g_k}{\partial x} + \frac{\partial g_k}{\partial y}y_1 + \cdots + \frac{\partial g_k}{\partial y_k}y_{k+1}\right),$$

$$k = 1, \ldots, r-1.$$

Analogues of Lemmas 5.6.1 and 5.6.2 are easily established by induction.

Lemma 5.6.4. *Let hypotheses* (I)–(III) *be satisfied. Then, for every* $k = 1, \ldots, r$, *the function* g_k *is of class* C^{r-k} *in* $\hat{\Omega} \times \mathbb{R}^k$ *and*

$$g_k(x, y, \ldots, y_k) = v_k(x, y, \ldots, y_{k-1}) + (f'(x))^{-k}\frac{\partial g}{\partial y}(x, y)y_k,$$

where $v_k: \hat{\Omega} \times \mathbb{R}^{k-1} \to \mathbb{R}$ *is a function of class* C^{r-k}.

Lemma 5.6.5. *Let hypotheses* (I)–(III) *be satisfied, and let* $\varphi: X \to \mathbb{R}$ *be a function of class* C^r *in* X. *If*

$$\psi(f(x)) = g(x, \varphi(x))$$

for $x \in X$, *then the derivatives of* φ *and* ψ *satisfy the relations*

$$\psi^{(k)}(f(x)) = g_k(x, \varphi(x), \varphi'(x), \ldots, \varphi^{(k)}(x)), \qquad k = 1, \ldots, r.$$

For an $x_0 \in X \setminus \{0\}$ write $X_0 = [f(x_0), x_0]$. The proof of the following extension theorem is straightforward.

Theorem 5.6.2. *Let hypotheses* (I)–(III) *be satisfied and let* $0 \notin X$. *Then every function* $\varphi_0: X_0 \to \mathbb{R}$ *of class* C^r *in* X_0 *and such that* $\varphi_0(x) \in \Omega_x$ *for* $x \in X_0$ *and*

$$\varphi_0(f(x_0)) = g(x_0, \varphi_0(x_0)), \tag{5.6.26}$$

$$\varphi_0^{(k)}(f(x_0)) = g_k(x_0, \varphi_0(x_0), \ldots, \varphi_0^{(k)}(x_0)), \qquad k = 1, \ldots, r, \tag{5.6.27}$$

can be uniquely extended onto X *to a* C^r *solution* $\varphi: X \to \mathbb{R}$ *of equation* (5.6.25). *Thus* (5.6.25) *has in* X *a* C^r *solution depending on an arbitrary function.*[1]

Now we pass to the case where $0 \in X$. Then the values

$$\varphi(0) = d, \qquad \varphi^{(k)}(0) = d_k, \qquad k = 1, \ldots, r,$$

must satisfy the equations

$$d = g(0, d), \tag{5.6.28}$$

$$d_k = g_k(0, d, d_1, \ldots, d_k), \qquad k = 1, \ldots, r. \tag{5.6.29}$$

Since by (II) $|f'(0)| \leq 1$, we may note that if the condition

$$\left| (f'(0))^{-k} \frac{\partial g}{\partial y} (0, d) \right| < 1 \tag{5.6.30}$$

holds for $k = r$, then it also holds for $k = 0, 1, \ldots, r-1$ and it follows from Lemma 5.6.4 that then there is a unique solution $(d_1, \ldots, d_r)$ of (5.6.29) depending on d that satisfies (5.6.28).

We shall prove the following.

Theorem 5.6.3. *Let hypotheses* (I)–(III) *be satisfied and let* $0 \in X$. *If there exist a solution* d *of* $d = g(0, d)$, $c \in (0, a)$, *and* $\delta > 0$ *such that* $R := [0, c] \times [d - \delta, d + \delta] \subset \Omega$ *and if the condition*

$$\left| (f'(0))^{-r} \frac{\partial g}{\partial y} (0, d) \right| < 1 \tag{5.6.31}$$

is satisfied, then equation (5.6.25) *has a* C^r *solution* $\varphi: X \to \mathbb{R}$ *such that* $\varphi(0) = d$ *which depends on an arbitrary function*[1].

Proof. It follows from Remark 5.3.2 that if c and δ are chosen small enough

[1] See Remark 5.3.1.

and $x_0 \in (0, c)$, then every continuous function $\varphi_0: X_0 \to \mathbb{R}$ satisfying (5.6.26) and the condition $|\varphi_0(x) - d| < \delta$ for $x \in X_0$ can be uniquely extended onto X to a continuous solution φ of equation (5.6.25) such that $\varphi(0) = d$. By Theorem 5.6.2 this φ is of class C^r in $X \setminus \{0\}$, provided that φ_0 was so in X_0 and fulfilled conditions (5.6.27). By Lemmas 5.6.4 and 5.6.5 the derivatives of φ satisfy in $X \setminus \{0\}$ the equations $(k = 1, \ldots, r)$

$$\varphi^{(k)}(f(x)) = (f'(x))^{-k} \frac{\partial g}{\partial y}(x, \varphi(x)) \varphi^{(k)}(x) + v_k(x, \varphi(x), \ldots, \varphi^{(k-1)}(x)).$$

(5.6.32)

In virtue of (5.6.31) the numbers d_k are uniquely determined by (5.6.29). Put $\varphi^{(k)}(0) = d_k$, $k = 1, \ldots, r$. Thus equations (5.6.32) are fulfilled also for $x = 0$. Using Theorem 3.1.8 we show by induction that $\varphi^{(k)}$, $k = 1, \ldots, r$, are continuous at $x = 0$. Thus φ is a C^r solution of (5.6.25) in X. ∎

Remark 5.6.4. In the case of $r = \infty$, of course, (5.6.30) is assumed to hold for all $k \in \mathbb{N}_0$. The proof above can then be applied to get C^∞ solutions of equation (5.6.25). It suffices to take a φ_0 of class C^∞ in X_0 and fulfilling conditions (5.6.26) and (5.6.27) for all $k \in \mathbb{N}$. The above argument shows that the extension φ of φ_0 is now of class C^r in X for all $r \in \mathbb{N}$, and hence of class C^∞ in X.

Comments. A weaker version of Theorem 5.6.1 first appeared in Choczewski [3]. The second part of the proof of the Theorem (for the case where condition (5.6.10) does not hold) is due to J. Matkowski [5], [6]. For Theorems 5.6.2 and 5.6.3 see Choczewski [2], [3].

The differentiability of C^r solutions of (5.6.1) with respect to a parameter has been studied by S. Czerwik [11].

Concerning solutions of (5.6.1) whose rth derivative is absolutely continuous see Sieczko [1].

5.7 Local analytic solutions

In this section x is a complex variable and we consider functions with values in $\mathbb{C}$. We shall be interested in solutions φ, analytic in a neighbourhood of the origin, of the equation

$$\varphi(x) = h(x, \varphi(f(x))).$$ (5.7.1)

The results we are going to discuss are due to W. Smajdor [1] and J. Matkowski [4], [7], [8].

5.7A. Unique solution

We make the following assumptions.

(i) $X \subset \mathbb{C}$ is an open set containing the origin.
(ii) $f: X \to X$ is analytic in X, $f(0) = 0$, $|f'(0)| < 1$.
(iii) $h: \Omega \to \mathbb{C}$ is analytic in Ω, where $\Omega \subset \mathbb{C}^2$ is an open set containing a point $(0, d)$ such that $d = h(0, d)$.

We put $\hat{\Omega} = (X \times \mathbb{C}) \cap \Omega$ and define the functions $h_k: \hat{\Omega} \times \mathbb{C}^k \to \mathbb{C}$, $k \in \mathbb{N}$, by recurrence (5.6.2). They are analytic in $\hat{\Omega} \times \mathbb{C}^k$ and formula (5.6.3) holds for $k \in \mathbb{N}$, where $u_k: \hat{\Omega} \times \mathbb{C}^{k-1} \to \mathbb{C}$ are analytic functions.

If $U \subset X$ is a neighbourhood of the origin and $f(U) \subset U$, and $\varphi: U \to \mathbb{C}$ is an analytic solution of equation (5.7.1) such that

$$\varphi(0) = d, \tag{5.7.2}$$

then the values

$$\varphi^{(k)}(0) = d_k, \qquad k \in \mathbb{N}, \tag{5.7.3}$$

satisfy the infinite system of equations

$$d_k = h_k(0, d, d_1, \ldots, d_k), \qquad k \in \mathbb{N}. \tag{5.7.4}$$

System (5.7.4) has a unique solution $(d_1, d_2, \ldots)$ provided

$$S(k) := (f'(0))^k \frac{\partial h}{\partial y} (0, d) \neq 1$$

for all $k \in \mathbb{N}$. In the other case, since $|f'(0)| < 1$, there may exist only one $K \in \mathbb{N}$ such that $S(K) = 1$ and if $u_K(0, d, d_1, \ldots, d_{K-1}) = 0$ (where $d_1, \ldots, d_{K-1}$ have been determined from the first $K - 1$ equations (5.7.4)), then d_K is arbitrary and system (5.7.4) has a family of solutions depending on the parameter d_K. Otherwise system (5.7.4) has no solution.

For any solution $(d_1, d_2, \ldots)$ of (5.7.4) the formal power series

$$\varphi(x) = \sum_{k=1}^{\infty} \frac{d_k}{k!} x^k \tag{5.7.5}$$

satisfies formally equation (5.7.1). We shall prove that (5.7.5) represents an analytic function in a neighbourhood of the origin (W. Smajdor [1]).

Theorem 5.7.1. *Let hypotheses* (i)–(iii) *be fulfilled, and let* $(d_1, d_2, \ldots)$ *be a solution of system* (5.7.4). *Then equation* (5.7.2) *has a unique analytic solution* φ *in a neighbourhood of the origin fulfilling conditions* (5.7.2) *and* (5.7.3). *In other words, series* (5.7.5) *has a positive radius of convergence.*

Proof. Since $|f'(0)| < 1$, there exists an $r \in \mathbb{N}$ such that $|S(r)| < 1$. We shall seek the solution φ in the form

$$\varphi(x) = P(x) + x^r \Phi(x), \qquad P(x) := d + \sum_{k=1}^{r} \frac{d_k}{k!} x^k,$$

where Φ is an analytic function in a neighbourhood of the origin. Such a φ is a solution of (5.7.1) if and only if Φ satisfies the equation

$$\Phi(x) = H(x, \Phi(f(x))), \tag{5.7.6}$$

where

$$H(x, y) = x^{-r}(h(x, P(f(x)) + y(f(x))^r) - P(x)). \tag{5.7.7}$$

We want to find analytic solutions of (5.7.6) with the aid of Banach's Theorem. To this end we should first know that the function H, given by (5.7.7), is analytic in a neighbourhood of $(0, 0)$. The proof of this property of H will be outlined only.

We start with the fact that f and h are analytic in some neighbourhoods of 0 and $(0, d)$, respectively. Fix a $D > 0$. There are $c > 0$ and $\delta > 0$ such that $|(P(f(x)) + f(x)^r y) - d| \leqslant \delta$ for $|x| \leqslant c$, $|y| \leqslant D$. Moreover, $f(x) = xF(x)$, where F is analytic for $|x| \leqslant c$, and $h(x, y) = \sum_{j=1}^{\infty} a_j(x)(y - d)^j$ for $|x| \leqslant c$, $|y - d| \leqslant \delta$, where a_j are analytic for $|x| \leqslant c$. Hence for $|x| \leqslant c$, $|y| \leqslant D$

$$x^r H(x, y) = \sum_{j=0}^{\infty} a_j(x)(P(f(x)) - d + F(x)^r(x^r y))^j - P(x) = \sum_{k=0}^{\infty} b_k(x)x^{kr}y^k,$$

where b_k are analytic for $|x| \leqslant c$.

Since φ formally satisfies (5.7.1), the first nonzero term of the expansion of b_0 is that with x^r. For, insert the formal power series (5.7.5) into (5.7.1) and extract from the equality obtained the formula

$$x^r \Phi(x) = b_0(x) + x^r \sum_{k=1}^{\infty} b_k(x)x^{(k-1)r}(\Phi(f(x)))^k.$$

This yields $b_0(x) = x^r B_0(x)$ with a function B_0 analytic for $|x| \leqslant c$. Consequently, H is analytic for $|x| \leqslant c$, $|y| \leqslant D$.

What is more, the function H fulfils $H(0, 0) = 0$ and

$$\left| \frac{\partial H}{\partial y}(0, 0) \right| = |S(r)| < 1. \tag{5.7.8}$$

Let Φ be the set of all functions Φ which are analytic for $|x| \leqslant c$, $\Phi(0) = 0$, $|\Phi(x)| \leqslant D$ for $|x| \leqslant c$. Endow this set with the metric

$$\rho(\Phi_1, \Phi_2) = \sup_{|x| \leqslant b} |\Phi_1(x) - \Phi_2(x)|,$$

where b is fixed, $0 < b < c$, to get a complete metric space. (Since analytic functions are uniquely extendable, ρ actually is a metric in Φ.)

Equation (5.7.6) induces in Φ the transform T:

$$T(\Phi)(x) = H(x, \Phi(f(x))).$$

This is a contractive self-mapping of Φ provided we have chosen c and D so that the following hold: $|f(x)| \leqslant |x|$ for $|x| \leqslant c$ (see (iii)), $|H(x, y) - H(x, \bar{y})| \leqslant \Theta|y - \bar{y}|$, $\Theta \leqslant 1$, for $|x| \leqslant c$, $|y| \leqslant D$, $|\bar{y}| \leqslant D$ (see (5.7.8)),

$|H(x, 0)| \leq (1 - \Theta)D$ for $|x| \leq c$ (by $H(0, 0) = 0$ and the continuity of H). We omit the detailed verification.

Banach's Theorem yields now the unique solution $\Phi \in \mathbf{\Phi}$ of equation (5.7.6). This implies the existence and uniqueness of a local analytic solution $\varphi(x) = P(x) + x^r\Phi(x)$ of equation (5.7.1). Of course, this φ satisfies (5.7.2) and (5.7.3) for $k = 1, \ldots, r$. But it has to satisfy (5.7.3) also for $k > r$, since, owing to (5.7.8) and the inequality $|f'(0)| < 1$, the d_k are uniquely determined for $k > r$ by (5.7.4). ∎

Remark 5.7.1. The solution φ of (5.7.1) we have found is given as the limit of the sequence of successive approximations $(\varphi_n)_{n \in \mathbb{N}_0}$, where $\varphi_0(x) = P(x) + x^r\Phi_0(x)$, Φ_0 is analytic in a neighbourhood of the origin, $\Phi_0(0) = 0$; and $\varphi_{n+1}(x) = h(x, \varphi_n(f(x)))$.

5.7B. Continuous dependence on the data

Now consider the sequence of equations

$$\varphi_m(x) = h_m(x, \varphi_m(f_m(x))), \qquad m \in \mathbb{N}_0, \tag{5.7.9}$$

and assume the following.

(iv) For every $m \in \mathbb{N}_0$ the function $f_m: X \to X$ is analytic in X, $f_m(0) = 0$, $|f'_m(0)| < 1$.

(v) For every $m \in \mathbb{N}_0$ the function $h_m: \Omega \to \mathbb{C}$ is analytic in Ω, where $\Omega \subset \mathbb{C}^2$ is an open set containing the point $(0, \eta_m)$ such that $\eta_m = h_m(0, \eta_m)$. Moreover,

$$\lim_{m \to \infty} \eta_m = \eta_0. \tag{5.7.10}$$

(vi) $\lim_{m \to \infty} f_m = f_0$ a.u. in X, and $\lim_{m \to \infty} h_m = h_0$ a.u. in Ω.

Let the functions $h_{mk}: \hat{\Omega} \times \mathbb{C}^k \to \mathbb{C}$ be constructed for h_m and f_m, $m \in \mathbb{N}_0$, $k \in \mathbb{N}$, analogously as h_k were for h and f (see (5.6.2)). We shall need solutions $(\eta_{m1}, \eta_{m2}, \ldots)$, for every $m \in \mathbb{N}_0$, of the infinite system

$$\eta_{mk} = h_{mk}(0, \eta_m, \eta_{m1}, \ldots, \eta_{mk}), \qquad k \in \mathbb{N}, \tag{5.7.11}$$

such that

$$\lim_{m \to \infty} \eta_{mk} = \eta_{0k}, \qquad k \in \mathbb{N}. \tag{5.7.12}$$

The question arises whether such solutions of (5.7.11) can actually be found. We shall use the abbreviation

$$S(m, k) := (f'_m(0))^k \frac{\partial h_m}{\partial y}(0, \eta_m). \tag{5.7.13}$$

Consider the case where $S(m, k) \neq 1$ for $m \in \mathbb{N}_0$, $k \in \mathbb{N}$. The solution of (5.7.11) is then unique –

$$\eta_{mk} = (1 - S(m, k))^{-1} u_{mk}(0, \eta_m, \eta_{m1}, \ldots, \eta_{m,k-1}),$$

for $m \in \mathbb{N}_0$, $k \in \mathbb{N}$ (see (5.6.2) and (5.6.3)) – and convergence (5.7.12) follows from (vi). If

$$S(m, K) = 1 \qquad \text{for } m \in \mathbb{N}_0, \tag{5.7.14}$$

where $K \in \mathbb{N}$ is independent of m and there exist η_{mk} satisfying (5.7.11), then (5.7.12) holds if and only if it holds for $k = K$. This is so since if η_{mK} does exist for an $m \in \mathbb{N}_0$ the η_{mk} for $k > K$ depend continuously on it as a parameter. Finally, if K in (5.7.14) may depend on m, $K = K(m)$, then $K(m) = K(0)$ for sufficiently large m. In fact, by (vi) there exist positive constants M and $s < 1$ such that

$$\left| \frac{\partial h_m}{\partial y} (0, \eta_m) \right| \leqslant M, \qquad |f'_m(0)| \leqslant s, \ m \in \mathbb{N}_0,$$

whence, by (5.7.13), (5.7.14),

$$1 = |S(m, K(m))| \leqslant M s^{K(m)}, \qquad m \in \mathbb{N}_0.$$

Thus $K(m)$ is bounded. If $K(m)$ were not ultimately constant, we would have $S(p_m, K_1) = S(q_m, K_2) = 1$ for $K_1 \neq K_2$ and for increasing sequences (p_m), (q_m) of positive integers. Letting $m \to \infty$, we would obtain $S(0, K_1) = S(0, K_2) = 1$, whence necessarily $f'_0(0) = 1$ (see (5.7.13)), a contradiction. Consequently $K(m) = K = \text{const}$ for sufficiently large m, and letting $m \to \infty$ in (5.7.14) we obtain, by (5.7.13), $K = K(0)$.

Now we are ready to formulate the following (Matkowski [4], [7]).

Theorem 5.7.2. *Let hypotheses* (i), (iv)–(vi) *be satisfied, and let the numbers* η_{mk}, $m \in \mathbb{N}_0$, $k \in \mathbb{N}$, *satisfy* (5.7.11) *and* (5.7.12). *Then there exists a* $c > 0$ *such that for every* $m \in \mathbb{N}_0$ *equation* (5.7.9) *has in the disc* $\{x \in \mathbb{C}: |x| < c\}$ *an analytic solution* φ_m *fulfilling the conditions*

$$\varphi_m(0) = \eta_m, \qquad \varphi_m^{(k)}(0) = \eta_{mk}, \qquad k \in \mathbb{N}.$$

Moreover, $\lim_{m \to \infty} \varphi_m = \varphi_0$ *a.u. in the disc.*

Proof. In view of what we have just been discussing, condition (5.7.12) implies that, omitting possibly a finite number of equations (5.7.9), we can find an $r \in \mathbb{N}$ (independent of m) such that $|S(m, r)| < 1$ for all m. Therefore, by a similar argument as in the proof of Theorem 5.7.1, we find, for every $m \in \mathbb{N}_0$, a unique local analytic solution φ_m of (5.7.9) which is given by the formula

$$\varphi_m(x) = P_m(x) + x^r \Phi_m(x), \qquad P_m(x) = \eta_m + \sum_{k=1}^{r} \frac{\eta_{mk}}{k!} x^k. \tag{5.7.15}$$

Every Φ_m in (5.7.15) is obtained as the unique fixed point of the contraction mapping T_m acting in the metric space $\boldsymbol{\Phi}$ we have introduced in that proof, and defined by

$$T_m(\Phi)(x) = H_m(x, \Phi(f_m(x))).$$

The function H_m is given by (5.7.7) with all the symbols h, P, f supplied with the index m. (By (vi), (5.7.10) and (5.7.11) the constants c and D occurring in the definition of Φ can be made independent of m.)

We shall show that $\lim_{m \to \infty} T_m = T_0$ uniformly in Φ, which space is compact in view of Vitali's Theorem.

Take an $\varepsilon > 0$ and a Θ independent of m, $|S(m, r)| < \Theta < 1$. Since $(\partial H_m / \partial y)(0, 0) = S(m, r)$ (see (5.7.8) and (5.7.13)) we can assume that c and D are so small that we have for $m \in \mathbb{N}_0$

$$|H_m(x, y) - H_m(x, \bar{y})| \leqslant \Theta |y - \bar{y}|, \qquad |x| \leqslant c, |y| \leqslant D, |\bar{y}| \leqslant D. \quad (5.7.16)$$

By (vi) the sequence (H_m) tends to H_0 uniformly for $|x| \leqslant c, |y| \leqslant D$. Thus there is an $N \in \mathbb{N}$ such that for $|x| \leqslant c, |y| \leqslant D$

$$|H_m(x, y) - H_0(x, y)| \leqslant (1 - \Theta)\varepsilon \qquad \text{for } m \geqslant N. \quad (5.7.17)$$

The space Φ is compact. Thus $L := \sup_{\Phi \in \Phi} \sup_{|x| \leqslant b} |\Phi'(x)|$ is finite (b occurs in the definition of the metric in Φ). Since f_m tends to f_0 uniformly for $|x| \leqslant b$, we have for $\Phi \in \Phi, |x| \leqslant b$

$$|\Phi(f_m(x)) - \Phi(f_0(x))| \leqslant L |f_m(x) - f_0(x)| \leqslant \varepsilon \qquad \text{for } m > N, \quad (5.7.18)$$

provided N is large enough.

Consequently, in view of (5.7.16), (5.7.17) and (5.7.18), we obtain for $\Phi \in \Phi$, $|x| \leqslant b$ and $m > N$

$$\begin{aligned} |T_m(\Phi)(x) - T_0(\Phi)(x)| &\leqslant |H_m(x, \Phi(f_m(x))) - H_0(x, \Phi(f_m(x)))| \\ &\quad + |H_0(x, \Phi(f_m(x))) - H_0(x, \Phi(f_0(x)))| \\ &\leqslant (1 - \Theta)\varepsilon + \Theta |\Phi(f_m(x)) - \Phi(f_0(x))| \leqslant \varepsilon, \end{aligned}$$

whence

$$\rho(T_m(\Phi), T_0(\Phi)) \leqslant \varepsilon \qquad \text{for } m > N \text{ and } \Phi \in \Phi.$$

By Theorem 1.6.4 the sequence Φ_m of the fixed points of T_m tends in the metric of Φ to the fixed point Φ_0 of the limit mapping T_0. In other words, $\lim_{m \to \infty} \Phi_m = \Phi_0$ uniformly in $\{x \in \mathbb{C} : |x| \leqslant b\}$, and hence, by Vitali's Theorem, a.u. in the disc $\{x \in \mathbb{C} : |x| < c\}$. By (5.7.15), in view of (5.7.10) and (5.7.12), (φ_m) converges to φ_0 a.u. in the latter disc. ∎

Comments. For the first version of Theorem 5.7.1 see Read [1]. Analytic solutions of (5.7.1) are also studied by G. P. Pelyukh [1], J. Ger [2]–[5], and continuous dependence in the nonuniqueness case by J. Matkowski [8].

5.8 Equations in measure spaces

We conclude this chapter with a study of integrable solutions φ of the equation

$$\varphi(x) = h(x, \varphi(f(x))). \quad (5.8.1)$$

We shall follow the notation and conventions adopted in Section 4.7, adding one more definition to them (see Šragin [1]).

Definition 5.8.1. Let (X, S, μ) be a measure space. A function $h: X \times \mathbb{R} \to \mathbb{R}$ is called *sup-measurable* iff the function $h(\cdot, \psi)$ is measurable for every measurable function $\psi: X \to \mathbb{R}$.

Remark 5.8.1. In the case where $X \subset \mathbb{R}^N$ and μ is the Lebesgue measure, h is sup-measurable whenever for every $y \in \mathbb{R}$ the function $h(\cdot, y)$ is measurable and for almost every $x \in X$ the function $h(x, \cdot)$ is continuous (see Carathéodory [1], Šragin [1]).

We assume that

(i) (X, S, μ) is a σ-finite measure space, $\mu(X) > 0$,
(ii) $f: X \to X$ is one-to-one, both f and f^{-1} are S-measurable and the measures μf and μf^{-1} (see Section 4.7) are absolutely continuous with respect to μ and

$$\mu\left(X \setminus \bigcup_{n=0}^{\infty} A_n\right) = 0, \qquad A_n := f^n(X) \setminus f^{n+1}(X), \ n \in \mathbb{N}_0,$$

(iii) $h: X \times \mathbb{R} \to \mathbb{R}$ is sup-measurable, and

$$|h(x, y) - h(x, \bar{y})| \leq g(x)|y - \bar{y}| \qquad \text{for } y, \bar{y} \in \mathbb{R}, \qquad (5.8.2)$$

where the function $g: X \to \bar{\mathbb{R}}$ is measurable and finite a.e. in X.

We may note that the continuity of $h(x, \cdot)$ (a.e. in X) results from condition (5.8.2).

As in the linear case (Section 4.7), the results we are going to present are due to J. Matkowski [13]. Consult also Baron–Ger [2].

5.8A. Existence and uniqueness of L^p solutions

Since by (5.8.2) the difference φ of two solutions φ_1, φ_2 of equation (5.8.1) satisfies inequality (4.7.10), $|\varphi(x)| \leq g(x)|\varphi(f(x))|$, the following uniqueness theorem is an immediate consequence of Theorems 4.7.1 and 4.7.2. Here the sequences (u_n) and (w_n) are defined by (4.7.8), and $L^p := L^p(X)$.

Theorem 5.8.1. *Let hypotheses* (i)–(iii) *be fulfilled. If either*

(a) *the series* $\sum_{n=0}^{\infty} \prod_{i=0}^{n-1} u_i$ *diverges,*

or

(b) *the sequence* $(\prod_{i=0}^{n-1} w_i)_{n \in \mathbb{N}}$ *is bounded,*

then every two solutions $\varphi_1, \varphi_2 \in L^p$ *of* (5.8.1) *are equal a.e. in* X.

To ensure the existence we have to make stronger assumptions.

Theorem 5.8.2. *Let hypotheses* (i)–(iii) *be fulfilled. If* $h(\cdot, 0) \in L^p$ *and there exists an* $s \in (0, 1)$ *such that*

$$(g)^p \leqslant s \frac{\mathrm{d}(\mu f)}{\mathrm{d}\mu} \qquad a.e.\ in\ X, \tag{5.8.3}$$

then equation (5.8.1) *has an essentially unique* L^p *solution* φ, *given by*

$$\varphi(x) = \lim_{n \to \infty} \varphi_n(x) \qquad a.e.\ in\ X, \tag{5.8.4}$$

where $\varphi_0 \in L^p$ *is arbitrary, and*

$$\varphi_{n+1}(x) = h(x, \varphi_n(f(x))), \qquad n \in \mathbb{N}_0. \tag{5.8.5}$$

Proof. The form of φ suggests a proof via Banach's Theorem. We take the space $\mathbf{L}^p = L^p / \approx$, where the relation $\approx$ denotes equality a.e. in X. When equipped with the norm $\|\cdot\|$ defined by (4.7.7) with (4.7.6), $\mathbf{L}^p$ is a Banach space. Let T be defined for $\varphi \in \mathbf{L}^p$ by

$$T(\varphi) = h(\cdot, \varphi \circ f) \qquad a.e.\ in\ X. \tag{5.8.6}$$

If $\varphi \in L^p$, then $\varphi \circ f$ is measurable (since f is S-measurable). Thus so is $T(\varphi)$ in view of (iii). The inequality

$$(a + b)^p \leqslant (2 \max(a, b))^p \leqslant 2^p(a^p + b^p), \qquad a \geqslant 0, b \geqslant 0,$$

will be useful in proving that $T(\varphi)$ is in L^p. Indeed, fix an $x \in X$ and take $a = |h(x, \varphi \circ f(x)) - h(x, 0)|$ and $b = |h(x, 0)|$ and estimate by (5.8.2)

$$|h(x, \varphi \circ f(x))|^p \leqslant 2^p((g(x))^p |\varphi \circ f(x)|^p + |h(x, 0)|^p).$$

Therefore, in view of (5.8.3), we have

$$|T(\varphi)|^p \leqslant 2^p \left(s \left(\frac{\mathrm{d}(\mu f)}{\mathrm{d}\mu} \right) |\varphi \circ f|^p + |h(\cdot, 0)|^p \right)$$

a.e. in X, whence

$$\int_X |T(\varphi)|^p \mathrm{d}\mu \leqslant 2^p \left(s \int_{f(X)} |\varphi|^p \mathrm{d}\mu + \int_X |h(\cdot, 0)|^p \mathrm{d}\mu \right) < \infty.$$

i.e. $T(\varphi) \in L^p$. Moreover, if $\varphi_1 = \varphi_2$ a.e. in X, then, by the absolute continuity of the measure μf^{-1} with respect to μ, $T(\varphi_1) = T(\varphi_2)$ a.e. in X. Thus $T(\mathbf{L}^p) \subset \mathbf{L}^p$.

Finally, T is a contraction. For, take $\varphi_1, \varphi_2 \in \mathbf{L}^p$ and use (5.8.2) and (5.8.3):

$$\| T(\varphi_1) - T(\varphi_2) \|^{1/\alpha(p)} = \int_X |h(\cdot, \varphi_1 \circ f) - h(\cdot, \varphi_2 \circ f)|^p \mathrm{d}\mu$$

$$\leqslant s \int_X |(\varphi_1 \circ f) - (\varphi_2 \circ f)|^p \frac{\mathrm{d}(\mu f)}{\mathrm{d}\mu} \mathrm{d}\mu \leqslant s \|\varphi_1 - \varphi_2\|^{1/\alpha(p)},$$

and the desired inequality results.

By Banach's Theorem there exists an essentially unique solution $\varphi \in \mathbf{L}^p$ of equation (5.8.1), and the sequence (φ_n) given by (5.8.5) converges to φ in measure. Now, since (5.8.5) can be written as $\varphi_{n+1} = T(\varphi_n)$, the above estimation yields by induction

$$\|\varphi_{n+1} - \varphi_n\| \leqslant s^{\alpha(p)n} \|\varphi_1 - \varphi_0\|, \qquad n \in \mathbb{N}_0,$$

and relation (5.8.4) results from Lemma 4.7.1. ∎

The next theorem resembles slightly Theorem 5.2.7 in which continuous solutions of (5.8.1) have been found with the help of some solutions of associated inequalities.

Theorem 5.8.3. *Let hypotheses* (i)–(iii) *be satisfied. If there exist* $\varphi_1^*, \varphi_2^* \in L^p$ *satisfying inequalities* (5.2.11) *and* (5.2.12), *respectively, a.e. in* X, $\varphi_1^* \leqslant \varphi_2^*$ *a.e. in* X, *and if, for almost every* $x \in X$, *the function* $h(x, \cdot)$ *is increasing in the interval* $[\varphi_1^*(x), \varphi_2^*(x)]$, *then equation* (5.8.1) *has a solution in* L^p.

If, moreover, condition (5.8.3) *holds with* $s = 1$, *then this solution is essentially unique.*

Proof. Put $\varphi_0 = \varphi_1^*$, and define the functions φ_n by (5.8.5). Then all φ_n are measurable, and for almost all $x \in X$ the sequence $(\varphi_n(x))$ is increasing. Moreover, $\varphi_1^* \leqslant \varphi_n \leqslant \varphi_2^*$ a.e. in X, $n \in \mathbb{N}_0$. Thus $\varphi_n \in L^p$, $n \in \mathbb{N}_0$, and $\varphi = \lim_{n \to \infty} \varphi_n$ exists a.e. in X and belongs to L^p. Since by (5.8.2) the function $h(x, \cdot)$ is continuous for almost all $x \in X$, this φ satisfies equation (5.8.1) a.e. in X. Since $s = 1$ in (5.8.3), we have $u_i \geqslant 1$, and the uniqueness assertion results from Theorem 5.8.1(a). ∎

Still another condition ensuring the existence of L^p solutions of (5.8.1), which we are going to present, is quite analogous to that in Theorem 4.7.5 for linear equations.

Theorem 5.8.4. *Let hypotheses* (i)–(iii) *be satisfied. If* $h(\cdot, 0) \in L^p$ *and the series*

$$\sum_{n=0}^{\infty} \prod_{i=0}^{n-1} w_i^{\alpha(p)} \left(\int_{f^n(X)} |h(\cdot, 0)|^p \, d\mu \right)^{\alpha(p)}$$

converges, then equation (5.8.1) *has a solution* $\varphi \in L^p$.

If, moreover, condition (b) *of Theorem 5.8.1 is fulfilled, then this solution is essentially unique.*

Proof. Take $\varphi_0 = 0$ and define the sequence (φ_n) by (5.8.5). Putting $\hat{h} := h(\cdot, 0)$ we obtain by induction, applying (5.8.2),

$$|\varphi_{n+1} - \varphi_n| \leqslant |\hat{h} \circ f^n| \prod_{i=0}^{n-1} g \circ f^i \qquad \text{a.e. in } X, n \in \mathbb{N}_0,$$

whence (see (4.7.5)),

$$\|\varphi_{n+1} - \varphi_n\|^{1/\alpha(p)} = \int_X |\varphi_{n+1} - \varphi_n|^p \, d\mu$$

$$\leq \int_X \prod_{i=0}^{n-1} \left((g \circ f^i)^p \bigg/ \left(\frac{d(\mu f)}{d\mu} \circ f^i \right) \right) |\hat{h} \circ f^n|^p \frac{d(\mu f^n)}{d\mu} \, d\mu$$

$$\leq \left(\prod_{i=0}^{n-1} w_i \right) \int_{f^n(X)} |\hat{h}|^p \, d\mu.$$

Thus $\varphi_n \in L^p$ for $n \in \mathbb{N}_0$ and the series $\sum_{n=0}^{\infty} \|\varphi_{n+1} - \varphi_n\|$ converges. By Lemma 4.7.1 the sequence $(\varphi_n)_{n \in \mathbb{N}_0}$ converges a.e. in X to a function $\varphi \in L^p$.

This φ satisfies (5.8.1) a.e. in X. For, let A be the set of those $x \in X$ at which $(\varphi_n(x))$ diverges, and let $B \subset \{x \in X : g(x) = \infty\}$ be the set of those $x \in X$ for which $h(x, \cdot)$ is not continuous. The sets A and B as well as the union $C := \bigcup_{n=-\infty}^{\infty} f^n(A \cup B)$ have measure zero. We have $f(C) = C$, whence $f(X \setminus C) = f(X) \setminus C$. Therefore $f(X \setminus C) \subset X \setminus C$, i.e., if $x \in X \setminus C$ then $f(x) \in X \setminus C$, so that the function $h(x, \cdot)$ is continuous and there exist the limits $\lim_{n \to \infty} \varphi_n(x) = \varphi(x)$ and $\lim_{n \to \infty} \varphi_n(f(x)) = \varphi(f(x))$. Hence by (5.8.4) and (5.8.5)

$$\varphi(x) = \lim_{n \to \infty} \varphi_n(x) = \lim_{n \to \infty} h(x, \varphi_{n-1}(f(x))) = h(x, \varphi(f(x))).$$

Thus φ satisfies equation (5.8.1) in $X \setminus C$, which means a.e. in X. The uniqueness assertion results from Theorem 5.8.1(b). ∎

5.8B. L^1 solutions

Before proceeding with a theorem on L^1 solutions of (5.8.1) we shall prove a lemma on Jensen's inequality (see e.g. Chapter V in Feller [1], Kuczma [51, p. 181]).

Lemma 5.8.1. *Let* (X, S, μ) *be a measure space with* $\mu(X) = 1$, *let* $I \subset \mathbb{R}$ *be an interval, and let* $\gamma : I \to \mathbb{R}$ *be a concave function. If* $\varphi : X \to I$ *belongs to* L^1, *then*

$$\int_X \gamma \circ \varphi \, d\mu \leq \gamma \left(\int_X \varphi \, d\mu \right). \tag{5.8.8}$$

Proof. Fix some $b \in \text{int } I$. There is a $k \in \mathbb{R}$ such that for $t \in I$

$$\gamma(t) - \gamma(b) \leq k(t - b).$$

For instance, any k between the right and left derivatives of (concave!) γ at b will do. Choosing $b = \int_X \varphi \, d\mu$, $t = \varphi(x)$, and integrating over X, we obtain (5.8.8), because the right-hand side of the inequality we get after integration equals zero. ∎

Remark 5.8.2. If γ is convex instead of concave, Lemma 5.8.1 remains valid, but the inequality sign in (5.8.8) is reversed.

Now we replace (iii) by the following assumption.

(iv) $h: X \times \mathbb{R} \to \mathbb{R}$ is sup-measurable, and

$$|h(x, y) - h(x, \bar{y})| \leq g(x)\gamma(|y - \bar{y}|) \qquad \text{for } y, \bar{y} \in \mathbb{R},$$

where $g: X \to \mathbb{R}$ is measurable and finite a.e. in X, $\gamma: \mathbb{R}^+ \to \mathbb{R}^+$ is concave and $\gamma(t) < t$ for $t > 0$.

Theorem 5.8.5. *Let hypotheses* (i), (ii), (iv) *be satisfied, and let* $\mu(X) = 1$. *If* $h(\cdot, 0) \in L^1$ *and* g *satisfies* (5.8.3) *with* $p = s = 1$, *then equation* (5.8.1) *has an essentially unique solution* $\varphi \in L^1$.

Proof. As in the proof of Theorem 5.8.2 we consider the Banach space L^1 of integrable functions on X modulo equality almost everywhere. The norm in L^1 is defined by (4.7.7) with $p = \alpha(p) = 1$.

Define the transform T on L^1 by (5.8.6). The same argument as in the proof of Theorem 5.8.2 when applied to the case $p = 1$ and $s = 1$ shows that T maps the space L^1 into itself.

Further we have for $\varphi_1, \varphi_2 \in L^1$

$$\|T(\varphi_1) - T(\varphi_2)\| \leq \int_X \gamma(|(\varphi_1 \circ f) - (\varphi_2 \circ f)|) \frac{\mathrm{d}(\mu f)}{\mathrm{d}\mu}\, \mathrm{d}\mu$$

$$= \int_{f(X)} \gamma(|\varphi_1 - \varphi_2|)\mathrm{d}\mu \leq \int_X \gamma(|\varphi_1 - \varphi_2|)\mathrm{d}\mu \leq \gamma(\|\varphi_1 - \varphi_2\|),$$

where the last inequality results from Lemma 5.8.1. Thus the theorem follows from the fixed-point theorem 1.5.2. ∎

Remark 5.8.3. It also follows from Theorem 1.5.2 that the sequence of successive approximations (5.8.5), with an arbitrary $\varphi_0 \in L^1$, converges to φ in measure.

5.8C. Extension theorems

In many cases the assumptions guaranteeing the existence and uniqueness of integrable solutions of equation (5.8.1) are not satisfied in the whole of X, but only in a certain subset of X. Then the preceding theorems furnish integrable solutions of (5.8.1) in this subset only. The question arises as to whether such a solution can be extended to one which is integrable on X. Below we prove two theorems to this effect.

Let X_0 be a subset of X. Write

$$X^* := X \setminus X_0.$$

We make the following assumptions.

(v) There exists a (finite or infinite) sequence of $X_n \in S$, $n \in \mathbb{N}$, such that $X = \bigcup X_n$, $X_n \subsetneqq X_{n+1}$.

(vi) $f: X \to X$ is one-to-one, f and f^{-1} are S-measurable, and the measures μf and μf^{-1} are absolutely continuous with respect to μ. Moreover

$$f(X_{n+1}) \subset X_n, \qquad \mu\left(X_0 \setminus \bigcup_{k=0}^{\infty} (f^k(X_0) \setminus f^{k+1}(X_0))\right) = 0.$$

(vii) $h: X \times \mathbb{R} \to \mathbb{R}$ is sup-measurable, $h(\cdot, 0) \in L^p(X^*)$. Moreover,

$$|h(x, y) - h(x, \bar{y})| \le g(x)|y - \bar{y}| \qquad \text{for } x \in X^*, \ y, \bar{y} \in \mathbb{R},$$

where $g: X^* \to \mathbb{R}$ is measurable and finite a.e. in X^*, and inequality (5.8.3) holds with an $s > 0$.

We start with the following.

Theorem 5.8.6. *Let hypotheses* (i) *and* (v)–(vii) *be satisfied, and let* $\varphi_0 \in L^p(X_0)$ *be a solution of equation* (5.8.1) *in* X_0. *Then there exists an essentially unique extension* φ *of* φ_0 *onto* X *satisfying equation* (5.8.1) *a.e. in* X. *Moreover,* $\varphi \in L^p(X_n)$ *for* $n \in \mathbb{N}_0$. *Hence* $\varphi \in L^p(X)$ *whenever the sequence* (X_n) *is finite.*

Proof. We define the functions φ_n, $n \in \mathbb{N}$, a.e. on X_n, by recurrence (5.8.5). It is easy to see that $\varphi_{n+1}|X_n = \varphi_n$ and that the function φ defined a.e. on X by the formula

$$\varphi(x) = \varphi_n(x) \qquad \text{for } x \in X_n, \ n \in \mathbb{N}_0,$$

is the desired extension which is essentially unique.

It remains to prove that $\varphi_n \in L^p(X_n)$, $n \in \mathbb{N}_0$. For $n = 0$ it is true by hypothesis. Assume that $\varphi_n \in L^p(X_n)$ for an $n \ge 0$ and write

$$Y_n := X_{n+1} \setminus X_n \tag{5.8.9}$$

so that $Y_n \subset X^*$. Similarly as in the proof of Theorem 5.8.2 we obtain

$$\int_{Y_n} |\varphi_{n+1}|^p \mathrm{d}\mu = \int_{Y_n} |h(\cdot, \varphi_n \circ f)|^p \mathrm{d}\mu \le \int_{Y_n} (g|\varphi_n \circ f| + |h(\cdot, 0)|)^p \mathrm{d}\mu$$

$$\le 2^p \int_{X_{n+1}} (g|\varphi_n \circ f|)^p \mathrm{d}\mu + 2^p \int_{X^*} |h(\cdot, 0)|^p \mathrm{d}\mu$$

$$\le 2^p s \int_{X_n} |\varphi_n|^p \mathrm{d}\mu + 2^p \int_{X^*} |h(\cdot, 0)|^p \mathrm{d}\mu < \infty.$$

Hence

$$\int_{X_{n+1}} |\varphi_{n+1}|^p \mathrm{d}\mu = \int_{Y_n} |\varphi_{n+1}|^p \mathrm{d}\mu + \int_{X_n} |\varphi_n|^p \mathrm{d}\mu < \infty,$$

since $\varphi_{n+1} = \varphi_n$ on X_n and $X_{n+1} = Y_n \cup X_n$. Induction completes the proof. ∎

Assuming more we can prove that $\varphi \in L^p(X)$ also in the case of some infinite (X_n).

Theorem 5.8.7. *Let the assumptions of Theorem 5.8.6 be satisfied, let the sequence (X_n) be infinite, and let $f(X_{n+1}) = X_n$ for $n \in \mathbb{N}_0$. If, moreover, there exists an $N \in \mathbb{N}$ and an $s \in (0, 1)$ such that (5.8.3) holds a.e. in $X \setminus X_N$, and if the series (with Y_n defined by (5.8.9))*

$$\sum_{n=0}^{\infty} \left(\int_{Y_n} |h(\cdot, 0)|^p d\mu \right)^{\alpha(p)}$$

converges, then $\varphi \in L^p(X)$.

Proof. We have $X \setminus X_N = \bigcup_{n \geq N} Y_n$. Let us put, for $n \geq N$,

$$a_n := \left(\int_{Y_{n+1}} |\varphi_n|^p d\mu \right)^{\alpha(p)}, \qquad b_n := \left(\int_{Y_n} |h(\cdot, 0)|^p d\mu \right)^{\alpha(p)}.$$

By estimates similar to those occurring in the proof of Theorem 5.8.6, using additionally Minkowski's inequality (see Note 5.9.16), we arrive at

$$a_{n+1} \leq s^{\alpha(p)} a_n + b_n, \qquad n > N,$$

where $s^{\alpha(p)} < 1$. In virtue of Theorem 6.7.1 or 6.7.2 the series $\sum a_n$ converges, whence again by Minkowski's inequality

$$\left(\int_{X \setminus X_N} |\varphi| d\mu \right)^{\alpha(p)} \leq \sum_{n=1}^{\infty} a_{N+n} < \infty.$$

By Theorem 5.8.6 we have $\varphi \in L^p(X_N)$, whence

$$\left(\int_X |\varphi|^p d\mu \right)^{\alpha(p)} \leq \left(\int_{X_N} |\varphi|^p d\mu \right)^{\alpha(p)} + \left(\int_{X \setminus X_N} |\varphi|^p d\mu \right)^{\alpha(p)} < \infty$$

which proves that $\varphi \in L^p$. ∎

5.8D. L^p solution depending on an arbitrary function

It is more convenient to study this problem for the equation

$$\varphi(f(x)) = g(x, \varphi(x)). \tag{5.8.10}$$

The function g will be subjected to the following condition.

(viii) $g: X \times \mathbb{R} \to \mathbb{R}$ is sup-measurable, and

$$|g(x, y) - g(x, \bar{y})| \leq q(x)|y - \bar{y}| \qquad \text{for } y, \bar{y} \in \mathbb{R},$$

where $q: X \to \mathbb{R}^+$ is measurable and finite a.e. in X.

Theorem 5.8.8. *Let hypotheses (i), (ii) and (viii) be satisfied. If there exists an $s \in (0, 1)$ such that*

$$(q \circ f^{-1})^p \leq s \frac{d(\mu f)}{d\mu} \qquad \text{a.e. in } X, \tag{5.8.11}$$

and the series (with A_n defined in (ii))

$$\sum_{n=1}^{\infty} \left(\int_{A_n} |g(f^{-1}, 0)|^p d\mu \right)^{\alpha(p)}$$

converges, then every measurable function $\varphi_0 \in L^p(A_0)$ can be uniquely extended to a solution $\varphi: X \to \mathbb{R}$ of equation (5.8.10), and $\varphi \in L^p(X)$.

Proof. It is easily checked that the required solution-extension is given by

$$\varphi(x) = \varphi_n(x) \qquad \text{for } x \in A_n, \ n \in \mathbb{N}_0,$$

where

$$\varphi_{n+1}(x) = g(f^{-1}(x), \varphi_n(f^{-1}(x))) \qquad \text{for } x \in A_{n+1}, \ n \in \mathbb{N}_0.$$

Hence

$$|\varphi_{n+1}| \le (q \circ f^{-1})|\varphi_n \circ f^{-1}| + |g(f^{-1}, 0)|,$$

whence we get by Minkowski's inequality and (5.8.11)

$$\left(\int_{A_{n+1}} |\varphi_{n+1}|^p d\mu \right)^{\alpha(p)} \le s^{\alpha(p)} \left(\int_{A_n} |\varphi_n|^p d\mu \right)^{\alpha(p)} + \left(\int_{A_{n+1}} |g(f^{-1}, 0)|^p d\mu \right)^{\alpha(p)}$$

which can be written in the form $\tilde{a}_{n+1} \le s^{\alpha(p)} \tilde{a}_n + \tilde{b}_n$. The proof then ends similarly to that of Theorem 5.8.7. ∎

5.9 Notes

5.9.1. We supply an example to illustrate situations we have discussed in Subsection 5.2B.

Example 5.9.1. Consider the equation

$$\varphi(x) = x + (1 + x^2)\varphi(f(x))/(1 + \varphi(f(x))), \tag{5.9.1}$$

with $f(x) = x/(1 + x)$, $X = \mathbb{R}^+$, $\Omega = \mathbb{R}^+ \times \mathbb{R}^+$, $d = 0$, $R = [0, 1] \times [0, 1]$. We have $h(x, y) = x + (1 + x^2)y/(y + 1)$, $\gamma(x) = 1 + x^2$, and

$$f^n(x) = \frac{x}{1 + nx}, \qquad G_n(x) = \prod_{i=0}^{n-1} \left(1 + \left(\frac{x}{1 + ix} \right)^2 \right).$$

By Theorem 5.2.8 equation (5.9.1) has a unique continuous solution $\varphi: \mathbb{R}^+ \to \mathbb{R}$ such that $\varphi(0) = 0$. On the other hand, Theorem 5.2.6 cannot be applied here, since series (5.2.8) with $H(x) = |h(x, 0)| = x$, $x \in \mathbb{R}^+$,

$$\sum_{n=0}^{\infty} G_n(x) H(f^n(x)) = \sum_{n=0}^{\infty} G_n(x) \frac{x}{1 + nx} \ge \sum_{n=0}^{\infty} \frac{x}{1 + nx},$$

diverges for $x > 0$.

5.9.2. We exhibit the example we have announced in Subsection 5.1A. It shows the behaviour of continuous solutions φ of a nonlinear equation, fulfilling $\varphi(0) = d$, for different values of d.

Example 5.9.2. Take the simple equation

$$\varphi(f(x)) = \varphi(x)^3 \qquad (5.9.2)$$

with $X = \mathbb{R}^+$, $\Omega = X^2$, and an arbitrary f fulfilling hypothesis (ii) in Section 5.3. Here $g(x, y) = y^3$ and the equation $d = g(0, d) = d^3$ has two solutions $d_1 = 0$ and $d_2 = 1$ belonging to $\Omega_0 = X$. In a neighbourhood of $(0, d_1)$ the conditions of Theorem 5.3.4 are fulfilled and equation (5.9.2) has a continuous solution $\varphi: X \to \mathbb{R}$ such that $\varphi(0) = 0$ depending on an arbitrary function.

On the other hand, if we write (5.9.2) in the equivalent form

$$\varphi(x) = \varphi(f(x))^{1/3},$$

i.e. in form (5.9.4) with $h(x, y) = y^{1/3}$, then in a neighbourhood of $(0, d_2)$ we have $|h(x, y) - h(x, \bar{y})| \leqslant \Theta |y - \bar{y}|$, $\Theta < 1$. Theorem 5.2.3 applies and thus equation (5.9.2) has a unique continuous solution $\varphi: X \to \mathbb{R}$ such that $\varphi(0) = 1$ (in fact it is the solution $\varphi = 1$).

5.9.3. Example 5.9.2 shows also that in the case which has been studied in Section 5.3, only functions φ_0 sufficiently close to d can be extended to a continuous solution of (5.9.4) on X fulfilling $\varphi(0) = d$. In the case of equation (5.9.2) every continuous $\varphi_0: X_0 \to \mathbb{R}^+$ satisfying $\varphi_0(f(x_0)) = (\varphi(x_0))^3$ can be extended to a continuous solution $\varphi: (0, \infty) \to \mathbb{R}^+$ of (5.9.2). But only if $\varphi_0(x) < 1$ in X_0, then φ_0 can be extended to a continuous solution of (5.9.2) on X, by putting $\varphi(0) = 0$.

Actually, the limit of φ at zero is $0, 1, \infty$ according to whether $\varphi_0(x)$ is less than, equal to, or greater than 1 in X_0. Moreover, if φ_0 takes in X_0 values both less and greater than 1, then all the points of $[0, \infty]$ are the cluster points of $\varphi(x)$ as $x \to 0$.

5.9.4. Consider the equation

$$\varphi(f(x)) = g(x, \varphi(x)). \qquad (5.9.3)$$

If in (iii) of Section 5.3 we replace the equality sign in $g(x, \Omega_x) = \Omega_{f(x)}$ by the inclusion '$\subset$' and release the invertibility condition on $g(x, \cdot)$, then in Theorems 5.3.2–5.3.4 we obtain the solution of (5.9.3) (depending on an arbitrary function) only locally, in a right neighbourhood of zero. See Remark 5.3.1.

5.9.5. The unique solution φ of the equation

$$\varphi(x) = h(x, \varphi(f(x))) \qquad (5.9.4)$$

we have found in Theorems 5.4.2 or 5.4.3 can be obtained as the limit of a sequence of successive approximations (5.4.9). The initial term φ_0 is to be taken as a function equal near zero to the polynomial $P(x)$ or $p(x)$ and such that $\varphi_0(f(x)) \in \Omega_x$.

5.9.6. The solutions of (5.9.4) occurring in Theorems 5.4.2 and 5.4.3 are increasing, or increasing and convex, provided we impose analogous conditions on f and h (see Theorem 5.1.1(c), (d)) and on the polynomials occurring in (5.4.8). Note also the following (Kuczma [38]).

Proposition 5.9.1. *Let the hypotheses of Theorem 5.4.3 be satisfied, and assume that f is increasing and convex in X, and h is increasing with respect to either variable and convex in Ω. Then equation (5.9.4) has a unique convex solution $\varphi: X \to \mathbb{R}$ such that $\varphi(x) = d + o(x)$, $x \to 0$.*

(Uniqueness follows from Theorem 5.4.3, since the asymptotic property and convexity of φ imply condition (5.4.13).)

5.9.7. The main Theorem 5.6.1 on C^r solutions has, among others, the following consequence (Choczewski [3]).

Proposition 5.9.2. *Let hypotheses (i)–(iii) of Section 5.6 be satisfied. If the condition*

$$\left| \frac{\partial h}{\partial y}(0, d) \right| < 1 \tag{5.9.5}$$

holds, then equation (5.9.4) has a unique continuous solution $\varphi: X \to \mathbb{R}$ such that $\varphi(0) = d$. This solution is in fact of class C^r in X.

(Such a φ does exist and is unique because of Theorem 5.2.3. On the other hand, by Theorem 5.6.1 equation (5.9.4) has a unique C^r solution $\hat{\varphi}: X \to \mathbb{R}$ satisfying $\hat{\varphi}(0) = d$, $\hat{\varphi}^{(k)}(0) = d_k$, $k = 1, \ldots, r$, with d_k uniquely determined by (5.6.6) (see (5.6.3)). Therefore $\hat{\varphi} = \varphi$, and $\varphi \in C^r(X)$.)

5.9.8. Change (5.9.5) into $|(f'(0))^p (\partial h/\partial y)(0, d)| < 1$ with a p, $0 < p < r$, and fix some $d_1, \ldots, d_p$ satisfying (5.6.6). An analogue of Proposition 5.9.3 yields the only C^p solution of (5.9.4) such that $\varphi(0) = d$, $\varphi^{(k)}(0) = d_k$, $k = 1, \ldots, p$, which, moreover, is of class C^r in X.

5.9.9. We formulate the theorem on the continuous dependence of C^r solutions on the data which we mentioned in Remark 5.6.3.

Consider the sequence of equations

$$\varphi_m(x) = h_m(x, \varphi_m(f_m(x))), \qquad m \in \mathbb{N}_0, \tag{5.9.6}$$

and define the functions h_{mk}, $k = 1, \ldots, r$, by formulae (5.6.2) with f and h replaced by f_m and h_m, respectively.

Proposition 5.9.3. *Let hypotheses (i), (ii) (with f_m in place of f) and (iii) (with h_m and η_m, $m \in \mathbb{N}_0$, in place of h and d, respectively) of Section 5.6 be satisfied. Let the numbers $\eta_{m1}, \eta_{m2}, \ldots, \eta_{mr}$, $m \in \mathbb{N}_0$, satisfy the system $\eta_{mk} = h_{mk}(0, \eta_m, \eta_{m1}, \ldots, \eta_{mk})$, $k = 1, \ldots, r$, and let $\lim \eta_m = \eta_0$, $\lim \eta_{mk} = \eta_{0k}$ $(m \to \infty)$. Further assume that f_m tends to f_0 and h_m tends to h_0 $(m \to \infty)$, together with all derivatives up to order r, a.u. on X and on Ω, respectively.*

If the conditions $f'_m(x) \leqslant 1$, $x \in [0, c]$, $m \in \mathbb{N}_0$, $c \in X \setminus \{0\}$, *and*

$$\left| (f'_m(0))^r \frac{\partial h_m}{\partial y} (0, \eta_m) \right| < 1, \qquad m \in \mathbb{N}_0,$$

hold, then for every $m \in \mathbb{N}_0$ *equation* (5.9.6) *has a unique* C^r *solution* $\varphi_m : X \to \mathbb{R}$ *such that* $\varphi_m(0) = \eta_m$, $\varphi_m^{(k)}(0) = \eta_{mk}$, $k = 1, \ldots, r$. *Moreover,* φ_m *tends to* φ_0 *as* $m \to \infty$, *together with all the derivatives, up to order* r, *a.u. in* X.

(For the proof see Matkowski [9].)

5.9.10. Equation (5.9.4) has been investigated by J. Jelonek [1] in the class Lip $C^r(X)$ of functions whose rth derivative satisfies a Lipschitz condition. In particular, the existence of a unique $\varphi \in$ Lip $C^r(X)$ has been proved under conditions like those of Theorem 5.6.1 but with Lipschitz conditions imposed on all the rth derivatives of h and with the inequality $|(f'(0))^{r+1}(\partial h/\partial y)(0, d)| < 1$ in place of (5.6.7).

Lipschitzian Nemitskiĭ operators acting in Lip $C^r(X)$ are also determined and they turn out to be affine. See the situation in Lip X we have described in Subsection 5.5C.

5.9.11. The assumption $f'(x) \neq 0$ in Theorem 5.6.2 is essential. A suitable example is to be found in Kuczma [26, p. 89]. However, we might postulate this condition only in a right vicinity of zero, say V. Taking $x_0 \in V$ we extend φ_0 from X_0 onto V, then rewrite equation (5.9.3) in form (5.9.4) and use Theorem 5.1.1 to obtain the extension to a solution on X.

5.9.12. Let f and g in equation (5.9.3) map open subsets of X and $X \times Y$, respectively, into Y; X and Y being Banach spaces. Differentiable solutions (in Fréchet's sense) of equation (5.9.3) in such a case have been investigated by M. Sablik [3]. In particular, the existence-and-uniqueness Theorem 5.6.1 and the lack-of-uniqueness Theorem 5.6.3 are generalized to the case of local solutions of (5.9.3) being of class C^r in a neighbourhood of $0 \in X$. The key assumptions (5.6.9), resp. (5.6.31), are replaced by some conditions concerning mutual positions of spectra of operators $f'(0)$ and $g'_y(0, d(0))$. Here d is a $C^\infty(X, Y)$ mapping that satisfies the conditions corresponding to (5.6.28) and (5.6.29) (we put $D(x) := (x, d(x))$:

$$(d \circ f)^{(k)}(0) = (g \circ D)^{(k)}(0), \qquad k = 0, \ldots, r$$

$$d^{(m)}(0) = 0, \qquad m > r.$$

5.9.13. The following version of Vitali's Theorem is used in the proof of Theorem 5.7.2.

Proposition 5.9.4. *If a sequence of functions analytic and equibounded in a disc* $D := \{x \in \mathbb{C} : |x| < c\}$, $c > 0$, *converges in a compact disc contained in* D, *then it converges a.u. in* D.

5.9.14. Conditions of Theorem 5.8.3 are fulfilled, in particular, if $h(\cdot, 0) \geqslant 0$ a.e. in X, $h(\cdot, 0) \in L^p$, for every $x \in X$ the function $h(x, \cdot)$ is increasing in $\mathbb{R}^+$ and there is a $t > 0$ such that

$$h(x, th(f(x), 0)) \leqslant th(x, 0) \qquad \text{a.e. in } X. \qquad (5.9.7)$$

Then we take $\varphi_1^* = 0$, $\varphi_2^* = th(\cdot, 0)$. We illustrate this by the following example; see Matkowski [13].

Example 5.9.3. Consider the equation

$$\varphi(x) = 2(x - x^2) \frac{\varphi(x^2)}{1 + |\varphi(x^2)|} + \frac{1}{\sqrt{x}}, \qquad x \in X := (0, 1), \qquad (5.9.8)$$

and let μ be the Lebesgue measure. Since $f(x) = x^2$, $\mathrm{d}(\mu f)/\mathrm{d}\mu = f' = 2\,\mathrm{id}$. Further, $h(x, y) = 2(x - x^2)y(1 + |y|)^{-1} + x^{-1/2}$. Condition (5.9.7) holds with $t = \frac{3}{2}$, $h(x, 0) = x^{-1/2}$ is in $L^1(X)$; $g(x) = 2(x - x^2)$, so that $g/f' < 1$, i.e. (5.8.3) holds with $p = s = 1$. Theorem 5.8.3 yields an essentially unique solution $\varphi_b \in L^1((0, b))$ of equation (5.9.8) for every $b < 1$. Letting $b \to 1$, we obtain an essentially unique common extension φ of φ_b which satisfies (5.9.8) a.e. in X. The function φ is in $L^1(X)$ since it obeys $0 \leqslant \varphi(x) \leqslant \varphi_2^*(x) = \frac{3}{2}x^{-1/2}$ a.e. in X.

(Note that Theorem 5.8.2 cannot be applied to equation (5.9.8).)

5.9.15. To exemplify the situation we have discussed in Subsection 5.8C assume that $X = (0, a)$ and $f: X \to X$ is strictly increasing, $f(x) < x$ in X. Let $X_0 = (0, x_0)$, $x_0 \in X$. Write $x_n := f^{-n}(x_0)$, $n \in \mathbb{N}_0$. The sequence (x_n) is infinite if $\lim_{x \to a^-} f(x) = a$, and it is finite if this limit is less than a. In the latter case, if x_{m-1} is the last existing x_n, we put additionally $x_m = a$. Then, with $X_n := (0, x_n)$, hypothesis (v) of Section 5.8 holds true.

5.9.16. Minkowski's inequality we used in the proofs of Theorems 5.8.7 and 5.8.8 is the triangle inequality for the norm (4.7.7) in $L^p(X)$:

$$\left(\int_X |\varphi_1 + \varphi_2|^p \,\mathrm{d}\mu \right)^{\alpha(p)} \leqslant \left(\int_X |\varphi_1|^p \,\mathrm{d}\mu \right)^{\alpha(p)} + \left(\int_X |\varphi_2|^p \,\mathrm{d}\mu \right)^{\alpha(p)},$$

where $\varphi_1, \varphi_2 \in L^p(X)$; see formula (4.7.6) for $\alpha(p)$.

5.9.17. The functional equation

$$\varphi(x)^2 + \varphi(ix)^2 = 1, \qquad x \in \mathbb{C} \setminus \{0\}, \qquad (5.9.7)$$

is satisfied by the Žukovski function $\Phi(x) = \frac{1}{2}(x + 1/x)$. The general solution of (5.9.7) in the class $\mathbf{\Phi}$ of functions that are analytic in $\mathbb{C} \setminus \{0\}$ and either are analytic or have a pole at the origin has been found by H. Haruki [2].

Proposition 5.9.4. *The general solution* $\varphi \in \mathbf{\Phi}$ *of equation* (5.9.7) *is given by*

$$\varphi = \Phi \circ p \qquad \text{or} \qquad \varphi = \Phi \circ \exp q = \cosh \circ q,$$

where Φ *is the Žukovski function;* $p(x) := x^3 \sum_{k=m}^{\infty} a_k x^{4k}$, $x \in \mathbb{C} \setminus \{0\}$, *is a* *meromorphic function, $p(x) \neq 0$ for $x \neq 0$, $q(x) := x^2 \sum_{n=0}^{\infty} b_n x^{4n}$, $x \in \mathbb{C}$, is an entire function, and $m \in \mathbb{Z}$, a_k, $b_n \in \mathbb{C}$.*

5.9.18. Theorem 5.5.3 when applied to the special case of equation (5.9.3),

$$\varphi(\sin x) = \cos \varphi(x), \qquad\qquad (5.9.8)$$

in the interval $X := [0, \pi/2]$ $(\Omega = X \times \mathbb{R})$ says that C^∞ solutions to (5.9.8) in X depend on an arbitrary function, and every such solution satisfies $\varphi(0) = c$, where $c = \cos c$, and $\varphi^{(k)}(0) = 0$, $k \in \mathbb{N}$. Thus, in particular, nonconstant solutions of (5.9.8) are not analytic at the origin.

By a direct examination of equation (5.9.8), T. L. McCoy [1] proved that there is a nonconstant solution to (5.9.8) which is in $C^\infty(\mathbb{R})$. Note that Theorem 5.5.3 does not work when $X = \mathbb{R}$ as the derivative of $f(x) = \sin x$ vanishes at some points of $\mathbb{R}$.

6

Equations of higher orders and systems of linear equations

6.0 Introduction

The objective of this chapter is to discuss linear equations of order N,

$$\sum_{i=0}^{N} a_i(x)\varphi(f_i(x)) = h(x), \qquad N \in \mathbb{N}, \tag{6.0.1}$$

as well as systems of linear equations, written in the matrix form,

$$\varphi(f(x)) = g(x)\varphi(x) + h(x), \tag{6.0.2}$$

where the values of φ lie in the space $\mathbb{R}^N$.

In fact, not much has been done specifically for these equations. Obviously, one way of considering them is to apply results for nonlinear equations presented in Chapter 7. We leave this task out of consideration and concentrate rather upon methods of solving several special cases of equation (6.0.1) or system (6.0.2). In Section 6.7 we describe how to reduce equation (6.0.1), where f_i are iterates f^i of a function f, to systems of form (6.0.2).

To begin with, we consider Cauchy's functional equations in the domain restricted to the graph of a function. We also show a characterization of the Gaussian normal distribution through equations of type (6.0.1). Next, special equations (6.0.1) are used in the ergodic theory and as a method of decomposition of two-place functions.

Passing from those examples of applications to samples of the theory we consider the following items: cyclic equations (6.0.1), system (6.0.2) with a constant matrix g, C^∞ solutions of (6.0.2) and some linear recurrence inequalities. Results on smooth solutions of (6.0.1) that may be obtained via Banach's Theorem are quoted in Section 6.6.

6.1 Particular solutions of some special equations

6.1A. Cauchy's functional equations on a curve

We consider the Cauchy equations

$$\varphi(x + y) = \varphi(x) + \varphi(y)$$

and

$$\varphi(x + y) = \varphi(x)\varphi(y)$$

(see Aczél [2], Aczél–Dhombres [1]) on the graph $\{(x, y) \in \mathbb{R}^2 : y = f(x)\}$ of a function f. We want to establish conditions for the resulting equations of second order

$$\varphi(x + f(x)) = \varphi(x) + \varphi(f(x)) \qquad (6.1.1)$$

and

$$\varphi(x + f(x)) = \varphi(x)\varphi(f(x)) \qquad (6.1.2)$$

to have a family of linear, resp. exponential, functions as the only solutions.

We assume that

(I) The function $f: \mathbb{R}^+ \to \mathbb{R}^+$ is continuous and increasing in $\mathbb{R}^+$, $f(0) = 0$, $f(x) > 0$ for $x \neq 0$.

Obviously, equation (6.1.2) goes over into (6.1.1) by taking logarithms of both sides. The following lemma shows that we actually can do this.

Lemma 6.1.1. *Under hypothesis* (I) *every solution* $\varphi: \mathbb{R}^+ \to \mathbb{R}$ *of equation* (6.1.2) *which is both continuous and positive at* $x = 0$ *is positive on* $\mathbb{R}^+$.

Proof. Put $g(x) = x + f(x)$. The function $g: \mathbb{R}^+ \to \mathbb{R}^+$ is a continuous bijection of $\mathbb{R}^+$ onto itself. Thus we may define the function $h: \mathbb{R}^+ \to \mathbb{R}^+$ as $h = g^{-1}$. Equation (6.1.2) is equivalent to

$$\varphi(x) = \varphi(h(x))\varphi(x - h(x)).$$

We shall show that $\varphi(x) > 0$ in $\mathbb{R}^+$. Suppose the contrary: $\varphi(x_0) \leqslant 0$ for an $x_0 > 0$. Then φ is nonpositive at one of the points $h(x_0)$, $x_0 - h(x_0)$. Denote this point by x_1. Thus we have $\varphi(x_1) \leqslant 0$ and $x_1 \leqslant \max(h(x_0), x_0 - h(x_0))$. Now we can repeat this argument with x_0 replaced by x_1. In this way a sequence $(x_n)_{n \in \mathbb{N}_0}$ is constructed such that

$$\varphi(x_n) \leqslant 0, \qquad 0 \leqslant x_{n+1} \leqslant \max(h(x_n), x_n - h(x_n)) < x_n, \quad n \in \mathbb{N}_0.$$

This sequence is convergent to a limit $b \in \mathbb{R}^+$, and either $b \leqslant h(b)$ or $h(b) \leqslant 0$, therefore $b = 0$. By the continuity of φ at zero we have $0 < \varphi(0) = \lim_{n \to \infty} \varphi(x_n) \leqslant 0$, a contradiction. ∎

The following theorem (Zdun [1]) will be proved in Subsection 6.1D.

Theorem 6.1.1. *Let hypothesis* (I) *be satisfied.*

(1) *The functions* $\varphi(x) = cx$, $c \in \mathbb{R}$, *are the only solutions* $\varphi: \mathbb{R}^+ \to \mathbb{R}$ *of equation* (6.1.1) *that are differentiable at zero.*

(2) *The functions* $\varphi(x) = e^{cx}$, $c \in \mathbb{R}$, *are the only solutions* $\varphi: \mathbb{R}^+ \to \mathbb{R}$ *of equation* (6.1.2) *that are differentiable at zero and such that* $\varphi(0) = 1$.

6.1B. The Gaussian normal distribution

It is easily seen that the function

$$\varphi(x) = \frac{1}{\sigma\sqrt{(2\pi)}} \exp(-x^2/2\sigma^2), \qquad \sigma > 0, \tag{6.1.3}$$

fulfils the condition (for every $a > 0$, $b > 0$ such that $a^2 + b^2 = 1$)

$$\varphi(x) = \sigma\sqrt{(2\pi)}\varphi(ax)\varphi(bx). \tag{6.1.4}$$

Following E. Vincze [2] (see also Laha–Lukacs–Rényi [1]) we shall prove later the following characterization theorem for function (6.1.3).

Theorem 6.1.2. *Let* a, b *be positive constants such that* $a^2 + b^2 = 1$. *The density* (6.1.3) *of the Gaussian normal distribution is the only solution* $\varphi: \mathbb{R} \to \mathbb{K}$ *of equation* (6.1.4) *which is twice differentiable at zero and fulfils the condition*

$$\int_{\mathbb{R}} \varphi(x)\,\mathrm{d}x = 1. \tag{6.1.5}$$

6.1C. Equation of *N*th order

The functional equations that occur in Subsections 6.1A and 6.1B are (or can be reduced to) particular cases of the following homogeneous equation of *N*th order:

$$\varphi(x) = \sum_{i=1}^{N} a_i(x)\varphi(f_i(x)). \tag{6.1.6}$$

We assume that

(i) $X = (0, a]$, $0 < a \leqslant \infty$, $a_i: X \to \mathbb{R}^+$ are arbitrary functions, $f_i: X \to X$ are continuous, $0 < f_i(x) < x$ in X, $i = 1, \ldots, N$.

We shall solve equation (6.1.6) in the case where

$$\sum_{i=0}^{N} a_i(x)(f_i(x)/x)^r = 1, \qquad x \in X, \tag{6.1.7}$$

for an $r \in \mathbb{N}_0$, and in the class of functions having the property

$$\varphi(x) = \sum_{k=0}^{r} \eta_k x^k + o(x^r), \qquad x \to 0, \tag{6.1.8}$$

with some $\eta_0, \ldots, \eta_r \in \mathbb{R}$.

Having assumed (i) and (6.1.7) we see, since $a_i(x) \geq 0$, $f_i(x)/x < 1$, that if $r \geq 1$ (by (6.1.7) at least one $a_i(x)$ does not vanish), then

$$\sum_{i=1}^{N} a_i(x)(f_i(x)/x)^k > 1 \qquad \text{for } x \in X, \ k = 0, \ldots, r-1.$$

Thus if φ satisfies (6.1.6) and has property (6.1.8), then $\eta_0 = \eta_1 = \cdots = \eta_{r-1} = 0$. We see that assumption (6.1.7) implies the following form of property (6.1.8):

$$\varphi(x) = \eta_r x^r + o(x^r), \qquad x \to 0. \tag{6.1.9}$$

We shall now prove the following.

Theorem 6.1.3. *Let hypothesis* (i) *and condition* (6.1.7) *be fulfilled, the latter with an* $r \in \mathbb{N}_0$. *Then the only solution* $\varphi : X \to \mathbb{R}$ *of equation* (6.1.6) *satisfying condition* (6.1.8) *is just* $\varphi(x) = \eta_r x^r$.

Proof. $r = 0$. Let φ be a solution of (6.1.6) satisfying (6.1.9) with $r = 0$. Thus

$$\lim_{x \to 0+} \varphi(x) = \eta_0, \tag{6.1.10}$$

and (6.1.7) becomes

$$\sum_{i=1}^{N} a_i(x) = 1. \tag{6.1.11}$$

We have to prove that $\varphi(x) \equiv \eta_0$.

Put $F(x) := \max_{1 \leq i \leq N} f_i(x)$. The function $F : X \to X$ is continuous and fulfils the condition $0 < F(x) < x$ in X. Fix an arbitrary $x_0 \in X$. It follows from (6.1.6) and (6.1.11) that the values of φ at the points $f_1(x_0), \ldots, f_N(x_0)$ cannot be all greater (smaller) than $\varphi(x_0)$. Thus among them there are points, say u_1 and u_2, which satisfy $\varphi(u_1) \leq \varphi(x_0) \leq \varphi(v_1)$. Moreover, putting $u_0 = v_0 = x_0$, we have $u_1 \leq F(u_0)$, $v_1 \leq F(v_0)$.

Replace in (6.1.6) x first by u_1 and then by v_1. We get, from (6.1.11), the points $u_2, v_2 \in X$ such that $\varphi(u_2) \leq \varphi(u_1)$, $\varphi(v_1) \leq \varphi(v_2)$, and $u_2 \leq F(u_1)$, $v_2 \leq F(v_1)$. Repeating this procedure, we obtain two sequences of points $u_n, v_n \in X$ such that $u_0 = v_0 = x_0$,

$$u_{n+1} \leq F(u_n), \qquad v_{n+1} \leq F(v_n), \qquad n \in \mathbb{N}_0, \tag{6.1.12}$$

and

$$\varphi(u_{n+1}) \leq \varphi(u_n) \leq \cdots \leq \varphi(x_0) \leq \cdots \leq \varphi(v_n) \leq \varphi(v_{n+1}), \qquad n \in \mathbb{N}_0. \tag{6.1.13}$$

It follows from (6.1.12) (see Theorem 1.2.1) that $\lim_{n \to \infty} u_n = \lim_{n \to \infty} v_n = 0$, whence, on letting $n \to \infty$ in (6.1.13), we obtain by (6.1.10) $\varphi(x_0) = \eta_0$. Since x_0 was arbitrary, $\varphi(x) = \eta_0$ in X. On the other hand, this function evidently satisfies equation (6.1.6) and condition (6.1.10).

$r \geq 1$. This case can be reduced to the preceding one. Take a solution $\varphi : X \to \mathbb{R}$ of (6.1.6) satisfying (6.1.8), note that then (6.1.9) holds and put

$$\hat{\varphi}(x) = x^{-r} \varphi(x). \tag{6.1.14}$$

This function satisfies (6.1.10) with η_r in place of η_0, and is a solution of the equation

$$\hat{\phi}(x) = \sum_{i=1}^{N} \hat{a}_i(x)\hat{\phi}(f_i(x)), \qquad \hat{a}_i(x) := a_i(x)(f_i(x)/x)^r.$$

Because of (6.1.7), the $\hat{a}_i$ satisfy (6.1.11). By the first part of the proof we get $\hat{\phi} = \eta_r$, i.e. $\varphi(x) = \eta_r x^r$, as required.

The converse is obvious. ∎

6.1D. Proofs

Theorems 6.1.1 and 6.1.2 follow from Theorem 6.1.3.

Proof of Theorem 6.1.1. (1) As in the proof of Lemma 6.1.1 we may write equation (6.1.1) in the equivalent form

$$\varphi(x) = \varphi(h(x)) + \varphi(x - h(x)). \tag{6.1.15}$$

We are looking for solutions of (6.1.1) that are differentiable at the origin, i.e. satisfy (set in (6.1.15) $x = 0$ to get $\varphi(0) = 0$)

$$\varphi(x) = cx + o(x), \qquad x \to 0, \; c := \varphi'(0).$$

Clearly, condition (6.1.7) is satisfied for equation (6.1.15), with $N = 2$, $r = 1$. By Theorem 6.1.3 $\varphi(x) = cx$ in $(0, \infty)$. Since $\varphi(0) = 0$, $\varphi(x) = cx$ in $\mathbb{R}^+$.

(2) By Lemma 6.1.1 we can take logarithms of both the sides of (6.1.2). Applying (1) of the theorem we get $\log \varphi(x) = cx$ in $\mathbb{R}^+$. ∎

Proof of Theorem 6.1.2. We need consider only the case where $\mathbb{K} = \mathbb{C}$. Let $\varphi: \mathbb{R} \to \mathbb{C}$ be a solution of (6.1.4), twice differentiable at $x = 0$. Write

$$\sigma\sqrt{(2\pi)} \cdot \varphi(x) = \rho(x)e^{i\alpha(x)}, \qquad \rho: \mathbb{R} \to \mathbb{R}^+, \; \alpha: \mathbb{R} \to \mathbb{R}. \tag{6.1.16}$$

The functions just introduced satisfy

$$\rho(x) = \rho(ax)\rho(bx), \tag{6.1.17}$$

$$\alpha(x) = \alpha(ax) + \alpha(bx) + 2k(x)\pi, \tag{6.1.18}$$

where $k: \mathbb{R} \to \mathbb{Z}$. Setting $x = 0$ in (6.1.17), we obtain $\rho(0) = 0$ or $\rho(0) = 1$. The ρ is continuous at zero, and if $\rho(0) = 1$ it is twice differentiable at zero. If $\rho(0) = 0$, then $\rho = 0$ (see Note 6.9.3) whence $\varphi = 0$, which is incompatible with (6.1.5). Thus $\rho(0) = 1$. Differentiating (6.1.17) and then setting $x = 0$ we obtain $\rho'(0) = 0$. Thus $\rho(x) = 1 + c_1x^2 + o(x^2)$, $x \to 0$, $2c_1 := \rho''(0)$. Moreover, $\rho(x) > 0$ in a neighbourhood U of zero. By taking logarithms of both the sides of (6.1.17) we reduce this equation to (6.1.6), where $N = 2$, $a_1(x) = a_2(x) = 1$, $f_1(x) = ax$, $f_2(x) = bx$. The condition $a^2 + b^2 = 1$ then yields (6.1.7) for $r = 2$. By Theorem 6.1.3[1], since $\log \rho(x) = c_1x^2 + o(x^2)$, $x \to 0$, we get $\log \rho(x) = c_1x^2$ and $\rho(x) = \exp(c_1x^2)$ in U. By Baron's Extension Theorem 7.1.2 this formula holds in $\mathbb{R}$.

---

[1] In fact, the form of the solutions of the equations considered is to be obtained first in $\mathbb{R}^+$, and then in $(-\infty, 0]$, transformed into $\mathbb{R}^+$ by the change of variables $x' = -x$.

By (6.1.16), $\alpha(x) = i \log \rho(x) - i \log \varphi(x)$ for small $|x|$, as $\sigma\sqrt{(2\pi)}\,\varphi(0) = 1$, where log is any branch of the logarithm on $\mathbb{C}$ cut along the negative real axis. Thus α also is twice differentiable at zero. By (6.1.18) the function k must be continuous at zero, and hence constant in a neighbourhood V of zero. Thus the function $\beta(x) := \alpha(x) + 2k\pi$ satisfies in V the functional equation

$$\beta(x) = \beta(ax) + \beta(bx),$$

so that $\beta'(0) = 0 = \beta(0)$. Therefore $\beta(x) = c_2 x^2 + o(x^2)$, $x \to 0$, and by Theorem 6.1.3 we get $\beta(x) = c_2 x^2$ for $x \in V$. Finally we obtain $\varphi(x) = (\sigma\sqrt{(2\pi)})^{-1} \exp[(c_1 + ic_2)x^2]$ in V. Formula (6.1.5) yields $c_1 = -\frac{1}{2}$ and $c_2 = 0$, and (6.1.3) in $\mathbb{R}^+$ follows from the Extension Theorem 7.1.2. ∎

Remark 6.1.1. Theorem 6.1.1 remains valid also if $X = \mathbb{R}$. For, it is enough to apply the argument described in the footnote on p. 239.

6.2 Further applications

Linear equations of higher orders appear in a natural way in the ergodic theory, and help to determine densities of measures that are invariant under some specific mappings of a compact interval. A problem that also leads to such equations is that of decomposition of a function of two variables into a sum of functions of separate variables.

6.2A. Invariant measures under piecewise linear transformations

Assume we are given a mapping $f: X \to X$, where $X := [0, 1]$.

Definition 6.2.1. A function $\varphi \in L^1(X)$, $\varphi \geq 0$, is said to be an *invariant density under f* if and only if the measure defined for every measurable $E \subset X$ as

$$\mu(E) := \int_E \varphi(x)\,dx$$

is invariant under the mapping f:

$$\mu(E) = \mu(f^{-1}(E)).$$

Invariant densities are fixed points of the *Frobenius–Perron operator* $P_f: L^1(X) \to L^1(X)$ (see e.g. Lasota–Yorke [1]):

$$(P_f \varphi)(x) = \frac{d}{dx} \int_{f^{-1}([0,x])} \varphi(s)\,ds.$$

Let X admit a partition, $X = \bigcup_{i=1}^N X_i$, into intervals X_i whose interiors do not intersect each other. If f is strictly monotonic on each interval X_i, then the Frobenius–Perron operator takes the form

$$P_f\varphi = \sum_{i=1}^{N} (\varphi \circ f_i)|f_i'|,$$

where $f_i := (f|X_i)^{-1}$, and $f(X_i) = X$ for $i = 1, \ldots, N$. Specifying further the function f to be piecewise linear, X being the range of each piece, we find that the invariant densities under such an f are the solutions of the functional equation (see Matkowski [18])

$$\varphi(x) = \sum_{i=1}^{N} a_i \varphi(f_i(x)), \tag{6.2.1}$$

where $f_i: X \to X$ are linear functions and $a_i := |f_i'(x)| = \text{const.}$

For instance, if we take an α, $0 < \alpha < 1$, and put

$$f(x) = \begin{cases} x/\alpha & \text{for } x \in [0, \alpha], \\ (x-1)/(\alpha-1) & \text{for } x \in (\alpha, 1], \end{cases}$$

then equation (6.2.1) becomes

$$\varphi(x) = \alpha\varphi(\alpha x) + (1-\alpha)\varphi(1 - (1-\alpha)x), \qquad x \in X. \tag{6.2.2}$$

From Kominek–Matkowski [2] we take two theorems on this equation: the first giving its general solution and the other showing that its Riemann-integrable solutions must be constant functions. We put

$$f_1(x) := \alpha x, \qquad f_2(x) := 1 - (1-\alpha)x, \qquad x \in X.$$

Theorem 6.2.1. *For every function $\varphi_0: [0, \alpha] \to \mathbb{R}$ such that $\varphi_0(0) = \varphi_0(\alpha)$ there exists a unique solution $\varphi: X \to \mathbb{R}$ of equation (6.2.2) such that $\varphi|_{[0,\alpha]} = \varphi_0$. Thus the solution of equation (6.2.1) depends on an arbitrary function.*

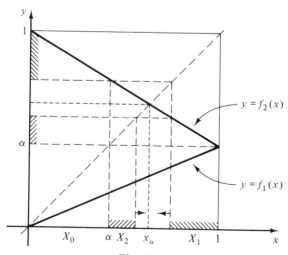

Fig. 6.1

Proof. Let $x_\alpha := (2 - \alpha)^{-1}$, so that $f_2(x_\alpha) = x_\alpha$. We put

$$X_n := f_2^n([0, \alpha]), \qquad n \in \mathbb{N}_0.$$

One verifies that $\mathrm{int}(X_i \cap X_j) = \varnothing$ for $i \neq j$ and $\bigcup_{n=1}^\infty X_n = [\alpha, 1] \setminus \{x_\alpha\}$ (see Fig. 6.1).

Inserting in (6.2.2) $x = 0$ and $x = 1$ we see that every solution of (6.2.2) has to fulfil the condition $\varphi(0) = \varphi(1) = \varphi(\alpha)$. Let $\varphi_0 \colon X_0 \to \mathbb{R}$, $\varphi_0(0) = \varphi_0(\alpha)$, be an arbitrary function. Define $\varphi_n \colon X_n \to \mathbb{R}$ inductively by

$$\varphi_{n+1}(x) = \frac{1}{1-\alpha} [\varphi_n(f_2^{-1}(x)) - \varphi_0(f_1(f_2^{-1}(x)))], \qquad x \in X_{n+1}, n \in \mathbb{N}_0.$$

(6.2.3)

The function

$$\hat{\phi}(x) = \varphi_n(x) \qquad \text{for } x \in X_n, n \in \mathbb{N}_0,$$

is well defined in $X \setminus \{x_\alpha\}$, as $f_2^{-1}(X_{n+1}) = X_n \subset [\alpha, 1]$, $n \in \mathbb{N}$, and $f_1([\alpha, 1]) \subset X_0$ (see Fig. 6.1). Substituting $x = x_\alpha$ in (6.2.2) we get $\varphi(\alpha x_\alpha) = \varphi(x_\alpha)$. Therefore the function

$$\varphi(x) = \begin{cases} \hat{\phi}(x), & x \in X \setminus \{x_\alpha\}, \\ \hat{\phi}(\alpha x_\alpha), & x = x_\alpha, \end{cases} \tag{6.2.4}$$

maps X into $\mathbb{R}$ and is a solution of equation (6.2.2). Obviously, it is the unique extension of φ_0 to a solution of (6.2.2) on X. ∎

Theorem 6.2.2. *If $\varphi \colon X \to \mathbb{R}$ is a Riemann-integrable solution of equation (6.2.2) then φ is a constant function.*

Proof. This will be outlined only. With the aid of the function f_j^n defined by

$$f_1^1 := f_1, \qquad f_2^1 := f_2,$$
$$f_k^{n+1} := f_1 \circ f_k^n \qquad \text{for } k = 1, 2, \ldots, 2^n,$$
$$f_{2^n+k}^{n+1} := f_2 \circ f_{2^{n+1}-k}^n \qquad \text{for } k = 1, \ldots, 2^n, n \in \mathbb{N},$$

we get the partition $X = \bigcup_{j=1}^{2^n} f_j^n(X)$, for every $n \in \mathbb{N}$.

Let A_j^n denote the length of the interval $f_j^n(X)$. The formulae that follow can be verified by induction.

$$\sum_{i=1}^{2^n} A_j^n = 1, \qquad n \in \mathbb{N}, \tag{6.2.5}$$

$$0 < A_j^n \leq (\max(\alpha, 1 - \alpha))^n, \qquad n \in \mathbb{N}, \tag{6.2.6}$$

$$\varphi(x) = \sum_{i=1}^{2^n} A_j^n \varphi(f_j^n(x)), \qquad n \in \mathbb{N}. \tag{6.2.7}$$

By (6.2.5) and (6.2.6) formula (6.2.7) is nothing other than a Riemann integral sum for φ in X. On letting $n \to \infty$ in (6.2.7) we obtain that $\varphi(x) = \int_0^1 \varphi(t)\,dt = \mathrm{const}$. ∎

Taking a continuous nonconstant $\varphi_0\colon X_0 \to \mathbb{R}$ we get the function φ continuous on $X \setminus \{x_\alpha\}$. By Theorem 2.6.2 the solution (6.2.4), as nonconstant, cannot be continuous at x_α. Thus we have the following.

Corollary 6.2.1. *There exist nonconstant solutions of equation (6.2.2) with the unique discontinuity at the fixed point $x = x_\alpha$ of f_2.*

Remark 6.2.1. The considerations above can be applied, without any essential changes, to equation (6.2.1), provided that $\sum_{i=1}^{N} a_i = 1$.

Remark 6.2.2. Since equation (6.2.2) has not been solved in the space L^1, we do not know whether there exist measures invariant under f which are different from the Lebesgue measure (which corresponds to constant densities).

6.2B. Decomposition of two-place functions

The problem we touch upon now belongs to the approximation theory. Let $G\colon X \times Y \to \mathbb{R}$ be a continuous function, X, Y being compact spaces. We ask for conditions under which there exist continuous functions $g_1\colon X \to \mathbb{R}$, $g_2\colon Y \to \mathbb{R}$ such that

$$G(x, y) = g_1(x) + g_2(y) \tag{6.2.8}$$

at least on a compact subspace Z of $X \times Y$. It turns out that in some instances the representation (6.2.8) exists if and only if a class of functional equations of type (6.0.1) has continuous solutions. This approach to the problem has been proposed by R. C. Buck [1], [2] (see also Baron [12]).

We are going to consider the simple case where $X = Y = [0, 1]$ and Z is the union of the graphs of two given continuous functions $f_1, f_2\colon X \to X$. We put $\Gamma := \{x \in X\colon f_1(x) = f_2(x)\}$.

Theorem 6.2.3. *The following assertions are equivalent.*

(1) *For every continuous function $G\colon X^2 \to \mathbb{R}$ there exist continuous functions $g_1, g_2\colon X \to \mathbb{R}$ such that (6.2.8) holds for $(x, y) \in Z$.*
(2) *The functional equation*

$$\varphi(f_1(x)) - \varphi(f_2(x)) = h(x), \qquad x \in X, \tag{6.2.9}$$

has a continuous solution $\varphi\colon X \to \mathbb{R}$ for every continuous $h\colon X \to \mathbb{R}$ such that $h(x) = 0$ for $x \in \Gamma$.

Proof. $(1) \Rightarrow (2)$. Take an $h \in C(X, \mathbb{R})$, $h|\Gamma = 0$. There exists a continuous function $G\colon X^2 \to \mathbb{R}$, whose restriction to the set Z is given by

$$G(x, y) = \begin{cases} h(x), & y = f_1(x), x \in X, \\ 0, & y = f_2(x), x \in X. \end{cases}$$

Relation (6.2.8) applied to this G yields
$$h(x) = g_1(x) + g_2(f_1(x)), \qquad 0 = g_1(x) + g_2(f_2(x)), \qquad x \in X,$$
so that $g_2(f_1(x)) - g_2(f_2(x)) = h(x)$, $x \in X$. Thus $\varphi = g_2$ is the required continuous solution of (6.2.9).

$(2) \Rightarrow (1)$. Let us fix a $G \in C(X^2, \mathbb{R})$. Equation (6.2.9) with the continuous $h: X \to \mathbb{R}$ given by
$$h(x) := G(x, f_1(x)) - G(x, f_2(x)), \qquad x \in X, \qquad (6.2.10)$$
has a continuous solution $\varphi: X \to \mathbb{R}$. Now we take
$$g_1(x) := G(x, f_2(x)) - \varphi(f_2(x)), \qquad g_2(y) := \varphi(y), \qquad x, y \in X.$$
Obviously,
$$g_1(x) + g_2(y) = G(x, y) \qquad \text{for } x \in X, \ y = f_2(x).$$
By (6.2.9) and (6.2.10), for $x \in X$, $y = f_1(x)$ we also get
$$g_1(x) + g_2(y) = G(x, f_2(x)) - \varphi(f_2(x)) + \varphi(f_1(x))$$
$$= G(x, f_2(x)) + h(x) = G(x, f_1(x)) = G(x, y).$$
Therefore (6.2.8) holds on the union Z of the graphs of f_1 and f_2. ∎

Comments. The first result on invariant densities, obtained via functional equations, seems to be that by A. Lasota–J. A. Yorke [1], where the density was invariant with respect to a self-mapping of the unit interval, piecewise of class C^2. The general solution of equation (6.2.2) with $\alpha = \frac{1}{2}$ has been found by A. Boyarsky–R. Ger [1], who have also proved that its continuous solutions are constant functions only (result unpublished).

However, from the point of view of the ergodic theory the results presented in Subsection 6.2A are rather unsatisfactory. As follows from Boyarsky–Haddad [1], solutions of (6.2.2) of unbounded variation would be particularly interesting for that theory.

The problem of determining all solutions of bounded variation of equation (6.2.1) has been reduced by J. Matkowski [18] to that of finding all monotonic solutions of this equation.

6.3 Cyclic equations

Particularly simple is the case where in equation (6.0.1) the functions f_i are the iterates of a function f –
$$\sum_{i=0}^{N} a_i(x)\varphi(f^i(x)) = h(x) \qquad (6.3.1)$$
– and the substitution f is cyclic, $f^{N+1} = \mathrm{id}$. Namely, we take (6.3.1), then replace x by $f(x)$, in turn by $f^2(x)$, and so on, up to $f^N(x)$. We obtain a system of $N + 1$ linear equations with $N + 1$ unknowns $\varphi(x)$,

$\varphi(f(x)), \ldots, \varphi(f^N(x))$. Thus the problem of solving (6.3.1) reduces to that of solving a system of algebraic linear equations, and is therefore settled.

6.3A. Homogeneous equation with a finite group of substitutions

The method described can be applied in the more general case of equation (6.0.1), where the functions f_i generate a finite group of substitutions. Writing also into the equation, if necessary, some summands with coefficients zero, we may assume that all functions from this group occur in the equation.

We shall be concerned with the homogeneous equation

$$\sum_{i=0}^{N} a_i(x)\varphi(f_i(x)) = 0, \qquad f_0 = \mathrm{id}. \tag{6.3.2}$$

The results we are going to present are due to S. B. Presić [1], [2]. Assume that

(i) X is an arbitrary set, and the functions $f_i: X \to X$, $i = 0, \ldots, N$, where $f_0 = \mathrm{id}_X$, form a group $\mathscr{F}$ under composition.

Thus f_0 is the neutral element of $\mathscr{F}$. The composition of any two $f_i, f_j \in \mathscr{F}$ yields an $f_p \in \mathscr{F}$. The index p, $0 \leqslant p \leqslant N$, is uniquely determined for every $i, j = 0, 1, \ldots, N$. We shall use the notation

$$p = i \circ j.$$

For every $j, k = 0, \ldots, N$ there is a unique i, $0 \leqslant i \leqslant N$, such that $i \circ j = k$, i.e. $f_i \circ f_j = f_k$.

Let $\mathscr{G}$ be the linear representation of the group $\mathscr{F}$. Thus $\mathscr{G}$ is the multiplicative group of matrices $C_k = (c_{ij}^k)$, $k = 0, \ldots, N$, whose entries are defined by

$$c_{ij}^k = \begin{cases} 1 & \text{if } j = i \circ k, \\ 0 & \text{otherwise}. \end{cases}$$

Inserting into (6.3.2), in place of x, in turn $f_0(x), f_1(x), \ldots, f_N(x)$, we obtain a system of $N + 1$ equations, which can be written in the form

$$A(x) \begin{pmatrix} \varphi(f_0(x)) \\ \vdots \\ \varphi(f_N(x)) \end{pmatrix} = 0 \in \mathbb{R}^{N+1}. \tag{6.3.3}$$

We are going to show that

$$A(f_k(x))C_k = C_k A(x), \qquad k = 0, \ldots, N. \tag{6.3.4}$$

Writing $A(x) = (a_{ij}(x))$, we have, by (6.3.3), $a_{ij}(x) = a_q(f_i(x))$, where q is determined by $q \circ i = j$. Thus the (i, j)th entry of the matrix $A(f_k(x))$ is $a_q(f_i \circ f_k(x)) = a_q(f_{i \circ k}(x))$. If in the equality $q \circ i = j$ we keep one of the indices q, i, j fixed, and let another run from 0 to N, then the third index will run over the whole set of $N + 1$ integers $\{0, 1, \ldots, N\}$. Consequently the (i, s)th entry of the matrix $A(f_k(x))C_k$ is

$$\sum_{q=0}^{N} a_q(f_{i \circ k}(x))c_{q \circ i,s}^k = a_m(f_{i \circ k}(x)),$$

where m is determined by $s = (m \circ i) \circ k$. Similarly, the (i, s)th entry of the matrix $C_k A(x)$ is

$$\sum_{q=0}^{N} c_{iq}^k a_r(f_q(x)),\tag{6.3.5}$$

where $r \circ q = s$. The only c_{iq}^k different from zero is that for which $q = i \circ k$. This yields $r \circ (i \circ k) = s = (m \circ i) \circ k = m \circ (i \circ k)$ by the associativity of the composition. Hence $r = m$ and (6.3.5) reduces to $a_m(f_{i \circ k}(x))$. Thus (6.3.4) is established.

6.3B. Compatibility

We accept the following definition.

Definition 6.3.1. A matrix function $B: X \to \mathbb{K}^{N \times N}$ is said to be *compatible with $\mathscr{F}$* iff

$$B(f_k(x)) = C_k B(x) C_k^{-1} \qquad \text{for } k = 0, \ldots, N.\tag{6.3.6}$$

Relation (6.3.4) we have just shown says that the matrix $A(x)$, occurring in (6.3.3), is compatible with $\mathscr{F}$. The role of compatibility is explained by the following.

Lemma 6.3.1. Let hypothesis (i) be satisfied. If $B: X \to \mathbb{K}^{N \times N}$ is compatible with $\mathscr{F}$, then for every function $g: X \to \mathbb{K}$ there exists a function $\varphi: X \to \mathbb{K}$ such that

$$\begin{pmatrix} \varphi(f_0(x)) \\ \vdots \\ \varphi(f_N(x)) \end{pmatrix} = B(x) \begin{pmatrix} g(f_0(x)) \\ \vdots \\ g(f_N(x)) \end{pmatrix}.\tag{6.3.7}$$

Proof. We take B and g having the stated properties and denote by $\varphi_k(x)$ the kth entry of the column matrix occurring on the right-hand side of (6.3.7). The thing to show is that $\varphi_k = \varphi_0 \circ f_k$, $k = 0, \ldots, N$. First observe that, since $0 \circ k = k$ and by the definition of c_{0j}^k,

$$\varphi_k(x) = \sum_{j=0}^{N} c_{0j}^k \varphi_j(x).$$

This, together with (6.3.6), yields

$$\begin{pmatrix} \varphi_k(x) \\ \vdots \end{pmatrix} = C_k \begin{pmatrix} \varphi_0(x) \\ \vdots \end{pmatrix} = C_k B(x) C_k^{-1} C_k \begin{pmatrix} g(f_0(x)) \\ \vdots \end{pmatrix} = B(f_k(x)) C_k \begin{pmatrix} g(f_0(x)) \\ \vdots \end{pmatrix}.$$

Hence, writing $B(x) = (b_{ij}(x))$,

$$\varphi_k(x) = \sum_{i=0}^{N} \sum_{j=0}^{N} b_{0i}(f_k(x))c_{ij}^k g(f_j(x)) = \sum_{i=0}^{N} b_{0i}(f_k(x))g(f_{i \circ k}(x)).$$

On the other hand, we have by the definition of $\varphi_0(x)$

$$\varphi_0(f_k(x)) = \sum_{i=0}^{N} b_{0i}(f_k(x))g(f_i \circ f_k(x)) = \sum_{i=0}^{N} b_{0i}(f_k(x))g(f_{i \circ k}(x)),$$

i.e., $\varphi_k = \varphi_0 \circ f_k$. Thus the function $\varphi = \varphi_0$ has property (6.3.7). ∎

6.3C. General solution

In order to solve system (6.3.3) we first find a matrix $B(x)$ compatible with $\mathscr{F}$ and such that

$$A(x)(B(x)A(x) + E) = 0, \tag{6.3.8}$$

where E is the unit matrix of order $N + 1$.

Let $r(x)$ be the rank of $A(x)$, and let $D(x)$ be the diagonal matrix of order $N + 1$ with $r(x)$ units and $N + 1 - r(x)$ zeros on the main diagonal. There exist nonsingular matrices $P(x)$ and $Q(x)$, of order $N + 1$, such that $A(x) = P(x)D(x)Q(x)$. We take

$$B(x) = \frac{1}{N+1} \sum_{i=0}^{N} C_i^{-1} B_0(f_i(x))C_i, \qquad B_0(x) := -Q(x)^{-1}D(x)P(x)^{-1}. \tag{6.3.9}$$

We check that this matrix is compatible with $\mathscr{F}$:

$$(N+1)B(f_k(x)) = \sum_{i=0}^{N} C_i^{-1}B_0(f_{i \circ k}(x))C_i$$

$$= \sum_{i=0}^{N} C_k C_{i \circ k}^{-1} B_0(f_{i \circ k}(x))C_{i \circ k}C_k^{-1}$$

$$= C_k \left(\sum_{j=0}^{N} C_j^{-1}B_0(f_j(x))C_j \right)C_k^{-1} = (N+1)C_k B(x)C_k^{-1}.$$

Here we have used the relation $C_{i \circ k} = C_i C_k$. We omit a routine calculation of (6.3.8) for $B(x)$ given by (6.3.9).

Determination of the general solutions to (6.3.2) now presents no difficulties.

Theorem 6.3.1. *Let hypothesis* (i) *be fulfilled and let* $a_i \colon X \to \mathbb{K}$ *be arbitrary functions,* $i = 0, \ldots, N$. *The general solution* $\varphi \colon X \to \mathbb{K}$ *of equation* (6.3.2) *is given by the formula*

$$\varphi(x) = \sum_{j=0}^{N} \beta_{0j}(x)g(f_j(x)), \tag{6.3.10}$$

where $(\beta_{ij}(x)) = B(x)A(x) + E$, $B(x)$ is defined by (6.3.9), $A(x)$ occurs in (6.3.3), and $g: X \to \mathbb{K}$ is an arbitrary function.

Proof. Let $g: X \to \mathbb{K}$ be an arbitrary function. We have to show that (6.3.10) yields a solution of (6.3.2).

Since the matrices $B(x)$ and $A(x)$ are compatible with $\mathscr{F}$, so is the matrix $B(x)A(x) + E$. By Lemma 6.3.1 there exists a function $\varphi: X \to \mathbb{K}$ such that relation (6.3.7) holds with that $B(x)$ replaced by $B(x)A(x) + E$. This, together with (6.3.8), implies (6.3.3), which is equivalent to (6.3.2). This φ is given by (6.3.10) as follows from (6.3.7) just referred to, since $f_0 = \mathrm{id}_X$.

On the other hand, let $\varphi: X \to \mathbb{K}$ be a solution of (6.3.2). Then (6.3.3) is satisfied and thus also

$$
\begin{pmatrix} \varphi(f_0(x)) \\ \vdots \\ \varphi(f_N(x)) \end{pmatrix} = (\beta_{ij}(x)) \begin{pmatrix} \varphi(f_0(x)) \\ \vdots \\ \varphi(f_N(x)) \end{pmatrix},
$$

i.e. (6.3.10) holds with $g = \varphi$. ∎

6.4 Matrix equation with constant g

In this section we turn to the matrix equation (6.0.2). We want to determine solutions $\Phi: X \to \mathbb{K}^N$ of the equation

$$\Phi(f(x)) = g\Phi(x) + h(x), \tag{6.4.1}$$

where X is an arbitrary set, $f: X \to X$, $h: X \to \mathbb{K}^N$ are arbitrary functions, and the matrix $g \in \mathbb{K}^{N \times N}$ is independent of x.

6.4A. Reduction to systems of scalar equations

Let $B \in \mathbb{K}^{N \times N}$ be a nonsingular matrix such that BgB^{-1} has the Jordan normal form

$$
BgB^{-1} = \begin{pmatrix} \Lambda_1 & & 0 \\ & \ddots & \\ 0 & & \Lambda_m \end{pmatrix}, \tag{6.4.2}
$$

with the boxes $\Lambda_k = \Lambda(\lambda_k)$, where

$$
\Lambda = \Lambda(\lambda) = \begin{pmatrix} \lambda & 1 & 0 & \cdots & 0 & 0 \\ 0 & \lambda & 1 & \cdots & 0 & 0 \\ \cdot & \cdot & \cdot & \cdots & \cdot & \cdot \\ 0 & 0 & 0 & \cdots & \lambda & 1 \\ 0 & 0 & 0 & \cdots & 0 & \lambda \end{pmatrix}. \tag{6.4.3}
$$

Here $\lambda_k \in \mathbb{K}$, $k = 1, \ldots, m$, are the characteristic roots of the matrix g (some of them may be equal). Setting $\hat{\Phi} = B\Phi$ and $\hat{h} = Bh$, we obtain from (6.4.1) the equation

$$\hat{\Phi}(f(x)) = (BgB^{-1})\hat{\Phi}(x) + \hat{h}(x). \tag{6.4.4}$$

By (6.4.2) this equation splits into m independent ones of the form

$$\Phi(f(x)) = \Lambda\Phi(x) + h(x), \tag{6.4.5}$$

where $\Phi, h: X \to \mathbb{K}^n$, $n \in \mathbb{N}$, and $\Lambda \in \mathbb{K}^{n \times n}$ is given by (6.4.3). In the kth of the m equations we mentioned we have Λ_k in place of Λ, and, instead of h, the kth component of the function $\hat{h}$, corresponding to this Λ_k. If n_k is the order of Λ_k, then $n_1 + \cdots + n_m = N$.

Thus we may assume that equation (6.4.1) has the simpler form (6.4.5). Let us put $\Phi = (\varphi_1, \ldots, \varphi_n)^T$, $h = (h_1, \ldots, h_n)^T$. Then equation (6.4.5) can be written as the system of scalar equations

$$\left.\begin{array}{l} \varphi_1(f(x)) = \lambda\varphi_1(x) + \varphi_2(x) + h_1(x), \\ \cdots\cdots\cdots\cdots\cdots\cdots\cdots\cdots\cdots\cdots\cdots\cdots\cdots \\ \varphi_{n+1}(f(x)) = \lambda\varphi_{n-1}(x) + \varphi_n(x) + h_{n-1}(x), \\ \varphi_n(f(x)) = \lambda\varphi_n(x) + h_n(x). \end{array}\right\} \tag{6.4.6}$$

Equations of system (6.4.6) can be solved successively, starting with that for φ_n – a scalar linear equation of first order. The theory developed in Chapters 2–4 may help to determine the function φ_n. After having inserted this φ_n into the last but one equation in (6.4.6), we get again a linear equation of order 1 with one unknown function φ_{n-1}. In this way we eventually solve all the equations of system (6.4.6), i.e. we solve equation (6.4.5). Getting then the solutions of the original equation (6.4.1) presents no difficulties.

6.4B. Real solutions when some characteristic roots of g are complex numbers

In this case the just described procedure leading to system (6.4.6) has to be modified. If $\lambda_k = \alpha_k + i\beta_k$, then instead of the two boxes Λ_k, Λ_{k+1}, corresponding to λ_k and its conjugate $\lambda_{k+1} = \bar{\lambda}_k$, we get a box of the form (we suppress the index k)

$$\Lambda = \begin{pmatrix} \alpha & -\beta & 1 & 0 & \cdots & 0 & 0 \\ \beta & \alpha & 0 & 1 & \cdots & 0 & 0 \\ & & & \ddots & & & \\ 0 & 0 & & & \cdots & 1 & 0 \\ 0 & 0 & & & \cdots & 0 & 1 \\ 0 & 0 & 0 & 0 & \cdots & \alpha & -\beta \\ 0 & 0 & 0 & 0 & \cdots & \beta & \alpha \end{pmatrix}. \tag{6.4.7}$$

Assuming that Λ in (6.4.5) has form (6.4.7), we get instead of (6.4.6) the following system of equations:

$$\left.\begin{aligned}
\varphi_1(f(x)) &= \alpha\varphi_1(x) - \beta\varphi_2(x) + \varphi_3(x) + h_1(x), \\
\varphi_2(f(x)) &= \beta\varphi_1(x) + \alpha\varphi_2(x) + \varphi_4(x) + h_2(x), \\
&\,\,\dots\dots\dots\dots\dots\dots\dots\dots\dots\dots\dots\dots\dots \\
\varphi_{n-3}(f(x)) &= \alpha\varphi_{n-3}(x) + \beta\varphi_{n-2}(x) + \varphi_{n-1}(x) + h_{n-3}(x), \\
\varphi_{n-2}(f(x)) &= \beta\varphi_{n-3}(x) + \alpha\varphi_{n-2}(x) + \varphi_n(x) + h_{n-2}(x), \\
\varphi_{n-1}(f(x)) &= \alpha\varphi_{n-1}(x) - \beta\varphi_n(x) + h_{n-1}(x), \\
\varphi_n(f(x)) &= \beta\varphi_{n-1}(x) + \alpha\varphi_n(x) + h_n(x).
\end{aligned}\right\} \quad (6.4.8)$$

Here even the last equation contains two unknowns: φ_n and φ_{n-1}. In this situation we pass to complex-valued functions and solve a scalar equation of the first order, whose solution furnishes the real-valued one of the last two equations in system (6.4.8). The idea is made precise in the following obvious lemma.

Lemma 6.4.1. *Let X be an arbitrary set, $f: X \to X$ and $h_1, h_2: X \to \mathbb{R}$ arbitrary functions, and let $\lambda = \alpha + i\beta$, $\alpha, \beta \in \mathbb{R}$. If $\varphi_1, \varphi_2: X \to \mathbb{R}$ satisfy the system of equations*

$$\left.\begin{aligned}
\varphi_1(f(x)) &= \alpha\varphi_1(x) - \beta\varphi_2(x) + h_1(x), \\
\varphi_2(f(x)) &= \beta\varphi_1(x) + \alpha\varphi_2(x) + h_2(x),
\end{aligned}\right\} \quad (6.4.9)$$

then the function $\varphi: X \to \mathbb{C}$ given by $\varphi = \varphi_1 + i\varphi_2$ satisfies the equation

$$\varphi(f(x)) = \lambda\varphi(x) + h(x), \quad (6.4.10)$$

where $h: X \to \mathbb{C}$ is defined as $h = h_1 + ih_2$. Conversely, the real, φ_1, and imaginary, φ_2, parts of a solution $\varphi: X \to \mathbb{C}$ of (6.4.10) satisfy system (6.4.9) with $h_1 := \operatorname{Re} h$, $h_2 := \operatorname{Im} h$.

Thus, having found φ_{n-1} and φ_n via Lemma 6.4.1, we insert them into the $(n-2)$nd and $(n-3)$rd equations in (6.4.8) and obtain again a system of form (6.4.9), Lemma 6.4.1 again works, and so on, till all the equations of system (6.4.8) will have been solved.

6.4C. Special system of two linear equations

As an example we solve system (6.4.6) in the particular case, where X is a real interval, the function f is linear, $f(x) = bx$, and $h = 0$. In accordance with the discussion in Subsection 6.4B, we need consider only the case $n = 2$. Thus we are going to deal with the system

$$\left.\begin{aligned}
\varphi_1(bx) &= \lambda\varphi_1(x) + \varphi_2(x), \\
\varphi_2(bx) &= \lambda\varphi_2(x).
\end{aligned}\right\} \quad (6.4.11)$$

We assume that

(i) $X = (0, a|, 0 < a \leqslant \infty, 1 \neq b > 0, \lambda \in \mathbb{K} \setminus \{0\}$.

We start by examining the second equation in (6.4.11). Write $\lambda = |\lambda| e^{i\alpha}$, $|\lambda| > 0$, and $c := \log_b |\lambda|$. Then $(bx)^c = |\lambda| x^c$ and for the function $\psi(x) := x^{-c} \varphi_2(x)$ we get the equation $\psi(bx) = e^{i\alpha} \psi(x)$. Writing further $\psi(x) = e^{i\alpha t} \omega(t)$, with $t = \log_b x$, we see that $e^{i(t+1)\alpha} \omega(t+1) = \psi(bx) = e^{i\alpha} \psi(x) = e^{i\alpha} e^{i\alpha t} \omega(t)$. Thus necessarily $\omega(t+1) = \omega(t)$, i.e. ω is periodic of period 1. In this way we arrive at the formula

$$\varphi_2(x) = x^{\log_b |\lambda|} e^{i\alpha \log_b x} \omega(\log_b x). \tag{6.4.12}$$

This is the general solution of the second equation in (6.4.11), $\omega : \mathbb{R} \to \mathbb{C}$ being an arbitrary periodic function of period 1.

Note that formula (6.4.12) applies also in the case of a real λ, and for $\alpha = 0$, i.e. $\lambda > 0$, reduces to

$$\varphi_2(x) = x^{\log_b \lambda} \omega(\log_b x). \tag{6.4.13}$$

However, if $\alpha = \pi$, i.e. $\lambda < 0$, then (6.4.12) yields a complex-valued function φ_2. If we want to remain within real-valued functions we have to replace (6.4.12) by the formula

$$\varphi_2(x) = x^{\log_b |\lambda|} \omega(\log_b x), \tag{6.4.14}$$

where $\omega : \mathbb{R} \to \mathbb{R}$ is now an arbitrary antiperiodic function of period 1, $\omega(t+1) = -\omega(t)$.

Now we pass to the first equation in (6.4.11). With the notation $c = \log_b |\lambda|$, $t = \log_b x$, this equation may be written, in view of (6.4.12), as

$$\varphi_1(bx) = \lambda \varphi_1(x) + x^c e^{i\alpha t} \omega(t). \tag{6.4.15}$$

Then the function $\psi(x) = x^{-c} e^{-i\alpha t} \varphi_1(x)$ must be a solution of the equation

$$\psi(bx) = \psi(x) + \lambda^{-1} \omega(t). \tag{6.4.16}$$

It is easy to guess a particular solution $\hat{\psi}$ of equation (6.4.16): $\hat{\psi}(x) = t\lambda^{-1} \omega(t) = \lambda^{-1} \omega(\log_b x) \log_b x$. Hence the function

$$\hat{\varphi}_1(x) := x^c e^{i\alpha \log_b x} \hat{\psi}(x)$$

is a particular solution of equation (6.4.15). The difference $\varphi_1 - \hat{\varphi}_1$ satisfies the homogeneous equation associated with (6.4.15) which is nothing else than the second equation in (6.4.11). Thus the general solution of equation (6.4.15) has the form

$$\varphi_1(x) = x^{\log_b |\lambda|} e^{i\alpha \log_b x} (\lambda^{-1} \omega(\log_b x) \log_b x + \omega_0(\log_b x)), \tag{6.4.17}$$

where $\omega_0 : X \to \mathbb{K}$ is an arbitrary periodic function of period 1.

In the case where $\lambda \in \mathbb{R}$ and $\lambda > 0$, (6.4.17) reduces to

$$\varphi_1(x) = x^{\log_b \lambda} (\lambda^{-1} \omega(\log_b x) \log_b x + \omega_0(\log_b x)). \tag{6.4.18}$$

However, if $\lambda < 0$ and $\mathbb{K} = \mathbb{R}$, then we get instead of (6.4.15) the equation

$$\varphi_1(bx) = \varphi_1(x) + x^c \omega(\log_b x), \qquad \omega(t+1) = -\omega(t),$$

whose general solution has the form

$$\varphi_1(x) = x^{\log_b |\lambda|}(\lambda^{-1}\omega(\log_b x)\log_b x + \omega_0(\log_b x)), \qquad (6.4.19)$$

where $\omega_0 \colon X \to \mathbb{R}$ is an arbitrary antiperiodic function of period 1, $\omega_0(t+1) = -\omega_0(t)$.

The above considerations lead to the following.

Theorem 6.4.1. *Let hypothesis* (i) *be fulfilled. The general solution $\varphi_1, \varphi_2 \colon X \to \mathbb{K}$ of system* (6.4.11) *is described in Table 6.1.*

Remark 6.4.1. It is easily seen that the functions φ_1 and φ_2 are continuous in X if and only if so are ω, ω_0 in $\mathbb{R}$. The same is true for solutions of class C^r.

Remark 6.4.2. If $X = [0, a|$, i.e. $0 \in X$, then we may assign the functions φ_1, φ_2 at $x = 0$ only the values $d_1 = d_2 = 0$ if $\lambda \neq 1$, and $d_2 = 0$, d_1 arbitrary, if $\lambda = 1$ (see (6.4.11)). It is easy to find out under what conditions the extended functions φ_1, φ_2 are continuous in X. This could also be derived from the general theorems in Section 6.3.2. See Note 6.9.7.

6.4D. Discussion of more general cases

If the function f in (6.4.1) is not of the form $f(x) = bx$, we may try to reduce it to such a form by a change of variables $x' = \sigma(x)$, $y' = \sigma(y)$. In order for us to have $y' = bx'$, σ must satisfy the Schröder equation

$$\sigma(f(x)) = b\sigma(x).$$

This equation has appeared in Chapters 2–4; see also Chapter 8. Effecting the substitution $x' = \sigma(x)$ in (6.4.1) we get an equation of the same form, but with $f(x') = bx'$.

If the function g in equation (6.0.2) is not constant, we may try a change of the unknown function $\hat{\Phi}(x) = S(x)\Phi(x)$ with an $S \colon X \to \mathbb{K}^{N \times N}$. Equation (6.0.2) becomes

$$\hat{\Phi}(f(x)) = S(f(x))g(x)S(x)^{-1}\hat{\Phi}(x) + S(f(x))h(x).$$

Table 6.1

Case	Formulae for φ_1 and φ_2	Conditions
$\mathbb{K} = \mathbb{C}$	(6.4.17) and (6.4.12)	$\alpha = \arg \lambda$; ω, ω_0 periodic of period 1
$\mathbb{K} = \mathbb{R}$, $\lambda > 0$	(6.4.18) and (6.4.13)	ω, ω_0 periodic of period 1
$\mathbb{K} = \mathbb{R}$, $\lambda < 0$	(6.4.19) and (6.4.14)	ω, ω_0 antiperiodic of period 1

This will have the form (6.4.5) if we choose the function S so as to satisfy

$$S(f(x))g(x) = \Lambda S(x). \tag{6.4.20}$$

Under suitable conditions regarding the given functions f and g and the constant matrix Λ, it can be proved that equation (6.4.20) has a solution S. This will follow from the general theorems in the next section and in Chapter 7.

Comments. For the material in this section see Pelyukh–Sarkovskiĭ [3]. The book contains also some information on solutions to equation (6.4.20).

6.5 Local C^∞ solutions of a matrix equation

In this section we shall present some theorems concerning local C^∞ solutions φ of the functional equation

$$\varphi(f(x)) = g(x)\varphi(x) + h(x) \tag{6.5.1}$$

which map a neighbourhood of the origin in $\mathbb{R}^K$ into $\mathbb{R}^N$ ($K, N \in \mathbb{N}$). Here the situation is particularly difficult, since the structure of (even very regular) functions on $\mathbb{R}^K$ is much more complicated than in the one-dimensional case.

All results that follow are due to L. P. Kučko [2].

6.5A. Preliminaries

We assume that

(i) $X \subset \mathbb{R}^K$, $K \geqslant 1$, is an open set and $0 \in X$.
(ii) The functions $f: X \to X$, $g: X \to \mathbb{R}^{N \times N}$ and $h: X \to \mathbb{R}^N$, $N \geqslant 1$, are of class C^∞, $f(0) = 0$, and the matrix $f'(0)$ is nonsingular.

If in equation (6.5.1) we make a linear change of variables $y = Tx$, with a nonsingular matrix $T \in \mathbb{R}^{K \times K}$, then (6.5.1) goes over into an equation of the same form with the new unknown function $\tilde\varphi = \varphi \circ T^{-1}$ and with $\tilde f := T \circ f \circ T^{-1}$, $\tilde g := g \circ T^{-1}$, $\tilde h := h \circ T^{-1}$ in place of the former f, g, h, respectively. The functions $\tilde f, \tilde g, \tilde h$ also satisfy (ii). and, moreover, we have $\tilde f'(0) = Tf'(0)T^{-1}$.

Further, multiplying equation (6.5.1) by a nonsingular matrix $S \in \mathbb{R}^{N \times N}$, we get another equation of form (6.5.1) with SgS^{-1} and Sh replacing the original g and h, respectively.

Making both the transformations one after another we can see that there is a one-to-one correspondence between the solutions φ of equation (6.5.1) and $\hat\varphi = S(\varphi \circ T^{-1})$ of the transformed one. In addition, φ is of class C^∞ if and only if $\hat\varphi$ is.

This discussion shows that we may assume that the matrices $f'(0) \in \mathbb{R}^{K \times K}$

and $g(0) \in \mathbb{R}^{N \times N}$ have convenient norms. For we can always replace these matrices by $Tf'(0)T^{-1}$, and $Sg(0)S^{-1}$, respectively, and apply Lemma 1.1.1.

The derivatives $\eta_k = \varphi^{(k)}(0)$ of φ at the origin, $k \in \mathbb{N}_0$, are multidimensional matrices. Differentiating equation (6.5.1) and then setting $x = 0$, we obtain an infinite system of linear equations for η_k

$$\eta_k = \sum_{i=0}^{k} Q_{ki}(0)\eta_i + h^{(k)}(0), \qquad k \in \mathbb{N}_0. \tag{6.5.2}$$

Here the functions Q_{ki}, whose values are multidimensional matrices, belong to class C^∞ in X.

In the sequel we assume that system (6.5.2) has a solution $(\eta_k)_{k \in \mathbb{N}_0}$.

We end the preparations with the following.

Definition 6.5.1. A function of class C^∞ whose value as well as the values of all its derivatives at a point x_0 are zero, is called *flat at* x_0. The abbreviation '*O-flat*' will stand for the property '*flat at the origin*'.

6.5B. Existence of a unique solution

We start with the following.

Theorem 6.5.1. *Let hypotheses* (i), (ii) *be fulfilled. Suppose all the characteristic roots of the matrix $f'(0)$ lie outside the unit circle. If (η_k) is a solution of system (6.5.2), then there exists a unique local C^∞ solution $\varphi: U \to \mathbb{R}^N$ of equation (6.5.1) in a neighbourhood $U \subset X$ of the origin, fulfilling the conditions*

$$\varphi^{(k)}(0) = \eta_k, \qquad k \in \mathbb{N}_0. \tag{6.5.3}$$

Proof. First we reduce the problem to that of finding O-flat solutions of an associated equation. This requires some preparation.

The characteristic roots of the matrix $(f^{-1})'(0)$ lie inside the unit circle. As a result of Lemma 1.1.1 we may assume that $\|(f^{-1})'(0)\| < 1$. Thus we may find a disc $U := \{x \in \mathbb{R}^K : |x| < \delta\}$ and a constant c, $0 < c < 1$, such that $f(\mathrm{cl}\, U) \subset X$, the inverse function f^{-1} exists and is of class C^∞ in U, and

$$|f^{-1}(x)| \leqslant c|x| \qquad \text{for } x \in U. \tag{6.5.4}$$

Let $\varphi_0: U \to \mathbb{R}^N$ be an arbitrary function of class C^∞ with $\varphi_0^{(k)}(0) = \eta_k$ for $k \in \mathbb{N}_0$. (Such a function always exists; see e.g. Whitney [1] or Pliś–Ważewski [1].)

Now we write

$$\varphi(x) = \varphi_0(x) + \psi(x).$$

A function $\varphi: U \to \mathbb{R}^N$ is a C^∞ solution of (6.5.1) fulfilling conditions (6.5.3) if and only if the function ψ is an O-flat solution of the equation

$$\psi(f(x)) = g(x)\psi(x) + h_0(x), \tag{6.5.5}$$

where $h_0(x) := h(x) - \varphi_0(f(x)) + g(x)\varphi_0(x)$. The function h_0 is O-flat, as follows from (6.5.2) and the properties of φ_0.

Replace x by $f^{-1}(x)$ in (6.5.5) and iterate the resulting equation. What is obtained suggests that a solution of equation (6.5.5) should have the form

$$\psi(x) = \sum_{n=0}^{\infty} \left(\prod_{i=1}^{n} g(f^{-i}(x)) \right) h_0(f^{-n-1}(x)). \tag{6.5.6}$$

We shall prove that the series in (6.5.6) converges a.u. in U.

The O-flatness of h_0 implies the existence of positive constants C_{kr} such that

$$|h_0^{(k)}(x)| \leqslant C_{kr}|x|^r \qquad \text{for } x \in U \text{ and } k, r \in \mathbb{N}_0. \tag{6.5.7}$$

Choose a positive integer r such that $Mc^r < 1$, where $M := \sup_U |g(x)|$, and c occurs in (6.5.4). Use (6.5.4) and (6.5.7) to majorize the series from (6.5.6) by the convergent one $\sum_{n=0}^{\infty} C_{0r} M^n c^{nr} \delta^r$. Hence ψ is defined and continuous in U, and clearly $\psi(0) = 0$. Differentiating series (6.5.6) we check in the same manner that all the derivatives of ψ exist, are continuous in U, and vanish at $x = 0$. Thus ψ defined by (6.5.6) is O-flat. Verification that ψ satisfies (6.5.5) in U is routine.

To prove the uniqueness, suppose that φ_1 and φ_2, both of class C^∞ in U, satisfy equation (6.5.1) and conditions (6.5.3). Then $\varphi := \varphi_1 - \varphi_2$ is a solution of the homogeneous equation

$$\varphi(f(x)) = g(x)\varphi(x), \tag{6.5.8}$$

and is O-flat. Thus relations (6.5.7) hold with φ in place of h_0 and with appropriate constants $\bar{C}_{kr} > 0$. We get from (6.5.8) for $x \in U$

$$|\varphi(x)| = \prod_{i=1}^{n} \|g(f^{-i}(x))\| \|\varphi(f^{-n}(x))\| \leqslant \bar{C}_{0r} M^n c^{nr} \delta^r, \qquad n, r \in \mathbb{N}.$$

It remains to take an $r \in \mathbb{N}$ such that $Mc^r < 1$ and let $n \to \infty$ to obtain hence $\varphi = 0$ in U, i.e. $\varphi_1 = \varphi_2$. ∎

Remark 6.5.1. Theorem 6.5.1 could also be proved via a fixed point theorem, whereas the above argument could also have been used to prove Theorem 3.4.2.

6.5C. Solution depending on an arbitrary function

In the case where all the characteristic roots of $f'(0)$ lie inside the unit circle, one can expect a large set of C^∞ solutions of (6.5.1). Indeed, we have the following result.

Theorem 6.5.2. *Let hypotheses* (i), (ii) *be fulfilled, and suppose that all the characteristic roots of $f'(0)$ lie inside the unit circle, whereas those of $g(0)$ are all zeros. If (η_k) is a solution of system* (6.5.2), *then equation* (6.5.1) *has a* *local C^∞ solution $\varphi: U \to \mathbb{R}^N$ in a neighbourhood U of the origin, fulfilling conditions* (6.5.3).

If, moreover, $\det g(x) \neq 0$ in a vicinity of the origin, then this solution depends on an arbitrary function.

Proof. As in the proof of Theorem 6.5.1, we shall show that formula (6.5.6), where h_0 is O-flat, defines an O-flat function $\psi: U \to \mathbb{R}^N$.

We may assume that $\|(f^{-1})'(0)\| > 1$ and we may change the values of f and g outside $U := \{x \in \mathbb{R}^K : |x| < \delta\}$ so that

$$g(x) = 0 \quad \text{for } |x| \geq \delta_0 > \delta, \qquad |f^{-1}(x)| \geq \delta_0 \quad \text{for } |x| \geq \delta.$$

Consequently, if δ and δ_0 are chosen small enough, we will have $|f^{-n}(x)| \geq \delta_0$ for $x \in U$ and $n \geq n(x)$, and $|f(x)| < |x|$ in U.

With such f and g in mind we see that for every $x \in U$ series (6.5.6) contains only a finite number of nonzero terms, and obviously converges. Moreover, ψ is of class C^∞ in $U \setminus \{0\}$. The thing to prove is

$$\lim_{x \to 0} \psi^{(k)}(x) = 0 \qquad \text{for } k \in \mathbb{N}_0. \tag{6.5.9}$$

Fix a positive $\varepsilon < 1$. In view of Lemma 1.1.1 we may assume that $\|g(x)\| < \varepsilon$ for $|x| \leq \delta_1 < \delta$. Using this in (6.5.5), we get, since $h_0(0) = 0$,

$$\limsup_{x \to 0} |\psi(x)| \leq \varepsilon \limsup_{x \to 0} |\psi(x)|.$$

To infer hence (6.5.9) we need the boundedness of ψ in U. Choose then a positive $\delta_2 < \inf_{|x| = \delta_1} |f(x)| < \delta_1$ and put $R_0 = \{x \in \mathbb{R}^K : \delta_2 \leq |x| \leq \delta_1\}$, $R_n = f^n(R_0)$, $n \in \mathbb{N}$. It follows from (6.5.5) that the suprema M_n of ψ on R_n fulfil the inequalities

$$M_{n+1} \leq \varepsilon M_n + L, \qquad n \in \mathbb{N}_0,$$

where $L := \sup_{|x| \leq \delta_1} |h_0(x)|$. In virtue of Theorems 6.8.1 or 6.8.2 the sequence (M_n) is bounded, and therefore so is ψ in U. This proves (6.5.9) for $k = 0$. Differentiating (6.5.5) and repeating this argument for the equations obtained we prove (6.5.9) for $k = 1, 2, \ldots$ as well.

Now assume that $g(x)$ is nonsingular just in $U \setminus \{0\}$. Take an $x_0 \in U \setminus \{0\}$ and let $V_0 := \{x \in \mathbb{R}^K : |x - x_0| < \rho\}$ with a $\rho > 0$ so small that $V_0 \subset U$ and $V_0 \cap f(V_0) = \varnothing$. Hence all the sets $V_n := f^n(V_0), n \in \mathbb{N}_0$, are mutually disjoint.

Let $\varphi_0: V_0 \to \mathbb{R}^N$ be an arbitrary function of class C^∞. We define φ_0 on V_n recursively,

$$\varphi_0(x) := g(f^{-1}(x)) \varphi_0(f^{-1}(x)), \qquad x \in V_n, n \in \mathbb{N}_0, \tag{6.5.10}$$

and put $\varphi_0(0) = 0$. One may verify, as for ψ above, that φ_0 is of class C^∞ in $V = \bigcup_{n=0}^\infty V_n \cup \{0\}$, and is O-flat. Thus we may extend φ_0 onto U to an O-flat function.

We want to prove that there is a solution $\hat{\varphi}: U \to \mathbb{R}^N$ of the homogeneous equation

$$\hat{\varphi}(f(x)) = g(x)\hat{\varphi}(x), \tag{6.5.11}$$

which is O-flat and coincides with φ_0 on V. Looking for such a $\hat{\varphi}$ in the form $\hat{\varphi}(x) = \varphi_0(x) + \psi(x)$, we see that ψ must be a solution of equation (6.5.5) with $h_0(x) := g(x)\varphi_0(x) - \varphi_0(f(x))$. Moreover, ψ has to vanish on V and be O-flat. By the first part of the proof the function defined by (6.5.6) has all the desired properties. In particular, ψ vanishes on V. Further, since

$$\det g(x) \neq 0$$

in $U \setminus \{0\}$, the function φ does not vanish on $V \setminus \{0\}$ provided $\varphi_0 \neq 0$ on V_0. Thus for an arbitrary C^∞ function $\varphi_0: V_0 \to \mathbb{R}^N$ the homogeneous equation (6.5.11) has a nontrivial C^∞ solution $\varphi: U \to \mathbb{R}^N$, O-flat and such that $\varphi|V_0 = \varphi_0$. This means that the C^∞ solution $\varphi: U \to \mathbb{R}^N$ of equation (6.5.1), satisfying (6.5.3), depends on an arbitrary function. ■

6.5D. Two existence theorems

Combining Theorems 6.5.1 and 6.5.2 we obtain the following.

Theorem 6.5.3. *Let hypotheses* (i), (i) *be fulfilled and let all the characteristic roots of $f'(0)$ lie inside the unit circle. Suppose that in a neighbourhood of the origin g has the form*

$$g(x) = \begin{pmatrix} g_1(x) & 0 \\ g_3(x) & g_2(x) \end{pmatrix},$$

where $g_1(x) \in \mathbb{R}^{p \times p}$, $g_2(x) \in \mathbb{R}^{q \times q}$, $p + q = N$, and the characteristic roots of $g_1(0)$ are all zero, whereas $g_1(0)$ is nonsingular. If (η_k) is a solution of system (6.5.2), then equation (6.5.1) has a local C^∞ solution $\varphi: U \to \mathbb{R}^N$ in a neighbourhood U of the origin, fulfilling conditions (6.5.3).

Proof. Write $\varphi = \varphi_0 + \psi$, with arbitrary φ_0 of class C^∞, fulfilling conditions (6.5.3). Then ψ must be a local C^∞ solution of equation (6.5.5), and must be O-flat. Let us decompose ψ and h_0 according to the form of g, i.e. let us put $\psi = (\psi_1, \psi_2)$, $h_0 = (h_1, h_2)$ $(\psi_1(x), h_1(x) \in \mathbb{R}^p)$. Then (6.5.5) becomes

$$\psi_1(f(x)) = g_1(x)\psi_1(x) + h_1(x),$$
$$\psi_2(f(x)) = g_2(x)\psi_2(x) + h_2(x) + g_3(x)\psi_1(x).$$

The existence of a suitable ψ_1 follows from Theorem 6.5.2. Since all the characteristic roots of $(f^{-1})'(0)$ lie outside the unit circle, we replace in the second equation x by $f^{-1}(x)$ and apply Theorem 6.5.1 to the resulting equation. In this way we get the desired ψ_2. ■

The last theorem in this section covers the case where the characteristic roots of $f'(0)$ lie both outside and inside the unit circle.

Theorem 6.5.4. *Let hypotheses* (i), (ii) *be fulfilled. Suppose that no characteristic root of $f'(0)$ lies on the unit circle and that g is as in Theorem 6.5.3. If (η_k) is a solution of system* (6.5.2), *then equation* (6.5.1) *has a local C^∞ solution $\varphi: U \to \mathbb{R}^N$ in a neighbourhood $U \subset X$ of the origin, fulfilling conditions* (6.5.3).

Proof. Assume that i characteristic roots of $f'(0)$ lie inside, and j outside the unit circle, $i + j = K$, $i > 0$, $j > 0$. In other cases Theorems 6.5.1 or 6.5.3 can be applied. Write every $x \in \mathbb{R}^K$ as $x = (x_+, x_-)$ with $x_+ \in \mathbb{R}^i$, $x_- \in \mathbb{R}^j$, i.e. $\mathbb{R}^K = \mathbb{R}^i \times \mathbb{R}^j$, where the subspaces correspond to the above introduced partition of the characteristic roots of $f'(0)$. In virtue of the Hadamard–Perron Theorem (see e.g. Anosov [1]), there is a neighbourhood $V \subset X$ of the origin, $V = V_1 \times V_2$, such that both $V_1 \subset \mathbb{R}^i$ and $V_2 \subset \mathbb{R}^j$ are invariant under f. Again we need consider only equation (6.5.5) and look for O-flat solutions $\psi: U \to \mathbb{R}^N$.

By Theorem 6.5.3 there exists a function $\tilde{\psi}: \tilde{U} \to \mathbb{R}^N$, $\tilde{U} \subset V_1$, which is O-flat (in $\mathbb{R}^i$) and satisfies the equation obtained from (6.5.5) by setting $x_- = 0$. Put $\psi_0(x) := \tilde{\psi}(x_+)$. Then ψ_0 is O-flat (in $\mathbb{R}^K$).

Write $\psi = \psi_0 + \psi_1$. Then ψ_1 has to satisfy the equation

$$\psi_1(f(x)) = g(x)\psi_1(x) + h_1(x), \tag{6.5.12}$$

where the function $h_1(x) := h_0(x) - \psi_0(f(x)) + g(x)\psi_0(x)$ is flat at every point of $\mathbb{R}^i$ lying in some neighbourhood, say U, of $x = 0 \in \mathbb{R}^K$. We may have in U the inequalities

$$\left|h_1^{(k)}(x)\right| \leqslant C_{kr}\left|x_-\right|^r, \quad k, r \in \mathbb{N}_0, \qquad \left|(f^{-1}(x))_-\right| \leqslant c\left|x_-\right|,$$

where $c < 1$. Making use of them we can prove, in the same way as in the proof of Theorem 6.5.1, the convergence of the series in (6.5.6) formed for equation (6.5.12), i.e. with h_1 in place of h_0. The sum of the series is a solution ψ_1 of equation (6.5.12), O-flat as needed. ■

Comments. Some further results on local C^∞ solutions of both the inhomogeneous and homogeneous equations ((6.5.1), resp. (6.5.8)) may be found in Kučko [1], [2].

6.6 Smooth solutions of the equation of Nth order

Existence and uniqueness of continuous, differentiable, and integrable solutions $\varphi: X \to \mathbb{R}$ of the linear functional equation of Nth order

$$\varphi(x) = \sum_{i=1}^N a_i(x)\varphi(f_i(x)) + h(x) \tag{6.6.1}$$

often can be proved by a use of Banach's Theorem. We are going to present, without proofs, some theorems of this sort.

In all cases the solution in question will be given by the formula

$$\varphi(x) = P(x) + \sum_{n=1}^{\infty} \sum_{i_1,\ldots,i_n=1}^{N} \left(\prod_{j=1}^{n} a_{i_j} \circ f_{i_j} \circ \cdots \circ f_{i_1}(x) \right) \hat{h} \circ f_{i_n} \circ \cdots \circ f_{i_1}(x), \quad (6.6.2)$$

where

$$\hat{h}(x) := h(x) + \sum_{i=1}^{N} a_i(x) P(f_i(x)) - P(x), \quad (6.6.3)$$

and P is a suitable polynomial.

Throughout this section the index i will run from 1 to N.

6.6A. Continuous and differentiable solutions

For continuous solutions of (6.6.1) we have the following.

Theorem 6.6.1. *Let* $X = [0, a]$, $0 < a < \infty$, *and let the functions* $f_i: X \to X$, $a_i: X \to \mathbb{R}$, $h: X \to \mathbb{R}$ *be continuous on* X. *If*

$$\sum_i |a_i(x)| < 1 \qquad \text{for } x \in X, \quad (6.6.4)$$

then equation (6.6.1) has a unique continuous solution $\varphi: X \to \mathbb{R}$. *This solution is given by formula (6.6.2) with (6.6.3), where* $P = 0$.

Passing to differentiable solutions we first state a lemma on the form of the kth derivative of a solution of (6.6.1); compare Lemma 3.4.1.

Lemma 6.6.1. *Let* $X = [0, a]$, $0 < a < \infty$, *and let* $f_i: X \to X$, $a_i: X \to \mathbb{R}$, $h: X \to \mathbb{R}$ *be of class* C^r *in* X, $1 < r \leqslant \infty$. *If* $\varphi: X \to \mathbb{R}$ *is of class* C^r *in* X *and satisfies (6.6.1), then for* $k = 1, \ldots, r$

$$\varphi^{(k)}(x) = \sum_i a_i(x)(f'_x(x))^k \varphi^{(k)}(f_i(x))$$

$$+ \sum_{j=0}^{k-1} \sum_i P_{kij}(x) \varphi^{(j)}(f_i(x)) + h^{(k)}(x), \quad (6.6.5)$$

where the functions $P_{kij}: X \to \mathbb{R}$ *are of class* C^{r-k} *in* X.

In particular, the values $d_k = \varphi^{(k)}(0)$, $k = 0, \ldots, r$, of a C^r solution of (6.6.1) have to satisfy the system of linear equations resulting from (6.6.5) on setting $x = 0$. Every solution $(d_0, d_1, \ldots, d_r)$ of this system will be called *admissible*.

Theorem 6.6.2. *Let the interval* X *and the functions* f_i, a_i, h *be as in Lemma 6.6.1. Further assume that* $f_i(0) = 0$, *and*

$$M + K \sum_{j=0}^{r-1} a^{r-j} < 1, \quad (6.6.6)$$

where

$$M := \sup_X \sum_i |a_i(x)(f'_i(x))^r|, \qquad K := \max_{0 \leqslant j \leqslant r-1} \sup_X \sum_i |P_{rij}(x)|.$$

Then for every admissible system $(d_0, d_1, \ldots, d_r)$ there exists a unique C^r solution $\varphi \colon X \to \mathbb{R}$ of equation (6.6.1) fulfilling the initial conditions $\varphi^{(k)}(0) = d_k, k = 0, \ldots, r$. This solution is given by formula (6.6.2) with (6.6.3), where $P(x) = \sum_{k=0}^{r} (d_k x^k / k!)$.

Remark 6.6.1. We may note that condition (6.6.6) implies that

$$\sum_i |a_i(x)(f'_i(x))'| < 1 \qquad \text{for } x \in X. \tag{6.6.7}$$

On the other hand, if condition (6.6.7) is fulfilled, then (6.6.6) holds provided the length a of the interval X is small enough.

6.6B. Integrable solutions

We assume, as usual (see Section 4.7), that

(i) $(X, \mathbf{S}, \mu)$ is a σ-finite measure space and $\mu(X) > 0$.
(ii) The functions $a_i \colon X \to \overline{\mathbb{R}}$ and $h \colon X \to \mathbb{R}$ are measurable.
(iii) The functions $f_i \colon X \to X$ are one-to-one and $\mathbf{S}$-measurable.
(iv) The measures μf_i and μf_i^{-1} (see Section 4.7 for the definitions) are absolutely continuous with respect to μ.
(v) The set $\bigcup_{n=0}^{\infty} (f_i^n(X) \setminus f_i^{n+1}(X))$ has measure zero.

Theorem 6.6.3. Let hypotheses (i)–(v) be satisfied. If there exist nonnegative numbers s_i such that

$$|a_i|^p \leq s_i^{1/\alpha(p)} \frac{d(\mu f_i)}{d\mu} \qquad \text{a.e. in } X,$$

where $\alpha(p)$ is defined in (4.7.6), and

$$\sum_i s_i < 1,$$

then equation (6.6.1) has an essentially unique solution $\varphi \in L^p$. This solution is given a.e. in X by formula (6.6.2) with (6.6.3), where $P = 0$.

Comments. Concerning results reported in this section and some related ones see Bajraktarević [5] (Theorem 6.6.1), Majcher [1], [3] (Theorem 6.6.2) and Matkowski [13] (Theorem 6.6.3).

6.7 Equation of Nth order with iterates of one function

In most cases it is convenient to reduce the linear equation in question, i.e.

$$\sum_{i=0}^{N} a_i(x) \varphi(f^i(x)) = h(x), \tag{6.7.1}$$

to a system of linear equations of order 1. In the present section we describe two methods for such a reduction, both based on ideas which have for long been exploited in the theory of differential equations.

6.7A. Reduction of order

Let X be an arbitrary set, and let $f: X \to X$ be an arbitrary function. We assume that a_i, h and φ occurring in equation (6.7.1) are functions of the type $X \to \mathbb{K}$ and that $a_N = 1$ (which can be achieved whenever $a_N(x) \neq 0$ in X). Write

$$\psi(x) := \varphi(f(x)) - \lambda(x)\phi(x), \tag{6.7.2}$$

where $\lambda(x)$ is to be determined. We get by induction for $i = 1, \ldots, N$

$$\varphi(f^i(x)) = \psi(f^{i-1}(x)) + \sum_{k=1}^{i-1} \left(\prod_{j=k}^{i-1} \lambda(f^j(x)) \right) \psi(f^{k-1}(x)) + \varphi(x) \prod_{j=0}^{i-1} \lambda(f^j(x)),$$

which inserted into (6.7.1) yields

$$\sum_{j=0}^{N-1} b_j(x)\psi(f^j(x)) + \left(\sum_{i=1}^{N} a_i(x) \prod_{j=0}^{i-1} \lambda(f^j(x)) + a_0(x) \right)\varphi(x) = h(x) \tag{6.7.3}$$

with $b_{N-1} = a_N = 1$ and some b_j, $j = 0, \ldots, N-2$, depending, among others, on λ. Making the coefficient of φ in (6.7.3) vanish we get a linear equation of order $N-1$ for the function ψ. Repeating this procedure $N-1$ times we eventually reduce (6.7.1) to a system of N linear equations of order 1, and of form (6.7.2),

$$\varphi_i(f(x)) - \lambda_i(x)\varphi_i(x) = \varphi_{i+1}(x), \qquad i = 0, \ldots, N-1, \tag{6.7.4}$$

where $\varphi_0 = \varphi$ and $\varphi_N = h$. System (6.7.4) is relatively easy to handle, since the rich theory of linear equations of the first order can be applied here; see Chapters 2–4.

The difficulty lies in finding a solution $\lambda: X \to \mathbb{K}$ of the equations of the type

$$\sum_{i=1}^{N} a_i(x) \prod_{j=0}^{i-1} \lambda(f^j(x)) + a_0(x) = 0 \tag{6.7.5}$$

which appear at each stage of the reducing procedure. This is a nonlinear equation of order $N-1$. Moreover, here the conditions corresponding to (5.1.2) or (5.3.2) are not fulfilled, and the theory presented in Chapter 5 does not work. For $N = 2$ equation (6.7.5) is discussed in Kuczma–Vopěnka [1]; see also Krzeszowiak–Dybiec [1]. For $N > 2$ there are no general results known.

However, the situation is not so hopeless, since it is enough to find a single particular solution of (6.7.5), and this may sometimes be guessed. See the example in Note 6.9.10.

Let us also note that it may happen that equation (6.7.5) has no solutions of the desired regularity, although equation (6.7.1) has such; see Kuczma–Vopěnka [1], Kuczma [26, pp. 260–1].

6.7B. Constant coefficients

Particularly simple is the case where the coefficients a_i are constant

$$\sum_{i=0}^{N} a_i \varphi(f^i(x)) = h(x). \tag{6.7.6}$$

Then we may seek a constant solution $\lambda(x) \equiv \lambda$ of equation (6.7.5), which now becomes an algebraic one

$$a_N \lambda^N + a_{N-1} \lambda^{N-1} + \cdots + a_0 = 0. \tag{6.7.7}$$

After choosing in (6.7.2) $\lambda(x) \equiv \lambda_0$, a root of (6.7.7), we obtain for ψ the equation

$$\sum_{i=0}^{N-1} b_i \psi(f^i(x)) = h(x),$$

where

$$(\lambda - \lambda_0)(b_{N-1} \lambda^{N-1} + \cdots + b_0) = a_N \lambda^N + a_{N-1} \lambda^{N-1} + \cdots + a_0.$$

Proceeding further in this way, after $N-1$ steps we arrive at system (6.7.4), where $\lambda_i(x) \equiv \lambda_i$ form the complete set of the roots of (6.7.7).

We will not go into further details here. Generally speaking, the method just described, if it works, allows one to reduce equation (6.7.6) to a particularly simple system of equations of order 1.

6.7C. Reduction to a matrix linear equation

This method may also be considered as classical, and has for long been used in the theory of differential equations. We put

$$\varphi_i(x) = \varphi(f^{i-1}(x)), \qquad i = 1, 2, \ldots, N. \tag{6.7.8}$$

Then equation (6.7.1) reduces to the system

$$\left.\begin{array}{l} \varphi_{i+1}(x) = \varphi_i(f(x)), \qquad i = 1, 2, \ldots, N-1, \\ a_N(x)\varphi_N(f(x)) + a_{N-1}(x)\varphi_N(x) + \cdots + a_0(x)\varphi_1(x) = h(x). \end{array}\right\} \tag{6.7.9}$$

System (6.7.9) may be written as a single matrix equation (we assume again $a_N(x) = 1$)

$$\Phi(f(x)) + g(x)\Phi(x) = H(x), \tag{6.7.10}$$

where $\Phi: X \to \mathbb{K}^N$ is given as $\Phi(x) = (\varphi_1(x), \ldots, \varphi_N(x))$, $H: X \to \mathbb{K}^N$, $H(x) = (0, \ldots, 0, h(x))$ and $g: X \to \mathbb{K}^{N \times N}$ is the matrix

$$g(x) = \begin{pmatrix} 0 & -1 & 0 & \cdots & 0 \\ 0 & 0 & -1 & \cdots & 0 \\ \cdot & \cdot & \cdot & \cdots & \cdot \\ 0 & 0 & 0 & \cdots & -1 \\ a_0(x) & a_1(x) & a_2(x) & \cdots & a_{N-1}(x) \end{pmatrix}.$$

This method always works, but the resulting system (6.7.10) is not nearly as simple as (6.7.4). Here the results of Chapters 2–4 are not applicable. In Sections 6.5 and 6.6 we have presented some results regarding the general N-dimensional linear equation (6.7.10).

Comments. Equation (6.7.6), of Nth order with constant coefficients, has been investigated in Kordylewski–Kuczma [4], Kuczma [26, pp. 262–5], Badura [1]; and also Gelfond [1], Berg [1], [2]. More on the method of reduction presented in Subsection 6.7C may be found in Kordylewski [3]. For still other approaches to the problem of reduction, also patterned on the theory of differential equations, see Turdiev–Šarifova [1].

6.8 Linear recurrence inequalities

We conclude this chapter with some theorems on sequences fulfilling linear inequalities. These theorems often serve in this book as a tool, but they have also an independent interest.

Also in this section the index i will always run through integers from 1 to N.

6.8A. System of inequalities

We consider the inequalities

$$a_{i,n+1} \leqslant \sum_{j=1}^{N} s_{ij} a_{jn} + b_{in}, \qquad n \in \mathbb{N}_0, \tag{6.8.1}$$

where a_{in} and b_{in}, $n \in \mathbb{N}_0$, are nonnegative numbers and $S = (s_{ij}) \in \mathbb{R}^{N \times N}$, $N \geqslant 1$, is a matrix with nonnegative entries. The following result is due to J. Matkowski [13].

Theorem 6.8.1. *Assume that all the characteristic roots of S lie inside the unit circle. Then the sequences $(a_{in})_{n \in \mathbb{N}}$ satisfying system (6.8.1) have the following properties.*

(a) *If $\sum_{n=0}^{\infty} b_{in}$ converge, then so do $\sum_{n=0}^{\infty} a_{in}$.*
(b) *If the sequences (b_{in}) are bounded, then so are (a_{in}).*
(c) *If $\lim_{n \to \infty} b_{in} = 0$, then also $\lim_{n \to \infty} a_{in} = 0$.*

Proof. (a) Let us introduce the multi-indices $J_k := (j_1, \dots, j_k)$. The components of J_k will run from 1 to N, independently from each other. By induction we get from (6.8.1)

$$a_{i,n+1} \leqslant \sum_{J_{n+1}} s_{ij_1} s_{j_1 j_2} \cdots s_{j_n j_{n+1}} a_{j_{n+1},0}$$

$$+ \sum_{m=1}^{n} \sum_{J_m} s_{ij_1} \cdots s_{j_{m-1} j_m} b_{j_m, n-m} + b_{in}, \qquad n \in \mathbb{N}_0. \tag{6.8.2}$$

By Lemmas 1.5.1 and 1.5.3 there exist positive numbers $s, r_1, \ldots, r_N$ such that $s < 1$ and

$$\sum_{j=1}^{N} s_{ij} r_j \leqslant s r_i. \tag{6.8.3}$$

Since inequalities (6.8.3) are homogeneous, we may also have

$$a_{i0} \leqslant r_i, \tag{6.8.4}$$

and $\sum_{n=0}^{\infty} b_{in} \leqslant r_i$. Hence

$$\sum_{J_{n+1}} s_{ij_1} \cdots s_{j_n j_{n+1}} a_{j_{n+1},0} \leqslant s^{n+1} r_i, \tag{6.8.5}$$

and

$$\sum_{n=1}^{\infty} \sum_{m=1}^{n} \sum_{J_m} s_{ij_1} \cdots s_{j_{m-1} j_m} b_{j_m, n-m} = \sum_{m=1}^{\infty} \sum_{n=m}^{\infty} \sum_{J_m} s_{ij_1} \cdots s_{j_{m-1} j_m} b_{j_m, n-m}$$

$$\leqslant \sum_{m=1}^{\infty} \sum_{J_m} s_{ij_1} \cdots s_{j_{m-1} j_m} r_{j_m} \leqslant \sum_{m=1}^{\infty} s^m r_i < \infty.$$

By this and (6.8.5) assertion (a) follows from (6.8.2).

(b) Choose positive $s, r_1, \ldots, r_N$ satisfying (6.8.3), (6.8.4) and $b_{in} \leqslant r_i$, $n \in \mathbb{N}_0$. We get by (6.8.2) and (6.8.5)

$$a_{i,n+1} \leqslant s^{n+1} r_i + \sum_{m=1}^{n} s^m r_i + r_i \leqslant r_i (1-s)^{-1},$$

i.e. the sequences (a_{in}) are bounded.

(c) By (b) the numbers $p_i := \limsup_{n \to \infty} a_{in}$ are finite. Clearly, they are nonnegative. It follows from (6.8.1), on letting $n \to \infty$ over suitable sequences of indices, that

$$p_i \leqslant \sum_{j=1}^{N} s_{ij} p_j.$$

Thus $p_i = 0$, by Lemma 1.5.2. ∎

6.8B. Consequences for single inequalities

First we consider the recurrence inequality of Nth order

$$a_{n+N} \leqslant s_1 a_{n+N-1} + \cdots + s_N a_n + b_n, \qquad n \in \mathbb{N}_0, \tag{6.8.6}$$

where s_i and $a_n, b_n, n \in \mathbb{N}_0$, are nonnegative numbers.

Theorem 6.8.2. *Assume that all the roots of the polynomial*

$$P(t) = t^N - s_1 t^{N-1} - \cdots - s_N$$

are in absolute value less than 1. If (a_n) is a solution of inequality (6.8.6), $a_n \geqslant 0$ for $n \in \mathbb{N}_0$, then

(a) *If $\sum_{n=0}^{\infty} b_n < \infty$, then $\sum_{n=0}^{\infty} a_n < \infty$.*
(b) *If the sequence (b_n) is bounded, then so is also (a_n).*
(c) *If $\lim_{n \to \infty} b_n = 0$, then also $\lim_{n \to \infty} a_n = 0$.*

Proof. Define s_{ij}, $j = 1, \ldots, N$, a_{in}, b_{in}, $n \in \mathbb{N}_0$, by

$$s_{1j} = s_j, \quad s_{i+1,i} = 1, \quad s_{ij} = 0 \qquad \text{for } i \neq 1 \text{ and } j \neq i - 1,$$

$$a_{in} = a_{n+N-i},$$

$$b_{1n} = b_n, \quad b_{in} = 0 \qquad \text{for } i \neq 1.$$

Inequality (6.8.6) implies that (6.8.1) holds. Moreover, $(-1)^N P(t)$ is just the characteristic polynomial of the matrix $S = (s_{ij})$. Thus the theorem follows from Theorem 6.8.1. ∎

Finally, we quote without proof a result on sequences satisfying (6.8.6) ($b_n = 0$) and having arbitrary terms. It is due to D. C. Russell [1], and belongs to the spirit 'boundedness + generalized monotonicity implies convergence'.

Theorem 6.8.3. *Let s_i be real numbers such that $\sum_i s_i = 1$. Then every bounded sequence (a_n) satisfying the inequality*

$$a_{n+N} \leqslant s_1 a_{n+N-1} + \cdots + s_N a_n, \qquad n \in \mathbb{N}_0,$$

converges if and only if the polynomial

$$Q(t) = 1 + q_1 t + \cdots + q_{N-1} t^{N-1}, \qquad q_i := 1 - \sum_{j=1}^{i} s_j,$$

has no roots in the set $\{t \in \mathbb{C}: |t| = 1, t \neq 1\}$.

Remark 6.8.1. The condition on the roots of $Q(t)$ is certainly fulfilled if all s_i are nonnegative; see Copson [1]. Concerning some related results see Copson [1] and Borwein [1].

6.9 Notes

6.9.1. Theorem 6.1.4 when applied to equation (6.1.2) with $f(x) = x$ yields the following characterization of the exponential function (see Zdun [1]).

Proposition 6.9.1. *A function $\varphi: \mathbb{R} \to \mathbb{R}$, differentiable at zero and such that $\varphi(0) = 1$, satisfies the functional equation*

$$\varphi(2x) = [\varphi(x)]^2$$

if and only if it is the exponential function $\varphi(x) = e^{cx}$, $c \in \mathbb{R}$.

6.9.2. The following result resembles Theorem 6.1.2.

Proposition 6.9.2. *Let a_i, b_i, $i = 1, \ldots, N$, be positive constants such that $\sum_{i=1}^{N} a_i^2 b_i = 1$. If a characteristic function $\varphi: \mathbb{R} \to \mathbb{C}$ satisfies the functional equation*

$$\varphi(x) = \sum_{i=1}^{N} [\varphi(a_i x)]^{b_i},$$

then φ is the characteristic function of a normal distribution.

(If we assume that there exists a (finite) second moment of the distribution in question, then the proposition follows from Theorem 6.1.3 (with $r = 2$, for then φ is twice differentiable at zero). Without any assumptions on the moments the proof is much more difficult; see Laha–Lukacs [1].)

6.9.3. The following argument is needed in the proof of Theorem 6.1.2. A solution $\rho: \mathbb{R} \to \mathbb{R}^+$ of equation (6.1.17) also satisfies

$$\rho(x) = \rho(a^{n+1}x) \prod_{j=0}^{n-1} \rho(a^j bx), \qquad x \in \mathbb{R}, \quad n \in \mathbb{N}.$$

On letting $n \to \infty$ we get hence $\rho = 0$ whenever ρ is continuous at zero and $\rho(0) = 0$.

6.9.3. The following argument is needed in the proof of Theorem 6.1.2. A solution $\rho: \mathbb{R} \to \mathbb{R}^+$ of equation (6.1.17) also satisfies

$$\varphi(x) = \alpha\varphi(x) + (1 - \alpha)\varphi(1 - (1 - \alpha)x), \qquad x \in [0, 1], \qquad (6.9.1)$$

is a constant function. The continuity assumption is essential: Dirichlet's function satisfies (6.9.1) with $\alpha = \frac{1}{2}$.

6.9.5. Consider Boyarsky's problem from ergodic theory.

Find an $r \in [3, 4]$ such that there exists an integrable function $\varphi: [0, 1] \to \mathbb{R}^+$ with $\operatorname{supp} \varphi \subset X_r := [r^2(4 - r)/16, r/4]$ and $\int_0^1 \varphi(x)dx = 1$ satisfying the equation

$$\varphi(\tfrac{1}{2} + x) + \varphi(\tfrac{1}{2} - x) = 2rx\phi(r(\tfrac{1}{4} - x^2)) \qquad \text{for } x \in X := [0, \tfrac{1}{2}]. \quad (6.9.2)$$

K. Baron [11] proved that if $[\frac{1}{2}, r/4] \not\subset X_r$ (which amounts to $r \leqslant 1 + \sqrt{5} = 3.236\ldots$), then there does not exist any function φ satisfying (6.9.2) and having all the desired properties.

6.9.6. If $\varphi: [0, 1] \to \mathbb{R}^+$ solves (6.9.2), then it also satisfies

$$\varphi(x) + \varphi(1 - x) = r|1 - 2x|\varphi(f_r(x)) \qquad \text{for } x \in [0, 1], \qquad (6.9.2')$$

where $f_r(x) = rx(1 - x), x \in [0, 1], r \in [3, 4]$. By using (6.9.2') and some results on the orbit structure of f_r (Collet–Eckmann [1, p. 117], Jakobson [1]), P. Alsholm [1] proved that there is a set $A \subset [3.57, 4]$ of positive Lebesgue measure such that when $r \in A$ equation (6.9.2) has no nontrivial Riemann integrable solutions.

Finally, after having observed that equation (6.9.2') is satisfied a.e. in $[0, 1]$ by the density function of a measure which is absolutely continuous with respect to Lebesgue measure and invariant under f_r, P. Alsholm [1, (II)] comes, by another result of M. V. Jakobson [1] to the following conclusion. For every $r \in [0, 4]$ equation (6.9.2') has no Riemann integrable solutions $\varphi \geqslant 0$ with $\operatorname{supp} \varphi$ of positive Lebesgue measure.

6.9.7. Let $X = [0, a|$. In Subsection 6.4C we considered the system

$$\varphi_1(bx) = \lambda\varphi_1(x) + \varphi_2(x), \qquad \varphi_2(bx) = \lambda\varphi_2(x) \qquad (6.9.3)$$

with $0 < b \neq 1$. The solutions of system (6.9.3) continuous on $X \setminus \{0\}$, determined by Theorem 6.4.1, which remain continuous at the origin are described in Table 6.2. (Similarly one can establish under what conditions the solutions of (6.9.3) are of class C^r in X.)

Table 6.2

Conditions	Continuous solutions on X
$\log_b |\lambda| > 0$	Whenever ω and φ_0, occurring in Table 6.1, are continuous in $\mathbb{R}$
$\log_b |\lambda| < 0$	$\varphi_1 = \varphi_2 = 0$
$|\lambda| = 1, \lambda \neq 1$	$\varphi_1 = \varphi_2 = 0$
$\lambda = 1$	$\varphi_1 = \text{const (arbitrary from } \mathbb{K}); \varphi_2 = 0$

6.9.8. If condition (6.6.4) or (6.6.7) is fulfilled only in a neighbourhood of zero, and $0 < f_i(x) < x$ in $X \setminus \{0\}$, $i = 1, \ldots, N$, then we can apply Theorem 6.6.1 or 6.6.2 in this neighbourhood and then extend the solution obtained via Extension Theorem 7.2.1. In the case of Theorem 6.6.2 the extension is of class C^r in X.

6.9.9. If the assumptions of Theorem 6.6.3 are fulfilled only on a subset X_0 of X, we can apply the theorem to obtain a solution $\varphi \in L^p(X_0)$ of equation (6.6.1), and then, if possible, to extend this solution onto X.

(For extension theorems for integrable solutions of (also nonlinear) equations see R. Węgrzyk [2].)

6.9.10. The following example illustrates the method of solving equations of type (6.7.1) we have discussed in Subsection 6.7A.

Example 6.9.1. Let us consider the equation

$$\varphi\left(\frac{x}{4}\right) - \frac{3x^2 + 19x - 10}{2x^2 + 17x + 8} \varphi\left(\frac{x}{2}\right) + \frac{x^2 + x - 6}{2x^2 + 9x + 4} \varphi(x) = \frac{5x}{4x + 2}. \quad (6.9.4)$$

Here $N = 2$, $X = \mathbb{R}^+$, $f(x) = x/2$. Equation (6.7.4) becomes

$$\lambda\left(\frac{x}{2}\right)\lambda(x) - a_1(x)\lambda(x) + a_0(x) = 0, \quad (6.9.5)$$

where $a_1(x)$ and $a_0(x)$ are the coefficients of $\varphi(x/2)$ and $\varphi(x)$ in (6.9.4), respectively. Rational coefficients in (6.9.5) suggest that this equation may have a rational solution. Indeed, the homography $\lambda(x) = (x + 3)/(x + 4)$ is a particular solution of equation (6.9.5). Thus (6.9.4) reduces to the system

$$\varphi\left(\frac{x}{2}\right) - \frac{x + 3}{x + 4} \varphi(x) = \psi(x), \qquad \psi\left(\frac{x}{2}\right) - \frac{x - 2}{2x + 1} \psi(x) = \frac{5x}{4x + 2}. \quad (6.9.6)$$

Smooth solutions $\varphi: X \to \mathbb{R}$, or local analytic solutions $\varphi: U \to \mathbb{C}$ in a neighbourhood $U \subset \mathbb{C}$ of the origin, of equation (6.9.4) can be determined from (6.9.6) with the aid of results contained in Chapter 3 or 4.

6.9.11. Let $X = (0, \infty)$ and let $A_i, a_i \in X, i = 1, \ldots, N$. M. Laczkovich [1] determined measurable solutions of the equation

$$\varphi(x) = \sum_{i=1}^{N} A_i \varphi(a_i x). \tag{6.9.7}$$

A more elementary proof, based on Baron–Jarczyk [2], of Laczkovich's result stated below has been supplied by W. Jarczyk [10].

Proposition 6.9.4. *Let $\lambda_1, \ldots, \lambda_k$ be the real roots of the equation $\sum_{i=1}^{N} A_i a_i^\lambda = 1$, where $A_i, a_i > 0, i = 1, \ldots, N$. Then for every measurable solution $\varphi: X \to X$ of equation (6.9.7) there are nonnegative measurable functions $s_j (j = 1, \ldots, k)$ on X such that $s_j(a_i x) = s_j(x)$ for every i and j and*

$$\varphi(x) = \sum_{j=1}^{k} s_j(x) x^{\lambda_j} \qquad a.e. \text{ in } X.$$

If $\log a_i / \log a_m$ is irrational for at least one pair $(i, m) \in \{1, \ldots, N\}^2$, then each s_j is a constant function.

Example 6.9.2. Consider the fundamental Cauchy equation (see Subsection 6.1A) restricted to the line $y = 2x$ and written in the form

$$\varphi(x) = \varphi\left(\frac{x}{3}\right) + \varphi\left(\frac{2x}{3}\right), \qquad x \in X. \tag{6.9.8}$$

The only real root of $(\frac{1}{3})^\lambda + (\frac{2}{3})^\lambda = 1$ is $\lambda = 1$ and $\log \frac{1}{3} / \log \frac{2}{3}$ is irrational. Proposition 6.9.4 yields the general measurable solution $\varphi: X \to X$ of (6.9.8) in the form $\varphi(x) = cx$ a.e. in X. $c \in \mathbb{R}^+$.

6.9.12. Let $0 < a_1 < \cdots < a_n$ be some constants. The equation

$$\varphi(x) + \sum_{i=1}^{N} \varphi(a_i x) = 0 \tag{6.9.9}$$

need not have nontrivial continuous solutions $\varphi: \mathbb{R} \to \mathbb{K}$. For instance, if $a_i = b^i, b > 1, i = 1, \ldots, N$, then it is seen by replacing x in (6.9.9) by bx and subtracting the resulting equation from (6.9.9) that $\varphi(x) = \varphi(b^{N+1} x)$. If 0 is in the domain of φ, the continuity implies $\varphi = \text{const}$, and (6.9.9) yields $\varphi = 0$.

The number of continuous solutions of (6.9.9) is not known. The existence of a nontrivial solution, continuous on $\mathbb{R}$, is proved by M. Newman–M. Sheingorn [1] (see also Wobst [1]). In particular, they showed that the equation

$$\varphi(x) + \varphi(ax) + \varphi(bx) = 0, \qquad 1 < a < b,$$

has a nontrivial continuous solution $\varphi: \mathbb{R} \to \mathbb{K}$ if and only if $b \neq a^2$.

6.9.13. When studying a physical problem, R. Schilling came across the following mathematical one. Let $q \in (0, 1)$ be fixed. Find solutions $\varphi: \mathbb{R} \to \mathbb{R}$

of the equation

$$\varphi(qx) = \frac{1}{4q}(\varphi(x-1) + \varphi(x+1) + 2\varphi(x)), \qquad (6.9.10)$$

satisfying further the condition

$$\varphi(x) = 0 \qquad \text{for } |x| > \frac{q}{1-q}. \qquad (6.9.11)$$

A partial solution to the problem is due to K. Baron [14].

Proposition 6.9.5. *If* $q \in (0, \sqrt{2} - 1)$ *then the zero function is the only solution* $\varphi: \mathbb{R} \to \mathbb{R}$ *of equation* (6.9.10) *satisfying* (6.9.11) *and bounded in a neighbourhood of the origin.*

Schilling himself knew this fact for $q \in (0, \frac{1}{3})$. In the case where $q \in (0, \frac{1}{3}]$ W. Förg-Rob [1] proved that every solution to (6.9.10) and (6.9.11) is zero a.e. in $\mathbb{R}$ and described the structure of all solutions to the problem.

If $q = \frac{1}{2}$, then (6.9.10) and (6.9.11) become

$$\varphi\left(\frac{x}{2}\right) = \frac{1}{2}(\varphi(x-1) + \varphi(x+1)) + \varphi(x), \quad \varphi(x) = 0 \qquad \text{for } |x| > 1. \quad (6.9.12)$$

Schilling's solution of (6.9.12) – $\varphi(x) = 1 - |x|$ for $|x| \le 1$, $\varphi(x) = 0$ otherwise – has been proved by K. Baron [14] to be the only one in the class of functions that have a continuous second derivative on $(-\delta, 1)$, $\delta > 0$, and the value unity at zero.

(Schilling's problem remains unsolved in the case where $q \in [\sqrt{2} - 1, 1) \setminus \{\frac{1}{2}\}$.)

6.9.14. Theorem 6.2.3 works, in particular, in the case where $f_1(x) = x/2$, $f_2(x) = (1 + x)/2$, $x \in X := [0, 1]$. The equation

$$\varphi(f_1(x)) - \varphi(f_2(x)) = h(x)$$

then has a continuous solution $\varphi: X \to \mathbb{R}$ for every continuous $h: X \to \mathbb{R}$ (Baron [12]).

6.9.15. A special case of equation (6.1.6), $\varphi(x) = \varphi(1 - x) + \varphi(x/(1 - x))$, $x \in [0, \frac{1}{2}]$, is considered by D. B. Small [1]. Theorem 6.1.3 does not work here.

7

Equations of infinite order and systems of nonlinear equations

7.0 Introduction

As was announced in Section 0.1, in this chapter we are going to consider functional equations of arbitrary order. An equation of countable order may simply be written as

$$F(x, \varphi(f_1(x)), \varphi(f_2(x)), \ldots) = 0.$$

To proceed to uncountable orders we slightly rearrange the notation. The following functional equation 'solved' with respect to $\varphi(x)$,

$$\varphi(x) = H(x, \varphi), \tag{7.0.1}$$

covers not only all the equations considered up to now but also those of infinite order. Many equations admitting infinitesimal operations like differentiation and integration performed upon the unknown function are also special cases of (7.0.1).

If $H(x, \varphi) := h(x, \varphi \circ f(\cdot, x))$ equation (7.0.1) reduces to the still fairly general functional equation

$$\varphi(x) = h(x, \varphi \circ f(\cdot, x)) \tag{7.0.2}$$

which we shall mainly be interested in. The factors S and X of the domain of the given function $f: S \times X \to X$ may be quite arbitrary here. Taking $S := \{1, \ldots, n\}$ and putting $f_i := f(i, \cdot)$, $i \in S$, we arrive at

$$\varphi(x) = h(x, \varphi(f_1(x)), \ldots, \varphi(f_n(x))), \tag{7.0.3}$$

the functional equation of nth order, which reduces to a quite familiar one in the case $n = 1$ (see Chapter 5).

To visualize the generality of equation (7.0.2), take for instance $S = X = \mathbb{R}$, $f(s, x) := s - x$, $(s, x) \in \mathbb{R}^2$, and $h(x, y) := -y''(\pi)$, $(x, y) \in \mathbb{R} \times C^2(\mathbb{R})$, to obtain the well-known differential equation with deviated argument

$$\varphi(x) = -\varphi''(\pi - x)$$

(see Fite [1] and Ševelo [1]). Taking $X = S = [0, 1]$, $f(s, x) := s$, $(s, x) \in [0, 1]^2$ and $h(x, y) := \int_0^1 K(x, s) y(s) \mathrm{d}s + g(x)$, $(x, y) \in [0, 1] \times L^1([0, 1])$ (with continuous kernel K and integrable function g) we obtain the Fredholm integral equation of the second kind:

$$\varphi(x) = \int_0^1 K(x, s) \varphi(s) \mathrm{d}s + g(x).$$

For $X = S = \mathbb{R}$, $f(s, x) = x + s$, $(s, x) \in \mathbb{R}^2$, and $h(x, y) = \inf\{x^2 + s^2 + y(s):$ $s \in \mathbb{R}\}$, $(x, y) \in \mathbb{R} \times \mathscr{F}(\mathbb{R})$, we arrive at

$$\varphi(x) = \inf\{x^2 + s^2 + \varphi(x + s): s \in \mathbb{R}\},$$

the functional equation which arises in the theory of multistage control process and was first introduced by R. Bellman [1] (see also Fischer [1] and Walter [1]).

The advantages of considering functional equations in such a general setting are evident especially in the situation where the finite order versions of the results concerning the existence, uniqueness, continuous dependence, etc. of (7.0.2) (or even (7.0.1)) are subtle enough, i.e. as good as those obtained while considering equation (7.0.3) directly. As we shall see, in many instances this is actually the case; all the results in that spirit we are going to present in the sequel are substantially due to K. Baron [9]. As previously, in many cases the solution of the equation in question is first obtained locally, in a subset of the set considered, and then the solution is extended onto a larger set. We begin our investigations with very important extension theorems obtained by K. Baron. Then we deal with his results on the existence, uniqueness and continuous dependence of solutions of equation (7.0.2) mainly in the class of continuous functions. Lyapunov stability and approximate solutions are also considered. The rest of the present chapter concerns more subtle results on finite order functional equations and their systems in different classes of functions: C^r, analytic, integrable or with prescribed asymptotic behaviour.

7.1 Extending solutions

7.1A. General extension theorem

Looking at equation (7.0.2) (resp. (7.0.1)) we realize at once that the domain of the given function h (resp. H) plays an important role. Some settlements are self-evident: if we look for solutions $\varphi: X \to Y$ and a map $f: S \times X \to X$ is given, then the situation is quite clear provided that h is defined on the whole product $X \times \mathscr{F}(S, Y)$, where $\mathscr{F}(S, Y)$ stands for the family of all functions mapping the set S of indices into Y. In such a case the following, purely set-theoretical, extension theorem holds true (see Baron [7], [9]).

Theorem 7.1.1. *Let* $h: X \times \mathscr{F}(S, Y) \to Y$ *and* $f: S \times X \to X$ *be given. If, for some set* $U \subset X$, *we have* $f(S \times U) \subset U$ *and for any* $x \in X$ *there exists an* $n \in \mathbb{N}$ *such that*

$$f(s_1, \cdot) \circ \cdots \circ f(s_n, \cdot)(x) \in U \qquad (7.1.1)$$

for all n-*tuples* $(s_1, \ldots, s_n) \in S^n$, *then each solution* $\varphi_0: U \to Y$ *of the equation*

$$\varphi(x) = h(x, \varphi \circ f(\cdot, x)) \qquad (7.1.2)$$

admits exactly one solution $\varphi: X \to Y$ *of* (7.1.2) *such that* $\varphi|_U = \varphi_0$.

Proof. Put

$$f_s := f(s, \cdot), \qquad s \in S,$$

and

$$U_0 := U, \qquad U_{k+1} := \bigcap_{s \in S} f_s^{-1}(U_k), \qquad k \in \mathbb{N}_0.$$

Obviously,

$$f_s(U_{k+1}) \subset U_k \qquad (7.1.3)$$

for all $s \in S$ and $k \in \mathbb{N}_0$.

The inclusion $U_k \subset U_{k+1}$ clearly holds for $k = 0$. Suppose it is true for a $k \in \mathbb{N}_0$ and take an $x \in U_{k+1}$. Then, for every $s \in S$, $f_s(x) \in U_k \subset U_{k+1}$, i.e. $x \in \bigcap_{s \in S} f_s^{-1}(U_{k+1}) = U_{k+2}$. Therefore, by induction, $U_k \subset U_{k+1}$ for all $k \in \mathbb{N}_0$. Moreover, by means of (7.1.1), one has $X = \bigcup_{k \in \mathbb{N}_0} U_k$.

Now, define a sequence of functions $\varphi_k: U_k \to Y$, $k \in \mathbb{N}_0$, by the formula

$$\varphi_{k+1}(x) := h(x, \varphi_k \circ f(\cdot, x)), \qquad x \in U_{k+1},$$

where $\varphi_0: U_0 \to Y$ is the given solution of (7.1.1). In view of (7.1.3), the definition is correct. We shall show that

$$\varphi_{k+1}|_{U_k} = \varphi_k \qquad (7.1.4)$$

for all $k \in \mathbb{N}_0$. In fact, for $k = 0$ equality (7.1.4) is true since φ_0 satisfies (7.1.2) in $U_0 = U$. Assume (7.1.4) to be satisfied for some $k \in \mathbb{N}_0$ and fix an $x \in U_{k+1}$. Then, for any $s \in S$, one has $f_s(x) \in U_k$ whence, by (7.1.4),

$$\varphi_{k+1}(f_s(x)) = \varphi_k(f_s(x)),$$

i.e.

$$\varphi_{k+1} \circ f(\cdot, x) = \varphi_k \circ f(\cdot, x)$$

and, consequently,

$$\varphi_{k+2}(x) = h(x, \varphi_{k+1} \circ f(\cdot, x)) = h(x, \varphi_k \circ f(\cdot, x)) = \varphi_{k+1}(x),$$

which finishes the induction. So, we may define the function $\varphi: X \to Y$ by the formula

$$\varphi(x) := \varphi_k(x) \qquad \text{for } x \in U_k, \ k \in \mathbb{N}_0,$$

which, obviously, has all the properties desired. ∎

To cover also some more delicate and interesting cases it turns out to be necessary to admit also some subsets Ω of the product $X \times \mathscr{F}(S, Y)$ as a possible domain of the function h. However, in such a case, without any additional assumptions, Theorem 7.1.1 is no longer valid (see Note 7.9.2). Keeping the notation of the proof of Theorem 7.1.1 we must assume the following hypothesis –

(H) Let $\Phi_n \subset \mathscr{F}(U_n, Y)$, $n \in \mathbb{N}_0$. For every $n \in \mathbb{N}_0$ and $\varphi \in \Phi_n$ the function
$\psi(x) := h(x, \varphi \circ f(\cdot, x))$, $x \in U_{n+1}$, belongs to Φ_{n+1} and the pair
$(x, \psi \circ f(\cdot, x))$ belongs to Ω for any $x \in U_{n+2}$

– to get the assertion of Theorem 7.1.1 in the case where h maps Ω into Y. The proof remains unchanged. Moreover, the extension $\varphi: X \to Y$ of the solution $\varphi_0: U_0 \to Y$ has the property that $\varphi|_{U_n} \in \Phi_n$ for all $n \in \mathbb{N}_0$ (see Baron [9]).

Corollary 7.1.1. *Under the assumptions of Theorem 7.1.1 if, moreover, X and Y are topological spaces, U is open, h, φ_0 and $f(s, \cdot)$ are continuous for all $s \in S$, then the extension $\varphi: X \to Y$ of the solution φ_0 is continuous provided that the intersection*

$$\bigcap_{s \in S} f(s, \cdot)^{-1}(V)$$

is open for any open set $V \subset X$ such that $U \subset V$.

Remark 7.1.1. The latter condition is certainly fulfilled in the case of finite set S, if the space $\mathscr{F}(S, Y)$ is endowed with the Tychonoff topology.

7.1B. Two sufficient conditions

Now, we supply conditions for the assumption (7.1.1) to be satisfied (see Baron [9]).

Theorem 7.1.2. *Let (E, ρ) be a metric space, S a nonempty set. Suppose that $X \subset E$, f maps the product $S \times X$ into X and $\xi \in \mathrm{cl}\, X$. Put $\|x\| := \rho(x, \xi)$, $x \in E$, and $a := \sup\{\|x\|: x \in X\}$. Let γ be an increasing right continuous real function defined on an interval $I = [0, a]$ or $[0, a)$ according as the supremum is attained or not, and let $\gamma(t) < t$ for $t \in I \setminus \{0\}$. If*

$$\|f(s, x)\| \leqslant \gamma(\|x\|) \qquad \text{for all pairs } (s, x) \in S \times X \tag{7.1.5}$$

then (7.1.1) holds true for every neighbourhood U of ξ.

Proof. Take an $x \in X$ and elements $s_1, \ldots, s_k$ from S; then

$$\|f(s_1, \cdot) \circ \cdots \circ f(s_k, \cdot)(x)\| \leqslant \gamma^k(\|x\|)$$

and since $\lim_{k \to \infty} \gamma^k(t) = 0$ for all $t \in I$ (see Theorem 1.2.1) the element $f(s_1, \cdot) \circ \cdots \circ f(s_k, \cdot)(x)$ falls into a given neighbourhood U of ξ provided that k is large enough. This is (7.1.1). ∎

Theorem 7.1.3. *Assume the hypotheses of the preceding theorem (except those for γ) and suppose that the set $\{\xi\} \cup \{x \in X: \|x\| \leqslant \|x_0\|\}$ is compact for all $x_0 \in X$. If the family $\{f(s,\cdot): s \in S\}$ is locally equicontinuous and $\sup_{s \in S} \|f(s, x)\| < \|x\|$ for all $x \in X \setminus \{\xi\}$, then (7.1.1) holds true for every neighbourhood U of ξ.*

Proof. First, observe that the function $\sigma: X \cup \{\xi\} \to \mathbb{R}$ given by the formula

$$\sigma(x) := \sup_{s \in S} \|f(s, x)\| \qquad \text{for } x \neq \xi,\ \sigma(\xi) = 0,$$

is continuous. Indeed, fix $\varepsilon > 0$, $x \in X$, and take any sequence $(x_n)_{n \in \mathbb{N}}$ of elements of $X \cup \{\xi\}$ tending to x. The local equicontinuity of the family $\{f(s,\cdot): s \in S\}$ ensures that $\rho(f(s, x_n), f(s, x)) < \varepsilon$ for all $s \in S$ and all sufficiently large $n \in \mathbb{N}$, whence

$$\|f(s, x_n)\| = \rho(f(s, x_n), \xi) \leqslant \varepsilon + \rho(f(s, x), \xi) = \varepsilon + \|f(s, x)\|$$

and, consequently,

$$\limsup_{n \to \infty} \sup_{s \in S} \|f(s, x_n)\| \leqslant \varepsilon + \sup_{s \in S} \|f(s, x)\|.$$

Therefore, on letting $\varepsilon \to 0$, we infer that σ is upper semicontinuous at x and hence upper semicontinuous in view of the arbitrariness of $x \in X$. On the other hand, $\sigma|_X$ is lower semicontinuous as a supremum of a family of continuous functions. Thus $\sigma|_X$ is continuous. The continuity of σ at ξ results directly from the hypotheses assumed. Put

$$\gamma(t) := \sup\{\sigma(x): x \in X \cup \{\xi\},\ \|x\| \leqslant t\}, \qquad t \in \mathbb{R}^+.$$

The monotonicity of γ is obvious and since

$$\|f(s, x)\| \leqslant \sigma(x) \leqslant \gamma(\|x\|),$$

condition (7.1.5) is fulfilled, in particular. To apply Theorem 7.1.2, it remains to show that γ is right continuous at any point $t_0 \in [0, a)$ where $a := \sup_{x \in X} \|x\|$ and that $\gamma(t) < t$ for $t \in (0, a)$.

To this aim, take any decreasing sequence $(t_n)_{n \in \mathbb{N}}$ from $(0, a)$ converging to t_0 and let $\bar{x}$ be an element of X such that $t_n \leqslant \|\bar{x}\|$ for all $n \in \mathbb{N}$ (disregarding, if necessary, a finite number of elements of $(t_n)_{n \in \mathbb{N}}$). Obviously, the limit $\alpha := \lim_{n \to \infty} \gamma(t_n)$ does exist and $\alpha \geqslant \gamma(t_0)$. If we had $\gamma(t_0) < \alpha$, then, for some $\varepsilon > 0$ and for any $n \in \mathbb{N}$, we would get $\gamma(t_0) + \varepsilon < \gamma(t_n)$, $n \in \mathbb{N}$, whence

$$\gamma(t_0) + \varepsilon \leqslant \sigma(x_n), \qquad n \in \mathbb{N},$$

with a sequence $(x_n)_{n \in \mathbb{N}}$ of elements of $X \cup \{\xi\}$ such that $\|x_n\| \leqslant t_n \leqslant \|\bar{x}\|$, $n \in \mathbb{N}$. Choosing a convergent subsequence $(x_{n_k})_{k \in \mathbb{N}}$ of $(x_n)_{n \in \mathbb{N}}$ (possible on account of the compactness of the set $\{\xi\} \cup \{x \in X: \|x\| \leqslant \|\bar{x}\|\}$) and making use of the continuity of σ, we infer that $\gamma(t_0) + \varepsilon \leqslant \sigma(x) \leqslant \gamma(t_0)$ (x being the limit of $(x_{n_k})_{k \in \mathbb{N}}$), a contradiction. Therefore, the right continuity of γ is proved. To show that $\gamma(t) < t$ for $t \in (0, a)$ fix such a t and choose an $x \in X$ such that $\gamma(t) = \sigma(x)$ and $\|x\| \leqslant t$ (for apply the continuity of σ and the appropriate compactness again). If $\gamma(t) = 0$, then $\gamma(t) < t$; if $\gamma(t) > 0$, then $\gamma(t) < \|x\| \leqslant t$. ∎

7.1C. Extending of continuous solutions

From Theorems 7.1.1, 7.1.3 and Corollary 7.1.1 we obtain the following.

Theorem 7.1.4. *Let (E, ρ) be a metric space, S a nonempty set and let Y be a topological space. Suppose that $X \subset E$, $\xi \in \mathrm{cl}\, X$, U is a neighbourhood of ξ, $h\colon X \times \mathscr{F}(S, Y) \to Y$ is continuous (with respect to the Tychonoff topology in $\mathscr{F}(S, Y)$) and $f\colon S \times X \to X$ is such that the family $\{f(s, \cdot)\colon s \in S\}$ is locally equicontinuous. Put $\|x\| := \rho(x, \xi)$, $x \in E$. If*

(a) *$\sup_{s \in S} \|f(s, x)\| < \|x\|$ for all $x \in X \setminus \{\xi\}$,*
(b) *the set $\{\xi\} \cup \{x \in X\colon \|x\| \leqslant \|x_0\|\}$ is compact for all $x_0 \in X$,*
(c) *for any open set $V \subset X$ such that $U \subset V$ the intersection $\bigcap_{s \in S} f(s, \cdot)^{-1}(V)$ is open,*

then, for every continuous function $\varphi_0\colon U \to Y$ such that

$$\varphi_0(x) = h(x, \varphi_0 \circ f(\cdot, x)) \qquad \text{for } x \in U,$$

there exists exactly one function $\varphi\colon X \to Y$ such that $\varphi|_U = \varphi_0$ and

$$\varphi(x) = h(x, \varphi \circ f(\cdot, x)) \qquad \text{for all } x \in X.$$

This extension φ is continuous on X.

In particular, taking $E = Y = \mathbb{R}$ with the usual metric, $S = \{1, \ldots, n\}$, $X = |0, a|$, $0 < a \leqslant \infty$, $\xi = 0$, $f(i, \cdot) := f_i$, $i \in S$, we immediately get from Theorem 7.1.4 the following.

Corollary 7.1.2. *Let $n \in \mathbb{N}$, $X = |0, a|$, $0 < a \leqslant \infty$, and let $f_i\colon X \to X$ be continuous and such that $0 < f_i(x) < x$ for $x \in X \setminus \{0\}$, $i \in \{1, \ldots, n\}$. If the function $h\colon X \times \mathbb{R}^n \to \mathbb{R}$ is continuous, then for every continuous function $\varphi_0\colon X \cap [0, \delta) \to \mathbb{R}$, $\delta \in (0, a)$, such that*

$$\varphi_0(x) = h(x, \varphi_0(f_1(x)), \ldots, \varphi_0(f_n(x))), \qquad x \in X \cap [0, \delta),$$

there exists exactly one function $\varphi\colon X \to \mathbb{R}$ such that $\varphi|_{X \cap [0, \delta)} = \varphi_0$ and

$$\varphi(x) = h(x, \varphi(f_1(x)), \ldots, \varphi(f_n(x))) \qquad \text{for all } x \in X.$$

This extension φ is continuous in X.

If we wish to rid ourselves of the assumption that the given function h is defined on the whole product $X \times \mathbb{R}^n$ (resp. $X \times \mathscr{F}(S, Y)$) we have to adopt additional hypotheses of the type (H) from Subsection 7.1A. Explicit statements become less readable, however, and so we omit them here, referring to Baron [9] for details.

7.2 Existence and uniqueness

In this section we accept the following general assumptions.

The space $(X, \mathcal{T})$ is a topological space and $\mathcal{G} \subset \mathcal{T}$ is a family of open sets fulfilling the following conditions:

(i) $\bigcup \mathcal{G} = X$;

(ii) cl U is compact for all $U \in \mathcal{G}$;

(iii) the intersection of any two members of $\mathcal{G}$ either is empty or belongs to $\mathcal{G}$.

The space (Y, σ) is a complete metric space and $\boldsymbol{\Phi} \subset C(X, Y)$ is a given nonempty family of transformations which has the following properties

(iv) for every $U \in \mathcal{G}$ the family $\boldsymbol{\Phi}_U := \{\varphi|_{\mathrm{cl}\, U} : \varphi \in \boldsymbol{\Phi}\}$ is closed with respect to the uniform convergence;

(v) every continuous map $\varphi : X \to Y$ having the property $\varphi|_{\mathrm{cl}\, U} \in \boldsymbol{\Phi}_U$, $U \in \mathcal{G}$, belongs to $\boldsymbol{\Phi}$.

The function $H : X \times \boldsymbol{\Phi} \to Y$ is such that

(vi) $H(\cdot, \varphi) \in \boldsymbol{\Phi}$ for any $\varphi \in \boldsymbol{\Phi}$;

(vii) for any $U \in \mathcal{G}$ and any two members φ and ψ of $\boldsymbol{\Phi}$, if $\varphi|_U = \psi|_U$, then $H(\cdot, \varphi)|_U = H(\cdot, \psi)|_U$.

7.2A. Basic result

Following K. Baron [9] (see also Baron [4]) we are going to prove the following fairly general result.

Theorem 7.2.1. *Let* $(X, \mathcal{T})$, (Y, σ), $\mathcal{G}$, $\boldsymbol{\Phi}$ *and* H *have the meaning described above i.e. let assumptions* (i)–(vii) *be satisfied. Put*

$$d_U(\varphi, \psi) := \sup_{x \in \mathrm{cl}\, U} \sigma(\varphi(x), \psi(x)), \qquad \varphi, \psi \in \boldsymbol{\Phi}, \; U \in \mathcal{G}.$$

If, for any $U \in \mathcal{G}$ *and any* $\varepsilon > 0$, *there exists a* $\delta > 0$ *such that*

$$\varepsilon \leqq d_U(\varphi, \psi) < \varepsilon + \delta \quad \text{implies} \quad \sigma(H(x, \varphi), H(x, \psi)) < \varepsilon, \qquad x \in \mathrm{cl}\, U, \; \varphi, \psi \in \boldsymbol{\Phi},$$

$$\tag{7.2.1}$$

then the functional equation

$$\varphi(x) = H(x, \varphi) \tag{7.2.2}$$

has exactly one solution $\varphi \in \boldsymbol{\Phi}$. *This solution is given as the limit of the sequence of successive approximations*

$$\varphi_{n+1} := H(\cdot, \varphi_n), \qquad n \in \mathbb{N}_0,$$

where φ_0 *is an arbitrary map from* $\boldsymbol{\Phi}$. *The convergence is almost uniform on* X.

Proof. Fix a $U \in \mathcal{G}$ and observe that for any member φ of $\boldsymbol{\Phi}_U$ there exists a function $\psi \in \Phi$ such that $\varphi = \psi|_{\mathrm{cl}\,U}$. This allows one to find a function $E_U \colon \Phi_U \to \Phi$ such that $E_U(\varphi)|_{\mathrm{cl}\,U} = \varphi$, $\varphi \in \Phi_U$; in particular, $E_U(\varphi)$ yields an extension of φ to a function from Φ.

On account of (vi), the formula

$$T(\varphi) := H(\cdot, E_U(\varphi))|_{\mathrm{cl}\,U}, \qquad \varphi \in \Phi_U,$$

establishes a transformation mapping the family Φ_U into itself. Putting

$$d(\varphi, \psi) := d_U(E_U(\varphi), E_U(\psi)), \qquad \varphi, \psi \in \Phi_U,$$

and using (iv) one may easily check that the pair (Φ_U, d) yields a complete metric space whereas (7.2.1) ensures that for every $\varepsilon > 0$ there exists a $\delta > 0$ such that for all $\varphi, \psi \in \Phi_U$ the implication

$$\varepsilon < d(\varphi, \psi) < \varepsilon + \delta \Rightarrow d(T(\varphi), T(\psi)) \leqslant \varepsilon \tag{7.2.3}$$

holds true. Moreover, for any $\varphi, \psi \in \Phi_U$ such that $\varphi \neq \psi$, the number $\varepsilon := d(\varphi, \psi) = d_U(E_U(\varphi), E_U(\psi))$ is positive whence, by means of the compactness of cl U and (7.2.1), there exists an $x_0 \in \mathrm{cl}\,U$ such that

$$d(T(\varphi), T(\psi)) = d(H(\cdot, E_U(\varphi))|_{\mathrm{cl}\,U}, H(\cdot, E_U(\psi))|_{\mathrm{cl}\,U})$$

$$= \sup_{x \in \mathrm{cl}\,U} \sigma(H(x, E_U(\varphi)), H(x, E_U(\psi)))$$

$$= \sigma(H(x_0, E_U(\varphi)), H(x_0, E_U(\psi))) < \varepsilon$$

for some $x_0 \in \mathrm{cl}\,U$, because $H(\cdot, \xi)$ is continuous for $\xi \in \Phi$. Therefore

$$d(T(\varphi), T(\psi)) < d(\varphi, \psi) \qquad \text{for all } \varphi, \psi \in \Phi_U, \ \varphi \neq \psi,$$

which jointly with (7.2.3) allows one to apply Matkowski's fixed point Theorem 1.5.1 (see also Meir–Keeler [1]). Consequently, there exists exactly one function $\varphi_U \in \Phi_U$ such that

$$\varphi_U = T(\varphi_U) = H(\cdot, E_U(\varphi_U))|_{\mathrm{cl}\,U}. \tag{7.2.4}$$

Moreover, φ_U is the uniform limit of a sequence $(\varphi_{U,n})_{n \in \mathbb{N}_0}$, where $\varphi_{U,0}$ is arbitrarily taken from Φ_U and

$$\varphi_{U,n+1} := H(\cdot, E_U(\varphi_{U,n}))|_{\mathrm{cl}\,U}, \qquad n \in \mathbb{N}_0.$$

The next step is to show that

for all $U, V \in \mathscr{G}$ and $x \in U \cap V$ we have $\varphi_U(x) = \varphi_V(x) = \varphi_{U \cap V}(x)$. (7.2.5)

To this end, observe that in view of (iii) the intersection $W := U \cap V$ belongs to $\mathscr{G}$ and that $\varphi_U|_{\mathrm{cl}\,W} \in \Phi_W$. Therefore

$$E_W(\varphi_U|_{\mathrm{cl}\,W})|_{\mathrm{cl}\,W} = \varphi_U|_{\mathrm{cl}\,W} = (E_U(\varphi_U)|_{\mathrm{cl}\,U})|_{\mathrm{cl}\,W} = E_U(\varphi_U)|_{\mathrm{cl}\,W}$$

whence, in particular,

$$E_W(\varphi_U|_{\mathrm{cl}\,W})|_W = E_U(\varphi_U)|_W$$

which jointly with (vii) implies that

$$H(\cdot, E_W(\varphi_U|_{\mathrm{cl}\,W}))|_W = H(\cdot, E_U(\varphi_U))|_W$$

and, by means of the continuity of $H(\cdot, \xi)$, $\xi \in \Phi$,

$$H(\cdot, E_W(\varphi_U|_{\mathrm{cl}\,W}))|_{\mathrm{cl}\,W} = H(\cdot, E_U(\varphi_U))|_{\mathrm{cl}\,W}$$
$$= (H(\cdot, E_U(\varphi_U))|_{\mathrm{cl}\,U})|_{\mathrm{cl}\,W} = \varphi_U|_{\mathrm{cl}\,W}$$

on account of (7.2.4). The uniqueness of the solution φ_W gives now the equality

$$\varphi_W = \varphi_U|_{\mathrm{cl}\,W}.$$

Analogously, $\varphi_W = \varphi_V|_{\mathrm{cl}\,W}$, i.e. (7.2.5) is proved. This allows one (see (i)) to define a map $\varphi: X \to Y$ by the formula

$$\varphi(x) := \varphi_U(x) \qquad \text{for } x \in U.$$

Obviously, φ is continuous whence, in particular, $\varphi|_{\mathrm{cl}\,U} = \varphi_U \in \Phi_U$, $U \in \mathscr{G}$; thus $\varphi \in \Phi$ because of (v). Moreover, since $E_U(\varphi_U)|_{\mathrm{cl}\,U} = \varphi_U = \varphi|_{\mathrm{cl}\,U}$, $U \in \mathscr{G}$, we have $E_U(\varphi_U)|_U = \varphi_U$, $U \in \mathscr{G}$, whence, by means of (vii),

$$H(\cdot, E_U(\varphi_U))|_U = H(\cdot, \varphi)|_U, \qquad U \in \mathscr{G}. \tag{7.2.6}$$

To show that φ is a solution of equation (7.2.2), fix an $x \in X$ and find a $U \in \mathscr{G}$ such that $x \in U$ (see (i)); then

$$\varphi(x) = \varphi_U(x) = H(x, E_U(\varphi_U)) = H(x, \varphi)$$

on account of (7.2.6).

To prove the uniqueness, assume that $\psi \in \Phi$ is also a solution of equation (7.2.2), i.e.

$$\psi = H(\cdot, \psi).$$

Then, in particular, $\psi|_{\mathrm{cl}\,U} \in \Phi_U$ for all $U \in \mathscr{G}$. Consequently

$$E_U(\psi|_{\mathrm{cl}\,U})|_{\mathrm{cl}\,U} = \psi|_{\mathrm{cl}\,U}, \qquad U \in \mathscr{G},$$

and hence

$$E_U(\psi|_{\mathrm{cl}\,U})|_U = \psi|_U \qquad \text{for all } U \in \mathscr{G}.$$

Applying (vii) again as well as the continuity of $H(\cdot, \xi)$, $\xi \in \Phi$, we infer that

$$H(\cdot, E_U(\psi|_{\mathrm{cl}\,U}))|_{\mathrm{cl}\,U} = H(\cdot, \psi)|_{\mathrm{cl}\,U} = \psi|_{\mathrm{cl}\,U}, \qquad U \in \mathscr{G}.$$

Now, the uniqueness of $\varphi_U \in \Phi_U$ satisfying (7.2.4) implies the equality $\varphi_U = \psi|_{\mathrm{cl}\,U}$, $U \in \mathscr{G}$. The definition of φ gives $\varphi = \psi$, as required.

It remains to show that taking an arbitrary map $\varphi_0 \in \Phi$ and putting

$$\varphi_{n+1} := H(\cdot, \varphi_n), \qquad n \in \mathbb{N}_0,$$

we have $\varphi = \lim \varphi_n$ uniformly on every compact subset of X. To this end put $\varphi_{U,0} := \varphi_0|_{\mathrm{cl}\,U}$, $U \in \mathscr{G}$, and $\varphi_{U,n+1} := H(\cdot, E_U(\varphi_{U,n}))|_{\mathrm{cl}\,U}$, $n \in \mathbb{N}_0$. We shall show that

$$\varphi_{U,n} = \varphi_n|_{\mathrm{cl}\,U} \tag{7.2.7}$$

for all $U \in \mathscr{G}$ and $n \in \mathbb{N}_0$. By definition, (7.2.7) holds for $n = 0$. Assuming (7.2.7) for an $n \in \mathbb{N}_0$, we have

$$\varphi_n|_U = \varphi_{U,n}|_U = E_U(\varphi_{U,n})|_U$$

whence by (vii) and the continuity of $H(\cdot, \xi)$, $\xi \in \Phi$, we get

$$H(\cdot, \varphi_n)|_{\mathrm{cl}\, U} = H(\cdot, E_U(\varphi_{U,n}))|_{\mathrm{cl}\, U} = \varphi_{U,n+1}.$$

Thus

$$\varphi_{n+1}|_{\mathrm{cl}\, U} = H(\cdot, \varphi_n)|_{\mathrm{cl}\, U} = \varphi_{U,n+1},$$

i.e. (7.2.7) is proved.

Now, take any compact $K \subset X$ and choose elements $U_1, \ldots, U_k$ from $\mathscr{G}$ such that $K \subset \bigcup_{i=1}^k U_i$ (see (i)). Fix an $\varepsilon > 0$; since $(\varphi_{U,n})_{n \in \mathbb{N}_0}$ tends uniformly to φ_U, $U \in \mathscr{G}$, we may find an $N \in \mathbb{N}$ such that

$$\sigma(\varphi_{U_i,n}(x), \varphi_{U_i}(x)) < \varepsilon \tag{7.2.8}$$

for all $x \in U_i$, $i \in \{1, \ldots, k\}$, and all $n \geq N$. To finish the proof it suffices to show that

$$\sigma(\varphi_n(x), \varphi(x)) < \varepsilon \qquad \text{for all } x \in K \text{ and all } n \geq N.$$

To this end, take an $x \in K$ and fix an $n \geq N$. There exists an $i \in \{1, \ldots, k\}$ such that $x \in U_i$ and, by the definition of φ jointly with (7.2.7), we obtain

$$\sigma(\varphi_n(x), \varphi(x)) = \sigma(\varphi_{U_i,n}(x), \varphi_{U_i}(x)) < \varepsilon$$

in view of (7.2.8). ∎

7.2B. Important special case

As to a particular case of (7.0.1), Theorem 7.2.1 may be applied to the functional equation

$$\varphi(x) = h(x, \varphi \circ f(\cdot, x)). \tag{7.2.9}$$

The point is to impose appropriate assumptions on the given functions h and f in order that the function H given by the formula

$$H(x, \varphi) := h(x, \varphi \circ f(\cdot, x)) \tag{7.2.10}$$

should satisfy hypotheses (vi), (vii) and (7.2.1) occurring in the statement of Theorem 7.2.1. The following theorem (see Baron [9]) seems to be the most natural result of such endeavours.

Theorem 7.2.2. Let $(X, \mathscr{T})$, (Y, σ), $\mathscr{G} \subset \mathscr{T}$ and $\Phi \subset C(X, Y)$ satisfy assumptions (i)–(v). Suppose that S is a given nonempty set and $f: S \times X \to X$ is such that $f(s, \cdot)(\mathrm{cl}\, U) \subset \mathrm{cl}\, U$ for all $s \in S$ and $U \in \mathscr{G}$. Put

$$\mathscr{F}_0(S, Y) := \{\varphi \circ f(\cdot, x): (x, \varphi) \in X \times \Phi\}.$$

If a function $h: X \times \mathscr{F}_0(S, Y) \to Y$ satisfies the conditions

> for any member φ of Φ the function $\psi: X \to Y$ given by
> $\psi(x) := h(x, \varphi \circ f(\cdot, x))$, $x \in X$, belongs to Φ, too, $\tag{7.2.11}$

> for any $U \in \mathscr{G}$ and any $\varepsilon > 0$ there exists a $\delta > 0$ such that,
> for all $y_1, y_2 \in \mathscr{F}_0(S, Y)$, $\sigma(y_1(s), y_2(s)) < \varepsilon + \delta$, $s \in S$, implies
> $\sigma(h(x, y_1), h(x, y_2)) < \varepsilon$, for all $x \in \mathrm{cl}\, U$, $\tag{7.2.12}$

then the functional equation (7.2.9) *has exactly one solution* $\varphi \in \Phi$. *This solution is given as the almost uniform limit of the sequence of successive approximations* $(\varphi_n)_{n \in \mathbb{N}_0}$ *where* φ_0 *is an arbitrary map from* Φ *and*

$$\varphi_{n+1}(x) := h(x, \varphi_n \circ f(\cdot, x)), \qquad x \in X, \ n \in \mathbb{N}_0.$$

Proof. Define a function $H: X \times \Phi \to Y$ by (7.2.10) and note that (vi) results directly from (7.2.11). To check (vii) take a $U \in \mathscr{G}$, an $x \in U$ and functions $\varphi, \psi \in \Phi$ such that $\varphi|_U = \psi|_U$. Then also $\varphi|_{\mathrm{cl}\, U} = \psi|_{\mathrm{cl}\, U}$ because φ and ψ are continuous, whence, in view of our assumption on f, we derive the equality

$$y_1(s) := \varphi(f(s, x)) = \psi(f(s, x)) =: y_2(s), \qquad s \in S.$$

Thus the antecedent of implication (7.2.12) is fulfilled for any $\varepsilon > 0$ which implies that $\sigma(h(x, y_1), h(x, y_2)) < \varepsilon$ for all $\varepsilon > 0$. Hence

$$H(x, \varphi) = h(x, \varphi \circ f(\cdot, x)) = h(x, \psi \circ f(\cdot, x)) = H(x, \psi),$$

i.e. (vii) holds true.

Finally, to verify (7.2.1), fix a $U \in \mathscr{G}$ and an $\varepsilon > 0$. Take a $\delta > 0$ assigned to ε according to (7.2.12) and fix any two members φ and ψ from Φ such that $d_U(\varphi, \psi) \in [\varepsilon, \varepsilon + \delta)$. For any $x \in \mathrm{cl}\, U$ we have $f(s, x) \in \mathrm{cl}\, U$, $s \in S$, and

$$\sigma(y_1(s), y_2(s)) = \sigma(\varphi(f(s, x)), \psi(f(s, x))) \leqslant d_U(\varphi, \psi) < \varepsilon + \delta, \qquad s \in S.$$

So, by (7.2.12),

$$\sigma(H(x, \varphi), H(x, \psi)) = \sigma(h(x, \varphi \circ f(\cdot, x)), h(x, \psi \circ f(\cdot, x)))$$
$$= \sigma(h(x, y_1), h(x, y_2)) < \varepsilon,$$

i.e. (7.2.1) is satisfied. It remains to apply Theorem 7.2.1. ∎

7.2C. An application

To illustrate the result just obtained we shall reproduce here one of the number of appropriate applications exhibited in Baron [9].

Example 7.2.1. Let $g: \mathbb{R} \to \mathbb{R}$ be a continuous function such that $g(x)/x \in (0, 1)$ for all $x \in \mathbb{R} \setminus \{0\}$ and let

$$f(s, x) := g(x - \xi(s)) + \xi(s), \qquad (s, x) \in \mathbb{R},$$

where $\xi: \mathbb{R} \to \mathbb{R}$ is a given bounded and continuous function, say $|\xi(s)| \leqslant M$ for $s \in \mathbb{R}$. Moreover, suppose that $F: \mathbb{R} \to \mathbb{R}$ is continuous and $(a_m)_{m \in \mathbb{N}}$ is a real sequence such that $\sum_{m=1}^{\infty} |a_m| \leqslant 1$ whereas $(A_m)_{m \in \mathbb{N}}$ is a sequence of nonempty bounded subsets of $\mathbb{R}$. Making use of Theorem 7.2.2 we shall show that the functional equation

$$\varphi(x) = \sum_{m=1}^{\infty} a_m \sin \sup_{s \in A_m} \varphi(f(s, x)) + F(x)$$

has exactly one continuous solution $\varphi: \mathbb{R} \to \mathbb{R}$ given as the almost uniform limit of the sequence of successive approximations. To this end take $(X, \mathscr{T})$

to be $\mathbb{R}$ with the usual topology, $\mathcal{G} := \{(-M - m, M + m): m \in \mathbb{N}\}$, (Y, σ) to be $\mathbb{R}$ with the natural metric and $\Phi := C(\mathbb{R})$. The function $h: \mathbb{R} \times \mathcal{F}_0(\mathbb{R}, \mathbb{R}) \to \mathbb{R}$ given by the formula

$$h(x, y) := \sum_{m=1}^{\infty} a_m \sin \sup_{s \in A_m} y(s) + F(x), \qquad (x, y) \in \mathbb{R} \times \mathcal{F}_0(\mathbb{R}, \mathbb{R})$$

(note that $\mathcal{F}_0(\mathbb{R}, \mathbb{R}) = \{\varphi \circ f(\cdot, x): (x, \varphi) \in \mathbb{R} \times C(\mathbb{R})\} \subset C(\mathbb{R})$ and so the definition is correct) satisfies conditions (7.2.11) and (7.2.12). In fact, fix a continuous function $\varphi: \mathbb{R} \to \mathbb{R}$ and put $\psi_m := \sup_{s \in A_m} \varphi \circ f(s, \cdot)$, $m \in \mathbb{N}$. To prove (7.2.11), it suffices to show that ψ_m, $m \in \mathbb{N}$, are continuous. However, to this end it is enough to state that the family $\{\varphi \circ f(s, \cdot): s \in \mathbb{R}\}$ is locally equicontinuous. So, fix arbitrarily a point $x_0 \in \mathbb{R}$ and an $\varepsilon > 0$. Let $m \in \mathbb{N}$ be such that $|x_0| < M + m$ and choose a $\delta_0 > 0$ such that for every $x \in [-M - m, M + m]$

$$|x - y| < \delta_0 \quad \text{implies} \quad |\varphi(x) - \varphi(y)| < \varepsilon.$$

Let $\delta > 0$ be small enough to have $(x_0 - \delta, x_0 + \delta) \subset [-M - m, M + m]$ and

$$|f(s, x) - f(s, x_0)| = |g(x - \xi(s)) - g(x_0 - \xi(s))| < \delta_0$$

provided that $|x - x_0| < \delta$ (this is possible in view of the uniform continuity of restrictions of g to compact intervals). Then, for any $x \in \mathbb{R}$,

$$|x - x_0| < \delta \quad \text{implies} \quad |\varphi(f(s, x)) - \varphi(f(s, x_0))| < \varepsilon$$

for all $s \in \mathbb{R}$.

To make sure that (7.2.12) holds true in the present case observe first that for a given $\varepsilon > 0$ there exists a $\delta > 0$ such that

$$|x - y| < \varepsilon + \delta \quad \text{implies} \quad |\sin x - \sin y| < \varepsilon \tag{7.2.13}$$

for all $x, y \in \mathbb{R}$. Without loss of generality we may assume that $\varepsilon \in (0, 2]$. Take $\delta := -\varepsilon + 2 \arcsin \frac{1}{2}\varepsilon > 0$ and numbers $x, y \in \mathbb{R}$ such that $|x - y| < \varepsilon + \delta$; then

$$|\sin x - \sin y| \leqslant 2 \left| \sin \frac{x - y}{2} \right| = 2 \left| \sin \frac{|x - y|}{2} \right| < 2 \sin \frac{\varepsilon + \delta}{2} = \varepsilon,$$

as claimed.

Now, fix two continuous functions $y_i: \mathbb{R} \to \mathbb{R}$, $i \in \{1, 2\}$, an $\varepsilon > 0$ and choose a $\delta > 0$ to have (7.2.13). Suppose that

$$|y_1(s) - y_2(s)| < \varepsilon + \delta, \qquad s \in \mathbb{R}. \tag{7.2.14}$$

In view of the boundedness of A_m and the continuity of y_i one can find points $s_{i,m} \in \mathrm{cl}\, A_m$ such that

$$\sup_{s \in A_m} y_i(s) = \max_{s \in \mathrm{cl}\, A_m} y_i(s) = y_i(s_{i,m}),$$

$i \in \{1, 2\}$, $m \in \mathbb{N}$. Moreover, we have

$$|y_1(s_{1,m}) - y_2(s_{2,m})| \in \{y_1(s_{1,m}) - y_2(s_{2,m}), y_2(s_{2,m}) - y_1(s_{1,m})\}$$

as well as $y_2(s_{2,m}) \geqslant y_2(s_{1,m})$ and $y_1(s_{1,m}) \geqslant y_1(s_{2,m})$, $m \in \mathbb{N}$, when

$$|y_1(s_{1,m}) - y_2(s_{2,m})| < \varepsilon + \delta, \qquad m \in \mathbb{N},$$

on account of (7.2.14). Consequently (disregarding the trivial case $a_m = 0$, $m \in \mathbb{N}$)

$$\left|h(x, y_1) - h(x, y_2)\right| \leqslant \sum_{m=1}^{\infty} |a_m| \left|\sin y_1(s_{1,m}) - \sin y_2(s_{2,m})\right|$$

$$< \varepsilon \sum_{m=1}^{\infty} |a_m| \leqslant \varepsilon,$$

for all $x \in \mathbb{R}$, i.e. (7.2.12) holds true.

Finally, for $U = (-M - m, M + m)$ and $s \in \mathbb{R}$ one has

$$f(s, \cdot)(\text{cl } U) \subset \text{cl } U = [-M - m, M + m].$$

In fact, fix an $x \in \mathbb{R}$ such that $|x| \leqslant M + m$ and $s \in \mathbb{R}$; then

$$|f(s, x)| = |g(x - \xi(s)) + \xi(s)| \leqslant M + m$$

because of the fact that $g(z)/z \in (0, 1)$, $z \in \mathbb{R} \setminus \{0\}$, and that $|\xi(s)| \leqslant M$, $s \in \mathbb{R}$. Now, it suffices to apply Theorem 7.2.2 to obtain the desired result.

It should be emphasized that assumption (iv) in Theorem 7.2.1 (resp. Theorem 7.2.2) imposed upon the given subfamily Φ of the collection $C(X, Y)$ of all continuous mappings from X into Y is somewhat restrictive and, in practice, one usually confines oneself to a few cases like $\Phi = C(X, Y)$ (the most important one) or $\Phi = C(X, \mathbb{R}^+)$ or $\Phi = C(X, [a, b])$. Monotonic, convex or analytic functions are also good positive examples of admissible subfamilies of $C(\mathbb{R})$ but the collections of Lipschitzian, differentiable or C^r-mappings are not.

7.2D. Lipschitzian solutions

Here we supply some general results on Lipschitzian solutions of equation (7.2.9). Following the ideas of K. Baron [9] (see also Baron–Matkowski [1]) we shall prove the following.

Theorem 7.2.3. *Let (X, ρ) be a metric space, $(Y, \|\cdot\|)$ a finite dimensional Banach space and S a nonempty set. Assume that $\xi \in X$ is a fixed point of all functions $f_s := f(s, \cdot)$, $s \in S$, where $f: S \times X \to X$ satisfies the condition*

$$\rho(f(s, x), f(s, y)) \leqq m(s)\rho(x, y), \qquad (s, x), (s, y) \in S \times X,$$

with some function $m: S \to \mathbb{R}^+$. Let $U \subset X$ be a compact set, $\xi \in U$, and $f_s(U) \subset U$ for all $s \in S$. Suppose that $h: X \times \mathscr{F}(S, Y) \to Y$ satisfies the equality $h(\xi, \eta) = \eta$ for some η and the Lipschitz type condition

$$\left\| h(x, y) - h(\bar{x}, \bar{y}) \right\| \leqq L\rho(x, \bar{x}) + \beta(\|y - \bar{y}\|)^1$$

---

[1] For $u \in \mathscr{F}(S, Y)$, $\|u\|$ stands here and in the sequel for the function $S \ni s \mapsto \|u(s)\| \in \mathbb{R}^+$ belonging to $\mathscr{F}(S, \mathbb{R}^+)$. The symbol η under the h sign denotes the constant function $S \ni s \mapsto \eta \in \mathbb{R}$.

for all $(x, y), (\bar{x}, \bar{y}) \in X \times \mathscr{F}(S, Y)$, *where* $\beta: \mathscr{F}(S, \mathbb{R}^+) \to \bar{\mathbb{R}}$ *is such that*

(a) $\beta(m) < 1$,
(b) $u \leq v$ *implies* $\beta(u) \leq \beta(v)$, $u, v \in \mathscr{F}(S, \mathbb{R}^+)$,
(c) $\beta(\alpha u) \leq \alpha \beta(u)$ *for* $\alpha \in \mathbb{R}^+$ *and* $u \in \mathscr{F}(S, \mathbb{R}^+)$,
(d) $\beta(1)$ *is finite*.

Then there exists a solution $\varphi_0: U \to Y$ *of equation* (7.2.9) *fulfilling the conditions* $\varphi_0(\xi) = \eta$ *and*

$$\|\varphi_0(x) - \varphi_0(y)\| \leq \frac{L}{1 - \beta(m)} \rho(x, y), \qquad x, y \in U.$$

If, moreover, for every $x \in X$, *there is an* $n \in \mathbb{N}$ *such that* $f_{s_1} \circ \cdots \circ f_{s_n}(x) \in U$ *for all* $(s_1, \ldots, s_n) \in S^n$, *then there is exactly one Lipschitzian solution* $\varphi: X \to Y$ *of that equation such that* $\varphi(\xi) = \eta$ *with the Lipschitz constant unchanged.*

Proof. Put $r := L/(1 - \beta(m))$ and define the space $\mathscr{L}$ of all functions $\varphi: U \to Y$ such that $\varphi(\xi) = \eta$ and $\|\varphi(x) - \varphi(y)\| \leq r\rho(x, y)$ for all $x, y \in U$. The set $\mathscr{L}$ is nonempty since the constant function $\varphi(x) = \eta$, $x \in U$, belongs to $\mathscr{L}$. Moreover, $\mathscr{L}$ forms a closed convex subset of the Banach space $C(U, Y)$ endowed with the uniform convergence norm. To prove that $\mathscr{L}$ is compact we need only show that the sets $B_x := \{\varphi(x): \varphi \in \mathscr{L}\}$, $x \in U$, have compact closures (observe that the elements of $\mathscr{L}$ being Lipschitzian (with the same constant) form a family of equicontinuous functions). To this end, put $d := \sup_{x \in U} \rho(x, \xi)$; owing to the compactness of U, d is finite, whence

$$\|\varphi(x) - \eta\| = \|\varphi(x) - \varphi(\xi)\| \leq r\rho(x, \xi) \leq rd, \qquad \varphi \in \mathscr{L}$$

i.e. B_x is bounded and hence precompact because of the finite dimension of Y.

Now, define a map $T: \mathscr{L} \to \mathscr{F}(U, Y)$ by the formula

$$T(\varphi)(x) := h(x, \varphi \circ f(\cdot, x)), \qquad x \in U.$$

Note that $T(\mathscr{L}) \subset \mathscr{L}$; in fact, using (b) and (c) among others, for any x and y from U and $\varphi \in \mathscr{L}$ we have $T(\varphi)(\xi) = h(\xi, \varphi \circ f(\cdot, \xi)) = h(\xi, \eta) = \eta$ and

$$\|T(\varphi)(x) - T(\varphi)(y)\| = \|h(x, \varphi \circ f(\cdot, x) - h(y, \varphi \circ f(\cdot, y))\|$$
$$\leq L\rho(x, y) + \beta(\|\varphi \circ f(\cdot, x) - \varphi \circ f(\cdot, y)\|$$
$$\leq L\rho(x, y) + \beta(r\rho(f(\cdot, x), f(\cdot, y)))$$
$$\leq L\rho(x, y) + \beta(r\rho(x, y)m) \leq L\rho(x, y) + r\rho(x, y)\beta(m)$$
$$= (L + r\beta(m))\rho(x, y) = r\rho(x, y),$$

i.e. $T(\varphi) \in \mathscr{L}$, as claimed.

On the other hand, T is Lipschitzian itself (and hence continuous). Indeed, fix φ and ψ from $\mathscr{L}$; then

$$\|T(\varphi) - T(\psi)\| = \sup_{x \in U} \|T(\varphi)(x) - T(\psi)(x)\|$$

$$= \sup_{x \in U} \|h(x, \varphi \circ f(\cdot, x)) - h(x, \psi \circ f(\cdot, x))\|$$

$$\leqslant \sup_{x \in U} \beta(\|\varphi \circ f(\cdot, x) - \psi \circ f(\cdot, x)\|)$$

$$\leqslant \sup_{x \in U} \beta\left(\sup_{z \in U} \|\varphi(z) - \psi(z)\|\right) = \beta(\|\varphi - \psi\|) \leqslant \beta(1)\|\varphi - \psi\|$$

(in the latter two terms the symbols $\|\varphi - \psi\|$ and 1 are understood as constant functions $S \ni s \mapsto \|\varphi - \psi\| \in \mathbb{R}^+$, resp. $S \ni s \mapsto 1 \in \mathbb{R}^+$; the finiteness of $\beta(1)$ is simply assumed (see (d))).

Applying Schauder's Theorem (see Section 1.5) we obtain the existence of a solution $\varphi_0 \in \mathscr{L}$ of equation (7.2.9).

By Baron's Extension Theorem 7.2.1, this solution can be uniquely extended onto X to a solution $\varphi: X \to Y$ of that equation. We shall show that

$$\|\varphi(x) - \varphi(y)\| \leqslant r\rho(x, y) \qquad \text{for all } x, y \in X. \tag{7.2.15}$$

To this end, we put $U_0 := U$, $U_{k+1} := \bigcap_{s \in S} f_s^{-1}(U_k)$, $k \in \mathbb{N}_0$; and we recall that $\bigcup_{k \in \mathbb{N}_0} U_k = X$ and $f_s(U_{k+1}) \subset U_k \subset U_{k+1}$, $k \in \mathbb{N}_0$ (see the proof of Theorem 7.1.1). Therefore, to prove (7.2.15), it suffices to show that

$$\|\varphi(x) - \varphi(y)\| \leqslant r\rho(x, y) \qquad \text{for all } x, y \in U_k \tag{7.2.16}$$

and for all $k \in \mathbb{N}_0$. This is certainly true for $k = 0$. Assume (7.2.16) to hold for some $k \in \mathbb{N}_0$ and take arbitrary $x, y \in U_{k+1}$. Then

$$\|\varphi(x) - \varphi(y)\| = \|h(x, \varphi \circ f(\cdot, x)) - h(x, \varphi \circ f(\cdot, y))\|$$
$$\leqslant \beta(\|\varphi \circ f(\cdot, x) - \varphi \circ f(\cdot, y)\|)$$

and, since $f_s(x), f_s(y) \in U_k$,

$$\|\varphi(f(s, x)) - \varphi(f(s, y))\| = \|\varphi(f_s(x)) - \varphi(f_s(y))\| \leqslant r\rho(f_s(x), f_s(y))$$
$$\leqslant rm(s)\rho(x, y),$$

for all $s \in S$, by means of (7.2.16). Consequently, by (c) and (a),

$$\|\varphi(x) - \varphi(y)\| \leqslant r\rho(x, y)\beta(m) \leqslant r\rho(x, y),$$

as desired.

Obviously, $\varphi(\xi) = \varphi_0(\xi) = \eta$. To prove the uniqueness, suppose that $\varphi, \psi: X \to Y$ are two Lipschitzian solutions of equation (7.2.9), and $\varphi(\xi) = \psi(\xi) = \eta$. Put $\lambda(x) := \|\varphi(x) - \psi(x)\|$, $x \in X$. Then

$$0 \leq \lambda(x) \leq \beta(\lambda \circ f(\cdot, x)), \qquad x \in X. \tag{7.2.17}$$

On the other hand, $\lambda(x) \leq \|\varphi(x) - \varphi(\xi)\| + \|\psi(\xi) - \psi(x)\| \leq c\rho(x, \xi)$, $x \in X$, for some constant $c \geq 0$. We shall show that

$$\lambda(x) \leq c\beta(m)^k \rho(x, \xi), \qquad x \in X, \qquad (7.2.18)$$

for all $k \in \mathbb{N}_0$. For $k = 0$ it has just been shown. Assuming (7.2.18) to hold for some $k \in \mathbb{N}_0$ we have

$$\lambda(f_s(x)) \leq c\beta(m)^k \rho(f_s(x), \xi) = c\beta(m)^k \rho(f_s(x), f_s(\xi))$$
$$\leq c\beta(m)^k m(s)\rho(x, \xi), \qquad x \in X, s \in S,$$

i.e.

$$\lambda \circ f(\cdot, x) \leq c\beta(m)^k \rho(x, \xi)m, \qquad x \in X.$$

Now, applying (7.2.17), (b) and (c), we get

$$\lambda(x) \leq \beta(\lambda \circ f(\cdot, x)) \leq c\beta(m)^k \rho(x, \xi)\beta(m) = c\beta(m)^{k+1}\rho(x, \xi), \qquad x \in X.$$

By induction, (7.2.18) holds for all $k \in \mathbb{N}_0$ whence, on letting $k \to \infty$, we get $\lambda(x) = 0$, $x \in X$; hence $\varphi = \psi$. ∎

7.2E. Denumerable order

Let us take $S = \mathbb{N}$ and $m = (s_n)_{n \in \mathbb{N}}$, $s_n \geq 0$, $n \in \mathbb{N}$. If $(q_n)_{n \in \mathbb{N}}$, $q_n \geq 0$, $n \in \mathbb{N}$, is such that

$$\sum_{n=1}^{\infty} q_n < \infty, \qquad \sum_{n=1}^{\infty} q_n s_n < 1,$$

and $\beta(u) := \sum_{n=1}^{\infty} q_n u_n$ for $u = (u_n)_{n \in \mathbb{N}} \in \mathscr{F}(\mathbb{N}, \mathbb{R}^+)$, then the assumptions (a)–(d) of Theorem 7.2.3 are satisfied. This leads immediately to the following.

Theorem 7.2.4. *Let (X, ρ) be a metric space and $(Y, \|\cdot\|)$ a finite-dimensional Banach space. Assume that $\xi \in X$ is a fixed point of all functions $f_n: X \to X$, $n \in \mathbb{N}$, such that*

$$\rho(f_n(x), f_n(y)) \leq s_n \rho(x, y), \qquad x, y \in X, n \in \mathbb{N}.$$

Let $U \subset X$ be a compact set containing ξ and such that $f_n(U) \subset U$ for all $n \in \mathbb{N}$. Suppose further that $h: X \times Y^N \to Y$ fulfils the Lipschitz condition

$$\|h(x, y_1, y_2, \ldots) - h(\bar{x}, \bar{y}_1, \bar{y}_2, \ldots)\| \leq L\rho(x, \bar{x}) + \sum_{n=1}^{\infty} q_n \|y_n - \bar{y}_n\|$$

for all $x, \bar{x} \in X$ and $y_n, \bar{y}_n \in Y$, $n \in \mathbb{N}$, where

$$\sum_{n=1}^{\infty} q_n < \infty \qquad and \qquad \sum_{n=1}^{\infty} s_n q_n < 1.$$

If, for some $\eta \in Y$, we have $h(\xi, \eta, \eta, \ldots) = \eta$, then there exists a Lipschitzian solution $\varphi_0: U \to Y$ of the functional equation

$$\varphi(x) = h(x, \varphi(f_1(x)), \varphi(f_2(x)), \ldots) \qquad (7.2.19)$$

such that $\varphi_0(\xi) = \eta$. If, moreover, for every $x \in X$, there exists a $k \in \mathbb{N}$ such that $f_{n_1} \circ \cdots \circ f_{n_k}(x) \in U$ for all $(n_1, \ldots, n_k) \in \mathbb{N}^k$, there exists exactly one Lipschitzian solution $\varphi: X \to Y$ of equation (7.2.19) satisfying the condition $\varphi(\xi) = \eta$.

Taking (in Theorem 7.2.3) $S = \{1, \ldots, k\}$, $m = (s_1, \ldots, s_k)$, $\beta(u) = \beta(u_1, \ldots, u_k) := \sum_{n=1}^{k} q_n s_n$ where $q_1, \ldots, q_k$ are nonnegative and such that $\sum_{n=1}^{k} q_n s_n < 1$, and $h: X \times Y^k \to Y$ such that

$$\|h(x, y_1, \ldots, y_k) - h(\bar{x}, \bar{y}_1, \ldots, \bar{y}_k)\| \leqslant L\rho(x, \bar{x}) + \sum_{n=1}^{k} q_n \|y_n - \bar{y}_n\|,$$

$x, \bar{x} \in X$, $y_1, \ldots, y_k$, $\bar{y}_1, \ldots, \bar{y}_k \in Y$, we get the result of K. Baron–J. Matkowski [1] (even better, their main result has a local character only).

For further examples (including uncountable orders) as well as for more detailed considerations concerning proper subsets of the product $X \times \mathscr{F}(S, Y)$ as possible domains of the functions h determining the right-hand side of equation (7.2.9) the reader is referred to Baron [9].

7.3 Stability

Among various notions of stability investigated in the theory of differential equations the oldest and perhaps the most important is that of the Lyapunov stability. Roughly speaking, a solution φ is stable (in the sense of Lyapunov) provided that any other solution sufficiently close to φ at the initial moment lies entirely in an *a priori* prescribed ε-hull of φ. The choice of the initial point is practically unrestricted.

However, if we consider an iterative functional equation, then practically everything is decided in a vicinity of a fixed point of the given 'inner' function(s) occurring in the equation. Thus this point should be chosen as the most favourable initial point. This makes the situation essentially different from that met in the theory of differential equations.

The stability theorems we present in this section do not then concern any rigorous definition of stability. The results are due to K. Baron–R. Ger [1], [2]; see also Baron [9].

7.3A. Main result

Let us start with the following.

Theorem 7.3.1. *Suppose that* $(X, \mathscr{T})$ *is a topological space and* (Y, σ) *is a metric space. Moreover, let* $f: S \times X \to X$, *where* S *is a nonempty set, satisfy for every neighbourhood* U *of a point* $\xi \in X$ *the condition*

for every $x \in X$ *there exists an* $n \in \mathbb{N}$ *such that for all n-tuples* $(s_1, \ldots, s_n) \in S^n$ *one has* $f(s_1, \cdot) \circ \cdots \circ f(s_n, \cdot)(x) \in U$.   (7.3.1)

Let $h: X \times \mathscr{F}(S, Y) \to Y$ *fulfil the condition*

$$\sigma(h(x, y_1), h(x, y_2)) \leq \beta(x, \sigma(y_1, y_2)), \qquad (x, y_1), (x, y_2) \in X \times \mathscr{F}(S, Y),$$

with a function $\beta: X \times \mathscr{F}(S, \mathbb{R}^+) \to \mathbb{R}$ such that

for every $\delta \in \mathbb{R}^+$, $u \in \mathscr{F}(S, \mathbb{R}^+)$ and $x \in X$ we have
$$u \leq \delta \Rightarrow \beta(x, u) \leq \delta. \tag{7.3.2}$$

Then for every two solutions (continuous at ξ) $\varphi, \psi: X \to Y$ of the equation

$$\varphi(x) = h(x, \varphi \circ f(\cdot, x)) \tag{7.3.3}$$

the estimate

$$\sigma(\varphi(x), \psi(x)) \leq \sigma(\varphi(\xi), \psi(\xi)) \tag{7.3.4}$$

is satisfied for all $x \in X$.

Proof. Put $\lambda(x) := \sigma(\varphi(x), \psi(x))$, $x \in X$, and fix an $\varepsilon > 0$. The continuity of φ and ψ at ξ implies the existence of a neighbourhood U of ξ such that

$$\sigma(\varphi(x), \varphi(\xi)) < \tfrac{1}{2}\varepsilon \qquad \text{and} \qquad \sigma(\psi(x), \psi(\xi)) < \tfrac{1}{2}\varepsilon$$

for all $x \in U$. Hence

$$\lambda(x) = \sigma(\varphi(x), \psi(x)) \leq \varepsilon + \sigma(\varphi(\xi), \psi(\xi)) =: \delta$$

for all $x \in U$. On the other hand,

$$\lambda(x) = \sigma(h(x, \varphi \circ f(\cdot, x)), h(x, \psi \circ f(\cdot, x)))$$
$$\leq \beta(x, \sigma(\varphi \circ f(\cdot, x), \psi \circ f(\cdot, x))) = \beta(x, \lambda \circ f(\cdot, x)),$$

i.e.

$$\lambda(x) \leq \beta(x, \lambda \circ f(\cdot, x)), \qquad x \in X.$$

Putting, as usual, $f_s := f(s, \cdot)$, $s \in S$, and $U_0 := U$, $U_{n+1} := \bigcap_{s \in S} f_s^{-1}(U_n)$, $n \in \mathbb{N}_0$, we have

$$f_s(U_{n+1}) \subset U_n \subset U_{n+1}, \quad n \in \mathbb{N}_0, \qquad \text{and} \qquad X = \bigcup_{n=0}^{\infty} U_n, \tag{7.3.5}$$

which easily results from (7.3.1) (see Theorem 7.1.1). This implies that

$$\lambda(x) \leq \delta \qquad \text{for } x \in U_n \tag{7.3.6}$$

for all $n \in \mathbb{N}_0$. In fact, (7.3.6) is obvious for $n = 0$; assuming (7.3.6) for some $n \in \mathbb{N}$ and taking an $x \in U_{n+1}$ we get, by (7.3.5), $f_s(x) \in U_n$, $s \in S$, whence $\lambda(f_s(x)) \leq \delta$, $s \in S$, and consequently

$$\lambda(x) \leq \beta(x, \lambda \circ f(\cdot, x)) \leq \delta$$

on account of (7.3.2). This proves that (7.3.3) holds for n replaced by $n + 1$. This jointly with the equality in (7.3.5) gives the estimation

$$\lambda(x) \leq \delta = \varepsilon + \sigma(\varphi(\xi), \psi(\xi)) \qquad \text{for all } x \in X.$$

Now, letting ε tend to zero, we obtain (7.3.4) for all $x \in X$. ∎

7.3B. Special results

Assumption (7.3.2) is rather restrictive. Taking, for instance, $S = \mathbb{N}$ and $\beta(x, (y_n)_{n \in \mathbb{N}}) := \sum_{n=1}^{\infty} g_n(x) y_n$, $(x, (y_n)_{n \in \mathbb{N}}) \in X \times Y^{\mathbb{N}}$ we arrive at the following particular case of Theorem 7.3.1 (see Baron–Ger [1]).

Theorem 7.3.2. *Let* $(X, \mathscr{T})$ *be a topological space and* (Y, σ) *a metric space,* $\xi \in X$. *If* $f_n \colon X \to X$ *are such that for every neighbourhood* U *of* ξ *and every* $x \in X$ *there exists a* $k \in \mathbb{N}$ *such that*

$$f_{n_1} \circ \cdots \circ f_{n_k}(x) \in U$$

for every $n_1, \ldots, n_k \in \mathbb{N}$, *and* $h \colon X \times Y^{\mathbb{N}} \to Y$ *fulfils the Lipschitz condition*

$$\sigma(h(x, y_1, y_2, \ldots), h(x, \bar{y}_1, \bar{y}_2, \ldots)) \leqslant \sum_{n=1}^{\infty} g_n(x)\sigma(y_n, \bar{y}_n)$$

for $x \in X$, $y_n, \bar{y}_n \in Y$ *with functions* $g_n \colon X \to \mathbb{R}^+$, $n \in \mathbb{N}$, *such that*

$$\sum_{n=1}^{\infty} g_n(x) \leqslant 1, \qquad x \in X, \tag{7.3.7}$$

then for any two solutions (continuous at ξ) $\varphi, \psi \colon X \to Y$ *of the functional equation* (7.3.3) *we have* (7.3.4) *for all* $x \in X$.

Assumption (7.3.7) is just the translation of (7.3.2) into the present case. A comparison with Theorem 7.2.4, for instance, would lead one to assume $\sum_{n=1}^{\infty} g_n(x) < \infty$, $x \in X$, rather than (7.3.7). To obtain such results we have to restrict the class of solutions considered and strengthen the conditions on the family $\{f(s, \cdot) \colon s \in S\}$ (resp. $\{f_n \colon n \in \mathbb{N}\}$). If (X, ρ) is a metric space then instead of the continuity of the solution $\varphi \colon X \to Y$ at ξ we shall now require that

$$\sigma(\varphi(x), \varphi(\xi)) \leqslant L_\varphi \rho(x, \xi)^\alpha, \qquad x \in X, \tag{7.3.8}$$

where L_φ is a positive constant (depending on φ) and α is a fixed nonnegative number. For brevity, denote by B_α the family of all mappings $\varphi \colon X \to Y$ fulfilling condition (7.3.8).

Theorem 7.3.3. *Let* (X, ρ) *and* (Y, σ) *be metric spaces,* diam $X < \infty$, $\xi \in X$, $L, \alpha \in \mathbb{R}^+$, *and let* S *be a nonempty set. Assume that a function* $f \colon S \times X \to X$ *fulfils the condition*

$$\rho(f(s, x), \xi) \leqslant m(s)\rho(x, \xi), \qquad (s, x) \in S \times X, \tag{7.3.9}$$

with some function $m \colon S \to \mathbb{R}^+$. *Let* $h \colon X \times \mathscr{F}(S, Y) \to Y$ *satisfy the Lipschitz-type condition*

$$\sigma(h(x, y), h(x, \bar{y})) \leqslant \beta(\sigma(y, \bar{y})) \tag{7.3.10}$$

for all $x \in X$ *and* $y, \bar{y} \in \mathscr{F}(S, Y)$, *where* $\beta \colon \mathscr{F}(S, \mathbb{R}^+) \to \bar{\mathbb{R}}$ *is such that*

(a) $\beta(m^\alpha) < 1$,
(b) $u \leqslant v$ *implies* $\beta(u) \leqslant \beta(v)$, $u, v \in \mathscr{F}(S, \mathbb{R}^+)$,
(c) $\beta(\omega u) \leqslant \omega \beta(u)$ *for* $\omega \in \mathbb{R}^+$ *and* $u \in \mathscr{F}(S, \mathbb{R}^+)$,
(d) $\beta(u + v) \leqslant \beta(u) + \beta(v)$, $u, v \in \mathscr{F}(S, \mathbb{R}^+)$,
(e) $\beta(1)$ *is finite.*

Then for any positive number ε there exists a positive δ such that for any two solutions $\varphi, \psi \in B_\alpha$ of the functional equation

$$\varphi(x) = h(x, \varphi \circ f(\cdot, x)) \tag{7.3.11}$$

for which $\varphi(\xi), \psi(\xi) \in Y_0 := \{\eta \in Y: \ \sigma(h(x, \eta), \eta) \leqslant L\rho(x, \xi)^\alpha, \ x \in X\}$, the implication

$$\sigma(\varphi(\xi), \psi(\xi)) \leqslant \delta \Rightarrow \sigma(\varphi(x), \psi(x)) < \varepsilon$$

is satisfied for all $x \in X$.

Proof. Fix a $\varphi \in B_\alpha$ such that $\varphi(\xi) \in Y_0$ and put

$$l_\varphi := \sup_{x \in X \setminus \{\xi\}} \frac{\sigma(\varphi(x), \varphi(\xi))}{\rho(x, \xi)^\alpha}.$$

First we shall show that (see (a))

$$l_\varphi \leqslant \frac{L}{1 - \beta(m^\alpha)} =: M. \tag{7.3.12}$$

In fact, observe that $h(\xi, \eta) = \eta$ for $\eta \in Y_0$ and that (7.3.9) implies the equality $f(s, \xi) = \xi$ for all $s \in S$; thus, with the aid of the subsequent use of (7.3.10), (7.3.9), (b) and (c), we get for $x \in X$:

$$\sigma(\varphi(x), \varphi(\xi)) = \sigma(h(x, \varphi \circ f(\cdot, x)), h(\xi, \varphi \circ f(\cdot, \xi)))$$
$$= \sigma(h(x, \varphi \circ f(\cdot, x)), h(\xi, \varphi(\xi)))$$
$$\leqslant \beta(\sigma(\varphi \circ f(\cdot, x), \varphi(\xi))) + \sigma(h(x, \varphi(\xi)), \varphi(\xi))$$
$$\leqslant \beta(l_\varphi m^\alpha \rho(x, \xi)^\alpha) + L\rho(x, \xi)^\alpha \leqslant (l_\varphi \beta(m^\alpha) + L)\rho(x, \xi)^\alpha,$$

whence, by the definition of l_φ, we get $l_\varphi \leqslant l_\varphi \beta(m^\alpha) + L$, which is just (7.3.12) because of (a).

Take any two members φ and ψ from B_α. With the use of (7.3.12) we get

$$\sigma(\varphi(x), \psi(x)) \leqslant \sigma(\varphi(x), \varphi(\xi)) + \sigma(\varphi(\xi), \psi(\xi)) + \sigma(\psi(\xi), \psi(x))$$
$$\leqslant \sigma(\varphi(\xi), \psi(\xi)) + 2M\rho(x, \xi)^\alpha, \qquad x \in X. \tag{7.3.13}$$

We shall show that

$$\sigma(\varphi(x), \psi(x)) \leqslant \beta(1)^k \sigma(\varphi(\xi), \psi(\xi)) + 2M\beta(m^\alpha)^k \rho(x, \xi)^\alpha, \qquad x \in X, \tag{7.3.14}$$

for all $k \in \mathbb{N}_0$. In fact, for $k = 0$ relation (7.3.14) reduces to (7.3.13). Assume (7.3.14) to hold for some $k \in \mathbb{N}_0$ and take an $x \in X$; then, on account of (7.3.10), (7.3.14), (d), (c) and (7.3.9),

$$\sigma(\varphi(x), \psi(x)) = \sigma(h(x, \varphi \circ f(\cdot, x)), h(x, \psi \circ f(\cdot, x)))$$
$$\leqslant \beta(\sigma(\varphi \circ f(\cdot, x), \psi \circ f(\cdot, x)))$$
$$\leqslant \beta(\beta(1)^k \sigma(\varphi(\xi), \psi(\xi)) + 2M\beta(m^\alpha)^k \rho(f(\cdot, x), \xi)^\alpha)$$
$$\leqslant \beta(1)^{k+1} \sigma(\varphi(\xi), \psi(\xi)) + 2M\beta(m^\alpha)^{k+1} \rho(x, \xi)^\alpha,$$

which finishes the inductive proof of (7.3.14).

Now, given an $\varepsilon > 0$, choose a $k \in \mathbb{N}_0$ large enough to have (see (a))
$c := 2M(\operatorname{diam} X)^\alpha \beta(m^\alpha)^k < \varepsilon$ and put

$$\delta := \frac{\varepsilon - c}{1 + \beta(1)^k}$$

to obtain the implication from the assertion of the theorem on the basis of
(7.3.14). ∎

Remark 7.3.1. In the case $\beta(1) \leqslant 1$ we get Theorem 7.3.1 (resp. Theorem
7.3.2) under the more restrictive assumption of the preceding Theorem 7.3.3
(it suffices to pass to infinity with $k \in \mathbb{N}_0$ in (7.3.14)). Actually, the case
$\beta(1) = 1$ is the only interesting one since, otherwise, we have the uniqueness
of the solutions considered.

Remark 7.3.2. A particular case $S = \mathbb{N}$, $\beta((y_n)_{n \in \mathbb{N}}) := \sum_{n=1}^{\infty} q_n y_n^\alpha$, $\beta(1) = \sum_{n=1}^{\infty} q_n < \infty$ and $m = (s_n)_{n \in \mathbb{N}}$ with $\beta(m^\alpha) = \sum_{n=1}^{\infty} q_n s_n^\alpha < 1$, coincides with
Theorem 7.3.3 from the paper of K. Baron–R. Ger [1]; we omit its
(obvious) detailed statement.

Remark 7.3.3. The function $\beta(y) := \sup_{s \in S} y(s)$, $y \in \mathscr{F}(S, \mathbb{R}^+)$, is the other
nontrivial realization of hypotheses (b)–(e). If $(S, \mathfrak{M}, \mu)$ is a finite measure
space and $\mathscr{F}_0(S, \mathbb{R}^+)$ is the family of all measurable members from $\mathscr{F}(S, \mathbb{R}^+)$,
then $\beta(y) := \int_S y \, d\mu$, $y \in \mathscr{F}_0(S, \mathbb{R}^+)$, stands for another example; slight
modifications (like those in Theorem 7.2.2, for instance) are needed regarding
the domain of h (see also the paper Baron–Ger [2] in connection with this).
More generally, positive linear functionals on suitable subspaces (cones) of
$\mathscr{F}(S, \mathbb{R}^+)$ may also be taken into account.

7.3C. Comments

The theorems of the present section say nothing about the existence of the
solutions considered. The existence is simply assumed. On the other hand,
these results imply the uniqueness of solutions in a given class of functions
fulfilling the same initial condition. Therefore, looking e.g. at the assumptions
of Theorem 7.3.3 we realize that from the stability point of view the situation
becomes interesting indeed provided that the cardinality of the set $Y_0 = \{\eta \in Y : \sigma(h(x, \eta), \eta) \leqslant L\rho(x, \xi)^\alpha, x \in X\}$ is large enough; in particular, a large set of
solutions of the 'numeric' equation $h(\xi, \eta) = \eta$ is to be desired in this context.

We omit here the analogues of Theorem 7.3.1–7.3.3 for systems of
equations of infinite orders. Such (routine) results are obtainable (see
Baron–Ger [1] and Note 7.9.5).

Finally, let us mention here that the word *stability* has also been used in
the theory of iterative functional equations in another meaning. We shall

explain it in the case of the equation

$$\varphi(f(x)) = g(x, \varphi(x)), \tag{7.3.15}$$

where $f: X \to X$ and $g: X \times Y \to Y$ are given functions. Put $g_1 := g$ and $g_{n+1}(x, y) := g(f^n(x), g_n(x, y))$, $(x, y) \in X \times Y$, $n \in \mathbb{N}$. After D. Brydak [4] we say that equation (7.3.15) is *iteratively stable* in the class $\mathbf{\Phi} \subset \mathscr{F}(X, Y)$ provided that there exists a positive number k such that for every $\varepsilon > 0$ and for every map $\psi \in \mathbf{\Phi}$ satisfying the inequalities

$$\sigma(\psi(f^n(x)), g_n(x, \psi(x))) < \varepsilon, \qquad x \in X, n \in \mathbb{N}, \tag{7.3.16}$$

there exists a solution $\varphi \in \mathbf{\Phi}$ of equation (7.3.15) such that

$$\sigma(\varphi(x), \psi(x)) \le k\varepsilon, \qquad x \in X.$$

This type of stability has been investigated by Brydak [4], Choczewski–Turdza–Węgrzyk [1] and Turdza [1]–[4]. However, the existence of a function $\psi \in \mathbf{\Phi}$ satisfying (7.3.16) is by no means obvious even for $n = 1$. It is a question of interest especially when a solution $\varphi \in \mathbf{\Phi}$ of the equation considered does not exist.

7.4 Approximate solutions

A function ψ nearly satisfying the equation in question (with a prescribed accuracy) may be considered as an approximate solution of this equation. In this section we discuss approximate solutions of the equation

$$\varphi(x) = h(x, \varphi \circ f(\cdot, x)),$$

in principle for the class $\mathbf{\Phi}$ of continuous functions.

7.4A. Two special equations

To visualize emphatically the effect we are going to deal with let us see what the behaviour of the two simple functional equations

$$\varphi(x) = -\varphi(x^2) + x(x - 1), \qquad x \in I = [0, 1], \tag{7.4.1}$$

and

$$\psi(x) = \psi\left(\frac{x}{x+1}\right) - x, \qquad x \in I, \tag{7.4.2}$$

looks like, if we consider them for functions $\varphi, \psi: I \to \mathbb{R}$. Both of them have no continuous solutions in I. To see this, we might use suitable results of Chapter 3; however, proceeding directly we get immediately (induction)

$$(-1)^n \varphi(x^{2^n}) = \varphi(x) - \sum_{k=0}^{n-1} (-1)^k q(x^{2^k}), \qquad x \in I, n \in \mathbb{N}, \tag{7.4.1'}$$

where $q(t) = t(1 - t)$, $t \in I$, and

$$\psi\left(\frac{1}{n+1}\right) = \psi(1) + \sum_{k=1}^{n} \frac{1}{k}, \qquad n \in \mathbb{N}, \tag{7.4.2'}$$

for any solutions $\varphi, \psi: I \to \mathbb{R}$ of equations (7.4.1) and (7.4.2), respectively. Thus, every solution φ of (7.4.1) in $[0, 1)$ continuous at zero is automatically analytic and

$$\varphi(x) = \sum_{k=0}^{\infty} (-1)^k q(x^{2^k}) = -x + 2 \sum_{k=0}^{\infty} (-1)^k x^{2^k}, \qquad x \in [0, 1);$$

the latter function, however, has no limit as $x \to 1-$ (see Steinhaus [1], for instance), which implies that any solution $\varphi: I \to \mathbb{R}$ of equation (7.4.1) has to be discontinuous either at 0 or at 1. Similarly, because of the divergence of the harmonic series we learn from (7.4.2') that any solution $\psi: I \to \mathbb{R}$ of equation (7.4.2) has to be discontinuous at zero.

In spite of the lack of continuous solutions of (7.4.1) and (7.4.2) in I, *for every $\varepsilon > 0$ there exist polynomials(!) φ_ε and ψ_ε such that*

$$|\varphi_\varepsilon(x) + \varphi_\varepsilon(x^2) - q(x)| < \varepsilon \qquad \text{for } x \in I \qquad (7.4.1'')$$

and

$$\left| \psi_\varepsilon(x) - \psi_\varepsilon\left(\frac{x}{x+1}\right) - x \right| < \varepsilon \qquad \text{for } x \in I. \qquad (7.4.2'')$$

The existence of the polynomial φ_ε may be derived in a way similar to that used by R. C. Buck [1]. To this aim, assign to any integer $p > 4$ a continuous function $q_p: I \to \mathbb{R}$, $0 \leq q_p \leq q$, vanishing on $[0, 1/p] \cup [1 - 1/p, 1]$ and coinciding with q on $[2/p, 1 - 2/p]$. Then, for every $x \in (0, 1)$ and $k \in \mathbb{N}$, the relation $q_p(x^{2^k}) \neq 0$ implies that $x^{2^k} \in (1/p, 1 - 1/p)$ or, equivalently, $k \in K(x) :=$ $(\alpha(x), \beta(x)) \cap \mathbb{N}_0$, where the number $\beta(x) - \alpha(x) = \log p/\log(p/(p-1))$ does not depend on $x \in (0, 1)$. Thus there exists an $N = N(p) \in \mathbb{N}$ such that for all $x \in (0, 1)$ the cardinality of $K(x)$ does not exceed N.

Now, fix an $\varepsilon > 0$ and take an integer $p > 4$ such that $q(2/p) < \varepsilon/6$. Let $\rho \in (0, 1)$ be sufficiently close to 1 to have $1 - \rho^{N+1} < \varepsilon/6$. Put

$$\tilde{\varphi}_\varepsilon(x) := \sum_{k=0}^{\infty} (-\rho)^k q_p(x^{2^k}) = \sum_{k \in K(x)} (-\rho)^k q_p(x^{2^k}), \qquad x \in I.$$

This function is continuous and one may easily check that

$$\tilde{\varphi}_\varepsilon(x) + \rho\tilde{\varphi}_\varepsilon(x^2) = q_p(x), \qquad x \in I. \qquad (7.4.3)$$

Moreover,

$$|\tilde{\varphi}_\varepsilon(x)| \leq \sum_{k \in K(x)} \rho^k \leq \frac{1 - \rho^{N+1}}{1 - \rho} < \frac{\varepsilon}{6(1-\rho)}, \qquad x \in I.$$

Hence, in view of (7.4.3) and by the choice of p and ρ, we obtain

$$|\tilde{\varphi}_\varepsilon(x) + \tilde{\varphi}_\varepsilon(x^2) - q(x)| \leq |q_p(x) - q(x)| + (1 - \rho)|\tilde{\varphi}_\varepsilon(x^2)|$$

$$\leq q\left(\frac{2}{p}\right) + (1 - \rho)\frac{\varepsilon}{6(1-\rho)} < \tfrac{1}{3}\varepsilon. \qquad (7.4.4)$$

Theorem 7.4.1 below guarantees (see Note 7.9.7) the existence of a continuous function $\tilde{\psi}_\varepsilon \colon I \to \mathbb{R}$ such that

$$\left| \tilde{\psi}_\varepsilon(x) - \tilde{\psi}_\varepsilon\left(\frac{x}{x+1}\right) - x \right| < \tfrac{1}{3}\varepsilon, \qquad x \in I. \tag{7.4.5}$$

Now, the classical Weierstrass Approximation Theorem allows one to find polynomials φ_ε and ψ_ε such that

$$\left| \tilde{\varphi}_\varepsilon(x) - \varphi_\varepsilon(x) \right| < \tfrac{1}{3}\varepsilon \qquad \text{and} \qquad \left| \tilde{\psi}_\varepsilon(x) - \psi_\varepsilon(x) \right| < \tfrac{1}{3}\varepsilon, \qquad x \in I. \tag{7.4.6}$$

Thus, for every $x \in I$, with the aid of (7.4.6) and (7.4.4) we get

$$|\varphi_\varepsilon(x) + \varphi_\varepsilon(x^2) - q(x)| \leq |\varphi_\varepsilon(x) - \tilde{\varphi}_\varepsilon(x)| + |\varphi_\varepsilon(x^2) - \tilde{\varphi}_\varepsilon(x^2)|$$
$$+ |\tilde{\varphi}_\varepsilon(x) + \tilde{\varphi}_\varepsilon(x^2) - q(x)| < \varepsilon,$$

i.e. relation (7.4.1″). Similarly, (7.4.6) and (7.4.5) imply (7.4.2″).

7.4B. Approximation in Buck's sense

This is the procedure we have just described. It was first applied by R. C. Buck [1], [2] to approximate a function of two variables by the sum of functions in one variable (see also Baron [8] and Baron–Ger [2]). The papers Głowacki [1], [2], Kwapisz [2] and Rożnowski [1] are also related. The word 'approximation' is perhaps not quite adequate; observe that the polynomials $(\varphi_{1/n})_{n \in \mathbb{N}}$ in the above two examples approximate nothing since continuous solutions of equations (7.4.1) and (7.4.2) do not exist in $[0, 1]$. This shows only that such sequences must necessarily be divergent. Below we reproduce here the main results of the last chapter of Baron [9].

Theorem 7.4.1. *Let* (X, ρ) *be a compact metric space,* $(Y, \|\cdot\|)$ *a finite-dimensional Banach space and* S *a nonempty set. Suppose that* $\Phi \subset \mathscr{F}(X, Y)$ *is a nonempty family closed with respect to uniform convergence and such that* $\{t\varphi \colon t \in [0, 1), \varphi \in \Phi\} \subset \Phi$. *Let two functions* $f \colon S \times X \to X$ *and* $h \colon X \times \mathscr{F}(S, Y) \to Y$ *be given such that*

$$\rho(f(s, x), f(s, y)) \leq m(s)\rho(x, y), \qquad (s, x), (s, y) \in S \times X, \tag{7.4.7}$$

relation (7.3.1) is satisfied for any neighbourhood U *of a point* $\xi \in X$ *(which becomes a common fixed point for all the transformations* $f_s := f(s, \cdot)$, $s \in S$*),* $h(\xi, 0) = 0$ *and*

$$\|h(x, y) - h(\bar{x}, \bar{y})\| \leq L\rho(x, \bar{x}) + \beta(\|y - \bar{y}\|), \qquad (x, y), (\bar{x}, \bar{y}) \in X \times \mathscr{F}(S, Y), \tag{7.4.8}$$

where $L \in \mathbb{R}^+$ *and* $\beta \colon \mathscr{F}(S, \mathbb{R}^+) \to \bar{\mathbb{R}}$ *satisfies the conditions*

(a) *for any constant function* $S \ni s \to c \in \mathbb{R}^+$ *one has* $\beta(c) \leq c$ *and* $\beta(cm) \leq c$,
(b) $u \leq v$ *implies* $\beta(u) \leq \beta(v)$, $u, v \in \mathscr{F}(S, \mathbb{R}^+)$,

(c) $\alpha\beta(u) \leqslant \beta(\alpha u)$, $\alpha \in (0, 1)$, $u \in \mathcal{F}(S, \mathbb{R}^+)$, and

(d) *for any sequence* $(u_n)_{n \in \mathbb{N}}$ *of elements of* $\mathcal{F}(S, \mathbb{R}^+)$ *uniformly convergent to a* $u \in \mathcal{F}(S, \mathbb{R}^+)$ *and such that* $u_n \leqslant mL \operatorname{diam} X$, $n \in \mathbb{N}$, *one has*

$$\liminf_{n \to \infty} \beta(u_n) \leqslant \beta(u).$$

If, for every $\varphi \in \mathbf{\Phi}$, *the function* $X \ni x \mapsto h(x, \varphi \circ f(\cdot, x)) \in Y$ *belongs to* $\mathbf{\Phi}$ *in turn, then, for every* $\varepsilon > 0$, *there exists a Lipschitzian function* $\varphi \in \mathbf{\Phi}$ *such that* $\varphi(\xi) = 0$ *and*

$$\|\varphi(x) - h(x, \varphi \circ f(\cdot, x))\| \leqslant \varepsilon \tag{7.4.9}$$

for all $x \in X$.

Proof. Fix a $\vartheta \in (0, 1)$ and put

$$\mathcal{L} := \left\{ \varphi \in \mathbf{\Phi} \colon \|\varphi(x) - \varphi(y)\| \leqslant \frac{1}{\vartheta} L\rho(x, y), \; x, y \in X, \text{ and } \varphi(\xi) = 0 \right\}.$$

Clearly $\mathcal{L} \neq \varnothing$ ($0 \in \mathcal{L}$) and $\mathcal{L}$ forms a closed subset of the space $C(X, Y)$ endowed with the usual uniform convergence norm. Moreover, the transformation $T \colon \mathcal{L} \to \mathcal{F}(X, Y)$ given by the formula

$$T(\varphi)(x) := h(x, (1 - \vartheta)\varphi \circ f(\cdot, x)), \qquad x \in X, \; \varphi \in \mathcal{L},$$

maps $\mathcal{L}$ into itself. To see this, fix a $\varphi \in \mathcal{L}$ and points $x, y \in X$. By assumption, $T(\varphi) \in \mathbf{\Phi}$. Moreover, $T(\varphi)(\xi) = h(\xi, (1 - \vartheta)\varphi \circ f(\cdot, \xi)) = h(\xi, 0) = 0$ and with the aid of the subsequent use of (7.4.8), (b), (7.4.7) and (a) we obtain

$$\|T(\varphi)(x) - T(\varphi)(y)\| = \|h(x, (1 - \vartheta)\varphi \circ f(\cdot, x)) - h(y, (1 - \vartheta)\varphi \circ f(\cdot, y))\|$$

$$\leqslant L\rho(x, y) + \beta((1 - \vartheta)\|\varphi \circ f(\cdot, x) - \varphi \circ f(\cdot, y)\|)$$

$$\leqslant L\rho(x, y) + \beta\left((1 - \vartheta)\frac{L}{\vartheta}\rho(f(\cdot, x), f(\cdot, y))\right)$$

$$\leqslant L\rho(x, y) + \beta\left((1 - \vartheta)\frac{L}{\vartheta}m\rho(x, y)\right)$$

$$\leqslant L\left(1 + \frac{1}{\vartheta}(1 - \vartheta)\right)\rho(x, y) = \frac{L}{\vartheta}\rho(x, y)$$

i.e. $T(\varphi) \in \mathcal{L}$ as desired.

To apply Banach's Theorem (see Section 1.5) it remains to check that T itself is contractive. To this end, fix any two members φ, ψ from $\mathcal{L}$; then again by (7.4.8), (b) and (a),

$$\|T(\varphi) - T(\psi)\| = \sup_{x \in X} \|T(\varphi)(x) - T(\psi)(x)\|$$

$$\leqslant \sup_{x \in X} \beta((1 - \vartheta)\|\varphi \circ f(\cdot, x) - \psi \circ f(\cdot, x)\|)$$

$$\leqslant \sup_{x \in X} \beta\left((1 - \vartheta) \sup_{z \in X} \|\varphi(z) - \psi(z)\|\right) \leqslant (1 - \vartheta)\|\varphi - \psi\|,$$

i.e. T is a contraction since $\vartheta \in (0, 1)$ implies that so does $1 - \vartheta$.

Consequently, T has (exactly one) fixed point $\varphi_\vartheta \in \Phi$ and, therefore, owing to the unrestricted choice of a $\vartheta \in (0, 1)$, we obtain a subfamily $\{\varphi_\vartheta : \vartheta \in (0, 1)\}$ of Φ such that

$$\|\varphi_\vartheta(x) - \varphi_\vartheta(y)\| \leqslant \frac{1}{\vartheta} L\rho(x, y), \qquad x, y \in X,$$

and

$$\varphi_\vartheta(x) = h(x, (1 - \vartheta)\varphi_\vartheta \circ f(\cdot, x)), \qquad x \in X,$$

for all $\vartheta \in (0, 1)$. In particular, the functions $\vartheta \varphi_\vartheta$, $\vartheta \in (0, 1)$, are equicontinuous and uniformly bounded (recall that X is compact whence $\operatorname{diam} X < \infty$). Since $(Y, \|\cdot\|)$ is of finite dimension, the classical version of the Arzelà–Ascoli Theorem assures us that there exists a sequence $(\vartheta_n)_{n \in \mathbb{N}}$ in $(0, 1)$ converging to zero and such that $(\vartheta_n \varphi_{\vartheta_n})_{n \in \mathbb{N}}$ is uniformly convergent on the whole of X. We shall show that

$$\lim_{n \to \infty} \vartheta_n \varphi_{\vartheta_n} =: \varphi = 0.$$

To this end, note that

$$\|\varphi_{\vartheta_n}(x)\| = \|\varphi_{\vartheta_n}(x) - \varphi_{\vartheta_n}(\xi)\|$$
$$= \|h(x, (1 - \vartheta_n)\varphi_{\vartheta_n} \circ f(\cdot, x)) - h(\xi, (1 - \vartheta_n)\varphi_{\vartheta_n} \circ f(\cdot, \xi))\|$$
$$\leqslant L\rho(x, \xi) + \beta((1 - \vartheta_n)\|\varphi_{\vartheta_n} \circ f(\cdot, x)\|), \qquad x \in X, n \in \mathbb{N},$$

whence by (c)

$$\|\vartheta_n \varphi_{\vartheta_n}(x)\| \leqslant \vartheta_n L\rho(x, \xi) + \beta((1 - \vartheta_n)\|\vartheta_n \varphi_{\vartheta_n} \circ f(\cdot, x)\|), \qquad x \in X, n \in \mathbb{N},$$

and, passing to the limit, we get

$$\|\varphi(x)\| \leqslant \beta(\|\varphi \circ f(\cdot, x)\|), \qquad x \in X, \tag{7.4.10}$$

on account of (7.4.7) (implying $\rho(f(\cdot, x), \xi) \leqslant m \operatorname{diam} X$) and (d).

Fix arbitrarily an $\eta > 0$ and a neighbourhood U of ξ such that $\|\varphi(x)\| = \|\varphi(x) - \varphi(\xi)\| \leqslant \eta$ for $x \in U$ (which is possible since φ, being a uniform limit of Lipschitzian functions vanishing at ξ, is continuous and vanishes at ξ). Now making use of (7.3.1) and (7.4.10) we extend this estimation onto the entire X (see the proof of Theorem 7.3.1, for instance) getting $\|\varphi(x)\| \leqslant \eta$, $x \in X$. Owing to the arbitrariness of η this proves that $\varphi = 0$, as claimed.

Finally, fix an $\varepsilon > 0$ and choose an $n \in \mathbb{N}$ such that

$$\vartheta_n \|\varphi_{\vartheta_n}(x)\| \leqslant \varepsilon \qquad \text{for all } x \in X$$

and put $\varphi(x) := h(x, \varphi_{\vartheta_n} \circ f(\cdot, x))$, $x \in X$. Then

$$\|\varphi_{\vartheta_n}(f_s(x)) - \varphi(f_s(x))\|$$
$$= \|h(f_s(x), (1 - \vartheta_n)\varphi_{\vartheta_n} \circ f(\cdot, f_s(x))) - h(f_s(x), \varphi_{\vartheta_n} \circ f(\cdot, f_s(x)))\|$$
$$\leqslant \beta(\vartheta_n \|\varphi_{\vartheta_n} \circ f(\cdot, f_s(x))\|) \leqslant \beta(\varepsilon) \leqslant \varepsilon$$

for all $(s, x) \in S \times X$ and consequently

$$\|\varphi(x) - h(x, \varphi \circ f(\cdot, x))\| \leqslant \beta(\|\varphi_{\vartheta_n} \circ f(\cdot, x) - \varphi \circ f(\cdot, x)\|) \leqslant \beta(\varepsilon) \leqslant \varepsilon, \qquad x \in X,$$

which proves (7.4.9). Obviously, $\varphi \in \Phi$. It remains to show that φ is Lipschitzian and $\varphi(\xi) = 0$. For that purpose, take any $x, y \in X$; then

$$\|\varphi(x) - \varphi(y)\| \leqslant L\rho(x, y) + \beta(\|\varphi_{\vartheta_n} \circ f(\cdot, x) - \varphi_{\vartheta_n} \circ f(\cdot, y)\|)$$

$$\leqslant L\rho(x, y) + \beta\left(\frac{1}{\vartheta_n} L\rho(x, y)m\right) \leqslant L\left(1 + \frac{1}{\vartheta_n}\right)\rho(x, y).$$

Lastly, $\varphi(\xi) = h(\xi, \varphi_{\vartheta_n} \circ f(\cdot, \xi)) = h(\xi, \varphi_{\vartheta_n}(\xi)) = h(\xi, 0) = 0.$ ∎

Theorem 7.4.2. *Under the assumptions of the preceding theorem, for every $\varepsilon > 0$ there exists a function $h^*: X \times \mathscr{F}(S, Y) \to Y$ such that*

$$\|h(x, y) - h^*(x, y)\| \leqslant \varepsilon \qquad \text{for all } (x, y) \in X \times \mathscr{F}(S, Y)$$

and the functional equation

$$\varphi(x) = h^*(x, \varphi \circ f(\cdot, x))$$

has a Lipschitzian solution $\varphi \in \Phi$ which vanishes at ξ.

Proof. Given $\varepsilon > 0$ it suffices to take $\varphi \in \Phi$ such that relation (7.4.9) holds true and to define

$$h^*(x, y) := h(x, y) + \varphi(x) - h(x, \varphi \circ f(\cdot, x)) \qquad (7.4.11)$$

for $(x, y) \in X \times \mathscr{F}(S, Y)$. ∎

Strengthening the assumptions on Φ a bit we are able to weaken those regarding h admitting its 'correction' suggested by (7.4.11). Such results are desirable indeed because the Lipschitz condition (7.4.8) is very restrictive in the presence of the strong assumptions upon the function β.

7.4C. Uniform approximation of a continuous mapping by a Lipschitzian one

We start with the following technical result.

Lemma 7.4.1. *Let (X, ρ) be a compact metric space and $(Y, \|\cdot\|)$ a finite-dimensional Banach space. Then for every continuous function $P: X \to Y$ and for every $\varepsilon > 0$ there exists a Lipschitzian mapping $Q: X \to Y$ such that*

$$\|P(x) - Q(x)\| \leqslant \varepsilon \qquad \text{for all } x \in X. \qquad (7.4.12)$$

Proof. Assume first $Y = \mathbb{R}$ and denote by Lip X the family of all Lipschitzian functions on X. Evidently, Lip $X \subset C(X, \mathbb{R})$ forms a real linear space closed with respect to pointwise multiplication of elements and containing all constant functions. Moreover, for any two distinct points $a, b \in X$ the function $p: X \to Y$ given by the formula

$$p(x) := \frac{\rho(x, a) - \rho(x, b)}{\rho(x, a) + \rho(x, b)}, \qquad x \in X,$$

is well defined, has the property $p(a) = -1 \neq 1 = p(b)$ and, as a simple calculation shows,

$$\left| p(x) - p(y) \right| \leqslant \frac{4 \operatorname{diam} X}{\rho(a, b)} \rho(x, y), \qquad x, y \in X,$$

i.e. $p \in \operatorname{Lip} X$. Now, it remains to apply the well-known Stone Approximation Theorem.

Proceeding to the general case, fix a base $\{e_1, \ldots, e_n\}$ of Y (over $\mathbb{R}$) and a function $P \in C(X, Y)$; then $P = p_1 e_1 + \cdots + p_n e_n$ where $p_i \colon X \to \mathbb{R}$ are continuous for $i \in \{1, \ldots, n\}$. Fix a positive ε arbitrarily and, applying the first part of the proof, find Lipschitzian functions $q_i \colon X \to \mathbb{R}$ such that

$$\left| p_i(x) - q_i(x) \right| \leqslant \frac{1}{\displaystyle\sum_{i=1}^{n} \|e_i\|} \, \varepsilon, \qquad x \in X, \, i \in \{1, \ldots, n\}.$$

Then the function $Q := \sum_{i=1}^{n} q_i e_i$ is Lipschitzian and satisfies (7.4.12). ∎

Theorem 7.4.3. *Assume that $(X, \rho), (Y, \|\cdot\|), \Phi, f, \xi$ and β satisfy the hypotheses of Theorem 7.4.1. If, moreover, the sum of any member of Φ and a function from $C(X, Y)$ is in Φ, and $h \colon X \times \mathscr{F}(S, Y) \to Y$ is such that*

(a) *for every $\varphi \in \Phi$ the function $X \ni x \mapsto h(x, \varphi \circ f(\cdot, x)) \in Y$ belongs to Φ in turn,*

(b) *$h(\xi, 0) = 0$,*

(c) *there exists a function $F \in C(X, Y)$ such that*

$$\left\| (h(x, y) + F(x)) - (h(\bar{x}, \bar{y}) + F(\bar{x})) \right\| \leqslant L\rho(x, \bar{x}) + \beta(\|y - \bar{y}\|)$$

for some $L \geqslant 0$ and all pairs $(x, y), (\bar{x}, \bar{y}) \in X \times \mathscr{F}(S, Y)$,

then for every $\varepsilon > 0$ there exists a Lipschitzian function $\varphi \in \Phi$ such that $\varphi(\xi) = 0$ and

$$\left\| \varphi(x) - h(x, \varphi \circ f(\cdot, x)) \right\| \leqslant \varepsilon$$

for all $x \in X$.

Proof. Given an $\varepsilon > 0$, apply Lemma 1 and choose a Lipschitzian function $G \colon X \to Y$ such that $\|F(x) - G(x)\| \leqslant \frac{1}{3}\varepsilon$ for all $x \in X$. One can easily check that the function

$$g(x, y) := h(x, y) + (F(x) - F(\xi)) - (G(x) - G(\xi)), \qquad (x, y) \in X \times \mathscr{F}(S, Y),$$

satisfies the Lipschitz condition

$$\left\| g(x, y) - g(\bar{x}, \bar{y}) \right\| \leqslant (L + L_G)\rho(x, \bar{x}) + \beta(\|y - \bar{y}\|),$$

$(x, y), (\bar{x}, \bar{y}) \in X \times \mathscr{F}(S, Y)$, where L_G denotes a Lipschitzian constant for G. Moreover, $g(\xi, 0) = h(\xi, 0) = 0$ and for any member φ of the class Φ the function

$$X \ni x \mapsto g(x, \varphi \circ f(\cdot, x)) = h(x, \varphi \circ f(\cdot, x)) + F(x) - G(x) + \operatorname{const} \in Y$$

belongs to $\boldsymbol{\Phi} + C(X, Y) \subset \boldsymbol{\Phi}$. Therefore, on account of Theorem 7.4.1, there exists a function $\varphi \in \boldsymbol{\Phi}$ such that

$$\|\varphi(x) - g(x, \varphi \circ f(\cdot, x))\| \leqslant \tfrac{1}{3}\varepsilon$$

for all $x \in X$; hence

$$\|\varphi(x) - h(x, \varphi \circ f(\cdot, x))\|$$
$$\leqslant \|\varphi(x) - g(x, \varphi \circ f(\cdot, x))\| + \|g(x, \varphi \circ f(\cdot, x) - h(x, \varphi \circ f(\cdot, x))\|$$
$$\leqslant \|\varphi(x) - g(x, \varphi \circ f(\cdot, x))\| + \|F(x) - G(x)\| + \|F(\xi) - G(\xi)\| \leqslant \varepsilon$$

for $x \in X$. ∎

7.4D. Polynomial approximate solutions

In Subsection 7.4A we got approximate solutions to (7.4.1) and (7.4.2) in the class of polynomials. We conclude the present section with a result going in this direction. We shall show, in particular, that close to an equation which (possibly) has no continuous solutions there are equations having solutions of the highest possible regularity.

Theorem 7.4.4. *Let $X \subset \mathbb{R}^n$ be a compact set and let all the assumptions of Theorem 7.4.1 (resp. Theorem 7.4.3) be satisfied with $Y = \mathbb{R}$. Then for every $\varepsilon > 0$ there exists a polynomial $\varphi_\varepsilon \colon \mathbb{R}^n \to \mathbb{R}$ such that $\varphi_\varepsilon(\xi) = 0$ and*

$$|\varphi_\varepsilon(x) - h(x. \varphi_\varepsilon \circ f(\cdot, x))| < \varepsilon \tag{7.4.13}$$

for all $x \in X$. Moreover, for every $\varepsilon > 0$ there exists a function $h^* \colon X \times \mathcal{F}(S, \mathbb{R}) \to \mathbb{R}$ such that $|h(x, y) - h^*(x, y)| \leqslant \varepsilon$ for all $(x, y) \in X \times \mathcal{F}(S, \mathbb{R})$ and the functional equation

$$\varphi(x) = h^*(x, \varphi \circ f(\cdot, x)) \tag{7.4.14}$$

has a solution $\varphi \colon X \to \mathbb{R}$ vanishing at ξ and being a restriction of a polynomial of n variables to the set X.

Proof. Theorem 7.4.1 (resp. Theorem 7.4.3) guarantees that for a given $\varepsilon > 0$ there exists a Lipschitzian function $\psi \in \boldsymbol{\Phi}$ such that $\psi(\xi) = 0$ and

$$|\psi(x) - h(x, \psi \circ f(\cdot, x))| \leqslant \tfrac{1}{3}\varepsilon \qquad \text{for all } x \in X.$$

With the aid of the classical Weierstrass Theorem we may find a polynomial $\varphi_\varepsilon \colon \mathbb{R}^n \to \mathbb{R}$ such that $\varphi_\varepsilon(\xi) = 0$ and

$$|\varphi_\varepsilon(x) - \psi(x)| \leqslant \tfrac{1}{3}\varepsilon \qquad \text{for } x \in X.$$

Then

$$|\varphi_\varepsilon(x) - h(x, \varphi_\varepsilon \circ f(\cdot, x))|$$
$$\leqslant |\varphi_\varepsilon(x) - \psi(x)| + |\psi(x) - h(x, \psi \circ f(\cdot, x))| + |h(x, \psi \circ f(\cdot, x)) - h(x, \varphi_\varepsilon \circ f(\cdot, x))|$$
$$\leqslant \tfrac{2}{3}\varepsilon + \beta(\|\psi \circ f(\cdot, x) - \varphi_\varepsilon \circ f(\cdot, x)\|) \leqslant \tfrac{2}{3}\varepsilon + \beta(\tfrac{1}{3}\varepsilon) \leqslant \tfrac{2}{3}\varepsilon + \tfrac{1}{3}\varepsilon = \varepsilon$$

for $x \in X$, as desired.

To prove the latter part of the theorem, take the polynomial φ_ε just obtained and put

$$h^*(x, y) := h(x, y) + \varphi_\varepsilon(x) - h(x, \varphi_\varepsilon \circ f(\cdot, x)), \qquad (x, y) \in X \times \mathscr{F}(S, \mathbb{R}).$$

Then $\varphi_\varepsilon|_X$ satisfies (7.4.14) in an obvious way whereas the estimate

$$|h - h^*| \leq \varepsilon$$

results immediately from (7.4.13). ∎

7.5 Continuous dependence

The effect that has been investigated in the preceding section might also be interpreted loosely as a sort of discontinuity in the behaviour of functional equations considered: each of the functional equations

$$\varphi = H_n(x, \varphi), \qquad (7.5.1)$$

$n \in \mathbb{N}$, has a solution (even very regular), the sequence $(H_n)_{n \in \mathbb{N}}$ tends uniformly to H and, nevertheless, equation

$$\varphi = H(x, \varphi) \qquad (7.5.2)$$

has no, say continuous, solutions.

The present considerations go into a positive direction: from the hypotheses assumed the existence and uniqueness of a solution φ_n of (7.5.1) follow (including the 'limit' equation (7.5.2)) and the point is to state that the (unique) solution φ of the limit equation (7.5.2) is just a limit of the sequence $(\varphi_n)_{n \in \mathbb{N}}$. Such a nice (and desired) behaviour is just called continuous dependence and has already been considered in the present book in connection with equations of finite orders (see, in particular, Chapter 5).

7.5A. A general result

Following K. Baron [9] again, we return to the general functional equation (7.5.2).

Theorem 7.5.1. Let $(X, \mathscr{T})$ be a topological space and let $\mathscr{G} \subset \mathscr{T}$ be a family of open sets fulfilling conditions (i)–(iii) of Section 7.2. Suppose further that (Y, σ) is a complete metric space and $\mathbf{\Phi}$ is a given subfamily of $C(X, Y)$ such that assumptions (iv) and (v) of Section 7.2 are satisfied.

Let $H_n: X \times \mathbf{\Phi} \to Y$, $n \in \mathbb{N}_0$, be such that

$$H_n(\cdot, \varphi) \in \mathbf{\Phi} \text{ for any } \varphi \in \mathbf{\Phi} \text{ and for all } n \in \mathbb{N}_0, \qquad (7.5.3)$$

for any $U \in \mathscr{G}$ and any two members φ and ψ of $\mathbf{\Phi}$ such that
$\varphi|_U = \psi|_U$, one has $H_n(\cdot, \varphi)|_U = H_n(\cdot, \psi)|_U$, $n \in \mathbb{N}_0$, $\qquad (7.5.4)$

for any $\varphi \in \mathbf{\Phi}$ the function $H_0(\cdot, \varphi)$ is the a.u. limit of the sequence $(H_n(\cdot, \varphi))_{n \in \mathbb{N}}$. $\qquad (7.5.5)$

Put $d_U(\varphi, \psi) := \sup_{x \in \mathrm{cl}\, U} \sigma(\varphi(x), \psi(x))$, $\varphi, \psi \in \Phi$, $U \in \mathscr{G}$. *If for any* $U \in \mathscr{G}$ *and any* $\varepsilon > 0$ *there exists a* $\delta > 0$ *such that for every* $n \in \mathbb{N}_0$

$$\varepsilon \leqslant d_U(\varphi, \psi) < \varepsilon + \delta \quad \text{implies} \quad \sigma(H_n(x, \varphi), H_n(x, \psi)) < \varepsilon, \quad x \in \mathrm{cl}\, U,\ \varphi, \psi \in \Phi,$$

$$(7.5.6)$$

then for every $n \in \mathbb{N}_0$ *the functional equation* (7.5.1) *has exactly one solution* $\varphi_n \in \Phi$. *Moreover, the sequence* $(\varphi_n)_{n \in \mathbb{N}}$ *tends almost uniformly to* φ_0.

Proof. In the light of Theorem 7.2.1 only the latter statement requires a motivation. Fix a $U \in \mathscr{G}$ and consider a function $E_U : \Phi_U \to \Phi$ such that $E_U(\varphi)|_{\mathrm{cl}\, U} = \varphi$, $\varphi \in \Phi_U$ (see the proof of Theorem 7.2.1). The transformations $T_n : \Phi_U \to \mathscr{F}(X, Y)$ defined by the formula

$$T_n(\varphi) := H_n(\cdot, E_U(\varphi))|_{\mathrm{cl}\, U}, \qquad \varphi \in \Phi_U,\ n \in \mathbb{N}_0,$$

map their common domain Φ_U into itself because of (7.5.3). Moreover, on account of (7.5.5), the transformation $T_0(\varphi)$ is the uniform limit of the sequence $(T_n(\varphi))_{n \in \mathbb{N}}$, $\varphi \in \Phi_U$. Further, for any two distinct members φ, ψ of the class Φ_U, we have $\varepsilon := d_U(\varphi, \psi) > 0$ whence, by (7.5.6), for all $n \in \mathbb{N}_0$,

$$d_U(T_n(\varphi), T_n(\psi)) = \sup_{x \in \mathrm{cl}\, U} \sigma(H_n(x, E_U(\varphi)), H_n(x, E_U(\psi)))$$

$$= \sigma(H_n(x_0, E_U(\varphi)), H_n(x_0, E_U(\psi))) < \varepsilon = d_U(\varphi, \psi)$$

(the existence of such an $x_0 \in \mathrm{cl}\, U$ results from the compactness of $\mathrm{cl}\, U$ and the fact that $H_n(\cdot, \xi) \in \Phi \subset C(X, Y)$ for all $n \in \mathbb{N}_0$ and $\xi \in \Phi$). Finally, if for some $\varphi, \psi \in \Phi_U$ and $\varepsilon > 0$ one has $d_U(\varphi, \psi) \in (\varepsilon, \varepsilon + \delta)$, where δ is assigned for ε in accordance with (7.5.6), then

$$d_U(T_n(\varphi), T_n(\psi)) = \sup_{x \in \mathrm{cl}\, U} \sigma(H_n(x, E_U(\varphi)), H_n(x, E_U(\psi))) \leqslant \varepsilon$$

on account of (7.5.6), again. Thus putting $d(\varphi, \psi) := d_U(E_U(\varphi), E_U(\psi))$, $\varphi, \psi \in \Phi_U$ we see that the complete metric space (Φ_U, d) and the sequence $(T_n)_{n \in \mathbb{N}_0}$ satisfy all the assumptions of the 'continuous dependence' type fixed-point theorem 1.6.1: if $\bar{\varphi}_n$ stands for the unique fixed point of T_n, $n \in \mathbb{N}_0$, then $\lim_{n \to \infty} \bar{\varphi}_n = \bar{\varphi}_0$. However,

$$\bar{\varphi}_n = \varphi_n|_{\mathrm{cl}\, U}, \qquad n \in \mathbb{N}_0, \tag{7.5.7}$$

where $\varphi_n = H_n(\cdot, \varphi_n)$ is the unique fixed point of H_n, $n \in \mathbb{N}_0$ (whose existence is guaranteed by Theorem 7.2.1), because the equality $E_U(\varphi_n|_{\mathrm{cl}\, U})|_{\mathrm{cl}\, U} = \varphi_n|_{\mathrm{cl}\, U}$ implies, in view of (7.5.4) and the continuity of $H(\cdot, \xi)$, $\xi \in \Phi$, that

$$\varphi_n|_{\mathrm{cl}\, U} = H_n(\cdot, \varphi_n)|_{\mathrm{cl}\, U} = H_n(\cdot, E_U(\varphi_n|_{\mathrm{cl}\, U}))|_{\mathrm{cl}\, U} = T_n(\varphi_n|_{\mathrm{cl}\, U})$$

whence (7.5.7) follows by means of the uniqueness of the fixed point $\bar{\varphi}_n$ of the transformation T_n, $n \in \mathbb{N}_0$. Eventually, we arrive at relation

$$\varphi_0|_{\mathrm{cl}\, U} = \bar{\varphi}_0 = \lim_{n \to \infty} \bar{\varphi}_n = \lim_{n \to \infty} \varphi_n|_{\mathrm{cl}\, U} \tag{7.5.8}$$

(the convergence being uniform) satisfied for all $U \in \mathscr{G}$.

To finish the proof, fix a compact set $K \subset X$; then there exists a finite subfamily $\{U_1, \ldots, U_n\}$ of $\mathcal{G}$ such that $K \subset \bigcup_{i=1}^n U_i$ and the uniform convergence (7.5.8) with U replaced by U_i, $i \in \{1, \ldots, n\}$, forces the sequence $(\varphi_n|_K)_{n \in \mathbb{N}}$ to tend uniformly to $\varphi_0|_K$. ∎

7.5B. Important special case

As in Section 7.2 we shall now present an interpretation of the result just obtained in the particular case

$$H_n(x, \varphi) := h_n(x, \varphi \circ f_n(\cdot, x)), \qquad n \in \mathbb{N}_0. \tag{7.5.9}$$

Theorem 7.5.2. *Let* $(X, \mathcal{T})$, (Y, σ), $\mathcal{G} \subset \mathcal{T}$ *and* $\Phi \subset C(X, Y)$ *satisfy the assumptions of the preceding theorem. Suppose that X is metrizable, S is a given set and $f_n \colon S \times X \to X$ are such that $f_n(s, \cdot)(\text{cl } U) \subset \text{cl } U$ for all $U \in \mathcal{G}$, $s \in S$ and $n \in \mathbb{N}_0$; moreover, assume that for every compact set $K \subset X$ the sequence $(f_n|_{S \times K})_{n \in \mathbb{N}}$ tends uniformly to $f_0|_{S \times K}$.*

Put $\mathcal{F}_0(S, Y) := \{\varphi \circ f_n(\cdot, x) \colon (x, \varphi) \in X \times \Phi, \ n \in \mathbb{N}_0\}$. *If the functions* $h_n \colon X \times \mathcal{F}_0(S, Y) \to Y$, $n \in \mathbb{N}_0$, *satisfy the conditions*

> *for every member φ of Φ the functions $\psi_n \colon X \to Y$ given by $\psi_n(x) := h_n(x, \varphi \circ f_n(\cdot, x))$, $n \in \mathbb{N}_0$, $x \in X$, belong to Φ, in turn,* (7.5.10)

> *for every $\varphi \in \Phi$ the sequence of functions $\omega_n \colon X \to Y$ defined by $\omega_n(x) := h_n(x, \varphi \circ f_0(\cdot, x))$, $x \in X$, $n \in \mathbb{N}_0$, tends almost uniformly to ω_0,* (7.5.11)

> *for any $U \in \mathcal{G}$ and any $\varepsilon > 0$ there exists a $\delta > 0$ such that for $y_1, y_2 \in \mathcal{F}_0(S, Y)$ $\sigma(y_1(s), y_2(s)) < \varepsilon + \delta$, $s \in S$, implies $\sigma(h_n(x, y_1), h_n(x, y_2)) < \varepsilon$ for $x \in \text{cl } U$ and $n \in \mathbb{N}_0$,* (7.5.12)

> *for any $n \in \mathbb{N}_0$, $x \in X$ and $y_1, y_2 \in \mathcal{F}_0(S, Y)$ there exists an $s \in S$ such that $\sigma(h_n(x, y_1), h_n(x, y_2)) \leq \sigma(y_1(s), y_2(s))$,* (7.5.13)

then the functional equation

$$\varphi(x) = h_n(x, \varphi \circ f_n(\cdot, x))$$

has exactly one solution $\varphi_n \in \Phi$, $n \in \mathbb{N}_0$. Moreover, the sequence $(\varphi_n)_{n \in \mathbb{N}}$ tends almost uniformly to φ_0.

Proof. Define the functions $H_n \colon X \times \Phi \to Y$, $n \in \mathbb{N}_0$, by formula (7.5.9). A direct verification (with the use of (7.5.10) and (7.5.12)) shows that all the functions H_n, $n \in \mathbb{N}_0$, satisfy the assumptions of Theorem 7.2.1. In the light of Theorem 7.5.1, the only thing to prove is the convergence requirement (7.5.5).

Fix a $\varphi \in \Phi$, a compact set $K \subset X$, an $\varepsilon > 0$ and choose a finite subfamily $\{U_1, \ldots, U_n\}$ of $\mathcal{G}$ such that

$$K \subset \bigcup_{i=1}^n U_i.$$

Let $K_1 := \bigcup_{i=1}^{n} \mathrm{cl}\, U_i$; our assumptions on the family $\mathcal{G}$ imply that K_1 is compact as well. Hence the continuous function $\varphi|_{K_1}$ is uniformly continuous; in particular, there exists a $\delta > 0$ such that for all $x, y \in K_1$

$$\rho(x, y) < \delta \quad \text{implies} \quad \sigma(\varphi(x), \varphi(y)) < \tfrac{1}{2}\varepsilon, \tag{7.5.14}$$

where ρ stands for a metric generating the topology $\mathcal{T}$.

Since $(f_n|_{S \times K})_{n \in \mathbb{N}}$ tends uniformly to $f_0|_{S \times K}$ we may find an $N_1 \in \mathbb{N}$ such that

$$\rho(f_n(s, x), f_0(s, x)) < \delta \quad \text{for all } (s, x) \in S \times K \text{ and all } n \geqslant N_1, \tag{7.5.15}$$

whereas (7.5.11) guarantees the existence of an $N_2 \in \mathbb{N}$ such that

$$\sigma(\omega_n(x), \omega_0(x)) < \tfrac{1}{2}\varepsilon \quad \text{for all } x \in K \text{ and all } n \geqslant N_2. \tag{7.5.16}$$

For an arbitrary pair $(s, x) \in S \times K$ and an $n \geqslant N := \max(N_1, N_2)$, relation (7.5.15) jointly with (7.5.14) implies

$$\sigma(\varphi(f_n(s, x)), \varphi(f_0(s, x))) < \tfrac{1}{2}\varepsilon. \tag{7.5.17}$$

Fix an $n \in \mathbb{N}$, $n \geqslant N$, and an $x \in K$. Putting $y_1 := \varphi \circ f_n(\cdot, x)$, $y_2 := \varphi \circ f_0(\cdot, x)$ and making use of (7.5.13) we derive the existence of an $s_0 \in S$ such that

$$\sigma(h_n(x, \varphi \circ f_n(\cdot, x)), h_n(x, \varphi \circ f_0(\cdot, x))) = \sigma(h_n(x, y_1), h_n(x, y_2))$$
$$\leqslant \sigma(y_1(s_0), y_2(s_0))$$
$$= \sigma(\varphi(f_n(s_0, x)), \varphi(f_0(s_0, x))) < \tfrac{1}{2}\varepsilon$$

by means of (7.5.17).

Finally,

$$\sigma(H_n(x, \varphi), H_0(x, \varphi)) = \sigma(h_n(x, \varphi \circ f_n(\cdot, x)), h_0(x, \varphi \circ f_0(\cdot, x)))$$
$$\leqslant \sigma(h_n(x, \varphi \circ f_n(\cdot, x)), h_n(x, \varphi \circ f_0(\cdot, x)))$$
$$+ \sigma(h_n(x, \varphi \circ f_0(\cdot, x)), h_0(x, \varphi \circ f_0(\cdot, x)))$$
$$< \tfrac{1}{2}\varepsilon + \sigma(\omega_n(x), \omega_0(x)) < \tfrac{1}{2}\varepsilon + \tfrac{1}{2}\varepsilon = \varepsilon,$$

in view of (7.5.18) and (7.5.16). ∎

Remark 7.5.1. The condition $f_n|_{S \times K} \Rightarrow f_0|_{S \times K}$ for each compact $K \subset X$ reduces to the almost uniform convergence of $(f_n(s, \cdot))_{n \in \mathbb{N}}$ to $f_0(s, \cdot)$, $s \in S$, in the case of finite set S.

Remark 7.5.2. Theorem 5.2.4 becomes a special case of Theorem 7.5.2 up to the assumptions upon the domain of functions h_n, $n \in \mathbb{N}_0$. More subtle, but also more involved, such assumptions are found in Baron [9].

7.6 A survey of results on systems of nonlinear equations of finite orders

We have already mentioned earlier that, in many instances, the results of Sections 7.1–7.5 admit their analogues for systems of equations of an arbitrary order. K. Baron considers, for example, a pretty general system

$$\varphi_i(x) = H_i(x, \varphi_1, \ldots, \varphi_m), \qquad i \in \{1, \ldots, m\}, \tag{S}$$

under assumptions as general as those dealt with in Section 7.1. Baron's results on (S) are collected under the symptomatic heading: 'The same for systems' (see Baron [9, Chapter 2]) which suggests that anybody familiar with the results and methods applied in our Sections 7.1–7.5 will find nothing especially exciting while dealing with (S). This is not far wrong. First of all, the results concerning (S) should reduce to those known for (7.2.2) before in the case $m = 1$ (in all probability no one would expect to get something deeper) although sometimes it might happen that such a special case leads eventually to a weaker assertion. On the other hand, getting suitable analogues for systems causes, in many cases, essential difficulties, mainly technical. As we shall see later on, Matkowski's fixed-point Theorem 1.5.4 turns out to be a pretty useful tool while solving systems of functional equations; although theoretically equivalent to the Banach contraction principle it usually helps considerably to overcome the difficulties just mentioned. But already the statement of Matkowski's theorem might give an unpleasant impression for someone who does not love too many indices. Trying to imagine what suitable analogues for infinite systems might look like[1], we have decided to confine ourselves to systems of two kinds only:

$$\varphi_i(x) = h_i(x; \varphi_1(f_1(x)), \ldots, \varphi_1(f_n(x)); \ldots; \varphi_m(f_1(x)), \ldots, \varphi_m(f_n(x))),$$
$$\tag{7.6.1}$$

$i \in \{1, \ldots, m\}$, for short, *h-systems* and

$$\varphi_i(f_0(x)) = g_i(x; \varphi_1(x), \varphi_1(f_1(x)), \ldots, \varphi_1(f_n(x)); \ldots;$$
$$\varphi_m(x), \varphi_m(f_1(x)), \ldots, \varphi_m(f_n(x))), \tag{7.6.2}$$

$i \in \{1, \ldots, m\}$ (*g-systems*), assuming permanently, unless explicitly stated otherwise, that n, m are fixed positive integers,

$$\varphi = (\varphi_1, \ldots, \varphi_m), \qquad f = (f_1, \ldots, f_n),$$

and

$$h = (h_1, \ldots, h_m) \qquad \text{and} \qquad g = (g_1, \ldots, g_m).$$

In particular, we restrict our further investigations to systems of equations of finite orders. For $n = m = 1$ system (7.6.1) reduces to the functional equation (5.0.1) whereas (7.6.2) coincides simply with equation (5.3.1), both studied widely in Chapter 5. This observation is not only formal: knowing already the behaviour of solutions of (5.0.1) (resp. 5.3.1) we can predict corresponding assertions for system (7.6.1) (resp. (7.6.2)). We have learned, for instance, that equation (5.3.1) is more convenient to handle in the case of lack of uniqueness of solutions and we shall see that so is system (7.6.2).

---

[1] Despite the discontents expressed here, a successful attempt in this direction has been made by W. Jarczyk [1].

In this section we are going to present various results on the behaviour of solutions of systems (7.6.1) and (7.6.2) in different function classes. The results will be presented without proofs. Some of the proofs lacking which are instructive in a sense will be presented in the next section.

7.6A. Continuous solutions of h-systems

The following result concerning the h-system (7.6.1) is due to K. Baron [6].

Theorem 7.6.1. *Let* $(X, \mathcal{T})$ *be a topological space and let* (Y_i, σ_i) *be complete metric spaces,* $i \in \{1, \dots, m\}$. *Suppose that functions* $f_k \colon X \to X$ *are continuous for* $k \in \{1, \dots, n\}$ *and* $h_i \colon X \times Y_1^n \times \cdots \times Y_m^n \to Y_i$ *satisfy the Lipschitz condition*

$$\sigma_i(h_i(x, y_{1,1}, \dots, y_{1,n}; \dots; y_{m,1}, \dots, y_{m,n}),$$
$$h_i(x, \bar{y}_{1,1}, \dots, \bar{y}_{1,n}; \dots; \bar{y}_{m,1}, \dots, \bar{y}_{m,n}))$$
$$\leqslant \sum_{j=1}^{m} \sum_{k=1}^{n} a_{i,j,k} \sigma_j(y_{j,k}, \bar{y}_{j,k}), \qquad i \in \{1, \dots, m\}, \qquad (7.6.3)$$

for all $x \in X$ *and* $y_{j,k}, \bar{y}_{j,k} \in Y_j, j \in \{1, \dots, m\}, k \in \{1, \dots, n\}$. *If there exists an open covering* $\{U_p \colon p \in \mathbb{N}\}$ *of* X *such that* $\operatorname{cl} U_p$ *is compact and* $\operatorname{cl} U_p \subset U_{p+1}$, $p \in \mathbb{N}$, *and if all the characteristic roots of the* $m \times m$ *matrix*

$$\left(\sum_{k=1}^{n} a_{i,j,k} \right)_{i,j \in \{1, \dots, m\}} \qquad (7.6.4)$$

are less than unity in absolute value, then system (7.6.1) *has a unique continuous solution* $\varphi \colon X \to Y_1 \times \cdots \times Y_m$.

In the next theorem, due to G. Majcher [1], we obtain only the existence of a continuous solution of system (7.6.1).

Theorem 7.6.2. *Let* (X, ρ) *be a metric space and let* $(B_i, \|\cdot\|_i)$ *be Banach spaces,* $i \in \{1, \dots, m\}$. *Suppose further that the self-mappings* ω_i *of* $\mathbb{R}^+$ *are increasing and vanishing at zero,* $i \in \{1, \dots, m\}$ *and the functions* $h_i \colon X \times Y_1^n \times \cdots \times Y_m^n \to Y_i$ *satisfy the Lipschitz conditions*

$$\|h_i(x, y_{1,1}, \dots, y_{m,n}) - h_i(\bar{x}, \bar{y}_{1,1}, \dots, \bar{y}_{m,n})\|_i \leqslant \omega_i(\rho(x, \bar{x}))$$

whenever $\|y_{j,k} - \bar{y}_{j,k}\| \leqslant \omega_j(\rho(f_k(x), f_k(\bar{x})))$, $k \in \{1, \dots, n\}$, $i, j \in \{1, \dots, m\}$. *Then for any continuous function* $f \colon X \to X^n$ *system* (7.6.1) *has a continuous solution* $\varphi \colon X \to Y_1 \times \cdots \times Y_m$.

We conclude the consideration of continuous solutions of system (7.6.1) by adding a result which formally does not belong to that subject matter. Namely, we shall impose no regularity conditions whatsoever upon the solution; nevertheless, we get the uniqueness of the solution. This is so because the values of φ are assumed to lie in a compact set; this is a sort of boundedness condition. The result reads as follows.

Theorem 7.6.3. *Let X be an arbitrary set and let (Y_i, σ_i) be compact spaces, $i \in \{1, \ldots, m\}$. If $f: X \to X^n$ is an arbitrary function and functions $h_i: X \times Y_1^n \times \cdots \times Y_m^n \to Y_i$ satisfy the Lipschitz conditions (7.6.3) for all $x \in X$, $y_{j,k}, \bar{y}_{j,k} \in Y_j$, $i, j \in \{1, \ldots, m\}$, $k \in \{1, \ldots, n\}$, then system (7.6.1) has a unique solution $\varphi: X \to Y_1 \times \cdots \times Y_m$ whenever all the characteristic roots of the matrix (7.6.4) are less than unity in absolute value.*

In the case where the h_i are real- or complex-valued functions such that
$$|y_{j,k}| \leqslant K, j \in \{1, \ldots, m\}, k \in \{1, \ldots, n\} \quad \text{implies} \quad |h(x, y_{1,1}, \ldots, y_{m,n})| \leqslant K,$$
the latter theorem gives the existence and uniqueness of solutions φ of (7.6.1) whose components φ_i are bounded by K, $i \in \{1, \ldots, m\}$.

In connection with this let us mention here a slightly more general question. Let $A: X \to \mathbb{K} \setminus \{0\}$ be an arbitrary function. The problem of determining solutions $\varphi: X \to \mathbb{K}^m$ of system (7.6.1) such that the components φ_i, $i \in \{1, \ldots, m\}$, are majorized by KA in absolute value, where K is a constant which may depend on φ, can be reduced to that of determining bounded solutions of (7.6.1) by introducing a new unknown function $\psi := (1/A)\varphi$. Such problems have been investigated for various functions A by H. Adamczyk [1] and S. Czerwik [14], [15].

Further results concerning continuous solutions of system (7.6.1) and of related functional equations can be found in Bajraktarević [3], Baron [2], [3], [6], [9], Baron–Ger [2], Choczewski [1], Kordylewski [1], [3], Kordylewski–Kuczma [3], Kuczma [26, Ch. XII], Kwapisz [1], Kwapisz–Turo [1], [3], [4] and Majcher [1].

7.6B. Solutions of h-systems with a prescribed asymptotic behaviour

Following Z. Kominek [2] we consider solutions of (7.6.1) which are smoothly approximated by polynomials near the origin (see also Matkowski [10]).

Theorem 7.6.4. *Let f_k be self-mappings of an interval $[0, a]$, $0 < a \leqslant \infty$, such that $0 < f_k(x) < x$ for $x \in (0, a]$, $f_k(0) = 0$, and $f_k(x) \leqslant s_k x$ for $x \in [0, b)$, $0 < b \leqslant a, s_k > 0, k \in \{1, \ldots, n\}$. Suppose that the functions $h_i: [0, a] \times \mathbb{R}^{mn} \to \mathbb{R}$ vanish at the origin and satisfy the Lipschitz conditions*

$$|h_i(x, y_{1,1}, \ldots, y_{m,n}) - h(x, \bar{y}_{1,1}, \ldots, \bar{y}_{m,n})| \leqslant \sum_{j=1}^{m} \sum_{k=1}^{n} a_{i,j,k} |y_{j,k} - \bar{y}_{j,k}|$$

for all $x \in [0, b)$ and $y_{j,k}, \bar{y}_{j,k} \in (-c, c)$ for $k \in \{1, \ldots, n\}$, $i, j \in \{1, \ldots, m\}$ and for some positive number c. Lastly, assume that we are given a function

$P: \mathbb{R} \to \mathbb{R}^m$ whose real components P_i are polynomials of degree at most $r - 1 \in \mathbb{N}$ and $P_i(0) = 0$, $i \in \{1, \dots, m\}$.

If the $m \times m$ matrix

$$\left(\sum_{k=1}^{n} a_{i,j,k} s_k^r \right)_{i,j \in \{1,\dots,m\}} \tag{7.6.5}$$

has all the characteristic roots less than unity in absolute value, then any two solutions $\varphi = (\varphi_1, \dots, \varphi_m)$ and $\psi = (\psi_1, \dots, \psi_m)$ of system (7.6.1) such that both $\varphi - P$ and $\psi - P$ are $O(x^r)$ as $x \to 0$ coincide in a right neighbourhood of zero. If, moreover, $f = (f_1, \dots, f_n)$ is continuous then $\varphi(x) = \psi(x)$ for all $x \in [0, a|$.

Obviously, given a map φ and looking for polynomials P_i such that $\varphi_i(x) = P_i(x) + O(x^r)$ as $x \to 0$, $i \in \{1, \dots, m\}$, we make use of Taylor's Formula provided that φ is smooth enough. In particular, Theorem 7.6.4 leads immediately to the following uniqueness result for C^r mappings.

Corollary 7.6.1. Let $f: [0, a| \to \mathbb{R}^n$ and $h: [0, a| \times \mathbb{R}^{mn} \to \mathbb{R}^m$ be C^r functions such that $0 < f_k(x) < x$, $x \in (0, a|$, $k \in \{1, \dots, n\}$, and $h(0, 0, \dots, 0) = 0$. If all the characteristic roots of the matrix

$$\left(\sum_{k=1}^{n} \left| \frac{\partial h_i}{\partial y_{j,k}} (0, \dots, 0) f_k'(0)^r \right| \right)_{i,j \in \{1,\dots,m\}} \tag{7.6.6}$$

are less than unity in absolute value, then for any m-tuple $(\eta_{1,p}, \dots, \eta_{m,p}) \in \mathbb{R}^m$, $p \in \{1, \dots, r - 1\}$, system (7.6.1) has at most one C^r solution φ such that $\varphi_i(0) = 0$ and $\varphi_i^{(p)}(0) = \eta_{i,p}$ for $p \in \{1, \dots, r - 1\}$ and $i \in \{1, \dots, m\}$.

What about the existence? This question is answered by the following (Kominek [2]).

Theorem 7.6.5. *Under the assumptions of Theorem 7.6.4, if all the characteristic roots of matrix (7.6.5) are less than unity in absolute value and if $h = (h_1, \dots, h_m)$ satisfies the asymptotic relation*

$$h(x, P_1(f_1(x)), \dots, P_m(f_n(x))) - P(x) = o(x^r), \qquad x \to 0,$$

then there exists a positive number d such that system (7.6.1) has a solution $\varphi: [0, d) \to \mathbb{R}^m$ having the asymptotic property

$$\varphi(x) - P(x) = O(x^r), \qquad x \to 0.$$

Remark 7.6.1. By Baron's extension Theorem 7.1.1 the solution φ just obtained may uniquely be extended onto the whole interval $[0, a|$ whenever $f = (f_1, \dots, f_n)$ is continuous.

7.6C. Differentiable solutions of h-systems

There is a visible connection of a suitable asymptotics and C^r solutions of the h-system of equations (7.6.1). After the appearance of the result we presented as Theorem 5.6.1, in connection with (7.6.1) for $n = m = 1$, the theory of C^r solutions (uniqueness, existence, continuous dependence) was developed by J. Matkowski [5], [6]. This inspired Z. Kominek [1], [3], [4] for further investigations regarding C^r solutions of systems of nonlinear equations. Below we present some of his results.

We start with a notational convention: for brevity, the symbol $\{y_{j,k}\}$ denotes in the sequel a member of $\mathbb{R}^{mn}$, namely

$$\{y_{j,k}\} := (y_{1,1}, \ldots, y_{1,n}; \ldots; y_{m,1}, \ldots, y_{m,n});$$

recall that in the whole section the indices i, j run over the set $\{1, \ldots, m\}$ while k belongs always to $\{1, \ldots, n\}$.

Moreover, given sufficiently smooth functions $h = (h_1, \ldots, h_m) : [0, a| \times \mathbb{R}^{mn} \to \mathbb{R}^m$ and $f : [0, a| \to \mathbb{R}^m$, we define functions $h_{i,p} : [0, a| \times \mathbb{R}^{mn} \times \mathbb{R}^{pmn} \to \mathbb{R}$, $i \in \{1, \ldots, m\}$, $p \in \{0, \ldots, r\}$, as follows:

$$h_{i,0} := h_i, \qquad y_{j,k,0} := y_{j,k}, \tag{7.6.7}$$

$$h_{i,p+1}(x, \{y_{j,k}\}, \{y_{j,k,1}\}, \ldots, \{y_{j,k,p+1}\}) := \frac{\partial h_{i,p}}{\partial x}(x, \{y_{j,k}\}, \ldots, \{y_{j,k,p}\})$$

$$+ \sum_{j=1}^{m} \sum_{k=1}^{n} \sum_{q=0}^{p} \frac{\partial h_{i,p}}{\partial y_{j,k,q}}(x, \{y_{j,k}\}, \ldots, \{y_{j,k,p}\}) y_{j,k,q+1} f'_k(x), \tag{7.6.8}$$

$i \in \{1, \ldots, m\}$, $p \in \{0, \ldots, r-1\}$ (compare Section 5.6).

Theorem 7.6.6. *Assume that all the hypotheses of Corollary 7.6.1 are satisfied. If $f'_k(x) \leqslant 1$ for $x \in [0, c] \subset [0, a|$ and if all the characteristic roots of matrix (7.6.6) are less than unity in absolute value, then for every solution $\{\eta_{i,p} : i = 1, \ldots, m; p = 1, \ldots, r\}$ of the system of 'numeric' equations*

$$\eta_{i,p} = h_{i,p}(0, \{0\}, \eta_{1,1}, \ldots, \eta_{1,1}; \ldots; \eta_{m,1}, \ldots, \eta_{m,1};$$

$$\ldots; \eta_{1,p}, \ldots, \eta_{1,p}; \ldots; \eta_{m,p}, \ldots, \eta_{m,p}) \tag{7.6.9}$$

(each of $\eta_{1,1}, \ldots, \eta_{m,p}$ is taken n times here) system (7.6.1) has exactly one C^r solution $\varphi = (\varphi_1, \ldots, \varphi_m) : [0, a| \to \mathbb{R}^m$ such that

$$\varphi_i(0) = 0 \quad \text{and} \quad \varphi_i^{(p)}(0) = \eta_{i,p}, \quad i \in \{1, \ldots, m\}, p \in \{1, \ldots, r\} \tag{7.6.10}$$

Remark 7.6.2. In the case $f'_k(0) = 1$, $k \in \{1, \ldots, n\}$ (not excluded by the assumptions of Theorem 7.6.6) one may even get rid of the assumption $f'_k(x) \leqslant 1$ for $x \in [0, c)$ with no violation of the assertion of Theorem 7.6.6.

Remark 7.6.3. It can also be proved (Kominek [3]) that the solution obtained in Theorem 7.6.6 depends on the given functions in a continuous manner.

7.6D. Analytic solutions of h-systems

The solutions will be investigated in the complex domain which is more natural for such type considerations. The result we are going to report on is contained in Baron–Ger–Matkowski [1] (see also W. Smajdor [5]). We shall assume that the given function f is now defined in a disc

$$U := \{x \in \mathbb{C} : |x| < \rho\}$$

and, motivated by the previous studies concerning the real case, we shall require that

$$|f_k(x)| \leqslant |x| \qquad \text{for } x \in U. \tag{7.6.11}$$

This, however, is satisfied only if $|f'_k(0)| < 1$ or $f_k(x) = s_k x, x \in U$, with $|s_k| = 1$, $k \in \{1, \ldots, n\}$. However, if among the f_k there is only one, say f_{k_0}, whose derivative at zero lies on the unit circle and $|f'_k(0)| < 1$ for the remaining f_k and if the Schröder equation

$$\sigma(f_{k_0}(x)) = f'_{k_0}(0)\sigma(x)$$

has an invertible solution $\sigma: U \to \mathbb{C}$, then the change of variables $y = \sigma(x)$ will transform f_{k_0} into the linear map $x \mapsto f'_{k_0}(0)x$. Nevertheless, we must be conscious that, unfortunately, the seemingly natural requirement (7.6.11) is much more restrictive than its real counterpart.

Theorem 7.6.7. *Suppose that functions $f_k: U \to \mathbb{C}^m$ are analytic and satisfy (7.6.11) for all $k \in \{1, \ldots, n\}$ and that $h_i: U \times \mathbb{C}^{mn} \to \mathbb{C}$ are analytic in $U \times \mathbb{C}^{mn}$ and vanishing at the origin for $i \in \{1, \ldots, m\}$. If, for some $r \in \mathbb{N}$, all the characteristic roots of matrix (7.6.6) are less than unity in absolute value then, for any solution $\{\eta_{i,p}: i = 1, \ldots, m; p = 1, \ldots, r\}$ of (7.6.9), system (7.6.1) has a unique analytic solution $\varphi = (\varphi_1, \ldots, \varphi_m): U \to \mathbb{C}^m$ satisfying (7.6.10) provided that the radius ρ of the disc U is small enough.*

Remark 7.6.4. The requirement that the characteristic roots of the matrix

$$\left(\sum_{n=1}^{n} \left| \frac{\partial h_i}{\partial y_{j,k}} (0, \{0\}) f'_k(0)^r \right| \right)_{i,j \in \{1, \ldots, m\}}$$

all lie in the open unit disc for some $r \in \mathbb{N}$ is by no means restrictive whenever all the derivatives $f'_k(0)$, $k \in \{1, \ldots, n\}$, do so. This is just the case where $|f'_k(0)| = 1$ for some $k \in \{1, \ldots, n\}$ which possibly causes inconvenience.

7.6E. Integrable solutions of h-systems

We shall modify slightly a result obtained by J. Matkowski [13] in Chapter 4 of his exhaustive treatment of integrable solutions of functional equations. Matkowski examines the following system

$$\varphi_i(x) = h_i(x, \varphi_1(f_{i,1}(x)), \ldots, \varphi_m(f_{i,m}(x)))$$

$i \in \{1, \ldots, m\}$, which for $m = 1$ reduces to a nonlinear functional equation of the first order. Nevertheless, his methods extend almost automatically to a more general system (7.6.1) considered in our present section.

Let $(X, \mathfrak{M}, \mu)$ be a measure space. We remind the reader that a function $F: X \times \mathbb{R}^q \to \mathbb{R}$ is called sup-measurable if and only if for any q-tuple $(\psi_1, \ldots, \psi_q)$ of real measurable functions defined on X the superposition $F(\cdot, \psi_1, \ldots, \psi_q)$ yields a measurable function in turn. If $X \subset \mathbb{R}^s$ and μ is the s-dimensional Lebesgue measure then the two conditions together

(a) $F(\cdot, u_1, \ldots, u_q)$ is Lebesgue measurable for any fixed q-tuple $(u_1, \ldots, u_q) \in \mathbb{R}^q$,
(b) $F(x, \cdot, \ldots, \cdot)$ is continuous for any fixed $x \in X$,

are sufficient for F to be sup-measurable (see Carathéodory [1] and Šragin [1]).

Theorem 7.6.8. *Let $(X, \mathfrak{M}, \mu)$ be a σ-finite measure space, $\mu(X) > 0$, and let f_k be $\mathfrak{M}$-measurable invertible self-mappings of X such that f_k^{-1} are $\mathfrak{M}$-measurable and the measures μf_k and μf_k^{-1} are absolutely continuous with respect to μ, $k \in \{1, \ldots, n\}$. Further, let $h_i: X \times \mathbb{R}^{mn} \to \mathbb{R}$ be sup-measurable and satisfy the Lipschitz condition*

$$\left| h_i(x, y_{1,1}, \ldots, y_{m,n}) - h_i(x, \bar{y}_{1,1}, \ldots, \bar{y}_{m,n}) \right| \leqslant \sum_{j=1}^{m} \sum_{k=1}^{n} g_{i,j,k}(x) \left| y_{j,k} - \bar{y}_{j,k} \right|$$

for all $x \in X$, $y_{j,k}, \bar{y}_{j,k} \in \mathbb{R}$, where $g_{i,j,k}: X \to \bar{\mathbb{R}}$ are given μ-almost everywhere finite $\mathfrak{M}$-measurable functions, $i, j \in \{1, \ldots, m\}$, $k \in \{1, \ldots, n\}$. Put (see (4.7.6))

$$\alpha(p) := \begin{cases} 1 & \text{for } p \in (0, 1), \\ 1/p & \text{for } p \in [1, \infty) \end{cases}$$

and denote by g_k the Radon–Nikodym derivative of the measure μf_k with respect to μ, $k \in \{1, \ldots, n\}$.

Assume that for some $p \in (0, \infty)$ there exist non-negative numbers $s_{i,j,k}$ such that

$$\left(\frac{1}{g_k(x)} g_{i,j,k}(x)^p \right)^{\alpha(p)} \leqslant s_{i,j,k} \qquad \mu\text{-a.e. in } X,$$

$i, j \in \{1, \ldots, m\}$, $k \in \{1, \ldots, n\}$. If all the characteristic roots of the matrix

$$\left(\sum_{k=1}^{n} s_{i,j,k} \right)_{i,j \in \{1, \ldots, m\}}$$

are less than unity in absolute value and if the functions $h_i(\cdot, 0, \ldots, 0)$ belong to $L^p(X, \mathfrak{M}, \mu)$, $i \in \{1, \ldots, m\}$, then system (7.6.1) has a μ-essentially unique solution $\varphi: X \to \mathbb{R}^m$ such that $\varphi_i \in L^p(X, \mathfrak{M}, \mu)$ for $i \in \{1, \ldots, m\}$. Moreover,

φ is a limit of a μ-almost everywhere convergent sequence of successive approximations starting with an arbitrary member of $L^p(X, \mathfrak{M}, \mu)$.

We conclude the list of results on system (7.6.1) with the following observation. In the particular but important case where the given functions f_k coincide with iterates f_0^k of a function f_0, $k \in \{1, \ldots, n\}$, i.e. $f = (f_1, f_2, \ldots, f_n) = (f_0, f_0^2, \ldots, f_0^n)$, system (7.6.1) may be reduced to a system of equations of the first order via the substitutions

$$\psi_{pm+q} := \varphi_q \circ f_0^p, \qquad q \in \{1, \ldots, m\}, \, p \in \{0, 1, \ldots, n-1\}$$

(Kordylewski [3]). Then system (7.6.1) takes the form

$$\psi_i(x) = h_i(x, \psi_1(f_0(x)), \ldots, \psi_{mn}(f_0(x))) \qquad \text{for } i \in \{1, \ldots, m\},$$
$$\psi_i(x) = \psi_{i-m}(f_0(x)) \qquad \text{for } i \in \{m+1, \ldots, mn\}.$$

The latter system can be written as a single equation

$$\psi(x) = h^*(x, \psi(f_0(x)))$$

with $\psi = (\psi_1, \ldots, \psi_{mn})$ and $h^* = (h_1^*, \ldots, h_{mn}^*)$, where

$$h_i^*(x, y_1, \ldots, y_{mn}) := \begin{cases} h_i(x, y_1, \ldots, y_{mn}) & \text{for } i \in \{1, \ldots, m\}, \\ y_{i-m} & \text{for } i \in \{m+1, \ldots, mn\}, \end{cases}$$

for all $(x, y_1, \ldots, y_{mn})$ from the domain of $h = (h_1, \ldots, h_m)$. This allows one to apply directly the results of Chapter 5, for instance, and if need be to translate the appropriate theorems back into the language of the system we were starting with.

7.6F. Continuous solutions of g-systems

In the case where $f_k = f_1^k$ and $f_0 = f_1^{n+1}$, $k \in \{1, \ldots, n\}$, we may proceed quite analogously as in Subsection 7.6E, i.e., we reduce (7.6.2) to a single functional equation of the first order

$$\psi(f_1^{n+1}(x)) = g^*(x, \psi(x))$$

and then make use of the theory described in Chapter 5.

Passing to continuous solutions of the (general) system (7.6.2) we shall present here two theorems which yield slight modifications of some of Z. Kominek's results devoted in principle to C^r solutions (see Kominek [4]). The index k is now allowed to run over some sets not necessarily equal to $\{1, \ldots, n\}$.

Theorem 7.6.9. Let f_k, $k \in \{0, 1, \ldots, n\}$, be continuous and strictly increasing self-mappings of an interval $(0, a]$, $0 < a \leqslant \infty$, such that

$$0 < f_0(x) < f_n(x) \leqslant f_k(x) < x \qquad \text{for } x \in (0, a] \text{ and } k \in \{1, \ldots, n-1\}. \quad (7.6.12)$$

Suppose further that continuous functions $g_i: (0, a] \times \mathbb{R}^{m(n+1)} \to \mathbb{R}, i \in \{1, \ldots, m\},$

are given such that for every fixed $x \in (0, a|$ and $y_{j,k} \in \mathbb{R}$, $j \in \{1, \ldots, m\}$, $k \in \{1, \ldots, n\}$, the map

$$g(x, \cdot, y_{1,1}, \ldots, y_{1,n}, \ldots, \cdot, y_{m,1}, \ldots, y_{m,n}), \qquad (7.6.13)$$

where $g = (g_1, \ldots, g_m)$, is a homeomorphism of $\mathbb{R}^m$ onto itself.

Then, for an arbitrarily taken point $x_0 \in (0, a|$ and an arbitrary continuous function $\hat{\phi} = (\hat{\phi}_1, \ldots, \hat{\phi}_m): [f_0(x_0), x_0] \to \mathbb{R}^m$ such that

$$\hat{\phi}(f_0(x_0)) = g(x; \hat{\phi}_1(x_0), \hat{\phi}_1(f_1(x_0)), \ldots, \hat{\phi}_1(f_n(x_0)); \ldots;$$

$$\hat{\phi}_m(x_0), \hat{\phi}_m(f_1(x_0)), \ldots, \hat{\phi}_m(f_n(x_0))) \quad (7.6.14)$$

there exists exactly one continuous solution $\varphi = (\varphi_1, \ldots, \varphi_m): (0, a| \to \mathbb{R}^m$ of system (7.6.2) such that $\varphi|_{[f_0(x_0), x_0]} = \hat{\phi}$.

Requiring the solution obtained to be extendable onto the interval $[0, a|$ we have to impose much more restrictive assumptions on the given functions. Namely, we have the following.

Theorem 7.6.10. *Under the assumptions of the preceding theorem if, moreover,*

(a) *the function g admits a continuous extension onto the product $[0, a| \times \mathbb{R}^{m(n+1)}$,*

(b) $|g(x, y_{0,1}, \ldots, y_{m,n}) - g(x, \bar{y}_{0,1}, \ldots, \bar{y}_{m,n})| \leqslant \sum\limits_{j=1}^{m} \sum\limits_{k=0}^{n} a_{i,j,k} |y_{j,k} - \bar{y}_{j,k}|$

for all $x \in [0, a|$ and $y_{j,k}, \bar{y}_{j,k} \in \mathbb{R}$, $j \in \{1, \ldots, m\}$, $k \in \{0, \ldots, n\}$,

(c) *all the characteristic roots of the matrix*

$$\left(\sum_{k=0}^{n} a_{i,j,k} \right)_{i,j \in \{1,\ldots,m\}}$$

are less than unity in absolute value,

then for an arbitrarily taken $x_0 \in (0, a|$ and an arbitrary continuous function $\hat{\phi}: [f_0(x_0), x_0] \to \mathbb{R}^m$ satisfying (7.6.14) there exists exactly one continuous solution $\varphi: [0, a| \to \mathbb{R}^m$ of system (7.6.2) such that $\varphi|_{[f_0(x_0), x_0]} = \hat{\phi}$.

Related results can also be found in Choczewski [1], Dyjak [1], Krzeszowiak–Dybiec [1] and Kuczma [26, Ch. XII].

7.6G. Differentiable solutions of g-systems

Roughly speaking, on replacing in the hypotheses of Theorem 7.6.9 the adjective 'continuous' by the phrase 'of class C^r' we get a suitable result saying on the dependence of C^r solutions of the g-system (7.6.2) on an arbitrary C^r function. To present the precise statement we have to introduce new functions $g_{i,p}: [0, a| \times \mathbb{R}^{m(n+1)} \times \mathbb{R}^{pm(n+1)} \to \mathbb{R}$ determined by g_i and their derivatives in a way analogous to that applied while examining C^r solutions of system (7.6.1) (see formulae (7.6.7) and (7.6.8)).

We put

$$g_{i,0} := g_i, \qquad y_{j,k,0} := y_{j,k}$$

and

$$g_{i,p+1}(x, \{y_{j,k}\}, \{y_{j,k,1}\}, \dots, \{y_{j,k,p+1}\}) := \frac{1}{f_0'(x)}\left[\frac{\partial g_{i,p}}{\partial x}(x, \{y_{j,k}\}, \dots, \{y_{j,k,p}\})\right.$$
$$\left. + \sum_{j=1}^{m}\sum_{k=0}^{n}\sum_{q=0}^{p}\frac{\partial g_{i,p}}{\partial y_{j,k,q}}(x, \{y_{j,k}\}, \dots, \{y_{j,k,p}\})y_{j,k,q+1}f_k'(x)\right],$$

$i \in \{1, \dots, m\}$, $p \in \{0, \dots, r-1\}$ (compare Subsection 5.6C).
The following result is due to Z. Kominek [4].

Theorem 7.6.11. *Given an* $r \in \mathbb{N} \cup \{\infty\}$, *let* f_k, $k \in \{0, \dots, n\}$, *be strictly increasing* C^r *self-mappings of the interval* $[0, a|, 0 < a \leqslant \infty$, *such that relations* (7.6.12) *are satisfied and* f_0' *never vanishes on* $[0, a|$. *Suppose further that* C^r *functions* $g_i: [0, a| \times \mathbb{R}^{m(n+1)} \to \mathbb{R}, i \in \{1, \dots, m\}$, *are given such that for every fixed* $x \in [0, a|$ *and* $y_{j,k} \in \mathbb{R}, j \in \{1, \dots, m\}, k \in \{1, \dots, n\}$ *the map* (7.6.13), *where* $g = (g_1, \dots, g_m)$, *is a* C^r *diffeomorphism of* $\mathbb{R}^m$ *onto itself.*

Then, for an arbitrarily taken point $x_0 \in (0, a|$ *and an arbitrary* C^r *function* $\hat{\phi} = (\hat{\phi}_1, \dots, \hat{\phi}_m): [f_0(x_0), x_0] \to \mathbb{R}^m$ *such that* (7.6.14) *holds true and*

$$\hat{\phi}_i^{(p)}(f_0(x_0)) = g_{i,p}(x_0; \hat{\phi}_1(x_0), \dots, \hat{\phi}_m(f_n(x_0)); \hat{\phi}_1'(x_0), \dots, \hat{\phi}_m'(f_n(x_0)),$$
$$\dots, \hat{\phi}_1^{(p)}(x_0), \dots, \hat{\phi}_m^{(p)}(f_n(x_0))) \quad (7.6.15)$$

for $i \in \{1, \dots, m\}$ *and* $p \in \{1, \dots, r\}$, *there exists exactly one* C^r *solution* $\varphi = (\varphi_1, \dots, \varphi_m): (0, a| \to \mathbb{R}^m$ *of system* (7.6.2) *such that* $\varphi|_{[f_0(x_0), x_0]} = \hat{\phi}$.

To get the C^r solution of (7.6.2) on the interval $[0, a|$ (containing the fixed point of all the f_k) we are under the necessity of imposing further strong assumptions on the given functions (quite analogously to what we had to do while dealing with continuous solutions of system (7.6.2) on $[0, a|$; see Theorem 7.6.10).

Theorem 7.6.12. *Under the assumptions of Theorem 7.6.11 if, moreover,* $g(0, \{0\}) = 0$, *there exist constants* $s_k > 1$ *such that*

$$\frac{1}{f_0'(x)} \leqslant s_0 \quad \text{and} \quad \frac{1}{f_0'(x)}f_k'(x) \leqslant s_k \quad \text{for } x \in [0, a| \text{ and } k \in \{1, \dots, n\}$$

and all the characteristic roots of the matrix

$$\left(\sum_{k=0}^{n}\left|\frac{\partial g_i}{\partial y_{j,k}}(0, \{0\})\right|s_k^r\right)_{i,j \in \{1,\dots,m\}}$$

are less than unity in absolute value, then for an arbitrarily taken point $x_0 \in (0, a|$ *and an arbitrary function* $\hat{\phi} = (\hat{\phi}_1, \dots, \hat{\phi}_m): [f_0(x_0), x_0] \to \mathbb{R}^m$ *fulfilling conditions* (7.6.14) *and* (7.6.15), *there exists exactly one* C^r *solution* $\varphi: [0, a| \to \mathbb{R}^m$ *of system* (7.6.2) *such that* $\varphi|_{[f_0(x_0), x_0]} = \hat{\phi}$ *and* $\varphi(0) = 0$.

The most striking feature of the 12 theorems we have listed in the section we are just finishing is the requirement that the characteristic roots of a suitable $m \times m$ matrix (with positive entries $s_{i,j}$, $i, j \in \{1, \ldots, m\}$) be less than unity in absolute value. This does not touch Theorems 7.6.9 and 7.6.11 only, but the price we had to pay for it was the phenomenon of (possible) nonexistence of the solutions considered in the whole of the domain we are interested in. We recall (see Theorems 1.5.4 and 1.5.5) that, due to Matkowski's fixed-point theorem, such assumptions may be effectively verified in practice: putting

$$a^0_{i,j} := \begin{cases} s_{i,j} & \text{for } i \neq j \\ 1 - s_{i,i} & \text{for } i = j \end{cases}, \qquad i, j \in \{1, \ldots, m\},$$

and

$$a^{p+1}_{i,j} := \begin{cases} a^p_{1,1} a^p_{i+1,j+1} + a^p_{i+1,1} a^p_{1,j+1} & \text{for } i \neq j, \\ a^p_{1,1} a^p_{i+1,j+1} - a^p_{i+1,1} a^p_{1,j+1} & \text{for } i = j, \end{cases}$$

$i, j \in \{1, \ldots, m - p - 1\}$, $p \in \{0, \ldots, m - 2\}$, it suffices to check that $a^p_{i,i}$ are positive for $i \in \{1, \ldots, m - p\}$ and $p \in \{0, \ldots, m - 1\}$.

7.7 Three sample proofs

We are going to present here the detailed proofs of Theorems 7.6.1, 7.6.7, and 7.6.8 from the preceding section trusting that the methods exhibited are sufficiently representative to equip the reader with the tools necessary to prove the remaining theorems of Section 7.6, if he should consider it advisable.

Proof of Theorem 7.6.1. Suppose first that the space X is compact and consider the complete metric spaces $C(X, Y_i)$ endowed with the usual uniform convergence metrics d_i, $i \in \{1, \ldots, m\}$. Define the transformations T_i: $\times^m_{j=1} C(X, Y_j) \to \mathcal{F}(X, Y_i)$ by the formula

$$T_i(\varphi_1, \ldots, \varphi_m)(x) := h_i(x; \{\varphi_i(f_j(x))\}), \qquad x \in X, \ \varphi_j \in C(X, Y_j),$$

$i, j \in \{1, \ldots, m\}$. Relation (7.6.3) implies the continuity of $h = (h_1, \ldots, h_m)$ which implies that

$$T_i\left(\times^m_{j=1} C(X, Y_j) \right) \subset C(X, Y_i), \qquad i \in \{1, \ldots, m\}.$$

Moreover, (7.6.3) leads immediately to

$$d_i(T_i(\varphi_1, \ldots, \varphi_m), T_i(\psi_1, \ldots, \psi_m)) \leqslant \sum_{j=1}^m s_{i,j} d_j(\varphi_j, \psi_j),$$

where

$$s_{i,j} := \sum_{k=1}^n a_{i,j,k}$$

for all $\varphi_i, \psi_i \in C(X, Y_i)$, $i, j \in \{1, \ldots, m\}$. Now, assumption (7.6.4) jointly with

Theorem 1.5.5 implies that the transformation $T = (T_1, \ldots, T_m)$ has exactly one fixed point, say $\varphi = (\varphi_1, \ldots, \varphi_m)$, in the space $\times_{j=1}^{m} C(X, Y_j)$, which finishes the first part of the proof.

Passing to the general case, we see that for every $p \in \mathbb{N}$ there exists exactly one continuous solution $\hat{\phi}_p = (\hat{\phi}_{p,1}, \ldots, \hat{\phi}_{p,m})$: cl $U_p \to \times_{j=1}^{m} Y_j$ of system (7.6.1). To complete the proof, it suffices to put

$$\varphi(x) := \hat{\phi}_p(x) \qquad \text{for } x \in U_p, \, p \in \mathbb{N}. \quad \blacksquare$$

Proof of Theorem 7.6.7. A tedious although quite elementary induction proves the following (compare Sections 5.6 and 5.7):

(a) the functions $h_{i,p} : U \times \mathbb{C}^{mn} \times \mathbb{C}^{pmn} \to \mathbb{C}$ defined by recurrence for formulae (7.6.7) and (7.6.8) are analytic, $(i, p) \in \{1, \ldots, m\} \times \mathbb{N}_0$;
(b) for any analytic functions $\varphi_1, \ldots, \varphi_m \in \mathcal{F}(U, \mathbb{C})$ one has

$$\frac{d^p}{dx^p} h_i(x, \varphi_1(f_1(x)), \ldots, \varphi_m(f_n(x)))$$

$$= h_{i,p}(x, \varphi_1(f_1(x)), \ldots, \varphi_m(f_m(x)); \varphi_1'(f_1(x)), \ldots, \varphi_m'(f_n(x)); \ldots;$$
$$\varphi_1^{(p)}(f_1(x)), \ldots, \varphi_m^{(p)}(f_n(x)))$$

for all $x \in U$ and all $(i, p) \in \{1, \ldots, m\} \times \mathbb{N}$.

Take any solution $\{\eta_{i,p} : i = 1, \ldots, m, \, p = 1, \ldots, r\}$ of system (7.6.9) and put

$$P_i(x) := \sum_{q=1}^{r} \frac{1}{q!} \eta_{i,q} x^q, \qquad x \in \mathbb{C}, \, i \in \{1, \ldots, m\},$$

$$H_i(x, \{y_{j,k}\}) := \frac{1}{x^r} [h_i(x, \{P_j(f_k(x)) + f_k(x)^r y_{j,k}\}) - P_i(x)],$$

$(x, \{y_{j,k}\}) \in (U \setminus \{0\}) \times \mathbb{C}^{mn}$ for $i \in \{1, \ldots, m\}$. We shall show that the functions H_i are restrictions of analytic functions defined in $U \times \mathbb{C}^{mn}$, $i \in \{1, \ldots, m\}$. To this end, fix an $i \in \{1, \ldots, m\}$ and consider the analytic function

$$U \ni x \mapsto u_i(x) := h_i(x, \{P_j(f_k(x))\}) - P_i(x) \in \mathbb{C}.$$

On account of (a), (b) and the analyticity of P_j, $j \in \{1, \ldots, m\}$, we get the equalities

$$u_i^{(p)}(x) = h_{i,p}(x, \{P_j(f_k(x))\}, \ldots, \{P_j^{(p)}(f_k(x))\})$$

$$- \left(\eta_{i,p} + \sum_{q=p+1}^{r} \frac{1}{(q-p)!} \eta_{i,q} x^{q-p} \right), \qquad x \in U, \, p \in \{1, \ldots, r\},$$

whence, by putting $x = 0$ and making use of (7.6.9), we infer that $u_i^{(p)}(0) = 0$ for all $p \in \{1, \ldots, m\}$. Moreover, since $h_i(0, \{0\}) = 0$ and $f_k(0) = 0$, $k \in \{1, \ldots, n\}$, we have also $u_i(0) = 0$. Therefore,

$$u_i(x) = \alpha_i x^{r+1} + \cdots, \qquad x \in U,$$

whence

$$H_i(x, \{0\}) = \frac{1}{x^r} u_i(x) = \alpha_i x + \cdots, \qquad x \in U \setminus \{0\},$$

i.e. $H_i(\cdot, \{0\})$ is the restriction to $U \setminus \{0\}$ of a function analytic in U. Further, observe that, for any $(x, \{y_{j,k}\}) \in (U \setminus \{0\}) \times \mathbb{C}^{mn}$,

$$\frac{\partial H_i}{\partial y_{j,k}} (x, \{y_{j,k}\}) = \left(\frac{f_k(x)}{x}\right)^r \frac{\partial h_i}{\partial y_{j,k}} (x, \{P_j(f_k(x)) + f_k(x)^r y_{j,k}\})$$

$$= (f'_k(0) + c_k x + \cdots)^r \frac{\partial h_i}{\partial y_{j,k}} (x, \{P_j(f_k(x)) + f_k(x)^r y_{j,k}\}),$$

(7.7.1)

which states that these partial derivatives are also restrictions of functions analytic in the product $U \times \mathbb{C}^{mn}$; for brevity, we shall use the same symbols for suitable extensions.

Obviously, the analytic function

$$U \times \mathbb{C} \ni (x, y_{1,1}) \mapsto \int_0^{y_{1,1}} \frac{\partial H_i}{\partial w} (x, w, 0, \ldots, 0) dw + H_i(x, \{0\})$$

coincides with the function

$$(U \setminus \{0\}) \times \mathbb{C} \ni (x, y_{1,1}) \mapsto H_i(x, y_{1,1}, 0, \ldots, 0) \in \mathbb{C}$$

on the set $(U \setminus \{0\}) \times \mathbb{C}$. Repeating this procedure mn times we get the existence of an analytic extension of H_i onto the product $U \times \mathbb{C}^{mn}$; denote this extension by H_i again.

It is now easy to check that analytic functions $\varphi_1, \ldots, \varphi_m \colon U \to \mathbb{C}^m$ such that $\varphi_i(0) = 0$, $\varphi_i^{(p)}(0) = \eta_{i,p}$, $p \in \{1, \ldots, r\}$, $i \in \{1, \ldots, m\}$ satisfy system (7.6.2) if and only if the (unique) analytic functions $\Phi_i \colon U \to \mathbb{C}$ satisfying the relations

$$\varphi_i(x) = P_i(x) + x^r \Phi_i(x), \qquad x \in U, \, i \in \{1, \ldots, m\}$$

yield a solution of the system

$$\left.\begin{array}{l} \Phi_i(x) = H_i(x, \Phi_1(f_1(x)), \ldots, \Phi_m(f_n(x))), \\ \Phi_i^{(p)}(0) = 0, \end{array}\right\} \tag{7.7.2}$$

$i \in \{1, \ldots, m\}$, $p \in \{0, 1, \ldots, r\}$. Therefore to prove our theorem it suffices to show that system (7.7.2) has a solution $\Phi = (\Phi_1, \ldots, \Phi_m)$ where Φ_i are analytic in a neighbourhood of zero.

Recall that we have assumed that the matrix (7.6.6) has all the characteristic roots less than unity in absolute value which jointly with (7.7.1) allows one to find a polydisc

$$U(\alpha, \beta) = \{x \in \mathbb{C} \colon |x| \leqslant \alpha\} \times \{y \in \mathbb{C} \colon |y| \leqslant \beta\}^{mn},$$

$\alpha \in (0, \rho)$, $\beta > 0$, such that the numbers

$$a_{i,j,k}(\alpha, \beta) := \max_{(x, \{y_{j,k}\}) \in U(\alpha, \beta)} \left|\frac{\partial H_i}{\partial y_{j,k}} (x, \{y_{j,k}\})\right|$$

$i, j \in \{1, \ldots, m\}, k \in \{1, \ldots, n\}$, are small enough to guarantee that the matrix $(a_{i,j}(\alpha, \beta))_{i,j \in \{1,\ldots,m\}}$ with non-negative entries $a_{i,j}(\alpha, \beta) := \sum_{k=1}^{n} a_{i,j,k}(\alpha, \beta)$ has all its characteristic roots less than unity in absolute value, too.

According to Lemmas 1.5.3 and 1.5.1 there exist positive numbers $\rho_i \leqslant \beta$, $i \in \{1, \ldots, m\}$, such that the differences

$$\varepsilon_i := \rho_i - \sum_{j=1}^{m} a_{i,j}(\alpha, \beta)\rho_j \qquad (7.7.3)$$

are positive for all $i \in \{1, \ldots, m\}$. Take $\alpha_0 \in (0, \alpha)$ small enough that

$$|H_i(x, \{0\})| \leqslant \varepsilon_i \qquad \text{for all } x \in U_0 := \{z \in \mathbb{C}: |z| < \alpha_0\}. \qquad (7.7.4)$$

Lastly, note that the functions H_i fulfil the Lipschitz condition

$$|H_i(x, \{y_{j,k}\}) - H_i(x, \{\bar{y}_{j,k}\})| \leqslant \sum_{j=1}^{m} \sum_{k=1}^{n} a_{i,j,k}(\alpha, \beta)|y_{j,k} - \bar{y}_{j,k}| \qquad (7.7.5)$$

for all $(x, y_{j,k}), (x, \bar{y}_{j,k}) \in U(\alpha, \beta)$, $i, j \in \{1, \ldots, m\}$, $k \in \{1, \ldots, n\}$.

The rest is standard. Put $\mathscr{F}_i := \{\Phi_i: U_0 \to \mathbb{C}: \Phi_i \text{ is analytic and } |\Phi_i(x)| \leqslant \rho_i, x \in U_0\}$ and

$$d_i(\Phi_i, \Psi_i) := \sup_{x \in U_0} |\Phi_i(x) - \Psi_i(x)|, \qquad \Phi_i, \Psi_i \in \mathscr{F}_i,$$

so that the pair $(\mathscr{F}_i, d_i)$ becomes a complete metric space, $i \in \{1, \ldots, m\}$. Let $\mathscr{F} := \times_{i=1}^{m} \mathscr{F}_i$ and let $T_i: \mathscr{F} \to \mathscr{F}(U_0, \mathbb{C})$ be defined by

$$T_i(\Phi)(x) := H_i(x, \{\Phi_j(f_k(x))\}), \qquad x \in U_0,$$

for $\Phi = (\Phi_1, \ldots, \Phi_m) \in \mathscr{F}$ and $i \in \{1, \ldots, m\}$. Then

(c) $T_i(\Phi) \in \mathscr{F}_i$, $i \in \{1, \ldots, m\}$.

In fact, take an $x \in U_0$ and $i \in \{1, \ldots, m\}$ to get

$$|T_i(\Phi)(x)| = |H_i(x, \{\Phi_j(f_k(x))\})| \leqslant |H_i(x, \{\Phi_j(f_k(x))\}) - H_i(x, \{0\})| + |H_i(x, \{0\})|$$

$$\leqslant \sum_{j=1}^{m} \sum_{k=1}^{n} a_{i,j,k}(\alpha, \beta)|\Phi_j(f_k(x))| + \varepsilon_i$$

$$\leqslant \sum_{j=1}^{m} \sum_{k=1}^{n} a_{i,j,k}(\alpha, \beta)\rho_j + \varepsilon_i = \sum_{j=1}^{m} a_{i,j}(\alpha, \beta)\rho_j + \varepsilon_i = \rho_i$$

by virtue of (7.7.5), (7.7.4) and (7.7.3).

(d) $\qquad d_i(T_i(\Phi), T_i(\Psi)) \leqslant \sum_{j=1}^{m} a_{i,j}(\alpha, \beta)d_i(\Phi_i, \Psi_i), \ i \in \{1, \ldots, m\}$

for all $\Phi = (\Phi_1, \ldots, \Phi_m)$ and $\Psi = (\Psi_1, \ldots, \Psi_m)$ belonging to $\mathscr{F}$.

Indeed,

$$d_i(T_i(\Phi), T_i(\Psi)) = \sup_{x \in U_0} \left| H_i(x, \{\Phi_j(f_k(x))\}) - H_i(x, \{\Psi_j(f_k(x))\}) \right|$$

$$\leqslant \sup_{x \in U_0} \sum_{j=1}^{m} \sum_{k=1}^{n} a_{i,j,k}(\alpha, \beta) \left| \Phi_j(f_k(x)) - \Psi_j(f_k(x)) \right|$$

$$\leqslant \sum_{j=1}^{m} \sum_{k=1}^{n} a_{i,j,k}(\alpha, \beta) \sup_{x \in U_0} \left| \Phi_j(x) - \Psi_j(x) \right|$$

$$= \sum_{j=1}^{m} a_{i,j}(\alpha, \beta) d_j(\Phi_j, \Psi_j).$$

To finish the proof it remains to apply Theorem 1.5.5. ∎

Proof of Theorem 7.6.8. Fix a $p > 0$ and consider the space $L^p(X, \mathfrak{M}, \mu)$ as a linear space endowed with the paranorm

$$\|\varphi\| := \left(\int_X |\varphi|^p \, d\mu \right)^{\alpha(p)}, \qquad \varphi \in L^p(X, \mathfrak{M}, \mu).$$

In particular, we have the Minkowski inequality (see Note 5.9.16)

$$\|\varphi + \psi\| \leqslant \|\varphi\| + \|\psi\|, \qquad \varphi, \psi \in L^p(X, \mathfrak{M}, \mu), \tag{7.7.6}$$

and the convergence in the sense of $\|\cdot\|$ coincides with the convergence in measure (see Section 4.7).

For $\varphi = (\varphi_1, \ldots, \varphi_m) \in \Omega := L^p(X, \mathfrak{M}, \mu)^m$ and $i \in \{1, \ldots, m\}$ define the transforms $T_i(\varphi)$ by the formula

$$T_i(\varphi)(x) := h_i(x, \{\varphi_i(f_k(x))\}), \qquad x \in X.$$

We shall show that T_i maps Ω into $L^p(X, \mathfrak{M}, \mu)$, $i \in \{1, \ldots, m\}$. To this end fix an $x \in X$ and a $\varphi \in \Omega$; then on account of the sup-measurability of h_i the map $T_i(\varphi)$ is measurable in turn, $i \in \{1, \ldots, n\}$, and with the use of the inequality $(a_1 + \cdots + a_s)^p/s^p \leqslant a_1^p + \cdots + a_s^p$, valid for all $a_1, \ldots, a_s \geqslant 0$, $s \in \mathbb{N}$, $p > 0$, we get

$$|T_i(\varphi)(x)|^p \leqslant \left(\sum_{j=1}^{m} \sum_{k=1}^{n} g_{i,j,k}(x) |\phi_j(f_k(x))| + |h_i(x, \{0\})| \right)^p$$

$$\leqslant (mn+1)^p \left(\sum_{j=1}^{m} \sum_{k=1}^{n} g_{i,j,k}(x)^p |\varphi_j(f_k(x))|^p + |h_i(x, \{0\})|^p \right)$$

$$\leqslant (mn+1)^p \left(\sum_{j=1}^{m} \sum_{k=1}^{n} s_{i,j,k}^{1/\alpha(p)} |\varphi_j(f_k(x))|^p g_k(x) + |h_i(x, \{0\})|^p \right)$$

whence

$$\int_X |T_i(\varphi)|^p d\mu \leqslant (mn+1)^p \left(\sum_{j=1}^m \sum_{k=1}^n s_{i,j,k}^{1/\alpha(p)} \int_X |\varphi_j \circ f_k|^p g_k d\mu + \int_X |h_i(\cdot, \{0\})|^p d\mu \right)$$

$$= (mn+1)^p \left(\sum_{j=1}^m \sum_{k=1}^n s_{i,j,k}^{1/\alpha(p)} \int_{f_k(X)} |\varphi_j|^p d\mu + \int_X |h_i(\cdot, \{0\})|^p d\mu \right)$$

$$< \infty$$

because we have assumed that $h_i(\cdot, \{0\})$ is in $L^p(X, \mathfrak{M}, \mu)$ and in view of the obvious inequality

$$\int_{f_k(X)} |\varphi_j|^p d\mu \leqslant \|\varphi_j\|^{1/\alpha(p)}, \qquad i,j \in \{1, \ldots, m\}, \, k \in \{1, \ldots, n\}.$$

It yet remains to note that for equivalent maps $\varphi, \psi \in \Omega$ the maps $T_i(\varphi)$ and $T_i(\psi)$ are equivalent, too. Indeed, for any $j \in \{1, \ldots, m\}$ and $k \in \{1, \ldots, n\}$, one has

$$\{x \in X : \varphi_j(f_k(x)) = \psi_j(f_k(x))\} = f_k^{-1}(\{x \in X : \varphi_j(x) \neq \psi_j(x)\}) \in \mathfrak{M}$$

because $E^j := \{x \in X : \varphi_j(x) \neq \psi_j(x)\} \in \mathfrak{M}$ and f_k is $\mathfrak{M}$-measurable; moreover, $\mu(E_j) = 0$ whence $\mu(f_k^{-1}(E_j)) = (\mu f_k^{-1})(E_j) = 0$ by means of the absolute continuity of the measure μf_k^{-1} with respect to μ.

On the other hand, Minkowski's inequality (7.7.6) implies that for any $\varphi, \psi \in \Omega$ and $i \in \{1, \ldots, m\}$

$$\|T_i(\varphi) - T_i(\psi)\| = \left(\int_X |h_i(x, \{\varphi_j(f_k(x))\}) - h_i(x, \{\psi_j(f_k(x))\})|^p d\mu \right)^{1/\alpha(p)}$$

$$\leqslant \left(\int_X \left(\sum_{j=1}^m \sum_{k=1}^n g_{i,j,k} |\varphi_j \circ f_k - \psi_j \circ f_k| \right)^p d\mu \right)^{\alpha(p)}$$

$$\leqslant \sum_{j=1}^m \sum_{k=1}^n \left(\int_X (g_{i,j,k} |\varphi_j \circ f_k - \psi_j \circ f_k|)^p d\mu \right)^{\alpha(p)}$$

$$\leqslant \sum_{j=1}^m \sum_{k=1}^n \left(\int_X s_{i,j,k}^{1/\alpha(p)} |\varphi_j \circ f_k - \psi_j \circ f_k|^p g_k d\mu \right)^{\alpha(p)}$$

$$= \sum_{j=1}^m \sum_{k=1}^n \left(\int_{f_k(X)} s_{i,j,k}^{1/\alpha(p)} |\varphi_j - \psi_j|^p d\mu \right)^{\alpha(p)}$$

$$\leqslant \sum_{j=1}^m \sum_{k=1}^n s_{i,j,k} \|\varphi_j - \psi_j\|.$$

Thus the first assertion of the theorem results now directly from Theorem 1.5.5. To prove the other take any map $\bar{\varphi}_0 = (\varphi_{0,1}, \ldots, \varphi_{0,m}) \in \Omega$ and define $\bar{\varphi}_{s+1} = (\varphi_{s+1,1}, \ldots, \varphi_{s+1,m})$ as follows:

$$\varphi_{s+1,i}(x) := h_i(x, \{\varphi_{s,j}(f_k(x))\}), \qquad x \in X, \, i \in \{1, \ldots, m\}, \, s \in \mathbb{N}.$$

By Lemmas 1.5.3 and 1.5.1 we derive the existence of positive numbers $r_1, \ldots, r_m$ and an $\alpha \in (0, 1)$ such that

$$\sum_{j=1}^{m} s_{i,j} r_j \leq \alpha r_i \tag{7.7.7}$$

and

$$\|\varphi_{1,i} - \varphi_{0,i}\| \leq r_i \tag{7.7.8}$$

where $s_{i,j} := \sum_{k=1}^{n} s_{i,j,k}$, $i, j \in \{1, \ldots, m\}$. By induction,

$$\|\varphi_{s+1,i} - \psi_{s,i}\| \leq \alpha^s r_i, \qquad s \in \mathbb{N}, \ i \in \{1, \ldots, m\}. \tag{7.7.9}$$

Indeed, for $s = 0$ this is just (7.7.8). Assuming (7.7.9) to hold for some $s \in \mathbb{N}_0$ and $i \in \{1, \ldots, m\}$, and proceeding along the same lines as during the verification of the contractiveness of T_i we get

$$\|\varphi_{s+2,i} - \varphi_{s+1,i}\| \leq \sum_{j=1}^{m} \sum_{k=1}^{n} s_{i,j,k} \|\varphi_{s+1,i} - \varphi_{s,i}\|$$

$$\leq \sum_{j=1}^{m} s_{i,j} \alpha^s r_i \leq \alpha^{s+1} r_i$$

in view of (7.7.7), which finishes the inductive proof of (7.7.9). Therefore, $\sum_{s \in \mathbb{N}_0} \|\varphi_{s+1,i} - \varphi_{s,i}\| < \infty$ for all $i \in \{1, \ldots, m\}$ and an appeal to Lemma 4.8.1 shows that the series $\sum_{s \in \mathbb{N}_0} (\varphi_{s+1,i} - \varphi_{s,i})$ is μ-almost everywhere convergent to a function $\psi_i \in L^p(X, \mathfrak{M}, \mu)$, whence

$$\varphi_{s,i} \xrightarrow[s \to \infty]{} \varphi_i := \varphi_{0,i} + \psi_i \in L^p(X, \mathfrak{M}, \mu) \qquad \mu\text{-almost everywhere,}$$

for all $i \in \{1, \ldots, m\}$, say,

$$\varphi_{s,i}(x) \xrightarrow[s \to \infty]{} \varphi_i(x) \qquad \text{for all } x \in X \setminus E, \ i \in \{1, \ldots, m\}$$

and $\mu(E) = 0$. Then also $F := \bigcup_{k=1}^{n} f_k^{-1}(E)$ is a null-set and for every $x \in X \setminus F$ we have

$$\varphi_{s,j}(f_k(x)) \xrightarrow[s \to \infty]{} \varphi_j(f_k(x)),$$

$j \in \{1, \ldots, m\}$, $k \in \{1, \ldots, n\}$. The continuity of h_i with respect to the last mn variables and the definition of the sequences $(\varphi_{s,i})_{s \in \mathbb{N}_0}$ lead now to the equalities

$$\varphi_i(x) = h_i(x, \{\varphi_j(f_k(x))\})$$

for all $x \in X \setminus F$ and $i \in \{1, \ldots, m\}$ which finishes the proof by virtue of the fact that $\mu(F) = 0$. ∎

7.8 Continuous solutions – deeper uniqueness conditions

A system of functional equations, say the h-system (7.6.1), may be reduced to a single equation of higher order,

$$\varphi(x) = h(x, \varphi(f_1(x)), \ldots, \varphi(f_n(x))), \tag{7.8.1}$$

which in turn is a very particular case of the equation

$$\varphi(x) = h(x, \varphi \circ f(\cdot, x)).\qquad (7.8.2)$$

Criteria for the existence and uniqueness of, in principle, continuous solutions of (7.8.2) are given in Section 7.2. However, Theorem 7.2.2, although pretty general and subtle enough (see Example 7.2.1), remains useless, for instance, for the equation (Baron–Sablik [1])

$$\varphi(x) = \frac{1}{n} \sqrt[3]{(1+x)} \cdot \sum_{k=1}^{n} \sin \varphi(f_n(x)),\qquad (7.8.3)$$

where all the f_k are continuous self-mappings of $\mathbb{R}^+$ which satisfy $0 < f_k(x) \leqslant x/(1+x)$, $x \in (0, \infty)$, $k \in \{1, \ldots, n\}$. The existence question is trivial: $\varphi = 0$ is a solution. Trying to apply Theorem 7.2.2 we realize that for the function

$$h(x, y_1, \ldots, y_n) = \frac{1}{n} \sqrt[3]{(1+x)} \cdot \sum_{k=1}^{n} \sin y_k, \qquad (x, y_1, \ldots, y_n) \in \mathbb{R}^+ \times \mathbb{R}^n$$

$$(7.8.4)$$

condition (7.2.12) fails to hold. Does equation (7.8.3) possess nontrivial continuous solutions? Theorem 7.6.1 does not work. The point is that the factors

$$a_k(x) := \frac{1}{n} \sqrt[3]{(1+x)}, \qquad x \in \mathbb{R}^+, k \in \{1, \ldots, n\},$$

exclude any kind of contraction or even just nonexpansion, as we have

$$\sum_{k=1}^{n} a_k(x) = \sqrt[3]{(1+x)} > 1 \qquad \text{for all } x \in (0, \infty).$$

The uniqueness type results of the present section will prove to be helpful in particular in the case of equation (7.8.3). It turns out that deep uniqueness tests are obtainable at the level of generality we dealt with in Sections 7.1–7.5. Below we report on K. Baron's results (Baron [9], also Baron–Sablik [1] and Baron [5]).

7.8A. A crucial inequality

When dealing with the problem of the uniqueness of solutions one usually proceeds as follows. Suppose that we are given a set X, a metric space (Y, σ), and that $\varphi, \psi: X \to Y$ are two solutions of equation (7.8.2) with the function $h: X \times \mathscr{F}(S, Y) \to Y$ fulfilling the Lipschitz-type condition

$$\sigma(h(x, y), h(x, z)) \leqslant \alpha(x)\beta(x, \sigma(y, z)), \qquad x \in X, y \in \mathscr{F}(S, Y), \quad (7.8.5)$$

with some functions α and β. Write

$$\lambda(x) := \sigma(\varphi(x), \psi(x)), \qquad x \in X.$$

Then

$$\lambda(x) = \sigma(h(x, \varphi \circ f(\cdot, x)), h(x, \psi \circ f(\cdot, x)))$$
$$\leqslant \alpha(x)\beta(x, \sigma(\varphi \circ f(\cdot, x), \psi \circ f(\cdot, x)))$$
$$= \alpha(x)\beta(x, \lambda \circ f(\cdot, x))$$

for all $x \in X$, which shows that in case (7.8.5) the problem of the uniqueness of solutions of equation (7.8.2) transforms itself to the question whether $\lambda = 0$ is the only nonnegative solution of inequality

$$\lambda(x) \leqslant \alpha(x)\beta(x, \lambda \circ f(\cdot, x)), \qquad x \in X. \tag{7.8.6}$$

Before stating the first result let us check what stands behind (7.8.6) in the case of our introductory testing functional equation (7.8.3). We have here $S = \{1, \ldots, n\}$, $X = \mathbb{R}^+$ and $Y = \mathbb{R}$ (with the usual topologies), $\mathscr{F}(S, Y) = \mathbb{R}^n$ and, by (7.8.4),

$$\left| h(x, y) - h(x, z) \right| = \frac{1}{n} \sqrt[3]{(1 + x)} \cdot \left| \sum_{k=1}^{n} (\sin y_k - \sin z_k) \right|$$

$$\leqslant \sqrt[3]{(1 + x)} \cdot \sum_{k=1}^{n} \frac{1}{n} \left| \sin y_k - \sin z_k \right|$$

for $x \in \mathbb{R}^+$ and $y = (y_1, \ldots, y_n)$, $z = (z_1, \ldots, z_n) \in \mathbb{R}^n$; therefore we get for $x \geqslant 0$, $y, z \in \mathbb{R}^n$,

$$\left| h(x, y) - h(x, z) \right| \leqslant \sqrt[3]{(1 + x)} \cdot \sum_{k=1}^{n} \frac{2}{n} \left| \sin \frac{|y_k - z_k|}{2} \right|, \tag{7.8.7}$$

i.e. (7.8.5) is satisfied with

$$\alpha(x) := \sqrt[3]{(1 + x)} \qquad \text{and} \qquad \beta(x, y) := \sum_{k=1}^{n} \frac{2}{n} \left| \sin \frac{y_k}{2} \right|, \tag{7.8.8}$$

for $x \in \mathbb{R}^+$, $y = (y_1, \ldots, y_n) \in \mathbb{R}^n$.

Consequently, inequality (7.8.6) becomes

$$\lambda(x) \leqslant \sqrt[3]{(1 + x)} \cdot \sum_{k=1}^{n} \frac{2}{n} \left| \sin \tfrac{1}{2}\lambda(f_k(x)) \right|, \qquad x \in \mathbb{R}^+. \tag{7.8.9}$$

The following general result describes the behaviour of solutions of the crucial inequality (7.8.6).

Theorem 7.8.1. *Let $(X, \mathscr{T})$ be a topological space, $\xi \in X$, and let S be a given nonempty set. Suppose that a function $f: S \times X \to X$ is given such that for every $x \in X$ and for every neighbourhood V of ξ there exists a subneighbourhood U of ξ and a $p \in \mathbb{N}$ for which*

$$f(S \times U) \subset U \subset V \tag{7.8.10}$$

and

$$f(s_1, \cdot) \circ \cdots \circ f(s_p, \cdot)(x) \in U \qquad \text{for all } s_1, \ldots, s_p \in S. \tag{7.8.11}$$

Further, let two functions $\alpha \in \mathscr{F}(X, \mathbb{R}^+)$ *and* $\beta \colon X \times \mathscr{F}(S, \mathbb{R}^+) \to \bar{\mathbb{R}}$ *be given and have the following properties:*

(1) *there exists a self-mapping g of X such that*

$$\alpha \circ g^p \circ f(\cdot, x) \leqslant \alpha(g^{p+1}(x))$$

for all $x \in X$ and $p \in \mathbb{N}_0$,

(2) *for every $\delta \in \bar{\mathbb{R}}$ and every pair $(x, u) \in X \times \mathscr{F}(S, \mathbb{R}^+)$ such that $u \leqslant \delta$ one has $\beta(x, u) \leqslant \gamma(\delta)$, where $\gamma \colon \bar{\mathbb{R}} \to \bar{\mathbb{R}}$ is such that*

$$\alpha \neq 1 \ \text{ implies } \ \gamma \geqslant 0$$

and

$$\gamma(\alpha(x)\delta) \leqslant \alpha(x)\gamma(\delta), \qquad x \in X, \ \delta \in \bar{\mathbb{R}}.$$

Given an $r \in \mathbb{R}$, if for every $x \in X \setminus \{\xi\}$ and $\varepsilon > 0$ there exists a $\delta \in (r, \infty)$ such that for every $p \in \mathbb{N}$ there exists a $q \in \mathbb{N} \cap [p, \infty)$ for which

$$\gamma^q(\delta) \sum_{i=0}^{q-1} \alpha(g^i(x)) \leqslant r + \varepsilon, \tag{7.8.12}$$

then every solution λ of inequality (7.8.6), continuous at ξ, and such that $\lambda(\xi) \leqslant r$, satisfies the global estimate $\lambda(x) \leqslant r$ for $x \in X$.

7.8B. Result for the testing equation

Before proceeding with the proof we shall show, with the aid of Theorem 7.8.1, that the zero function is the only nonnegative solution of inequality (7.8.9) continuous at $\xi = 0$, which proves that the only solution of equation (7.8.3) continuous at zero is just $\varphi = 0$. For note that in the light of Baron's extension Theorem 7.1.1 it suffices to prove that there exists a $b > 0$ such that $\varphi(x) = 0$, $x \in [0, b)$, is the only solution of (7.8.3) in $[0, b)$ continuous at zero. To this end, take $b := (\pi/2)^3 - 1$ and put

$$g(x) := \frac{x}{1+x}, \quad x \in [0, b) \qquad \text{and} \qquad \gamma(\delta) := \begin{cases} 0 & \text{for } \delta < 0, \\ 2 \sin \dfrac{\delta}{2} & \text{for } \delta \in [0, \pi], \\ 2 & \text{for } \delta > \pi. \end{cases} \tag{7.8.13}$$

Relations (7.8.10) and (7.8.11) are evidently satisfied with $f(k, x) := f_k(x)$, $x \in [0, b)$, $k \in \{1, \dots, n\}$, since $f_k(U) \subset U$ for all $k \in \{1, \dots, n\}$ and all neighbourhoods U of 0 in $[0, b)$ and $f_k^p(x) \to 0$ as $p \to \infty$, $x \in [0, b)$, $k \in \{1, \dots, n\}$ (see Theorem 1.2.1). Relation (1) assumes the form

$$\sqrt[3]{\left(1 + \frac{f_k(x)}{1 + p f_k(x)}\right)} \leqslant \sqrt[3]{\left(1 + \frac{x}{1 + (p+1)x}\right)}, \qquad x \in [0, b),$$

equivalent to

$$f_k(x) \leqslant \frac{x}{1+x}, \qquad x \in [0, b),$$

the latter being assumed for $k \in \{1, \ldots, n\}$. To check (2), take a $\delta \in \mathbb{R}$, an $x \in [0, b)$ and $y = (y_1, \ldots, y_n) \in \mathbb{R}^n$ such that $0 \leqslant y_k \leqslant \delta$ for all $k \in \{1, \ldots, n\}$; then

$$2 \left| \sin \frac{y_k}{2} \right| \leqslant \gamma(\delta) \qquad \text{for } k \in \{1, \ldots, n\}.$$

Indeed, the estimates are trivial for $\delta > \pi$ and for $\delta \in [0, \pi]$ they result from the fact that $\sin|_{[0,\pi/2]}$ is a nonnegative and increasing function. Consequently (see (7.8.8)),

$$\beta(x, y) = \sum_{k=1}^{n} \frac{2}{n} \left| \sin \frac{y_k}{2} \right| \leqslant \sum_{k=1}^{n} \frac{1}{n} \gamma(\delta) = \gamma(\delta),$$

as desired. Then, we have to derive the inequality

$$\gamma(\sqrt[3]{(1 + x)} \cdot \delta) \leqslant \sqrt[3]{(1 + x)} \cdot \gamma(\delta) \qquad \text{for } x \in [0, b) \text{ and } \delta \in \mathbb{R},$$

which, by means of the definition of γ, is self-evident for $\delta \in \mathbb{R} \setminus [0, \pi]$. For $\delta \in [0, \pi]$ we shall consider two subcases.

(a) $\sqrt[3]{(1 + x)} \cdot \delta \leqslant \pi$; then

$$\gamma(\sqrt[3]{(1 + x)} \cdot \delta) = 2 \sin(\tfrac{1}{2}\sqrt[3]{(1 + x)} \cdot \delta) = \sqrt[3]{(1 + x)} \cdot \delta \, \frac{\sin(\tfrac{1}{2}\sqrt[3]{(1 + x)} \cdot \delta)}{\tfrac{1}{2}\sqrt[3]{(1 + x)} \cdot \delta}$$

$$\leqslant \sqrt[3]{(1 + x)} \cdot \delta \, \frac{\sin \tfrac{1}{2}\delta}{\tfrac{1}{2}\delta} = \sqrt[3]{(1 + x)} \cdot \gamma(\delta)$$

since the function $(0, \pi/2) \ni t \mapsto \sin t / t$ is decreasing.

(b) $\sqrt[3]{(1 + x)} \cdot \delta > \pi$; then, because of $x \in [0, b) = [0, (\pi/2)^3 - 1)$ we infer that $\delta \in (2, \pi]$ whereupon

$$\sqrt[3]{(1 + x)} \cdot \gamma(\delta) = 2 \cdot \sqrt[3]{(1 + x)} \cdot \sin \tfrac{1}{2}\delta \geqslant \frac{2}{\delta} \pi \sin \tfrac{1}{2}\delta = \pi \, \frac{\sin \tfrac{1}{2}\delta}{\tfrac{1}{2}\delta} \geqslant \pi \, \frac{\sin \tfrac{1}{2}\pi}{\tfrac{1}{2}\pi} = 2$$

$$= \gamma(\sqrt[3]{(1 + x)} \cdot \delta)$$

which was to be shown.

To apply Theorem 7.8.1 with $r = 0$, it remains to check that for a fixed $x \in (0, b)$ and an $\varepsilon > 0$ there exists a $\delta > 0$ such that for every $p \in \mathbb{N}$ the inequality (7.8.12) holds for some $q \geqslant p$. For note that

$$\sum_{i=0}^{q-1} \alpha(g^i(x)) = \sum_{i=0}^{q-1} \sqrt[3]{\frac{1 + (i + 1)x}{1 + ix}} = \sqrt[3]{(1 + qx)}$$

and taking $\delta = 1$ it suffices to show that

$$\gamma^q(1) \sqrt[3]{(1 + qx)} \leqslant \varepsilon$$

for almost all $q \in \mathbb{N}$; the latter will certainly be satisfied provided that the sequence $(\gamma^q(1))_{q \in \mathbb{N}}$ behaves like $O(1/\sqrt{q})$ as q tends to infinity. This is actually the case; to become convinced let us appeal to Thron's Theorem 1.3.5. In fact, we have

$$2 \sin \tfrac{1}{2}\delta = \delta[1 + p(\delta)] \qquad \text{where } p(\delta) = -\frac{\delta^2}{24} + \frac{\delta^4}{384} - \cdots$$

whence $\lim_{\delta \to 0} p(\delta) = 0$, $1 + p(\delta) = \sin \tfrac{1}{2}\delta / \tfrac{1}{2}\delta \in (0, 1)$ for $\delta \in (0, \pi)$ and $c := \lim_{\delta \to 0}(1/\delta^2)p(\delta) = -\tfrac{1}{24} < 0$. Therefore, $\lim_{q \to 0}(\sqrt{q} \cdot \gamma^q(1)) = 2\sqrt{3}$; in particular $\gamma^q(1) = O(1/\sqrt{q})$, $q \to \infty$, as claimed.

7.8C. Proof of Theorem 7.8.1

Suppose that $\lambda: X \to \mathbb{R}$ is a solution of inequality (7.8.6), λ is continuous at ξ and $\lambda(\xi) \leqslant r$. Fix arbitrarily an $\varepsilon > 0$ and an $x \in X \setminus \{\xi\}$ and take a $\delta \in (r, \infty)$ such that for every $p \in \mathbb{N}$ there exists a positive integer $q \geqslant p$ for which (7.8.12) holds true. Let V be a neighbourhood of ξ such that

$$\lambda(z) \leqslant \delta \qquad \text{for } z \in V. \tag{7.8.14}$$

Choose a neighbourhood U of ξ and $p \in \mathbb{N}$ to satisfy (7.8.10) and (7.8.11) and fix a $q \in \mathbb{N}$, $q \geqslant p$, such that

$$\gamma^q(\delta) \sum_{i=0}^{q-1} \alpha(g^i(x)) \leqslant r + \varepsilon. \tag{7.8.15}$$

Putting $U_0 := U$ and $U_{n+1} := \bigcap_{s \in S} f(s, \cdot)^{-1}(U_n)$, $n \in \mathbb{N}$, we have (7.3.5) whence $x \in U_p \subset U_q$. In view of (7.8.15) and the unrestricted choice of ε and x, to prove that $\lambda \leqslant r$ it remains to show that

$$\lambda(z) \leqslant \gamma^n(\delta) \sum_{i=0}^{n-1} \alpha(g^i(z)) \qquad \text{for } z \in U_n, \tag{7.8.16}$$

for all $n \in \mathbb{N}$. For observe that $\lambda(t) \leqslant \delta$ for $t \in U_0 = U \subset V$ by virtue of (7.8.14). So, taking a $z \in U_1$ we have $f(s, z) \in U_0$ whence $\lambda(f(s, z)) \leqslant \delta$ for $s \in S$ and, consequently, $\beta(z, \lambda \circ f(\cdot, z)) \leqslant \gamma(\delta)$ by means of (2); this jointly with (7.8.6) implies

$$\lambda(z) \leqslant \alpha(z)\beta(z, \lambda \circ f(\cdot, z)) \leqslant \gamma(\delta)\alpha(z),$$

which coincides with (7.8.16) for $n = 1$. Assume (7.8.16) to hold for an $n \in \mathbb{N}$ and take a $z \in U_{n+1}$; then $f(s, z) \in U_n$, $s \in S$, whence

$$\lambda(f(s, z)) \leqslant \gamma^n(\delta) \sum_{i=0}^{n-1} \alpha(g^i(f(s, z))) \leqslant \gamma^n(\delta) \sum_{i=0}^{n-1} \alpha(g^{i+1}(z)) = \gamma^n(\delta) \sum_{i=1}^{n} \alpha(g^i(z)),$$

on account of (1). Therefore, by a repeated use of (2), we get

$$\lambda(z) \leqslant \alpha(z)\beta(z, \lambda \circ f(\cdot, z)) \leqslant \alpha(z)\gamma\left(\gamma^n(\delta) \sum_{i=1}^{n} \alpha(g^i(z))\right)$$

$$\leqslant \alpha(z)\gamma^{n+1}(\delta) \sum_{i=1}^{n} \alpha(g^i(z)) = \gamma^{n+1}(\delta) \sum_{i=0}^{n} \alpha(g^i(z)),$$

which yields (7.8.16) for n replaced by $n + 1$. Induction completes the proof. ∎

7.8D. Uniqueness implies existence

By Theorem 7.8.1 the equation (see Section 7.4)

$$\varphi(x) = \varphi\left(\frac{x}{x+1}\right) - x \qquad (7.8.17)$$

has at most one solution $\varphi: [0, 1] \to \mathbb{R}$ continuous at zero. But we have seen that (7.8.17) admits no such solutions. Consequently in this case uniqueness does not imply existence. But surprisingly enough, it turns out that uniqueness type theorems may sometimes be used in order to obtain the existence of solutions. We present a result to this effect.

Theorem 7.8.2. Let $(X, \mathcal{T})$ be a topological space, S a nonempty set, $\xi \in X$, and let $f: S \times X \to X$ be such that ξ is a fixed point of all transformations $f_s := f(s, \cdot)$, $s \in S$. Let $\Phi \subset C(X, \mathbb{R})$ be a class of functions closed with respect to uniform convergence, $a, b \in \Phi$, $a \leq b$, and $\Phi \subset \{\varphi \in C(X, \mathbb{R}): a \leq \varphi \leq b\}$. Suppose that we are given a function $h: X \times \mathcal{F}(S, \mathbb{R}) \to \mathbb{R}$ such that

(a) for every monotonic (hence convergent) sequence $(\varphi_n)_{n \in \mathbb{N}}$ of elements of Φ one has

$$h\left(\lim_{n \to \infty} \varphi_n \circ f(\cdot, x)\right) = \lim_{n \to \infty} h(x, \varphi_n \circ f(\cdot, x)), \qquad x \in X,$$

(b) for any two functions $y_1, y_2 \in \mathcal{F}(S, \mathbb{R})$ and any $x \in X$ the inequalities $a \circ f(\cdot, x) \leq y_1 \leq y_2 \leq b \circ f(\cdot, x)$ imply that $h(x, y_1) \leq h(x, y_2)$,
(c) for any $\varphi \in \Phi$ the map $X \ni x \mapsto h(x, \varphi \circ f(\cdot, x)) \in \mathbb{R}$ is in Φ.

If the functional equation

$$\varphi(x) = h(x, \varphi \circ f(\cdot, x)) \qquad (7.8.18)$$

has at most one solution φ continuous at ξ such that

$$a \leq \varphi \leq b \qquad (7.8.19)$$

and $\varphi(\xi) = \eta$ where η is the only solution of equation $h(\xi, \eta) = \eta$ in the interval $[a(\xi), b(\xi)]$ then equation (7.8.18) has exactly one solution $\varphi: X \to \mathbb{R}$ such that estimates (7.8.19) hold true. This solution belongs to Φ.

Proof. Write

$$\varphi_0 := a, \qquad \varphi_{n+1}(x) := h(x, \varphi_n \circ f(\cdot, x))$$

and

$$\psi_0 := b, \qquad \psi_{n+1}(x) := h(x, \psi_n \circ f(\cdot, x))$$

for $x \in X$ and $n \in \mathbb{N}_0$. Assumption (c) ensures that both the families $\{\varphi_n: n \in \mathbb{N}_0\}$ and $\{\psi_n: n \in \mathbb{N}_0\}$ are contained in Φ, whereas (b) forces the sequence $(\varphi_n)_{n \in \mathbb{N}_0}$, resp. $(\psi_n)_{n \in \mathbb{N}_0}$, to be increasing, resp. decreasing. Moreover,

$$a \leq \varphi_n \leq \psi_n \leq b \qquad \text{for all } n \in \mathbb{N}_0.$$

Put $\varphi := \lim \varphi_n$ and $\psi := \lim \psi_n$. Plainly, we have

$$a \leq \varphi \leq \psi \leq b \tag{7.8.20}$$

as well as

$$\varphi(x) = h(x, \varphi \circ f(\cdot, x)), \qquad \psi(x) = h(x, \psi \circ f(\cdot, x)), \qquad x \in X,$$

the latter two equalities resulting from (a). In particular, the (assumed) uniqueness of solutions of the 'numeric' equation $h(\xi, \eta) = \eta$ in the interval $[a(\xi), b(\xi)]$, implies that

$$\varphi(\xi) = \eta = \psi(\xi). \tag{7.8.21}$$

The lower semicontinuity of φ and the upper semicontinuity of ψ jointly with (7.8.20) and (7.8.21) imply that both φ and ψ are continuous at ξ and, since equation (7.8.18) has at most one solution lying between a and b and continuous at ξ, we infer that

$$\varphi = \psi$$

whence, in particular, φ is continuous. It remains to show that $\varphi \in \Phi$. To this end, take any compact set $K \subset X$; then the sequence $(\varphi_n|_K)_{n \in \mathbb{N}}$ and the function $\varphi|_K$ satisfy all the hypotheses of Dini's Uniform Convergence Theorem (see Engelking [1], for instance). The assertion desired results now from the assumption that Φ is closed under almost uniform convergence. ■

Remark 7.8.1. The hypotheses assumed for the class Φ force a and b to be solutions of the functional inequalities

$$\varphi(x) \leq h(x, \varphi \circ f(\cdot, x))$$

and

$$h(x, \varphi \circ f(\cdot, x)) \leq \varphi(x),$$

respectively.

Remark 7.8.2. In Theorem 7.8.2 the assumption of the continuity of solutions at ξ is essential. To see this, take any continuous function $A: \mathbb{R} \to \mathbb{R}$ such that $A(x) = -\sin x$ for $x \in [-\frac{1}{2}, \frac{1}{2}]$ and $A(x) = 1$ for $x \in [1, 0)$ and consider the following equation:

$$\varphi(x) = \frac{1}{n} \sum_{k=1}^{n} A(\varphi(f_k(x))), \tag{7.8.22}$$

where f_k are continuous self-mappings of $\mathbb{R}^+$ satisfying $0 < f_k(x) \leq x/(1 + x)$, $x \in (0, \infty)$, $k \in \{1, \dots, n\}$. A reasoning similar to that applied at the beginning of the present section shows that the zero function is the only solution of (7.8.22) continuous at zero. Nevertheless, the (discontinuous at zero) function

$$\varphi(x) := \begin{cases} 0 & \text{for } x = 0 \\ \sqrt[3]{}(1 + x) & \text{for } x \in (0, \infty) \end{cases}$$

yields a solution of equation (7.8.22) on the halfline $\mathbb{R}^+$.

7.9 Notes

7.9.1. The most general functional equation we have examined in the present chapter as well as in the whole book is

$$\varphi(x) = H(x, \varphi). \tag{7.9.1}$$

In Section 7.0 we pointed out that it covers numerous equations of various kinds considered in mathematics. The problem of existence and uniqueness of continuous solutions of (7.9.1) as well as that of the convergence of successive approximations for this equation were investigated by M. Kwapisz–J. Turo [2] (see also [4]). They considered also (see Kwapisz–Turo [3]) such problems regarding directly the equation

$$\varphi(x) = h(x, \varphi \circ f(\cdot, x)) \tag{7.9.2}$$

studied most widely in the present chapter. K. Baron, whose results were extensively reproduced here (Baron [9]) writes in this paper: 'In what concerns these problems ... the results obtained ... are crossing those of the authors mentioned above'. The main difference seems to lie in a radically different distribution of emphasis. In the present book we are interested in topics stemming naturally from the theory of iterative functional equations (see the Preface).

7.9.2. Extension theorems like those presented in Section 7.1 are of great importance in the theory of iterative functional equations. We add an example throwing some light onto the role played by very frequently occurring assertion

$$f(s_1, \cdot) \circ \cdots \circ f(s_n, \cdot)(x) \in U, \qquad x \in X, \tag{7.9.3}$$

where $U \subset X$ is a given set such that $f(S \times U) \subset U$. Recalling our considerations on the domains of attraction (Section 1.2) we see at once a close connection of that subject matter with requirement (7.9.3). One might be inclined to replace (7.9.3) by a more natural and agreeable looking condition

for every $x \in X$ there exists an $n \in \mathbb{N}$ such that $f_s^n(x) \in U, s \in S$ (7.9.4)

(f_s stands here, as usual, for the map $f(s, \cdot)$, $s \in S$).

Unfortunately, assumption (7.9.4) is too weak to yield the uniqueness of a suitable extension (see Theorem 7.2.1).

Example 7.9.1. Fix a $\vartheta \in (0, 1)$, put

$$f_1(x) := \vartheta x \text{ for } x \in (-1, 1), \, f_1(-1) = 1, \, f_1(1) = 0,$$

$$f_2(x) := \vartheta x \text{ for } x \in (-1, 1), \, f_2(-1) = 0, \, f_2(1) = -1,$$

and consider the equation

$$\varphi(x) = \tfrac{1}{2}\varphi(f_1(x)) + 2\varphi(f_2(x)).\tag{7.9.5}$$

The function $\varphi_0 \colon (-1, 1) \to \mathbb{R}$ vanishing identically is a solution of (7.9.5) admitting for every $y \in \mathbb{R}$ an extension of the form

$$\varphi(x) = 0 \text{ for } x \in (-1, 1), \ \varphi(-1) = y, \ \varphi(1) = 2y,$$

onto the closed interval $[-1, 1]$. On the other hand, for any neighbourhood U of zero and any $x \in [-1, 1]$ both $f_1^n(x)$ and $f_2^n(x)$ are in U provided that n is large enough.

The next example again shows that the question of the domain of the given function h in (7.9.2) is very important while dealing with extension problems.

Example 7.9.2. Take $S = \{1\}$, $f_1(x) = f(x) = \tfrac{1}{2}x$, $x \in \mathbb{R}$, and $\Omega := \mathbb{R} \times \{0\} \subsetneqq \mathbb{R} \times \mathcal{F}(\{1\}, \mathbb{R}^+)$. Let $h \colon \Omega \to \mathbb{R}$ be given by

$$h(x, 0) := \begin{cases} 0 & \text{for } x \in (-1, 1), \\ x & \text{for } x \in \mathbb{R} \setminus (-1, 1). \end{cases}$$

Then the zero function is (a unique) solution of the equation

$$\varphi(x) = h(x, \varphi(f(x)))$$

in $(-1, 1)$ but there exists no solution in $\mathbb{R}$.

7.9.3. Formally, J. Kordylewski's extension theorem (Kordylewski [3]) does not become a special case of Theorem 7.1.1. Assuming (H), however (see Section 7.1), by a careful selection of Φ_n, $n \in \mathbb{N}_0$, one is able to derive Kordylewski's theorem from such a generalized version of Theorem 7.1.1. The details are described precisely in Baron [9, Remark 3.5]. The same touches J. Matkowski's lemma on extending meromorphic solutions of a linear equation (see Matkowski [2]).

7.9.4. At the beginning of Section 7.6 we have mentioned that keeping the level of generality of Section 7.2 one may obtain existence and uniqueness results regarding the systems of infinite orders. For instance, we have the following analogue of Theorem 7.2.1 (Baron [9]).

Proposition 7.9.1. *Let* $(X, \mathcal{T})$ *be a topological space and let* $\mathcal{G}$ *be an open covering of* X *fulfilling conditions* (i) *and* (iii) *of Section 7.2 and let* (Y_i, σ_i), $i \in \{1, \ldots, m\}$, *be complete metric spaces. Suppose that* $\Phi_i \subset C(X, Y_i)$ *are nonempty,* $\Phi_{U,i} := \{\varphi|_{\mathrm{cl}\, U} \colon \varphi \in \Phi\}$ *are closed under uniform convergence and such that every* $\varphi \in C(X, Y_i)$ *having the property* $\varphi|_{\mathrm{cl}\, U} \in \Phi_{U,i}$, $U \in \mathcal{G}$, *belongs to* Φ_i, $i \in \{1, \ldots, m\}$.

Assume further that the functions $H_i \colon X \times \Phi \to Y_i$ *where* $\Phi := \Phi_1 \times \cdots \times \Phi_m$ *are such that*

(a) $H_i(\,\cdot\,, \varphi_1, \ldots, \varphi_m) \in \Phi_i$ *for all* $(\varphi_1, \ldots, \varphi_m) \in \Phi$,

(b) *for any $U \in \mathcal{G}$ and any two members $\varphi, \psi \in \Phi$ such that $\varphi|_U = \psi|_U$ one has $H_i(\cdot, \varphi)|_U = H_i(\cdot, \psi)|_U$ for all $i \in \{1, \ldots, m\}$,*

(c) *for every $U \in \mathcal{G}$ there exists a matrix $(a_{i,j})_{i,j \in \{1,\ldots,m\}}$ such that for all $i \in \{1, \ldots, m\}$ and $\varphi = (\varphi_1, \ldots, \varphi_m), \psi = (\psi_1, \ldots, \psi_m) \in \Phi$ we have*

$$\sigma_i(H_i(x, \varphi), H_i(x, \psi)) \leq \sum_{j=1}^{m} a_{i,j} \sup_{z \in \mathrm{cl}\, U} \sigma_j(\varphi_j(z), \psi_j(z))$$

for all $x \in U$.

Then the system

$$\varphi_i(x) = H_i(x, \varphi_1, \ldots, \varphi_m), \qquad i \in \{1, \ldots, m\},$$

has exactly one solution $\varphi: X \to Y_1 \times \cdots \times Y_m$ belonging to Φ. This solution is given as the almost uniform limit of a sequence of successive approximations starting from an arbitrary member of Φ.

7.9.5. It would definitely be desirable to get rid of the assumption of the boundedness of the metric space (X, ρ) in Theorem 7.3.3 (diam $X < \infty$). Unfortunately, the only known result in this direction comprises only some very special functions h (close to linear with respect to the second variable); see Theorem 4 in Baron–Ger [1].

7.9.6. Apart from Lyapunov stability and that described at the end of Section 7.3 the so-called *interval (set) stability* has also been introduced and studied (Czerni [2]–[4]; see also Choczewski [14], Turdza [10]).

7.9.7. Theorem 7.4.1 actually applies to equation (7.4.2) (see Section 7.4A). For take $X = I = [0, 1]$, $Y = \mathbb{R}$, $\Phi = C(I, \mathbb{R})$, $S = \{1\}$, $f(1, x) = x/(x + 1)$, $x \in I$, $h(x, y) := y - x$, $(x, y) \in I \times \mathbb{R}$. Then $f(1, \cdot)$ satisfies the following conditions: (7.4.7) with $m(1) = 1$ and (7.3.1) with $\xi = 0$, and h has property (7.4.8) with $L = 1$ and $\beta = \mathrm{id}_{\mathbb{R}^+}$.

7.9.8. Continuing a study of approximate solutions (see Section 7.4) K. Baron has examined under what conditions there exist Lipschitzian mappings φ_i from a metric space (X, ρ) into normed space $(Y, \|\cdot\|_i)$, $i \in \{1, \ldots, m\}$, satisfying the system of inequalities

$$\|\varphi_i(x) - h_i(x, \varphi_1 \circ f(\cdot, x), \ldots, \varphi_m \circ f(\cdot, x))\|_i \leq \varepsilon$$

for a given $\varepsilon > 0$, a function $f: S \times X \to X$ (S a given set of indices) and $h_i: X \times \mathcal{F}(S, Y_1) \times \cdots \times \mathcal{F}(S, Y_m) \to Y_i$, $i \in \{1, \ldots, m\}$. The character of the results obtained does not differ essentially from those reported on in Section 7.4 (see Baron [10] and also Baron–Jarczyk [1], Dankiewicz [2]).

7.9.9. The choice of initial condition in Kominek's Theorem 7.6.4, $\varphi_i(0) = 0$, $i \in \{1, \ldots, m\}$, is in no circumstance essential; the same applies to other results on systems of equations: instead of the conditions $h_i(0, \{0\}) = 0$ and $\varphi_i(0) = 0$ one may consider the conditions $\eta_i = h_i(0; \eta_1, \ldots, \eta_1; \ldots; \eta_m, \ldots, \eta_m)$ and $\varphi_i(\xi) = \eta_i$, respectively, $i \in \{1, \ldots, m\}$.

Moreover, the assertion of Theorem 7.6.4 states that any two solutions from the class considered coincide in a right neighbourhood of zero. It may happen, however, that there is no common interval of equality for all solutions (see Kuczma–Matkowski [1] for a suitable example).

7.9.10. Results concerning the g-system

$$\varphi_i(f_0(x)) = g_i(x; \varphi_1(x), \varphi_1(f_1(x)), \ldots, \varphi_m(f_m(x))) \qquad (7.9.6)$$

are not too numerous; this is caused by the fact that in principle the fixed-point theorem approach becomes useless. In general, we expect to deal with the abundance of solutions. Section 7.6 presents two results touching continuous and differentiable solutions. The following theorem (Matkowski [13]) yields a counterpart of Theorem 7.6.8 with regard to the following special case of system (7.9.6)

$$\varphi_i(f(x)) = g_i(x, \varphi_1(x), \ldots, \varphi_m(x)), \qquad i \in \{1, \ldots, m\}. \qquad (7.9.7)$$

Proposition 7.9.2. Let $(X, \mathfrak{M}, \mu)$, f and p satisfy the assumption spoken of in Theorem 7.6.8 (with f_k replaced by f). Assume that functions $g_i: X \times \mathbb{R}^m \to \mathbb{R}$ are sup-measurable and

$$|g_i(x, y) - g_i(x, z)| \leq \sum_{j=1}^m a_{i,j}(x)|y_j - z_j|$$

for $x \in X$, $y = (y_1, \ldots, y_m)$, $z = (z_1, \ldots, z_m) \in \mathbb{R}^m$, $i \in \{1, \ldots, m\}$, where the functions $a_{i,j}: X \to \mathbb{R}$ are measurable and μ-a.e. finite in X, $i, j \in \{1, \ldots, m\}$. If

(a) the complement of the union $\bigcup_{n=0} A_n$, where $A_n := f^n(X) \setminus f^{n+1}(X)$, $n \in \mathbb{N}_0$, is a null set,

(b) there exist numbers $s_{i,j} \geq 0$, $i, j \in \{1, \ldots, m\}$, such that

$$\left[\frac{(a_{i,j} \circ f^{-1})^p}{d(\mu f^{-1})/d\mu} \right]^{\alpha(p)} \leq s_{i,j} \qquad a.e. \text{ in } X, \ i, j \in \{1, \ldots, m\}$$

and the matrix $(s_{i,j})_{i,j \in \{1,\ldots,m\}}$ has all its characteristic roots less than unity in absolute value,

(c) $\displaystyle\sum_{n=1}^{\infty} \left(\int_{A_n} |g_i(f^{-1}(x), 0, \ldots, 0)|^p d\mu \right)^{\alpha(p)} < \infty \qquad for \ i \in \{1, \ldots, m\}$,

then every system $(\varphi_{0,1}, \ldots, \varphi_{0,m}): A_0 \to \mathbb{R}^m$ of m measurable functions such that

$$\int_{A_0} |\varphi_{0,i}|^p d\mu < \infty, \qquad i \in \{1, \ldots, m\},$$

admits a unique extension onto X to a solution $(\varphi_1, \ldots, \varphi_m)$ of system (7.9.7); all the functions φ_i, $i \in \{1, \ldots, m\}$ belong to $L^p(X, \mathfrak{M}, \mu)$.

7.9.11. The simple equation

$$\varphi(x) = h(x, \varphi(f_1(x)), \ldots, \varphi(f_n(x))) \qquad (7.9.8)$$

may be considered as the special case $(m = 1)$ of system (7.6.1) and we may apply Theorems 7.6.1–7.6.10 to equation (7.9.8). All these theorems require that the characteristic roots of the matrix

$$\left(\sum_{k=1}^{n} a_{i,j,k} b_k \right)_{i,j\in\{1,\ldots,m\}} \tag{7.9.9}$$

(where $a_{i,j,k}$ are the Lipschitz coefficients of h_i and b_k are suitable constants connected with functions f_k) are less than unity in absolute value. In the present case $(m = 1)$ matrix (7.9.9) reduces to a number

$$s = \sum_{k=1}^{n} s_k b_k \tag{7.9.10}$$

and thus this crucial condition states that $|s| < 1$. If, however, the functions f_k occurring in (7.9.8) are successive iterates of the same function f: $f_k = f^k$ for $k \in \{1, \ldots, n\}$ then we can reduce (7.9.8) to the system

$$\varphi_k(x) = h_k(x, \varphi_1(f(x)), \ldots, \varphi_n(f(x))), \qquad k \in \{1, \ldots, n\},$$

where $h_1 = h$, $h_k(x, y_1, \ldots, y_n) := y_{k-1}$, $k \in \{2, \ldots, n\}$, and $\varphi_k := \varphi \circ f^{k-1}$, $k \in \{1, \ldots, n\}$. The Lipschitz coefficients of h_i are now $a_{1,j} = s_j$, $j \in \{1, \ldots, n\}$, $a_{i,i-1} = 1$ for $i \in \{2, \ldots, n\}$ and $a_{i,j} = 0$ for the remaining indices. The matrix (7.9.9) takes now the form $(a_{i,j}b)_{i,j\in\{1,\ldots,n\}}$, where b is a constant connected with f, and its characteristic polynomial is (up to the factor $(-1)^n$)

$$p(\lambda) = \lambda^n - s_1 b\lambda^{n-1} - \cdots - s_{n-1} b\lambda - s_n.$$

Thus instead of number (7.9.10) we obtain the roots of the above polynomial and now the crucial condition says that all the roots of p should have the absolute values less than unity. This, in general, does not change too much except for the cases where the numbers $a_{i,j}$ are replaced by the partial derivatives $\partial h_i / \partial y_j$ (calculated at the initial point $(0, \eta)$). Then the condition that all the roots of polynomial p should be less than unity in absolute value is replaced by the condition that all the roots of the polynomial

$$\lambda^n - \hat{s}_1 \lambda^{n-1} - \cdots - \hat{s}_{n-1} \lambda - \hat{s}_n \tag{7.9.11}$$

should have the absolute values less than unity, where $\hat{s}_k := (\partial h / \partial y_k)(0, \eta)$, $k \in \{1, \ldots, n\}$.

The important point is that $\hat{s}_k$ may now be negative. For instance, consider the equation

$$\varphi(x) = \tfrac{5}{4}\varphi\left(\frac{x}{2}\right) - \tfrac{3}{8}\varphi\left(\frac{x}{4}\right)$$

for functions $\varphi\colon \mathbb{R}^+ \to \mathbb{R}$. Here $s_1 = \hat{s}_1 = \tfrac{5}{4}$, $s_2 = \tfrac{3}{8}$ but $\hat{s}_2 = -\tfrac{3}{8}$. The roots of polynomial p $(b=1)$ are $\tfrac{3}{2} > 1$ and $-\tfrac{1}{4}$. On the other hand, the roots of polynomial (7.9.11) are $\tfrac{3}{4}$ and $\tfrac{1}{2}$ and hence both lie in $(0, 1)$.

8

On conjugacy

8.0 Introduction

In this chapter we deal with the conjugacy equation

$$\varphi(f(x)) = g(\varphi(x)). \tag{8.0.1}$$

The mappings $f: X \to X$ and $g: Y \to Y$ are said to be *conjugate* iff there exists a bijective solution $\varphi: X \to Y$ of equation (8.0.1).

There are important cases of (8.0.1) (when X and Y are subsets of $\mathbb{K}^N$): the Schröder equation

$$\sigma(f(x)) = S\sigma(x), \tag{8.0.2}$$

where $S \in \mathbb{K}^{N \times N}$; and the Abel equation

$$\alpha(f(x)) = \alpha(x) + A, \qquad A \in \mathbb{K}^N, \tag{8.0.3}$$

expressing the fact that f is conjugate with a linear g, $g(y) = Sy$ or with $g(y) = y + A$, respectively; and also the Böttcher equation, corresponding to $g(y) = y^p$ and $N = 1$,

$$\beta(f(x)) = [\beta(x)]^p. \tag{8.0.4}$$

Equations (8.0.2) and (8.0.4), as well as that of permutable functions

$$\varphi(f(x)) = f(\varphi(x)) \tag{8.0.5}$$

are dealt with in Sections 8.2, 8.3 and 8.6. Conjugacy itself of functions is discussed in Sections 8.1 and 8.4. A part of the theory of conjugate (resp. commuting) formal power series and analytic functions is presented in Section 8.5 (resp. Section 8.7). There is also a result in Section 8.5 concerning a particularly interesting case of Abel's equation (8.0.3) ($N = 1$, $A = 1$). Section 8.8 contains some examples and supplementary results.

8.1 Conjugacy

8.1A. Change of variables

Let X be an arbitrary set, and let f and g be some self-mappings of X. We are looking for a change of variables $x' = \varphi(x)$ that transforms f into g. In other words, the relationship $y = f(x)$ in new coordinates should take the form $y' = g(x')$. Replacing y' by $\varphi(y)$ and x' by $\varphi(x)$ we obtain hence $\varphi(y) = g(\varphi(x))$, or else, since $y = f(x)$,

$$\varphi(f(x)) = g(\varphi(x)). \tag{8.1.1}$$

Usually we require that the function φ looked for should have some additional properties. Since we want to return to old variables, we should have an invertible φ. Often we need φ to be sufficiently regular. We will be particularly interested in determining whether two given functions are conjugate via a function φ of class C^r, if X is a subset of $\mathbb{R}^N$, or via an analytic function φ if X is a subset of $\mathbb{C}^N$.

Change of variables has been used several times in earlier chapters of this book. Suppose that we study the equation

$$F(x, \varphi(x), \varphi(f(x))) = 0 \tag{8.1.2}$$

in a set X. Assume there is an invertible solution $\hat\varphi: X \to X$ of equation (8.1.1). Then, with $x' = \hat\varphi(x)$ and $\psi = \varphi \circ \hat\varphi^{-1}$, equation (8.1.2) goes over into

$$F(\hat\varphi^{-1}(x'), \psi(x'), \psi(g(x'))) = 0, \tag{8.1.3}$$

i.e., again an equation of form (8.1.2), but with f replaced by g. If g is simpler than f, equation (8.1.3) may be easier to solve than (8.1.2).

8.1B. Properties of the conjugacy relation

In many cases it is sufficient to effect the transformation of f into g only locally, in a neighbourhood of a point $\xi \in X$, usually a fixed point of f. Then we are looking for local, or locally invertible, solutions φ of (8.1.1). Assume X to be a neighbourhood of $\xi = 0 \in \mathbb{K}^N$. (Without loss of generality we can always place ξ at the origin.) Then $\varphi: X \to X$ will certainly be locally invertible around 0 if

$$\det \varphi'(0) \neq 0. \tag{8.1.4}$$

Moreover, in order to preserve the fixed point $\xi = 0$ of f, we postulate that

$$\varphi(0) = 0. \tag{8.1.5}$$

We introduce the notion of smooth conjugacy.

Definition 8.1.1. The functions $f: X \to X$ and $g: X \to X$, where X is a neighbourhood of the origin in $\mathbb{K}^N$, are said to be C^r-conjugate (resp. A-conjugate) iff there exists a function φ, defined and having continuous derivatives up to order $r \geqslant 1$ (resp. defined and analytic) in a neighbourhood of the origin in $\mathbb{R}^N$ (resp. $\mathbb{C}^N$), satisfying (8.1.4) and (8.1.5) and such that (8.1.1) holds in a neighbourhood of the origin.

Each of the function classes occurring in Definition 8.1.1 forms a group under the operation of composition of functions[1]. It follows that each of the conjugacy relations is transitive. Therefore, if f and g are conjugate with the same function $h: X \to Y$, they are themselves conjugate. It then becomes natural to investigate equation (8.1.1) first for some particular functions g. We do this in Sections 8.2 and 8.3.

We conclude the section with two simple general results.

Theorem 8.1.1. *Let X be a neighbourhood of the origin in $\mathbb{K}^N$ and let the functions $f: X \to X$ and $g: X \to X$ be differentiable at 0, $f(0) = 0$, $g(0) = 0$. If f and g are either C^r-conjugate or A-conjugate, then the matrices $f'(0)$ and $g'(0)$ are conjugate, and hence they have the same Jordan normal form.*

Proof. Differentiating (8.1.1) we obtain
$$\varphi'(f(x))f'(x) = g'(\varphi(x))\varphi'(x),$$
whence on setting $x = 0$ we get, by (8.1.4), $f'(0) = C^{-1}g'(0)C$, where $C := \varphi'(0)$, which means that the matrices $f'(0)$ and $g'(0)$ are conjugate. ∎

Connections between conjugacy and orbit structure (see Ulam [1, Ch. VI, § 2], Rice–Schweizer–Sklar [1], Targoński [8]) show the following.

Theorem 8.1.2. *The functions $f: X \to X$ and $g: X \to X$ are conjugate if and only if they are orbit isomorphic, i.e. there is a bijection $\varphi: X \to Y$ such that*
$$f^m(x) = f^n(y) \Leftrightarrow g^m(\varphi(x)) = g^n(\varphi(y)) \qquad (8.1.6)$$
for $n, m \in \mathbb{N}_0$ and every $x, y \in X$.

Proof. If f and g are conjugate, then (8.1.1) holds with a bijective $\varphi: X \to Y$. Thus $\varphi \circ f^i = g^i \circ \varphi$, $i \in \mathbb{N}_0$ (induction), and consequently
$$f^m(x) = f^n(y) \Leftrightarrow \varphi(f^m(x)) = \varphi(f^n(y)) \Leftrightarrow g^m(\varphi(x)) = g^n(\varphi(y)),$$
Conversely, if (8.1.6) holds with a bijective $\varphi: X \to Y$, then for $m = 1$, $n = 0$ we have $g(\varphi(x)) = \varphi(y)$ whenever $y = f(x)$, i.e., φ is just the function conjugating f and g. ∎

8.2 Linearization

If the function g in (8.1.1) is linear, the procedure of transforming f into g is referred to as *linearization*.

The linearization method is of importance in, among others, the theory

---

[1] More exactly, the functions from a suitable class should be replaced by the germs of functions coinciding in a neighbourhood of the origin.

of dynamical systems. It allows one to reduce the study of the nonlinear dynamical system ($\dot{x} := dx/dt$)

(I) $\dot{x} = F(x)$, $x \in \mathbb{R}^N$,

to the investigation of the linear one

(II) $\dot{z} = F'(0)z$,

provided that F fulfils suitable conditions. Let $x(t, y)$ (resp. $z(t, y)$) be the solution of (I) (resp. (II)), say, in $[0, 1]$, such that $x(0, y) = y$ (resp. $z(0, y) = y$). So we have, in particular,

$$x(t, y) = [\exp(F'(0)t)]y + x^*(t, y), \qquad \frac{\partial x^*}{\partial y}(t, 0) = 0.$$

We write

$$f(y) := x(1, y), \qquad S = f'(0) = \exp(F'(0)).$$

Suppose that σ is an invertible (and smooth enough) solution of the Schröder equation

$$\sigma(f(y)) = S\sigma(y)$$

in a neighbourhood of the origin, and set

$$T(y) := \int_0^1 z(-s, y)\sigma(y)x(s, y)ds.$$

Then the relation (e.g. Sternberg [3])

$$T(x(t, T^{-1}(y))) = z(t, y)$$

connects the solutions of (I) with those of (II). For more details see Hartman [4, Ch. 9]. The idea of linearization goes back to H. Poincaré [1].

8.2A. The Schröder equation

This is the fundamental equation of linearization. Let S be a matrix, $S \in \mathbb{K}^{N \times N}$. We denote the unknown function by σ. We have

$$\sigma(f(x)) = S\sigma(x). \tag{8.2.1}$$

If the function f has a fixed point at the origin and is differentiable there, then, by Theorem 8.1.1, equation (8.2.1) can have a smooth solution only if the matrices S and $f'(0)$ are conjugate: $f'(0) = CSC^{-1}$. Setting $\hat{\sigma} = C\sigma$, we obtain for $\hat{\sigma}$ the equation $\hat{\sigma}(f(x)) = f'(0)\hat{\sigma}(x)$. Therefore in the sequel we shall assume that

$$f'(0) = S. \tag{8.2.2}$$

To find local analytic solutions of the Schröder equation in $\mathbb{C}^N$ we assume that

(i) X is a neighbourhood of zero in $\mathbb{C}^N$, and $f: X \to \mathbb{C}^N$ is an analytic function, $f(0) = 0$, $f'(0) = S$, $\det S \neq 0$.

Differentiating (8.2.1) and then setting $x = 0$ we obtain

$$\sigma'(0)S = S\sigma'(0),$$

which is satisfied, e.g., by $\eta_1 := \sigma'(0) = E$, the unit matrix of order N. In general, the multidimensional matrices $\eta_p := \sigma^{(p)}(0)$, $p = 2, 3, \ldots$, must satisfy an infinite system of equations resulting on differentiating (8.2.1) p times and then setting $x = 0$. For $p \geqslant 2$, such η_p may exist or not. However, as follows from a result by W. Smajdor [4] on formal solutions of nonlinear equations, all the η_p for $p \geqslant 2$ can be uniquely determined whenever

$$s_{k_1} \cdots s_{k_p} \neq s_i \qquad \text{for } i, k_1, \ldots, k_p = 1, \ldots, N, \ p = 2, 3, \ldots \quad (8.2.3)$$

where s_k are the characteristic roots of S. Condition (8.2.3) can equivalently be written as

$$s_1^{q_1} \cdots s_N^{q_N} \neq s_i \qquad \text{for } i = 1, \ldots, N, \ q_1, \ldots, q_N \in \mathbb{N}_0, \quad (8.2.4)$$

where $\sum_{j=1}^{N} q_j = p$, $p = 2, 3, \ldots$. The following theorem on local analytic solutions to (8.2.1) is also a consequence of a result by W. Smajdor [2].

Theorem 8.2.1. *Let hypothesis* (i) *be satisfied. If condition* (8.2.4) *and*

$$|s_i| < 1 \qquad \text{for } i = 1, \ldots, N$$

are fulfilled by the characteristic roots $s_1, \ldots, s_N$ *of* S, *then equation* (8.2.1) *has a unique local analytic solution* $\sigma: U \to \mathbb{C}^N$ *in a neighbourhood* U *of the origin, fulfilling the condition* $\sigma'(0) = E$.

8.2B. Unique local C^r solution in $\mathbb{R}^N$

For the result presented, see Kuczma [41]. We assume that

(ii) $X \subset \mathbb{R}^N$ is a neighbourhood of the origin, and $f: X \to \mathbb{R}^N$ is of class C^r, $r \geqslant 1$, $f(0) = 0$, $f'(0) = S$, $\det S \neq 0$,

(iii) $f^{(r)}(x) = f^{(r)}(0) + O(|x|^\delta)$, $|x| \to 0$, $0 \leqslant \delta \leqslant 1$.

Theorem 8.2.2. *Let hypotheses* (ii), (iii) *be satisfied, and let the characteristic roots* $s_1, \ldots, s_N$ *of* S *satisfy*

$$0 < |s_1| \leqslant \cdots \leqslant |s_N| < 1.$$

If condition (8.2.4) *is fulfilled for* $p = 2, \ldots, r$ *whenever* $r \geqslant 2$, *and if*

$$|s_N|^{r+\delta} < |s_1|,$$

then equation (8.2.1) *has a unique* C^r *solution* $\sigma: U \to \mathbb{R}^N$ *in a neighbourhood* U *of the origin, such that* $\sigma(0) = 0$, $\sigma'(0) = E$ *and*

$$\sigma^{(r)}(x) = \sigma^{(r)}(0) + O(|x|^\delta), \qquad |x| \to 0. \quad (8.2.5)$$

Proof. We are going to apply Banach's Theorem. This requires some preparations.

Given $\varepsilon > 0$, by Lemma 1.1.1, we can find a nonsingular matrix A such that simultaneously

$$\|A^{-1}SA\| < |s_N| + \varepsilon, \qquad \|A^{-1}S^{-1}A\| < |s_1|^{-1} + \varepsilon.$$

Then, with $\tilde{f}(x) = A^{-1}f(Ax)$, $\hat{\sigma}(x) = A^{-1}\sigma(Ax)$, equation (8.2.1) is equivalent to $\hat{\sigma}(\tilde{f}(x)) = A^{-1}SA\hat{\sigma}(x)$. Thus we may assume that

$$\|S\|^{r+\delta}\|S^{-1}\| < 1,$$

whence, in particular, $\|S\| < 1$. Consequently, we can find a $\Theta < 1$ and a neighbourhood $U := \{x \in \mathbb{R}^N : |x| \leqslant b\}$, $b > 0$, of the origin such that $0 < |f(x)| < \Theta|x|$ in $U \setminus \{0\}$ and $\|f'(x)\| < \Theta$ in U; $\Theta^{r+\delta}\|S^{-1}\| < 1$

Let $\boldsymbol{\Phi}$ be the space of all functions $\sigma : U \to \mathbb{R}^N$ that are of class C^r in U and satisfy (8.2.5) and the conditions

$$\sigma(0) = 0, \quad \sigma'(0) = E, \quad \sigma^{(q)}(0) = \eta_q, \quad q = 2, \ldots, r,$$

where η_q are the (uniquely determined) values at zero of the derivatives of a solution of (8.2.1). In $\boldsymbol{\Phi}$ we introduce the metric

$$\rho(\sigma_1, \sigma_2) = \sup_{0 < |x| \leqslant b} |x|^{-\delta}\|\sigma_1^{(r)}(x) - \sigma_2^{(r)}(x)\|_r,$$

where the norm $\|y\|_q$ is defined for $y = (y_{k_1 \ldots k_q}^i)$ by

$$\|y\|_q = \left(\sum_{i=1}^N \sum_{k_1=1}^N \cdots \sum_{k_q=1}^N |y_{k_1 \ldots k_q}^i|^2\right)^{1/2}, \quad q = 1, \ldots, r. \tag{8.2.6}$$

We have for $0 < |x| \leqslant b$, $\sigma_1, \sigma_2 \in \boldsymbol{\Phi}$, and $q = 0, \ldots, r-1$

$$\|\sigma_1^{(q)}(x) - \sigma_2^{(q)}(x)\|_q \leqslant |x| \sup_{0 < |t| \leqslant |x|} \|\sigma_1^{(q+1)}(t) - \sigma_2^{(q+1)}(t)\|_{q+1}$$

$$\leqslant |x| \sup_{0 < |t| \leqslant |x|} (|t|/|x|)^{-\delta}\|\sigma_1^{(q+1)}(t) - \sigma_2^{(q+1)}(t)\|_{q+1},$$

whence

$$\sup_{0 < |x| \leqslant b} |x|^{-\delta}\|\sigma_1^{(q)}(x) - \sigma_2^{(q)}(x)\|_q \leqslant b \sup_{0 < |x| \leqslant b} |x|^{-\delta}\|\sigma_1^{(q+1)}(x) - \sigma_2^{(q+1)}(x)\|_{q+1}.$$

By a repeated use of this estimate we get

$$\sup_{0 < |x| \leqslant b} |x|^{-\delta}\|\sigma_1^{(q)}(x) - \sigma_2^{(q)}(x)\|_q \leqslant b^{r-q}\rho(\sigma_1, \sigma_2), \tag{8.2.7}$$

for $q = 0, \ldots, r-1$. This implies that $(\boldsymbol{\Phi}, \rho)$ is a complete metric space, since the convergence in $\boldsymbol{\Phi}$ is equivalent to the uniform convergence in U of the sequence of functions together with sequences of their derivatives up to order r.

For $\sigma \in \boldsymbol{\Phi}$ we define the mapping T by

$$T(\sigma)(x) = S^{-1}\sigma(f(x)).$$

Since $|f(x)| < \Theta|x| < |x|$, $T(\sigma)$ is well defined in U. It is in $C^r(U, \mathbb{R}^N)$, and fulfils the conditions $T(\sigma)(0) = 0$, $T(\sigma)'(0) = E$, $T(\sigma)^{(q)}(0) = \eta_q$, $q = 2, \ldots, r$.

We have also (consult Lemma 3.4.1)

$$T(\sigma)^{(r)}(x) = S^{-1}\sigma^{(r)}(f(x))f'(x)^r + \sum_{q=1}^{r-1} S^{-1}\sigma^{(q)}(f(x))B_q(x). \quad (8.2.8)$$

Here $\sigma^{(p)}$, $(f')^r$ (see Note 8.8.5) and B_q are multi-dimensional matrices, and the entries of B_q are of class C^{q-1} in U, $B_1 = f^{(r)}$.

To prove (8.2.5) for $T(\sigma)$ first put $x = 0$ in (8.2.8) to obtain

$$\eta_r = T(\sigma)^{(r)}(0) = S^{-1}\eta_r S^r + \sum_{q=1}^{r-1} S^{-1}\eta_q B_q(0).$$

Using this we may write

$$T(\sigma)^{(r)}(x) - T(\sigma)^{(r)}(0) = S^{-1}(\sigma^{(r)}(f(x)) - \eta_r)f'(x)^r + S^{-1}\eta_r[f'(x)^r - S^r]$$
$$+ \sum_{q=1}^{r-1} [S^{-1}\eta^{(q)}(f(x))B_q(x) - S^{-1}\eta_q B_q(0)]$$
$$+ S^{-1}[\sigma'(f(x)) - E]f^{(r)}(x) + S^{-1}(f^{(r)}(x) - f^{(r)}(0)).$$

Observe that in each bracket there appears a difference of the values, at x and at 0, of a function which is at least of class C^1 on U. Therefore, every term containing an expression in brackets is $O(|x|)$. Next, the first term becomes $O(|f(x)|^\delta)$, by (8.2.5); and the last one is $O(|x|^\delta)$, because of (iii). Since $O(|x|) \subset O(|x|^\delta)$ (because $0 \leqslant \delta \leqslant 1$) and $O(|f(x)|^\delta) \subset O(|x|^\delta)$ (because $|f(x)| \leqslant \Theta|x|$), we actually get (8.2.5) with $T(\sigma)$ in place of σ.

We have shown that T maps Φ into itself. Moreover, we have for $\sigma_1, \sigma_2 \in \Phi$, again by (8.2.8)

$$\rho(T(\sigma_1), T(\sigma_2)) \leqslant \sup_{0<|x|\leqslant b} |x|^{-\delta}\{\|S^{-1}(\sigma_1^{(r)}(f(x)) - \sigma_2^{(r)}(f(x)))(f'(x))^r\|_r$$

$$+ \sum_{q=1}^{r-1} \|S^{-1}(\sigma_1^{(q)}(f(x)) - \sigma_2^{(q)}(f(x)))B_q(x)\|_r\}.$$

For $q < r$ we have the estimate (with $y := f(x)$, $t := |y|/|x|$, $x \neq 0$)

$$|x|^{-\delta}\|S^{-1}(\sigma_1^{(q)}(y) - \sigma_2^{(q)}(y))B_q(x)\|_r \leqslant C_q t^\delta(|y|^{-\delta}\|\sigma_1^{(q)}(y) - \sigma_2^{(q)}(y)\|_q),$$

where C_q depend on f; whereas for $q = r$ we have (see Note 8.8.5)

$$|x|^{-\delta}\|S^{-1}(\sigma_1^{(q)}(y) - \sigma_2^{(r)}(y))f'(x)^r\|_r$$
$$\leqslant \|S^{-1}\|\|f'(x)\|^r t^\delta(|y|^{-\delta}\|\sigma_1^{(r)}(y) - \sigma_2^{(r)}(y)\|_r). \quad (8.2.9)$$

Hence we get by (8.2.7) (note that both $\|f'(x)\|$ and t are less than Θ)

$$\rho(T(\sigma_1), T(\sigma_2)) \leqslant \left(\Theta^{r+\delta}\|S^{-1}\| + \sum_{q=1}^{r-1} C_q b^{r-q}\Theta^\delta\|S^{-1}\|\right)\rho(\sigma_1, \sigma_2).$$

Thus T is a contraction provided b has been chosen small enough. Banach's Theorem yields the unique solution $\sigma \in \Phi$ of (8.2.1). ∎

Remark 8.2.1. For $\delta = 0$ Theorem 8.2.2 yields the existence of a unique local C^r solution σ of (8.2.1) such that $\sigma(0) = 0$, $\sigma'(0) = E$, under the assumptions that f is of class C^r in X, $f(0) = 0$, $f'(0) = S$, $|s_N|^r/|s_1| < 1$, and condition (8.2.4) holds for $p = 2, \ldots, r$, $r \geqslant 2$. (This result is due to S. Sternberg [3]; see also Hartman [1], [4].)

8.2C. Further results on smooth solutions

These concern solutions of (8.2.1) which are at least continuously differentiable. We present them without proofs. Note that, in general, there is no uniqueness attached to the solutions in question.

Since $\eta_1 = E$ always exists, we can get rid of condition (8.2.4) if we weaken the regularity requirement on σ.

Theorem 8.2.3. *Let hypothesis* (ii) *be satisfied with* $r \geqslant 2$, *and let* $|s_k| < 1$ *for* $k = 1, \ldots, N$, *where* s_k *are the characteristic roots of* S. *Then equation* (8.2.1) *has a* C^1 *solution* $\sigma: U \to \mathbb{R}^N$ *in a neighbourhood* U *of the origin, fulfilling the condition* $\sigma'(0) = E$.

Regarding the proof of Theorem 8.2.3 the reader is referred to Hartman [1], [4].

If not all characteristic roots of S lie inside the unit circle, we can prove similarly to Theorem 6.5.4 the following.

Theorem 8.2.4. *Let hypothesis* (ii) *be satisfied with* $r \geqslant 2$, *and let* $|s_k| \neq 1$ *for* $k = 1, \ldots, N$, *where* s_k *are the characteristic roots of* S. *If condition* (8.2.4) *is fulfilled for* $p = 2, \ldots, r$, *then equation* (8.2.1) *has a solution* $\sigma: U \to \mathbb{R}^N$ *in a neighbourhood* U *of the origin, of a class* C^ρ, $1 \leqslant \rho \leqslant r$, *and such that* $\sigma'(0) = E$. *Moreover* ρ *tends to infinity when* r *does, and* $\rho = \infty$ *whenever* $r = \infty$.

For a detailed proof see Sternberg [4], Hartman [4].

8.3 The Böttcher equation

In the case where $f(x) \sim x^p$ as $x \to 0$ (the fixed point of f) we may ask for conditions under which f is conjugate just with $g(x) = x^p$. We assume that $p > 1$.

The conjugacy equation now reads

$$\beta(f(x)) = (\beta(x))^p, \tag{8.3.1}$$

and is called the *Böttcher equation* (Böttcher [1], [2]). We shall confine ourselves to the one-dimensional case.

8.3A. Complex case

We start with some results concerning local analytic solutions (LAS) of

(8.3.1) in a neighbourhood of the origin on the complex plane. We assume that

(i) $X \subset \mathbb{C}$ is a neighbourhood of the origin,
(ii) $f: X \to \mathbb{C}$ is an analytic function, $f(x) = x^p F(x)$, where $p > 1$ is an integer and $F(0) \neq 0$.

Theorem 8.3.1. *Let hypotheses* (i), (ii) *be satisfied. Then equation* (8.3.1) *has an LAS* β_0 *such that* $\beta_0(0) = 0$ *and* $\beta_0'(0) \neq 0$.

Proof. Fix a c such that $c^{p-1} = F(0)$, and consider the equation

$$\varphi(x) = \sqrt[p]{[F(x)\varphi(f(x))]}, \tag{8.3.2}$$

where $\sqrt[p]{u}$ denotes this branch of the root in a neighbourhood of $u = c^p$ for which $\sqrt[p]{c^p} = c$. By Theorem 5.7.1 equation (8.3.2) has a unique LAS φ such that $\varphi(0) = c$. The function $\beta_0(x) = x\varphi(x)$ meets the conditions of the theorem. ∎

In order to determine the general LAS of (8.3.1) we first prove the following.

Lemma 8.3.1. *Let hypotheses* (i), (ii) *be satisfied. Then* $\beta = 1$ *is the only LAS of equation* (8.3.1) *satisfying the condition* $\beta(0) = 1$.

Proof. Let an analytic function $\beta \neq 1$ satisfy (8.3.1) and $\beta(0) = 1$. Then $\beta(x) = 1 + x^r\varphi(x)$ with an $r \geq 1$ and $\varphi(0) \neq 0$. Inserting this into (8.3.1) we get $\varphi(0) = 0$, a contradiction. ∎

The next lemma yields LASs of the special case of (8.3.1):

$$\hat{\beta}(x^p) = (\hat{\beta}(x))^p. \tag{8.3.3}$$

Lemma 8.3.2. *If* $\hat{\beta} \neq 0$ *is an LAS of* (8.3.3) *and hypothesis* (i) *is satisfied, then*

$$\hat{\beta}(x) = \varepsilon x^r, \tag{8.3.4}$$

where ε *is a* $(p-1)$*st root of unity and* $r \in \mathbb{N}_0$.

Proof. Let $\hat{\beta}$ be an LAS of (8.3.3), $\hat{\beta} \neq 0$. We may write it as $\hat{\beta}(x) = x^r\varphi(x)$, $r \geq 0$, $\varphi(0) \neq 0$. Then φ satisfies (8.3.3) whenever $\hat{\beta}$ does, and $\varphi(0)$ is a $(p-1)$st root of unity, say ε. Hence $\varepsilon^{-1}\varphi$ also satisfies (8.3.3) and equals 1 at zero. By Lemma 8.3.1 we have $\varphi = \varepsilon$, and $\hat{\beta}(x) = \varepsilon x^r$. ∎

This lemma helps to prove the following.

Theorem 8.3.2. *Let hypotheses* (i), (ii) *be satisfied. Then the only LASs* β *of equation* (8.3.1) *are the functions* $\beta = 0$ *and*

$$\beta(x) = \varepsilon_j(\beta_0(x))^r, \qquad j = 1, \dots, p-1, \; r \in \mathbb{N}_0, \tag{8.3.5}$$

where the ε_j *are the* $(p-1)$*st roots of unity, and* β_0 *is a fixed LAS of* (8.3.1) *such that* $\beta_0(0) = 0$ *and* $\beta_0'(0) \neq 0$.

Proof. The zero function clearly is a solution, so assume that an analytic function $\beta \neq 0$ satisfies (8.3.1). Then $\hat{\beta} = \beta \circ \beta_0^{-1} \neq 0$ satisfies equation (8.3.3), is thus of form (8.3.4), and (8.3.5) follows. ∎

8.3B. Asymptotic behaviour and regularity of real solutions

Now we assume the following.

(iii) $X = |0, a|, 0 < a \leqslant \infty$.
(iv) $f: X \to X$ is continuous, $0 < f(x) < x$ in $X \setminus \{0\}$. Moreover, $f(x) = x^p F(x)$, where F approaches a finite and positive limit as $x \to 0$, and $p > 1$ (not necessarily an integer).

Theorem 8.3.3. *Let hypotheses* (iii), (iv) *be satisfied and let* $0 \notin X$. *Then, for every* $r \in \mathbb{R}$, *equation* (8.3.1) *has a unique solution* $\beta: X \to \mathbb{R}$ *such that* $\lim_{x \to 0} x^{-r} \beta(x)$ *exists, is finite and positive. This solution is continuous in X and is given by the formula*

$$\beta(x) = \lim_{n \to \infty} (f^n(x))^{rp^{-n}}. \tag{8.3.6}$$

(a) *If F is strictly increasing in X, then so are the* β *for* $r > 0$.
(b) *If F is defined and of class* C^s, $1 \leqslant s < \infty$, *in* $X \cup \{0\}$, *then so are the* β *in X.*

Proof. Extend f and F onto $X \cup \{0\}$ by putting $f(0) := 0$, $F(0) := \lim_{x \to 0} F(x)$, to get them continuous in $X \cup \{0\}$. In virtue of Theorem 5.2.3 the equation

$$\varphi(x) = (F(x)^r \varphi(f(x)))^{1/p} \tag{8.3.7}$$

has a unique continuous solution $\varphi: X \cup \{0\} \to \mathbb{R}^+$, $r \in \mathbb{R}$ being fixed. (Take $\Omega = (X \cup \{0\}) \times \mathbb{R}^+$ in Theorem 5.2.3 and $d \neq 0$ determined by $d^p = F(0)d$.) This solution is given as the limit of the sequence (φ_n) of functions defined recursively by

$$\varphi_{n+1}(x) = [F(x)^r \varphi_n(f(x))]^{1/p}, \qquad n \in \mathbb{N}_0, \tag{8.3.8}$$

and starting with, e.g., $\varphi_0 = d := \varphi(0) = F(0)^{r/(p-1)}$. One gets from (8.3.8) by induction that

$$\varphi_n(x) = x^{-r} d^{p^{-n}} (f^n(x))^{rp^{-n}}, \qquad n \in \mathbb{N}_0.$$

The function $\beta(x) = x^r \varphi(x)$ fulfils the conditions of the first part of the theorem.

To prove the uniqueness, assume that $\beta_i = x^r \varphi_i(x)$, $i = 1, 2$, are two solutions of (8.3.1) such that φ_1 and φ_2 approach finite and positive limits as $x \to 0$. Then $\varphi = \varphi_1/\varphi_2$ satisfies (8.3.7) with $r = 0$, whence $\varphi(x) = (\varphi(f^n(x)))^{p^{-n}}$ for $n \in \mathbb{N}$. On letting $n \to \infty$ we obtain hence $\varphi = 1$, because of $0 < \lim_{x \to 0} \varphi(x) < \infty$.

To prove (a), observe that, proceeding as in the proof of Theorem 5.2.2(c), we get φ strictly increasing in X. Assertion (b) follows from Proposition 5.9.3 when applied to (8.3.7). ∎

Theorem 8.3.3 is found in Kuczma [21]. It implies part (a) of the next one which is essentially due to G. Szekeres [1], and concerns the case where $0 \in X$.

Theorem 8.3.4. *Let hypotheses* (iii) *(with* $0 \in X$), (iv) *be satisfied and let F be of class* C^s, $1 \leqslant s < \infty$, *in X.*

(a) *Equation* (8.3.1) *has a unique solution* $\beta: X \to \mathbb{R}$ *of class* C^1 *in X and such that* $\beta'(0) > 0$. *This solution is given by formula* (8.3.6) *with* $r = 1$ *and is in fact of class* C^s *in X.*

(b) *This solution is strictly increasing in X if f is.*

(c) *If, moreover,*

$$f'(x) = pbx^{p-1} + O(x^{p-1+\delta}), \qquad x \to 0, \tag{8.3.9}$$

where b and δ *are positive constants, then*

$$\log \beta(x) = \frac{1}{p-1} \log b + \log x + O(x^\delta), \qquad x \to 0. \tag{8.3.10}$$

Proof. (b) By $\beta'(0) > 0$, the function β is strictly increasing in a neighbourhood of zero, and by (8.3.1) this property is preserved in X.

(c) Condition (8.3.9) implies that $\log(f(x)/bx^p) = O(x^\delta)$, $x \to 0$. Thus there exists an $M > 0$ such that

$$\left| \log f(x) - \log b - p \log x \right| \leqslant Mx^\delta$$

for $x \in (0, x_0]$, with an $x_0 \in X \setminus \{0\}$. Hence it follows by induction that

$$\left| \log f^n(x) - \frac{p^n - 1}{p-1} \log b - p^n \log x \right| \leqslant M \frac{p^n - 1}{p-1} x^\delta$$

for $x \in (0, x_0]$ and $n \in \mathbb{N}$. Since $\log \beta(x) = \lim_{n \to \infty} p^{-n} \log f^n(x)$, this inequality yields (8.3.10) on letting $n \to \infty$. ∎

Remark 8.3.1. Since $\beta'(0) > 0$, $\beta(0) = 0$, the function β is positive in a vicinity of zero, and hence, by (8.3.1), in $X \setminus \{0\}$. Thus the function $\sigma = \log \beta$ satisfies in $X \setminus \{0\}$ the Schröder equation $\sigma(f(x)) = p\sigma(x)$.

8.4 Conjugate functions

We continue to deal with the conjugacy problems announced in Section 8.1. We want to find function classes in which the necessary condition of A-conjugacy or C^r-conjugacy (see Definition 8.1.1), established in Theorem 8.1.1, becomes also sufficient.

8.4A. N-dimensional case

Let us recall the conjugacy equation:

$$\varphi(f(x)) = g(\varphi(x)). \tag{8.4.1}$$

The functions f and g are assumed to have the following properties.

(i) $f, g : X \to \mathbb{K}^N$, where $X \subset \mathbb{K}^N$ is a neighbourhood of the origin; $f(0) = g(0) = 0$. Both functions are either analytic (if $\mathbb{K} = \mathbb{C}$) or of class $C^r, r \geq 2$ (if $\mathbb{K} = \mathbb{R}$), in X. Moreover, $\det f'(0) \neq 0$, and the characteristic roots $s_1, \ldots, s_N$ of $f'(0)$ are all less than unity in absolute value.

As a consequence of Theorems 8.1.1 and 8.2.1 we obtain in view of the transitivity of the conjugacy relation (see Section 8.1) the following.

Theorem 8.4.1. *Assume* (i) *with* $\mathbb{K} = \mathbb{C}$ *and let* s_i, $i = 1, \ldots, N$, *satisfy*

$$s_1^{q_1}, \ldots, s_N^{q_N} \neq s_i \qquad \text{for } q_1, \ldots, q_N \in \mathbb{N}_0, \qquad \sum_{j=1}^{N} q_j = p, \, p = 2, 3, \ldots \quad (8.4.2)$$

The functions f *and* g *are* A-conjugate *if and only if the matrices* $f'(0)$ *and* $g'(0)$ *are conjugate.*

Similarly, from Theorems 8.1.1 and 8.2.2 (see also Remark 2.1) we get the following.

Theorem 8.4.2. *Assume* (i) *with* $\mathbb{K} = \mathbb{R}$ *and let* s_i, $i = 1, \ldots, N$, *satisfy* (8.4.2) *for* $p = 2, \ldots, r$ *and the condition*

$$\left(\max_i |s_i| \right)^r \Big/ \min_i |s_i| < 1.$$

Then the functions f *and* g *are* C^r-conjugate *if and only if the matrices* $f'(0)$ *and* $g'(0)$ *are conjugate.*

Remark 8.4.1. As follows from Theorem 8.2.3, if $\mathbb{K} = \mathbb{R}$, then the functions f and g, satisfying (i) (so that $r \geq 2$) are C^1-conjugate if and only if the matrices $f'(0)$ and $g'(0)$ are conjugate (no additional conditions on s_i are necessary).

8.4B. One-dimensional case

The case where $0 < |f'(0)| < 1$ is covered by Theorems 8.4.1 and 8.4.2 for $N = 1$ (Note 8.8.11). Let $f'(0) = 0$. For A-conjugacy we have

Theorem 8.4.3. *Let* $X \subset \mathbb{C}$ *be a neighbourhood of zero, and let* $f : X \to \mathbb{C}$ *be an analytic function,* $f(0) = f'(0) = 0$. *Let* $g : X \to \mathbb{C}$ *be an analytic function,* $g(0) = 0$. *Then the functions* f *and* g *are* A-conjugate *if and only if at the origin they have zeros of the same order.*

Proof. If f and g are A-conjugate by a function φ, then φ is analytic and $\varphi(x) = x\Phi(x)$, $\Phi(0) \neq 0$ (as $\varphi'(0) \neq 0$). Write $f(x) = x^p F(x)$, $g(x) = x^q G(x)$, $F(0) \neq 0$, $G(0) \neq 0$. Inserting this into (8.4.1) we obtain $p = q$, which shows that the condition is necessary. The sufficiency results from Theorem 8.3.1. ∎

The above proof works also for the C^r-conjugacy in the real case (however, the sufficiency results now from Theorem 8.3.4).

Theorem 8.4.4. *Let* $X = [0, a|, 0 < a \leq \infty$, *and let* $f: X \to \mathbb{R}$, $g: X \to \mathbb{R}$ *have the form* $f(x) = x^p F(x)$, $g(x) = x^q G(x)$, $p > 1$, $q > 1$, *with* F *and* G *of class* C^r $(r \geq 1)$ *in* X, $F(0) \neq 0$, $G(0) \neq 0$. *Then* f *and* g *are* C^r-*conjugate if and only if* $p = q$.

Remark 8.4.2. We might also handle the case of a full neighbourhood of zero, by splitting it into two one-sided neighbourhoods.

The case where $f'(0) = 1$ is most difficult. Here we supply a necessary and sufficient condition for C^1-conjugacy of asymptotically comparable functions (Szekeres [1]). Our main tool will be Theorem 3.5.6 on the Abel equation.

Theorem 8.4.5. *Let* $X = [0, a|, 0 < a \leq \infty$, *and let* $f: X \to \mathbb{R}$, $g: X \to \mathbb{R}$ *be of class* C^1 *in* X *and such that* $f(0) = g(0) = 0$ *and*

$$f'(x) = 1 - b(p+1)x^p + O(x^{p+\delta}), \qquad x \to 0, \tag{8.4.3}$$

$$g'(x) = 1 - d(q+1)x^q + O(x^{q+\eta}), \qquad x \to 0, \tag{8.4.4}$$

where all the constants occurring are positive. Then f *and* g *are* C^1-*conjugate if and only if* $p = q$ (p *and* q *not necessarily integers*).

Proof. Conditions (8.4.3) and (8.4.4) imply that $0 < f(x) < x$, $f'(x) \neq 0$ and $0 < g(x) < x$, $g'(x) \neq 0$ in a right vicinity of the origin, which may be assumed to be $X \setminus \{0\}$, since the conjugacy is a local property. Moreover (8.4.3) and (8.4.4) imply that, for $x \to 0$,

$$f(x) = x - x^{p+1}(b + O(x^\delta)), \qquad g(x) = x - x^{q+1}(d + O(x^\eta)). \tag{8.4.5}$$

Suppose that f and g are C^1-conjugate, i.e. $g = \varphi \circ f \circ \varphi^{-1}$ with φ of class C^1 in X, $\varphi(0) = 0$, $\varphi'(0) \neq 0$ (see Definition 8.1.1). We have $(z := \varphi^{-1}(x))$

$$g(x) - x = \varphi(f(z)) - \varphi(z) = \varphi'(\Theta_x)(f(z) - z) = -\varphi'(\Theta_x)(bz^{p+1} + O(z^{p+1+\delta}))$$
$$= -b\varphi'(0)^{-p}x^{p+1}(1 + o(1)),$$

as $x \to 0$, where Θ_x is a point between $f(z)$ and z, whence $\lim_{x \to 0} \Theta_x = 0$. This together with (8.4.5) for g implies that $p = q$.

Now suppose that $p = q$, and let $\alpha_i: X \setminus \{0\} \to \mathbb{R}$, $i = 1, 2$, be C^1 solutions of the corresponding Abel equations

$$\alpha_1(f(x)) = \alpha_1(x) + 1, \qquad \alpha_2(g(x)) = \alpha_2(x) + 1,$$

with the properties, as $x \to 0$ (see Theorem 3.5.6),

$$\alpha_1'(x) = -b^{-1}x^{-p-1}(1 + O(x^\tau)), \tag{8.4.6}$$

$$\alpha_2'(x) = -d^{-1}x^{-p-1}(1 + O(x^\sigma)), \tag{8.4.7}$$

where $\tau = \min(p, \delta)$, $\sigma = \min(p, \eta)$. Define $\varphi: X \to \mathbb{R}$ by

$$\varphi = \alpha_2^{-1} \circ \alpha_1 \qquad \text{in } X \setminus \{0\}, \ \varphi(0) = 0. \tag{8.4.8}$$

Then φ is continuous on X, of class C^1 on $X \setminus \{0\}$, and (8.4.1) holds. So it is enough to prove that the limit of φ' at zero is finite and different from zero. This will follow from the relation

$$\varphi'(x) = \left(\frac{b}{d}\right)^{1/p} + O(x^\rho), \qquad x \to 0, \ \rho = \min(\tau, \sigma) = \min(p, \delta, \eta) > 0. \quad (8.4.9)$$

To prove (8.4.9), note that (8.4.8) implies $\varphi'(x) = \alpha_1'(x)/\alpha_2'(\varphi(x))$, $x \in X \setminus \{0\}$. Let us estimate $\varphi(x)$ as $x \to 0$. We see that relation (8.4.6), resp. (8.4.7), yields

$$\alpha_1(x) = (bp)^{-1} x^{-p} (1 + O(x^\sigma)), \qquad (8.4.10)$$

$$\alpha_2(x) = (dp)^{-1} x^{-p} (1 + O(x^\sigma)). \qquad (8.4.11)$$

Let $x := \alpha_2^{-1}(y)$. Since $y \to \infty$ if and only if $x \to 0$, relation (8.4.11) implies

$$x = (dpy)^{-1/p} (1 + O(x^\sigma)), \qquad y \to \infty. \qquad (8.4.12)$$

This means that $x = O(y^{-1/p})$, and (8.4.12) becomes

$$\alpha_2^{-1}(y) = (dp)^{-1/p} y^{-1/p} (1 + O(y^{-\sigma/p})), \qquad y \to \infty. \qquad (8.4.13)$$

Now, use (8.4.10) to get

$$\alpha_1(x)^{-1/p} = (bp)^{1/p} x (1 + O(x^\tau)), \qquad x \to 0,$$

and put $y = \alpha_1(x)$ in (8.4.13). Thus $O(y^{-\sigma/p}) = O(x^\sigma)$, $x \to 0$, and, according to (8.4.8), we obtain

$$\varphi(x) = \alpha_2^{-1}(\alpha_1(x)) = (dp)^{-1/p}(bp)^{1/p} x (1 + O(x^\tau))(1 + O(x^\sigma))$$

$$= \left(\frac{b}{d}\right)^{1/p} x (1 + O(x^\rho)), \qquad x \to 0,$$

with ρ as in (8.4.9). Finally, by (8.4.7) we arrive at

$$\alpha_2'(\varphi(x)) = -d^{-1} \left(\frac{d}{b}\right)^{(p+1)/p} x^{-p-1} (1 + O(x^\rho))^{-p-1} (1 + O(x^\sigma))$$

$$= -b^{-1} \left(\frac{d}{b}\right)^{1/p} x^{-p-1} (1 + O(x^\rho)), \qquad x \to 0,$$

which, together with (8.4.6), yields

$$\varphi'(x) = \frac{\alpha_1'(x)}{\alpha_2'(\varphi(x))} = \left(\frac{b}{d}\right)^{1/p} \frac{1 + O(x^\tau)}{1 + O(x^\sigma)}, \qquad x \to 0,$$

and (8.4.9) follows. ∎

8.5 Conjugate formal series and analytic functions

Two complex *formal power series*, short FPSs, f and g are said to be *formally conjugate* iff there exists an invertible FPS φ such that $\varphi \circ f = g \circ f$ (in other words $f = \varphi^{-1} \circ g \circ \varphi$). If the three series have positive radii of convergence, then f and g are said to be *analytically conjugate*.

8.5A. Julia's equation and the iterative logarithm

The theory of conjugate FPSs is based on some notions and results by J. Ecalle [2] concerning the Julia equation

$$\lambda(f(x)) = f'(x)\lambda(x). \tag{8.5.1}$$

Let f be an FPS of the form

$$f(x) = x + \sum_{n=m}^{\infty} b_n x^n, \qquad b_m \neq 0, \, m \geq 2. \tag{8.5.2}$$

Suppose that an FPS

$$\lambda(x) = \sum_{n=0}^{\infty} c_n x^n \tag{8.5.3}$$

formally satisfies equation (8.5.1). Insert (8.5.2) and (8.5.3) into (8.5.1) to get

$$\sum_{n=1}^{\infty} \sum_{i=1}^{n} \binom{n}{i} c_n x^{n-i} \left(\sum_{k=m}^{\infty} b_k x^k \right)^i = \sum_{n=0}^{\infty} \sum_{i=0}^{n} (i+m) b_{i+m} c_{n-i} x^{n+m-1}.$$

Equating the coefficients of x^{m+j-1}, $j \in \mathbb{N}_0$, we obtain

$$jc_j b_m + A_j = mb_m c_j + B_j, \qquad A_0 = B_0 = 0, \tag{8.5.4}$$

where the terms A_j and B_j contain only c_i with $i < j$, and are homogeneous in the c_i. For $j = 0, \ldots, m-1$ relation (8.5.4) yields $c_j = 0$. Thus $A_m = B_m = 0$, c_m may be arbitrary, and then the c_j for $j > m$ can be uniquely determined from (8.5.4) and are homogeneous functions of c_m. Thus we have the following.

Theorem 8.5.1. *If f is an FPS of the form (8.5.2), then equation (8.5.1) has a unique formal solution λ_0 of the form*

$$\lambda_0(x) = b_m x^m + \sum_{n=m+1}^{\infty} c_n x^n. \tag{8.5.5}$$

The general formal solution λ of (8.5.1) is given by $\lambda(x) = c\lambda_0(x)$, where c is an arbitrary constant (a parameter).

Theorem 8.5.1 gives rise to the following definitions which are due to J. Ecalle [2].

Definition 8.5.1. (1) The FPS λ_0 given by (8.5.5) is said to be the *iterative logarithm* (logit f) of the FPS f given by (8.5.2) and is denoted by f_*.

(2) The number $m - 1$ (see (8.5.5)) is called the *iterative valuation* (valit f) of f.

(3) By the *iterative residuum* (resit f) of f we mean the coefficient of x^{-1} in the formal Laurent series $1/f_*$, or, which is the same, the coefficient of x^{m-1} in the FPS $1/f_0$, where $f_*(x) = x^m f_0(x)$.

There is a one-to-one correspondence between f and logit f.

Theorem 8.5.2. *To every FPS (8.5.5) there corresponds a unique FPS (8.5.2) such that $\lambda_0 = \operatorname{logit} f$.*

Proof. This follows on inserting (8.5.2) and (8.5.5) into (8.5.1) and equating the coefficients of x^{2m+j-1}. ∎

The convergence of the series (8.5.5), in the case where the series (8.5.2) converges in a neighbourhood of the origin, is a difficult problem. What is more, it is rather a rare situation, as may be seen from the theorem below whose proof will not be given here.

Theorem 8.5.3. *Let f be a meromorphic function, regular at the origin and having the expansion (8.5.2). If the FPS f_* has a positive radius of convergence, then*

$$f(x) = \frac{x}{1 + bx}, \qquad b \in \mathbb{C}.$$

Theorem 8.5.3 is implied by the results of I. N. Baker [9] (see also Szekeres [4]) and Erdős–Jabotinsky [1].

8.5B. Formally conjugate power series

We start with necessary conditions for two FPSs

$$\left.\begin{aligned} f(x) &= x + \sum_{n=m}^{\infty} b_n x^n, & b_m \neq 0, m \geq 2, \\ g(x) &= x + \sum_{n=k}^{\infty} a_n x^n, & a_k \neq 0, k \geq 2, \end{aligned}\right\} \tag{8.5.6}$$

to be conjugate. Regarding the theorem that follows see Ecalle [1], [2], Erdős–Jabotinsky [1], Muckenhoupt [1].

Theorem 8.5.4. *Let f and g, of form (8.5.6), be formally conjugate FPSs, i.e.*

$$\varphi(f(x)) = g(\varphi(x)) \tag{8.5.7}$$

holds with an invertible FPS φ, the inverse of which has the form

$$\varphi^{-1}(x) = \sum_{n=1}^{\infty} d_n x^n, \qquad d_1 \neq 0. \tag{8.5.8}$$

Then, for $f_* = \operatorname{logit} f$ and $g_* = \operatorname{logit} g$, we have

$$f_* = (g_* \circ \varphi)/\varphi', \tag{8.5.9}$$

$$\operatorname{valit} f = \operatorname{valit} g, \tag{8.5.10}$$

$$\operatorname{resit} f = \operatorname{resit} g. \tag{8.5.11}$$

Proof. In view of Theorem 8.5.1 the series f_* and g_* are uniquely determined.

Inserting (8.5.6) and (8.5.8) into the relation $\varphi^{-1} \circ g = f \circ \varphi^{-1}$ we obtain

$$\sum_{n=1}^{\infty} \sum_{i=1}^{n} \binom{n}{i} d_n x^{n-i} \left(\sum_{j=k}^{\infty} a_j x^j \right)^i = \sum_{n=m}^{\infty} b_n \left(\sum_{i=1}^{\infty} d_i x^i \right)^n$$

whence it follows immediately that $m = k$, yielding (8.5.10); and also $b_m = a_m d_1^{1-m}$. Hence the series on the right-hand side of (8.5.9) has the form $b_m x^m + \cdots$. Thus it is logit f, if it satisfies the Julia equation (8.5.1). We check that (note that $g_* \circ g = g' \circ g_*$, by the definition of g_*)

$$\frac{g_* \circ (\varphi \circ f)}{\varphi' \circ f} = \frac{g_* \circ (g \circ \varphi)}{\varphi' \circ f} = \frac{(g_* \circ g) \circ \varphi}{\varphi' \circ f} = \frac{g' \circ \varphi}{\varphi' \circ f} g_* \circ \varphi = \frac{(g' \circ \varphi)\varphi'}{\varphi' \circ f} \frac{g_* \circ \varphi}{\varphi'}.$$

When (8.5.7) is differentiated, the first ratio here becomes f', i.e.

$$\frac{g_* \circ \varphi}{\varphi'} \circ f = f' \cdot \frac{g_* \circ \varphi}{\varphi'}$$

and we get (8.5.9) in view of Theorem 8.5.1.

In order to calculate the iterative residua of g and f we need to find the coefficient of x^{-1} in $1/g_*$ and $1/f_*$, respectively. This coefficient depends only on a finite number of the coefficients in the FPSs g_* and φ (respectively in the FPSs f_* and φ). Thus there exist polynomials G and Φ (curtailments of g_* and φ) such that the coefficients of x^{-1} in $1/G$ and in $\Phi'/(G \circ \Phi)$ are equal to those in $1/g_*$ and in $1/f_* = \varphi'/(g_* \circ \varphi)$, respectively. But for polynomials these coefficients are the usual residua of the corresponding functions, and can be expressed by the known integral formulae. Thus we have

$$\text{resit } f = \frac{1}{2\pi i} \int \frac{\Phi'(x)dx}{G(\Phi(x))} = \frac{1}{2\pi i} \int \frac{du}{G(u)} = \text{resit } g,$$

where the path of integration is a (positively oriented) closed contour around the origin. This proves (8.5.11). ■

The relation of conjugacy is transitive. Thus we may first look for conditions for f to be conjugate with a specific (sample) FPS. We attempt to find a $D \in \mathbb{C}$ for f to be formally conjugate with

$$g(x) = x + x^m + Dx^{2m-1}. \tag{8.5.12}$$

Supposing that (8.5.7) holds with an FPS φ, the inverse of which has expansion (8.5.8), we obtain

$$\sum_{n=m}^{\infty} b_n \left(\sum_{k=1}^{\infty} d_k x^k \right)^n = \sum_{p=1}^{\infty} \sum_{i=1}^{p} d_p \binom{p}{i} x^{p-i+mi}(1 + Dx^{m-1})^i. \tag{8.5.13}$$

Comparing in (8.5.13) the coefficients of x^m yields $b_m d_1^m = d_1$, whence d_1 may be any of the $m-1$ values of the $(m-1)$st root of b_m. Equating then the coefficients of x^{m+j-1} for $j \geq 2$ we find

$$mb_m d_1^{m-1} d_j + A_j = jd_j + B_j, \tag{8.5.14}$$

where A_j and B_j depend only on d_i with $i < j$, and, for $j \geq m$, B_j may depend also on D. Since $b_m d_1^{m-1} = 1$, relation (8.5.14) yields unique d_j for $j = 2, \ldots, m - 1$. For $j = m$ we have $B_m = B_m^* + d_1 D$, where B_m^* is independent of D. Since (8.5.14) now becomes $A_m = B_m$, we get $D = d_1^{-1}(A_m - B_m^*)$. For $j > m$ we can again determine the d_j from (8.5.14), and d_m is arbitrary.

To proceed further we need the following lemma, which results from comparing the coefficients in the formula that we get from (8.5.1) with $\lambda = f_*$ (see (8.5.4)).

Lemma 8.5.1. *If*

$$f(x) = x + x^m + dx^{2m-1} + \sum_{n=2m}^{\infty} b_n x^n$$

then

$$f_*(x) = x^m + (D - \tfrac{1}{2}m)x^{2m-1} + \sum_{n=2m+2}^{\infty} c_n x^n.$$

Now, we can prove that D does not depend on the choice of the root d_1. For suppose that f given by (8.5.2) is formally conjugate with g defined by (8.5.12). By Lemma 8.5.1 $g_*(x) = x^m + (D - \tfrac{1}{2}m)x^{2m-1} + \cdots$, i.e. $(1/g_*)(x) = x^{-m} - (D - \tfrac{1}{2}m)x^{-1} + \cdots$, Theorem 8.5.4 says that resit $f =$ resit $g = -D + \tfrac{1}{2}m$, and $D = \tfrac{1}{2}m -$ resit f depends only on f.

It follows from what has been said so far that f is conjugate with g of form (8.5.12) if and only if $D = \tfrac{1}{2}m -$ resit f. Thus two series of form (8.5.6) are conjugate with the same g given by (8.5.12) if they have the same iterative valuation and iterative residuum. These considerations, together with Theorem 8.5.4, imply the following.

Theorem 8.5.5. *Let f and g be FPSs with $f(0) = g(0) = 0$, $f'(0) = g'(0) = 1$. Series f and g are formally conjugate if and only if valit $f =$ valit g and resit $f =$ resit g.*

There is one more fact that can be derived from our discussion based on examining relation (8.5.13). Namely, if we stop after having determined from (8.5.14) the d_j for $j = 1, \ldots, m - 1$, then we get a polynomial $P(x) = d_1 x + \cdots + d_{m-1}x^{m-1}$ such that $\bar{g} := P^{-1} \circ f \circ P$ has a form $\bar{g}(x) = x + x^m + dx^{2m-1} +$ terms of higher orders, since the latter bear only on c_j with $j > m$. We shall formulate this fact as a lemma, which will be useful in the next subsection.

Lemma 8.5.2. *If f is an FPS of form (8.5.2), then there exists a polynomial P such that*

$$P^{-1} \circ f \circ P(x) = x + x^m + dx^{2m-1} + \cdots,$$

where $d = \tfrac{1}{2}m -$ resit f.

8.5C. Conjugate analytic functions

The conditions expressed in Theorem 8.5.5 are also necessary, but not sufficient for the analytic conjugacy of f and g if they actually represent analytic functions. In fact, let f be an analytic function with expansion (8.5.2) such that f_* has a positive radius of convergence, and let g be given by (8.5.12), where $D = \frac{1}{2}m - \text{resit } f$. Then f and g are formally conjugate in virtue of Theorem 8.5.5. If f and g were analytically conjugate, then we would have $g = \varphi^{-1} \circ f \circ g$ with an analytic function φ, and by Theorem 8.5.4 $g_* = (f_* \circ \varphi)/\varphi'$ would have a positive radius of convergence. But this is impossible in view of Theorem 8.5.3, as $g(x) \neq x/(1 + bx)$.

Necessary and sufficient conditions for analytic conjugacy of functions f, g with $f(0) = g(0) = 0$, $f'(0) = g'(0) = 1$ have been given by J. Ecalle [3], but they are too complicated to be reproduced here. However, after B. Muckenhoupt [1], we shall prove the following.

Theorem 8.5.6. *Let $X \subset \mathbb{C}$ be a neighbourhood of the origin, and let $f: X \to \mathbb{C}$ and $g: X \to \mathbb{C}$ be analytic functions such that $f(0) = g(0) = 0$, $f'(0) = g'(0) = 1$. If their iterative logarithms f_* and g_* have positive radii of convergence and f, g are formally conjugate, then they are analytically conjugate.*

Proof. By Lemma 8.5.2 and Theorem 8.5.5 it suffices to consider f and g of the form

$$f(x) = x + x^m + dx^{2m-1} + \sum_{n=2m}^{\infty} b_n x^n, \quad g(x) = x + x^m + dx^{2m-1} + \sum_{n=2m}^{\infty} a_n x^n,$$

where $d = \frac{1}{2}m + r$, $m = 1 + \text{valit } f = 1 + \text{valit } g$, $r = \text{resit } f = \text{resit } g$, for the series above are formally conjugate with the corresponding original ones.

By Lemma 8.5.1 we have

$$f_*(x) = x^m + (d - \tfrac{1}{2}m)x^{2m-1} + \cdots, \quad g_*(x) = x^m + (d - \tfrac{1}{2}m)x^{2m-1} + \cdots.$$
$$(8.5.15)$$

Consider the differential equation

$$y' = f_*(y)/g_*(x). \tag{8.5.16}$$

We want to find its solutions of the form $y(x) = x + x^m z(x)$. Then z should satisfy

$$z' = \frac{f_*(x + x^m z) - g_*(x) - mzx^{m-1}g_*(x)}{x^m g_*(x)}. \tag{8.5.17}$$

By (8.5.15) the right-hand side of (8.5.17) is an analytic function of (x, z) in a neighbourhood of $(0, 0)$. By Cauchy's Existence Theorem for differential equations, (8.5.17) has an analytic solution z in a neighbourhood of the origin. The corresponding function y is analytic and invertible in a neighbourhood of the origin. Write

$$h = y^{-1} \circ f \circ y. \tag{8.5.18}$$

Treating h as an FPS we find similarly as in the first lines of the proof of Theorem 8.5.4 that

$$h(x) = x + x^m + \sum_{n=m+1}^{\infty} h_n x^n$$

(observe that now $b_m = 1$). By Theorem 8.5.1 the FPS $h_* = x^m + \cdots$ makes sense. We want to prove that $h_* = g_*$.

Indeed, (8.5.16), (8.5.18), the relation $f_* \circ f = f' \cdot f_*$ and again (8.5.16) imply

$$(y' \circ h)(g_* \circ h) = f_* \circ (y \circ h) = (f_* \circ f) \circ y = (f' \circ y)(f_* \circ y) = (f' \circ y)y'g_*.$$

But (8.5.18) yields $(y' \circ h)h' = (f' \circ y)y'$ so that

$$g_*(h(x)) = h'(x)g_*(x).$$

In view of (8.5.15) this means that $g_* = h_*$. By Theorem 8.5.2 also $g = h$ and (8.5.18) shows that g is analytically conjugate with f. ∎

Now we pass to the case where $|f'(0)| = 1$ but $f'(0) \neq 1$. Observe that the condition $f'(0) = g'(0)$ is necessary for functions f and g (both analytic in a neighbourhood of the origin) to be analytically conjugate. Theorem 8.5.7, whose part (b) is due to B. Muckenhoupt [1] contains conditions equivalent to analytical conjugacy. (If $f'(0)^p = 1$, $f^p \neq \mathrm{id}$, then Theorem 8.7.6 applies.)

Theorem 8.5.7. *Let $X \subset \mathbb{C}$ be a neighbourhood of the origin and let $f \colon X \to \mathbb{C}$, $g \colon X \to \mathbb{C}$ be analytic functions such that $f(0) = g(0) = 0$. Consider two cases:*

(a) *$f'(0) \in S$, where S is the Siegel set (see Definition 4.3.1),*
(b) *$f'(0) \neq 1$ is a pth root of unity and $f^p = \mathrm{id}$.*

Necessary and sufficient conditions for g to be analytically conjugate with f are

in case (a): $g'(0) = f'(0)$,
in case (b): $g'(0) = f'(0)$ and $g^p = \mathrm{id}$.

Proof. Part (a) follows from Theorem 8.3.1. In case (b) the necessity of the condition $g^p = \mathrm{id}$ is obvious, and the sufficiency is a consequence of Theorem 4.6.2. ∎

8.5D. Abel's equation

Results of Subsection 8.5A may also be used to prove a theorem on complex solutions to the Abel equation

$$\alpha(f(x)) = \alpha(x) + 1 \tag{8.5.19}$$

in the case where $s = f'(0) = 1$ (Ecalle [2]; for $s \neq 1$ see Subsection 4.6B).

Theorem 8.5.8. *Let f having expansion (8.5.2) be analytic in a neighbourhood*

*of the origin. Then equation (8.5.19) has solutions α defined in a vicinity of
the origin and such that*

$$\alpha(x) = c_0 \log x + \sum_{i=1}^{m-1} c_i x^{-i} + \varphi(x), \qquad c_0, \ldots, c_{m-1} \in \mathbb{C}, \quad (8.5.20)$$

*where φ is an analytic function in a neighbourhood of the origin, if and only
if $f_* = \text{logit } f$ has a positive radius of convergence. If f_* actually is a function
analytic at the origin, then α is determined uniquely up to an additive constant
and is given by the formula*

$$\alpha(x) = c + \int_{x_0}^{x} \left[\int_{x_0}^{f(x_0)} \frac{dt}{f_*(t)} \right]^{-1} \frac{dt}{f_*(t)}, \qquad (8.5.21)$$

*where c is an arbitrary constant (a parameter), x_0 is a point arbitrarily fixed
in a vicinity V of the origin, and the integration is over an arbitrary path in
V joining x_0 with x, respectively $f(x_0)$.*

Proof. Let α, defined in a vicinity of the origin, be a solution of (8.5.19) with
property (8.5.20). Since (8.5.19) cannot have a solution analytic at the origin,
not all the c_i are zero. Thus $\lambda(x) = 1/\alpha'(x)$ is analytic at the origin and satisfies
equation (8.5.1). In view of Theorem 8.5.1, we have $\lambda(x) = \gamma f_*(x)$ with a
$\gamma \neq 0$. Since $f_*(x) = x^m \psi(x)$ with $\psi(0) = b_m \neq 0$, the function $1/\lambda$ is analytic
in a vicinity V of the origin and has a pole of order m at 0. Consequently,
in V we have

$$\alpha(x) = c + \int_{x_0}^{x} \frac{dt}{\lambda(t)} = c + \gamma^{-1} \int_{x_0}^{x} \frac{dt}{f_*(t)}, \qquad (8.5.22)$$

where $x_0 \in V$ is arbitrary, and we integrate over an arbitrary path in V joining
x_0 with x. Since α satisfies (8.5.19), we have

$$\int_{x_0}^{f(x_0)} \frac{dt}{f_*(t)} = \int_{x_0}^{f(x_0)} \alpha'(t) dt = \gamma [a(f(x_0)) - \alpha(x_0)] = \gamma,$$

and so (8.5.21) results from (8.5.22).

Now assume that f_* is an analytic function in a neighbourhood of the
origin. Then in a vicinity of $x = 0$ we have $f_*(x) \neq 0$, $f(x) \neq -x$ and

$$\lim_{x \to 0} \int_{x}^{f(x)} \frac{dt}{f_*(t)} = \lim_{x \to 0} \frac{f(x)^{1-m} - x^{1-m}}{(1-m)b_m} = 1, \qquad (8.5.23)$$

where the integral is taken over the segment joining x and $f(x)$. So we can
find a vicinity V of the origin such that both f_* and the integral occurring
in (8.5.23) do not vanish in V. Define $\alpha(x)$ for $x \in V$ by (8.5.21) with $x_0 \in V$
arbitrarily fixed. Then it is easily seen that α has property (8.5.20) and satisfies
equation (8.5.19). ■

Comments. Regarding conjugacy problems we discussed in this and the
preceding sections see, in particular, Belickiĭ [1]–[9], Bratman [1],

Ditor [1], Ecalle [1]–[3], Fine–Kostant [1], Herman [1], Julia [3], Kneser [1], Kuczma [11], [23], Muckenhoupt [1], Nitecki [1], Sternberg [2], [6], Venti [1], Weitkämper [2].

8.6 Permutable functions

Let $g = f$ in the conjugacy equation (8.0.1). Then we obtain the equation

$$\varphi(f(x)) = f(\varphi(x)). \tag{8.6.1}$$

If (8.6.1) is satisfied, then we say that f and φ *commute* or are *permutable*.

Commuting functions have been extensively studied by many authors (see Comments at the end of this section). Here we confine ourselves to the one-dimensional case only and to presenting some results which can be obtained with the aid of solutions of the Schröder, Abel and Böttcher equations.

Permutability is a rare property of functions. In the Cartesian square of the space $\mathbf{C} = C([0,1], \mathbb{R})$ of continuous functions with the usual sup norm, the pairs of commuting functions form a nowhere dense set (Kuczma [27]). On the other hand, it can be deduced from Theorem 5.3.1 that for every strictly monotonic $f \in \mathbf{C}$ the solution $\varphi \in \mathbf{C}$ of equation (8.6.1) depends on an arbitrary function (Lipiński [1], Kuczma [26, pp. 213–14]).

However, if f and φ have a higher regularity, then often we are able to prove the uniqueness of φ satisfying (8.6.1) with a given f. The theorems to this effect we present below are based on the following scheme.

Suppose that f is conjugate to a function g, $\varphi_0 \circ f = g \circ \varphi_0$, where φ_0 is an invertible function smooth enough. If a function φ is permutable with f, then $\hat{\varphi} := \varphi_0 \circ \varphi$ satisfies equation (8.0.1). In a class of sufficiently smooth functions this equation may happen to have a unique solution, up to a parameter. In such a case, since both $\hat{\varphi}$ and φ_0 satisfy equation (8.0.1), they must be related in a form $\hat{\varphi} = G(c, \varphi_0)$, where c is the parameter. Hence $\varphi = \varphi_0^{-1} \circ G(c, \varphi_0)$, and we obtain a one-parameter family of solutions of (8.6.1).

Now, results on smooth permutable functions are obtained by the argument just described with $g(x) = sx$, $g(x) = x + 1$ and $g(x) = x^p$. Equation (8.0.1) then becomes the equation of Schröder (8.0.2), Abel (8.0.3) and Böttcher (8.0.4), respectively.

In the following, when we say 'all φ' we mean 'all functions φ commuting with f and satisfying $\varphi(0) = 0$'. Moreover, whenever we write $U \subset X$, we mean by U a neighbourhood of the origin. Finally, we put

$$s := f'(0).$$

In the case of analytic functions, if $0 < |s| \leqslant 1$, we obtain from Theorems 4.3.1, 4.6.1 and 4.6.3 the following.

Theorem 8.6.1. *Let $X \subset \mathbb{C}$ be a neighbourhood of the origin, and let $f: X \to \mathbb{C}$ be a function analytic on X, $f(0) = 0$. Further assume that*

$$0 < |s| < 1 \qquad \text{or} \qquad s \in S$$

(where S is the Siegel set – see Definition 4.3.1). Then all φ analytic in a $U \subset X$ are given by

$$\varphi(x) = \sigma^{-1}(c\sigma(x)), \qquad x \in X, \tag{8.6.2}$$

where

$$\sigma(x) = \lim_{n \to \infty} s^{-n} f^n(x), \tag{8.6.3}$$

and $c \in \mathbb{C}$ is an arbitrary constant.

In the real case Theorem 3.5.1 and Proposition 3.9.3 yield the following.

Theorem 8.6.2. *Let $X \subset \mathbb{R}$ be a neighbourhood of the origin, and let $f: X \to \mathbb{R}$ be a function of class C^1 on X, $f(0) = 0$, such that*

$$f'(x) = s + O(x^\delta), \qquad x \to 0, \, \delta > 0, \, 0 < |s| < 1.$$

Then all φ of class C^1 in a $U \subset X$ are given by formulae (8.6.2)–(8.6.3).

Similarly, if $s = 0$, we obtain from Theorems 8.3.1 and 8.3.2 the following.

Theorem 8.6.3. *Let X and f be as in Theorem 8.6.1 and, moreover, $f(x) = x^p F(x)$, $F(0) \neq 0$, $p > 1$ (an integer). Then the only φ analytic in a $U \subset X$ are $\varphi = 0$ and*

$$\varphi(x) = \beta_0^{-1}(\varepsilon_j(\beta_0(x))^r), \qquad x \in X, \, j = 1, \ldots, p-1, \, r \in \mathbb{N},$$

where β_0 is a function occurring in Theorem 8.3.1, and ε_j are $(p-1)$st roots of unity.

In the real case we have by Theorem 8.3.3 the following.

Theorem 8.6.4. *Let $X = [0, a|, \, 0 < a \leqslant \infty$, and let $f: X \to \mathbb{R}$ have the form $f(x) = x^p F(x)$, where F is continuous and strictly increasing in X, $F(0) > 0$, $p > 1$ (not necessarily an integer). Then all φ such that $\lim_{x \to 0} x^{-q} \varphi(x)$ exists, is finite and positive (where $q > 0$) are given by*

$$\varphi(x) = \beta^{-1}((\beta(x))^q), \qquad x \in X,$$

where β is given by formula (8.3.6) with $r = 1$.

In the case where $s = 1$ we have by Theorem 3.5.6 (also Theorem 8.4.4) the following.

Theorem 8.6.5. *Let $X = [0, a|, \, 0 < a \leqslant \infty$, and let $f: X \to \mathbb{R}$ be of class C^1, $f(0) = 0$, and fulfil condition (8.4.2). Then all φ of class C^1 in a $U \subset X$ such* *that $\varphi'(0) > 0$ are given by*

$$\varphi(x) = \begin{cases} \alpha^{-1}(\alpha(x) + c), & x \in X \setminus \{0\}, \\ 0, & x = 0, \end{cases} \tag{8.6.4}$$

where

$$\alpha(x) = \lim_{n \to \infty} \frac{f^n(x) - f^n(x_0)}{f^{n+1}(x_0) - f^n(x_0)}, \tag{8.6.5}$$

and $c \in \mathbb{R}$ is an arbitrary constant.

Remark 8.6.1. In the course of the proof of Theorem 8.4.5 we have shown that the function $\varphi := \alpha_2^{-1} \circ \alpha_1$ in $U \setminus \{0\}$, $\varphi(0) := 0$, where α_i, $i = 1, 2$, are C^1 solutions of suitable Abel equations, has a continuous derivative at zero. Taking here $\alpha_2 = \alpha$ (given by (8.6.5)) and $\alpha_1 = \alpha + c$, we get (8.6.4). Thus that φ is of class C^1 in the whole U, as claimed.

Comments. The references we give here cover also the problems of permutable formal power series, resp. analytic functions, which are discussed in the next section. Consult, among others, I. N. Baker [7], [8], Boyce [1], Chen [1], [2], Coifman [3], Ecalle [2], Erdős–Jabotinsky [1], Fatou [4], Huneke [1], Julia [2], Kopell [1], Krüppel [1], Kuczma [11], [27], Lipiński [2], Muckenhoupt [1], Pelczar [5], Matkowski [1], Zdun [18]. See also Kuczma [26, Ch. X] and references quoted therein.

8.7 Commuting formal series and analytic functions

Finally, we consider the permutability equation

$$\varphi(f(x)) = f(\varphi(x)) \tag{8.7.1}$$

first in the class of formal power series and then in that of analytic functions. The most interesting, but also the most difficult, is the case where $f'(0) = 1$.

8.7A. Formal power series that commute with a given one

Let f and φ_t be the formal power series (abbreviated FPSs)

$$f(x) = x + \sum_{n=m}^{\infty} b_n x^n, \qquad b_m \neq 0, \, m \leqslant 2, \tag{8.7.2}$$

$$\varphi_t(x) = x + t b_m x^m + \sum_{n=m+1}^{\infty} c_n(t) x^n. \tag{8.7.3}$$

We are looking for FPSs (8.7.3) that formally commute with f.

Inserting (8.7.2) and (8.7.3) into (8.7.1) and equating the coefficients of the resulting series we arrive at the following result (see Erdős–Jabotinsky [1], I. N. Baker [7] for details).

Theorem 8.7.1. *For every $t \in \mathbb{C}$ there exists a unique FPS $\varphi_t[f]$ of form (8.7.3) that formally commutes with f given by (8.7.2). For every $t, s \in \mathbb{C}$ the FPSs $\varphi_t := \varphi_t[f]$ and $\varphi_s := \varphi_s[f]$ formally commute as they satisfy*

$$\varphi_t \circ \varphi_s = \varphi_{t+s}. \tag{8.7.4}$$

The coefficients $c_n(t)$ of t^n in $\varphi_t[f]$ are polynomials in t.

We want to determine the form of all FPSs commuting with f. First we prove the following (Baker [7]).

Lemma 8.7.1. *For every FPS (8.7.2) and every primitive $(m-1)$st root of unity ε there exists an FPS γ such that*

$$\gamma(0) = 0, \quad \gamma'(0) = \varepsilon, \quad \gamma^{m-1}(x) = x, \tag{8.7.5}$$

and γ commutes with all $\varphi_t := \varphi_t[f]$ of form (8.7.3) associated with f.

Proof. By Theorem 8.5.4 the FPSs f and $g(x) := x + x^m + Dx^{2m-1}$, where $D = \frac{1}{2}m - \text{resit } f$ (see Definition 8.5.1) are conjugate. In other words, there is an FPS φ such that $g = \varphi \circ f \circ \varphi^{-1}$. Put $h(x) := \varepsilon x$. Clearly, h commutes with g, thus also with $g^n = f^n$. This means that h commutes with $\varphi \circ \varphi_n \circ \varphi^{-1}$ for all $n \in \mathbb{N}$. Let $d_j = d_j(n)$ be the coefficient of x^j in the FPS $\varphi \circ \varphi_n \circ \varphi^{-1}$. It follows that $\varepsilon d_j = d_j \varepsilon^j$, i.e. $d_j = 0$ except in the case where $j = 1 + (m-1)k$, $k \in \mathbb{N}_0$. Since $c_j(t)$ in $\varphi_t[f]$ are polynomials in t, each FPS $\varphi \circ \varphi_t \circ \varphi^{-1}$ also has all the coefficients zero except those of $x^{1+(m-1)k}$. Consequently h commutes with all the series $\varphi \circ \varphi_t \circ \varphi^{-1}$, $t \in \mathbb{C}$. Now it is easily seen that $\gamma := \varphi^{-1} \circ h \circ \varphi$ has all the desired properties. ∎

The following theorem and lemma are due to B. Muckenhoupt [1].

Theorem 8.7.2. *Let f be an FPS of form (8.7.2), and let φ be an FPS*

$$\varphi(x) = \sum_{n=1}^{\infty} c_n x^n. \tag{8.7.6}$$

If φ commutes formally with f, then either $\varphi = 0$ or there exist numbers $j \in \mathbb{N}$ and $t \in \mathbb{C}$ such that

$$\varphi = \gamma^j \circ \varphi_t,$$

where $\varphi_t = \varphi_t[f]$ has form (8.7.3), and γ is an FPS occurring in Lemma 8.7.1.

Proof. Fix a primitive $(m-1)$st root of unity ε. Inserting (8.7.2) and (8.7.6) into (8.7.1) and comparing the coefficients of x^m we obtain $c_m + b_m c_1^m = c_1 b_m + c_m$, whence either $c_1 = 0$, or $c_1 = \varepsilon^j$ with a $j \in \mathbb{N}$. Further, the coefficients of x^{m+k-1} for $k = 2, 3, \ldots$ are

$$c_{m+k-1} + mb_m c_1^{m-1} c_k + A_k = c_{m+k-1} + kc_k b_m + B_k, \tag{8.7.7}$$

where A_k and B_k contain only c_is with $i < k$ and $A_k = B_k = 0$ if $c_1 = \cdots = c_{k-1} = 0$. If $c_1 = 0$, then we get from (8.7.7) $c_k = 0$ for all k, i.e.,

$\varphi = 0$. If $c_1 = \varepsilon^j$, then we can find a unique c_k from (8.7.7) except for c_m, which can be arbitrary. Thus all the coefficients of an FPS commuting with f are uniquely determined by the first and the mth ones.

Let $c_1 = \varepsilon^j$. By Lemma 8.7.1 there exists an FPS γ commuting with all φ_t, $t \in \mathbb{C}$, and such that $\gamma^j(x) = c_1 x + \cdots$. Thus for any $t \in \mathbb{C}$ the first coefficient in $\gamma^j \circ \varphi_t$ is c_1. By a proper choice of t we can make the coefficient of x^m equal to c_m. Since both γ^j and φ_t commute with f, so does $\gamma^j \circ \varphi_t$, and thus it must be identical with φ. ∎

Lemma 8.7.2. *The FPS γ from Lemma 8.7.1 is unique.*

Proof. Suppose that γ_1 and γ_2 satisfy conditions (8.7.5) and are permutable with all $\varphi_t = \varphi_t[f]$, $t \in \mathbb{C}$. Thus, in particular, they commute with $\varphi_1 = f$. By Theorem 8.7.2 they are permutable with each other. Let $\chi := \gamma_1 \circ \gamma_2^{-1}$. Since γ_1 and γ_2 commute, we have $\chi^{m-1}(x) = x$. On the other hand, χ also commutes with f, whence, by Theorem 8.7.2, there are numbers $j \in \mathbb{N}$ and $t \in \mathbb{C}$ such that $\chi = \gamma_1^j \circ \varphi_t$. By (8.7.4) $\chi^{m-1} = \gamma_1^{(m-1)j} \circ \varphi_{(m-1)t}$. Thus $\varphi_{(m-1)t}(x) = x$, in turn $t = 0$ and $\chi = \gamma_1^j$. By (8.7.5) we have $\chi'(0) = 1$, whence $j \equiv 0 \pmod{(m-1)}$. Therefore $\chi(x) = x$, which implies $\gamma_1 = \gamma_2$. ∎

8.7B. Convergence of formal power series having iterative logarithm

Let us return to Theorem 8.7.1 and ask the question when the FPS $\varphi_t := \varphi_t[f]$ converges.

It follows from the uniqueness of φ_t that for integral t it is the expansion at zero of f^t, and so, if series (8.7.2) converges, φ_t has a positive radius of convergence. But for other values of t this fails to be true. We quote here without proof a corresponding result which is due to I. N. Baker [7] and J. Ecalle [2].

Theorem 8.7.3. *Let f be an analytic function in a neighbourhood of the origin, with expansion (8.7.2). The FPS $\varphi_t[f]$ has a positive radius of convergence either for all $t \in \mathbb{C}$, or only for t being integral multiples of a real rational number (depending on f).*

The following theorem (Erdős–Jabotinsky [1]) says which of the above described possibilities is actually the case.

Theorem 8.7.4. *Let f be an FPS of form (8.7.2). The FPS $\varphi_t[f]$ has a positive radius of convergence for all $t \in \mathbb{C}$ if and only if $f_* = \text{logit } f$ has a positive radius of convergence.*

Proof. Suppose that f_* has a positive radius of convergence. As in the proof of Theorem 8.5.6 we deduce that for every $t \in \mathbb{C}$ the differential equation

$y' = f_*(y)/f_*(x)$ has a unique local analytic solution y_t of the form $y_t(x) = x + tb_m x^m +$ higher order (h.o.) terms. Let us fix a $t \in \mathbb{C}$.

Put $u = y_t \circ f$. Then $u(x) = x + (t+1)b_m x^m +$ h.o. terms, and

$$u' = (y_t' \circ f)f' = f'(x)f_*(u)/f_*(f(x)) = f_*(u)/f_*(x),$$

since $\hat{\lambda}(x) := f_*(x)$ satisfies the Julia equation (8.5.1). By the uniqueness of y_t we obtain hence $u = y_{t+1}$.

Now put $v = f \circ y_t$. Again $v(x) = x + (t+1)b_m x^m +$ h.o. terms and

$$v' = (f' \circ y_t)y_t' = f'(y_t(x))f_*(y_t(x)/f_*(x)) = f_*(v)/f_*(x),$$

whence, again by the uniqueness, $v = y_{t+1}$. Thus y_t commutes with f, and since $\varphi_t := \varphi_t[f]$ is unique we obtain $y_t = \varphi_t$. Consequently φ_t is an analytic function in a neighbourhood of the origin.

Now suppose that all φ_t have a positive radius of convergence. Then $\varphi_t(x)$ is an analytic function of (x, t). We put

$$\lambda(x) := \frac{\partial}{\partial t} \varphi_t(x) \Big|_{t=0}.$$

Thus λ is analytic with an expansion $\lambda(x) = b_m x^m +$ h.o. terms. Setting $\varphi = \varphi_t$ in (8.7.1) and differentiating (8.7.1) with respect to t we obtain

$$\frac{\partial}{\partial t} \varphi_t(f(x)) = f'(\varphi_t(x)) \frac{\partial}{\partial t} \varphi_t(x).$$

Put here $t = 0$ and note that $\varphi_0(x) = f^0(x) = x$. We obtain $\lambda(f(x)) = f'(x)\lambda(x)$ and $\lambda = f_*$ in virtue of Theorem 8.5.1. ∎

8.7C. Permutable analytic functions

To determine analytic functions that commute with an analytic f first we need the following lemma (I. N. Baker [7]).

Lemma 8.7.3. *Let f be an analytic function in a neighbourhood of the origin, with expansion (8.7.2). If f_* has a positive radius of convergence, then the FPSs γ occurring in Lemma 8.7.1 also have positive radiii of convergence.*

Proof. First note that for a fixed ε, $\varepsilon^{m-1} = 1$, the series γ occurring in Lemma 8.7.1 is, by Lemma 8.7.2, unique. We put

$$\lambda(x) = x^m + Dx^{2m-1}, \qquad D := \tfrac{1}{2}m - \text{resit } f.$$

It follows from Theorem 8.5.1 that there is a unique FPS $g(x) = x + x^m +$ h.o. terms such that $\lambda = g_*$, i.e.

$$\lambda(g(x)) = g'(x)\lambda(x). \tag{8.7.8}$$

Since λ is a polynomial, it follows by Theorem 8.7.4 that all FPSs $\varphi_t := \varphi_t[g]$ of form (8.7.3) commuting with g have a positive radius of convergence. Thus $g = \varphi_1$ is an analytic function.

All the coefficients in the series g are zero except those of $x^{1+(m-1)k}$, $k \in \mathbb{N}_0$. For suppose the contrary, i.e. let a_q be the first nonzero coefficient with q not of the form $1 + (m-1)k$. Comparing the coefficients of x^{m+q-1} in both sides of (8.7.8) we get $ma_q = qa_q$, a contradiction. Thus g commutes with $h(x) := \varepsilon x$.

By Theorem 8.5.6 there exists an analytic function φ conjugating f and g, $\varphi \circ f \circ \varphi^{-1} = g$. The same argument as in the proof of Lemma 8.7.1 shows that $\hat{\gamma} := \varphi^{-1} \circ h \circ \varphi$ has all the properties of the series γ. Therefore $\gamma = \hat{\gamma}$, by the uniqueness of γ, and γ has a positive radius of convergence. ∎

The same argument we have used in the proof of Theorem 8.7.2 yields the following.

Theorem 8.7.5. *Let f be as in Lemma 8.7.3, and assume that f_* has a positive radius of convergence. Then the only analytic functions (8.7.6) commuting with f are $\varphi = 0$ and $\varphi = \gamma^j \circ \varphi_t$, $j \in \mathbb{N}$, $t \in \mathbb{C}$, where γ occur in Lemma 8.7.3 and $\varphi_t = \varphi_t[f]$.*

8.7D. Conjugacy again

Finally, after B. Muckenhoupt [1] we show how results on commutable functions may be used to attack the conjugacy problem for functions f and g with $f'(0)$ $(= g'(0))$ being a primitive pth root of unity.

Theorem 8.7.6. *Let f and g be FPSs (resp. analytic functions in a neighbourhood of the origin) such that $f(0) = g(0) = 0$ and $f'(0)$ is a primitive pth root of unity , but $f^p \neq \mathrm{id}$. Then f and g are formally (resp. analytically) conjugate if and only if $f'(0) = g'(0)$ and f^p is formally (resp. analytically) conjugate with g^p.*

Proof. Necessity is obvious. So let $f'(0) = g'(0)$ and let f^p and g^p be formally (resp. analytically) conjugate. Put $h = g^p$, so there exists an FPS (resp. analytic function) φ such that $\varphi \circ f^p \circ \varphi^{-1} = h$. Write $\tilde{g} := \varphi \circ f \circ \varphi^{-1}$. Thus $\tilde{g}^p = h$ and $\tilde{g}$ commutes with h. By Theorem 8.7.2 there exist a $j \in \mathbb{N}$ and the FPS $\tilde{\varphi}_t := \varphi_t[h]$ (see Theorem 8.7.1) such that $\tilde{g} = \tilde{\gamma}^j \circ \tilde{\varphi}_t$, where $\tilde{\gamma}$ is an FPS occurring in Lemma 8.7.1 (see Lemma 8.7.2), corresponding to the function h. Since g also commutes with (its iterate) h, we have, again by Theorem 8.7.2, $g = \tilde{\gamma}^i \circ \tilde{\varphi}_u$ with some $i \in \mathbb{N}$, $u \in \mathbb{C}$. Since $g'(0) = f'(0) = \tilde{g}'(0)$, we have to have $i \equiv j \pmod{(m-1)}$, whence $\tilde{\gamma}^i = \tilde{\gamma}^j$. Further, since $\tilde{g}^p = g^p$ and $\tilde{\gamma}$ commutes with $\tilde{\varphi}_t$ and $\tilde{\varphi}_u$, we have $\tilde{\gamma}^{jp} \circ \tilde{\varphi}_t^p = \tilde{\gamma}^{jp} \circ \tilde{\varphi}_u^p$, implying $\tilde{\varphi}_t^p = \tilde{\varphi}_u^p$. By (8.7.4), $\tilde{\varphi}_{pt} = \tilde{\varphi}_{pu}$, which implies $pt = pu$, whence $t = u$. Thus $\tilde{g} = \tilde{\gamma}^j \circ \tilde{\varphi}_t = \tilde{\gamma}^j \circ \tilde{\varphi}_u = g$. This means that $g = \varphi \circ f \circ \varphi^{-1}$, i.e., f and g are formally (resp. analytically) conjugate. ∎

Remark 8.7.1. Note that in the case considered $f'(0)^p = (f^p)'(0) = (g^p)'(0) = 1$ so that Theorems 8.5.5 and 8.5.6 apply.

8.8 Notes

8.8.1. The trapezoid functions (Fig. 8.1)

$$t_a: [0, 1] \to [0, 1], \qquad a \in (0, \tfrac{1}{2}),$$

$$t_a(x) = \begin{cases} x/a, & x \in [0, a], \\ 1, & x \in [a, 1-a], \\ (1-x)/a, & x \in [1-a, 1], \end{cases}$$

are conjugate with each other by a continuous function. Conjugacy itself may be derived from Theorem 8.1.2 (see Schweizer–Sklar [4]). The function which conjugates t_a with t_b is constructed as linear on each component of the complement of a Cantor set (see Schweizer–Sklar [6]).

8.8.2. The hat functions (Fig. 8.1)

$$h_{u,v}: [0, 1] \to [0, 1], \qquad u, v \in (0, 1], u \neq 1,$$

$$h_{u,v}(x) = \begin{cases} \dfrac{v}{u} x, & x \in [0, u], \\ \dfrac{v}{1-u} (1-x), & x \in [u, 1], \end{cases}$$

are conjugate if and only if they are conjugate by a continuous function (Weitkämper [1]). The hat functions that belong to the following classes are conjugate with each other: (1) $u > v$, (2) $u = v$, (3) $u < v$ and $1 - u > v$, (4) $u < v$ and $u = 1 - v$. If $1 - v < u < v$, two hat functions are conjugate only then, when the vertices of their graphs lie on some curves (Frank [1]).

8.8.3. The functions

$$f_p: [-1, 1] \to [-1, 1], \qquad p > 0,$$
$$f_p(x) = 2|x|^p - 1,$$

are conjugate with f_1 when $p \in [\tfrac{1}{2}, 2]$ and are not when $p \in (0, \tfrac{1}{2})$ (Mycielski [1]). The latter assertion follows from Theorem 8.1.2. Indeed, each f_p,

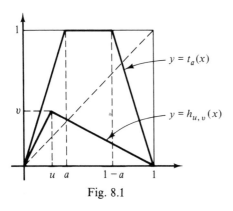

Fig. 8.1

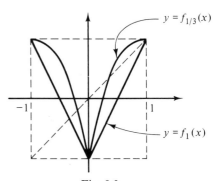

Fig. 8.2

$p \in (0, \frac{1}{2})$, has three fixed points whereas f_1 has only two (Fig. 8.2). The first assertion is implied by the following (Corollary 5 in Mycielski [1]).

Proposition 8.8.1. *Let* $f = f_1 \circ h$, *where* h *is a continuous strictly increasing map of* $[-1, 1]$ *onto itself. If* $f(a) = -1$, f_-^{-1} *is the inverse function to* $f|[-1, a]$, f_+^{-1} *is the inverse function to* $f|[a, 1]$ *and* $(f_-^{-1})'(x) > -1$, $(f_+^{-1})'(x) < 1$ *for almost all* $x \in [-1, 1]$, *then* f *is conjugate with* f_1.

(In the case where $1 < p \leqslant 2$ we may check that the function $f = h^{-1} \circ f_p \circ h$, where $h(x) = \sin \frac{1}{2}\pi x$, satisfies the conditions of Proposition 8.8.1. Thus f is conjugate with f_1. By its definition, f is conjugate with f_p, hence so is f_p with f_1. The function h has first been used by S. Ulam [1, Ch. VI, § 2.4], for a direct proof that f_2 is conjugate with f_1.)

8.8.4. The condition (8.2.4) of Theorem 8.2.2 for the matrix $S = f'(0)$ occurring in the Schröder equation

$$\sigma(f(x)) = S\sigma(x) \tag{8.8.1}$$

may be dropped if we weaken the regularity requirement on σ.

Proposition 8.8.2. *Let hypothesis* (ii) *in Section 8.2 be satisfied and let* $|s_k| \neq 1$ *for* $k = 1, \ldots, N$, *where* s_k *are the characteristic roots of* S. *Then equation* (8.8.1) *has a continuous and invertible solution* $\sigma: U \to \mathbb{R}^N$ *in a neighbourhood* U *of the origin in* $\mathbb{R}^N$.

(See Grobman [1], Hartman [2]–[4], Nitecki [1].)

8.8.5. The following facts are needed in the proof of Theorem 8.2.1 on p. 338.

Let $B = (b_k^j) \in \mathbb{K}^{N \times N}$ be a matrix. By B^r, $r \in \mathbb{N}$, we mean the r-fold tensor product of B, i.e. the multidimensional matrix with the valence (r, r):

$$B^r := (b_{k_1}^{j_1} \cdots b_{k_r}^{j_r}); \qquad j_m, k_m \in \{1, \ldots, N\}, m \in \{1, \ldots, r\}.$$

Further, by $u = (u_{k_1 \cdots k_N}^i)$ we denote the elements of the Banach space Y_r of all N^{r+1}-tuples of reals, endowed with the norm $\|u\|_r$ defined by (8.2.6) ($y = u$, $q = r$).

Now, with $A = (a_i^j) \in \mathbb{K}^{N \times N}$, u and B as above, we define $v = AuB^r \in Y_r$ by the formula

$$v = (v_{k_1 \cdots k_r}^j) = \left(\sum_{i=1}^N \sum_{j_1=1}^N \cdots \sum_{j_r=1}^N a_i^j u_{j_1 \cdots j_r}^i b_{k_1}^{j_1} \cdots b_{k_r}^{j_r} \right).$$

To obtain the first component in (8.2.8) we need to take here

$$A = S^{-1}, \qquad u = \sigma^{(r)}(f(x)), \qquad \sigma^{(r)} = \left(\frac{\partial^r \sigma_i}{\partial x_{j_1} \cdots \partial x_{j_r}} \right),$$

$B = f'(x) = (\partial f_j / \partial x_k(x))$, where the indexed σ_i, f_j are the components of $\sigma, f: U \to \mathbb{R}^N$, $U \subset \mathbb{R}^N$.

Finally, the estimates

$$\|AuB^r\|_r \le \|A\| \|u\|_r \|B\|^r, \qquad r \in \mathbb{N},$$

may be verified by induction. With suitable substitutions they become (8.2.9).

8.8.6. The Julia functional equation

$$\lambda(f(x)) = f'(x)\lambda(x) \tag{8.8.2}$$

may have nontrivial solutions although the corresponding Schröder equation $\sigma(f(x)) = f'(0)\sigma(x)$ has not. This may be seen from the following.

Example 8.8.1. Take $f(x) = x/(1 + x)$, $x \in (-1, 1)$. Thus $f'(0) = 1$. The Julia equation becomes

$$\lambda(x/(1 + x)) = (1 + x)^{-2}\lambda(x), \qquad x \in (-1, 1). \tag{8.8.3}$$

The functions $\lambda(x) = cx^2$, $x \in (-1, 1)$, satisfy (8.8.3), but the corresponding Schröder equation $\sigma(f(x)) = \sigma(x)$ has $\sigma = 0$ as the only differentiable or analytic (or even continuous) solution in a neighbourhood of the origin satisfying $\sigma(0) = 0$. To see this, let $n \to \pm \infty$ in $\sigma(f^n(x)) = \sigma(x)$.

8.8.7. Solutions of the Schröder equation (8.8.1) may yield solutions of the Julia equation (8.8.2). For instance, we have the following.

Proposition 8.8.3. *Let* $X \subset \mathbb{K}^N$ *be a neighbourhood of the origin, and let* $f: X \to \mathbb{K}^N$ *be continuously differentiable in* X, $f(0) = 0$, $f'(0) = S$, $\det S \ne 0$. *If the Schröder equation (8.8.1) has a* C^r *($r \ge 1$,* $\mathbb{K} = \mathbb{R}$*), resp. analytic* *(*$\mathbb{K} = \mathbb{C}$*), solution* $\sigma: U \to \mathbb{K}^N$, $\sigma(0) = 0$, $\det \sigma'(0) \ne 0$, *where* $U \subset X$ *is a neighbourhood of the origin, then equation (8.8.2) has a solution* $\lambda: U \to \mathbb{K}^N$, *of class* C^{r-1}, *resp. analytic, in* U *and such that* $\det \lambda'(0) \ne 0$. *(If* $\mathbb{K} = \mathbb{R}$, *the latter condition holds provided* $r > 1$.)

(It is readily checked that the function $\lambda: U \to \mathbb{K}^N$. $\lambda(x) = \sigma'(x)^{-1}\sigma(x)$ is the desired solution of (8.8.2).)

8.8.8. Consider the Böttcher equation

$$\beta(f(x)) = [\beta(x)]^p \tag{8.8.4}$$

and assume hypotheses (iii) and (iv) of Section 8.3. If $s < p$, then C^s solution of equation (8.8.4) in $X \setminus \{0\}$ depends on an arbitrary function (Anczyk [1]).

8.8.9. The existence of a unique solution of the Böttcher equation (8.8.4) may be also established in the class of functions that are conjugate with a convex or concave function via exponential function. We have the following (see Kuczma [22]).

Proposition 8.8.4. *Suppose that* $X = (0, a|, 0 < a \leqslant \infty$, $f: X \to X$ *is continuous*, $0 < f(x) < x$ *in* X, $f(x) = x^p F(x)$, $p > 1$, *and that* $\lim_{x \to 0}(\log F(x)/\log x) = 0$. *If the function* $\log \circ f \circ \exp$ *is convex or concave in* $(-\infty, \log a)$, *then equation* (8.8.4) *has a unique one-parameter family of solutions* $\beta: X \to \mathbb{R}$ *such that* $\log \circ \beta \circ \exp$ *is convex or concave in* $(-\infty, \log a)$. *These solutions are given by the formula*

$$\beta(x) = \lim_{n \to \infty} (f^n(x))^{r/\log f^n(x_0)},$$

where $x_0 \in X$ *is arbitrarily fixed and* $r \in \mathbb{R}$ *is the parameter.*

(The change of variables $\sigma = \log \circ \beta \circ \exp$, $\hat{f} = \log \circ f \circ \exp$ transforms (8.8.4) into the Schröder equation $\sigma(\hat{f}(x)) = p\sigma(x)$, to which Proposition 2.9.4 applies.)

8.8.10. The condition 'log $\circ f \circ \exp$ is convex (or concave)' of Proposition 8.8.4 may be expressed in terms of convexity (or concavity) of f with respect to the interpolatory family of functions $g(x) = cx^d$, $c \in (0, \infty)$, $d \in \mathbb{R}$ (see Beckenbach [1], Moldovan [1], and also Kuczma [22]).

8.8.11. In the case $N = 1$, Theorems 8.4.1 and 8.4.2 may be reformulated as follows.

Proposition 8.8.5. *Let* $X \subset \mathbb{K}$ *be a neighbourhood of zero, and let* $f: X \to \mathbb{K}$ *be a function analytic (resp. of class* C^r, $r \geqslant 2$*) in* X, $f(0) = 0$, $0 < |f'(0)| < 1$. *Let* $g: X \to \mathbb{K}$ *be a function analytic (resp. of class* C^r*) in* X, $g(0) = 0$. *The functions* f *and* g *are A-conjugate (resp.* C^r-*conjugate) if and only if* $f'(0) = g'(0)$.

(This remains true for $\mathbb{K} = \mathbb{R}$ and $r = 1$ if we assume additionally that $f'(x) = s + O(x^\delta)$, $\delta > 0$, as $x \to 0$, and that $g'(x)$ behaves likewise (see Theorem 3.5.1.)

8.8.12. L. P. Kučko [3] considers the problem of local C^∞-equivalence of two functional equations, $\varphi(f(x)) = G(x, \varphi(x))$ and $\Phi(f(x)) = H(x, \Phi(x))$, under a locally C^∞ mapping Ψ, $\varphi(x) \mapsto \Psi(x, \Phi(x))$ that transforms one equation into another. Such a mapping Ψ has to satisfy the equation $\Psi(f(x), H(x, y)) = G(x, \Psi(x, y))$. Note that if two equations are locally C^∞-equivalent, then there is a one-to-one correspondence between the sets of solutions to each of them, with the regularity of solutions preserved.

This technique when used to linearize the conjugacy equation (8.0.1), $\varphi(f(x)) = g(\varphi(x))$, yields, by a result of S. Sternberg [5], the following (Kučko [3]).

Proposition 8.8.6. *Let* f *and* g *be locally* C^∞ *mappings of* $\mathbb{R}^N$ *into itself*, $f(x) = f'(0)x + O(|x|^2)$ *and* $g(x) = g'(0)x + O(|x|^2)$ *as* $|x| \to 0$. *If* $f'(0) = g'(0)$ *and the characteristic roots* s_i, $i = 1, \ldots, N$, *of the matrix* $f'(0)$ *satisfy the* condition $s_i \neq s_1^{k_1} \cdots s_N^{k_N}$ for $i = 1, \ldots, N$ and $k_j \in \mathbb{N}_0$, then the conjugacy equation (8.0.1) is locally C^∞-equivalent to the Schröder equation (8.8.1) $(S = f'(0))$.

(The C^∞ mapping Ψ that transforms one equation into another now satisfies $\Psi(f'(0)y) = g(\Psi(y))$, i.e., it is a function that conjugates g with its differential at the origin.)

8.8.13. M. Boshernitzan–L. A. Rubel [1] consider coherent families of formal power series (or functions). An FPS given by

$$f(x) = \sum_{k=1}^{\infty} a_k x^k, \qquad a_k \in \mathbb{C}, \tag{8.8.5}$$

is said to be *compositionally coherent* (c.c.) iff the family $\{f^n : n \in \mathbb{N}\}$ is *coherent*, i.e., its members $y = f^n$ satisfy (formally) one and the same differential equation

$$P(x, y, y', \ldots, y^{(n)}) = 0, \tag{8.8.6}$$

where P is a (complex) polynomial in its $n + 2$ variables. We say that f is *totally compositionally incoherent* (t.c.i.) iff no infinite subfamily of $\{f^n : n \in \mathbb{N}\}$ is coherent. Any y satisfying *an* equation (8.8.6) is said to be *differentially algebraic*.

The following connection between the Schröder and Julia equations and the compositional coherency is found in Boshernitzan–Rubel [1].

Proposition 8.8.5. *The FPS (8.8.5) is c.c. or t.c.i. exactly when the FPS φ is differentially algebraic or not, where*

(1) *if $a_1 \neq 0$ and a_1 is not a root of unity, then $\varphi(x) = x + \sum_{k=2}^{\infty} b_k x^k$ is the (unique) FPS solution of the Schröder equation*

$$\varphi(f(x)) = a_1 \varphi(x), \tag{8.8.7}$$

(2) *if $a_1 = 1$, then $\varphi = \text{logit } f$ (see Definition 8.5.1).*

For instance, if f is a finite Blaschke product $(k > 1)$

$$f(x) = x \prod_{i=1}^{k} \frac{x - x_i}{1 - \bar{x}_i x}, \qquad 0 < |x_i| < 1, \, x_i \in \mathbb{C}, \, i = 1, \ldots, k,$$

then the solution φ of (8.8.7) is transcendentally transcendental (Carroll [1]), i.e., not differentially algebraic, hence f is t.c.i. On the other hand, polynomials x^m, $m \in \mathbb{N}$, and the Čebyšev polynomials $T_m(x)$ are c.c. A. Eremenko showed (F. Carroll, private communication) that a polynomial of nth order is c.c. if and only if it is conjugate by an affine function with x^m or $T_m(x)$. Thus $f(x) = cx + x^2$ are c.c. for $c = 0$ and $c = 2, 4, -2$, as they are conjugate by an affine function either to x^2 or to $T_2(x) = 2x^2 - 1$. Since $x + x^2$ is not c.c., $\text{logit}(x + x^2)$ is not differentially algebraic.

9

More on the Schröder and Abel equations

9.0 Introduction

In this chapter we are going to discuss some further problems concerning Schröder's,

$$\sigma(f(x)) = s\sigma(x), \tag{9.0.1}$$

and Abel's,

$$\alpha(f(x)) = \alpha(x) + 1, \tag{9.0.2}$$

equations and their applications. We start by introducing the notion of principal solutions of equations (9.0.1) and (9.0.2) and of the Koenigs and Lévy algorithms. In Section 9.2 we consider relations between the pre-Schröder system

$$\sigma[f(x)]^n = \sigma(f^n(x))[\sigma(x)]^{n-1}, \qquad n \in \mathbb{N},$$

and the Schröder equation (9.0.1).

Abel and Schröder equations jointly are shown in Section 9.3 to be useful for determining associative functions. Systems of form (9.0.2) yield transformations of differential equations with deviating argument (Section 9.4) and those of form (9.0.1) serve as a tool for characterization of norms in some sequence and function spaces (Section 9.5). Several further applications are reported in the Notes section.

Both the Schröder and the Abel equations play an important role throughout this book. Here we want to indicate problems for and applications of equations (9.0.1) and (9.0.2) which are presented in other chapters of the book. For the sake of brevity, function classes for solutions and domains of applications are itemized and tabulated (Tables 9.1 and 9.2).

9.1 Principal solutions

In most applications of the Schröder and Abel equations there appear their unique solutions defined by formulae involving iterates of the function f.

Following G. Szekeres [1] we distinguish special solutions of this kind which are called the principal solutions to Schröder's and Abel's equations.

In the sequel we take $X = [0, a|$, $0 < a \leqslant \infty$, and $f: X \to X$ to be a continuous strictly increasing function, $0 < f(x) < x$ in $X \setminus \{0\}$.

Definition 9.1.1. Let us assume that the limit

$$s := \lim_{n \to \infty} \frac{f^{n+1}(x)}{f^n(x)}, \qquad x \in X \setminus \{0\},$$

exists and it does not depend on x. Take an $x_0 \in X \setminus \{0\}$. If the limit

$$\sigma(x) = \lim_{n \to \infty} \frac{f^n(x)}{f^n(x_0)}, \qquad x \in X, \tag{9.1.1}$$

exists and is positive and finite in $X \setminus \{0\}$, then it satisfies

$$\sigma(f(x)) = s\sigma(x) \tag{9.1.2}$$

and is called the *principal solution of the Schröder equation* (9.1.2).

Table 9.1. *Schröder's equation*

Function class	Subsection	Applications	Section, etc.
Monotonic	2.3F	Branching processes (BP): limit distributions	2.1B, 2.6A
Convex	2.4D	BP: restricted stationary measures	2.6C
		Fractional iteration	11.4
Regularly varying	2.5B	BP: norming sequences	Note 2.8.14
$C^r, r \geqslant 1$	3.5A	Linearization	8.2
	8.2B	Permutability	8.6
		Fractional iteration	11.4
		Cross ratio	3.5D
		An integral equation	Note 3.8.14
Analytic	4.6A	Conjugacy	8.4A
	4.3A	Permutability	8.6
	8.2A	Fractional iteration	11.6
		Characterization of functions	10.2A, D
		Second-order iterative inequality	12.7

Table 9.2. *Abel's equation*

Function class	Subsection	Applications	Section, etc.
Convex	2.4D	BP: limit distributions	2.1B, 2.6A
		BP: restricted stationary measures	2.1D, 2.6C
$C^r, r \geqslant 1$	3.5C	Second-order differential equations	3.7C
		Conjugacy	8.4B
		Permutability	8.6
Analytic	4.6B		
	8.5D		

If we took in (9.1.1) an $x_1 \in X \setminus \{0\}$ in place of x_0, then the limit would differ from the previous one only by a constant multiple. Thus the principal solution of the Schröder equation is unique up to a constant factor.

In general, the principal solution behaves better near the origin than other solutions of (9.1.2). Conditions for the existence of limit (9.1.1) are contained in Theorems 2.3.12 and 2.4.4 and also in Lemma 2.5.1.

Remark 9.1.1. If there exists the limit

$$\tilde{\sigma}(x) = \lim_{n \to \infty} s^{-n} f^n(x), \qquad x \in X, \tag{9.1.3}$$

then limit (9.1.1) exists as well. The converse is not true. E. Seneta [5] proved that the key condition for the existence of $\tilde{\sigma}$, $0 < \tilde{\sigma} < \infty$, is the convergence of the improper integral

$$\int_0^\delta (f(x) - sx) x^{-2} dx$$

for $\delta \in X \setminus \{0\}$ (see Theorem 1.3.2). Formula (9.1.3) is called the *Koenigs algorithm* (Koenigs [1]); see Theorem 3.5.1.

Now let $(d_n)_{n \in \mathbb{N}}$ be any sequence of reals for which

$$\lim_{n \to \infty} \frac{1}{d_n} (f^{n+1}(x) - f^n(x)) = 1, \qquad x \in X \setminus \{0\}. \tag{9.1.4}$$

Definition 9.1.2. Let a sequence (d_n) have property (9.1.4). If there exists the limit

$$\alpha(x) := \lim_{n \to \infty} \frac{1}{d_n} (f^n(x) - f^n(x_0)), \qquad x \in X \setminus \{0\}, \tag{9.1.5}$$

where $x_0 \in X \setminus \{0\}$, then it satisfies

$$\alpha(f(x)) = \alpha(x) + 1 \tag{9.1.6}$$

and is called the *principal solution of the Abel equation* (9.1.6).

It may be easily shown that limit (9.1.5) does not depend on the choice of a sequence (d_n) satisfying (9.1.4), and that by replacing x_0 in (9.1.5) by another $x_1 \in X \setminus \{0\}$ we obtain a limit which differs from (9.1.5) by a constant summand. Thus the principal solution of the Abel equation is determined up to an additive constant.

Remark 9.1.2. If (9.1.4) holds for $d_n = f^{n+1}(x_0) - f^n(x_0)$, where $x_0 \in X \setminus \{0\}$ and if there exists the limit

$$\tilde{\alpha}(x) = \lim_{n \to \infty} \frac{f^n(x) - f^n(x_0)}{f^{n+1}(x_0) - f^n(x_0)}, \qquad x \in X \setminus \{0\}, \tag{9.1.7}$$

then it is the principal solution of (9.1.6). Formula (9.1.7) is called the

Lévy algorithm (Lévy [1]). Theorems 2.4.3 and 3.5.6 give conditions for the effectiveness of the Lévy algorithm.

Comments. Further results on the Schröder equation may be found in I. N. Baker [10], Carroll [1], Coifman [1]–[3], Diamond [2], Ecalle [2], Fatou [1], [4], Hartman [1]–[4], Julia [1], [2], Kneser [1], Kuczma [11], [24], [42], [44], Lundberg [1], Montel [1], Myrberg [1], Nitecki [1], Peschl [1], Peschl–Reich [1], L. Reich [1], Schubert [1], Seneta [5], W. Smajdor [6], [7], Śmiałkówna [1], Sternberg [3]–[5], Szekeres [1], Targoński [2]–[4], Töpfer [1]. See also Bellman [2, Ch. 14] and the references in Kuczma [11], [26, Ch. VI], [44].

For further results regarding the Abel equations in the one-dimensional case see Bödewadt [1], Burkart [2], Coifman [1]–[3], Dubuc [1], Ecalle [2], Fatou [1]–[3], [5], Julia [2], Szekeres [1]–[5], Tambs Lyche [1], Targoński [6], Töpfer [1], Zdun [12]. See also the references in Kuczma [26, Ch. VIII]. In the multi-dimensional case the Abel equation is not presented in this book, thus see, in particular, Diamond [1], Hoppe [1].

9.2 The pre-Schröder system

Let X be a set, and let $f: X \to X$ be a function. Solutions $\sigma: X \to \mathbb{K}$ of the Schröder equation

$$\sigma(f(x)) = s\sigma(x) \qquad (9.2.1)$$

are eigenfunctions (corresponding to the eigenvalue $s \in \mathbb{K}$) of the operator of substitution $T(\varphi) = \varphi \circ f$, acting in the space of mappings $\varphi: X \to \mathbb{K}$.

We want to eliminate the scalar factor s from (9.2.1). To this end we first iterate (9.2.1) n times, next raise both sides of (9.2.1) to the nth power, and compare the resulting equations. We arrive at the system of equations

$$[\sigma(f(x))]^n = \sigma(f^n(x))[\sigma(x)]^{n-1}, \qquad n \in \mathbb{N}, \qquad (9.2.2)$$

which has been introduced by Gy. Targoński [5] under the name of the *pre-Schröder system*.

9.2A. An equivalent system and automorphic functions

The pre-Schröder system (9.2.2) is interesting only for $n \geqslant 2$ (for $n = 1$ it is an identity). Every solution $\sigma: X \to \mathbb{K}$ of equation (9.2.1) satisfies all equations (9.2.2), $n \in \mathbb{N}$. The converse implication is not true, as has been shown by Z. Moszner [1] ($n = 2$) and R. Burkart [2] ($n > 2$).

Theorem 9.2.1. *A function $\sigma: X \to \mathbb{K}$ satisfies equation (9.2.2) for a given $n \geqslant 2$ if and only if there exists a function $\omega: X \to \mathbb{K}$ such that σ and ω satisfy the* *system of equations*

$$\sigma(f(x)) = \omega(x)\sigma(x) \tag{9.2.3}$$

and

$$[\omega(x)]^{n-1} = \prod_{i=1}^{n-1} \omega(f^i(x)). \tag{9.2.4}$$

Proof. Fix an $n \geq 2$. It is readily seen that (9.2.3) and (9.2.4) imply (9.2.2). Conversely, suppose that a function $\sigma: X \to \mathbb{K}$ satisfies (9.2.2). Put $\omega(x) = \sigma(f(x))/\sigma(x)$ if $\sigma(x) \neq 0$, and $\omega(x) = 0$ otherwise. It follows by (9.2.2) that $\sigma(x) = 0$ implies $\sigma(f(x)) = 0$. Thus (9.2.3) holds by the definition of $\omega: X \to \mathbb{K}$. Furthermore, if $\sigma(x) = 0$, then $\omega(x) = \omega(f(x)) = 0$, and (9.2.4) holds. And if $\sigma(x) \neq 0$ then

$$\omega(x)^n = \frac{\sigma(f(x))^n}{\sigma(x)^n} = \frac{\sigma(f^n(x))}{\sigma(x)}.$$

But (9.2.3) implies $\sigma(f^n(x)) = \omega(x) \cdots \omega(f^{n-1}(x))\sigma(x)$ and (9.2.4) follows. ∎

For $n = 2$ equation (9.2.4) takes the form

$$\omega(x) = \omega(f(x)) \tag{9.2.5}$$

which implies (9.2.4) for all $n \geq 2$. Thus, by Theorem 9.2.1, if (9.2.2) is satisfied for $n = 2$, then it is satisfied for all $n \in \mathbb{N}$.

Solutions of (9.2.5) are called *automorphic functions* (see, e.g., Ghermănescu [1]). Their structure is determined by the orbit structure of X under f, as we have the following.

Theorem 9.2.2. *Let X, Y be arbitrary sets and $f: X \to X$ an arbitrary function. A function $\omega: X \to \mathbb{K}$ satisfies equation (9.2.5) if and only if it is constant on the orbits under f.*

Proof. Let $x, y \in X$, $x \sim_f y$ (see Section 1.1). Then there exist $p, q \in \mathbb{N}_0$ such that $f^p(x) = f^q(y)$. Thus we get by (9.2.5) $\omega(x) = \omega(f^p(x)) = \omega(f^q(y)) = \omega(y)$. ∎

Remark 9.2.1. Theorems 9.2.1 and 9.2.2 remain valid if $\mathbb{K}$ is replaced by a (commutative) field $\mathbf{F}$. They imply that if X reduces to a single orbit under f, then system (9.2.2) and equation (9.2.1) are equivalent. But if X contains more than one orbit then (9.2.1) and (9.2.2) are not equivalent for a very large class of functions (Kuczma [39]). It is the case, in particular, if char $\mathbf{F} \neq 2$ and X contains an orbit $C \neq X$ without fixed points or with a fixed point of an even order, see Note 9.6.1.

9.2B. Equivalence of the Schröder equation and the pre-Schröder system

If we restrict the class of functions in which system (9.2.2) is investigated, it may happen that (9.2.2) and (9.2.1) are equivalent. Below we give two theorems to this effect (Kuczma–Targoński [1]). Assume that

(i) $X \subset \mathbb{K}$ is a neighbourhood of the origin,
(ii) $f: X \to X$ is a function such that $f(0) = 0$ and $f'(0)$ exists.

First we consider the class of functions that are differentiable at the origin.

Theorem 9.2.3. *Let hypotheses* (i), (ii) *be satisfied and let* $f(x) \neq 0$ *for* $x \in X \setminus \{0\}$, $\lim_{k \to \infty} f^k(x) = 0$ *for all* $x \in X$. *If a function* $\sigma: X \to \mathbb{K}$ *satisfies system* (9.2.2) *and has a finite derivative* $\sigma'(0) \neq 0$, *then* σ *satisfies* (9.2.1) *with* $s = f'(0)$.

Proof. We have $\sigma(x) \neq 0$ in $X \setminus \{0\}$. For $\sigma(x_0) = 0$ (with an $x_0 \in X \setminus \{0\}$) implies $\sigma(f^k(x_0)) = 0$ for $k \in \mathbb{N}_0$ and

$$\sigma'(0) = \lim_{k \to \infty} \frac{\sigma(f^k(x_0)) - \sigma(0)}{f^k(x_0)} = -\sigma(0) \Big/ \lim_{k \to \infty} f^k(x_0)$$

would be either zero or infinite, according as $\sigma(0) = 0$ or not, contrary to the assumption.

By Theorem 9.2.1 the function $\omega(x) = \sigma(f(x))/\sigma(x)$ for $x \in X \setminus \{0\}$ satisfies equation (9.2.5) (i.e. (9.2.4) for $n = 2$). Thus

$$\omega(x) = \omega(f^k(x)), \qquad x \in X \setminus \{0\}, \; k \in \mathbb{N}. \tag{9.2.6}$$

We have for $x \in X \setminus \{0\}$

$$\omega(f^k(x)) = \frac{\sigma(f^{k+1}(x))}{\sigma(f^k(x))} = \frac{\sigma(f^{k+1}(x))}{f^{k+1}(x)} \frac{f^k(x)}{\sigma(f^k(x))} \frac{f^{k+1}(x)}{f^k(x)}. \tag{9.2.7}$$

If $\sigma(0) \neq 0$, then $\lim_{k \to \infty} \omega(f^k(x)) = 1$, according to the first equality in (9.2.7). If $\sigma(0) = 0$, we forget the first equality in (9.2.7) and deduce from the other that $\lim_{k \to \infty} \omega(f^k(x)) = f'(0)$. By (9.2.6) in both cases ω is constant in $X \setminus \{0\}$, and σ satisfies (9.2.1) there with $s = 1$ or $s = f'(0)$. By the continuity of σ at zero, (9.2.1) holds in the whole of X. If $s = 1$, σ would be an automorphic function continuous at zero, thus constant in X. This would imply $\sigma'(0) = 0$, contrary to the assumption. Thus $s = f'(0)$. ∎

Now we pass to the class of functions analytic on X.

Theorem 9.2.4. *Let hypotheses* (i), (ii) *be satisfied and let* f *be analytic in* X, $f(x) \neq 0$ $(x \neq 0)$, $|f'(0)| < 1$. *If an analytic function* $\sigma: X \to \mathbb{K}$ *satisfies system* (9.2.2), *then it satisfies* (9.2.1) *with* $s = f'(0)^q$, *where* q *is the order of zero of* σ *at the origin. If* $\sigma = 0$, *then* s *may be arbitrary.*

Proof. Choose an $r > 0$ so that $U = \{x \in \mathbb{K}: |x| < r\} \subset X$ and $0 < |f(x)| < |x|$ in $U \setminus \{0\}$. If $\sigma(x_0) = 0$ for an $x_0 \in U \setminus \{0\}$, then $\sigma(f^k(x_0)) = 0$ for $k \in \mathbb{N}$ and, according to Theorem 1.2.5, $\sigma = 0$ in X. Obviously, (9.2.1) holds with any s.

Now assume that $\sigma(x) \neq 0$ in $U \setminus \{0\}$ and write $\sigma(x) = x^q \psi(x)$, $q \in \mathbb{N}_0$, $\psi(0) \neq 0$. Again $\omega(x) = \sigma(f(x))/\sigma(x)$ in $U \setminus \{0\}$ satisfies (9.2.5). We have for $x \in U \setminus \{0\}$

$$\omega(x) = \lim_{k \to \infty} \omega(f^k(x)) = \lim_{k \to \infty} \left[\left(\frac{f^{k+1}(x)}{f^k(x)} \right)^q \frac{\psi(f^{k+1}(x))}{\psi(f^k(x))} \right] = f'(0)^q.$$

Hence σ satisfies (9.2.1) with $s = f'(0)^q$ in $U \setminus \{0\}$. Since it is analytic, (9.2.1) holds in the whole X. ■

Comments. The equivalence of (9.2.2) and (9.2.1) and of particular equations (9.2.2) to each other (also in the case of a more general structure than a field) has been thoroughly investigated in Burkart [2], Drewniak–Kalinowski [1], Kuczma [34], [39], Kuczma–Targoński [1], Moszner [2], [3]. Similar problems corresponding to Abel's equation are dealt with in Burkart [2], Targoński [6].

9.3 Abel–Schröder systems and the associativity equation

Investigation of systems of Abel and/or Schröder equations has recently been motivated by problems from other fields of mathematics. In this section Abel and Schröder equations jointly appear in connection with the associativity equation

$$T(T(x, y), z) = T(x, T(y, z)), \qquad x, y, z \in [0, 1]. \tag{9.3.1}$$

We are interested in questions of the existence and uniqueness of solutions T of (9.3.1) which map the unit square into the unit interval, whose restrictions to some segments of the square are prescribed. All the results we present here are due to W. F. Darsow–M. J. Frank [1].

9.3A. Strict Archimedean associative functions

We restrict the considerations to a special class of solutions of (9.3.1). We write

$$X := [0, 1] \quad \text{and} \quad A^* := A \setminus \{0\} \quad \text{for } A \subset X.$$

Definition 9.3.1. By an *Archimedean* $T: X^2 \to X$ we mean a solution of (9.3.1) that is continuous, increasing with respect to each variable and satisfies the conditions $T(0, x) = T(x, 0) = 0$, $T(1, x) = T(x, 1) = x$ for $x \in X$ and $T(x, x) < x$ for $x \in (0, 1)$. An Archimedean T is said to be *strict* iff it is strictly increasing with respect to each variable on X^*. Strict Archimedean associative functions T will be abbreviated as 's.A.'.

It is known (see Aczél [2, § 6.2.2]) that all s.A. T are given by the formula

$$\left. \begin{array}{ll} T(x, y) = \varphi^{-1}(\varphi(x) + \varphi(y)), & x, y \in X^* \\ T(x, 0) = T(0, x) = 0, & x \in X, \end{array} \right\} \tag{9.3.2}$$

where $\varphi: X^* \to \mathbb{R}^+$ is an arbitrary decreasing bijection, which is called the *generator* of *T*.

Remark 9.3.1. A generator φ of an s.A. *T* is continuous in X^*.

The generator φ is determined up to a positive multiple. For, let *T* be s.A. such that (9.3.2) holds with the generators φ_1 and φ_2. Write $\Phi = \varphi_2 \circ \varphi_1^{-1}$. Therefore $\Phi: \mathbb{R}^+ \to \mathbb{R}^+$ is strictly increasing and satisfies the Cauchy equation

$$\Phi(u + v) = \Phi(u) + \Phi(v), \qquad u, v \in \mathbb{R}^+.$$

Thus $\Phi(u) = ku$, $k > 0$ (see Aczél [2, § 2.3.4], Kuczma [51, § 13.5]) and we get $\varphi_2 = k\varphi_1$.

Now we introduce the notions of the diagonal and the *u*-sections of the Archimedean associative function *T*. Let us put

$$f(x) := T(x, x), \tag{9.3.3}$$

and call *f* the *diagonal* of *T*. As a result of (9.3.2) with $y = x$ we obtain the Schröder equation

$$\varphi(f(x)) = 2\varphi(x), \qquad x \in X^*. \tag{9.3.4}$$

Similarly, let us fix a $u \in (0, 1)$, write

$$g(x) := T(x, u), \tag{9.3.5}$$

and call *g* the *u-section* of *T*. Assuming, without loss of generality, that $\varphi(u) = 1$ we get the Abel equation (put $y = u$ in (9.3.2))

$$\varphi(g(x)) = \varphi(x) + 1, \qquad x \in X^*. \tag{9.3.6}$$

Let us keep *u* fixed. Diagonals and *u*-sections of s.A. *T*s are characterized by the following.

Lemma 9.3.1. (1) *A self-mapping f of X is the diagonal of an s.A. T if and only if it is strictly increasing on X, maps X onto X, and $0 < f(x) < x$ in $(0, 1)$, $f(0) = 0$, $f(1) = 1$.*

(2) *A self-mapping g of X is the u-section of an s.A. T if and only if it is strictly increasing on X, maps X onto $[0, u]$, and $0 < g(x) < x$ in X^*, $g(0) = 0$, $g(1) = u$.*

Proof. The necessity in both cases is obvious; see (9.3.3), (9.3.5) and Definition 9.3.1. Given an *f* as specified we apply the extension method from Note 9.6.3 to get a strictly decreasing solution $\varphi: X^* \to \mathbb{R}^+$ (onto) of (9.3.4), which generates *T*. Similarly we get the sufficiency for case (2). ∎

Let *V* be a subset of *X*. All s.A. *T*s whose diagonals (resp. *u*-sections) coincide with each other on *V* can be determined through the following.

Theorem 9.3.1. *Let T_0 and T be s.A. with the generators φ_0 and φ, respectively.*

(a) *The restrictions to V of the diagonals of T_0 and T are identical if and only if $\varphi_0 = \sigma \circ \varphi$, where*

$$\sigma(2t) = 2\sigma(t), \qquad t \in \varphi(V^*). \tag{9.3.7}$$

(b) *For any fixed $u \in (0, 1)$, the restrictions to V of the u-sections of T_0 and T are identical if and only if $\varphi_0(u) = \varphi(u) = 1$ and $\varphi_0 = \alpha \circ \varphi$, where*

$$\alpha(t + 1) = \alpha(t) + 1, \qquad t \in \varphi(V^*). \tag{9.3.8}$$

Proof. It is enough to apply the formulae for the diagonal of T, $f = \varphi^{-1} \circ (2\varphi)$ (see (9.3.4)), and for the u-section of T, $g = \varphi^{-1} \circ (\varphi + 1)$ (see (9.3.6)). For T_0 the function φ is to be replaced by φ_0. ∎

9.3B. Abel and Schröder equations jointly

Now we may ask what can be said on a strict Archimedean T when both its diagonal f and u-section g are specified. A generator of such a T has to satisfy simultaneously the Schröder and Abel equations (9.3.4) and (9.3.6). Here we want to establish, in terms of generators, the conditions for the uniqueness of s.A. T, whose restrictions f and g are given.

Fix a $u \in X^*$ and assume we are given s.A. T_0 and T with the generators φ_0 and φ respectively, $\varphi_0(u) = \varphi(u) = 1$. Let the diagonals of T_0 and T coincide on a set $A \subset X$, and the u-sections on a set $B \subset X$. In view of Theorem 9.3.1, the function

$$\Phi := \varphi_0 \circ \varphi^{-1} \tag{9.3.9}$$

has to satisfy simultaneously the equations

$$\Phi(2x) = 2\Phi(x), \qquad x \in \varphi(A^*), \tag{9.3.10}$$

$$\Phi(x + 1) = \Phi(x) + 1, \qquad x \in \varphi(B^*). \tag{9.3.11}$$

For certain A and B the only common solution Φ to (9.3.10) and (9.3.11) is the identity function. If Φ were $\mathrm{id}_{\mathbb{R}^+}$, then necessarily $\varphi_0 = \varphi$, i.e., $T_0 = T$, and T would be uniquely determined by its restriction to the set $\{(x, x) \in X^2 : x \in A\} \cup (B \times \{u\})$.

We prove a uniqueness theorem referring to the case where either $A = X$ or $B = X$. Denote by f_0 and f the diagonals of T_0 and T. The u-sections are g_0 and g, respectively.

Theorem 9.3.2. *If T_0 and T are s.A., then each of the two conditions implies $T_0 = T$:*

(a) $g_0 = g$ *and* $f_0(x) = f(x)$ *for* $x \in A := [0, a], 0 < a \leqslant 1$.
(b) $f_0 = f$ *and* $g_0(x) = g(x)$ *for* $x \in B := [0, b], 0 < b \leqslant 1$.

Proof. (a) The function (9.3.9) satisfies equation (9.3.10) in $\varphi(A^*) = [\varphi(a), \infty)$

and equation (9.3.11) in $\varphi(X^*) = \mathbb{R}^+$. Equation (9.3.11) yields by induction $\Phi(x + n) = \Phi(x) + n$, $x \geqslant 0$, $n \in \mathbb{N}_0$, and since $\Phi(1) = \varphi_0(\varphi^{-1}(1)) = \varphi_0(u) = 1$, we get hence $\Phi(n) = n$ for $n \in \mathbb{N}$. From (9.3.10) we have by induction

$$\Phi(x) = 2^{-n}\Phi(2^n x) \qquad (9.3.12)$$

for $x \geqslant \varphi(a)$, $n \in \mathbb{N}$. Let us fix an integer $k \geqslant \varphi(a)$. For any positive dyadic rational $t = m \cdot 2^{-n}$ we have $\Phi(t) = \Phi(t + k) - k = 2^{-n}\Phi(2^n(t + k)) - k = 2^{-n}\Phi(m + k \cdot 2^n) - k = 2^{-n}(m + k \cdot 2^n) - k = t$, since $t + k \geqslant \varphi(a)$ and $m + k \cdot 2^n$ is an integer. Thus $\Phi(t) = t$ on a set dense in $\mathbb{R}^+$. The conclusion follows by the continuity of Φ (see Remark 9.3.1).

(b) Now (9.3.10) is satisfied in $\mathbb{R}^+$ and (9.3.11) in $[\varphi(b), \infty)$. Choose a positive integer k such that $2^k \geqslant \varphi(b)$. Use (9.3.12) (which now holds for $x \geqslant 0$, $n \in \mathbb{N}$) and $\varphi(1) = 1$ to get from (9.3.11) the relation $\Phi(m) = m$ valid for all integers $m \geqslant 2^k$. Let t be any positive dyadic rational. Take an integer $n \geqslant k$ for which $2^n t \in \mathbb{N}$ and $2^n t \geqslant 2^k$. Hence, $\Phi(t) = 2^{-n}\Phi(2^n t) = 2^{-n}(2^n t) = t$. The continuity of Φ again implies $\Phi(x) = x$ for $x \geqslant 0$. ∎

9.3C. Existence of generators

We pass to the problem of the existence of a strict Archimedean T in the following situation.

Let A and B be closed subintervals of X and let $u \in (0, 1)$ be fixed. Assume we are given continuous and strictly increasing functions $f: A \to X$ and $g: B \to X$. Under what conditions does there exist an s.a. T_0 such that f and g are the restrictions of its diagonal f_0 and its u-section g_0?

To answer this question we proceed as follows. First we find a continuous strictly decreasing φ mapping the set $A^* \cup f(A^*)$ into $\mathbb{R}^+$ and satisfying equation (9.3.4) in A^*. Next we need an increasing bijection σ of the set $\varphi(A^* \cup f(A^*))$ onto itself which satisfies equation (9.3.7) in $\varphi(A^*)$. Now, according to Theorem 9.3.1, those T_0 that have the given piece f of their diagonals f_0 are generated by the functions φ_0 whose restrictions to $A^* \cup f(A^*)$ have the form $\sigma \circ \varphi$.

Similarly, g is a piece of the u-section g_0 of T_0 if and only if the generator φ_0 of T_0 is a strictly decreasing extension onto X^* of the function $\alpha \circ \tilde{\varphi}$, where $\tilde{\varphi}: B^* \cup g(B^*) \to \mathbb{R}^+$ is a continuous strictly decreasing function that satisfies (9.3.6) in B^* and α is a strictly increasing self-mapping of the set $\tilde{\varphi}(B^* \cup g(B^*))$ satisfying (9.3.8) in $\tilde{\varphi}(B^*)$.

If the domains of φ and $\tilde{\varphi}$ do not overlap, the functions $\sigma \circ \varphi$ and $\alpha \circ \tilde{\varphi}$ can be extended to create a generator φ_0 of the sort wanted. But if the domains overlap the problem is difficult and in general remains unsolved. We describe a solution to the problem in a special case of interest, where

$$A = B = [u, 1].$$

(The only other case treated in Darsow–Frank [1] is $A = B = [0, u]$.)

The (continuous and strictly increasing) functions $f, g: [u, 1] \to X$ are to be restrictions of the diagonal, resp. the u-section of some s.A. T. Thus (see Definition 9.3.1) they should necessarily fulfil the conditions

$$0 < g(x) < f(x) < x \quad \text{in } (u, 1), \qquad 0 < f(u) = g(u) < u, \quad f(1) = 1, g(1) = u.$$

$$(9.3.13)$$

Now $A^* \cup f(A^*) = [f(u), 1]$. By the argument from Note 9.6.3 we find a continuous, strictly decreasing function φ defined on $[f(u), 1] = [g(u), 1]$ and satisfying (9.3.4) in $[u, 1]$, $\varphi(u) = 1$. Setting $x = u$ and next $x = 1$ in (9.3.4) we get, by (9.3.13), $\varphi(g(u)) = 2$ and $\varphi(1) = 0$. Thus we see that φ maps the interval $[g(u), 1]$ onto the whole of $[0, 2]$, and the interval $[u, 1]$ onto $X = \varphi(A)$.

Take the family of all compositions $\sigma \circ \varphi$ with continuous and strictly increasing functions $\sigma: [0, 2] \to [0, 2]$ (onto) satisfying (9.3.7) on $\varphi(A) = X$. We require that one of such compositions coincide with a solution of equation (9.3.6), i.e.,

$$(\sigma \circ \varphi)(g(x)) = (\sigma \circ \varphi)(x) + 1, \qquad x \in [u, 1],$$

or, what is the same thing,

$$\sigma(\varphi \circ g \circ \varphi^{-1}(t)) = \sigma(t) + 1, \qquad t = \varphi(x) \in X.$$

Therefore σ has to be a strictly increasing solution, from $[0, 2]$ onto itself, of the Abel–Schröder system

$$\sigma(2x) = 2\sigma(x), \qquad x \in X, \tag{9.3.14}$$

$$\sigma(h(x)) = \sigma(x) + 1, \qquad x \in X, \tag{9.3.15}$$

where h is the function conjugate with g via φ:

$$h(x) = (\varphi \circ g \circ \varphi^{-1})(x), \qquad x \in X. \tag{9.3.16}$$

Note that h is continuous, strictly increasing and $h(0) = \varphi(g(1)) = \varphi(u) = 1$, whereas $h(1) = \varphi(g(u)) = 2$. Thus h maps X onto $[1, 2]$.

In this way we have proved the following.

Theorem 9.3.3. *Let us fix a $u \in (0, 1)$ and let $f, g: [u, 1] \to X$ be continuous and strictly increasing functions subject to conditions* (9.3.13). *The following assertions are equivalent.*

(a) *There exists an s.A. T whose diagonal and u-section, when restricted to $[u, 1]$, are equal to f and g, respectively.*

(b) *There is a continuous and strictly increasing function σ from $[0, 2]$ onto itself, satisfying simultaneously equations* (9.3.14) *and* (9.3.15) *in X, where h is given by* (9.3.16) *with a strictly decreasing φ that maps $[g(u), 1]$ onto $[0, 2]$ and satisfies* (9.3.4) *in $A = [u, 1]$.*

Moreover, if (b) *is satisfied, then every extension of $\sigma \circ \varphi: g(A) \to [0, 2]$ to* *a continuous and decreasing bijection $\varphi_0: X^* \to \mathbb{R}^+$ is a generator of a strict Archimedean extension of f and g simultaneously.*

9.3D. A solution of the Abel–Schröder system

We are led to the study of system (9.3.14)–(9.3.15). It will be convenient to put

$$Y := [1, 2], \qquad Z := [0, 2].$$

We assume that

(i) h is a continuous and strictly increasing map of X onto Y.

We look for continuous and strictly increasing σs from Z onto Z satisfying both (9.3.14) and (9.3.15) in X. It is convenient to replace (9.3.14)–(9.3.15) by the equivalent infinite system

$$\psi(h(2^{-n}y)) = 1 + 2^{-n}\psi(y), \qquad n \in \mathbb{N}_0, \ y \in Y. \qquad (9.3.17)$$

Lemma 9.3.2. *Assume* (i) *to hold. If* $\sigma: Z \to Z$ *is a simultaneous solution of* (9.3.14) *and* (9.3.15) *in* X, *then* $\psi := \sigma|Y$ *satisfies* (9.3.17). *Conversely, if* $\psi: Y \to Y$ *satisfies* (9.3.17) *and* $\psi(1) = 1$, $\psi(2) = 2$, *then* $\sigma: Z \to Z$ *defined by*

$$\sigma(0) = 0,$$

$$\sigma(x) = 2^{-n}\psi(2^n x), \qquad 2^{-n} < x \leqslant 2^{-n+1}, \ n \in \mathbb{N}_0,$$

satisfies (9.3.14) *and* (9.3.15) *in* X. *The function* σ *is continuous or increasing or strictly increasing or surjective on* Z *precisely when so is (respectively)* ψ *on* Y.

Proof. Straightforward verification. ∎

We shall construct an appropriate solution of (9.3.17). The sequence

$$t_n := h(2^{-n}), \qquad n \in \mathbb{N}_0,$$

is strictly decreasing and $\lim_{n \to \infty} t_n = h(0) = 1$. Let $\psi_0: Y \to \mathbb{R}$ be any function with $\psi_0(1) = 1$, $\psi_0(2) = 2$. Define the sequence of functions $\psi_m: Y \to \mathbb{R}$ for $m \in \mathbb{N}$ by

$$\left.\begin{array}{l} \psi_m(1) := 1, \\ \psi_m(y) := 1 + 2^{-n}\psi_{m-1}(2^n h^{-1}(y)), \qquad t_{n+1} < y \leqslant t_n, \, n \in \mathbb{N}_0. \end{array}\right\} \quad (9.3.18)$$

Theorem 9.3.4. *For every bounded* $\psi_0: Y \to \mathbb{R}$ *with* $\psi_0(1) = 1$, $\psi_0(2) = 2$, *sequence* (9.3.18) *converges uniformly in* Y *to a bounded function* $\psi: Y \to \mathbb{R}$ *with* $\psi(1) = 1$, $\psi(2) = 2$ *that satisfies system* (9.3.17) *on* Y *(for all* $n \in \mathbb{N}_0$*). This solution* ψ *is continuous or increasing or onto* Y *precisely when so is (respectively)* ψ_0.

Proof. To each $y \in Y \setminus \{1\}$ there corresponds a unique $n \in \mathbb{N}_0$ such that $t_{n+1} < y \leqslant t_n$, i.e. $y_1 := 2^n h^{-1}(y) \in Y \setminus \{1\}$. Thus in view of (9.3.18)

$$|\psi_{m+1}(y) - \psi_m(y)| = 2^{-n}|\psi_m(y_1) - \psi_{m-1}(y_1)| \leqslant \tfrac{1}{2}|\psi_m(y_1) - \psi_{m-1}(y_1)|.$$

Arguing this way again for y_1, and so on, we get eventually a $y_m \in Y \setminus \{1\}$ such that

$$|\psi_{m+1}(y) - \psi_m(y)| \leq 2^{-m} |\psi_1(y_m) - \psi_0(y_m)| \leq M \cdot 2^{-m}$$

for $y \in Y$, $m \in \mathbb{N}$, the last inequality resulting from the boundedness of ψ_0. This yields the uniform convergence of ψ_m on Y to a function, say $\psi : Y \to \mathbb{R}$. By (9.3.18) we have for fixed $n \in \mathbb{N}_0$

$$\psi_m(h(2^{-n}y)) = 1 + 2^{-n} \psi_{m-1}(y), \qquad y \in Y, \; m \in \mathbb{N},$$

and so letting $m \to \infty$ we see that ψ satisfies (9.3.17) in Y.

Now, if ψ_0 is continuous or increasing or maps Y onto itself, then the same property holds for all functions ψ_m, $m \in \mathbb{N}$. The limit process preserves these properties as well.   ∎

Now we consider the question of uniqueness of continuous and increasing solutions $\psi : Y \to Y$ of (9.3.17). First we observe their behaviour on the set D of all dyadic rationals in $(1, 2)$, that is on

$$D := \left\{ d \in (1, 2) : d = 1 + \sum_{j=1}^{k} 2^{-n_1 - \cdots - n_j}, \; n_j \in \mathbb{N}, \; j = 1, \ldots, k, \; k \in \mathbb{N} \right\}.$$

It is useful to introduce the functions $h_n : Y \to [t_n, t_{n-1}]$ by

$$h_n(y) := h(2^{-n}y), \quad n \in \mathbb{N}.$$

By induction we get for ψ satisfying (9.3.17) and $d \in D$

$$\psi(h_{n_1} \circ \cdots \circ h_{n_k}(y)) = d + 2^{-n_1 - \cdots - n_k}(\psi(y) - 1), \qquad k \in \mathbb{N}. \quad (9.3.19)$$

Let us put here $y = 1$ and write

$$H(d) := h_{n_1} \circ \cdots \circ h_{n_k}(1) \qquad \text{for } d = 1 + \sum_{j=1}^{k} 2^{-n_1 - \cdots - n_k} \in D.$$

Let C be the range of H, $C := H(D) \subset (1, 2)$. One proves that H is a one-to-one, increasing mapping of D onto C.

Since $\psi(1) = 1$, we get from (9.3.19)

$$\psi(H(d)) = d, \qquad d \in D. \quad (9.3.20)$$

Thus ψ is uniquely determined on C.

Let the cl C be a proper subset of Y. Take any open interval $(a, b) \subset Y \setminus \text{cl } C$ contiguous to cl C. We claim that $\psi(a) = \psi(b)$, i.e. ψ is constant on $[a, b]$. Indeed $\psi(a) \leq \psi(b)$ since ψ is increasing. If $\psi(a) < \psi(b)$, then there would be a dyadic rational $d \in D$ such that $\psi(a) < d < \psi(b)$. Further, $d = \psi(c)$, where $c = H(d) \in C$ and $a < c < b$, since by (9.3.20) ψ is a strictly increasing bijection of C onto D. But (a, b) cannot contain points of C, a contradiction.

Finally, if C is dense in Y, then the monotonicity and continuity of ψ and (9.3.20) imply the uniqueness of the solution $\psi : Y \to Y$ (onto) of (9.3.17). Again by (9.3.20), ψ is necessarily strictly increasing in Y.

Upon applying Theorem 9.3.4 and Lemma 9.3.2, these facts give the proof of the following.

Theorem 9.3.5. *Let hypothesis* (i) *be satisfied. There is a unique continuous and increasing function* $\sigma: Z \to Z$ *satisfying* (9.3.14) *and* (9.3.15) *simultaneously in* X. *Moreover,* σ *is strictly increasing if and only if the set* $C = H(D)$ *is dense in* Y.

Remark 9.3.2. Theorem 9.3.5 implies that the s.A. T spoken of in Theorem 9.3.3(a) exist if and only if the set C is dense in $[1, 2]$. The elements of C can be expressed in terms of f and g, but checking the density is a difficult problem. See Darsow–Frank [1] for more details.

9.4 Abel systems and differential equations with deviations

To indicate the problem, let us take a differential equation of the first order with one deviating argument $f(x)$:

$$A(f) := A(x, y(x), y(f(x)), y'(x), y'(f(x))) = 0.$$

By a change of the variable $x \mapsto t = \varphi(x)$, the equation $A(f) = 0$ is transformed into another one with the deviating argument $g(t)$, where $g = \varphi \circ f \circ \varphi^{-1}$ (see Section 8.1). In this way we are led to the conjugacy equation (8.0.1), where, moreover, the function φ should be invertible and smooth enough. If we want to have a constant deviation $g(t) = t + c$, then (8.0.1) goes over into the Abel equation

$$\alpha(f(x)) = \alpha(x) + c.$$

To find an appropriate α, we may apply the theory of differentiable solutions of the equations in question (see Sections 5.6 and 3.4).

Now, consider an equation of nth order with several deviations $f_1(x), \ldots, f_k(x)$, denoted by $A_n(f_1, \ldots, f_k) = 0$. The existence of a transformation α that converts this equation into an equation $B_n(g_1, \ldots, g_n) = 0$ with constant deviations

$$g_i(t) = t + c_i, \qquad i = 1, \ldots, k, \; c_i \neq 0,$$

is equivalent to the existence of a common solution α to the system of the simultaneous Abel equations

$$\alpha(f_i(x)) = \alpha(x) + c_i, \qquad i = 1, \ldots, k, \; c_i \neq 0. \tag{9.4.1}$$

System (9.4.1) will be dealt with under the following assumptions.

(i) $X \subset \mathbb{R}$ is an open, finite or infinite interval.
(ii) $f_i: X \to X$ are of class C^n on X, $f'_x(x) > 0$ in X, $f_i(x) \neq x$ in X, $f_i(X) = X$, $i = 1, \ldots, k$, $n \in \mathbb{N}$.

9.4A. A group of transformations

We assume additionally

(iii) There exists a common solution $\alpha: X \to \mathbb{R}$ of (9.4.1) that is
of class C^n on X, $\alpha'(x) > 0$ on X.

From (iii) it follows that $\alpha(X) = \mathbb{R}$. For take an $x_0 \in X$, a $j \in \{1, \ldots, k\}$,
calculate

$$\lim_{m \to \pm \infty} \alpha(f_j^m(x_0)) = \lim_{m \to \pm \infty} (\alpha(x_0) + mc_j) = \pm \infty$$

and observe that α is continuous and strictly increasing in X. Equations
(9.4.1) then say that the functions f_i can be embedded into the one-parameter
group of transformations given by the formula $\alpha^{-1}(\alpha(x) + c)$, $c \in \mathbb{R}$. Every
finite composition of f_i and their inverses f_i^{-1} also belongs to this group.
Let us denote by $\mathscr{F}$ the set of all such compositions:

$$\mathscr{F} := \{ f_1^{s_1} \circ f_2^{s_2} \circ \cdots \circ f_k^{s_k}: s_i \in \mathbb{Z}, i = 1, \ldots, k \}.$$

The following properties of $\mathscr{F}$ are easily established.

Lemma 9.4.1. *Let hypotheses* (i)–(iii) *be satisfied.*
  (a) *Any two members of* $\mathscr{F}$ *commute and* $\mathscr{F}$ *is a group.*
  (b) *Every two functions from* $\mathscr{F}$ *that coincide at a point of* X *are identical.*

An important question is whether the set of all points lying on graphs of
functions from $\mathscr{F}$ is dense in X^2 or not. Write

$$G = \{(x, y) \in X^2: \text{there is an } f \in \mathscr{F} \text{ such that } f(x) = y\}.$$

Let us recall the density condition for some subsets of $\mathbb{R}$ (see Halmos [1]).

Property 9.4.1. Assume that $k \in \mathbb{N}$, $k \geqslant 2$ and $c_i \in \mathbb{R} \setminus \{0\}$, $i = 1, \ldots, k$. The set

$$M := \{ x \in \mathbb{R}: x = s_1 c_1 + \cdots + s_k c_k, \ s_i \in \mathbb{Z} \} \tag{9.4.2}$$

is dense in $\mathbb{R}$ if and only if at least one quotient c_i/c_j is irrational.
  We have the following (Neuman [9]).

Lemma 9.4.2. *Let hypotheses* (i)–(iii) *be satisfied.*
  (a) *The set* G *is dense in* X^2 *if and only if there is at least one pair* (i, j),
$i, j \in \{1, \ldots, k\}$ *such that the quotient* c_i/c_j *is irrational.*
  (b) *If* G *is not dense in* X^2, *then there is a function* $g: X \to X$, *of class* C^n
on X *and such that* $g'(x) > 0$, $g(x) > x$ *on* X *and* f_i *are iterates of* $g: f_i = g^{m_i}$,
$m_i \in \mathbb{Z}$, $i = 1, \ldots, k$.

Proof. (a) The transformation $T: X^2 \to \mathbb{R}^2$, $(x, y) \mapsto (\alpha(x), \alpha(y))$ is a diffeomorphism since $\alpha'(x) > 0$ on X. Thus the set $T(G)$ is dense in $\mathbb{R}^2$ precisely
when G is dense in X^2. But $T(G) = \{(t, t + \sum_{i=1}^k s_i c_i): t \in \mathbb{R}, s_i \in \mathbb{Z}\}$ is dense

in $\mathbb{R}^2$ if and only if the set M defined by (9.4.2) is dense in $\mathbb{R}$. Property 9.4.1 completes the proof.

(b) If G is not dense in $\mathbb{R}^2$, then, by (a), there is a $d > 0$ such that $c_i = m_i d$ with $m_i \in \mathbb{Z}$. The function $g \colon X \to X$, $g(x) = \alpha^{-1}(\alpha(x) + d)$ has the required properties. ∎

Owing to the assertion (b) of Lemma 9.4.1 the function $H \colon G \to \mathbb{R}$ given by

$$H(x, y) = f'(x), \qquad \text{where } f \in \mathscr{F} \text{ and } f(x) = y, \qquad (9.4.3)$$

is well defined. The function (9.4.3) has the following properties.

Lemma 9.4.3. *Assume* (i)–(iii) *to hold.*

(a) *The function* $H \colon G \to \mathbb{R}$ *defined by* (9.4.3) *is positive on* G *and satisfies the functional equation*

$$H(x, y)H(y, z) = H(x, z) \qquad (9.4.4)$$

whenever (x, y) *and* (y, z) *are in* G.

(b) *There exists an extension* H^* *of* H *onto* X^2 *which is positive, of class* C^{n-1} *and satisfies* (9.4.4) *on* X^2. *Moreover,* H *is uniquely extendable onto* X^2 *provided the set* G *is dense in* X^2.

Proof. (a) The positivity of H follows from (ii) and its definition. Now, if (x, y), $(y, z) \in G$ then there exist $h_1, h_2 \in \mathscr{F}$ such that $h_2(x) = y$, $h_1(y) = z$. Thus $(h_1 \circ h_2)(x) = z$ and $h_1 \circ h_2 \in \mathscr{F}$, whence $(x, z) \in G$. In virtue of $(h_1 \circ h_2)' = (h_1' \circ h_2)h_2'$ and (9.4.3) we have (9.4.4).

(b) We put

$$H^*(x, y) = \alpha'(x)/\alpha'(y) \qquad \text{on } X^2,$$

where α is the function from (iii). Obviously, H^* is positive, of class C^{n-1} and satisfies (9.4.4) on X^2. Let $(x, y) \in G$ so that $f(x) = y$ for an $f \in \mathscr{F}$. Thus f is a composition of iterates of f_i, $i = 1, \ldots, k$, and by (iii) we have $\alpha(f(x)) = \alpha(x) + \text{const}$. Consequently, $H^*(x, y) = \alpha'(x)/\alpha'(f(x)) = f'(x) = H(x, y)$, i.e., $H^* | G = H$. The uniqueness assertion is obvious. ∎

9.4B. Systems of simultaneous Abel equations

Lemmas 9.4.2 and 9.4.3 suggest the following existence theorem for system (9.4.1) (Neuman [10]).

Theorem 9.4.1. *Let hypotheses* (i), (ii) *be satisfied. In the following two cases there exist constant* $c_i \neq 0$, $i = 1, \ldots, k$, *such that system* (9.4.1) *has a solution* $\alpha \colon X \to \mathbb{R}$, *of class* C^n *on* X *and satisfying* $\alpha'(x) > 0$ *on* X (*i.e., assertion* (iii) *is true*).

(a) *There is a function* $g: X \to X$, *of class* C^n *on* X, *satisfying* $g'(x) > 0$, $g(x) > x$ *on* X, *and there are integers* $m_i \neq 0$ *such that* $f_i = g^{m_i}$, $i = 1, \ldots, k$.

(b) *The set* G *is dense in* X^2, *the function* $H: G \to \mathbb{R}$ *is well defined by* (9.4.3) *and has the extension* $H^*: X^2 \to \mathbb{R}$ *that is of class* C^{n-1} *on* X^2 *and satisfies equation* (9.4.4) *in* X^2.

Proof. (a) From Theorem 3.4.1 we know that the Abel equation

$$\alpha(g(x)) = \alpha(x) + 1$$

has a solution $\alpha: X \to \mathbb{R}$, of class C^n on X, satisfying $\alpha'(x) > 0$ in X. (This solution even involves an arbitrary function). The function α satisfies system (9.4.1) with $c_i = m_i$ since we have

$$\alpha(f_i(x)) = \alpha(g^{m_i}(x)) = \alpha(x) + m_i, \qquad i = 1, \ldots, k.$$

(b) First note that $H^* > 0$. Indeed, $H^* | G = H > 0$, by (ii) and (9.4.3). If H^* were not positive in X^2, then by the continuity there would exist a point $(u, v) \in X^2$ such that $H^*(u, v) = 0$. But then we would have by (9.4.4) for arbitrary $(x, y) \in X^2$

$$H^*(x, y) = H^*(x, u)H^*(u, v)H^*(v, y) = 0,$$

a contradiction.

Now we fix $x_0, y_0 \in X$ and define the function $\varphi: X \to \mathbb{R}$ as

$$\alpha(x) = \int_{x_0}^{x} H^*(s, y_0) ds, \qquad x \in X.$$

Then α is of class C^n on X, $\alpha'(x) > 0$ on X. We have

$$(\alpha \circ f_i \circ \alpha^{-1})'(t) = H^*(f_i(x), y_0) f_i'(x) / H^*(x, y_0), \qquad x = \alpha^{-1}(t). \quad (9.4.5)$$

Observe that $f_i \in \mathscr{F}$, whence by (9.4.3),

$$f_i'(x) = H(x, f_i(x)) = H^*(x, f_i(x)),$$

the function H^* being an extension of H. Thus we obtain from (9.4.5), since H^* satisfies (9.4.4) (we again put $x = \alpha^{-1}(t)$)

$$(\alpha \circ f_i \circ \alpha^{-1})'(t) = H^*(f_i(x), y_0) H^*(x, f_i(x)) / H^*(x, y_0) = 1,$$

$i = 1, \ldots, k$. Hence

$$(\alpha \circ f_i \circ \alpha^{-1})(t) = t + d_i,$$

where $d_i = \alpha(f_i(x_0)) \neq 0$ since $\alpha(x_0) = 0$, $f_i(x_0) \neq x_0$ and α is strictly increasing. Thus assertion (iii) holds for system (9.4.1) with $c_i = d_i$. ∎

Remark 9.4.1. Theorem 9.4.1 contains conditions under which a differential equation with variable deviations can be transformed into another one with constant deviations.

Comments. The method of transforming differential equations with deviations to simpler ones was proposed by F. Neuman [9], [10]. His method of finding smooth solutions of simultaneous Abel equations (Neuman [10]) is based on O. Borůvka's result concerning one-parameter groups of transformations (Borůvka [2]). For results on functional equations which are needed in this section see Bödewadt [1], and also Barvínek [1], Choczewski [2].

9.5 A Schröder system and a characterization of norms

The characterization of norms via functional equations and all other results in this section are due to J. Matkowski [16].

9.5A. Characterization of norms

Let us consider the space l^p of all real or complex sequences summable in the pth power ($p \geqslant 1$):

$$l^p = \left\{ x = (x_n)_{n \in \mathbb{N}} : x_n \in \mathbb{K}, \sum_{i=1}^{\infty} |x_i|^p < \infty \right\}.$$

This space, endowed with the norm

$$\|x\| = \left(\sum_{i=1}^{\infty} |x_i|^p \right)^{1/p},$$

is a Banach space. If we introduce

$$\varphi(t) = ct^p$$

the above formula becomes

$$\|x\|_\varphi = \varphi^{-1} \left(\sum_{i=1}^{\infty} \varphi(|x_i|) \right). \tag{9.5.1}$$

One may ask if there exists any strictly increasing φ from $\mathbb{R}^+$ onto itself, $\varphi(0) = 0$, different from the power function and such that the function (9.5.1) represents a norm in the space l_φ consisting of all φ-summable real or complex sequences x, i.e. those satisfying

$$\sum_{i=1}^{\infty} \varphi(|x_i|) < \infty.$$

We shall show that the answer to this question is negative.

Using the homogeneity condition for the norm (9.5.1) we shall reduce the problem to that of finding a common strictly increasing solution of two Schröder equations. We proceed as follows.

Take first the sequence $x = (1, 1, 0, 0, \ldots)$ and next $x = (1, 1, 1, 0, \ldots)$, that is the terms $x_i = 0$ whenever $i \geqslant 4$ in both cases. Since $\|tx\|_\varphi = t\|x\|_\varphi$ for $t > 0$, we get from (9.5.1)

$$\varphi^{-1}(2\varphi(t)) = t\alpha,$$
$$\varphi^{-1}(3\varphi(t)) = t\beta,$$

where $\alpha := \varphi^{-1}(2\varphi(1))$, $\beta = \varphi^{-1}(3\varphi(1))$. Thus φ^{-1} has to satisfy in $(0, \infty)$ the simultaneous Schröder equations

$$\left.\begin{aligned} \varphi^{-1}(2t) &= \alpha\varphi^{-1}(t), \\ \varphi^{-1}(3t) &= \beta\varphi^{-1}(t). \end{aligned}\right\} \tag{9.5.2}$$

We shall see that system (9.5.2) has power functions as the only solutions in question. In this way we shall obtain the characterization of the norm of l^p.

9.5B. A system of simultaneous Schröder equations

We are going to solve the system

$$\left.\begin{aligned} \sigma(at) &= \alpha\sigma(t), \\ \sigma(bt) &= \beta\sigma(t), \end{aligned}\right\} \quad t > 0, \tag{9.5.3}$$

where a, b, α, β are positive real constants, $a \neq 1$, $b \neq 1$.

Theorem 9.5.1. (a) *Let* $\log b/\log a$ *be irrational. If a strictly monotonic function* $\sigma: (0, \infty) \to \mathbb{R}$ *satisfies system (9.5.3), then there is a* $c \in \mathbb{R} \setminus \{0\}$ *such that*

$$\sigma(t) = ct^p \quad \text{for } t > 0, \tag{9.5.4}$$

where

$$p = \frac{\log \alpha}{\log a}. \tag{9.5.5}$$

Moreover,

$$\frac{\log \alpha}{\log a} = \frac{\log \beta}{\log b}. \tag{9.5.6}$$

(b) *If (9.5.6) holds, then functions (9.5.4) with p given by (9.5.5) and any* $c \in \mathbb{R}$ *satisfy system (9.5.3) (and are strictly monotonic whenever* $c \neq 0$ *and* $p \neq 0$).

Proof. Assertion (b) is obvious, so let us assume that $\log b/\log a$ is irrational and $\sigma: (0, \infty) \to \mathbb{R}$ is a strictly monotonic solution of equations (9.5.3). Replacing, if necessary, the first of these equations by the equivalent one $\sigma(a^{-1}t) = \alpha^{-1}\sigma(t)$ we may assume that $a > 1$; and similarly $b > 1$.

If we had $\sigma(t_0) = 0$ for a $t_0 \in (0, \infty)$, then by (9.5.3) also $\sigma(at_0) = 0 = \sigma(t_0)$, which is incompatible with the invertibility of σ. Thus, in particular, $\sigma(1) \neq 0$ and we may assume that $\sigma(1) = 1$ (otherwise we would consider the solution $\sigma/\sigma(1)$ instead of σ).

Let σ be strictly increasing in $(0, \infty)$. Then $\sigma > 0$. For $\sigma(t) \geqslant \sigma(1) = 1$ for $t \geqslant 1$, and if $t_0 \in (0, 1)$, then $a^n t_0 > 1$ for large $n \in \mathbb{N}$, so that, by (9.5.3), $\sigma(t_0) = \alpha^{-n}\sigma(a^n t_0) > \alpha^{-n} > 0$.

By (9.5.3) we obtain $\alpha = \sigma(a) > \sigma(1) = 1$ and $\beta = \sigma(b) > 1$, since $a > 1$ and $b > 1$. Iterating (9.5.3) and combining the resulting equations we get

$$\sigma(a^n b^m t) = \alpha^n \beta^m \sigma(t), \quad t > 0, n, m \in \mathbb{Z},$$

whence on setting $t = 1$

$$\sigma(a^n b^m) = \sigma(a)^n \sigma(b)^m, \qquad n, m \in \mathbb{Z}. \tag{9.5.7}$$

The quotient $\log b / \log a$ is irrational. Thus the set

$$D := \{x \in (0, \infty): x = a^n b^m, \ n, m \in \mathbb{Z}\}$$

is dense in $(0, \infty)$. (Observe that $a^n b^m = \exp(n \log a + m \log b)$ and recall Property 9.4.1.)

Let p be given by (9.5.5). Thus

$$\alpha = \sigma(a) = a^p. \tag{9.5.8}$$

The proof will be over if we show that $\sigma(b) = b^p$. For (9.5.7) then yields $\sigma(t) = t^p$ on D, and the density of D in $(0, \infty)$ together with the monotonicity of σ imply (9.5.4) with $c = 1$.

To get our result we take two sequences of rationals that approximate $\log b / \log a$ from both sides:

$$\lim_{n \to \infty} \frac{p_n}{q_n} = \lim_{n \to \infty} \frac{r_n}{s_n} = \frac{\log b}{\log a} \tag{9.5.9}$$

and

$$a^{p_n/q_n} < b < a^{r_n/s_n}, \qquad n \in \mathbb{N}, \tag{9.5.10}$$

where p_n, q_n, r_n, s_n are positive integers. The following inequality results from (9.5.8), (9.5.7), (9.5.10) in view of the monotonicity of σ:

$$(a^p)^{p_n} = \sigma(a)^{p_n} = \sigma(a^{p_n}) < \sigma(b^{q_n}) = \sigma(b)^{q_n}.$$

Similarly, $(a^p)^{r_n} > \sigma(b)^{s_n}$. Consequently

$$(a^p)^{p_n/q_n} < \sigma(b) < (a^p)^{r_n/s_n}, \qquad n \in \mathbb{N}.$$

This and (9.5.9) imply that

$$\sigma(b) = (a^p)^{\log_a b} = b^p,$$

as required. Since $\alpha = a^p$ and $\beta = b^p$, relation (9.5.6) holds as well.

In the case where σ is strictly decreasing the argument is similar. Finally, (9.5.4) follows from the homogeneity of equations (9.5.3). ∎

Now we are ready to prove the main theorem.

Theorem 9.5.2. *Suppose that* $\varphi: \mathbb{R}^+ \to \mathbb{R}^+$ *is strictly increasing and* $\varphi(0) = 0$. *If* $(l_\varphi, \|\cdot\|_\varphi)$ *is a normed space (with norm (9.5.1)), then necessarily there are* $c > 0$ *and* $p \geqslant 1$ *such that*

$$\varphi(t) = ct^p \qquad for \ t \geqslant 0.$$

Proof. The discussion in Subsection 9.5D shows that φ^{-1} has to satisfy system (9.5.2) in $\mathbb{R}^+$. This system is nothing else but (9.5.3) with $a = 1, b = 3$, $\sigma = \varphi^{-1}$. Since $\log 2 / \log 3$ is irrational, Theorem 9.5.1 applies. We obtain $\varphi^{-1}(u) = du^q$, $u \in (0, \infty)$, where $d > 0$ and $q \neq 0$. Since $\varphi(0) = 0$, $\varphi(t) = ct^p$

in $\mathbb{R}^+$, where $c = d^{-1/q}$ and $p = 1/q$. Consequently (9.5.1) reduces (for each c) to the usual norm in l^p, whence $p \geqslant 1$. Thus finally $\varphi(t) = ct^p$ in $\mathbb{R}^+$, where $c > 0$ and $p \geqslant 1$. ∎

9.6 Notes

9.6.1. Solutions of the pre-Schröder system (9.2.2) may not satisfy any Schröder equation (9.2.1). To see this take a set X, a function $f: X \to X$ and an orbit $C \subsetneq X$ which contains either no fixed points of f or a cycle of even order. We define the function $\sigma: X \to \mathbb{R}$

$$\sigma(x) = (-1)^{p-q} \quad \text{for } x \in C \quad \text{and} \quad \sigma(x) = 1 \quad \text{for } x \in X \setminus C,$$

where p and q are minimal positive integers such that $f^p(x) = f^q(x_0)$ for a certain fixed $x_0 \in C$.

Equations (9.2.3) and (9.2.4) are satisfied with $\omega(x)$ defined as -1 on C and 1 on $X \setminus C$. By Theorem 9.2.1 σ is a solution to (9.2.2). But it does not satisfy (9.2.1) with any s.

9.6.2. The following problem for automorphic functions is raised and partially solved by M. Laczkovich–Sz. Révész [1].

Let $f_1, \ldots, f_n$ be commuting maps of an arbitrary set A into itself: $f_i \circ f_j = f_j \circ f_i$. For a $\varphi: A \to \mathbb{R}$ write $(\Delta_f \varphi)(x) = \varphi(f(x)) - \varphi(x)$. Does the relation

$$\Delta_{f_1} \cdots \Delta_{f_n} \varphi = 0$$

imply the existence of functions $\varphi_i: A \to \mathbb{R}$ automorphic with respect to f_i: $\Delta_{f_i} \varphi_i = 0$ and such that $\varphi = \varphi_1 + \cdots + \varphi_n$?

The authors give a positive answer to this question in the following cases: (1) φ is a bounded function and f_i are arbitrary maps, (2) $\varphi \in L^p$, $1 \leqslant p < \infty$, and f_i are measurable maps which do not decrease measure, (3) $\varphi \in L^\infty$ (with σ-finite measure space $(X, \mathbf{S}, \mu)$) and f_i are measurable maps which do not map sets of positive measure into sets of measure zero.

9.6.3. Let $u \in [0, 1)$ and let $f: [u, 1] \to [0, 1]$ be strictly increasing and such that $0 < f(x) < x$ in $(u, 1)$, $f(1) = 1$, $f(0) = 0$ (if $u = 0$).

When $u = 0$, the extension method we referred to in the proof of Theorem 9.3.1 works for the Schröder quation (9.3.4),

$$\varphi(f(x)) = 2\varphi(x), \tag{9.6.1}$$

as follows. Take an $x_0 \in (0, 1)$ and put $x_k = f^k(x_0)$, $X_k = [x_{k+1}, x_k)$, $k \in \mathbb{Z}$. Then $(0, 1) = \bigcup_{k=-\infty}^\infty X_k$. Choose any strictly decreasing $\varphi_0: X_0 \to [1, 2]$ (onto) and extend it over $(0, 1]$ by defining $\varphi(x) = 2^k \varphi_0(f^{-k}(x))$, $x \in X_k$, $k \in \mathbb{Z}$, $\varphi(1) = 0$. The (strictly decreasing!) $\varphi: (0, 1] \to \mathbb{R}^+$ (onto) satisfies (9.6.1) and is the generator of a s.A. T whose diagonal is f.

When $u > 0$, the same method yields a strictly decreasing continuous solution of (9.6.1) on $[u, 1)$ (see p. 375). What we need is to take $x_0 = u$ and to observe that $[u, 1) = \bigcup_{n=1}^{\infty} X_{-n}$.

(In both cases the solution obtained depends on an arbitrary function. Compare e.g. Theorem 3.1.3).

9.6.4. To illustrate the procedure of reducing differential equations with deviating argument (see Subsection 9.4A) we propose the following example which is due to F. Neuman [9].

Example 9.6.1. The differential equation

$$y'(x) = ky(x^p), \qquad x \in (1, \infty)$$

where $k \neq 0$, $p \in (0, 1)$ is transformed to $z'(t) = g(t)z(t+1)$ with the aid of the transformation $t = \alpha(x)$, $y(x) = z(t) = (z \circ \alpha)(x)$, where $\alpha(x) = -(\log \log x)/\log p$ is a solution of the equation

$$\alpha(x^p) = \alpha(x) + 1.$$

Here $g(t) = -k \log p \exp(\exp(-t \log p) - t \log p)$.

9.6.5. We exhibit an example in which Theorem 9.4.1 is applicable (see Neuman [10]).

Example 9.6.2. (a) The differential equation

$$y'(x) = y(x^{1/2}) + y(x^4), \qquad x \in X := (1, \infty),$$

goes over into an equation with constant deviations $t - 1$, $t + 2$, by a change of variables $t = \alpha(x)$. This results from Theorem 9.4.2(a) since for $f_1(x) = x^{1/2}$ and $f_2(x) = x^4$ we have $f_1 = g^{-1}$ and $f_2 = g^2$, where $g: X \to X$, $g(x) = x^2$. The function α may be any solution of the equation $\alpha(x^{1/2}) = \alpha(x) + 1$, being of class C^2 on X, $\alpha'(x) > 0$.

(b) The differential equation

$$y'(x) = y(x^{1/2}) + y(x^3), \qquad x \in X,$$

can also be transformed into another one with constant deviations $t - K_1 \log 2$, $t + K_1 \log 3$, $K_1 > 0$. Now the set $G = \{(x, x^a): a = 2^p 3^q, p, q \in \mathbb{Z}, x \in X\}$ is dense in X^2. The function H (see (9.4.3)) is given by $H(x, y) = H(x, x^a) = (x^a)' = ax^{a-1}$ on G. It admits a smooth extension H^* onto X^2 which is defined as $H^*(x, y) = (y \log y)/(x \log x)$, $(x, y) \in X^2$. Integrating H^* with respect to the first variable we obtain $\alpha(x) = K_1 \log \log x + K_2$, $K_1 > 0$, which, by Theorem 9.4.2(b), is the desired transformation.

9.6.6. The assertion of Theorem 9.5.2 is true also for the space L_φ of all φ-integrable functions $x: [0, 1] \to \mathbb{R}$ endowed with the norm

$$\|x\|_\varphi = \varphi^{-1}\left(\int_0^1 \varphi(|x(u)|)\,du \right).$$

Thus the space L_φ is in fact an L^p space, with the usual p-norm.

Moreover, if a norm $\| \cdot \|$ in $\mathbb{R}^N$, $N \geq 4$, is given by

$$\|x\| = \psi^{-1}\left(\sum_{i=1}^{N} \varphi(|x_i|)\right)$$

with $\varphi, \psi: \mathbb{R}^+ \to \mathbb{R}^+$ and ψ strictly increasing and onto, then necessarily $\varphi(t) = c_1 t^p$, $\psi(t) = c_2 t^p$, $c_1, c_2 > 0$, $p > 1$.

(See Matkowski [15]. The homogeneity of the norm implies the formulae for φ and ψ with $p > 0$.)

9.6.7. The system of Schröder equations

$$\left.\begin{array}{l} \varphi(x + p) = a\varphi(x), \\ \varphi(x + q) = b\varphi(x) \end{array}\right\} \tag{9.6.2}$$

has been considered by W. E. Clark–A. Mukherjea [1]. The constants a and b are positive, and p/q is irrational.

Proposition 9.6.1. *If there is a nonzero solution $\varphi: R \to \mathbb{R}$ of system (9.6.2) which is positive at a point and bounded above on an interval, then $a^p = b^q$. Conversely, if $a^p = b^q$, then every Lebesgue measurable solution of (9.6.2) is equal a.e. to $\varphi(x) = ca^{x/p}$.*

(Compare Theorem 9.5.1. In Clark–Mukherjea [1] an example may be found of a solution of (9.6.2) which is not equivalent to any continuous function.)

9.6.8. Schröder and Abel equations are associated with nonautonomous difference equations

$$x(n + 1) = f_n(x(n)), \qquad x(0) = x,$$

as has been shown by P. Diamond [3].

In the proposition below we shall use the sequence of generalized iterates, with the terms

$$F_0(x) := x, \qquad F_{n+1}(x) := F_n(f_{n+1}(x)), \qquad n \in \mathbb{N}, \tag{9.6.3}$$

whenever $f_{n+1}(x)$ is in the domain of F_n. Write $U_\delta := \{x \in \mathbb{C}: |x| < \delta\}$, $\delta > 0$.

Proposition 9.6.2. *Let $f_n: U_\delta \to \mathbb{C}$ be analytic in U_δ, $|f_n(x)| < |x|$ for $x \in U_\delta \setminus \{0\}$, with the expansions*

$$f_n(x) = s(n)x + \sum_{i=2}^{\infty} a_i(n)x^i$$

and $0 < \rho \leq |s(n)| \leq r < 1$, $n \in \mathbb{N}$, with some $\rho \leq r$. Suppose that the sequence $(f_n(x))_{n \in \mathbb{N}}$ converges uniformly in U_δ to a function f. Then there exists the limit

$$\sigma(x) = \lim_{n \to \infty} F_n(x) \prod_{j=1}^{n} s(j)^{-1} \tag{9.6.4}$$

the convergence is uniform in a $U_\eta \subset U_\delta$ *and* σ *is an analytic solution on* U_η *of the Schröder equation* $\sigma(f(x)) = s\sigma(x)$, *where* $s = f'(0) = \lim_{n \to \infty} s(n)$.

(Algorithm (9.6.3) has been used by J. D. Church [1] for probability generating functions. If $f_n = f, n \in \mathbb{N}$, then (9.6.3) represents the usual sequence of iterates of f, limit (9.6.4) becomes the Koenigs limit (9.1.3), and Proposition 9.6.2 reduces to Theorem 4.6.1.)

9.6.9. The Schröder equation $\sigma(f(x)) = \frac{1}{2}\sigma(x)$, $f(x) = \frac{1}{2}(x + x^2)$, $x \in X := [0, 1)$, admits the principal solution $\sigma(x) = \lim_{n \to \infty} 2^n f^n(x)$, $x \in X$, which is strictly increasing in X. Similarly, the function $\pi(y) = \lim_{n \to \infty} f^n(1 - (\frac{2}{3})^n y)$ is defined and strictly decreasing in $\mathbb{R}^+$ and satisfies the Poincaré equation $\pi(\frac{3}{2}y) = f(\pi(y))$, $y \in \mathbb{R}^+$. Both functions can be used to define semigroups of continuous iterations (see Section 1.7) containing the discrete semigroup $\{f^n, n \in \mathbb{N}\}$, viz. $F_1^t = \sigma^{-1} \circ (2^{-t}\sigma)$ and $F_2^t = \pi \circ ((\frac{3}{2})^t \pi^{-1})$, $t \in \mathbb{R}^+$. This embedding problem is considered in Karlin–McGregor [1].

The two semigroups would coincide in $(0, 1)$ if and only if the function

$$K(y) = y^p \sigma(\pi(y)), \qquad p := \log 2 / \log \tfrac{3}{2}, \qquad (9.6.5)$$

were constant. However, although $K(y) = 1.213\ 205\ 784\ 311 \ldots$ for all $y > 0$, the value changes with y from the 13th decimal on.

The properties of K in the case of a general complex f are thoroughly investigated by S. Dubuc [2], who called K the Karlin–McGregor function and explained the phenomenon of the small variability of function (9.6.5).

9.6.10. M. C. Zdun [19] considered the simultaneous Abel equations

$$\left.\begin{aligned}
\varphi(f(x)) &= \varphi(x) + 1 \\
\varphi(g(x)) &= \varphi(x) + s
\end{aligned}\right\} \qquad (9.6.6)$$

for $x \in X := (0, a)$, $0 < a \leq \infty$. The most interesting case is the following.

(i) f and g are commuting homeomorphisms of X onto X and $f^m(x) \neq g^n(x)$ in X whenever $(m, n) \in \mathbb{Z}^2 \setminus \{(0, 0)\}$.

Let L be the set of limit points of the set $C(x) := \{f^m \circ g^n(x): n, m \in \mathbb{Z}\}$, $x \in X$. In Zdun [19] it is proved that L truly does not depend on x and either L is a perfect nowhere dense set or $L = [0, a]$.

Proposition 9.6.3. *Assume* (i) *to hold. There is only one* $s \in \mathbb{R}$ *such that system* (9.6.6) *has a continuous solution* $\varphi: X \to \mathbb{R}$. *This* s *is irrational, the continuous solution of* (9.6.6) *is unique up to an additive constant, is monotonic and maps* $L \cap X$ *onto* $\mathbb{R}$. *Moreover, it is invertible if and only if* $L = [0, a]$.

10

Characterization of functions

10.0 Introduction

There is a number of results showing how some functions, in particular elementary ones, can be characterized by means of iterative functional equations. Theorems we present in this chapter contain conditions under which a function is the only solution of 'its' functional equation(s).

From among the elementary functions we consider, in Sections 10.1–10.3, power, exponential, logarithmic and trigonometric functions. We determine two improper integrals related to the Gauss probability distribution function.

A separate Section 10.4 is devoted to Euler's gamma function to which several functional equations may be attributed. Though the section is lengthy enough, we could only give an incomplete survey of results concerning this function.

Finally, to show that iterative functional equations may bring about peculiar functions, we discuss continuous nowhere differentiable functions.

The material of this chapter is organized in a somewhat different manner from hitherto. In particular, we attach concluding remarks to each section instead of having them collected in the Notes section which now contains a few supplementary pieces of information only.

10.1 Power functions

10.1A. Identity

Characterizing linear functions without appealing to the fundamental Cauchy equation

$$\varphi(x + y) = \varphi(x) + \varphi(y) \qquad (C)$$

(see Aczél [2, § 2.1]) is rather a hopeless task. But, as we have seen in Section 6.1, we may postulate the Cauchy equation to be satisfied on a curve only and still get the functions $\varphi(x) = cx$ as the only solutions of the resulting equation in suitable function classes (see Theorem 6.1.1).

In this section we present a result of P. Volkmann [1], which is related to the Darboux theorem on automorphisms of $\mathbb{R}$ (see, e.g., Kuczma [51, § 14.4]). We suppose that equation (C) is fulfilled on the line $y = 1$, and another Cauchy equation $\varphi(xy) = \varphi(x)\varphi(y)$ on the line $y = x$, and thus we get Böttcher and Abel equations.

Theorem 10.1.1. *The only joint solution $\varphi: \mathbb{R} \to \mathbb{R}$ of the equations*

$$\varphi(x^2) = (\varphi(x))^2, \tag{10.1.1}$$

$$\varphi(x + 1) = \varphi(x) + 1 \tag{10.1.2}$$

is the identity function, $\varphi(x) = x$ for $x \in \mathbb{R}$.

Proof. Equations (10.1.1) and (10.1.2) imply $\varphi(x) \geq 0$ for $x \geq 0$ and

$$\varphi(m + x) = m + \varphi(x), \qquad m \in \mathbb{Z}, x \in \mathbb{R}. \tag{10.1.3}$$

The equality $(\varphi(1 - x))^2 = (\varphi(x - 1))^2$ yields $\varphi(-x) = -\varphi(x), x \in \mathbb{R}$. Now we want to get the estimate

$$|x - \varphi(x)| \leq n - m, \qquad x \in [m, n], m, n \in \mathbb{Z}, m \leq n. \tag{10.1.4}$$

In fact, $x - m \geq 0$ yields $\varphi(x) - m = \varphi(x - m) \geq 0$, i.e. $\varphi(x) \geq m$. Similarly, $0 \leq \varphi(n - x) = n + \varphi(-x) = n - \varphi(x)$. Hence $\varphi(x) \in [m, n]$ whenever x does, and (10.1.4) follows.

In particular, with $m = [x]$, $n = [x] + 1$ in (10.1.4), we have $|x - \varphi(x)| \leq 1$, $x \in \mathbb{R}$. Therefore, for $k = 2^n$, $n \in \mathbb{N}$, we get

$$1 \geq |x^k - \varphi(x^k)| = |x^k - \varphi(x)^k|$$
$$= |x - \varphi(x)||x^{k-1} + x^{k-2}\varphi(x) + \cdots + \varphi(x)^{k+1}|$$
$$\geq |x - \varphi(x)|x^{k-1},$$

provided $x \geq 0$, so that $\varphi(x) \geq 0$. Finally, we have

$$|x - \varphi(x)| \leq x^{-k+1}.$$

On letting $k \to \infty$ we get $\varphi(x) = x$ for $x > 1$. Our assertion follows now from (10.1.3). ∎

10.1B. Reciprocal

The condition of convexity is sufficient for the uniqueness of solutions of the equation

$$\varphi(x + 1) = \frac{\varphi(x)}{\varphi(x) + 1}. \tag{10.1.5}$$

Namely, we have the following characterization theorem for reciprocals, which is due to M. Kuczma [15].

Theorem 10.1.2. *The only convex or concave solutions* $\varphi: (0, \infty) \to \mathbb{R}$ *of equation (10.1.5) are* $\varphi = 0$ *and* $\varphi(x) = 1/(x + d)$, *where* $d \in \mathbb{R}^+$ *is an arbitrary constant (a parameter).*

Proof. Let $\varphi: (0, \infty) \to \mathbb{R}$ be a convex or concave solution of (10.1.5). If we had $\varphi(x_0) < 0$ for an $x_0 \in (0, \infty)$, then, because of

$$\varphi(x_0 + n) = \varphi(x_0)/(1 + n\varphi(x_0)), \qquad n \in \mathbb{N}, \tag{10.1.6}$$

the φ could not be convex nor concave. If $\varphi(x_0) = 0$, then by (10.1.6) $\varphi(x_0 + n) = 0$ for $n \in \mathbb{N}$, and $\varphi = 0$. If φ admits positive values only, then it follows from (10.1.6) that φ is convex and strictly decreasing. Thus $\alpha = \varphi^{-1}$ is convex and satisfies the Abel equation

$$\alpha\left(\frac{x}{x + 1}\right) = \alpha(x) + 1.$$

Since $\alpha_0(x) = 1/x$ is a convex solution of this equation, by Theorem 2.4.3 (see also Note 2.9.8) α and α_0 differ by a constant: $\alpha_0(x) - \alpha(x) = d$. Hence $\varphi^{-1}(x) = 1/x - d$, and $\varphi(x) = 1/(x + d)$, where $d \in \mathbb{R}^+$, as the domain of φ is to be $(0, \infty)$. ∎

Corollary 10.1.1. *The reciprocal* $\varphi(x) = 1/x$ *is the only convex solution* $\varphi: (0, \infty) \to (0, \infty)$ *of (10.1.5) such that* $\varphi(1) = 1$.

10.1C. Roots

Let $p \geqslant 2$ be an integer. We want to use the equation

$$\varphi(x^{p+1}) = x\varphi(x), \tag{10.1.7}$$

to characterize the function

$$\varphi(x) = \sqrt[p]{x}, \qquad x \in X,$$

where X is the complex plane cut along the negative real axis, and the branch of the root function on X is taken so that $\varphi(1) = 1$. We follow M. Kuczma [28].

Theorem 10.1.3. *The only solution* $\varphi: X \to \mathbb{C}$ *of (10.1.7) satisfying the condition*

$$\lim_{x \to 1} \varphi(x) = 1 \tag{10.1.8}$$

is the above-described root function $\varphi(x) = \sqrt[p]{x}$, $x \in X$.

Proof. Let $x_{n+1} = \sqrt[p+1]{x_n}$, $n \in \mathbb{N}_0$, $x_0 \in X$, where the branch of the $(p + 1)$st root is taken so that $\sqrt[p+1]{1} = 1$. Observe that the x_n are in X and $\lim_{n \to \infty} x_n = 1$, whenever $x_0 \neq 0$. For let $x_n = |x_n|e^{i\theta_n}$. If $\theta_n \in (-\pi, \pi)$ then so does $\theta_{n+1} = \theta_n/(p + 1)$. Moreover, $\lim_{n \to \infty} \theta_n = 0$, whereas $\lim_{n \to \infty} |x_n| = 1$.

Now we take a solution $\varphi: X \to \mathbb{C}$ of (10.1.7) satisfying (10.1.8) and define the function $\psi(x) := \varphi(x)/\sqrt[p]{x}$, $x \in X \setminus \{0\}$. This function is automorphic,

$$\psi(x^{p+1}) = \psi(x), \tag{10.1.9}$$

$\lim_{x \to 1} \psi(x) = 1$. Iterating (10.1.9), we get for the sequence (x_n) the relation $\psi(x_n) = \psi(x_0)$, $n \in \mathbb{N}$, $x_0 \in X \setminus \{0\}$. Passing to the limit yields $\psi(x_0) = 1$ for $x_0 \in X \setminus \{0\}$. The resulting equality $\varphi(x) = \sqrt[p]{x}$ remains valid for $x = 0$. ∎

Remark 10.1.1. Theorem 4.2.4 on analytic solutions of linear homogeneous equations yields $\varphi(x) = c\sqrt[p]{x}$, $c \in \mathbb{C}$, as the only analytic solutions of (10.1.7) in X.

10.1D. Comments

10.1D1. The equation of second order

$$\varphi(x) = \tfrac{1}{2}\varphi\left(\frac{x}{2} - 1\right) + \tfrac{1}{2}\varphi\left(\frac{x}{2} + 1\right), \qquad x \in (-2, 2),$$

has constant functions as the only Riemann integrable solutions $\varphi: (-2, 2) \to \mathbb{R}$ (see Mohr [1]). The same is true for the more general equation

$$\varphi(x) = p\varphi(px - 1) + q\varphi(qx + 1), \qquad p, q > 0, \ p + q = 1, \ x \in \left(-\frac{1}{q}, \frac{1}{p}\right),$$

$$\tag{10.1.10}$$

in the class of bounded φ which are continuous at least at one point. This follows from a result by N. Steinmetz–P. Volkmann [1] for an equation of form (10.1.10) but with nonconstant p and q and some nonlinear functions under the sign of φ.

By integrating (10.1.10) we get the Cauchy equation (C) on two lines: $y = px - 1$ and $y = qx + 1$, whose absolutely continuous solutions on $[-1/q, 1/p]$ are the functions $\varphi(x) = cx$, only. See Steinmetz [1].

10.1D2. Let $f: \mathbb{R} \to \mathbb{R}$ be a function with the property

$$\limsup_{n \to \infty} |f^n(x) - f^n(y)| = \infty, \qquad x, y > a, \ x \neq y.$$

Then the identity on $\mathbb{R}$ is the unique solution of equation (10.1.2) which is permutable with f, $\varphi \circ f = f \circ \varphi$, and bounded on an interval $[b, b + 1]$. This result is due to L. Volkmann–P. Volkmann [1] (see Theorem 10.1.1, which corresponds to the case where $f(x) = x^2$).

10.1D3. The function $\varphi(x) = 1/x$, $\varphi: (0, \infty) \to (0, \infty)$, can also be characterized as the only involution satisfying (10.1.5) and being continuous on a subinterval of $(0, \infty)$. Compare Theorem 11.8.1 which has been proved by L. Dubikajtis [2].

10.1D4. Theorem 2.4.2 when applied to the equation

$$\varphi(x + 1) - \varphi(x) = -\frac{1}{x(x + 1)}, \qquad x \in (0, \infty)$$

yields $\varphi(x) = 1/x$ as the only solution $\varphi: (0, \infty) \to (0, \infty)$ which is convex and such that $\varphi(1) = 1$.

10.1D5. Another characterization of the root may be obtained from a result by A. Ostrowski [3], who has proved that if f and g are rational functions, and the sequence of iterates $f^n(x)$ is not periodic, then algebraic functions φ satisfying the equation

$$\varphi(f(x)) = g(x)\varphi(x)$$

are kth roots of rational functions, $k \in \mathbb{N}$. It follows that the functions $\varphi(x) = c\sqrt[p]{x}$ are the only algebraic solutions of equation (10.1.7) $(c \in \mathbb{C}, p \geqslant 2)$. See Kuczma [28], where the real case is also considered.

10.1D6. The following theorems are pertinent to the characterization problems: Theorem 6.2.2 for constant functions, Theorem 8.8.2 for general power functions, and Theorem 11.7.1 for the identity.

10.1D7. Regarding a characterization of polynomials see Section 2.7, and, in particular, Theorem 2.7.3 (Kuczma [14]). For a characterization of the Bernoulli polynomials see Dickey–Kairies–Shank [1] and Kairies [1].

10.1D8. A recent result by L. Volkmann–P. Volkmann [1, (II)] reads as follows. If $\varphi: \mathbb{R} \to \mathbb{R}$ satisfies (10.1.7) (on $\mathbb{R}$) and $\varphi(|x|^{-1}) = |\varphi(x)|^{-1}$ (when $x \neq 0$ and $\varphi(x) \neq 0$), then $\varphi(x) = x$, $x \in \mathbb{R}$.

10.2 Logarithmic and exponential functions

The functional equations associated to logarithmic φ, respectively exponential ψ, are clearly those of Cauchy:

$$\varphi(xy) = \varphi(x) + \varphi(y)$$

and

$$\psi(x + y) = \psi(x)\psi(y)$$

(see Aczél [2, § 2.1], Aczél–Dhombres [1], Kuczma [51, §13.1]). Setting here $y = x$ we obtain the equations of Schröder

$$\varphi(x^2) = 2\varphi(x), \tag{10.2.1}$$

and of Böttcher

$$\psi(2x) = (\psi(x))^2,$$

as the simplest iterative functional equations which may characterize the functions φ and ψ, respectively.

From among several results showing that the logarithm is the unique solution of (10.2.1) in a class of functions we choose that by G. Fubini [1] (see also Skof [1]), both for historical reasons and for the originality of the proof. We are grateful to R. B. Burckell (private communication) for calling our attention to Fubini's paper.

10.2A. Logarithm

In 1899 G. Fubini [1] proved the following.

Theorem 10.2.1. *The only function* $\varphi: (1, \infty) \to \mathbb{R}$ *that satisfies equation* (10.2.1) *and the condition*

$$\lim_{x \to 1+} \frac{\varphi(x)}{x - 1} = 1 \tag{10.2.2}$$

is the logarithmic function, $\varphi(x) = \log x$, $x \in (1, \infty)$.

Proof (after Fubini). Let φ obey the conditions of the theorem. We put

$$\left.\begin{aligned} u_n &:= \frac{a_n \sqrt{(a_n^2 - b_n^2)}}{\varphi(x_n)}, \\ x_n &:= \frac{a_n + \sqrt{(a_n^2 - b_n^2)}}{b_n}, \end{aligned}\right\} \tag{10.2.3}$$

where a_n and b_n are defined recursively by

$$\left.\begin{aligned} a_{n+1} &= \sqrt{[a_n \cdot \tfrac{1}{2}(a_n + b_n)]}, \\ b_{n+1} &= \sqrt{(a_n b_n)}, \end{aligned}\right\} \qquad a_1 > b_1 > 0, \; n \in \mathbb{N}.$$

Thus $x_n > 1$, $n \in \mathbb{N}$. Both sequences (a_n) and (b_n) are monotonic and bounded, therefore convergent. Clearly,

$$\lim_{n \to \infty} a_n = \lim_{n \to \infty} b_n = K = K(a_1, b_1)$$

and

$$K(a_1 x, b_1 x) = x K(a_1, b_1), \qquad x > 0. \tag{10.2.4}$$

Thus (10.2.3) implies $\lim_{n \to \infty} x_n = 1$, and by (10.2.2) we get

$$\lim_{n \to \infty} u_n = \lim_{n \to \infty} \frac{a_n}{x_n - 1} \sqrt{(a_n^2 - b_n^2)} = \lim_{n \to \infty} \frac{a_n b_n \sqrt{(a_n + b_n)}}{\sqrt{(a_n - b_n)} + \sqrt{(a_n + b_n)}} = K^2.$$

Further, one checks that (u_n) is a constant sequence. Thus $K^2 = u_1$. Now, take arbitrary $x > 1$ and $t > 1$ and put $a_1 = \tfrac{1}{2}x(1 + t^2)$, $b_1 = xt$. Then $x_1 = t$ and $u_1 = \tfrac{1}{4}x^2(t^4 - 1)/\varphi(t)$. In turn, owing to (10.2.4), $K(a_1, b_1)^2 = x^2 K(\tfrac{1}{2}(1 + t^2), t)^2$. This yields $\varphi(t) = \tfrac{1}{4}(t^4 - 1)K(\tfrac{1}{2}(1 + t^2), t)^{-2}$. We see that the function $\varphi: (1, \infty) \to \mathbb{R}$ satisfying (10.2.1) and (10.2.2) is uniquely determined. Thus $\varphi(t) = \log t$, as the logarithm also satisfies (10.2.1) and (10.2.2). ∎

Remark 10.2.1. The simplest method of proving Theorem 10.2.1 has been shown to us by J. Aczél. Write equation (10.2.1) in the form

$$\varphi(\sqrt{x}) = \tfrac{1}{2}\varphi(x), \qquad x \in (0, \infty). \tag{10.2.5}$$

With $t_n := x^{2^{-n}}$, $n \in \mathbb{N}$, on iterating (10.2.5), we obtain

$$\varphi(x) = \frac{\varphi(t_n)}{t_n - 1} \frac{t_n - 1}{\log t_n} \log x, \qquad x \in (0, \infty), \; n \in \mathbb{N}.$$

Since $t_n \to 1$ as $n \to \infty$, by condition (10.2.2) with $(\log)'(1) = 1$ we get $\varphi(x) = \log x$.

One can also replace equation (10.2.5) by the equivalent one

$$\psi(\sqrt{x}) = (\sqrt{x} + 1)\psi(x), \qquad x \in (0, \infty), \tag{10.2.6}$$

where $\varphi(x) = (x - 1)\psi(x)$ for $0 < x \neq 1$, $\psi(1) = 1$. Theorem 3.1.2 says that equation (10.2.6) has the unique solution $\psi : (0, \infty) \to \mathbb{R}$ continuous at $x = 1$, $\psi(1) = 1$, and Theorem 10.2.1 follows.

10.2B. Exponential functions

W. Sierpiński [5, § 37] proved that there exists exactly one real number a such that

$$a^x \geq 1 + x$$

holds for $x \in \mathbb{R}$. This number is denoted by e and is given by the formula

$$e = \lim_{n \to \infty} \left(1 + \frac{1}{n}\right)^n.$$

By this definition of e, M. Kuczma [25] was led to the characterization of the exponential function that we are now going to discuss. We start at the point when the power is defined for positive integer exponents only and we do not assume previous knowledge of exponential functions. We do need, however, the notion, and some properties, of convergent real sequences. We have the following.

Theorem 10.2.2. *There exists exactly one function* $\varphi : \mathbb{R} \to (0, \infty)$ *that fulfils the conditions (for all* $x \in \mathbb{R}$*)*

$$\varphi(2x) = (\varphi(x))^2, \tag{10.2.7}$$

$$\varphi(-x) = \frac{1}{\varphi(x)}, \tag{10.2.8}$$

$$\varphi(x) \geq 1 + x. \tag{10.2.9}$$

Proof. We shall prove that the function φ is given by the formula

$$\varphi(x) = \lim_{n \to \infty} \left(1 + \frac{x}{n}\right)^n. \tag{10.2.10}$$

The sequence in (10.2.10) is increasing and positive, whenever $n > -x$, so that $\varphi(x) > 0$ for $x \in \mathbb{R}$. Next, we use the relation

$$\lim u_n = \lim v_n \quad \text{implies} \quad \lim\left(1 + \frac{u_n}{n}\right)^n = \lim\left(1 + \frac{v_n}{n}\right)^n, \tag{10.2.11}$$

as $n \to \infty$, to establish $\varphi(x + y) = \varphi(x)\varphi(y)$ for $x, y \in \mathbb{R}$. Thus φ satisfies (10.2.7) (put $y = x$) and (10.2.8) (put $y = -x$, use $\varphi(0) = 1$). Inequality (10.2.9) is obviously satisfied when $x \leqslant -1$. For $x > -1$ we have $n > -x$ from $n = 1$ on, so that the sequence in (10.2.10) is increasing from its first term on which is equal $1 + x$, and (10.2.9) holds as well. Thus function (10.2.10) actually fulfils the requirements of the theorem.

It remains to prove the uniqueness. Let now $\hat{\varphi}: \mathbb{R} \to (0, \infty)$ be another function satisfying (10.2.7)–(10.2.9). From (10.2.7) we get $\hat{\varphi}(x) = (\hat{\varphi}(2^{-n}x))^{2^n}$, and, taking n sufficiently large that the expressions in parentheses below are positive,

$$\hat{\varphi}(x) \geqslant (1 + 2^{-n}x)^{2^n}, \qquad \hat{\varphi}(-x) \geqslant (1 - 2^{-n}x)^{2^n}.$$

In view of this and (10.2.8) we obtain

$$(1 + 2^{-n}x)^{2^n} \leqslant \hat{\varphi}(x) = \frac{1}{\hat{\varphi}(-x)} \leqslant (1 + 2^{-n}x)^{2^n}(1 - 4^{-n}x^2)^{-2^n}. \quad (10.2.12)$$

Taking $u_n = -x^2/n$, $v_n = 0$ in (10.2.11) we see that the sequence $(p_n) := ((1 - x^2/n^2)^n)$ tends to 1, whence so does its sub-sequence (p_{2^n}). The other sequence in (10.2.12) is a sub-sequence of that in (10.2.10). Therefore $\hat{\varphi} = \varphi$. ∎

Remark 10.2.3. Let us write $e := \varphi(1)$. Then $\varphi(k) = (\varphi(1))^k = e^k$ for $k \in \mathbb{N}$. We may call φ the exponential function and write

$$e^x := \varphi(x), \qquad x \in \mathbb{R}.$$

It can be derived from (10.2.7)–(10.2.9) that the function so defined is strictly increasing and maps $\mathbb{R}$ onto $(0, \infty)$ and is analytic in $\mathbb{R}$. Exponential functions with an arbitrary base $a > 0$ are now defined by

$$a^x = \varphi_a(x) := \varphi(\varphi^{-1}(a)x).$$

This function is uniquely determined by (10.2.7), (10.2.8) and the inequality

$$\varphi_a(x) \geqslant 1 + ax.$$

10.2C. An improper integral

Following M. C. Zdun [2] we are going to evaluate the integral

$$\varphi(t) = \frac{2}{\sqrt{\pi}} \int_{-\infty}^{+\infty} \exp(-x^2) \cos tx \, dx, \quad (10.2.13)$$

by using the following corollary from Theorem 6.1.2 (see Vincze [2]).

Theorem 10.2.3. Let $\beta: \mathbb{R} \to \mathbb{R}$ be a function twice differentiable at zero, $\beta(0) = 1$, $\beta''(0) = 2c < 0$, which satisfies in $\mathbb{R}$ the Böttcher equation

$$\beta(\sqrt{2} \cdot t) = (\beta(t))^2. \quad (10.2.14)$$

Then necessarily $\beta(t) = \exp(ct^2)$.

We are going to show that φ given by (10.2.13) satisfies (10.2.14). We have

$$(\varphi(t))^2 = \frac{1}{\pi} \int_{-\infty}^{+\infty} \int_{-\infty}^{+\infty} \exp(-x^2 - y^2) \cos tx \cos ty \, dx \, dy$$

$$= \frac{1}{2\pi} (I_1 + I_2), \tag{10.2.15}$$

where

$$I_1 = \int_{-\infty}^{+\infty} \int_{-\infty}^{+\infty} \exp(-x^2 - y^2) \cos t(x + y) \, dx \, dy,$$

and I_2 differs from I_1 by '$x - y$' occurring in place of '$x + y$'. Changing the variables in I_1 ($\sqrt{2} \cdot x = u - v$, $\sqrt{2} \cdot y = u + v$) we have (Jacobian $= 1$, $x^2 + y^2 = u^2 + v^2$)

$$I_1 = \int_{-\infty}^{+\infty} \int_{-\infty}^{+\infty} \exp(-u^2 - v^2) \cos(\sqrt{2} \cdot tu) \, du \, dv$$

$$= \int_{-\infty}^{+\infty} \exp(-v^2) \, dv \int_{-\infty}^{+\infty} \exp(-u^2) \cos(\sqrt{2} \cdot tu) \, du = \pi \varphi(\sqrt{2} \cdot t),$$

since

$$\int_{-\infty}^{+\infty} \exp(-x^2) \, dx = \sqrt{\pi}. \tag{10.2.16}$$

Similarly, $I_2 = \pi \varphi(\sqrt{2} \cdot t)$. By (10.2.15) φ satisfies (10.2.14). In addition, (10.2.13) and (10.2.16) yield $\varphi(0) = 1$, and, since the function φ is twice differentiable in $\mathbb{R}$, we also have

$$\varphi''(0) = -\frac{2}{\sqrt{\pi}} \int_{-\infty}^{+\infty} x^2 \exp(-x^2) \, dx = -\tfrac{1}{2}.$$

Applying Theorem 10.2.3 with $c = -\tfrac{1}{4}$ we get the formula

$$\varphi(t) = \exp(-t^2/4).$$

It is clear that all the operations performed above on improper integrals are legitimate because of the absolute and uniform convergence of the integral (10.2.13).

Remark 10.2.4. Theorem 10.2.3 is a special case of a result by M. C. Zdun [2] saying that the function $\beta(x) = \exp(cx^r)$, $c \in \mathbb{R}$, $r > 0$, is the unique solution of the Böttcher equation

$$\beta(ax) = (\beta(x))^r, \qquad a^r > 0, \ a \neq 1,$$

that has the property

$$\beta(x) = 1 + cx^r + o(x^r), \qquad x \to 0.$$

10.2D. Comments

10.2D1. One might introduce logarithmic functions in a way similar to that we have shown in Subsection 10.2B for exponential functions. The Schröder equation

$$\sigma(\sqrt{x}) = \tfrac{1}{2}\sigma(x) \tag{10.2.17}$$

has a one-parameter family of solutions $\sigma: [1, \infty) \to \mathbb{R}$ (resp. $\sigma: (0, 1] \to \mathbb{R}$) which are convex in $[1, \infty)$ (resp. in $(0, 1]$). They are given by the formula for the principal solution of (10.2.17):

$$\sigma_a(x) = \lim_{n \to \infty} \frac{x^{2^{-n}} - 1}{a^{2^{-n}} - 1}, \qquad 0 < a \neq 1 \tag{10.2.18}$$

(see Theorem 2.4.4 and Section 9.1).

We call the function $\sigma_a: (0, \infty) \to \mathbb{R}$ the logarithmic function with the base a. Note that we cannot define logarithmic functions as convex solutions of (10.2.17) in $(0, \infty)$, as the latter are given by

$$\sigma(x) = \begin{cases} \sigma_a(x), & x \in (0, 1), \\ \sigma_b(x), & x \in [1, \infty), \end{cases}$$

with $1 < a < b$ or $0 < a < b < 1$. All the fundamental algebraic properties of logarithms can be derived from (10.2.18) (Kuczma [24]).

10.2D2. Theorem 3.5.1 offers us another definition of the logarithms on the whole of $(0, \infty)$. We may define them as the nontrivial C^1 solutions $\sigma: (0, \infty) \to \mathbb{R}$ of (10.2.17). They are given by the Koenigs algorithm (see Section 9.1)

$$\sigma(x) = c \lim_{n \to \infty} 2^n(x^{2^{-n}} - 1), \qquad c \neq 0. \tag{10.2.19}$$

It turns out (see Kuczma [24]) that the family of functions (10.2.19) is identical with that of functions (10.2.18). Definition (10.2.19) gives rise to the definition of the natural logarithm, corresponding to $c = 1$, and allows one to derive analytic properties of logarithms.

10.2D3. Regarding Comments 10.2D1 and 10.2D2 see also Kuczma [26, §6.12]. Both exponential and logarithmic functions have been characterized in Kuczma [10]. See also Theorem 6.1.1. Some results related to the contents of Subsection 10.2A may be found in Skof [1].

10.2D4. The Laplace integral

$$\varphi(t) = \frac{1}{\pi} \int_{-\infty}^{+\infty} \frac{\cos tx}{x^2 + 1} \, dx, \qquad t \in \mathbb{R}^+,$$

may be shown (see Haruki [1]) to satisfy equation (10.2.7), so that by Theorem 6.1.1 ($f(x) = x$, $c = -1$) we get $\varphi(t) = e^{-t}$, $t \in \mathbb{R}^+$.

This result may also be derived from Remark 10.2.4.

10.3 Trigonometric and hyperbolic functions

Our fundamental tool will be the equation (see also Heathcote–Pitman [1])

$$\varphi(2x) = 2(\varphi(x))^2 - 1, \tag{10.3.1}$$

which expresses the duplication formula for the cosine.

If φ satisfies (10.3.1), then the value $d = \varphi(0)$ must satisfy the equation $d = 2d^2 - 1$, whence $d = -\frac{1}{2}$ or $d = 1$. Similarly, differentiating (10.3.1) and then setting $x = 0$ we see that $\varphi'(0) = 0$.

10.3A. Cosine and hyperbolic cosine

The following result on solutions of (10.3.1) that are smooth at zero is due to G. Forder [1].

Theorem 10.3.1. *Let* $\varphi: \mathbb{R} \to \mathbb{R}$ *be a solution of equation* (10.3.1).

(a) *If* $\varphi(0) = -\frac{1}{2}$ *and* φ *is differentiable at zero, then* $\varphi(x) = -\frac{1}{2}$ *in* $\mathbb{R}$.

(b) *If* $\varphi(0) = 1$ *and* φ *is twice differentiable at zero, then* $\varphi(x) = \cos cx$ *or* $\varphi(x) = \cosh cx$, *depending on whether* $\varphi''(0) \leqslant 0$ *or* $\varphi''(0) \geqslant 0$, *where* $c = |\varphi''(0)|^{1/2}$.

Proof. (a) The function $\psi(x) := (\varphi(x) + \frac{1}{2})/x$ (defined as $0 \ (= \varphi'(0))$ at $x = 0$) is continuous at zero and satisfies the equation

$$\psi(x) = \tfrac{1}{2}x(\psi(\tfrac{1}{2}x))^2 - \psi(\tfrac{1}{2}x).$$

This is an equation of form (5.2.1) with $h(x, y) = \frac{1}{2}xy^2 - y$, $f(x) = \frac{1}{2}x$. In a sufficiently small neighbourhood $U \times V$ of $(0, \frac{1}{2})$ we have $|h(x, y) - h(x, \bar{y})| \leqslant (1 - x)|y - \bar{y}|$, and by Theorem 5.2.1 ψ is unique in U. Clearly $\psi(x) = 0$ in U, and by Baron's Extension Theorem 7.1.1 also in $\mathbb{R}$, which yields $\varphi = -\frac{1}{2}$.

(b) Now we have $\varphi(x) = 1 + \bar{d}x^2 + o(x^2)$, $x \to 0$, with $2\bar{d} = \varphi''(0)$. Thus the function $\psi(x) := (\varphi(x) - 1)/x^2$ for $x \neq 0$, $\psi(0) = \bar{d}$, is continuous at zero and satisfies the equation

$$\psi(x) = \psi(\tfrac{1}{2}x) + \tfrac{1}{8}x^2(\psi(\tfrac{1}{2}x))^2.$$

By Theorems 5.2.1 and 7.1.1 such a solution of this equation is unique for every $\bar{d} \in \mathbb{R}$. Thus equation (10.3.1) may have at most one solution, twice differentiable at zero and satisfying $\varphi(0) = 1$, $\varphi''(0) = 2\bar{d}$. Evidently $\varphi(x) = \cos cx$ or $\varphi(x) = \cosh cx$, according as $\bar{d} \leqslant 0$ or $\bar{d} \geqslant 0$, $c = \sqrt{(2|\bar{d}|)}$. ∎

10.3B. Periodic solutions of the cosine equation

The regularity condition on φ can be weakened if we add the assumption of periodicity. The following theorem and its proof are due to A. N. Šarkovskiĭ [6].

Theorem 10.3.2. *Let* $\varphi: \mathbb{R} \to \mathbb{R}$ *be a periodic solution of equation* (10.3.1), *continuous in a neighbourhood of zero. Then* $\varphi(x) = -\frac{1}{2}$ *or* $\varphi(x) = \cos cx$, *where* $c \in \mathbb{R}$ *is a constant.*

Proof. Let $\varphi: \mathbb{R} \to \mathbb{R}$ be a solution of (10.3.1) which is continuous in a neighbourhood of zero and for a certain $p > 0$ satisfies

$$\varphi(x + p) = \varphi(x). \qquad (10.3.2)$$

The range of such a φ is contained in $[-1, 1]$. Indeed, φ is bounded in an interval $[0, x_0]$, $x_0 > 0$. By (10.3.1) it is so on $[0, 2^i x_0]$ for every $i \in \mathbb{N}$, as $|\varphi(2x)| \leqslant 2\varphi(x)^2 + 1$. Hence φ is bounded on $[0, p]$, and by (10.3.2) on $\mathbb{R}$ as well. If we had $|\varphi(x_0)| > 1$ for an $x_0 \in \mathbb{R}$, then by (10.3.1) we would have $y := \varphi(2x_0) - 1 > 0$, next $\varphi(4x_0) - 1 = 2(\varphi(2x_0) + 1)y > 4y$, and by induction $\varphi(2^n x_0) - 1 > 2^n y$, $n \geqslant 2$. Thus $\lim_{n \to \infty} (\varphi(2^n x_0)) = \infty$, which contradicts the boundedness of φ. Hence $|\varphi(x)| \leqslant 1$ in $\mathbb{R}$.

The value of φ at zero can be either 1 or $-\frac{1}{2}$. First assume that $\varphi(0) = 1$. Then φ is positive on an interval $[0, x_0]$, $x_0 > 0$. Put

$$\beta_0(x) = \frac{1}{2\pi} \arccos \varphi(px) \qquad \text{for } x \in [0, x_0/p].$$

The function β_0 is continuous, nonnegative, and by (10.3.1) (with x replaced by $\frac{1}{2}x$) satisfies $\beta_0(x) = 2\beta_0(\frac{1}{2}x)$. By Baron's Extension Theorem 7.1.1 this β_0 can be uniquely extended to a solution $\beta: \mathbb{R}^+ \to \mathbb{R}$ of the Schröder equation

$$\beta(x) = 2\beta(\tfrac{1}{2}x), \qquad (10.3.3)$$

and β is continuous in $\mathbb{R}^+$. Obviously, $\beta(x) \geqslant 0$ in $\mathbb{R}^+$. The functions φ and $x \mapsto \cos 2\pi\beta(x/p)$, $x \geqslant 0$, coincide on $[0, x_0/p]$ and they both satisfy (10.3.1) in $\mathbb{R}^+$. Again by Theorem 7.1.1, they are equal, i.e.

$$\varphi(x) = \cos 2\pi\beta\left(\frac{x}{p}\right), \qquad x \in \mathbb{R}^+. \qquad (10.3.4)$$

We will show that $\beta(x) = kx$ (with a $k \in \mathbb{Z}$). This does hold when $\beta = 0$, thus from now on this case is excluded.

Because of (10.3.4) and the periodicity (10.3.2), β satisfies, besides (10.3.3), the equation

$$\beta(x + 1) = \varepsilon(x)\beta(x) + k(x), \qquad (10.3.5)$$

where $\varepsilon(x) = +1$ or -1 and $k(x)$ is an integer.

Assume for a moment that equation (10.3.5) reduces to

$$\beta(x + 1) = \beta(x) + k, \qquad (10.3.6)$$

with a fixed k. In such a case $\psi(x) = \beta(x) - kx$ satisfies simultaneously

$$\psi(2x) = 2\psi(x), \qquad \psi(x + 1) = \psi(x).$$

This implies that $\psi = 0$, since ψ is bounded being continuous and periodic, and by induction we have $\psi(2^n x) = 2^n \psi(x)$ for $n \in \mathbb{N}$. Hence $\beta(x) = kx$. Relation (10.3.4) now yields $\varphi(x) = \cos cx$, $c := 2k\pi/p$, $x \geqslant 0$. By (10.3.2) this remains true for $x < 0$, too.

Similarly we can treat the case where $\varphi(0) = -\frac{1}{2}$. First we deduce that

$$\varphi(x) = \cos\left[\frac{2\pi}{3}\left(1 + \alpha\left(\frac{x}{p}\right)\right)\right],$$

where $\alpha: (0, \infty) \to \mathbb{R}$ is continuous and satisfies

$$\alpha(x) = -2\alpha\left(\frac{x}{2}\right), \qquad \alpha(x+1) = \varepsilon(x)\alpha(x) + 3k(x) + \varepsilon(x) - 1,$$

where $\varepsilon(x) \in \{-1, 1\}$ and $k(x) \in \mathbb{Z}$. By an equally complicated argument as that below one arrives at $\varepsilon = 1$ and $k = 0$, i.e. $\alpha = 0$, whence $\varphi = -\frac{1}{2}$.

Reducing equation (10.3.5) to (10.3.6) is the most difficult task. We proceed as follows.

Write formula (10.3.5) for y and then for x, both in $(0, \infty)$, subtract one from the other and let $y \to x$. Since β is continuous at x, we get

$$\lim_{y \to x} [(\varepsilon(y) - \varepsilon(x))\beta(x) + (k(y) - k(x))] = 0.$$

We know that $\varepsilon(y) - \varepsilon(x) \in \{-2, 0, 2\}$ and $k(y) - k(x) \in \mathbb{Z}$. Thus, if $2\beta(x)$ is not an integer, then ε has to be constant in a neighbourhood of x. Hence $\varepsilon(y) = \varepsilon(x)$ in this neighbourhood, which implies $k(y) = k(x)$ there. This observation suggests the investigation of intervals on which $2\beta(x) \notin \mathbb{N}_0$.

An interval $(x_1, x_2) \subset (0, \infty)$ is called an *L-interval* iff $2\beta(x_1) \neq 2\beta(x_2)$, both are integers, and $2\beta(x)$ is not an integer for $x \in (x_1, x_2)$.

L-intervals do exist. For, since $\beta(0) = 0$ and β is continuous, there is a smallest x_2 such that $\beta(x_2) = \frac{1}{2}$ (otherwise we would have $0 \leqslant \beta(x) < \frac{1}{2}$ in $(0, \infty)$, contrary to (10.3.3)). Take now the largest $x_1 < x_2$ such that $\beta(x_1) = 0$ to get an *L*-interval (x_1, x_2).

The functions ε and k are constant on any *L*-interval. Further, by (10.3.5) we see that (x_1, x_2) is an *L*-interval if and only if $(x_1 + 1, x_2 + 1)$ is. On the other hand, since $\beta(0) = 0$ and by (10.3.5), $\beta(x)$ is an integer for integral x. Thus every *L*-interval is contained in an interval $(k, k+1)$, $k \in \mathbb{N}_0$; in every such interval there is the same number r of *L*-intervals; and in every interval

$$I_k := (2^k, 2^{k+1}), \qquad k \in \mathbb{N}_0,$$

there are exactly $2^k r$ *L*-intervals. Finally, $r \neq 0$, as we have constructed an *L*-interval contained in $(0, 1)$.

Let $(x_1, x_2) \subset I_k$ be an *L*-interval. Then $(2x_1, 2x_2) \subset I_{k+1}$, $2\beta(2x_1) = 4\beta(x_1)$, $2\beta(2x_2) = 4\beta(x_2)$, and these values differ by $+2$ or -2. Therefore $(2x_1, 2x_2)$ contains at least two *L*-intervals lying in I_{k+1}. But the number of *L*-intervals in I_{k+1} is twice the number of those in I_k. Consequently, to every *L*-interval J in I_k there correspond exactly two *L*-intervals contained in $2J \subset I_{k+1}$, and their images by the function β are disjoint.

We say that an *L*-interval $J_1 \subset I_k$ *generates* an *L*-interval $J_2 \subset I_{k+m}$, $m \in \mathbb{N}$, iff $J_2 \subset 2^m J_1$. From the above discussion we infer that every *L*-interval in I_k

generates exactly 2^m ones in I_{k+m}, whose images by the function β are pairwise disjoint. In particular, every L-interval in I_n, $n \in \mathbb{N}$, is generated by an L-interval in I_0.

The function β cannot be constant on any interval, for otherwise, by (10.3.1), φ would be constant on an interval of arbitrarily large length, in particular, exceeding p. By (10.3.2) φ would be constant on $\mathbb{R}$, whence $\beta = 0$, and this case has been excluded.

We shall show that β is injective. For an indirect proof suppose that $\beta(x') = \beta(x'')$ for some $x' \neq x''$. By the continuity of β, and since $\beta \neq \text{const}$ on $I := [x', x'']$, we may assume that $\beta(x) > 0$ in I. We can find an $s \in \mathbb{N}$ such that $2^{-s}I \subset (0, 1)$ and $0 < \beta(x) < \frac{1}{2}$ in $2^{-s}I$ (see (10.3.3)). Moreover, we have $\beta(2^{-s}x') = \beta(2^{-s}x'')$. Take $x_1 := 2^{-s}x' + 1$, $x_2 := 2^{-s}x'' + 1$. Thus $\beta(x_1) = \beta(x_2)$, $2\beta(x) \notin \mathbb{N}_0$ when $x \in [x_1, x_2]$ and $(x_1, x_2) \subset I_0 = (1, 2)$. Evidently (x_1, x_2) either is contained in an L-interval, or disjoint from any L-interval.

Let $\beta(x_3) \neq \beta(x_1)$, $x_3 \in (x_1, x_2)$. We may choose an $m \in \mathbb{N}$ so that $2^m |\beta(x_3) - \beta(x_1)| = |\beta(2^m x_3) - \beta(2^m x_1)| > 1$. Therefore between $2\beta(2^m x_3)$ and $2\beta(2^m x_1)$ there are at least two integers, say q and $q + 1$. By the continuity of β in $[x_1, x_2]$ and (10.3.3), we may find x_1', x_2' so that $x_1 \leqslant x_1' \leqslant x_3 \leqslant x_2' \leqslant x_2$ and $2\beta(2^m x_1') = 2\beta(2^m x_2') = q$. Starting with the points $2^m x_1'$ and $2^m x_2'$, we can construct, again by the continuity of β, two L-intervals J_1 and J_2, $J_1 \subset (2^m x_1, 2^m x_3) \subset I_m$, $J_2 \subset (2^m x_3, 2^m x_2) \subset I_m$ such that $2\beta(J_1) = 2\beta(J_2) = (q, q+1)$.

Let $J_0 \subset I_0 = (1, 2)$ be the L-interval that generates J_1. It is evident that $[x_1, x_2] \subset J_0$ as $2\beta(x)$ is not an integer when $x \in [x_1, x_2]$ and $J_1 \subset 2^m J_0$ implies $[x_1, x_2] \cap J_0 \neq \varnothing$. But $J_2 \subset (2^m x_3, 2^m x_2) \subset 2^m J_0$, and J_0 generates J_2 as well. Since $J_1 \cup J_2 \subset I_m$, the images $\beta(J_1)$ and $\beta(J_2)$ by β are disjoint, a contradiction.

We have proved that β is injective. It is also continuous and nonnegative, $\beta(0) = 0$, whence strictly increasing. This implies in view of (10.3.5) that $\varepsilon = +1$ on every interval on which it is constant. At the remaining points of $(0, \infty)$, where $2\beta(x)$ takes on integral values, the values of ε and k can be adjusted so that $\varepsilon = +1$. Then we shall have $\varepsilon = 1$, whence $k(x) = \beta(x+1) - \beta(x)$ is continuous and integral-valued, and hence constant.

Equation (10.3.5) actually becomes (10.3.6). ∎

Theorem 10.3.2 obviously implies (Šarkovskiĭ [6]) the following.

Corollary 10.3.1. *The only solution* $\varphi: \mathbb{R} \to \mathbb{R}$ *of equation* (10.3.1) *that is continuous in a neighbourhood of zero and fulfils the conditions*

$$\varphi(x + 2\pi) = \varphi(x), \qquad \varphi(x + p) \neq \varphi(x) \qquad \text{for } 0 < p < 2\pi,$$

is $\varphi(x) = \cos x$.

10.3C. Sine

The function $\varphi(x) = \sin x$ can be characterized by the equation

$$2(\varphi(x))^2 + \varphi\left(\frac{\pi}{2} - 2x\right) = 1 \tag{10.3.7}$$

(Dubikajtis [1]). Namely, we have the following.

Lemma 10.3.1. *If* $\varphi: \mathbb{R} \to \mathbb{R}$ *is an odd solution of equation* (10.3.7), *then*

$$\hat{\varphi}(x) = \varphi\left(\frac{\pi}{2} - x\right) \tag{10.3.8}$$

satisfies equation (10.3.1) *and is periodic of period* 2π. *Moreover,* φ *and* $\hat{\varphi}$ *have the same regularity in a neighbourhood of zero.*

Proof. We have by (10.3.7)

$$\varphi(x) = \varphi\left(\frac{\pi}{2} - 2\left(\frac{\pi}{4} - \frac{x}{2}\right)\right) = 1 - 2\left(\varphi\left(\frac{\pi}{4} - \frac{x}{2}\right)\right)^2$$

and

$$\varphi(\pi - x) = \varphi\left(\frac{\pi}{2} - 2\left(\frac{x}{2} - \frac{\pi}{4}\right)\right) = 1 - 2\left(\varphi\left(\frac{x}{2} - \frac{\pi}{4}\right)\right)^2,$$

whence

$$\varphi(\pi - x) = \varphi(x), \tag{10.3.9}$$

since φ is odd. This implies the periodicity of φ:

$$\varphi(2\pi + x) = \varphi(\pi - (2\pi + x)) = -\varphi(\pi + x) = -\varphi(\pi - (\pi + x)) = \varphi(x).$$

By (10.3.8) $\hat{\varphi}$ also is periodic of period 2π. Further, we have by (10.3.7)

$$2(\hat{\varphi}(x))^2 = 2\left(\varphi\left(\frac{\pi}{2} - x\right)\right)^2 = 1 - \varphi\left(2x - \frac{\pi}{2}\right) = 1 + \hat{\varphi}(2x),$$

i.e., $\hat{\varphi}$ satisfies (10.3.1). The assertion on regularity follows from (10.3.7) and (10.3.8). ∎

Lemma 10.3.1 allows us to translate results for equation (10.3.1) into ones for equation (10.3.7). In this way we obtain from Corollary 10.3.1 the following characterization of the sine function.

Theorem 10.3.3. *The only solution* $\varphi: \mathbb{R} \to \mathbb{R}$ *of equation* (10.3.7) *that is odd, continuous in a neighbourhood of zero, and fulfils the condition*

$$\varphi(x) > 0 \qquad for\ 0 < x < \frac{\pi}{2}, \tag{10.3.10}$$

is $\varphi(x) = \sin x$.

Proof. It follows from (10.3.9) and (10.3.10), since φ is odd, that $\varphi(x) > 0$

in $(0, \pi)$, $\varphi(x) < 0$ in $(\pi, 2\pi)$. Hence $\varphi(x + p) \not\equiv \varphi(x)$ for $0 < p < 2\pi$. Lemma 10.3.1 and Corollary 10.3.1 apply.   ∎

The direct proof of Theorem 10.3.3, however, is much simpler than that of Theorem 10.3.2 (see Dubikajtis [1], Kuczma [26, Ch. XI, § 2]).

10.3D. An improper integral

Following H. Haruki [4] we shall apply Theorem 10.3.1 and Proposition 6.9.1 to evaluate the integral

$$K(t) = \frac{2}{\sqrt{\pi}} \int_0^\infty \exp(-x^2)\cos(t^2/x^2)dx, \qquad t \geqslant 0. \qquad (10.3.11)$$

We proceed as in Subsection 10.2C, splitting $K(t)^2$ into two integrals

$$K(t)^2 = \tfrac{1}{2}(L(t) + M(t)), \qquad (10.3.12)$$

where

$$\frac{\pi}{4} L(t) := \int_0^\infty \int_0^\infty \exp(-x^2 - y^2)\cos t^2(x^{-2} + y^{-2})dxdy \quad (10.3.13)$$

and

$$\frac{\pi}{4} M(t) := \int_0^\infty \int_0^\infty \exp(-x^2 - y^2)\cos t^2(x^{-2} - y^{-2})dxdy. \quad (10.3.14)$$

In (10.3.13) we use polar coordinates $x = r\cos\tfrac{1}{2}\theta$, $y = r\sin\tfrac{1}{2}\theta$, and obtain (Jacobian $= r/2$, $x^{-2} + y^{-2} = 4(r\sin\theta)^{-2}$)

$$L(t) = \frac{2}{\pi} \int_0^\pi \int_0^\infty \exp(-r^2)\cos\frac{4t^2}{r^2\sin^2\theta} r\,dr\,d\theta.$$

We see that

$$L(t) = \frac{2}{\pi} \int_{-\infty}^{+\infty} \left(\int_0^\infty \exp(-y^2)\cos(4t^2/y^2)dy \right) \exp(-x^2)dx.$$

We arrive at (see (10.2.16))

$$L(t) = K(2t), \qquad t \geqslant 0. \qquad (10.3.15)$$

For the integral (10.3.14) we want to obtain the relation

$$M(t)^2 = M(2t), \qquad t \geqslant 0. \qquad (10.3.16)$$

This may be achieved by first using under the integral representing $M(t)^2$ the formula for the product of cosines,

$$M(t)^2 = \frac{16}{\pi^2} \int_0^\infty \int_0^\infty \int_0^\infty \int_0^\infty \exp(-x^2 - y^2 - u^2 - v^2)$$

$$\times \cos t^2(x^{-2} - y^{-2} + u^{-2} - v^{-2})dxdydudv,$$

next applying the transformation $x = r_1\cos\tfrac{1}{2}\theta_1$, $y = r_2\cos\tfrac{1}{2}\theta_2$, $u = r_1\sin\tfrac{1}{2}\theta_1$, $v = r_2\sin\tfrac{1}{2}\theta_2$ and manipulating as for $L(t)$ before.

By (10.3.16) and Proposition 6.9.1 we have

$$M(t) = \exp kt, \tag{10.3.17}$$

provided $M'(0) = k$. To determine k we first calculate

$$K'(t) = -\frac{2}{\sqrt{\pi}} \int_0^\infty 2(t/x^2) \exp(-x^2) \sin(t^2/x^2) dx$$

$$= -\frac{4}{\sqrt{\pi}} \int_0^\infty \exp(-t^2/y^2) \sin y^2 dy, \tag{10.3.18}$$

which for $t = 0$ reduces to the Fresnel integral

$$K'(0) = -\frac{4}{\sqrt{\pi}} \int_0^\infty \sin y^2 dy = -\sqrt{2}.$$

Further, since

$$M(t) = \frac{4}{\pi} \int_0^\infty \int_0^\infty \exp(-x^2 - y^2) \left(\cos \frac{t^2}{x^2} \cos \frac{t^2}{y^2} + \sin \frac{t^2}{x^2} \sin \frac{t^2}{y^2} \right) dx dy,$$

we obtain

$$M(t) = K(t)^2 + \frac{4}{\pi} \left(\int_0^\infty \exp(-x^2) \sin(t^2/x^2) dx \right)^2.$$

Differentiate this, and then set $t = 0$, to get (recall that $K(0) = 1$)

$$M'(0) = 2K(0)K'(0) = -2\sqrt{2}.$$

Thus $k = -2\sqrt{2}$ by (10.3.17) and $M(t) = \exp(-2\sqrt{2} \cdot t)$. Substituting this and (10.3.15) into (10.3.12) we have

$$K(2t) = 2K(t)^2 - \exp(-2\sqrt{2} \cdot t). \tag{10.3.19}$$

Putting

$$\varphi(t) := K(t) \exp(\sqrt{2} \cdot t) \tag{10.3.20}$$

we conclude that $\varphi(0) = 1$ and

$$\varphi(2t) = 2\varphi(t)^2 - 1, \qquad t \geqslant 0. \tag{10.3.21}$$

What we need in order to be able to apply Theorem 10.3.1 is the value of φ'' at zero. By (10.3.20) one gets

$$\varphi''(0) = 2K(0) + 2\sqrt{2} \cdot K'(0) + K''(0) = -2,$$

as $K''(0) = 0$, which can be seen from (10.3.18). Consequently $\varphi(t) = \cos(\sqrt{2} \cdot t)$, and relation (10.3.20) shows that

$$K(t) = \exp(-\sqrt{2} \cdot t) \cos(\sqrt{2} \cdot t), \qquad t \geqslant 0.$$

Remark 10.3.1. As in Subsection 10.2C, the operations involved are legitimate on the basis of the absolute and uniform convergence of the integral $K(t)$.

10.3E. Comments

10.3E1. Smooth solutions $\varphi: \mathbb{R} \to \mathbb{R}$ of equation (10.3.1), written in the equivalent form

$$\varphi(x) = 2\varphi(\tfrac{1}{2}x)^2 - 1, \tag{10.3.22}$$

may be derived directly from Theorem 5.6.1, applied to the interval $X = \mathbb{R}^+$, with $\Omega = X \times \mathbb{R}$, $f(x) = x/2$, $h(x, y) = 2y^2 - 1$. The solutions in $(-\infty, 0]$ are obtained by observing that $\hat{\varphi}(x) := \varphi(-x)$ also satisfies equation (10.3.22) on X. Both branches, when put together, furnish solutions φ on $\mathbb{R}$.

The value $d = \varphi(0)$ may be 1 or $-\tfrac{1}{2}$.

For $d = -\tfrac{1}{2}$ we apply Theorem 5.6.1 with $r = 2$, as

$$f'(0)^2 \left| \frac{\partial h}{\partial y}(0, d) \right| = \tfrac{1}{4}|4d| = \tfrac{1}{2} < 1.$$

Since $0 = d_1 = \varphi'(0) = \varphi''(0) = d_2$ (see (5.6.6)), we get $\varphi = -\tfrac{1}{2}$ as the only C^2 solution $\varphi: \mathbb{R} \to \mathbb{R}$ of equation (10.3.22) with $\varphi(0) = -\tfrac{1}{2}$.

For $d = 1$ we must take $r = 3$. Since now $d_1 = 0$, d_2 is arbitrary, $d_3 = 0$, we obtain the functions $\varphi(x) = \cos ax$ when $d_2 = -a^2$, and $\varphi(x) = \cosh ax$ when $d_2 = a^2$ as the only C^3 solutions of (10.3.22) in $\mathbb{R}$ satisfying $\varphi(0) = 1$.

Theorem 10.3.1 improves on both these results.

10.3E2. The function $h(x, y) = 2y^2 - 1$ in (10.3.22) is nothing but the Čebyšev polynomial T_2, where T_n are defined by $T_0(x) = 1$, $T_1(x) = x$, $T_{n+1}(x) = 2xT_n(x) - T_{n-1}(x)$ for $n \in \mathbb{N}$. Cz. Dyjak [2] proved that cosine and hyperbolic cosine can be characterized with the help of the equation (for a fixed $n \geqslant 2$)

$$\varphi(nx) = T_n(\varphi(x)). \tag{10.3.23}$$

More exactly, the assertion (b) of Theorem 10.3.1 holds true for equation (10.3.23) as well.

10.3E3. It follows from Theorem 5.6.3 applied to the equation

$$\varphi(\tfrac{1}{2}x) = \sqrt{[\tfrac{1}{2}(\varphi(x) + 1)]}, \tag{10.3.24}$$

and from Theorem 7.1.1 that the C^1 solution $\varphi: \mathbb{R} \to [1, \infty)$ of (10.3.22) satisfying $\varphi(0) = 1$ depends on an arbitrary function. The uniqueness cannot be assured even by the additional requirement of the convexity of φ in a neighbourhood of zero (Cooper [1]). However, $\varphi(x) = \cos x$ is the only solution $\varphi: [-\pi/2, \pi/2] \to \mathbb{R}$ of (10.3.24) that satisfies the condition $\lim_{x \to 0}\{\sqrt{[1 - \varphi(x)^2]}/x\} = 1$ (Fubini [1]).

10.3E4. Characterizations of cosine, similar to those we have presented in Subsections 10.3A and 10.3B, can also be found in Kannappan [1], Kuczma [16], Robbins [1]. By Lemma 10.3.1 they may be used to obtain theorems on sine.

10.3E5. The function $\varphi: [-\pi/2, \pi/2] \to [-1, 1]$, $\varphi(x) = \sin x$, is the only, differentiable at zero, solution of the equation

$$\varphi(2x) = 2\varphi(x)\sqrt{[1 - \varphi(x)^2]}$$

such that $\varphi'(0) = 1$, whereas the function $\varphi: [0, 2\pi] \to \mathbb{R}$, $\varphi(x) = \cos x$, is the unique continuous function that satisfies $\varphi(x_0) = \cos x_0$ and the equation

$$\varphi(f(x)) = 2\varphi(x)^2 - 1,$$

where $f(x) = 2x$ for $x \in [0, \pi]$ and $f(x) = 2(2\pi - x)$ for $x \in [\pi, 2\pi]$ and $x_0 \in (0, 2\pi)$ is such that the set $\{f''(x_0): n \in \mathbb{N}\}$ is dense in $[0, 2\pi]$.

Both results are found in Pelyukh–A. N. Šarkovskiĭ [1], where also other functions are characterized via iterative functional equations.

10.3E6. Periodic solutions of more general functional equations than (10.3.1) have also been studied, in particular, in Kuczma–Szymiczek [1]. See also Kuczma [26, Ch. XI, § 2] and the references quoted therein.

10.4 Euler's gamma function

In this section we present some results on various characterizations of Euler's gamma function by functional equations. We follow to a considerable extent the article by H.-H. Kairies [6]. Throughout this section we use the notation

$$X := (0, \infty).$$

The Euler gamma function $\Gamma: X \to X$ is defined by the relation

$$\Gamma(x) = \lim_{n \to \infty} \frac{n! \, n^x}{x(x+1)\cdots(x+n)}. \tag{10.4.1}$$

10.4A. The fundamental functional equation

It follows from (10.4.1) that $\Gamma(1) = 1$ and that $\varphi = \Gamma$ satisfies in X the equation

$$\varphi(x + 1) = x\varphi(x), \tag{10.4.2}$$

which will be called *fundamental*. We see that $\Gamma(k) = (k - 1)!$ for $k \in \mathbb{N}$.

Examining (10.4.1) again, we find that $\log \Gamma$ is convex. In other words, the gamma function is logarithmically convex. It follows from Theorem 2.4.2 that the properties just described completely characterize the function Γ. This result is due to H. Bohr–J. Mollerup [1] (see also Artin [1]).

Theorem 10.4.1. *If a function $\varphi: X \to X$ satisfies equation* (10.4.2), *is logarithmically convex, and $\varphi(1) = 1$, then $\varphi = \Gamma$.*

Remark 10.4.1. In Artin's proof of Theorem 10.4.1 (Artin [1]), formula (10.4.1) is deduced directly from the log-convexity of φ. It also follows from Theorem 2.4.2 that all the derivatives of $\log \Gamma$ (and hence of Γ) exist in X and

$$(-1)^k \frac{d^k}{dx^k} \log \Gamma(x) = (k - 1)! \sum_{n=0}^{\infty} \frac{1}{(x+n)^k} > 0, \qquad x \in X,$$

for $k = 2, 3, \ldots$, i.e., $(\log \Gamma)''$ is completely monotonic. It follows that Γ is analytic in X.

In Theorem 10.4.1 the condition of the logarithmic convexity cannot be replaced by that of convexity (Mayer [1], Kairies [1], [6]). In particular (Kairies [6]), if the positive constant γ is small enough, the function

$$\varphi(x) = \Gamma(x) \exp(\gamma \sin 2\pi x) \tag{10.4.3}$$

is analytic in X, satisfies (10.4.2) and $\varphi(1) = 1$, is convex in X, and even logarithmically convex in a right vicinity of zero; nevertheless $\varphi \neq \Gamma$.

On the other hand, the only function $\varphi: X \to X$, such that $(-1)^k(\log \varphi)^{(k)} \geq 0$ in X for $k = 2, 3, \ldots$ and that $\varphi(k) = (k-1)!$ for $k \in \mathbb{N}$ is $\varphi = \Gamma$ (Muldoon [1]). Note that equation (10.4.2) is not needed here.

The condition of the logarithmic convexity of φ can also be replaced by conditions of another kind (see Muldoon [2], Anastassiadis [3], [7], Eagle [1], Kuczma [26, Ch. V, §10]). We show two theorems to this effect.

Theorem 10.4.2. *If a function $\varphi: X \to X$ satisfies equation (10.4.2), $\varphi(1) = 1$, and, for some $\alpha \geq 0$, the function $\hat{\varphi}: X \to X$,*

$$\hat{\varphi}(x) = x^{\alpha}(e/x)^x \varphi(x) \tag{10.4.4}$$

is monotonic for large x, then $\varphi = \Gamma$.

Proof. We are going to apply Proposition 2.7.3 on monotonic solutions of the equation

$$\hat{\varphi}(x+1) = g(x)\hat{\varphi}(x), \tag{10.4.5}$$

which is actually satisfied by the function (10.4.4) when

$$g(x) = e(x/(x+1))^{x-\alpha+1}. \tag{10.4.6}$$

Indeed, we have for $x > 1$

$$\left(\log \frac{g(x)}{e}\right)' = \frac{x-\alpha+1}{x(x+1)} + \sum_{n=1}^{\infty}(-1)^n \frac{x^{-n}}{n} = \frac{1}{x}\left(\frac{-\alpha}{x+1} + \frac{1}{2x} - \frac{1}{3x^2} + \cdots\right).$$

This expression has a constant sign for large x, which shows that the function g is monotonic for large x.

By Proposition 2.7.3 we have, with $x_0 = 1$,

$$\hat{\varphi}(x) = c \lim_{n \to \infty} \frac{x^{\alpha-1}(1+n)^{\alpha-1}(x+n)^{x+n}}{x^x(x+n)^{\alpha-1}(1+n)^{1+n}} \prod_{i=1}^{n} \frac{1+i}{x+i}$$

$$= c \lim_{n \to \infty} \left[x^{\alpha-x}\left(\frac{1+n}{x+n}\right)^{\alpha-1}\left(\frac{x+n}{n}\right)^x\left(\frac{x+n}{1+n}\right)^n \frac{n!\, n^x}{x(x+1)\cdots(x+n)}\right]$$

$$= cx^{\alpha}e^{x-1}x^{-x}\Gamma(x).$$

Setting $x = 1$ we obtain $c = \hat{\varphi}(1) = e\varphi(1) = e$, by (10.4.4). Therefore $\hat{\varphi}(x) = x^{\alpha}(e/x)^x\Gamma(x)$, whence $\varphi = \Gamma$. ∎

Remark 10.4.2. Incidentally, this proof shows that, for every $\alpha \in \mathbb{R}^+$, the function (10.4.4) (with $\varphi = \Gamma$) actually is monotonic for large x (Muldoon [2]).

Theorem 10.4.3. *If a function* $\varphi:X \to X$ *satisfies equation* (10.4.2), $\varphi(1) = 1$, *and if*

$$\lim_{x \to \infty} x^{1/2}(e/x)^x \varphi(x) < \infty, \tag{10.4.7}$$

then $\varphi = \Gamma$.

Proof. With $\hat{\varphi}$ and g defined by (10.4.4) and (10.4.6) with $\alpha = \frac{1}{2}$, we obtain equation (10.4.5), which yields

$$\hat{\varphi}(x+n) = \hat{\varphi}(x) \prod_{i=0}^{n-1} g(x+i), \qquad n \in \mathbb{N}.$$

Hence, if c denotes the limit (10.4.7),

$$\hat{\varphi}(x) = c \lim_{n \to \infty} \left(\prod_{i=0}^{n-1} g(x+i) \right)^{-1}$$

$$= c \lim_{n \to \infty} \left(e^{-n} x^{1/2-x} (x+n)^{x+n+1/2} \prod_{i=0}^{n} \frac{1}{x+i} \right)$$

$$= c \lim_{n \to \infty} \left(x^{1/2-x} \left(\frac{x+n}{n} \right)^{x+n+1/2} \frac{1}{\sqrt{(2\pi)}} \frac{n^n e^{-n} \sqrt{(2\pi n)}}{n!} \frac{n! \, n^x}{x(x+1)\cdots(x+n)} \right).$$

In view of Stirling's formula

$$n! = n^n e^{-n} \sqrt{(2\pi n)} \cdot (1 + o(1)), \qquad n \to \infty,$$

we get hence $\hat{\varphi}(x) = c(2\pi)^{-1/2} x^{1/2}(e/x)^x \Gamma(x)$. Since $\varphi(1) = 1$, we obtain $\hat{\varphi}(1) = e$, whence $c = (2\pi)^{1/2}$ and $\varphi = \Gamma$. ∎

Remark 10.4.3. The condition $\varphi(1) = 1$ of Theorem 10.4.3 can be replaced by the requirement that the limit (10.4.7) be equal to $(2\pi)^{1/2}$.

10.4B. Riemann integrable solutions of an auxiliary equation

Our further considerations are based on a result which is due to H.-H. Kairies [6] and concerns the equation

$$\psi(x) = \frac{1}{2} \left(\psi \left(\frac{x}{2} \right) + \psi \left(\frac{x+1}{2} \right) \right). \tag{10.4.8}$$

Lemma 10.4.1. *Let* $\psi: X \to \mathbb{R}$ *be Riemann integrable on every compact interval contained in X, and suppose that there exists a limit, finite or not, $c := \lim_{x \to 0} \int_x^1 \psi(t)dt$. If ψ satisfies (10.4.8), then c is finite and $\psi(x) = c$ a.e. in X.*

Proof. We get by a repeated use of (10.4.8)

$$\psi(x) = 2^{-p} \sum_{i=0}^{2^p - 1} \psi\left(\frac{x+i}{2^p}\right), \qquad p \in \mathbb{N}. \tag{10.4.9}$$

Replacing here x by $2^p x$ and letting $p \to \infty$ we obtain

$$I(x) := \lim_{p \to \infty} \psi(2^p x) = \int_x^{x+1} \psi(t)\,dt, \tag{10.4.10}$$

as the sum in (10.4.9) is a Riemann sum for the integral in (10.4.10). Clearly, $I(x) = I(2^{-q}x)$ for $q \in \mathbb{N}$, whence

$$\int_1^{x2^{-q}+1} \psi(t)\,dt - \int_1^{x2^{-q}} \psi(t)\,dt = I(x).$$

Letting $q \to \infty$ we get $I(x) = c$, that is, c is finite.

Since I is constant, we get differentiating (10.4.10) that $\psi(x+1) = \psi(x)$ a.e. in X. Extend ψ onto $X \cup \{0\}$ by putting $\psi(0) = \psi(1)$, and let $A \subset \mathbb{R}^+$ be the set (of full measure) of those x for which $\psi(x+1) = \psi(x)$. Then we get for $x \in \bigcup_{p,i=0}^{\infty} (2^p A - i)$, on letting $p \to \infty$ in (10.4.9),

$$\psi(x) = \int_0^1 \psi(t)\,dt = c,$$

i.e., ψ is constant a.e. in X. ∎

10.4C. Gauss' multiplication theorem

This is the formula

$$\prod_{i=0}^{p-1} \varphi\left(\frac{x+i}{p}\right) = d_p p^{-x} \varphi(x), \qquad p = 2, 3, \ldots, \tag{10.4.11}$$

where

$$\frac{1}{p} d_p = \prod_{i=0}^{p-1} \Gamma\left(\frac{1+i}{p}\right), \qquad p = 2, 3, \ldots. \tag{10.4.12}$$

We shall prove that the gamma function satisfies (10.4.11) with (10.4.12). Let us fix an integer $p > 1$ and write

$$\Phi_p(x) = p^x \prod_{i=0}^{p-1} \Gamma\left(\frac{x+i}{p}\right).$$

Since Γ satisfies (10.4.2), so does Φ_p. Moreover, Φ_p is logarithmically convex. By Theorem 10.4.1 we have $\Phi_p(x) = \Phi_p(1)\Gamma(x)$ and (10.4.11) with (10.4.12) holds for $\varphi = \Gamma$ in view of $d_p = \Phi_p(1)$.

Now, the question arises, under what conditions equations (10.4.2) and (10.4.11) have $\varphi = \Gamma$ as the only solution.

The first result to be mentioned is due to E. Artin [1] and says that if $\varphi : X \to X$ is continuous and satisfies (10.4.2) and (10.4.11) for all $p = 2, 3, \ldots$, then $\varphi = \Gamma$.

Actually, it is enough to consider (10.4.11) for primes, $p \in \mathbb{P}$, only, since if a function φ satisfies (10.4.11) for two positive integers, it does so for their product as well. Now, it turns out (see Lucht [1], [2]) that if $\varphi: X \to X$ is continuous and satisfies (10.4.2) and (10.4.11) for all but a finite number of $p \in \mathbb{P}$, then $\varphi = \Gamma$.

On the other hand, a continuous function $\varphi: X \to X$ different from Γ but satisfying equations (10.4.2) and (10.4.11) for all p from a set $P \subset \mathbb{P}$ may be constructed in the case where the set P satisfies one of the following conditions:

(a) $\sum_{p \in P} 1/p < \infty$ (see Lucht [1]),
(b) $\sum_{p \notin P} 1/p = \infty$ and there is an increasing function $g: X \to X$, $\lim_{x \to \infty} g(x) = \infty$, such that for large x,

$$\sum_{\substack{p \notin P \\ p \leqslant x}} \frac{1}{p} \leqslant g(x),$$

(see Lucht–Wolke [2]).

However, if φ is more regular, then (10.4.2) and (10.4.11) for $p = 2$ alone already imply that $\varphi = \Gamma$ (see Theorem 10.4.6 below).

Before proceeding further we calculate the numbers d_p given by (10.4.12). They are

$$d_p = (2\pi)^{(p-1)/2} p^{1/2}, \qquad p \in \mathbb{N}. \tag{10.4.12'}$$

We prove this for $p = 2$.

Let $D(\mathbb{K})$ be the complement in $\mathbb{K}$ of the set of nonpositive integers. By a use of equation (10.4.2) the function Γ may be extended onto the whole $D(\mathbb{R})$ and this extension also satisfies (10.4.2) and (10.4.11).

Let us introduce the function $\omega: \mathbb{R} \setminus \mathbb{Z} \to \mathbb{R}$ by

$$\omega(x) = \Gamma(x)\Gamma(1-x) \sin \pi x. \tag{10.4.13}$$

The function ω is periodic of period 1, and, by (10.4.2) and $\Gamma(1) = 1$,

$$\lim_{x \to 0} \omega(x) = \lim_{x \to 0} \Gamma(1+x)\Gamma(1-x) \frac{\sin \pi x}{x} = \pi. \tag{10.4.14}$$

Since $\lim_{x \to 0} \omega'(x) = 0$, ω can be extended onto $\mathbb{R}$, to a function of class C^1 in $\mathbb{R}$. Because of (10.4.11) for $p = 2$ and $\varphi = \Gamma$ we have

$$\omega\left(\frac{x}{2}\right)\omega\left(\frac{x+1}{2}\right) = \tfrac{1}{4}d_2^2\omega(x). \tag{10.4.15}$$

Clearly, $\omega(x) > 0$ for $x \in (0, 1)$ whence and by (10.4.14) ω is positive in the whole $\mathbb{R}$. By (10.4.15) the function $\psi(x) = (d/dx)\log(\omega(x))$ satisfies equation (10.4.8), and being continuous and periodic is constant in virtue of Lemma 10.4.1. Hence $\log \omega(x)$ is a linear function, therefore constant, as it is periodic just like ω. So ω is constant, equal π, in view of (10.4.14). Thus we obtain from (10.4.15) that $d_2 = 2\sqrt{\pi}$, as claimed.

10.4D. Complex gamma function

If x is a complex variable, then formula (10.4.1) defines the gamma function in $D(\mathbb{C})$. (Now $n^x := \exp(x \log n)$.) This may be seen from the fact that the limit in (10.4.1) is the reciprocal of

$$x \exp(Cx) \sum_{n=1}^{\infty} \left(1 + \frac{x}{n}\right) \exp\left(-\frac{x}{n}\right),$$

and this is an entire function with simple zeros at $x = 0, -1, -2, \ldots$. Here C denotes the Euler constant

$$C := \lim_{n \to \infty} \left(\sum_{k=1}^{n} \frac{1}{k} - \log n\right).$$

The complex gamma function is then meromorphic in $\mathbb{C}$, with simple poles at $x = 0, -1, -2, \ldots$. Thus it obeys in $D(\mathbb{C})$ all identities known in the real case. We consider the Gauss multiplication formula (10.4.11) rewriting it in the form (see (10.4.12'))

$$p^{x-1/2} \sum_{k=0}^{p-1} \varphi\left(\frac{x+k}{p}\right) = (2\pi)^{(p-1)/2} \varphi(x), \qquad (10.4.16)$$

where $p^{x-1/2}$ stands for $\exp((x - \frac{1}{2}) \log p)$. The following characterization theorem, which we quote here without proof, is due to H. Haruki [3].

Theorem 10.4.4. *Let φ be a function meromorphic in $\mathbb{C}$ but analytic and never vanishing in $D(\mathbb{C})$. If φ satisfies (10.4.16) for a $p \in \mathbb{N} \setminus \{1\}$, then*

$$\varphi(x) = \exp\left(a(x - \tfrac{1}{2}) + \frac{2m\pi i}{p-1}\right) \Gamma(x),$$

where $a \in \mathbb{C}$ and $m \in \mathbb{Z}$.

Moreover, if $\varphi(1) = 1$, $\varphi'(1) = -C$ (the Euler constant), then $\varphi = \Gamma$.

Remark 10.4.4. The condition 'φ is analytic and never vanishes in $D(\mathbb{C})$' in Theorem 10.4.4 cannot be replaced by 'φ is analytic and does not identically vanish in $D(\mathbb{C})$'. The entire function given by $\varphi(x) = (\sin \pi x) \Gamma(x)$ serves as a counterexample (see Haruki [3]).

10.4E. Legendre's duplication formula and Euler's functional equation

We return to reals. For $p = 2$ equation (10.4.11) becomes (see (10.4.12'))

$$\varphi\left(\frac{x}{2}\right) \varphi\left(\frac{x+1}{2}\right) = 2\sqrt{\pi} \cdot 2^{-x} \varphi(x), \qquad (10.4.17)$$

which is known as *Legendre's duplication formula.*

Moreover, since we have shown in Subsection 10.4C that the function $\omega = \pi$, we see that $\varphi = \Gamma$ satisfies the equation

$$\varphi(x)\varphi(1-x) = \frac{\pi}{\sin \pi x}, \qquad x \in (0, 1), \tag{10.4.18}$$

known as *Euler's functional equation*.

First we shall prove that the Legendre equation (10.4.17) may characterize the gamma function (see Kairies [6], [1]). We are interested in the class of functions satisfying the conditions

(i) $\varphi : X \to X$ is either convex or of class C^1, and $\varphi(1) = 1$.

Theorem 10.4.5. *If φ satisfies* (i) *and equation* (10.4.17), *and there exists a finite or infinite limit* $\lim_{x \to 0} x\varphi(x)$, *then necessarily $\varphi = \Gamma$.*

Proof. Write $\hat{\phi}(x) = \log \varphi(x) - \log \Gamma(x)$ and $\psi(x) = \hat{\phi}'_+(x)$ (the right-sided derivative). Since $\hat{\phi}(1) = 0$, we have $\hat{\phi}(x) = \int_1^x \psi(t)\mathrm{d}t$. Further

$$\lim_{x \to 0} \hat{\phi}(x) = \lim_{x \to 0} (\log(x\varphi(x)) - \log(x\Gamma(x))) = \lim_{x \to 0} \log(x\varphi(x)).$$

The function ψ is Riemann integrable on every compact subinterval of X and, by (10.4.17), satisfies equation (10.4.8). By Lemma 10.4.1 it is constant a.e. in X, whence $\hat{\phi}$ is affine: $\hat{\phi}(x) = c(x-1)$, as $\hat{\phi}(1) = 0$. Setting $x = 1$ in (10.4.17) we obtain $\varphi(\frac{1}{2}) = \sqrt{\pi} = \Gamma(\frac{1}{2})$, whence $\hat{\phi}(\frac{1}{2}) = 0$. Thus $\hat{\phi} = 0$, i.e. $\varphi = \Gamma$. ∎

As an easy consequence of Theorem 10.4.5 we get a characterization of the gamma function as the simultaneous solution of (1) Legendre's and fundamental equations or (2) Legendre's and Euler's ones. The results are due to H.-H. Kairies [6]. If φ is of class C^1, part (1) of the theorem has already been proved by E. Artin [1].

Theorem 10.4.6. *If φ satisfies conditions* (i) *and either* (1) *equations* (10.4.17) *and* (10.4.2) *or* (2) *equations* (10.4.17) *and* (10.4.18), *then $\varphi = \Gamma$.*

Proof. (1) By (10.4.2) and the continuity of φ we have

$$\lim_{x \to 0} x\varphi(x) = \lim_{x \to 0} \varphi(x + 1) = \varphi(1).$$

(2) Continuity of φ and (10.4.18) imply

$$\lim_{x \to 0} x\varphi(x) = \lim_{x \to 0} \frac{\pi x}{\sin \pi x} (\varphi(1 - x))^{-1} = \frac{1}{\varphi(1)}.$$

Thus in both cases the theorem results from Theorem 10.4.5. ∎

Remark 10.4.5. We may observe that condition (i) alone does not imply that $\varphi = \Gamma$ is the only solution of the Legendre equation (10.4.17), even if φ is assumed of class C^r (Kairies [6]). Analogously, there are solutions φ of the Euler equation (10.4.18) other than Γ, even if φ is assumed of class C^1 and logarithmically convex, $\varphi(1) = 1$ (Kairies [1]). Finally, since function (10.4.3) satisfies (10.4.18) and (10.4.2) and $\varphi(1) = 1$, these requirements do not suffice to have $\varphi = \Gamma$, even in the class of convex and analytic functions.

10.4F. Logarithmic derivative of the gamma function

The digamma function $\varphi = \psi := \Gamma'/\Gamma$ satisfies the equation

$$\varphi(x + 1) - \varphi(x) = \frac{1}{x}. \tag{10.4.19}$$

This function may be characterized in the following way (Muldoon [2], Kairies [2], Artin [1]).

Theorem 10.4.7. *Assume that* $\varphi:X \to X$ *is of class* C^{p+1} *in* X *($p \in \mathbb{N}$ fixed),* $(-1)^p \varphi^{(p+1)}(x) \geqslant 0$ *in* X *and* $\varphi(1) = -C$ *(C is the Euler constant). If* φ *satisfies equation* (10.4.19), *then* $\varphi = \psi$.

Proof. The functions φ and ψ both satisfy equation (10.4.19) and the condition involving $(p + 1)$st derivative. The latter condition means that $(-1)^p \varphi$ and $(-1)^p \psi$ are convex functions of order p. Moreover, the function $x \mapsto (-1)^p x^{-1}$ is concave of order p, since its $(p + 1)$st derivative is nonpositive. The condition $\varphi(1) = \psi(1) = -C$ being fulfilled, it follows from Theorem 2.7.1 that $\varphi = \psi$. ∎

Remark 10.4.6. Theorem 10.4.1, with the logarithmic convexity condition for φ replaced by the assumption that $(-1)^p (\log \varphi)^{(p)} \geqslant 0$ for a $p \geqslant 2$ (Muldoon [2]) may be proved in the same way as Theorem 10.4.7.

Many further results on the Euler gamma function and related functions and functional equations may be found, for instance, in Artin [1], Anastassiadis [1]–[7], Bajraktarević [1], Eagle [1], Kairies [1]–[6], Kuczma [9], Lucht [1], [2], Lucht–Wolke [1], [2], Mayer [1], Muldoon [1], [2], Schmidt [1]. See also Kuczma [26, Ch. V, § 10] and the references quoted therein.

10.5 Continuous nowhere differentiable functions

We shall write for short c.n.d. functions. We owe the contents of Subsections 10.5B and 10.5C to H.-H. Kairies [7] (see also Kairies [8]).

10.5A. The Weierstrass c.n.d. functions

Let us consider the series

$$H(x) := \sum_{n=0}^{\infty} a^n h(b^n x), \qquad 0 < a < 1, \, x \in \mathbb{R}. \tag{10.5.1}$$

It represents a continuous function provided $h: \mathbb{R} \to \mathbb{R}$ is continuous and bounded. With $h(t) = \sin 2\pi t$, resp. $h(t) = \cos 2\pi t$, the function H becomes the Weierstrass function $S_{a,b}$, resp. $C_{a,b}$. The most far-reaching result concerning these functions is due to G. H. Hardy [1] and says that they are nowhere differentiable if $ab \geqslant 1$.

It is our aim to show that c.n.d. functions satisfy certain functional equations and may be characterized by some combinations of them. We restrict ourselves to the case $a = 1/p$, $b = p$ in (10.5.1), $p \in \mathbb{P}$ (p is a prime) and $h(t) = \sin 2\pi t$. The resulting function is denoted by S_p:

$$S_p(x) = \sum_{n=0}^{\infty} p^{-n} \sin 2\pi p^n x, \qquad x \in \mathbb{R}. \tag{10.5.2}$$

The function S_p is odd and periodic with period 1, but it satisfies also more interesting functional equations. By Hardy's result, S_p is a c.n.d. function.

10.5B. A characterization of S_p by homogeneous equations

Let $p \in \mathbb{P}$. We denote by $M(p)$ the following set of powers of p:

$$M(p) = \{p^i : i \in \mathbb{N}_0\}.$$

It is readily seen that the function $\varphi = S_p$ satisfies in $\mathbb{R}$ the equations

$$\varphi(x) = \sum_{n=0}^{p-1} \varphi\left(\frac{x+n}{p}\right) \tag{10.5.3}$$

(and also with p replaced by any number from $M(p)$) and

$$\sum_{m=0}^{q-1} \varphi\left(\frac{x+m}{q}\right) = 0, \tag{10.5.4}$$

for all $q \notin M(p)$. As the normalizing condition we take $\varphi(1/2p) = \sin(\pi/p)$.

Moreover, it is seen from (10.5.2) that S_p coincides term by term with its Fourier sine series. The following theorem is connected with this observation.

Theorem 10.5.1. *Let $p \in \mathbb{P}$. The only solution $\varphi: \mathbb{R} \to \mathbb{R}$ of equation (10.5.3) and equation (10.5.4) for all $q \in \mathbb{P} \setminus \{p\}$, which is continuous, odd, periodic of period 1 and such that*

$$b_1(\varphi) := 2 \int_0^1 \varphi(t) \sin 2\pi t \, dt = 1, \tag{10.5.5}$$

is $\varphi = S_p$.

Proof. We are going to find the Fourier series for φ. By a repeated use of (10.5.3) we get

$$\sum_{n=0}^{r-1} \varphi\left(\frac{x+n}{r}\right) = \varphi(x) \tag{10.5.6}$$

for all $r \in M(p)$, i.e. for all powers of p. Similarly, iteration of (10.5.4) shows that

$$\sum_{m=0}^{s-1} \varphi\left(\frac{x+m}{s}\right) = 0 \qquad (10.5.7)$$

for all $s \in \mathbb{N} \setminus \{1\}$ not divisible by p. Finally, (10.5.7) holds for any positive integer which is not a power of p. For any such k is a product rs, where r is a power of p and s is relatively prime with p, and we may use (10.5.6) and (10.5.7) to obtain

$$\sum_{j=0}^{k-1} \varphi\left(\frac{x+j}{k}\right) = \sum_{m=0}^{s-1} \sum_{n=0}^{r-1} \varphi\left(\frac{x+m+sn}{rs}\right)$$

$$= \sum_{m=0}^{s-1} \sum_{n=0}^{r-1} \varphi\left(\frac{1}{r}\left(\frac{x+m}{s}+n\right)\right)$$

$$= \sum_{m=0}^{s-1} \varphi\left(\frac{x+m}{s}\right) = 0.$$

Since φ is odd and 1-periodic, its Fourier series becomes

$$\tilde{\varphi}(x) = \sum_{k=1}^{\infty} b_k(\varphi) \sin 2\pi k x.$$

We calculate

$$b_k(\varphi) = 2 \int_0^1 \varphi(u) \sin 2\pi k u \, du$$

$$= 2 \sum_{j=0}^{k-1} \int_{j/k}^{(j+1)/k} \varphi(u) \sin 2\pi k u \, du.$$

Let us put $t = ku - j$ in the jth integral. We get

$$b_k(\varphi) = \frac{2}{k} \int_0^1 \sum_{j=0}^{k-1} \varphi\left(\frac{t+j}{k}\right) \sin 2\pi t \, dt.$$

By (10.5.6) and (10.5.5) we get $b_k(\varphi) = 1/k$ for $k \in M(p)$, whereas, by (10.5.7), $b_k(\varphi) = 0$ for $k \notin M(p)$. Thus

$$\tilde{\varphi}(x) = \sum_{k \in M(p)} \frac{1}{k} \sin 2\pi k x = S_p(x).$$

Now, Fejér's Theorem yields $\varphi = \tilde{\varphi}$ so that $\varphi = S_p$. ∎

Remark 10.5.1. The proof shows that if we replace (10.5.5) by the condition $b_1(\varphi) = c \neq 0$, then $\varphi = c S_p$. Thus the normalization $\varphi(1/2p) = \sin(\pi/p)$ instead of (10.5.5) would work as well.

10.5C. An inhomogeneous equation for S_p

The equation (10.5.4) may be given an inhomogeneous form. Namely, $\varphi = S_p$ satisfies, for any $k \in \mathbb{N} \setminus M(p)$, the equation

$$\varphi(x) = \sum_{j=0}^{k-1} \varphi\left(\frac{x+j}{k}\right) + S_p(x). \tag{10.5.8}$$

On the other hand, we have the following.

Theorem 10.5.2. *If a function $\varphi \colon \mathbb{R} \to \mathbb{R}$ is continuous, odd, 1-periodic and satisfies equation (10.5.3) and equation (10.5.8) for all $k \in \mathbb{P} \setminus \{p\}$, then $\varphi = S_p$.*

Proof. Put $\psi(x) = \varphi(x) - S_p(x)$, check that it satisfies (10.5.6) for any $r \in \mathbb{N}$, and that its Fourier series is

$$\tilde{\psi}(x) = \sum_{n=1}^{\infty} \frac{b_1(\psi)}{n} \sin 2n\pi x. \tag{10.5.9}$$

By Fejér's Theorem we have $\psi(x) = b_1(\psi) \cdot \frac{1}{2}\pi(1 - 2x)$ for $x \in (0, 1)$, the latter formula representing the sum of the series (10.5.9). Continuity and periodicity of ψ imply $b_1(\psi) = 0$, whence also $\psi = 0$. ∎

A disadvantage of using equation (10.5.8) to characterize S_p is the occurrence of S_p itself in it. Thus replace (10.5.8) by

$$\varphi(x) - \sum_{j=0}^{k-1} \varphi\left(\frac{x+j}{k}\right) = h_k(x), \qquad x \in \mathbb{R}, \tag{10.5.10}$$

where $h_k \colon \mathbb{R} \to \mathbb{R}$ is a function. If $\varphi \colon \mathbb{R} \to \mathbb{R}$ is continuous, then so is h_k when defined by (10.5.10). If $\varphi = S_p$, then $h_k = S_p$ for $k \in \mathbb{N} \setminus M(p)$, and, of course, $b_1(h_k) = 1$ for precisely the same k.

Upon these observations we may anticipate the following result.

Theorem 10.5.3. *Assume that $\varphi \colon X \to \mathbb{R}$ is continuous, of period 1 and odd. Moreover, let φ satisfy equation (10.5.3) and equation (10.5.10) for $k \in \mathbb{N} \setminus M(p)$, where $h_k \colon \mathbb{R} \to \mathbb{R}$ are any functions. Then $\varphi = S_p$ if and only if $b_1(h_k) = 1$ for all $k \in \mathbb{N} \setminus M(p)$.*

Proof. Only the sufficiency requires an argument. Let φ and h_k satisfy our assumptions. From (10.5.10) and the periodicity of φ we find

$$h_k(x + 1) - h_k(x) = \varphi(x + 1) - \varphi(x) + \varphi\left(\frac{x}{k}\right) - \varphi\left(\frac{x}{k} + 1\right) = 0,$$

for any $x \in \mathbb{R}$ and $k \in \mathbb{N} \setminus M(p)$. Thus h_k has period 1. Since by (10.5.10) h_k is continuous, as is φ, the Fourier sine coefficients $b_j(h_k)$ make sense. Again we find the Fourier series for φ.

For $k \in M(p)$ we get, by the same argument as in the proof of Theorem

10.5.1, $b_k(\varphi) = k^{-1}b_1(\varphi)$. For $k \in \mathbb{N} \setminus M(p)$ we have, by (10.5.10), the relations

$$b_1(\varphi) = 2 \int_0^1 \left[\sum_{j=0}^{k-1} \varphi\left(\frac{t+j}{k}\right) + h_k(t) \right] \sin 2\pi t \, dt$$

$$= 2k \int_0^1 \varphi(t) \sin 2k\pi t \, dt + 2 \int_0^1 h_k(t) \sin 2\pi t \, dt$$

$$= kb_k(\varphi) + b_1(h_k) = kb_k(\varphi) + 1.$$

Hence the Fourier series $\tilde{\varphi}$ of φ is given by

$$\tilde{\varphi}(x) = b_1(\varphi)S_p(x) + (b_1(\varphi) - 1) \sum_{k \in \mathbb{N} \setminus M(p)} \frac{1}{k} \sin 2k\pi x$$

$$= S_p(x) + (b_1(\varphi) - 1) \sum_{k=1}^{\infty} \frac{1}{k} \sin 2k\pi x$$

$$= S_p(x) + (b_1(\varphi) - 1)g(x),$$

where $g: \mathbb{R} \to \mathbb{R}$ is given by $g(x) = \frac{1}{2}\pi(1 - 2x)$ for $x \in (0, 1)$, $g(0) = 0$, and $g(x + 1) = g(x)$. By Fejér's Theorem we have $\varphi = \tilde{\varphi}$. Therefore the second summand above has to be a continuous function, which is the case only if $b_1(\varphi) = 1$. Consequently $\varphi = S_p$. ∎

10.5D. Comments

10.5D1. In Note 3.9.12 we have characterized the original Weierstrass c.n.d. function $C_{a,b}$ as the only bounded solution of the equation

$$\varphi(x) = a\varphi(bx) + h(x) \tag{10.5.11}$$

$(a \in (0, 1), b \text{ odd}, ab \geqslant 1 + 3\pi/2, h(x) = \cos x)$. Such a simple characterization is also possible for S_p which turns out to be the only solution $\varphi: \mathbb{R} \to \mathbb{R}$ of (10.5.11) $(a = p^{-N}, b = p^N, h(x) = \sum_{n=0}^{N-1} p^{-n} \sin 2\pi p^n x, p \in \mathbb{P};$ with some $N \in \mathbb{N})$ that satisfies $\varphi(x) = O(1)$ as $x \to \infty$ (Kairies [7]).

Similarly, the van der Waerden c.n.d. function $w: [0, 1] \to \mathbb{R}$ is the unique bounded solution of (10.5.11) $(a = \frac{1}{2}, b = 2, h(x)$ is the distance from x to the nearest integer); see de Rham [4]. The nowhere differentiability of w has been proved in Darsow–Frank–Kairies [1] by using elementary functional equations for w only (e.g., $w(x) - 2w(\frac{1}{2}x) = -x, w(\frac{1}{2}(x + 1)) - w(\frac{1}{2}x) = -x + \frac{1}{2})$.

10.5D2. Assume that the conditions of Theorem 10.5.2 are fulfilled except that equation (10.5.8) is postulated not for all primes k different from p but only for $k \in T \setminus \{p\}$, where $T \subset \mathbb{P}$. Then the following assertions hold true:

(a) if $\mathbb{P} \setminus T$ is finite, then $\varphi = S_p$,
(b) if $\sum_{q \in T} 1/q < \infty$, then φ is not necessarily S_p,
(c) if $T = \{q \in \mathbb{P}: q \not\equiv 1 \pmod{N}\}$ for an $N \in \mathbb{N}$, then also φ may be different from S_p.

The proof of (a) and the counterexample in the case (c) are based on results of L. Lucht [1] and L. Lucht–D. Wolke [1], respectively. For the proofs of all assertions the reader is referred to Kairies [7]. The question how to characterize the set T on which we postulate (10.5.8) in order to have $\varphi = S_p$ remains unsettled. The largest set T on which (10.5.8) may hold and nevertheless $\varphi \neq S_p$ is not known either.

10.5D3. The continuity condition imposed on φ in Theorems 10.5.1 and 10.5.3 cannot be replaced by $\varphi \in L^r$ for all $r > 0$. Counterexamples may be obtained as follows.

Define $S_p^*(x)$ to be $S_p(x)$, if x is irrational, and to be 0, if x is rational. Clearly, $S_p^* \neq S_p$, as $S_p^*(1/2p) = 0 \neq \sin(\pi/p) = S_p(1/2p)$.

Now, $S_p^* \in L^r[0, 1]$ for any $r > 0$, and $\varphi = S_p^*$ satisfies all the conditions of Theorem 10.5.1 except the continuity.

In the case of Theorem 10.5.3, consider a family $\{h_k: k \in \mathbb{N} \setminus M(p)\}$ of integrable functions with all $b_1(h_k) = 1$, and change their values on the rationals x by putting $h_k^*(x) = 0$. We have $b_1(h_k^*) = b_1(h_k)$, but (10.5.10) (with h_k^* in place of h_k) has the solution $\varphi = S_p^* \neq S_p$.

10.6 Notes

10.6.1. Other peculiar functions can also be characterized as the only solutions of, usually systems of, functional equations. We shall mention two such functions, without going into details.

(a) Sierpiński's carpet.

Proposition 10.6.1. *There is a unique even and bounded simultaneous solution* $\varphi: \mathbb{R} \to \mathbb{R}$ *of the equations*

$$\left.\begin{array}{ll} \varphi(t) + \varphi(t + \tfrac{1}{2}) = 0, & t \in \mathbb{R}, \\ 2\varphi(\tfrac{1}{4}t) + \varphi(t + \tfrac{1}{8}) = 1, & t \in [0, 1]. \end{array}\right\} \tag{10.6.1}$$

The function φ is continuous in $\mathbb{R}$, and the equations

$$x = \varphi(t), \qquad y = \varphi(t - \tfrac{1}{4}), \qquad t \in [0, 1],$$

define a continuous curve filling the square $[-1, 1]^2$.

(The function φ satisfies a single equation to which Theorem 4.1.1 applies and yields the unique (bounded) φ. The result itself is due to W. Sierpiński [2]. Regarding the detailed proof see Kuczma [26, Ch. XI, § 4].)

(b) Cantor's singular function.

Proposition 10.6.2. *There is a unique function* $\varphi: [0, 1] \to [0, 1]$ *which is bounded in* $[0, 1]$ *and satisfies in* $[0, 1]$ *the equations*

$$\varphi(\tfrac{1}{3}x) = \tfrac{1}{2}\varphi(x), \qquad \varphi(\tfrac{1}{3}(x + 1)) = \tfrac{1}{2}, \qquad \varphi(\tfrac{1}{3}(x + 2)) = \tfrac{1}{2} + \tfrac{1}{2}\varphi(x).$$

This function is continuous, increasing and is constant on a set of intervals which lie densely in $(0, 1)$.

(This is a result of W. Sierpiński [1]. The proof is similar to that of Proposition 10.6.1, and is presented e.g. in Kuczma [26, Ch. XI, § 5].)

10.6.2. For results concerning other characterizations of peculiar, mainly c.n.d., functions, in most cases by systems of simultaneous equations see Andreoli [1], de Rham [1]–[5], Dubuc [3], Julia [4], Ruziewicz [1], Sierpiński [3], [4], Wunderlich [1]. Regarding systems of simultaneous iterative functional equations see also Howroyd [1]–[3].

10.6.3. The cotangent can be characterized in the following ways:

(a) (Gupta [1], private communication) The unique function $\varphi: (0, 1) \to \mathbb{R}$ that is continuous, satisfies $\lim_{x \to 0}(\varphi(x) - 1/x) = 0$ and solves the equations

$$\varphi(x) = \tfrac{1}{2}[\varphi(\tfrac{1}{2}x) + \varphi(\tfrac{1}{2}(x + 1))], \qquad x \in (0, 1), \tag{10.6.2}$$

and $\varphi(x) + \varphi(1 - x) = 0$ (thus $\varphi(\tfrac{1}{2}) = 0$) is $\varphi(x) = \pi \cot \pi x$.

(b) (Jäger [1]) If $\varphi: (0, 1) \to \mathbb{R}$ is continuous and satisfies the equations

$$\varphi(x) = \frac{1}{n} \sum_{i=0}^{n-1} \varphi\left(\frac{x + i}{n}\right), \qquad x \in (0, 1), \tag{10.6.3}$$

for all $n \in \mathbb{N}$, then $\varphi(x) = a \cot \pi x + b$, $a, b \in \mathbb{R}$.

The main tool to prove (a) is a particular case of Lemma 10.4.1 due to E. Mohr [1] (see also Walter [1]) that concerns Riemann integrable solutions (on $[0, 1]$) of (10.6.2). Using it for the (continuous in $[0, 1]$) solution $\tilde{\varphi}$ of (10.6.2), $\tilde{\varphi}(x) := \varphi(x) - \pi \cot \pi x$, $x \in (0, 1)$, $\tilde{\varphi}(0) = \tilde{\varphi}(1) = 0$, one finds $\tilde{\varphi} = 0$ by $\tilde{\varphi}(\tfrac{1}{2}) = \varphi(\tfrac{1}{2}) = 0$.

That the cotangent satisfies (10.6.3) for $n \in \mathbb{N}$ is seen, for instance, from its resolution into partial fractions

$$\pi \cot \pi x = \frac{1}{x} + \sum_{k=1}^{\infty} \left(\frac{1}{x + k} - \frac{1}{x - k}\right), \qquad x \in \mathbb{Z}. \tag{10.6.4}$$

This formula is derived (Mohr [1], see also Walter [1]) as in (a), by showing that the function $\hat{\varphi}(x) := \pi \cot \pi x - \lim_{n \to \infty} \sum_{j=-n}^{n} 1/(x + j)$ for $x \in (0, 1)$, $\hat{\varphi}(0) = \hat{\varphi}(1) = 0$ shares all the properties of $\tilde{\varphi}$. In turn, (10.6.4) follows directly from Euler's identity for the sine,

$$\sin \pi x = \pi x \prod_{k=1}^{\infty} \left(1 - \frac{x^2}{k^2}\right), \qquad x \in \mathbb{R}, \tag{10.6.5}$$

by differentiation. To prove (10.6.5) H. Haruki [5] uses the equation

$$\varphi(x) = \varphi(\tfrac{1}{2}x)\varphi(\tfrac{1}{2}(1 - x)), \qquad x \in \mathbb{C}, \tag{10.6.6}$$

and shows that it has the unique entire solution $\varphi = 1$ and that the quotient of both sides of (10.6.5) is also an entire function (its singularities at the integers are removable) that satisfies (10.6.6).

Iterative roots and invariant curves

11.0 Introduction

Dealing with a continuous iteration semigroup $\{f^s: s > 0\}$ (see Section 1.7) and putting $f := f^1$ we get, in particular, the existence of solutions of the functional equations

$$\varphi^N = f \qquad (11.0.1)$$

for any positive integer N; indeed, it suffices to take $\varphi := f^{1/N}$. Having, however, a self-mapping f which, *a priori*, is not embeddable into any continuous (or even only rational) iteration semigroup, we may go on trying to find solutions of (11.0.1), at least for a given $N \in \mathbb{N}$. This is just the central idea of the present chapter.

For obvious intuitive reasons any solution of equation (11.0.1) is called the *Nth iterative root* of the function f. The symbol $f^{1/N}$ used above suggests also an alternative term: (the *N*th) *fractional iteration*, actually occurring in many papers.

The material of this chapter is organized as follows. Starting with the iterative roots of arbitrary mappings we proceed to discuss special aspects of this problem for functions from a subset of the real line (usually an interval) into itself. Next, we investigate local analytic solutions of equation (11.0.1) in a neighbourhood of the origin at the complex plane. Then we deal with iterative roots of identity, i.e. we seek for solutions of the so-called *Babbage equation*

$$\varphi^N = \mathrm{id} \qquad (11.0.2)$$

which in fact is one of the oldest iterative functional equations ever discussed. We shall be concerned mainly with continuous solutions of (11.0.2) and of its special case $N = 2$ defining the *involutions*.

The last part of the chapter is loosely related with the preceding ones. The goal is to find two-dimensional manifolds invariant with respect to a given transformation. This interesting geometrical problem leads, in a particular case, to the following functional equation:

$$\varphi(f(x, \varphi(x))) = g(x, \varphi(x)). \qquad (11.0.3)$$

The problem of existence and uniqueness of local Lipschitzian solutions of (11.0.3) in a neighbourhood of the origin is considered. The reason for associating such studies with finding iterative roots is just the following common feature of (11.0.1) and (11.0.3): both of them contain superpositions of the unknown function.

As previously, supplementary facts and information are collected in the final Notes section.

11.1 Purely set-theoretical case

In the whole of this short section, X denotes an arbitrary set, f is a self-mapping of X and N is a fixed positive integer.

We start with the following.

Theorem 11.1.1. *Let $\varphi: X \to X$ be a solution of the equation $\varphi^N = f$. Then φ is surjective (resp. injective, bijective) if and only if f is surjective (resp. injective, bijective).*

Proof. $\varphi(X) = X$ implies $f(X) = \varphi^N(X) = X$. Conversely, assume f to be surjective; if the set $X \setminus \varphi(X)$ were not empty, say $x_0 \in X \setminus \varphi(X)$ then, for any $x \in X$, we would have $\varphi(x) \neq x_0$ whence $f(x) = \varphi^N(x) = \varphi(\varphi^{N-1}(x)) \neq x_0$, which contradicts the surjectivity of f.

Since the composition of injections is an injection again we see that the injectivity of φ implies that of f. Conversely, assume f is injective; if φ were not injective then, for some $x, y \in X, x \neq y$, we would have $\varphi(x) = \varphi(y)$ whence

$$f(x) = \varphi^N(x) = \varphi^{N-1}(\varphi(x)) = \varphi^{N-1}(\varphi(y)) = \varphi^N(y) = f(y),$$

which contradicts the injectivity of f.

The appropriate bijectivity equivalence follows now immediately. ∎

For bijective mappings f the equation

$$\varphi^N = f \qquad (11.1.1)$$

was solved by S. Łojasiewicz [1] (see also Bajraktarević [9], Haĭdukov [1] and Kuczma [26]). The general solution in the case $N = 2$ has been described by R. Isaacs [1]. Extending the notions and methods of Isaacs (see also Sklar [1]), G. Zimmermann-Riggert [1], [2] gave a solution in the general case by reducing the problem of the existence of the Nth iterative roots to the problem of the so-called N-mateability. The whole procedure is described in detail in G. Targoński's monograph [8]. We shall confine ourselves here to the statement of the main result to give a flavour of such studies only.

For let n be a divisor of $N \geq 2$ and let $\Omega_1, \ldots, \Omega_n$ be orbits of f. Put $\Omega := \bigcup_{i=1}^n \Omega_i$. We say that the orbits $\Omega_1, \ldots, \Omega_n$ are N-mateable provided there exists a map $\varphi_\Omega: \Omega \to \Omega$ such that $\varphi_\Omega^N = f|\Omega$ and φ_Ω has one single orbit.

Theorem 11.1.2. *Equation* (11.1.1) *has a solution if and only if the family of all orbits of* F *admits a decomposition into a disjoint union of classes such that the cardinality of each class is a divisor of* N *and the elements of each class are* N-mateable.

Various necessary and sufficient conditions for N-mateability may be found in Targoński's book [8], too.

For further purposes we shall end this section with the proof of the following lemma which is a particular case of a result of R. Isaacs [1].

Lemma 11.1.1. *Suppose that* $f(a) = b$ *and* $f(b) = a$ *for some* $a, b \in X$, $a \neq b$. *If, for any* $x \in X$, *the equality* $f^2(x) = x$ *implies that* $x \in \{a, b, f(x)\}$, *then the equation*

$$\varphi^2 = f \tag{11.1.2}$$

has no solutions.

Proof. Suppose that $\varphi: X \to X$ is a solution of (11.1.2) and put $c := \varphi(a)$. Then $f^2(c) = \varphi^5(a) = \varphi(f^2(a)) = \varphi(a) = c$ and therefore $c \in \{a, b, f(c)\}$. If $c = a$, then $b = f(a) = \varphi^2(a) = \varphi(c) = \varphi(a) = c = a$, a contradiction. If $c = b$, then $b = f(a) = \varphi^2(a) = \varphi(c) = \varphi(b)$ whence $a = f(b) = \varphi^2(b) = b$, a contradiction again. Finally, if we had $f(c) = c$, then $\varphi(b) = \varphi(f(a)) = \varphi^3(a) = f(\varphi(a)) = f(c) = c$ whence $\varphi(a) = \varphi(b)$ and, consequently, we would get $b = f(a) = \varphi^2(a) = \varphi^2(b) = f(b) = a$, contrary to our assumption. ∎

11.2 Continuous and monotonic solutions

Such a situation as described in Lemma 11.1.1 cannot happen for any increasing function $f: \mathbb{R} \to \mathbb{R}$. Actually, as we shall see later, any continuous and strictly increasing function on the real line possesses continuous and strictly increasing iterative roots of all orders. But first we shall prove some preliminary results.

Lemma 11.2.1. *Let* $X \subset \mathbb{R}$ *be an arbitrary set and let* $f: X \to X$ *be strictly monotonic. Assume* $\varphi: X \to X$ *to be a monotonic iterative root of* f *and fix a point* $x_0 \in X$.

(a) *If* φ *is increasing, then the following conditions are equivalent:*

(1) $f(x_0) = x_0$;
(2) $\varphi(x_0) = x_0$;
(3) $\varphi(x_0) = f(x_0)$.

(b) *If* φ *and* f *are decreasing, then* $\varphi(x_0) = f(x_0)$ *if and only if* $f^2(x_0) = x_0$.

Proof. Suppose that φ is an iterative root of the Nth order. By Theorem 11.1.1 φ is strictly monotonic.

To prove (a) assume (1) and suppose that $\varphi(x_0) > x_0$. Then $\varphi^{k+1}(x_0) > \varphi^k(x_0)$ for all $k \in \mathbb{N}$ whence $f(x_0) = \varphi^N(x_0) > \varphi(x_0) > x_0$, a contradiction. Similarly, $\varphi(x_0) < x_0$ would imply $f(x_0) < x_0$. Therefore, implication (1) $\Rightarrow$ (2) has been proved. Assume (2); then $f(x_0) = \varphi^N(x_0) = x_0 = \varphi(x_0)$, i.e. (3) holds true. Now, suppose (3) to be satisfied and put $g := \varphi^{N-1}$. Then g is strictly increasing and $g(x_0) = \varphi^{N-1}(x_0) = x_0$; applying the first part of the proof with f replaced by g we infer that $f(x_0) = \varphi(x_0) = x_0$.

To prove (b) observe that N has to be odd, $\psi := \varphi^2$, $g := f^2$ are strictly increasing and $\psi^N = g$. On account of (a), relation $g(x_0) = x_0$ is equivalent to the equality $\psi(x_0) = x_0$, i.e. $f^2(x_0) = x_0$ is equivalent to $\varphi^2(x_0) = x_0$. The latter equality is satisfied if and only if $\varphi^{N-1}(x_0) = x_0$ (recall that N is odd) being equivalent to $f(x_0) = \varphi(x_0)$. ∎

Lemma 11.2.2. *Let $X \subset \mathbb{R}$ be an arbitrary set and let $f : X \to X$ be strictly increasing. Assume $\varphi : X \to X$ to be an increasing iterative root of f. Then the functions $(f - \mathrm{id}_X)(\varphi - \mathrm{id}_X)$ and $(\varphi - \mathrm{id}_X)(f - \varphi)$ are both nonnegative.*

Proof. Suppose that φ is an iterative root of the Nth order. By Theorem 11.1.1 φ is strictly increasing. Fix an $x_0 \in X$ and assume that $f(x_0) \geq x_0$. If we had $\varphi(x_0) < x_0$, then $\varphi^{k+1}(x_0) < \varphi^k(x_0)$ for all $k \in \mathbb{N}$ and we would get $f(x_0) = \varphi^N(x_0) < \varphi(x_0) < x_0$, a contradiction. Similarly, $f(x_0) \leq x_0$ implies $\varphi(x_0) \leq x_0$. This proves that the function $(f - \mathrm{id}_X)(\varphi - \mathrm{id}_X)$ is nonnegative. The remaining part of the proof is similar. ∎

Theorem 11.2.1. *Let $X \subset \mathbb{R}$ be an interval and let $f : X \to X$ be a strictly monotonic surjection. Assume $\varphi : X \to X$ to be an iterative root of f. Then φ is continuous if and only if φ is strictly monotonic.*

Proof. By Theorem 11.1.1 φ is bijective. Continuous bijection on an interval has to be strictly monotonic. The converse results from the fact that any monotonic surjection on an interval is continuous. ∎

11.2A. Strictly increasing continuous iterative roots

Now we can prove the following.

Theorem 11.2.2. *Let $X \subset \mathbb{R}$ be an interval and let f be a strictly increasing and continuous self-mapping of X. Then f possesses strictly increasing and continuous iterative roots of all orders. More precisely, for any positive integer $N \geq 2$, the strictly increasing and continuous solution of the functional equation*

$$\varphi^N(x) = f(x), \qquad x \in X,$$

depends on an arbitrary function.

Proof. Without loss of generality we may assume that $X = (a, b|$, $-\infty \leqslant a < b \leqslant \infty$, and $a < f(x) < x$ for all $x \in X$. Indeed, otherwise, setting $F := \{x \in X : f(x) = x\}$ we have $X = F \cup \bigcup_k X_k$ where X_k are pairwise disjoint intervals of the form $(\alpha, \beta|$ or $[\alpha, \beta)$ with $\alpha, \beta \in F$ or $\alpha = a$ or $\beta = b$. If φ is a (necessarily strictly) increasing iterative root of f, then Lemma 11.2.1 gives $\varphi(x) = x$ for $x \in F$, whence $\varphi(X_k) \subset X_k$ for each k; conversely, if for each k a self-mapping φ_k of X_k is a continuous Nth iterative root of $f|_{X_k}$, then the function $\varphi : X \to X$ defined by the formula

$$\varphi(x) := \begin{cases} \varphi_k(x) & \text{for } x \in X_k, \\ x & \text{for } x \in F, \end{cases}$$

yields a continuous and strictly increasing Nth iterative root of f (see Lemma 11.2.2).

Thus, to proceed, fix arbitrarily a point $x_0 \in X$, choose any points $x_1 > x_2 > \cdots > x_{N-1}$ from the interval $(f(x_0), x_0)$ and put

$$x_{k+N} := f(x_k)$$

(which is equivalent to $x_k = f^{-1}(x_{k+N})$) for all those $k \in \mathbb{Z}$ for which the recurrence procedure is performable. Put $J := \mathbb{Z} \cap [-k_0, \infty)$ provided that either $x_{-k_0} = b \in X$ or x_{-k_0+1} does exist but x_{-k_0} does not; otherwise, put $J := \mathbb{Z}$. Let $I_k := [x_{k+1}, x_k], k \in \mathbb{Z}$, if $J = \mathbb{Z}$; if, however, $J = \mathbb{Z} \cap (-k_0, \infty)$, then

$$I_k := \begin{cases} [x_{k+1}, x_k] & \text{for } k \in \mathbb{Z} \cap (-k_0, \infty), \\ [x_{-k_0}, b] \cap X & \text{for } k = k_0. \end{cases}$$

Finally, given arbitrary increasing homeomorphisms φ_k of I_k onto I_{k+1}, $k \in \{0, \ldots, N-2\}$, we put

$$\varphi_k(x) := f \circ \varphi_{k-N+1}^{-1} \circ \cdots \circ \varphi_{k-1}^{-1}(x), \qquad x \in I_k,$$

for $k \in \mathbb{N}, k \geq N - 1$, and

$$\varphi_k(x) := \varphi_{k+1}^{-1} \circ \cdots \circ \varphi_{k+N+1}^{-1} \circ f(x), \qquad x \in I_k,$$

for $k \in y \cap (-\infty, -1]$. Now, it is easy to check that the formula

$$\varphi(x) := \varphi_k(x) \qquad \text{for } x \in I_k, k \in J,$$

defines a continuous and strictly increasing Nth iterative root of f. ∎

Remark 11.2.1. Each continuous and increasing iterative root of a continuous and strictly increasing self-mapping of a real interval may be obtained in the manner just described. We omit the obvious detailed calculation.

11.2B. Strictly decreasing roots of strictly decreasing functions

If f is strictly decreasing, then, by Theorem 11.2.1, any continuous iterative root of f has to be strictly decreasing, too (because a composition of increasing functions remains increasing). Hence, the order of the root must be odd; there is no continuous even iterative root of a strictly decreasing function. Generally, the behaviour of iterative roots for decreasing functions gives less satisfaction than that for increasing ones.

What we are able to prove is the following.

Theorem 11.2.3. *Let X, Y be two disjoint real intervals and let Z be their union. Assume f to be a strictly decreasing and continuous self-mapping of Z such that*

$$f(X) = Y \qquad and \qquad f(Y) = X.$$

Then, for every odd $N \in \mathbb{N}$ and any strictly increasing solution $\psi \colon X \to X$ of the equation

$$\psi^N = f^2, \tag{11.2.1}$$

the formula

$$\varphi(x) := \begin{cases} f \circ \psi^{-\frac{1}{2}(N-1)}(x) & for \ x \in X, \\ \psi^{\frac{1}{2}(N+1)} \circ f^{-1}(x) & for \ x \in Y, \end{cases} \tag{11.2.2}$$

defines a continuous and strictly decreasing Nth iterative root of f. Conversely, each Nth iterative root of f has the above representation.

Proof. Without loss of generality we may assume that $N \geqslant 3$. Let $\psi \colon X \to X$ be an arbitrary continuous and (necessarily strictly) increasing solution of (11.2.1). Then the map $\varphi \colon Z \to Z$ given by formula (11.2.2) is continuous and strictly decreasing. Moreover, for any $x \in X$ one has

$$\varphi^2(x) = \psi^{\frac{1}{2}(N+1)} \circ f^{-1} \circ f \circ \psi^{-\frac{1}{2}(N-1)}(x) = \psi(x)$$

whereas, for any $x \in Y$, one gets

$$\varphi^2(x) = f \circ \psi^{-\frac{1}{2}(N+1)} \circ \psi^{\frac{1}{2}(N+1)} \circ f^{-1}(x) = f \circ \psi \circ f^{-1}(x).$$

Consequently, owing to the oddness of N, we obtain respectively

$$\varphi^N(x) = \varphi \circ \varphi^{N-1}(x) = \varphi \circ \psi^{\frac{1}{2}(N-1)}(x) = f \circ \psi^{-\frac{1}{2}(N-1)} \circ \psi^{\frac{1}{2}(N-1)}(x) = f(x),$$

$x \in X$, whereas

$$\varphi^N(x) = \varphi \circ \varphi^{N-1}(x) = \varphi \circ f \circ \psi^{\frac{1}{2}(N-1)} \circ f^{-1}(x)$$
$$= \psi^{\frac{1}{2}(N+1)} \circ f^{-1} \circ f \circ \psi^{\frac{1}{2}(N-1)} \circ f^{-1}(x) = \psi^N \circ f^{-1}(x) = f(x), \qquad x \in Y,$$

on account of (11.2.1).

To prove the converse it suffices to take $\psi := \varphi^2$. ∎

Theorem 11.2.3 covers more cases than it might seem at the first glance, and its seemingly artificial assumptions are quite adequate. In fact, observe first that, according to Theorem 11.2.2, equation (11.2.1) has always a strictly increasing and (necessarily) continuous solution. Take any surjective and, strictly decreasing and continuous self-mapping f of an interval $X \subset \mathbb{R}$. Put $F := \{x \in X \colon f^2(x) = x\}$; plainly, F is closed in X. Let intervals $X_k, k \in K \subset \mathbb{N}$, denote the components of $X \setminus F$. The function f has exactly one fixed point $x_0 \in X$ which, obviously, belongs to F. Let K^- be the set of all $k \in K$ such that X_k lies to the left of x_0. Put $Y_k := f(X_k)$, $k \in K^-$. Clearly, $Y_k \cap X_k = \varnothing$ for $k \in K^-$. Moreover, $f(Y_k) = f^2(X_k) \setminus X_k$ since the endpoints of X_k are fixed points of f^2 (or, possibly, the left point of X_k coincides with the left point of X). Consequently, f maps the union $Z_k := X_k \cup Y_k$, $k \in K^-$, onto itself. This reduces the situation to the case considered in Theorem 11.2.3. Therefore, for each $k \in K^-$, there exists a strictly decreasing and continuous solution φ_k of the equation $\varphi^N = f|_{Z_k}$. If x_k is an endpoint of X_k, then $\lim_{x \to x_k} \varphi_k^2(x) = x_k$ by means of Lemma 11.2.2 whence $\lim_{x \to x_k} \varphi_k(x) = \lim_{x \to x_k} f \circ \varphi_k^{1-N}(x) = f(x_k)$, $k \in K$. Thus, the function

$$\varphi(x) := \begin{cases} \varphi_k(x) & \text{for } x \in Z_k, \\ f(x) & \text{for } x \in F, \end{cases}$$

yields a strictly decreasing and continuous solution of the functional equation $\varphi^N(x) = f(x)$, $x \in X$. Consequently, we have proved the following.

Theorem 11.2.4. *Let* $X \subset \mathbb{R}$ *be an interval and let* f *be a strictly decreasing and continuous function from* X *onto* X. *For each odd* $N \in \mathbb{N}$ *there exists a strictly decreasing and continuous* N*th iterative root of* f.

11.2C. Strictly decreasing roots of strictly increasing functions

If N is an even number, then a strictly increasing and continuous self-mapping f of a real interval X can have also continuous and strictly decreasing Nth iterative roots. In fact, the problem reduces itself to a solution of the system

$$\varphi^2 = \psi \qquad \text{and} \qquad \psi^{\frac{1}{2}N} = f,$$

where ψ is increasing. In the light of Theorem 11.2.2, we may confine ourselves to the case $N = 2$ of equation (11.0.1). We begin with

Theorem 11.2.5. *Let* f *be an increasing homeomorphism of an open or closed real interval* X *onto* X. *Assume that* $\xi \in F := \{x \in X : f(x) = x\}$ *and put* $F^- := \{x \in F : x \leqslant \xi\}$, $F^+ := \{x \in F : x \geqslant \xi\}$. *Consider* $(X \setminus F^-) \cap (-\infty, \xi)$ *as the union of all members of the family* $\mathcal{F}$ *of disjoint open intervals with endpoints in* F^-. *Let* α *be a strictly decreasing map of* F^- *onto* F^+ *such that for every* $I = (a, b) \in \mathcal{F}$ *and* $J := (\alpha(b), \alpha(a))$ *one has* $(f(x) - x)(f(y) - y) < 0$ *for all* $(x, y) \in I \times J$. *Then for any such* I *and* J, *the continuous and strictly decreasing square iterative root of the function* $f|_{I \cup J}$ *depends on an arbitrary function.*

Proof. Without loss of generality we may assume that $f(x) < x$ for $x \in I$. Fix an arbitrary point $x_0 \in I$ and an arbitrary continuous and strictly decreasing function $\varphi_0 : [f(x_0), x_0] \to J$ such that $\varphi_0 \circ f(x_0) = f \circ \varphi_0(x_0)$. With the aid of Theorem 5.3.1 we may extend φ_0 to a continuous and strictly decreasing solution $\gamma : I \to J$ of the equation

$$\gamma \circ f(x) = f \circ \gamma(x), \qquad x \in I; \tag{11.2.3}$$

moreover, $\lim_{x \to a} \gamma(x) = \alpha(a)$ and $\lim_{x \to b} \gamma(x) = \alpha(b)$. Put $Z := I \cup J$. Then the function $\varphi : Z \to Z$ given by the formula

$$\varphi(x) := \begin{cases} \gamma(x) & \text{for } x \in I, \\ \gamma^{-1} \circ f(x) & \text{for } x \in J, \end{cases} \tag{11.2.4}$$

is strictly decreasing and continuous. Moreover, for $x \in X$, we have

$$\varphi^2(x) = \gamma^{-1} \circ f \circ \gamma(x) = f(x)$$

in view of (11.2.3) whereas, for $x \in J$,

$$\varphi^2(x) = \gamma \circ \gamma^{-1} \circ f(x) = f(x),$$

i.e. φ is a desired square iterative root of $f|_Z$. ∎

Remark 11.2.2. In case φ is a (necessarily strictly) decreasing and continuous square iterative root of a surjective, continuous and strictly increasing self-mapping f of an open or closed real interval X, the assumptions of Theorem 11.2.5 are actually satisfied. In fact, let ξ be the unique fixed point of φ. Obviously, $\xi \in F$. Let $I = (a, b) \in \mathscr{F}$ and suppose that $f(x) < x$ for $x \in I$. Put $\alpha := \varphi|_{F^-}$ and take a $y \in (\alpha(b), \alpha(a)) = \varphi(I)$. Then $y = \varphi(x)$ for an $x \in I$ whence $f(y) - y = \varphi^2(\varphi(x)) - \varphi(x) = \varphi(f(x)) - \varphi(x) > 0$.

Remark 11.2.3. The general strictly decreasing and continuous solution $\varphi : X \to X$ of equation (11.0.1) for $N = 2$ with a surjective, continuous and strictly increasing map $f : X \to X$ may be obtained as follows. We take all points $\xi \in F$ and bijections $\alpha : F^- \to F^+$ fulfilling the conditions described in Theorem 11.2.5 on intervals $I \in \mathscr{F}$ and we construct a suitable solution φ_I on $I \cup J$. We extend α onto F by taking $\bar\alpha : F \to F$ given by $\bar\alpha|_{F^-} = \alpha$ and $\bar\alpha|_{F^+} = \alpha^{-1}$, so that $\bar\alpha = \bar\alpha^{-1}$. Then setting

$$\varphi(x) = \varphi_I(x) \text{ for } x \in I \cup J, \ I \in \mathscr{F}, \qquad \varphi(x) = \bar\alpha(x) \text{ for } x \in F,$$

we check easily that φ is a strictly decreasing square iterative root of f. The continuity of φ results now from Theorem 11.2.1.

Comments. The description of continuous iterative roots we have presented in this section is due to P. I. Haĭdukov [1] and M. Kuczma [5, (b)].

11.3 Monotonic C^r solutions

We have seen (Theorem 11.2.2) that strictly increasing and continuous self-mappings of a real interval always possess iterative roots (of all orders) having the same regularity properties. Unfortunately, this is no longer valid for C^r mappings, in general. The surprise disappears if we realize that the extending procedure applied in the proof of Theorem 11.2.2 preserves the regularity only in the interiors of the intervals considered and there is no reason to expect higher regularity at the sticking points. Indeed, in many cases there exists no iterative root even of class C^1, although the given mapping is of class C^r and with strictly positive first derivative in the whole of its domain. We shall deal with C^1 iterative roots problem later on.

To establish any positive result, assume at first that φ is a C^r solution of the equation

$$\varphi^N = f \qquad (11.3.1)$$

where f is a C^r function mapping a real interval X into itself. Differentiating both sides of equation (11.3.1) p times, $p \in \{1, \dots, r\}$, we come to an equality of the form

$$f^{(p)}(x) = P_p(\varphi'(x), \dots, \varphi'(\varphi^{N-1}(x)); \dots; \varphi^{(p)}(x), \dots, \varphi^{(p)}(\varphi^{N-1}(x))),$$

$x \in X$, where P_p is a (uniquely determined) polynomial of $p \cdot N$ variables. This observation allows us to establish the following (Kuczma [31].

Theorem 11.3.1. *Assume $r \in \mathbb{N}$ and $N \in \mathbb{N} \setminus \{1\}$ to be fixed and let f be a C^r self-mapping of a real interval $X = (a, b|, -\infty \leqslant a < b \leqslant \infty$ such that $a < f(x) < x$ and $f'(x) > 0$ for $x \in X$. Fix arbitrarily a point $x_0 \in X$, put $x_N := f(x_0)$ and choose arbitrary points $x_1 > x_2 > \cdots > x_{N-1}$ from the interval (x_N, x_0). Then, for every strictly increasing C^r surjection $\varphi_i : [x_{i+1}, x_i] \to [x_{i+2}, x_{i+1}],$ $i \in \{0, \dots, N-2\}$ such that*

$$\varphi_i^{(p)}(x_{i+1}) = \varphi_{i+1}^{(p)}(x_{i+1}) \qquad \text{for } i \in \{0, \dots, N-3\} \text{ and } p \in \{1, \dots, r\} \quad (11.3.2)$$

(for $N = 2$ this condition disappears) and

$$P_p(\varphi'_0(x_0), \dots, \varphi'_{N-2}(x_{N-2}), \varphi'_{N-2}(x_{N-1}); \dots; \varphi_0^{(p)}(x_0), \dots,$$
$$\varphi_{N-2}^{(p)}(x_{N-2}), \varphi_{N-2}^{(p)}(x_{N-1})) = f^{(p)}(x_0) \qquad \text{for } p \in \{1, \dots, r\}, \quad (11.3.3)$$

there exists a unique function $\varphi : X \to X$ satisfying (11.3.1) and such that $\varphi|_{[x_{i+1}, x_i]} = \varphi_i$ for $i \in \{0, \dots, N-2\}$. This function is strictly increasing and of class C^r in X.

The proof is literally the same as that of Theorem 11.2.2. Assumptions (11.3.2) and (11.3.3) assure the C^r regularity of a solution constructed in that way. We omit the tedious although almost evident calculations.

Equally straightforward are the following two results.

Theorem 11.3.2. *Under the assumptions of Theorem 11.2.3, if, moreover, f is of class C^r, $1 \leqslant r \leqslant \infty$, in Z with nowhere vanishing first derivative and if ψ is of class C^r in X, then so is the solution φ given by formula (11.2.2).*

Theorem 11.3.3. *Under the assumptions of Theorem 11.2.5, if, moreover, f is of class C^r, $1 \leqslant r \leqslant \infty$, in $Z := I \cup J$ with nowhere vanishing first derivative in Z and if γ is a C^r solution of equation (11.2.3) in I, then the function $\varphi : Z \to Z$ given by (11.2.4) yields a C^r and strictly decreasing square iterative root of $f|_Z$.*

Somewhat deeper results will be presented in the next two sections with
regard to increasing and C^1 iterative roots. It turns out that their behaviour
depends essentially on whether the multiplier of the given function (i.e. its
derivative at the fixed point) vanishes or not. For that reason we shall
emphatically distinguish between these two cases.

11.4 C^1 iterative roots with nonzero multiplier

11.4A. A function with no smooth convex square roots

By means of the example formerly announced (J. Ger [1], see also Kuczma
[31]) we wish to exhibit two phenomena: (a) diffeomorphism with no smooth
(square) iterative roots; (b) a strictly increasing convex mapping with no
convex (square) iterative roots.

One function will serve for both purposes; that means that even a junction
of these two regularity requirements helps nothing.

Example 11.4.1. Fix an $s \in (0, 1)$ and points $x_0 < x_1$ from $(0, 1)$ such that
$x_1 < \sqrt{s \cdot x_0/s}$. Take any convex mapping $f \in C^1(\mathbb{R})$ and such that

$$f(x) = sx \quad \text{for } x \in (-\infty, x_0] \quad \text{and} \quad f(x) = \alpha x + \beta \quad \text{for } x \in [x_1, \infty),$$

where $\alpha := (1 - \sqrt{s \cdot x_0})/(1 - x_1)$, $\beta := (\sqrt{s \cdot x_0} - x_1)/(1 - x_1)$.

Suppose that a function $\varphi: \mathbb{R} \to \mathbb{R}$ satisfies the equation $\varphi^2 = f$ and φ is
a C^1 function or φ is convex. From Theorem 11.4.2 (resp. 11.4.3) below we
infer that $\varphi(x) = \sqrt{s \cdot x}$ for all $x \in [0, x_0]$. In particular,

$$\varphi(x_0) = \sqrt{s \cdot x_0} = f(x_1) = \varphi^2(x_1). \tag{11.4.1}$$

Moreover, in the case where φ is convex, Theorem 11.3.1 implies that $\varphi|_{(0,1)}$
is a C^1 function as well. In both cases there exists a finite limit $\lim_{x \to 1^-} \varphi'(x)$.
This leads to a contradiction. To see this, observe that φ has to be strictly
increasing since so is f and Theorem 11.1.1 holds. Thus the sequence
$x_n := \varphi^{-n}(x_0)$, $n \in \mathbb{N}_0$, is well defined and the 'new' x_1 coincides with the 'old'
one because of (11.4.1).

Further, $(x_n)_{n \in \mathbb{N}_0}$ tends to the nearest (and unique) fixed point
of φ at the right of x_0, i.e. $x_n \to 1$ as $n \to \infty$ (note that $\varphi(1) = 1$ on account
of Lemma 11.1.1 and there are no other fixed points of φ except that at
zero). And yet, the sequence $(\varphi'(x_n))_{n \in \mathbb{N}}$ diverges. To see this, observe that

$$\varphi'(x)\varphi'(\varphi(x)) = f'(x) \geq s > 0, \qquad x \in (0, 1), \tag{11.4.2}$$

whence

$$\varphi'(x_{n+2}) = \frac{f'(x_{n+2})}{\varphi'(\varphi(x_{n+2}))} = \frac{f'(x_{n+2})}{\varphi'(x_{n+1})} = \frac{f'(x_{n+2})}{f'(x_{n+1})} \varphi'(x_n), \qquad n \in \mathbb{N}_0,$$

and since $x_{n+2} > x_{n+1} > x_1$, we have $f'(x_{n+2}) = f'(x_{n+1}) = \alpha$, $n \in \mathbb{N}_0$, and

$$\varphi'(x_{n+2}) = \varphi'(x_n), \qquad n \in \mathbb{N}_0.$$

On the other hand, since (11.4.1) holds, we have $\varphi(x_1) = x_0$, whence by (11.4.2) $\varphi'(x_1) = f'(x_1)/\varphi'(x_0) = \alpha/\sqrt{s} \neq \varphi'(x_0)$. Thus the sequence $(\varphi'(x_n))_{n \in \mathbb{N}_0}$ is oscillating and hence divergent. This proves that φ cannot be of class C^1 in any neighbourhood of 1. Obviously, φ is neither convex nor concave in consequence of the nonmonotonic behaviour of the sequence $(\varphi'(x_n))_{n \in \mathbb{N}_0}$.

11.4B. Necessary conditions

We shall place the unique fixed point of a given mapping at zero. Such a step has obviously a technical meaning only. We proceed with the following.

Theorem 11.4.1. *Let f be a self-mapping of a real interval $X = [0, a|$, $0 < a \leqslant \infty$, satisfying the conditions*

$$0 < f(x) < x \quad for \ x \in X \setminus \{0\} \qquad and \qquad f'(x) > 0 \quad for \ x \in X.$$

If $\varphi: X \to X$ is a C^1 solution of the functional equation

$$\varphi^N = f, \tag{11.4.3}$$

then the infinite product

$$G(x, y) := \prod_{n=0}^{\infty} \frac{f'(f^n(x))}{f'(f^n(y))} \tag{11.4.4}$$

is convergent for all pairs $(x, \varphi(x))$ from the graph of φ whereas φ itself satisfies the differential equation

$$\varphi'(x) = s^{1/N} G(x, \varphi(x)), \qquad x \in X; \tag{11.4.5}$$

here $s := f'(0)$ is the (positive!) multiplier of f.

Proof. φ has to be strictly increasing. Indeed, otherwise, according to Theorem 11.1.1, φ would be strictly decreasing; this implies the existence of a (unique) fixed point $\xi \in X \setminus \{0\}$ of φ whence $f(\xi) = \xi$ which contradicts our assumption on f. Consequently, $\varphi'(x) \geqslant 0$ for $x \in X$. However, relation (11.4.3) implies that

$$\varphi'(x) \prod_{i=1}^{N-1} \varphi'(\varphi^i(x)) = f'(x), \qquad x \in X, \tag{11.4.6}$$

whence, in particular, it follows that, in fact, $\varphi'(x) > 0$ for all $x \in X$. Moreover, Lemma 11.2.1(a) implies the equality $\varphi(0) = 0$ whence $\varphi^i(0) = 0$ for $i \in \mathbb{N}$. Therefore, putting $x = 0$ in (11.4.6), we get the equality $\varphi'(0) = s^{1/N}$.

On the other hand, formula (11.4.6) implies the relation

$$\frac{f'(x)}{f'(\varphi(x))} = \frac{\varphi'(x)}{\varphi'(\varphi^N(x))} = \frac{\varphi'(x)}{\varphi'(f(x))}, \qquad x \in X,$$

and hence, for any $k \in \mathbb{N}$, we get

$$\prod_{n=0}^{k-1} \frac{f'(f^n(x))}{f'(f^n(\varphi(x)))} = \frac{\varphi'(x)}{\varphi'(f^k(x))}, \qquad x \in X,$$

since f and φ are permutable. We complete the proof by letting k tend to infinity. ∎

11.4C. Main existence theorem

Theorem 11.4.1 suggests that a C^1 iterative root (if it does exist) has to be unique. Actually, this has been proved, together with the existence, under some supplementary conditions on the given function (Bratman [1], Crum [1], Kuczma–A. Smajdor [3] and Zdun [9]). These results are collected in the following theorem whose proof yields a perfect occasion for a survey of natural applications of the results concerning the existence and uniqueness of solutions of the crucial iterative functional equations (i.e. that of Abel, Schröder and, more generally, linear equations) in various function classes.

Theorem 11.4.2. *Assume the hypotheses of Theorem 11.4.1 and put* $s := f'(0)$.
  (a) *If* $s = 1$ *and either*

(1) f *is concave*

or

(2) $f'(x) = 1 - b(m+1)x^m + O(x^{m+\delta})$, $x \to 0$,

then equation (11.4.3) has a unique C^1 *solution in* X*; this solution is given by the formula*

$$\varphi(x) = \alpha^{-1}(\alpha(x) + 1/N) \qquad \text{for } x \in X \setminus \{0\}, \ \varphi(0) = 0 \qquad (11.4.7)$$

where α *is a principal solution of the Abel functional equation:*

$$\alpha(f(x)) = \alpha(x) + 1, \qquad x \in X \setminus \{0\}, \tag{A}$$

  (b) *If* $s \in (0, 1)$ *and either*

(3) f *is convex or* f *is concave or*

(4) $f'(x) = s + O(x^\delta)$, $x \to 0$,

then equation (11.4.3) has a unique C^1 *solution in* X*; this solution is given by the formula*

$$\varphi(x) = \sigma^{-1}(s^{1/N}\sigma(x)), \qquad x \in X, \tag{11.4.8}$$

where σ *is a nontrivial principal solution of the Schröder functional equation:*

$$\sigma(f(x)) = s\sigma(x), \qquad x \in X. \tag{S}$$

Here m, b *and* δ *denote some positive constants.*

Proof. A straightforward verification shows that formulae (11.4.7) and (11.4.8) both give a solution of equation (11.4.3).

Part (a). Assume (1). Then Theorem 2.4.3 guarantees that the Abel equation (A) possesses the family of strictly decreasing and convex (principal; see Section 9.1) solutions given by the formula

$$\alpha(x) = c + \lim_{n \to \infty} \frac{f^n(x) - f^n(x_0)}{f^{n+1}(x_0) - f^n(x_0)}, \qquad x \in X \setminus \{0\}, \qquad (11.4.9)$$

where $c \in \mathbb{R}$ and $x_0 \in X \setminus \{0\}$. The left and right derivatives of α do exist at each point of the interior of X. Both of them are increasing and satisfy the linear functional equation

$$\beta(f(x)) = \frac{1}{f'(x)} \beta(x), \qquad x \in \text{int } X. \qquad (11.4.10)$$

Since $g := 1/f'$ is positive and increasing (f being concave has a decreasing derivative) and since, consequently $\inf_{x \in \text{int } X} g(x) = \lim_{x \to 0} g(x) = s = 1$, we may apply Theorem 2.3.1; therefore, our one-sided derivatives differ by a constant factor only. But convex mappings have differentiability points! Let $z \in \text{int } X$ be a differentiability point of α. If we had $\alpha'(z) = 0$, then we would have $\alpha'(x) = 0$ for all $x \geqslant z$, $x \in \text{int } X$ (recall that α is convex and decreasing) contradicting the strict monotonicity of α. Thus the proportionality factor has to be 1, i.e. α is differentiable on the whole of int X and $\alpha'(x) < 0$ for all $x \in \text{int } X$. Differentiable convex mapping is necessarily of class C^1. Hence α and, consequently, φ is of class C^1 in $X \setminus \{0\}$. It remains to prove that

$$\lim_{x \to 0} \varphi'(x) = 1. \qquad (11.4.11)$$

For observe that (11.4.7) implies

$$f(x) < \varphi(x) < x \qquad \text{for } x \in \text{int } X \qquad (11.4.12)$$

whence

$$1 \leqslant \frac{\alpha'(\varphi(x))}{\alpha'(x)} \leqslant \frac{\alpha'(f(x))}{\alpha'(x)} = \frac{1}{f'(x)}$$

for any $x \in \text{int } X$. Referring to (11.4.7) again we get $\varphi'(x) = \alpha'(x)/\alpha'(\varphi(x))$, $x \in \text{int } X$, whence

$$f'(x) < \varphi'(x) \leqslant 1, \qquad x \in \text{int } X,$$

from which we get (11.4.11) by letting x tend to zero.

To prove formula (11.4.7) and hence the uniqueness of φ observe that α' is a monotonic solution of equation (11.4.10) for which the function $g := 1/f'$

satisfies the condition

$$\lim_{n \to \infty} g(f^n(x)) = \frac{1}{f'(0)} = 1, \qquad x \in X.$$

Thus, we may apply Theorem 2.3.2 getting

$$0 > \alpha'(x) = c \prod_{n=0}^{\infty} \frac{g(f^n(x_0))}{g(f^n(x))} = c \prod_{n=0}^{\infty} \frac{f'(f^n(x))}{f'(f^n(x_0))}, \qquad x \in \text{int } X,$$

whence (see (11.4.4))

$$G(x, y) := \prod_{n=0}^{\infty} \frac{f'(f^n(x))}{f'(f^n(y))} = \frac{\alpha'(x)}{\alpha'(y)} \qquad \text{for all } x, y \in \text{int } X. \quad (11.4.13)$$

Now Theorem 11.4.1 gives $(s = 1)$

$$\varphi'(x) = \frac{\alpha'(x)}{\alpha'(\varphi(x))}, \qquad x \in \text{int } X, \qquad\qquad (11.4.14)$$

for any C^1 solution of equation (11.4.3). This says that taking such a solution, we have $(\alpha \circ \varphi)' = \alpha'$ and hence $\alpha \circ \varphi = \alpha + c_0$ for some $c_0 \in \mathbb{R}$ whence

$$\alpha(f(x)) = \alpha(\varphi^N(x)) = \alpha(x) + Nc_0, \qquad x \in \text{int } X,$$

and since α satisfies Abel's functional equation (A) we infer that $c_0 = 1/N$ and, finally, $\varphi(x) = \alpha^{-1}(\alpha(x) + 1/N)$ for all $x \in X \setminus \{0\}$. That $\varphi(0) = 0$ results from Lemma 11.2.1(a).

Assume (2). Then Theorem 3.5.6 guarantees directly that α, given, again, by formula (11.4.9) is strictly decreasing and of class C^1 in $X \setminus \{0\}$. Moreover, from relation (3.5.26) $(r = -b^{-1})$ we learn that

$$\alpha'(x) = -\lim_{n \to \infty} b^{-1}(nbm)^{1 + 1/m}(f^n)'(x), \qquad x \in X \setminus \{0\},$$

whence

$$\frac{\alpha'(x)}{\alpha'(y)} = \lim_{n \to \infty} \frac{(f^n)'(x)}{(f^n)'(y)} = \lim_{n \to \infty} \prod_{i=0}^{n-1} \frac{f'(f^i(x))}{f'(f^i(y))} = G(x, y)$$

for all $x, y \in X \setminus \{0\}$, i.e. (11.4.13) remains valid; this, as previously, implies the uniqueness of φ. To show that φ is of class C^1 at the origin, recall that (11.4.14) holds true whence on account of (3.5.23)

$$\varphi'(x) = \frac{\alpha'(x)}{\alpha'(\varphi(x))} = \frac{x^{m+1}\alpha'(x)}{\varphi(x)^{m+1}\alpha'(\varphi(x))}\left(\frac{\varphi(x)}{x}\right)^{m+1}$$

$$= \frac{-b^{-1} + O(x^\tau)}{-b^{-1} + O(\varphi(x)^\tau)}\left(\frac{\varphi(x)}{x}\right)^{m+1}, \qquad x \to 0, \qquad (11.4.14')$$

where $\tau := \min(m, \delta)$. On the other hand, estimates (11.4.12) are satisfied leading to the relation

$$\frac{f(x)}{x} < \frac{\varphi(x)}{x} < 1, \qquad x \in X \setminus \{0\},$$

and therefore, since $f'(0) = 1$, we have

$$\varphi'(0) = \lim_{x \to 0} \frac{\varphi(x)}{x} = 1.$$

Now, letting $x \to 0$ in (11.4.14'), we get (11.4.11) which yields the the desired result.

Part (b). Assume (3). Then Theorem 2.4.4 guarantees that the Schröder equation (S) possesses the family of monotonic and convex (resp. concave) principal solutions (see Section 9.1) given by the formula

$$\sigma(x) = c \lim_{n \to \infty} \frac{f^n(x)}{f^n(x_0)}, \qquad x \in X, \tag{11.4.15}$$

where $c \in \mathbb{R}^+$ and $x_0 \in X \setminus \{0\}$. Except the trivial case $\sigma = 0$ (corresponding to the choice $c = 0$) these solutions are actually strictly monotonic. Indeed, if σ were (nonzero) constant on an interval $I = [u, v] \subset X$, it would be constant on each $f^n(I)$, $n \in \mathbb{N}$, violating the convexity (concavity) of σ because by (S) the constant mappings $\sigma|_I$ and $\sigma|_{f^n(I)}$ are different.

To prove the C^1 regularity of the function φ given by (11.4.8) on the whole of X (no matter which nontrivial solution (11.4.15) has been taken) it suffices to reproduce the reasoning applied for the proof in case (1); this time

$$\lim_{x \to 0} \varphi'(x) = s^{1/N},$$

whereas any nontrivial σ given by (11.4.15) is of class C^1 in $X \setminus \{0\}$ with $\sigma'(x) \neq 0$, $x \in X \setminus \{0\}$, and satisfies the functional equation

$$\sigma'(f(x)) = \frac{s}{f'(x)} \sigma'(x), \qquad x \in X \setminus \{0\}. \tag{11.4.16}$$

Making use of Theorem 2.3.2 once more we deduce that

$$\sigma'(x) = \tilde{c} \prod_{n=0}^{\infty} \frac{f'(f^n(x))}{f'(f^n(x_0))}, \qquad x \in X \setminus \{0\},$$

with some $\tilde{c} \neq 0$, whence (see (11.4.4))

$$G(x, y) = \frac{\sigma'(x)}{\sigma'(y)} \qquad \text{for all } x, y \in X \setminus \{0\}. \tag{11.4.17}$$

Now, Theorem 11.4.1 gives

$$\varphi'(x) = s^{1/N} \frac{\sigma'(x)}{\sigma'(\varphi(x))}, \qquad x \in X \setminus \{0\},$$

for any C^1 solution of equation (11.4.1). This says that taking such a solution we have $(\sigma \circ \varphi)' = (s^{1/N}\sigma)'$ and hence

$$\sigma(\varphi(x)) = s^{1/N}\sigma(x) + c_0$$

for some $c_0 \in \mathbb{R}$ and all $x \in X \setminus \{0\}$. Letting here $x \to 0$ we get $c_0 = 0$ in view of continuity of σ at 0, since obviously $\sigma(0) = \varphi(0) = 0$. Therefore formula (11.4.8) holds true.

Assume (4). Then Theorem 3.5.1 guarantees directly that the Schröder equation (S) has a C^1 solution σ in X given by the formula

$$\sigma(x) = \lim_{n \to \infty} \frac{f^n(x)}{s^n}, \qquad x \in X.$$

Since σ' is a continuous solution of equation (11.4.16) we may apply Theorem 3.1.4 (which assures that 'case (A)' occurs) and, subsequently, Theorem 3.1.2, getting

$$\sigma'(x) = \sigma'(0) \lim_{n \to \infty} \prod_{i=0}^{n-1} \frac{f'(f^i(x))}{s}, \qquad x \in X.$$

This leads again to formula (11.4.17) and the rest follows along the same lines as in the preceding case. ∎

It is noteworthy to observe that a powerful corollary follows.

Corollary 11.4.1. *In the circumstances described in Theorem 11.4.2 the function f is embeddable into an iteration semigroup (see Section 1.7) whose members are of class C^1.*

This result is just at hand: it suffices to replace the number $1/N$ in formulae (11.4.7) and (11.4.8) by a 'continuous' parameter $t \in \mathbb{R}^+$. This should come as no surprise. Our regularity assumptions imposed on f turned out to be strong enough to produce elegant and smooth iteration semigroups (see also Kuczma [26, Ch. IX]).

Remark 11.4.1. Assumptions (2) and (4) are of a local character. This is not the case for (1) and (3) and this might seem somewhat restrictive. As a matter of fact, the problem is apparent only. Actually, there is no need to assume the concavity (convexity) of f on the whole of its domain; it suffices to assume it merely in a right vicinity of zero. Then we obtain local C^1 iterative roots which, by Theorem 11.3.1, can be uniquely extended onto the whole interval considered.

11.4D. Convex and concave iterative roots

In Subsection 11.4A (Example 11.4.1) we were faced by a situation where a convex and smooth map admits no convex square iterative roots. Nevertheless, if an iterative root does exist (which, for instance, actually takes place if we assume the concavity of both f and f' (Theorem 11.4.4 below, Zdun [13]), then it has a representation (11.4.7) or (11.4.8) depending on whether $s = 1$ or $s \in (0, 1)$. More precisely, we have the following result of M. Kuczma–A. Smajdor [2].

Theorem 11.4.3. *Let f be a convex or concave self-mapping of an open interval*

$(0, a) \subset \mathbb{R}$. Suppose that $(1/x)f(x) \to s \in (0, 1]$ as $x \to 0$, $f(x) < x$ for $x \in (0, a)$ and $f' > 0$ in the domain of its existence. If a convex or concave Nth iterative root φ of f does exist, then

$$s = 1 \quad implies \quad \varphi(x) = \alpha^{-1}(\alpha(x) + 1/N), \qquad x \in (0, a),$$

where α is a principal solution of Abel's equation (A), whereas

$$s \in (0, 1) \quad implies \quad \varphi(x) = \sigma^{-1}(\sigma(x) + 1/N), \qquad x \in (0, a),$$

where σ stands for a nontrivial principal solution of Schröder's equation (S).

The proof is similar to suitable parts of the proof of Theorem 11.4.2 and so we will omit it passing to Zdun's result announced previous to that.

Theorem 11.4.4. *Assume f to satisfy the assumptions of Theorem 11.4.1 and suppose additionally that both f and f' are concave. Then the unique C^1 iterative root (of a given order) of f is concave and this is the only concave iterative root (of that order) of f.*

Proof. Let $\varphi: X \to X$ be the unique C^1 solution of equation (11.4.3) (see Theorem 11.4.2). In particular, we have $\varphi'(x) > 0$ and $f(x) \leqslant \varphi(x) \leqslant x$ for all $x \in X$. As f and φ are permutable we get

$$\frac{\varphi'(f(x))}{\varphi'(x)} = \frac{f'(\varphi(x))}{f'(x)} =: \psi(x), \qquad x \in X,$$

and, since f' decreases, ψ is greater than or equal to 1 whence

$$\varphi'(f(x)) \geqslant \varphi'(x) \qquad \text{for all } x \in X.$$

Inductively, $\varphi'(f^n(x)) \geqslant \varphi'(x)$, $x \in X$, $n \in \mathbb{N}$; letting n tend to infinity one obtains $\varphi'(x) \leqslant \varphi'(0) \in [s, 1]$, i.e.

$$\varphi'(x) \leqslant 1 \qquad \text{for all } x \in X. \tag{11.4.18}$$

The function $\lambda := \log f'$ is evidently concave and decreasing (since, by assumption, f' is). Therefore, for $y \in X$ arbitrarily fixed and for any $x \in (\varphi(y), y)$ we have $\varphi(x) < \varphi(y) < x < y$ and hence

$$\frac{\lambda(y) - \lambda(x)}{y - x} \leqslant \frac{\lambda(x) - \lambda(\varphi(y))}{x - \varphi(y)} \leqslant \frac{\lambda(\varphi(y)) - \lambda(\varphi(x))}{\varphi(y) - \varphi(x)};$$

consequently, for some $\xi \in (x, y)$, we get

$$\varphi'(\xi)(\lambda(y) - \lambda(x)) = \frac{\varphi(y) - \varphi(x)}{y - x}(\lambda(y) - \lambda(x)) \leqslant \lambda(\varphi(y)) - \lambda(\varphi(x)).$$

In view of the fact that λ is decreasing and (11.4.18) holds true we come to the inequality

$$\lambda(y) - \lambda(x) \leqslant \lambda(\varphi(y)) - \lambda(\varphi(x))$$

which says that

$$\frac{f'(y)}{f'(x)} \leq \frac{f'(\varphi(y))}{f'(\varphi(x))},$$

i.e. $\psi(x) \leq \psi(y)$. Finally, bearing the continuity of ψ in mind, what we have proved is the implication

$$x \in [\varphi(y), y] \quad \text{implies} \quad \psi(x) \leq \psi(y) \qquad (11.4.19)$$

for any $y \in X$. Now, take any $u, v \in X$, $u < v$. Then, for some $n \in \mathbb{N}_0$, one has $u \in [\varphi^{n+1}(v), \varphi^n(v))$ whence, by (11.4.19),

$$\psi(u) \leq \psi(\varphi^n(v)) \leq \psi(\varphi^{n-1}(v)) \leq \cdots \leq \psi(v).$$

Thus ψ is increasing and, consequently, for any $x, y \in X$, $x < y$, we get

$$\frac{\varphi'(f(x))}{\varphi'(f(y))} \leq \frac{\varphi'(x)}{\varphi'(y)}$$

and, inductively,

$$\frac{\varphi'(f^n(x))}{\varphi'(f^n(y))} \leq \frac{\varphi'(x)}{\varphi'(y)} \qquad \text{for all } n \in \mathbb{N},$$

whence, as $n \to \infty$, we obtain $\varphi'(y) \leq \varphi'(x)$, i.e., φ' is decreasing and thus φ is concave. The uniqueness statement results from Theorem 11.4.3. ∎

11.5 C^1 iterative roots with zero multiplier

As we have pointed out several times (see Chapters 0, 1, 8) the Böttcher functional equation

$$\beta(f(x)) = \beta(x)^p \qquad \text{(B)}$$

occurs naturally and usually serves well when linearization via the Schröder (resp. the Abel) equation is not possible; the idea is to replace the linear mapping by a power function. The content of the present section may also be regarded as a step-by-step verification of the appropriateness of such an approach. All the results are due to M. Kuczma [33], [40].

11.5A. Abundance of solutions

Contrary to the case of mappings with nonvanishing multiplier, the C^1 iterative roots for functions with multiplier zero depend on an arbitrary function! (See Theorem 11.5.1 below.) So, the problem arises to choose a function class assuring the uniqueness. That one which fits well while we are considering Böttcher's equation is that ensuring a suitable asymptotic behaviour of the roots at the fixed point of the given map. This is described by Theorem 11.5.3 in detail.

We begin with a technical lemma whose standard proof will be omitted.

Lemma 11.5.1. *Let f be a C^1 self-mapping of an interval $X = [0, a|$, $0 < a \leqslant \infty$, such that $0 < f(x) < x$ and $f'(x) > 0$ for all $x \in (0, a|$ and $f'(0) = 0$ (multiplier zero). Then, for any constant $p > 1$, the function*

$$F(x) := \begin{cases} x^{-p} f(x) & \text{for } x \in X \setminus \{0\}, \\ b & \text{for } x = 0, \end{cases}$$

is of class C^1 if and only if

$$f'(x) = bpx^{p-1} + c(p+1)x^p + o(x^p), \qquad x \to 0, \qquad (11.5.1)$$

for some constants $b > 0$ and $c \in \mathbb{R}$.

Suppose now that f satisfies the assumptions of Lemma 11.5.1 and $\varphi : X \to X$ is a C^1 solution of the equation

$$\varphi^N = f. \qquad (11.5.2)$$

Then $\varphi'(x) > 0$ for $x \in X \setminus \{0\}$ and $\varphi'(0) = 0$; moreover, from the proof of Theorem 11.4.1 we learn that

$$\varphi'(f^k(x)) = \varphi'(x) \prod_{n=0}^{k-1} \frac{f'(f^n(\varphi(x)))}{f'(f^n(x))}, \qquad x \in X, \qquad (11.5.3)$$

for any $k \in \mathbb{N}$. Since the sequence $(f^k)_{k \in \mathbb{N}}$ of iterates of f tends to zero almost uniformly on X (see Theorem 1.2.2), the convergence

$$\lim_{k \to \infty} \prod_{n=0}^{k-1} \frac{f'(f^n(y))}{f'(f^n(x))} = 0 \qquad (11.5.4)$$

is almost uniform in (x, y) from the graph of the map $\varphi|_{(0,a|}$.

A (partially) converse result holds true.

Theorem 11.5.1. *Assume the hypotheses of Lemma 11.5.1 and fix an $x_0 \in (0, a|$. If the convergence (11.5.4) is almost uniform in (x, y) from the set*

$$\{(x, y) \in [f(x_0), x_0] \times \mathbb{R} : f(x) < y < x\}, \qquad (11.5.5)$$

then the C^1 iterative roots of f on $[0, a|$ form a family depending on an arbitrary function.

Proof. In view of Theorem 3.1, the assertion is obviously true on the interval $(0, a|$. The point is to extend any C^1 iterative root φ of $f|_{(0,a|}$ to a C^1 solution of equation (11.5.2) on the whole of $[0, a)$. To this aim observe that for any $x \in [f(x_0), x_0]$ the pair $(x, \varphi(x))$ belongs to set (11.5.5) (see Lemmas 11.2.2 and 11.2.1) which jointly with relation (11.5.3) and the assumed a.u. convergence (11.5.4) implies

$$\lim_{k \to \infty} \varphi'(f^k(x)) = 0$$

uniformly on the interval $[f(x_0), x_0]$. Consider any sequence $(y_n)_{n \in \mathbb{N}}$

of elements of $(0, x_0)$, $y_n \to 0$, and fix an $\varepsilon > 0$. Since

$$\bigcup_{m=0}^{\infty} [f^{m+1}(x_0), f^m(x_0)) = (0, x_0),$$

we have

$$y_n = f^{m_n}(z_n) \qquad \text{for some } z_n \in [f(x_0), x_0] \text{ and } m_n \geqslant n \in \mathbb{N};$$

consequently,

$$|\varphi'(y_n)| = |\varphi'(f^{m_n}(z_n))| < \varepsilon$$

for sufficiently large $n \in \mathbb{N}$, because $(\varphi' \circ f^{m_n})_{n \in \mathbb{N}}$ converges uniformly on $[f(x_0), x_0]$. This proves that

$$\lim_{x \to 0} \varphi'(x) = 0. \qquad (11.5.6)$$

Now an appeal to Lemma 11.2.2 shows that putting $\varphi(0) = 0$ we obtain a continuous extension of φ to the solution of (11.5.2) on $[0, a]$. Now, the relation

$$\frac{\varphi(x)}{x} = \varphi'(\xi) \qquad \text{for some } \xi \in (0, x), \ x \in (0, a],$$

and (11.5.6) give $\varphi'(0) = 0$ and again prove that the extension obtained is of class C^1 on $[0, a]$. ∎

11.5B. Convergence problem

But when does (11.5.4) hold a.u. on the set (11.5.5)? Fortunately, this is the case for quite a large class of functions as seen from the following.

Theorem 11.5.2. *Assume the hypotheses of Lemma 11.5.1 to be satisfied. If, moreover,*

$$f'(x) = bpx^{p-1} + o(x^{p-1}), \qquad x \to 0 \qquad (11.5.7)$$

for some positive constant b and some $p > 1$, then, for any $x_0 \in (0, a)$, each of the following two conditions is sufficient for the convergence (11.5.4) to be almost uniform in the set (11.5.5):

(i) *the function $F := \dfrac{1}{(\cdot)^p} f$ is increasing on $(0, a]$;*

(ii) *the inequality*

$$\left| \log \frac{F(f^n(x))}{F(f^n(y))} \right| < (p-1) \left| \log \frac{x}{y} \right|$$

holds true for any (x, y) from the set (11.5.5) and any $n \in \mathbb{N}_0$.

Proof. Write $g(x) := \dfrac{1}{x^{p-1}} f'(x)$, $x \in (0, a]$. Then (11.5.7) says that

$$\lim_{x \to 0} g(x) = bp \in (0, \infty).$$

On the other hand, for any $x, y \in (0, a|$, one has

$$\frac{f'(f^n(y))}{f'(f^n(x))} = \left(\frac{f^n(y)}{f^n(x)}\right)^{p-1} \frac{g(f^n(y))}{g(f^n(x))}.$$

Thus, it suffices to prove that both of conditions (i) and (ii) imply

$$\lim_{n \to \infty} \frac{f^n(y)}{f^n(x)} = 0 \tag{11.5.8}$$

almost uniformly in (x, y) from set (11.5.5).

Assume (i). We have (induction)

$$f^n(x) = x^{p^n} \prod_{i=0}^{n-1} F(f^i(x))^{p^{n-i-1}}, \qquad x \in (0, a|, \ n \in \mathbb{N}. \tag{11.5.9}$$

From the assumed monotonicity of F, in view of the fact that the iterates f^n are strictly increasing, we infer that

$$y < x \quad \text{implies} \quad F(f^i(y))^{p^{n-i-1}} \leqslant F(f^i(x))^{p^{n-i-1}}, \ i \in \{0, \ldots, n-1\}, \ n \in \mathbb{N},$$

which together with (11.5.9) implies that

$$0 < \frac{f^n(y)}{f^n(x)} \leqslant \left(\frac{y}{x}\right)^{p^n} \qquad \text{for } y < x, \ n \in \mathbb{N}.$$

Hence (11.5.8) holds a.u. in the set (11.5.5).

Assume (ii) and fix any compact subset A of (11.5.5). Note that for all n large enough we have

$$\left|\log \frac{F(f^n(x))}{F(f^n(y))}\right| \leqslant \tfrac{1}{2}(p-1)\left|\log \frac{x}{y}\right|, \qquad (x, y) \in A;$$

indeed, the left-hand side converges to zero (uniformly in (x, y) from A) whereas the right-hand side is bounded away from zero on A. Therefore, we may find a positive constant $\Theta < 1$ such that

$$\left|\frac{\log F(f^n(x)) - \log F(f^n(y))}{\log x - \log y}\right| \leqslant \Theta(p-1) \tag{11.5.10}$$

for all $(x, y) \in A$ and all $n \in \mathbb{N}_0$. Now, relation (11.5.9) gives

$$\log \frac{f^n(y)}{f^n(x)} = p^n(\log y - \log x) + \sum_{i=0}^{n-1} p^{n-i-1}(\log F(f^i(y)) - \log F(f^i(x)))$$

whence, by (11.5.10),

$$\left|\log \frac{f^n(y)}{f^n(x)}\right| \geqslant p^n\left|\log \frac{y}{x}\right| - \sum_{i=0}^{n-1} p^{n-i-1}\Theta(p-1)\left|\log \frac{y}{x}\right|$$

$$= \left(p^n - \Theta(p-1)\frac{p^n-1}{p-1}\right)\left|\log \frac{y}{x}\right| = ((1-\Theta)p^n + \Theta)\left|\log \frac{y}{x}\right|$$

and, since $y < x$ for $(x, y) \in A$, we arrive at

$$0 < \frac{f''(y)}{f''(x)} \leqslant \left(\frac{y}{x}\right)^{(1-\Theta)p^n + \Theta} \qquad \text{for all } (x, y) \in A.$$

This in turn implies that (11.5.8) holds a.u. in the set (11.5.5). ∎

11.5C. Uniqueness conditions

As pointed out in Subsection 11.5A, for eliminating the abundance of C^1 iterative roots of functions with multiplier zero we have to impose some requirements on the asymptotic behaviour of them at the fixed point of f. The following result is just to this effect.

Theorem 11.5.3. *Let f be a continuous self-mapping of an interval $X = [0, a|$. Suppose that*

(1) $0 < f(x) < x$ *for* $x \in X \setminus \{0\}$;
(2) $f(x) = bx^p + o(x^p)$, $x \to 0$, *for some* $b > 0$ *and* $p > 1$;
(3) $f(x) = x^p F(x)$, $x \in X$, *where* $F: X \to X$ *is a strictly increasing function.*

Then, for any $N \in \mathbb{N}, f$ has exactly one Nth iterative root $\varphi: X \to X$ having the property that for some positive real number r the limit

$$\lim_{x \to 0} \frac{1}{x^r} \varphi(x) \text{ exists and belongs to } (0, \infty). \qquad (11.5.11)$$

Moreover, φ is given by the formula

$$\varphi(x) = \beta^{-1}(\beta(x)^{p^{1/N}}), \qquad x \in X, \qquad (11.5.12)$$

where

$$\beta(x) := \lim_{n \to \infty} (f^n(x))^{p^{-n}}, \qquad x \in X, \qquad (11.5.13)$$

yields a solution of the Böttcher equation (B).

Proof. Armed with the tools of Sections 8.3 and 8.5 we proceed as follows. Assume that φ is any Nth iterative root of f satisfying (11.5.11). Obviously, φ commutes with f and an appeal to Theorem 8.5.4 assures us that

$$\varphi(x) = \beta^{-1}(\beta(x)^r), \qquad x \in X,$$

where β is given by (11.5.13) (note that (3) forces f to be strictly increasing). We have

$$\beta(f(x)) = \lim_{n \to \infty} (f^{n+1}(x))^{p^{-n}} = \left(\lim_{k \to \infty} (f^k(x))^{p^{-k}}\right)^p = \beta(x)^p$$

and, on the other hand,

$$f(x) = \varphi^N(x) = \beta^{-1}(\beta(x)^{r^N})$$

whence
$$\beta(x)^{r^N} = \beta(x)^p,$$
i.e. $r = p^{1/N}$. Thus φ has representation (11.5.12) and hence is unique.

To prove the existence, take the (unique) solution β of the Böttcher equation (B) fulfilling the condition
$$\lim_{x \to 0} \frac{1}{x} \beta(x) \in (0, \infty);$$
such a function does exist on account of Theorem 8.3.3. This solution is strictly increasing and is given by formula (11.5.13). Put $\varphi := \beta^{-1} \circ \beta^{p^{1/N}}$. Then $\varphi^N = f$ and, moreover, since $\beta(x) = dx + o(x)$, $x \to 0$, for some real constant $d > 0$, we deduce that
$$\beta^{-1}(x) = \frac{1}{d} x + o(x), \qquad x \to 0.$$
Therefore,
$$\varphi(x) = \beta^{-1}(\beta(x)^{p^{1/N}}) = \beta^{-1}((dx + o(x))^{p^{1/N}}) = \beta^{-1}(d^{p^{1/N}} x^{p^{1/N}} + o(x^{p^{1/N}}))$$
$$= d^{p^{1/N}-1} x^{p^{1/N}} + o(x^{p^{1/N}})$$
as $x \to 0$, i.e.
$$\lim_{x \to 0} x^{-p^{1/N}} \varphi(x) = d^{p^{1/N}-1} \in (0, \infty),$$
which shows that φ fulfils (11.5.11) with $r = p^{1/N}$. ∎

With the aid of Lemma 11.5.1 and applying Theorem 8.3.4 instead of Theorem 8.5.4 in the above proof one gets analogously the following.

Theorem 11.5.4. *Assume the hypotheses of Lemma 11.5.1 and suppose further that the derivative f' of f has the asymptotic property (11.5.1). Then the unique iterative root (of a given order) of f in the class of maps φ satisfying the asymptotic relation (11.5.11) has to be of class C^1.*

This might be expected. Working in the class guaranteeing the existence and uniqueness of iterative roots in case of multiplier zero we have got an additional smoothness property as a result of higher regularity requirement on the given function.

11.6 Complex domain and local analytic solutions

The goal now is to study the iterative roots of analytic mappings in a neighbourhood of the origin in the complex plane. Most interesting is the case where the multiplier of the given map is a primitive root of unity and the majority of the contents of this section will actually be devoted to that case. Similarly to the real case, we shall make an extensive use of suitable results concerning the analytic solutions of the basic functional equations, i.e. those of Schröder and Böttcher depending on whether the multiplier vanishes or not.

We shall be interested in local analytic solutions of the iterative roots equation

$$\varphi^N = f \tag{11.6.1}$$

where f stands for a given nonzero analytic function mapping a neighbourhood $X \subset \mathbb{C}$ of the origin into $\mathbb{C}$.

We may confine ourselves to the case where the multiplier $s := f'(0)$ does not exceed unity in absolute value. Such a restriction does not violate the generality by any means; in fact, if $|s| > 1$ then f (and hence also φ) is locally invertible and φ is a local analytic solution of (11.6.1) if and only if $\psi := \varphi^{-1}$ is a local analytic solution of the equation $\varphi^N = f^{-1}$; clearly, now the multiplier of f^{-1} is as desired.

11.6A. Existence and uniqueness

We shall distinguish the following three cases:

  (i) $|s| \in (0, 1)$;
  (ii) $|s| = 1$ and s belongs to the Siegel set S (see Definition 4.3.1);
  (iii) $s = 0$.

We begin with a theorem linking various aspects of iterative functional equation theory in a complex domain.

Theorem 11.6.1. *Assume that $X \subset \mathbb{C}$ is a neighbourhood of zero, $f : X \to \mathbb{C}$ is a nontrivial analytic function such that $f(0) = 0$ and $f'(0) = s$. Let N be a fixed positive integer.*

  (a) In case (i) or (ii) the map f has exactly N local analytic Nth iterative roots; all of them are given by the formula

$$\varphi(x) = \sigma^{-1}(s^{1/N}\sigma(x)), \qquad x \in U \subset X,$$

where $s^{1/N}$ stands for any of the N possible values of the complex root of s, U is a neighbourhood of zero and

$$\sigma(x) = \lim_{n \to \infty} \frac{f^n(x)}{s^n}, \qquad x \in U,$$

is a local analytic solution of the Schröder equation (S) $\sigma \circ f = s\sigma$.

  (b) In case (iii) f admits a representation of the form $f(x) = x^p F(x), x \in X$, where $p \in \mathbb{N}\setminus\{1\}$, F is analytic in X and nonvanishing at zero; if $p \notin \{r^N : r \in \mathbb{N}\setminus\{1\}\}$ then f has no local analytic iterative roots of Nth order whereas in the case where $p = r^N$ for some $r \in \mathbb{N}\setminus\{1\}$ there exist exactly $(r^N - 1)/(r - 1)$

local analytic solutions of equation (11.6.1) *and all of them are of the form*

$$\varphi(x) = \beta_0^{-1}(\varepsilon_j \beta_0(x)^r), \qquad x \in U \subset X, \tag{11.6.2}$$

where $\beta_0: U \to \mathbb{C}$ *is a local analytic solution of the Böttcher equation* (B) $\beta \circ f = (\beta)^p$ *such that* $\beta_0(0) = 0 \neq \beta_0'(0)$ *(see Theorem 8.3.1) and* ε_j *are the distinct* $[(r^N - 1)/(r - 1)]$*st roots of unity,* $j \in \{1, \ldots, (r^N - 1)/(r - 1)\}$.

Proof. Part (a) results directly from Theorem 8.5.1 since any iterative root of f commutes with f.

Part (b). Clearly, f is of the form asserted. Let φ be a local analytic Nth iterative root of f. Since φ is permutable with f an appeal to Theorem 8.6.3 proves that in some neighbourhood U of the origin one gets a representation (11.6.2) where β_0 is a local analytic solution of the Böttcher equation $\beta \circ f = (\beta)^p$ such that $\beta_0(0) = 0 \neq \beta_0'(0)$ (see Theorem 8.3.1), $\varepsilon_j =: \varepsilon$ is a $(p-1)$st root of unity for some $j \in \{1, \ldots, p-1\}$ and $r \in \mathbb{N}$. Consequently,

$$f(x) = \varphi^N(x) = \beta_0^{-1}(\varepsilon^{1+r+\cdots+r^{N-1}} \beta_0(x)^{r^N}), \qquad x \in U,$$

whence

$$\beta_0(x)^p = \beta_0(f(x)) = \varepsilon^{1+\cdots+r^{N-1}} \beta_0(x)^{r^N}, \qquad x \in U, \tag{11.6.3}$$

This implies $|\beta_0(x)|^p = |\beta_0(x)|^{r^N}, x \in U$, i.e. $p = r^N$ since $|\beta_0|$ cannot be constant. Thus p has to be a power of a positive integer $r > 1$ ($r = 1$ is excluded by $p > 1$). Relation (11.6.3) reads now as follows:

$$\beta_0(x)^p = \varepsilon^q \beta_0(x)^p, \qquad x \in U,$$

where $p = r^N$ and $q := (p-1)/(r-1)$. Since β_0 does not vanish identically on U we infer that

$$\varepsilon^q = 1,$$

and since, evidently, each qth root of unity yields simultaneously its $(p-1)$st root, the number of admissible ε_js amounts to exactly q, all being the qth complex roots of unity. Conversely, if ε is any such a root and $\varphi(x) := \beta_0^{-1}(\varepsilon \beta_0(x)^r), x \in U$, then

$$\varphi^N(x) = \beta_0^{-1}(\varepsilon^{1+\cdots+r^{N-1}} \beta_0(x)^{r^N}) = \beta_0^{-1}(\varepsilon^q \beta_0(x)^p) = \beta_0^{-1}(\beta_0(x)^p)$$
$$= \beta_0^{-1}(\beta_0(f(x))) = f(x), \qquad x \in U,$$

i.e. φ yields an Nth iterative root of f. ∎

11.6B. Multiplier 1

It remains to investigate the case where the multiplier s is a primitive root of unity. The idea is to reduce the whole thing to the case where s is simply 1 and that is why before treating the general case we establish some results referring to $s = 1$, i.e. to

$$f(x) = x + \sum_{n=m}^{\infty} b_n x^n, \qquad b_n \in \mathbb{C} \text{ for } n \geqslant m, \; b_m \neq 0, \; m \geqslant 2. \tag{11.6.4}$$

To prove the following theorem (see Muckenhoupt [1], Ecalle [2] and Brown–Kuczma [1]) no convergence requirements are needed.

Theorem 11.6.2. *Let f be a formal power series* (11.6.4). *Then equation* (11.6.1) *has exactly k formal solutions*

$$\varphi = \gamma^j \circ \varphi_{1/N}, \tag{11.6.5}$$

where k stands for the greatest common divisor of N and $m-1$, γ is described in Lemma 8.7.1 and $\varphi_{1/N} = \varphi_{1/N}[f]$ is the unique (permutable with f) formal power series occurring in Theorem 8.7.1; finally, j is any positive integer for which $m-1$ divides Nj.

Proof. Let φ be a formal Nth iterative root of f. On account of Theorem 8.7.1, $\varphi = \gamma^j \circ \varphi_t$ for some $j \in \mathbb{N}$ and $t \in \mathbb{C}$. Hence

$$f(x) = \varphi^N(x) = \gamma^{Nj} \circ \varphi_{Nt}(x) = \varepsilon^{Nj} x + \cdots, \tag{11.6.6}$$

where ε is primitive $(m-1)$st root of unity occurring in Lemma 8.7.1. Therefore, Nj must be a multiple of $m-1$ and consequently γ^{Nj} is just the identity. Now (11.6.6) implies $f = \varphi_{Nt}$ which jointly with the facts that $f = \varphi_1$ and that $\varphi_t = \varphi_t[f]$ is unique (Theorem 8.7.1) gives $t = 1/N$, as claimed.

To prove that actually there exist exactly k formal Nth iterative roots of f, note that $m-1 = kl$ and $N = kl'$ for some l and l' in $\mathbb{N}$ and since k is just the greatest common divisor of $m-1$ and N, we deduce that the only positive integers such that Nj is a multiple of $m-1$ are $j_i = li$, $i \in \mathbb{N}$. We have $j_{i+k} - j_i = kl = m-1$, whence $\gamma^{j_{i+k}} = \gamma^{j_i}$, $i \in \mathbb{N}$. Thus, among the γ^j admitted in the statement of our theorem, there are exactly k distinct ones, corresponding to $j \in \{j_1, \ldots, j_k\}$. ∎

Assuming additionally that the iterative logarithm of f (see Definition 8.5.1) has a positive radius of convergence we are able to prove that the formal iterative roots (11.6.5) of an analytic function $f: X \to \mathbb{C}$ are actual indeed. More precisely, we have the following.

Theorem 11.6.3. *Let f be an analytic function mapping a neighbourhood $X \subset \mathbb{C}$ of the origin into $\mathbb{C}$ and having the series expansion* (11.6.4) *for $x \in X$. If the iterative logarithm of f has a positive radius of convergence, then the cardinality of local analytic Nth iterative roots of f coincides with the greatest common divisor of N and $m-1$; all these iterative roots are given by formula* (11.6.5) *with γ, j and $\varphi_{1/N}$ having the same meanings as in Theorem 11.6.2.*

Proof. Apply Theorem 8.7.3 and repeat the method used in the proof of Theorem 11.6.2. ∎

The following theorem shows that the assumption of Theorem 11.6.3 on the iterative logarithm of f is necessary.

Theorem 11.6.4. *Assume the hypotheses of the preceding theorem and suppose that f admits local analytic iterative roots φ (of all orders) such that $\varphi'(0) = 1$. Then the iterative logarithm of f has a positive radius of convergence.*

Proof. It follows from Theorem 11.6.2 that the only formal solution of (11.6.1) fulfilling the condition $\varphi'(0) = 1$ is just $\varphi_{1/N}$ and henceforward $\varphi_{1/N}$ is convergent at the points of some neighbourhood of zero. By Theorem 8.7.3 so are all the series $\varphi_t[f]$ (see (8.7.3)). On account of Theorem 8.7.4 this forces the iterative logarithm of f to have a positive radius of convergence. ∎

11.6C. Multiplier being a primitive root of unity

Passing to this case, fix a positive integer N again and consider the formal power series

$$f(x) = sx + \sum_{n=2}^{\infty} B_n x^n, \qquad s, B_n \in \mathbb{C}, \, n \geqslant 2,$$

where s is a primitive pth root of unity. Assume that

$$\hat{f}(x) := f^p(x) = x + b_m x^m + \cdots, \qquad b_m \neq 0, \, m \geqslant 2,$$

and consider the formal power series $\hat{\phi}_t$ and $\hat{\gamma}$ associated with $\hat{f}$ in a way analogous to that described in Theorem 11.6.2. If

$$\hat{\gamma}(x) = \varepsilon x + \cdots$$

with $\varepsilon^{m-1} = 1$, then Theorem 11.6.2 guarantees that

$$f = \hat{\gamma}^j \circ \hat{\phi}_{1/p}$$

for some $j \in \mathbb{N}$. Consequently, $f(x) = \varepsilon^j x + \cdots$ whence

$$\varepsilon^j = s \tag{11.6.7}$$

and p divides $m - 1$. Finally, let k denote the greatest common divisor of N and $m - 1$. Now, we are in a position to formulate the following.

Theorem 11.6.5. *Under the circumstances prescribed above f has no formal Nth iterative roots unless*

$$p \text{ divides } m - 1 \qquad and \qquad k \text{ divides } \frac{1}{p}(m - 1). \tag{11.6.8}$$

If conditions (11.6.8) are met, then f has exactly k formal iterative roots of Nth order; all of them are given by the formula

$$\varphi = \hat{\gamma}^i \circ \hat{\phi}_{1/(Np)}$$

where i is an arbitrary solution of the congruence

$$Ni \equiv j \pmod{m-1} \tag{11.6.9}$$

and j is any positive integer satisfying (11.6.7).

Proof. Let φ be a formal solution of (11.6.1). Then φ commutes with $\hat{f} = f^p$ whence $\varphi = \hat{\gamma}^i \circ \hat{\phi}_t$ for some $i \in \mathbb{N}$ and $t \in \mathbb{C}$ (see Theorem 8.7.2 and Lemma 8.7.1). We have

$$\hat{\gamma}^{iNp} \circ \hat{\phi}_{Npt} = \hat{f}$$

which is possible only if $\hat{\gamma}^{iNp}$ coincides with the identity and $\hat{\phi}_{Npt} = \hat{f} = \hat{\phi}_1$, so that $t = 1/Np$. Consequently,

$$f = \varphi^N = \hat{\gamma}^{iN} \circ \hat{\phi}_{1/p} \qquad \text{and} \qquad \hat{\gamma}^{iN} \circ \hat{\phi}_{1/p} = \hat{\gamma}^j \circ \hat{\phi}_{1/p}.$$

This implies that $\hat{\gamma}^{iN} = \hat{\gamma}^j$ or, what amounts to the same thing, congruence (11.6.9) holds true. Put

$$d := \text{greatest common divisor of the numbers } N, \ m-1 \text{ and } p,$$

$$b := \frac{1}{d}p, \quad c := \frac{1}{d}k, \quad a := \frac{1}{bcd}(m-1), \quad u := \frac{1}{m-1}pj \quad \text{and} \quad v := \frac{1}{k}N.$$

Then a, b, c, d, u and v are positive integers and, since s is a primitive pth root of unity, the numbers u and bd are relatively prime. With the aid of this notation congruence (11.6.9) may equivalently be written as

$$ivcd = uac = labcd \tag{11.6.10}$$

for some $l \in \mathbb{Z}$. Since u and bd are relatively prime, d has to divide a and consequently

$$k \text{ divides } \frac{1}{p}(m-1).$$

Setting $z := (1/d)a$ we deduce from (11.6.10) that

$$iv = uz + lab, \tag{11.6.11}$$

which is equivalent to (11.6.10). Since v and ab are relatively prime, there exist integers i' and l' such that $i'v - l'ab = 1$ whence $i_0 := uzi'$ and $l_0 := uzl'$ satisfy (11.6.11). Thus (11.6.11) has an integral solution (i_0, l_0) and if (i, l) is another integral solution of (11.6.11) then $(i - i_0)v = (l - l_0)ab$ and an easy calculation shows that

$$i = i_0 + l''ab \tag{11.6.12}$$

with some $l'' \in \mathbb{Z}$. Conversely, every i of form (11.6.12) is a solution of (11.6.11) (with a suitable integer l) and hence it yields a solution of (11.6.9). Since every two i differing by a multiple of $m-1$ generate the same $\hat{\gamma}^i$, congruence (11.6.9) has exactly $cd = k$ solutions yielding different $\hat{\gamma}^i \circ \hat{\phi}_{1/(Np)}$. ∎

Similarly to the case $s = 1$ (Theorems 11.6.2 and 11.6.3) the result just obtained leads immediately to the existence of local analytic Nth iterative roots under the circumstances considered provided the iterative logarithm of the analytic function f^p has a positive radius of convergence. This goes smoothly due to Lemma 8.7.3 and Theorem 8.7.5 forcing the associated formal transformations $\hat{\gamma}$ and $\hat{\phi}_t$ to be actual. This yields the following.

Theorem 11.6.6. *Let* f *be an analytic function mapping a neighbourhood* $X \subset \mathbb{C}$ *of the origin into* $\mathbb{C}$ *and having the series expansion*

$$f(x) = sx + \sum_{n=2}^{\infty} B_n x^n, \qquad x \in X, \tag{11.6.13}$$

where s *is a* p*th primitive root of unity. If*

$$f^p(x) = x + b_m x^m + \cdots, \qquad b_m \neq 0, \; m \geqslant 2, \; x \in X,$$

then f *has no local analytic* N*th iterative roots unless conditions* (11.6.8) *are satisfied, where* k *denotes the greatest common divisor of* N *and* $m - 1$*. If conditions* (11.6.8) *are met and* $\text{logit}\, f^p$ *has a positive radius of convergence, then* f *has exactly* k *local analytic* N*th iterative roots* φ *given by*

$$\varphi = \hat{\gamma}^i \circ \hat{\phi}_{1/(Np)}$$

where i*,* $\hat{\gamma}$ *and* $\hat{\phi}_{1/(Np)}$ *have the same meaning as in Theorem* 11.6.5.

11.6D. Fractional iterates of the roots of identity function

Finally, we consider the situation where the pth iterate of f coincides with the identity. We do not exclude the possibility $p = 1$, i.e. we allow f to be the identity itself.

Theorem 11.6.7. *Let* f *be an analytic function mapping a neighbourhood* $X \subset \mathbb{C}$ *of the origin into* $\mathbb{C}$ *and having the series expansion* (11.6.13) *where* s *is a* p*th primitive root of unity for some* $p \in \mathbb{N}$ *and let* $f^p = \text{id}_X$*. Then the local analytic iterative roots of* f *depend on an arbitrary analytic function. More precisely, for any analytic function* g *in a neighbourhood of the origin such that* $g(0) = 0 \neq g'(0)$ *and any* N*th complex root* τ *of* s *the function*

$$\varphi := \sigma^{-1} \circ (\tau\sigma) \qquad \text{where } \sigma := \sum_{i=0}^{p-1} \frac{1}{s^i} g \circ f^i \tag{11.6.14}$$

yields a local analytic N*th iterative root of* f*. Conversely, each local analytic* N*th iterative root of* f *may be represented in that way.*

Proof. Let σ be given by (11.6.14). Note that

$$\sigma' = g' + \sum_{i=1}^{p-1} \frac{1}{s^i} (g' \circ f^i) \prod_{j=0}^{i-1} (f' \circ f^j)$$

whence $\sigma'(0) = pg'(0) \neq 0$ so that σ^{-1} actually exists. Moreover,

$$\sigma(f(x)) = \sum_{i=0}^{p-1} \frac{1}{s^i} g(f^{i+1}(x)) = s \sum_{i=1}^{p} \frac{1}{s^i} g(f^i(x))$$

$$= s\left(\sum_{i=0}^{p-1} \frac{1}{s^i} g(f^i(x)) - g(x) + \frac{1}{s^p} g(f^p(x)) \right)$$

$$= s\sigma(x)$$

since $s^p = 1$ and $f^p = \mathrm{id}_X$. Consequently,

$$\varphi^N(x) = \sigma^{-1}(\tau^N \sigma(x)) = \sigma^{-1}(s\sigma(x)) = f(x).$$

Conversely, assume that φ is a local analytic Nth iterative root of f. Put $\tau := \varphi'(0)$; then, on account of the equality

$$f' = \prod_{i=0}^{N-1} \varphi' \circ \varphi^i,$$

one gets $s = f'(0) = \varphi'(0)^N = \tau^N$, i.e. τ is an Nth complex root of s. Write

$$g := \sum_{j=0}^{N-1} \frac{1}{\tau^j} \varphi^j.$$

It remains to define σ by formula (11.6.14) which, in turn, implies $\sigma \circ \varphi = \tau\sigma$ and finishes the proof. ∎

Plainly, a perfect analogue of Theorem 11.6.7 in terms of formal power series may be formulated and proved along almost literally the same lines. We omit the detailed and self-evident arguments.

11.7 Babbage equation and involutions

A striking similarity of the problem of finding iterative roots to the operation of taking real or complex roots was most visible in former sections. Those natural analogies find also their expression in suggestive notational, conceptual as well as language (intentional) similarities. More generally, the phenomenon extrapolates itself obviously into analogies between 'continuous' iteration semigroups (flows) and the ordinary powers with 'continuous' exponents. In the complex domain while considering complex roots we arrive almost immediately at the question of complex roots of unity and then we realize at once that it suffices to confine ourselves to primitive roots only. Needless to say that the iterative analogue of the subject at hand is the question of iterative roots of the identity mapping, i.e. search for solutions of the Babbage functional equation (Babbage [1]–[3])

$$\varphi^N = \mathrm{id}. \tag{11.7.1}$$

The solutions of its special case $N = 2$ (corresponding to square roots of unity) appear usually under the name *involutory functions* or *involutions* for short. In the sequel we deal mainly with continuous solutions of (11.7.1). Some facts are just direct consequences of the results established in Section 11.2 with regard to a nonspecified right-hand side of the equation considered. For convenience, we shall formulate them explicitly here starting with the following theorem (see Vincze [1] and McShane [1]).

Theorem 11.7.1. *Let a self-mapping φ of a real interval be a continuous solution of equation (11.7.1). Then either φ itself is the identity mapping or N has to be even and φ is a strictly decreasing involution.*

Proof. This results immediately from Theorem 11.2.1 and Lemma 11.2.1; in case φ is strictly decreasing N must obviously be even, say $N = 2M$, and putting $\psi := \varphi^2$ we apply the results just quoted to the increasing solutions of the equation $\psi^M = \text{id}$. ∎

11.7A. Decreasing involutions

We are led to this problem by Theorem 11.7.1. Remark 11.2.3, jointly with the fact that in the present case $F = \{x: f(x) = x\}$ coincides with the domain of f, allows one to proceed smoothly. Noteworthy is the fact that any continuous and (necessarily strictly) decreasing involution on a real interval X yields a homeomorphism of the set $\{x \in X: x \leqslant \xi\}$ onto the set $\{x \in X: x \geqslant \xi\}$ where $\xi \in \text{int } X$ stands for the unique fixed point of φ. This forces the interval X to be either compact or open. Remark 11.2.3 with the additional observation just made leads easily to the following.

Theorem 11.7.2. *Let $X \subset \mathbb{R}$ be a compact or open interval. Fix any point $\xi \in \text{int } X$ and a strictly decreasing function α from $X^- := X \cap (-\infty, \xi]$ onto $X^+ := X \cap [\xi, \infty)$. Then the function $\varphi : X \to X$ given by the formula*

$$\varphi(x) := \begin{cases} \alpha(x) & \text{for } x \in X^-, \\ \alpha^{-1}(x) & \text{for } x \in X^+ \end{cases}$$

defines a continuous and strictly decreasing involution on X. Conversely, each decreasing involution on X admits a representation of this form.

Decreasing involutions (nontrivial increasing ones do not exist) on an interval have simple geometric interpretation: their graph has to be symmetric with respect to the diagonal $\{(x, y) \in \mathbb{R}^2 : x = y\}$. In fact, the relation

$$\varphi^2 = \text{id} \qquad (11.7.2)$$

forces φ to be invertible and says that $\varphi = \varphi^{-1}$. Thus it is not surprising that they may be constructed in ways other than that described in Theorem 11.7.2 (see, in particular, Aczél [1], Bogdanov [1], Haĭdukov [1], Schwerdtfeger [1] and Aczél–Radó [1]). To give a bit of the flavour of the other methods we shall present here two proofs of the following.

Theorem 11.7.3. *Consider two open intervals (a, b) and $(-c, c)$ in $\mathbb{R}$ with $c > 0$ and $a < b$, $a, b, c \in \mathbb{R} \cup \{-\infty, \infty\}$. Let σ be an arbitrary homeomorphism of (a, b) onto $(-c, c)$. Then the formula*

$$\varphi(x) := \sigma^{-1}(-\sigma(x)), \qquad x \in (a, b), \qquad (11.7.3)$$

defines a strictly decreasing square iterative root of identity. Conversely, each decreasing involution on (a, b) admits a representation of form (11.7.3).

Proof. Obviously, only the latter assertion requires an argument.

First method. Let $\varphi: (a, b) \to (a, b)$ be a decreasing solution of (11.7.2). Put $\sigma_0(x):=x - \varphi(x)$, $x \in (a, b)$. Clearly, σ_0 is an order preserving homeomorphism of (a, b) onto $(a - b, b - a)$ (see Theorems 11.1.1 and 11.2.1). Moreover, $\sigma_0 \circ \varphi = -\sigma_0$. Write $\sigma:=g \circ \sigma_0$ where g is an arbitrary odd homeomorphism of $(a - b, b - a)$ onto $(-c, c)$. Now,

$$\sigma \circ \varphi(x) = g \circ \sigma_0 \circ \varphi(x) = g(\varphi(x) - \varphi^2(x)) = g(\varphi(x) - x) = -g(x - \varphi(x))$$
$$= -g \circ \sigma_0(x) = -\sigma(x),$$

$x \in (a, b)$, i.e. σ is the desired homeomorphism.

Second method (Aczél [1]) refers to the symmetry of a graph with respect to the diagonal. Consider the rotation $r: \mathbb{R}^2 \to \mathbb{R}^2$ given by

$$r(x, y):=\frac{\sqrt{2}}{2}(x - y, x + y), \qquad (x, y) \in \mathbb{R}^2.$$

Transformation r maps the graph of a decreasing solution $\varphi: (a, b) \to (a, b)$ of (11.7.2) onto the set

$$\frac{\sqrt{2}}{2}\{(x - \varphi(x), x + \varphi(x)): x \in (a, b)\}. \tag{11.7.4}$$

Let, as previously, $\sigma_0(x):=x - \varphi(x)$, $x \in (a, b)$; σ_0 is a homeomorphism of (a, b) onto $(a - b, b - a)$. Then (11.7.4) is just a graph of the map $\omega: (-d, d) \to \mathbb{R}$, where $d:=(b - a)/\sqrt{2}$, given by the formula

$$\omega(t):=\sqrt{2} \cdot \sigma_0^{-1}(\sqrt{2} \cdot t) - t, \qquad t \in (-d, d),$$

or, equivalently,

$$\omega(t) = \frac{\sqrt{2}}{2}(\sigma_0^{-1}(\sqrt{2} \cdot t) + \sigma_0^{-1}(-\sqrt{2} \cdot t)), \qquad t \in (-d, d),$$

on account of the equality $\varphi = \sigma_0^{-1}(-\sigma_0)$. Therefore, the rotation r transforming the graph of φ onto the graph of an even continuous function ω such that $\eta(t):=t + \omega(t)$, $t \in (-d, d)$ is strictly increasing.

Conversely, let $\omega: (-d, d) \to \mathbb{R}$ (d as above) be an even continuous function such that $\eta:=\mathrm{id} + \omega$ is strictly increasing. By the rotation r^{-1} the graph of ω is transformed into the graph of a function $\varphi: (a, b) \to (a, b)$ whose analytic representation reads as follows:

$$\varphi(x) = \frac{\sqrt{2}}{2}\eta(-\eta^{-1}(\sqrt{2} \cdot x)), \qquad x \in (a, b).$$

With $\sigma(x):=\eta^{-1}(\sqrt{2} \cdot x)$, $x \in (a, b)$, this is identical to (11.7.3). ∎

Remark 11.7.1. In case a, b and c are finite, Theorem 11.7.3 remains valid if we replace the open intervals (a, b) and $(-c, c)$ by their closures. No change in the proof is needed regardless of whether the first or the second method is applied.

11.7B. Primitive iterative roots, homographies

The analogies between algebraic and iterative roots suggest introducing a 'primitive' iterative root of the identity. The following definition seems to be adequate: φ is a *primitive Nth iterative root of the identity* provided φ satisfies (11.7.1) but $\varphi^M \neq \text{id}$ for any $M \in \{1, \ldots, N-1\}$. As we have seen, in the real case continuous primitive iterative roots of identity on an interval do not exist for $N > 2$. It turns out that the property which actually is decisive to that effect is the connectedness of the domain. In the case where we admit disconnected domains (two components do suffice!) the situation changes diametrically: already among homographic mappings

$$\varphi(x) = \frac{ax + b}{cx + d}, \qquad ad - bc \neq 0, \tag{11.7.5}$$

we can find primitive iterative roots of all orders regardless of whether the domain is real or complex. To visualize this let us start with the following.

Theorem 11.7.4. *The only homographic (real or complex) Nth iterative roots of identity are of the form*

$$\varphi = A \circ \hat{\varphi} \circ A^{-1} \tag{11.7.6}$$

where $A: \mathbb{K} \to \mathbb{K}$ is an arbitrary similitude[1] and $\hat{\varphi}$ has one of the following three representations:

(i) $\hat{\varphi}(x) = \varepsilon x$, $x \in \mathbb{K}$, *where ε is any Nth root of unity;*

(ii) $\hat{\varphi}(x) = \delta \dfrac{1}{x}$, $x \in \mathbb{K} \setminus \{0\}$; *this solution occurs only in the case where N is even;*

(iii) $\hat{\varphi}(x) = \gamma - \gamma^2 \dfrac{\varepsilon}{(1 + \varepsilon)^2} \dfrac{1}{x}$, $x \in \mathbb{K} \setminus \{0\}$, *where $\gamma \in \mathbb{K} \setminus \{0\}$ and ε is an Nth root of unity, $\varepsilon \notin \{-1, 1\}$.*

Proof. It is well known that any homographic map φ defined by (11.7.5) admits a similitude $A: \mathbb{K} \to \mathbb{K}$ such that $\hat{\varphi} := A^{-1} \circ \varphi \circ A$ has one of the following three forms:

(I) $\hat{\varphi}(x) = \gamma x$, $x \in \mathbb{K}$,

or

(II) $\hat{\varphi}(x) = x + \delta$, $x \in \mathbb{K}$,

---

[1] That is, $A(x) = \alpha x + \beta$, $x \in \mathbb{K}$, for some $\alpha \neq 0$ and β from $\mathbb{K}$.

or

(III) $\hat{\phi}(x) = \gamma + \delta \dfrac{1}{x}, \quad x \in \mathbb{K} \setminus \{0\},$

$\gamma \in \mathbb{K}, \delta \in \mathbb{K} \setminus \{0\}$. A simple calculation shows that φ satisfies equation (11.7.1) if and only if $\hat{\phi}$ does.

Assume a homography φ to be an Nth iterative root of the identity. In case (I) the associated map $\hat{\phi}$ satisfies (11.7.1) if and only if γ is an Nth primitive root of unity, i.e. (i) holds.

Assume (II). One may immediately check that $\hat{\phi}(x) = x + \delta$, $x \in \mathbb{K}$, satisfies (11.7.1) if and only if $\delta = 0$ which is excluded.

Assume (III). Then (induction)

$$\hat{\phi}^n(x) = \frac{k_n x + k_{n-1}\delta}{k_{n-1}x + k_{n-2}\delta}, \qquad n \in \mathbb{N}, \tag{11.7.7}$$

where the coefficients k_n, $n \in \mathbb{N}_0 \cup \{-1\}$, are given by the recurrence

$$k_{-1} := 0, \quad k_0 := 1, \quad k_n := \gamma k_{n-1} + \delta k_{n-2}, \quad n \in \mathbb{N}. \tag{11.7.8}$$

Hence, again by induction,

$$k_n = \begin{cases} \dfrac{1}{2^{n+1}\Delta}\left((\gamma+\Delta)^{n+1} - (\gamma-\Delta)^{n+1}\right) & \text{if } \Delta := \sqrt{(\gamma^2 + 4\delta)} \neq 0, \\ (n+1)(\tfrac{1}{2}\gamma)^n & \text{if } \Delta = 0, \end{cases} \tag{11.7.9}$$

for all $n \in \mathbb{N}$; the square complex root here denotes either of the two possible values.

Inserting (11.7.7) into (11.7.1) and making use of (11.7.8) we obtain $k_{N-1}(x^2 - \gamma x - \delta) = 0$ for all $x \in \mathbb{K}$ except, possibly, a finite set of arguments. Obviously, this is equivalent to $k_{N-1} = 0$ which excludes the second possibility occurring in (11.7.9). Therefore, the first expression in (11.7.9) disappears for $n = N - 1$ whence

$$\gamma + \Delta = \varepsilon(\gamma - \Delta)$$

where ε is an Nth root of unity; $\varepsilon = 1$ gives $\Delta = 0$, a contradiction, whereas $\varepsilon = -1$ implies $\gamma = 0$ whence

$$\hat{\phi}(x) = \delta \frac{1}{x}, \qquad x \in \mathbb{K} \setminus \{0\},$$

is an involution (ii). If ε is a nonreal Nth root of unity, then $\Delta = \gamma(\varepsilon - 1)/(\varepsilon + 1)$ whence $\delta = -\varepsilon\gamma^2/(1 + \varepsilon)^2$ and

$$\hat{\phi}(x) = \gamma - \gamma^2 \frac{\varepsilon}{(1+\varepsilon)^2} \frac{1}{x}, \qquad x \in \mathbb{K} \setminus \{0\},$$

i.e. (iii) occurs. We omit the detailed calculations showing that the functions obtained are actually the Nth iterative roots of identity. ∎

Since homographies are the only invertible rational (meromorphic) functions we obtain immediately the following.

Theorem 11.7.5. *The only rational (resp. meromorphic) iterative roots of identity are of the form described in Theorem 11.7.4.*

Now, we are able to convince the reader that in the case of disconnected domains continuous primitive iterative roots of the identity (of all orders!) actually exist even on the real line. For, denote by ε a primitive Nth nonreal root of unity and observe that the function φ given by (11.7.6) with $\hat\varphi$ defined by (III) is an Nth primitive iterative root of the identity. Finally, since the number $(1+\varepsilon)^2/\varepsilon = \varepsilon + 2 + 1/\varepsilon$ is real ($|\varepsilon| = 1$ implies $1/\varepsilon = \bar\varepsilon$), so is its reciprocal. Thus the function

$$\hat\varphi(x) = \gamma - \gamma^2 \frac{\varepsilon}{(1+\varepsilon)^2} \frac{1}{x}, \qquad x \in \mathbb{R} \setminus \{0\},$$

yields, for any $\gamma \in \mathbb{R} \setminus \{0\}$, an Nth primitive iterative root of the identity for any $N \in \mathbb{N}$.

11.8 Another characterization of reciprocals

We have seen that the family of continuous decreasing involutions is pretty large. An appeal to Theorem 11.7.2 shows emphatically that higher regularity requirements (smoothness of any order) help nothing. What about analyticity? Even such a strong regularity assumption does not change the situation: there is no uniqueness of analytic square iterative roots of identity. In fact, take, for instance,

$$\varphi(x) := \frac{1}{x}, \qquad \psi(x) := -x + \tfrac{5}{2}, \qquad x \in (\tfrac{1}{2}, 2).$$

Since both of these two analytic involutions are also convex, even conditions of this type do not force the uniqueness of square iterative roots of identity. All this shows expressively that in order to obtain the uniqueness of solutions of the equation

$$\varphi^2 = \mathrm{id} \tag{11.8.1}$$

we must impose upon φ conditions of quite a different nature.

The contents of Chapter 10 yield a good prognosis for the following idea: just postulate another functional equation to be satisfied.

11.8A. Dubikajtis' theorem

The characterization of reciprocals we reproduce here after L. Dubikajtis [2] has already been announced in Note 10.1D3.

Theorem 11.8.1. *The only continuous function* $\varphi\colon (0, \infty) \to (0, \infty)$ *satisfying simultaneously the equations* (11.8.1) *and*

$$\varphi(x+1) = \frac{\varphi(x)}{\varphi(x)+1}, \qquad x \in (0, \infty), \tag{11.8.2}$$

is given by $\varphi(x) = 1/x$, $x \in (0, \infty)$.

To prove this we start with a lemma.

Lemma 11.8.1. *The set of all positive rationals is the smallest one in* $\mathbb{R}$ *among those containing* 1, *closed under translations on* 1 *and taking reciprocals; in other words, if* $Q \subset \mathbb{R}$ *is any set such that* $1 \in Q$ *and* $(Q+1) \cup 1/Q \subset Q$, *then* $\mathbb{Q} \cap (0, \infty) \subset Q$.

Proof. Obviously $\mathbb{N} \subset Q$. Suppose that

$$\bigcup_{k=1}^{m} \frac{1}{k} \mathbb{N} \subset Q \qquad \text{for some } m \in \mathbb{N}, \tag{11.8.3}$$

and take $x \in [1/(m+1)]\mathbb{N}$. Then $x = p + q/(m+1)$ where $p \in \mathbb{N}_0$ and $q \in \{0, \ldots, m\}$. Excluding the trivial case $q = 0$, we infer that $(m+1)/q \in Q$, by means of (11.8.3). Hence $q/(m+1)$ as well as $p + q/(m+1)$ belongs to Q, i.e. $\bigcup_{k=1}^{m+1} (1/k)\mathbb{N} \subset Q$. Induction shows that Q contains all positive rationals. ■

Proof of Theorem 11.8.1. Assume that $\varphi\colon (0, \infty) \to (0, \infty)$ is a continuous solution of both (11.8.1) and (11.8.2). Let Q be the set of all those points $x \in (0, \infty)$ for which there exists a sequence $(x_n)_{n \in \mathbb{N}}$ of positive numbers and such that $x_n \to x$ as $n \to \infty$ as well as

$$\lim_{n \to \infty} \varphi(x_n) = \frac{1}{x}.$$

Write $c := \varphi(1)$. Equation (11.8.2) and a straightforward induction show that

$$\varphi(n+1) = \frac{c}{1+nc} \qquad \text{for all } n \in \mathbb{N}_0,$$

whence, in view of (11.8.1),

$$n+1 = \varphi^2(n+1) = \varphi\left(\frac{c}{1+nc}\right), \qquad n \in \mathbb{N}_0.$$

Now, making use of (11.8.2) again,

$$\varphi\left(1 + \frac{c}{1+nc}\right) = \varphi\left(\frac{c}{1+nc}\right) \frac{1}{1 + \varphi\left(\dfrac{c}{1+nc}\right)} = \frac{n+1}{n+2} \xrightarrow[n \to \infty]{} 1,$$

which proves that $1 \in Q$; indeed, it suffices to take $x_n := 1 + c/(1+nc)$, $n \in \mathbb{N}$. To check the remaining assumptions of Lemma 11.8.1, assume that $x \in Q$

with $(x_n)_{n \in \mathbb{N}}$ as a corresponding sequence. Put

$$y_n := x_n + 1 \quad \text{and} \quad z_n := \varphi(x_n), \qquad n \in \mathbb{N}.$$

Then $y_n \to x + 1$ as $n \to \infty$ and, by (11.8.2),

$$\varphi(y_n) = \varphi(x_n + 1) = \frac{\varphi(x_n)}{1 + \varphi(x_n)} \xrightarrow[n \to \infty]{} \frac{1}{x} \frac{1}{1 + 1/x} = \frac{1}{x + 1},$$

i.e. $x + 1 \in Q$. Finally, $z_n \to 1/x$ as $n \to \infty$ and

$$\varphi(z_n) = \varphi^2(x_n) = x_n \xrightarrow[n \to \infty]{} x = \left(\frac{1}{x}\right)^{-1},$$

i.e. $1/x \in Q$. On account of Lemma 11.8.1 we get the inclusion

$$Q \cap (0, \infty) \subset Q.$$

This and the continuity of φ imply that $\varphi(x) = 1/x$, $x \in (0, \infty)$. $\blacksquare$

11.8B. Volkmann's theorem

In fact, L. Dubikajtis [2] has obtained a somewhat stronger result assuming the continuity of φ on some subinterval (a, b) of the positive half-line. He conjectured that some additional regularity assumption is indispensable to characterize reciprocals via system (11.8.1), (11.8.2). However, as P. Volkmann has recently shown, system (11.8.1), (11.8.2) alone admits $\varphi(x) = 1/x$, $x \in (0, \infty)$, as the only solution! Below we present this result in detail (see Volkmann [2]). Theorem 11.8.1 follows immediately from Volkmann's but we have decided to present its proof to illustrate different methods of possible approach.

Theorem 11.8.2. *The only function* $\varphi \colon (0, \infty) \to (0, \infty)$ *satisfying simultaneously equations (11.8.1) and (11.8.2) is given by* $\varphi(x) = 1/x$, $x \in (0, \infty)$.

Proof. Assume $\varphi \colon (0, \infty) \to (0, \infty)$ to be a solution of system (11.8.1), (11.8.2) and put $\psi := 1/\varphi$. Then the system considered assumes the form

$$\psi\left(\frac{1}{\psi(x)}\right) = \frac{1}{x}, \qquad x \in (0, \infty), \tag{11.8.4}$$

$$\psi(x + 1) = \psi(x) + 1, \qquad x \in (0, \infty). \tag{11.8.5}$$

Equation (11.8.4) alone implies the bijectivity of ψ. Therefore the inverse function ψ^{-1} maps the positive half-line onto itself. Moreover, one may easily check that ψ^{-1} is also a solution of system (11.8.4), (11.8.5). Consequently, any information on ψ furnished by system (11.8.4), (11.8.5) applies for ψ^{-1} as well. In particular, the property

$$\psi(x + m) = \psi(x) + m, \qquad x \in (0, \infty), \ m \in \mathbb{N}, \tag{11.8.6}$$

resulting immediately from (11.8.5), is shared automatically by ψ^{-1}, i.e.

$$\psi^{-1}(x+m) = \psi^{-1}(x)+m, \qquad x\in(0,\infty),\, m\in\mathbb{N}. \qquad (11.8.7)$$

Observe that

$$\psi((0,m]) \subset (0,m] \qquad \text{for all } m\in\mathbb{N}. \qquad (11.8.8)$$

Indeed, otherwise we would have $\psi(x) = y+m$ for some $m\in\mathbb{N}$, $x\in(0,m]$ and $y\in(0,\infty)$ whence, in view of (11.8.7),

$$x = \psi^{-1}(\psi(x)) = \psi^{-1}(y+m) = \psi^{-1}(y)+m > m,$$

a contradiction. This proves also that $\psi^{-1}((0,m]) \subset (0,m]$, $m\in\mathbb{N}$, which jointly with (11.8.8) gives the equality

$$\psi((0,m]) = (0,m], \qquad m\in\mathbb{N}. \qquad (11.8.9)$$

Now, we shall show that $\psi(1) = 1$. For, assuming the contrary and applying (11.8.9) with $m=1$ we would get $\psi(1)<1$ whence $1/\psi(1)>1$ and consequently, in view of (11.8.4) and (11.8.9),

$$1 = \psi\left(\frac{1}{\psi(1)}\right) > 1,$$

a contradiction. Thus $\psi(1) = 1$ which jointly with (11.8.5) implies

$$\psi(m) = m \qquad \text{for all } m\in\mathbb{N}. \qquad (11.8.10)$$

Let K_n denote the set of all continued fractions

$$r = u_0 + \cfrac{1}{u_1 + \cfrac{1}{u_2 + \cdots + \cfrac{1}{u_n}}}$$

where $u_0\in\mathbb{N}_0$ and $u_1,\ldots,u_n\in\mathbb{N}$, $n\in\mathbb{N}$. Put also $K_0:=\mathbb{N}$. It is well known that

$$\mathbb{Q}\cap(0,\infty) = \bigcup_{n\in\mathbb{N}_0} K_n. \qquad (11.8.11)$$

We shall inductively show that, whenever $r\in K_n$, we have

$$\psi(r) = r, \qquad \psi((0,r)) = (0,r) \qquad \text{and} \qquad \psi((r,\infty)) = (r,\infty), \qquad (11.8.12)$$

for all $n\in\mathbb{N}_0$. In fact, for $n=0$ (11.8.12) reduces to (11.8.10) and (11.8.9) because of the bijectivity of ψ and the equality $\psi((0,\infty)) = (0,\infty)$. Assume (11.8.12) to hold for some $n\in\mathbb{N}_0$ and fix arbitrarily an $s\in K_{n+1}$. Then

$$s = m + \frac{1}{t} \qquad \text{for some } m\in\mathbb{N}_n \text{ and } t\in K_n,$$

whence by (11.8.6), (11.8.4) and (11.8.12) applied for ψ^{-1} instead of ψ we obtain

$$t(s) = \psi\left(m+\frac{1}{t}\right) = m + \psi\left(\frac{1}{t}\right) = m + \frac{1}{\psi^{-1}(t)} = m + \frac{1}{t} = s.$$

Fix an $x \geqslant s = m + 1/t$; then (11.8.4) implies

$$\psi\left(\frac{1}{\psi(x-m)}\right) = \frac{1}{x-m} \leqslant t \in K_n$$

whence, by (11.8.12), $1/\psi(x-m) \leqslant t$. Now,

$$\psi(x) = \psi((x-m)+m) = \psi(x-m)+m \geqslant \frac{1}{t}+m = s,$$

i.e.

$$\psi([s, \infty)) \subset [s, \infty).$$

Replacing here ψ by ψ^{-1} we infer that $\psi([s, \infty)) = [s, \infty)$. Consequently, we have also $\psi((0, s)) = (0, s)$, which finishes the inductive proof of (11.8.12). Finally, note that for any $r, s \in \mathbb{Q} \cap (0, \infty)$, $r \leqslant s$, one has

$$\psi([r, s]) = \psi((0, s] \setminus (0, r)) = \psi((0, s]) \setminus \psi((0, r))$$
$$= (0, s] \setminus (0, r) = [r, s]$$

because of the bijectivity of ψ, (11.8.11) and (11.8.12). This implies that for any $x \in (0, \infty)$ and any $r, s \in \mathbb{Q} \cap (0, \infty)$ such that $r \leq x \leq s$ the value $\psi(x)$ belongs to $[r, s]$. Letting the difference $s - r$ tend to zero one obtains $\psi(x) = x$ for all $x \in (0, \infty)$. This means that $\varphi(x) = 1/x$, $x \in (0, \infty)$, which was to be proved. ∎

Remark 11.8.1. Equation (11.8.2) alone considered in the class of convex self-mappings of $(0, \infty)$ having a fixed point at 1 admits

$$\varphi(x) = \frac{1}{x}, \qquad x \in (0, \infty),$$

as the only solution (see Theorem 10.1.2).

11.9 Invariant curves

Various geometrical considerations yield especially good breeding grounds for the occurrence of interesting functional equations. In the present section we deal with one of such problems leading to a functional equation with a superposition of the unknown function.

Equations of this kind are in a sense close to those defining iterative roots and that is the reason we have decided to include investigations concerning the problem of invariant curves in the present chapter.

The question is to find k-dimensional manifolds $M \subset \mathbb{R}^n$, $k < n$, invariant with respect to a given transformation F of the space $\mathbb{R}^n$ into itself, i.e. such that the inclusion $F(M) \subset M$ holds true.

As it stands, the problem is pretty general and it still remains a vast area of research. A more exhaustive treatment of this problem can be found in Nitecki [1] (see also the references therein). Concerning further results see also Anosov [1], Brydak [2], [5], [6], [11], Dhombres [1], Fort [1], Hartman [4], Hirsch–Pugh [1], Hirsch–Pugh–Shub [1], Kuczma [3], [26], Lattès [1]–[4], Lush [1], Montel [1], Nabeya [1], Sternberg [1], Urabe [1] as well as the references in Kuczma [26, Ch. XIV].

11.9A. Simplifications and assumptions

In its full generality the problem is too hard for the tools we are equipped with in trying to keep the contents of this book at a reasonably elementary level. We shall content ourselves with a much more modest approach under restrictive assumptions.

First of all we shall confine ourselves to the case of curves on the real plane, i.e. we deal with $k = 1$, $n = 2$. The transformation F is supposed to be of class C^1 in a neighbourhood $U \subset \mathbb{R}^2$ of the origin and to have a fixed point at zero. The characteristic roots of the derivative $F'(0)$ are assumed to be nonzero and distinct in absolute value (in particular, such a requirement forces them to be real). On the other hand, we shall consider only those invariant curves that form a graph of a map φ such that $\varphi(0) = 0$.

If the function F is linear, then the one-dimensional eigenspaces determined by the eigenvalues of the matrix $F'(0)$ yield (very special) invariant manifolds passing through the origin. Thus, in the general case, we may expect the invariant curves to be close to the eigenspaces of $F'(0)$. This turns out to be actually the case (see Theorem 11.9.3 below).

In order to simplify our considerations we choose the eigenspaces of $F'(0)$ as the axes of the coordinate system. Such an assumption might be regarded as technical only since, analytically, the change (if necessary) can be done easily by a linear transformation $T: \mathbb{R}^2 \to \mathbb{R}^2$ represented by a 2×2 nonsingular matrix T_0 such that $T_0 F'(0) T_0^{-1}$ is diagonal. Finally, we assume that $F: U \to \mathbb{R}^2$ is of class C^1, and has the asymptotic property

$$F(z) = Az + o(|z|), \qquad z \to 0, \tag{11.9.1}$$

where $A = \mathrm{diag}(\lambda, \mu) := \begin{pmatrix} \lambda & 0 \\ 0 & \mu \end{pmatrix} \in \mathbb{R}^{2 \times 2}$, and $|\cdot|$ is the norm in $\mathbb{R}^2$. In the language of the components (f, g) of F this means that

$$f(z) = (\lambda, 0) \cdot z + o(|z|), \qquad g(z) = (0, \mu) \cdot z + o(|z|), \qquad z \to 0, \tag{11.9.2}$$

with $B := \mathrm{diag}(1/\lambda, 1/\mu)$, whereas are of class C^1.

11.9B. Equation of invariant curves and its Lipschitzian solutions

The condition that the curve $\{(x, \varphi(x)) \in \mathbb{R}^2 : x \in [-c, c]\}$ (the graph of a function $\varphi: [-c, c] \to \mathbb{R}$) should be invariant under F reads as follows:

$$\varphi(f(x, \varphi(x))) = g(x, \varphi(x)), \qquad x \in [-c, c]. \tag{11.9.3}$$

We shall prove the existence and, in some cases, the uniqueness of local Lipschitzian solutions of equation (11.9.3) in a neighbourhood of the origin (see Hadamard [1], Lattès [1] and Montel [1]).

Theorem 11.9.1. *Let* $F = (f, g)$ *be a* C^1 *transformation mapping a neighbourhood* $U \subset \mathbb{R}^2$ *of the origin into* $\mathbb{R}^2$ *and satisfying* (11.9.1) *with* $A = \operatorname{diag}(\lambda, \mu)$ *such that either*

$$0 < |\lambda| < |\mu| \quad \text{and} \quad |\lambda| < 1 \quad\quad\quad (11.9.4)$$

or

$$0 < |\mu| < |\lambda| \quad \text{and} \quad |\lambda| > 1. \quad\quad\quad (11.9.5)$$

Then for every positive real number L *there exist a* $c > 0$ *and exactly one local solution* $\varphi: [-c, c] \to \mathbb{R}$ *of equation* (11.9.3) *fulfilling the Lipschitz condition*

$$|\varphi(s) - \varphi(t)| \le L|s - t| \quad \text{for all } s, t \in [-c, c],$$

and such that $\varphi(0) = 0$. *This solution is differentiable at zero and we have* $\varphi'(0) = 0$.

Proof. Without loss of generality we may assume (11.9.4) since otherwise replacing F by its inverse we transform (11.9.5) into (11.9.4). In fact, $F^{-1} = (f^*, g^*)$ actually exists in a neighbourhood W of zero, remains C^1-smooth and satisfies the asymptotical condition

$$F^{-1}(z) = Bz + o(|z|), \quad z \to 0,$$

with $B := \operatorname{diag}(1/\lambda, 1/\mu)$, whereas

$$f^*(z) = \left(\frac{1}{\lambda}, 0\right) \cdot z + o(|z|) \quad \text{and} \quad g^*(z) = \left(\frac{1}{\mu}, 0\right) \cdot z + o(|z|) \quad \text{as } z \to 0.$$

There exists a C^1 mapping h from a neighbourhood $V \subset \mathbb{R}^2$ of the origin into $\mathbb{R}^2$ inverse to g with respect to the second variable, i.e. satisfying the relation

$$h(x, g(x, y)) = y$$

for (x, y) from a neighbourhood of zero. Moreover,

$$h(z) = \left(0, \frac{1}{\mu}\right) \cdot z + o(|z|), \quad z \to 0. \quad\quad\quad (11.9.6)$$

Fix any positive real number L and take any $c > 0$ small enough to have

$$D := \{(x, y) \in \mathbb{R}^2 : |x| \le c, |y| \le L|x|\} \subset U \cap V.$$

Write $I := [-c, c]$ and consider the class $\mathbf{\Phi}$ of all functions $\varphi: I \to \mathbb{R}$ such that $\varphi(0) = 0$ and

$$|\varphi(s) - \varphi(t)| \le L|s - t| \quad \text{for all } s, t \in I.$$

Endow $\mathbf{\Phi}$ with the metric

$$\rho(\varphi, \psi) := \sup_{x \in I \setminus \{0\}} \left| \frac{\varphi(x) - \psi(x)}{x} \right|, \quad \varphi, \psi \in \mathbf{\Phi};$$

then the pair $(\mathbf{\Phi}, \rho)$ becomes a complete metric space.

Since $|\lambda| < 1$ we may, diminishing c if necessary, make sure about the relation

$$f(x, \varphi(x)) \in I \qquad \text{for } x \in I \text{ and } \varphi \in \Phi$$

which allows us to define a transformation $T: \Phi \to C(I, \mathbb{R})$ by the formula

$$T(\varphi)(x) := h(x, \varphi(f(x, \varphi(x)))), \qquad x \in I.$$

We wish to show that, in fact, T is a self-mapping of Φ provided c is taken small enough. For observe that obviously $T(\varphi)(0) = 0$ for $\varphi \in \Phi$. Fix an $\varepsilon > 0$; it follows from (11.9.2) and (11.9.6) as well as from the regularity of f and h that

$$|f(x, y) - f(u, v)| \leqslant (|\lambda| + \varepsilon)|x - u| + \varepsilon|y - v| \tag{11.9.7}$$

and

$$|h(x, y) - h(u, v)| \leqslant \varepsilon|x - u| + \left(\frac{1}{|\mu|} + \varepsilon\right)|y - v| \tag{11.9.8}$$

for all $(x, y), (u, v) \in D$. Now for any $\varphi \in \Phi$ and any $(s, t) \in I^2$ one gets immediately the following estimate:

$$|T(\varphi)(s) - T(\varphi)(t)| \leqslant \left(\varepsilon + \left(\frac{1}{|\mu|} + \varepsilon\right)L((|\lambda| + \varepsilon) + L\varepsilon)\right)|s - t|.$$

The coefficient at the right-hand side tends as $\varepsilon \to 0$ to $|\lambda/\mu| \cdot L < L$ which proves that

$$|T(\varphi)(s) - T(\varphi)(t)| \leqslant L|s - t|, \qquad s, t \in I$$

whenever $\varepsilon > 0$ is small. This proves that T does transform Φ into itself.

To apply Banach's Theorem (see Section 1.5) it remains to show that T is a contraction.

With the aid of (11.9.7) and (11.9.8) we arrive easily at

$$\rho(T(\varphi), T(\psi)) \leqslant \left(\frac{1}{|\mu|} + \varepsilon\right)(|\lambda| + \varepsilon + 2L\varepsilon)\rho(\varphi, \psi)$$

for any $\varphi, \psi \in \Phi$. Diminishing ε (and hence c) if necessary we force the coefficient on the right-hand side to be less than 1 getting the existence and uniqueness of a solution φ of the equation

$$\varphi(x) = h(x, \varphi(f(x, \varphi(x)))) = T(\varphi(x)), \qquad x \in I,$$

which is equivalent to (11.9.3).

To prove that φ is differentiable at zero and $\varphi'(0) = 0$ observe that what we have proved so far yields, for each positive $K > 0$, a unique Lipschitzian solution φ_K such that

$$|\varphi_K(s) - \varphi_K(t)| \leqslant K|s - t|$$

on an interval I_K centred at zero. In particular, $I = I_L$ and $\varphi = \varphi_L$. The uniqueness of solutions implies however that $\varphi|_{I \cap I_K} = \varphi_L|_{I \cap I_K}$. Therefore $|\varphi(s)/s| \leqslant K$ for an arbitrarily small constant $K > 0$ and all s in $I \cap I_k$. This shows that $\varphi'(0) = 0$. ∎

11.9C. Lack of uniqueness of Lipschitzian solutions

The theorem just proved guarantees both existence and uniqueness of suitable invariant curves (graphs). It turns out that keeping the assumptions $0 \neq |\lambda| \neq |\mu| \neq 0$ untouched we are faced with the phenomenon of the abundance of invariant curves provided the conditions (11.9.4) or (11.9.5) are no longer valid. Actually, if they are replaced by

$$0 < |\mu| < |\lambda| < 1 \qquad (11.9.9)$$

or

$$1 < |\lambda| < |\mu| \qquad (11.9.10)$$

then equation (11.9.3) has a local Lipschitzian solution depending on an arbitrary function. More precisely, we have the following.

Theorem 11.9.2. *Let* $F = (f, g)$ *be a* C^1 *transformation mapping a neighbourhood* $U \subset \mathbb{R}^2$ *of the origin into* $\mathbb{R}^2$ *and satisfying* (11.9.1) *with* $A = \mathrm{diag}(\lambda, \mu)$ *such that either* (11.9.9) *or* (11.9.10) *holds true. Then for every positive real number* L *there exist a* $c > 0$ *and a local solution* $\varphi : [-c, c] \to \mathbb{R}$ *of equation* (11.9.3) *fulfilling the Lipschitz condition*

$$|\varphi(s) - \varphi(t)| \leqslant L|s - t| \qquad \text{for all } s, t \in [-c, c],$$

and such that $\varphi(0) = 0$. *This solution depends on an arbitrary* L-*Lipschitzian function and is differentiable at zero with* $\varphi'(0) = 0$.

Proof. Replacing, if necessary, F by F^{-1} we may assume (11.9.9). First, suppose additionally that $\lambda > 0$. Then, in view of (11.9.1) and the fact $\|A\| \leqslant \lambda < 1$, there exists a neighbourhood U_0 of zero such that $F(U_0) \subset U_0$. On account of Theorem 1.1.5 the origin is an attractive fixed point of $F|U_0$ and its domain of attraction $A_F(0)$ is also a neighbourhood of zero (see Theorem 1.1.4(b)). Fix a positive real number L and take $c > 0$ small enough to have

$$D := \{(x, y) \in \mathbb{R}^2 : x \in [0, c], |y| \leqslant Lx\} \subset U_0 \cap A_F(0)$$

as well as

$$0 < f(x, y) < x, \qquad |g(x, y)| \leqslant Lf(x, y) \qquad \text{for all } (x, y) \in D, x > 0. \quad (11.9.11)$$

Take an arbitrary point $(x_0, y_0) \in D$, $x_0 > 0$, and put $(x_n, y_n) := F^n(x_0, y_0)$, $n \in \mathbb{N}_0$. Then, in particular, $(x_0, y_0) \in A_F(0)$ whence $(x_n, y_n) \to (0, 0)$ as $n \to \infty$. The first of the relations (11.9.12) implies also that the sequence $(x_n)_{n \in \mathbb{N}}$ is strictly decreasing.

Now, let $\varphi_0 : [x_1, x_0] \to \mathbb{R}$ be an arbitrary L-Lipschitzian function fulfilling the boundary conditions $\varphi_0(x_0) = y_0$ and $\varphi_0(x_1) = y_1$ as well as the estimate

$$|\varphi_0(x)| \leqslant Lx \qquad \text{for all } x \in [x_1, x_0].$$

Write $\hat{f}(x):=f(x,\varphi_0(x))$, $x\in[x_1,x_0]$. The continuity of $\hat{f}$ jointly with the equalities $\hat{f}(x_0)=x_1$ and $\hat{f}(x_1)=x_2$ ensures that $\hat{f}([x_1,x_0])=[x_2,x_1]$. This application is one-to-one provided c has been chosen small enough. Indeed, suppose for the indirect proof that $\hat{f}(s)=\hat{f}(t)$ for two distinct points $s,t\in[x_1,x_0]$. Then, taking an $\varepsilon>0$ we get from (11.9.2) that

$$0=\hat{f}(s)-\hat{f}(t)=\lambda(s-t)+r(s,\varphi_0(s))-r(t,\varphi_0(t))$$

where $r(x,y)$ stands for the 'remainder', $f(x,y)-\lambda x$ being locally ε-Lipschitzian. Thus, diminishing c again (if necessary), we get the inequality

$$\lambda|s-t|\leqslant\varepsilon(|s-t|+|\varphi_0(s)-\varphi_0(t)|)\leqslant\varepsilon(1+L)|s-t|,$$

impossible for $\varepsilon<\lambda/(1+L)$.

Thus the formula

$$\varphi_1(f(x,\varphi_0(x)))=g(x,\varphi_0(x))$$

defines unambiguously a function $\varphi_1:[x_2,x_1]\to\mathbb{R}$ such that $\varphi_1(x_1)=y_1$, $\varphi_1(x_2)=y_2$ and $|\varphi_1(x)|\leqslant Lx$ for $x\in[x_2,x_1]$ on account of (11.9.11). Moreover, for any $\tilde{s},\tilde{t}\in[x_2,x_1]$, $\tilde{s}\neq\tilde{t}$, one has

$$\frac{|\varphi_1(\tilde{s})-\varphi_1(\tilde{t})|}{|\tilde{s}-\tilde{t}|}=\frac{|g(s,\varphi_0(s))-g(t,\varphi_0(t))|}{|\hat{f}(s)-\hat{f}(t)|}\leqslant\frac{\varepsilon+(|\mu|+\varepsilon)L}{\lambda-\varepsilon-L\varepsilon}\leqslant L$$

provided ε is small enough; here $s:=\hat{f}^{-1}(\tilde{s})$ and $t:=\hat{f}^{-1}(\tilde{t})$. Thus the function φ_1 has all the properties of φ_0 and we can inductively continue this procedure. The common extension of the sequence $\varphi_n:[x_{n+1},x_n]\to\mathbb{R}$, $n\in\mathbb{N}_0$, obtained in this way, i.e. the function $\varphi:[0,x_0]\to\mathbb{R}$ given by

$$\varphi(x):=\begin{cases}\varphi_n(x) & \text{for } x\in[x_{n+1},x_n],n\in\mathbb{N}_0,\\ 0 & \text{for } x=0\end{cases}$$

is an L-Lipschitzian solution of (11.9.3) in $[0,x_0]$; it remains to put finally $c:=x_0$.

Now, we are going to show that φ is differentiable at zero with $\varphi'(0)=0$. To this end, write $\tilde{f}(x):=f(x,\varphi(x))$, $x\in I$. Then $\tilde{f}$ is continuous in I and by (11.9.11) we have $0<\tilde{f}(x)<x$, $x\in I\setminus\{0\}$. By Theorem 1.2.4, $\tilde{f}^n$ tends to zero uniformly in I. Fix a $d\in(0,c]$. Taking (11.9.2) into account we get for $x\in(0,d]$

$$\frac{|\varphi(\tilde{f}(x))|}{|\tilde{f}(x)|}=\frac{|g(x,\varphi(x))|}{|f(x,\varphi(x))|}\leqslant\frac{\varepsilon x+(|\mu|+\varepsilon)|\varphi(x)|}{(\lambda-\varepsilon)x-\varepsilon|\varphi(x)|}$$

$$\leqslant\frac{\varepsilon x+(|\mu|+\varepsilon)|\varphi(x)|}{(\lambda-\varepsilon)x-\varepsilon Lx}=\frac{\varepsilon}{\lambda-\varepsilon-\varepsilon L}+\frac{|\mu|+\varepsilon}{\lambda-\varepsilon-\varepsilon L}\frac{|\varphi(x)|}{x},$$

where ε depends on d and $\varepsilon\to0$ as $d\to0$. The latter is also true for $\alpha(d):=\varepsilon/(\lambda-\varepsilon-\varepsilon L)$. Thus in view of (11.9.9) we may choose d sufficiently small that for a $\Theta<1$ we have

$$\frac{|\varphi(\tilde{f}(x))|}{\tilde{f}(x)}\leqslant\alpha(d)+\Theta\frac{|\varphi(x)|}{x}\qquad\text{for all }x\in(0,d].$$

An easy induction gives

$$\frac{\left|\varphi(\tilde{f}^n(x))\right|}{\tilde{f}^n(x)} \leqslant \frac{\alpha(d)}{1-\Theta} + \Theta^n \frac{|\varphi(x)|}{x}, \qquad x \in (0, d]. \qquad (11.9.12)$$

Now, fix an $\eta > 0$ arbitrarily and let d be such that

$$\frac{\alpha(d)}{1-\Theta} < \frac{\eta}{2}. \qquad (11.9.13)$$

Furthermore, let $N \in \mathbb{N}$ be such that $\Theta^n L < \eta/2$ for $n \geqslant N$. If $x \in (0, \tilde{f}^N(d)]$ then $x = \tilde{f}^n(\bar{x})$ for some $n > N$ and $\bar{x} \in (\tilde{f}(d), d]$. Thus (11.9.12) and (11.9.13) give

$$\frac{|\varphi(x)|}{x} = \frac{\left|\varphi(\tilde{f}^n(\bar{x}))\right|}{\tilde{f}^n(\bar{x})} \leqslant \frac{\alpha(d)}{1-\Theta} + \Theta^n \frac{|\varphi(\bar{x})|}{\bar{x}} \leqslant \frac{\eta}{2} + \Theta^n L < \eta,$$

for every $x \in (0, \tilde{f}^N(d)]$. Since η was arbitrary, this proves the desired differentiability of φ at 0.

Repeating the whole construction on an interval $[-c, 0]$ we finish the proof in the case where $\lambda > 0$.

If $\lambda < 0$, then we start with $c > 0$ such that $-c < f(x, y) < 0$ for $(x, y) \in D$, $x > 0$. We may assume that $F(U) \subset U$. Let the second iterate F^2 of F be $F^2 = (f_2, g_2)$. By the first part of the proof we can find a Lipschitzian solution $\varphi : [0, c] \to \mathbb{R}$ of the equation

$$\varphi(f_2(x, \varphi(x))) = g_2(x, \varphi(x)).$$

This solution depends on an arbitrary function. Now use equation (11.9.3) to define φ for negative arguments. It is not hard to check that φ so defined in a (closed) neighbourhood of zero yields a Lipschitzian solution of equation (11.9.3), differentiable at zero and such that $\varphi'(0) = 0$. ∎

11.9D. Comments

Recall that we have confined ourselves to the case of curves yielding a graph of a locally Lipschitzian mapping and passing through the origin. The graphs of all invariant curves we have determined in Subsections 11.9B and 11.9C turned out to be tangent to the one-dimensional eigenspace coinciding with the x-axis. What about the y-axis? The (affirmative) answer is trivial whenever we realize that if a curve is invariant under the transformation F then it remains invariant under the inverse transformation F^{-1}. This leads directly to the following conclusion.

Theorem 11.9.3. *Let $F = (f, g)$ be a C^1 transformation mapping a neighbourhood $U \subset \mathbb{R}^2$ of the origin into $\mathbb{R}^2$ and satisfying (11.9.1) with $A = \mathrm{diag}(\lambda, \mu)$ such that $0 < |\lambda| \neq |\mu| > 0$.*

If $|\lambda| < 1 < |\mu|$ or $|\mu| < 1 < |\lambda|$, then there is exactly one invariant curve tangent to the x-axis and exactly one tangent to the y-axis.

If $|\lambda| < |\mu| < 1$ or $1 < |\mu| < |\lambda|$, then there exists exactly one invariant curve tangent to the x-axis and those tangent to the y-axis depend on an arbitrary function.

If $|\mu| < |\lambda| < 1$ or $1 < |\lambda| < |\mu|$, then the situation is analogous to the former with the interchanged role of the axes.

Remark 11.9.1. In the case where $F = (f, g)$ is linear, S. Nabeya [1] determined all continuous solutions $\varphi \colon \mathbb{R} \to \mathbb{R}$ of equation (11.9.3) with no assumptions on the characteristic roots of the derivative. This is particularly interesting in view of the fact that if there exists a continuous and locally invertible solution σ of the Schröder equation $\sigma \circ F = F'(0)\sigma$ in a neighbourhood of the origin then under this transformation the function F gets reduced to the linear one. Thus, with the aid of Nabeya's result, we can find (locally) all continuous curves invariant under F. The solution σ of the suitable Schröder equation with the properties desired actually exists if the characteristic roots of $F'(0)$ omit 0 and 1. Nevertheless, in Theorem 11.9.1 the possibility $|\mu| = 1$ is quite admissible and the linearization in this case is extremely difficult.

11.9E. Euler's and other special equations

When (f, g) in (11.9.3) is the linear map of $\mathbb{R}^2$ into itself, given by $f(x, y) = x + y$, $g(x, y) = \gamma y$ $(x, y) \in \mathbb{R}^2$, we are led to the functional equation

$$\varphi(x + \varphi(x)) = \gamma\varphi(x) \tag{11.9.14}$$

which aroused considerable interest. If $\gamma = 1$, this equation is usually named *Euler's functional equation*, and for $\gamma = 2$ this is just the equation of a function additive on its own graph in the class of idempotent mappings, intensively studied nowadays (see Dhombres [3], Zdun [1], Forti [1], Sablik [2] and Jarczyk [5], [9], [11]; see also Section 6.1). Putting $\psi(x) := x + \varphi(x)$, $x \in \mathbb{R}$, we arrive at another interesting functional equation

$$\psi^2(x) = (\gamma + 1)\psi(x) - \gamma x \tag{11.9.15}$$

which reduces to the equation of *idempotence* in the case $\gamma = 0$, whereas for $\gamma = -1$ equation (11.9.15) becomes the functional equation of involutions: $\psi^2 = \mathrm{id}$, which we are already quite familiar with. In general, equation (11.9.15) is called an *equation of linear iteration of order 2* as it links ψ^2 linearly with ψ and identity. If the linking is not necessarily linear we are faced with the equation

$$\psi^2(x) = g(x, \psi(x))$$

investigated by M. Kuczma [3], [26] and M. Fort [1]. A study of equation (11.9.15) is found in the book of J. Dhombres [3, Chapter 6].

We shall restrict ourselves here to Euler's functional equation

$$\varphi(x + \varphi(x)) = \varphi(x) \tag{11.9.16}$$

on the real line (Euler [1], Kuratowski [1], Doksum [1], Lüssy [1], Wagner [1]). Following K. Kuratowski [1] we are going to prove the following.

Theorem 11.9.4. *Constant functions are the only solutions of Euler's equation* (11.9.16) *in the class of all functions* $\varphi: \mathbb{R} \to \mathbb{R}$ *such that* $\varphi + \mathrm{id}$ *has the Darboux property.*

Proof. As previously, we put $\psi := \mathrm{id} + \varphi$ whence

$$\psi^n = \mathrm{id} + n\varphi \qquad \text{for all } n \in \mathbb{N}_0, \tag{11.9.17}$$

which jointly with (11.9.16) gives

$$\varphi \circ \psi^n = \varphi, \qquad n \in \mathbb{N}_0. \tag{11.9.18}$$

If $\varphi = 0$, there is nothing to prove; so, assume that, e.g., $\varphi(a) > 0$ for some $a \in \mathbb{R}$. Obviously, the sequence $(\psi^n(a))_{n \in \mathbb{N}}$ increases to infinity. We shall show that the inequalities

$$\psi^n(a) \leqslant \psi^n(x) \leqslant \psi^{n+1}(a), \qquad n \in \mathbb{N}, \tag{11.9.19}$$

hold true provided $x \in [a, \psi(a)]$. For supposing (11.9.19) false, we get either

$$\psi^n(b) < \psi^n(a) \tag{11.9.20}$$

or

$$\psi^n(b) > \psi^{n+1}(a) \tag{11.9.21}$$

for some $b \in (a, \psi(a))$, $n \in \mathbb{N}$. The iterate ψ^n itself has the Darboux property as one may easily check that the superposition of any two functions having the Darboux property must have it, too. Therefore $\psi^n(c) = \psi^n(a)$ for a $c \in (b, \psi(a))$ if (11.9.20) occurs or $\psi^n(c) = \psi^{n+1}(a)$ for a $c \in (a, b)$ in the case (11.9.21). The first possibility jointly with (11.9.18) implies $\varphi(c) = \varphi(a)$ which says that $c = a$ in view of (11.9.17), a contradiction. Similarly, $\psi^n(c) = \psi^{n+1}(a)$ implies in turn that $\varphi(c) = \varphi(a)$ and consequently $c = \psi(a)$, a contradiction again. Thus (11.9.19) has been proved.

Jointly with (11.9.17), (11.9.19) says that

$$a - x \leqslant n(\varphi(x) - \varphi(a)) \leqslant a - x + \varphi(a), \qquad n \in \mathbb{N},$$

for all $x \in [a, \psi(a)]$ which is possible only if $\varphi(x) = \varphi(a)$ for all $x \in [a, \psi(a)]$. For any $x > a$ there exist a $y \in [a, \psi(a)]$ and an $n \in \mathbb{N}_0$ such that $x = \psi^n(y)$ whence $\varphi(x) = \varphi(\psi^n(y)) = \varphi(y) = \varphi(a)$. Thus $\varphi(x) = \varphi(a)$ for all $x \in [a, \infty)$ and for every $a \in \mathbb{R}$ such that $\varphi(a) > 0$.

Let $a_0 := \inf\{a \in \mathbb{R}: \varphi(a) > 0\}$. If we had $a_0 > -\infty$ then we would get $\varphi(x) \leqslant 0$ for all $x < a_0$ and $\varphi|_{(a_0, \infty)} = \text{const} > 0$ which contradicts the Darboux property of the function $\psi = \mathrm{id} + \varphi$. Thus $a_0 = -\infty$ which implies that φ is constant on the entire real line as claimed. ∎

11.10 Notes

11.10.1. Functional equations with superpositions of the unknown functions originate usually on the ground of different problems stemming from geometry, algebra or functional analysis. Regardless of whether the equation obtained is of iterative type or belongs to functional equations in several variables, their treatment requires definitely more sophisticated methods. The name 'iterative functional equations' occurring in the title of the present book refers to the iteration procedure unavoidably performed on the given function in the equation in a single variable in order to solve it. So, iteration usually is just a method and not the problem, except for the chapter to hand. Here iteration explicitly appears in the formulation of the problem to be dealt with.

11.10.2. Dealing with iterations we are often faced with somewhat unexpected situations. To give a bit of the flavour of such nonintuitive effects let us quote here an interesting (although negative) result of R. E. Rice–B. Schweizer–A. Sklar [1]. They are asking 'when is $\varphi(\varphi(z)) = az^2 + bz + c$?' and the surprising answer is 'never $(a \neq 0)$'. They prove even more: a quadratic polynomial has no iterative roots whatever on the entire complex plane. Their proof is purely combinatorial.

In the real case, however, the function $\mathbb{R} \ni x \mapsto |x|^{\sqrt{2}}$ for example, yields a square iterative root (even continuous) of the quadratic monomial $\mathbb{R} \ni x \mapsto x^2$. More generally, the continuous solution of the functional equation $\varphi^2(x) = x^2$, $x \in \mathbb{R}$, depends on an arbitrary function; namely, $\varphi(x) = \psi(|x|)$, $x \in \mathbb{R}$, where $\psi: \mathbb{R}^+ \to \mathbb{R}^+$ satisfies the equation $\psi^2(x) = x^2$, $x \in \mathbb{R}^+$ (see Theorem 11.2.2).

11.10.3. Even in the real case the behaviour of iterative roots may sharply contrast with what one might expect.

For instance, in the light of Theorem 11.2.2 (continuous and strictly increasing functions on the real line have infinitely many iterative roots of all orders) the following proposition appears to be extremely interesting (Dikof–Graw [1], Targoński [8]).

Proposition 11.10.1. *The strictly increasing function $\varphi: \mathbb{R} \to \mathbb{R}$, $x \mapsto \varphi(x) = \frac{1}{2}(x + [x]) + \frac{5}{4}$, has no iterative roots of any order.*

So, if we drop the assumption of continuity (observe, however, that the set of discontinuity points is denumerable here) the situation changes from the abundance of solutions to their nonexistence. Many further examples (less striking perhaps) of such diametrically opposite behaviour were presented in the present chapter.

11.10.4. Sometimes even restrictive regularity requirements did not guarantee any kind of uniqueness of iterative roots. On the other hand, there exist no continuous iterative roots of even order of a strictly decreasing function. As we might learn from other parts of this book it often happens that in the case where even analyticity does not force the solution to be unique, convexity does. In the case of decreasing involutions the situation regarding the possible criteria for uniqueness appeared almost hopeless: convexity did not help either. In fact, the additional functional equation had to be imposed to obtain the effect desired. Nevertheless, convexity played an essential role in various other considerations.

11.10.5. The study of convex iteration groups commenced with the works of A. Smajdor [3], [8] (see also Smajdor [6] and Ger–Smajdor [1] for related results). The first example of a convex function with convex (square) iterative roots has also been presented by A. Smajdor [1] in response to a problem raised by M. Kuczma [12]. In Example 11.4.1 essentially due to J. Ger [1], which plays an instructive role in Section 11.4, we showed another function with this property.

On the other hand, we have the following result due to R. Durret–T. M. Ligett [1] (private communication, unpublished).

Proposition 11.10.2. *The power function* $(0, 1) \ni x \mapsto x^{\alpha}$, $\alpha > 1$, *possesses convex iterative roots depending on an arbitrary function.*

The proof, though quite elementary, is too long to be reproduced here.

The question of convex (concave) iterative roots was also examined by A. Smajdor [3], [8] and M. C. Zdun [9], [14]. We mention here the following.

Proposition 11.10.3. *Let* f *be a strictly increasing convex self-mapping of an interval* $(0, a] \subset \mathbb{R}$ *such that* $0 < f(x) < x$ *for* $x \in (0, a]$. *If* f *is of class* C^1 *and the Julia functional equation*

$$\lambda(f(x)) = f'(x)\lambda(x)$$

admits a nontrivial convex solution, then f *has convex iterative roots of any order* $N \geqslant 2$. *Conversely, if*

$$\lim_{x \to 0} \frac{f(x)}{x} > 0$$

and f *has convex iterative roots of any order* $N \geqslant 2$, *then* f *is of class* C^1 *in* $(0, a]$ *and the function* $\lambda: (0, a] \to \mathbb{R}$ *given by the formula*

$$\lambda(x) := \lim_{N \to \infty} N(f^{1/N}(x) - x), \qquad x \in (0, a],$$

yields a nontrivial convex solution of Julia's equation; $f^{1/N}$ *stands here for a convex Nth iterative root of* f.

11.10.6. The following question was raised by Z. Moszner [1] as connected with various definitions of the notion of biscalar.

(1) If f is a C^r self-mapping of $\mathbb{R}^k$, $r \geq 1$, with $\det f'(x) > 0$ for all $x \in \mathbb{R}^k$, does there exist a square iterative root $\varphi \colon \mathbb{R}^k \to \mathbb{R}^k$ of f, of class C^r in $\mathbb{R}^k$ and having the property $\det \varphi'(x) > 0$, $x \in \mathbb{R}^k$?

(2) Does any function f as in (1) admit a factorization $f = f_1 \circ \cdots \circ f_m$ such that f_i are of class C^r in $\mathbb{R}^k$ with $\det f'_i(x) > 0$ for $x \in \mathbb{R}^k$, $i \in \{1, \ldots, m\}$, and every equation $\varphi_i^2 = f_i$ has a C^r-solution fulfilling the condition $\det \varphi'_i(x) > 0$, $x \in \mathbb{R}^k$, $i \in \{1, \ldots, m\}$?

Our Example 11.4.1 shows that the answer to (1) is negative even in the case $k = 1$. Another counterexample (Kuczma [31]) corresponding to the case $k = 2$ reads as follows.

Example 11.10.1. Take the map $f \colon \mathbb{R}^2 \to \mathbb{R}^2$ defined by the formula

$$f(x, y) := (-\tfrac{1}{2}(x^2 + y^2 + 1)(x + y), -\tfrac{1}{2}(x^2 + y^2 + 1)y), \qquad (x, y) \in \mathbb{R}^2.$$

An elementary computation shows that the equation $f^2(z) = z$ has three solutions: $a = (1, 0)$, $b = (-1, 0)$ and $c = (0, 0)$. Moreover, $f(a) = b$, $f(b) = a$ and $f(c) = c$. Lemma 11.1.1 then says that f has no square iterative roots. Noteworthy is the fact that in this example the components are simply polynomials and moreover the determinant $\det f'$ is not only positive but bounded away from zero: actually $\det f'(x, y) \geq \tfrac{1}{4}$ for all $(x, y) \in \mathbb{R}^2$.

For $k = 1$ the answer to (1) turns out to be affirmative with $m \leq 4$ (Kuczma [32]). To the best of our knowledge for $k > 1$ the question (2) remains unanswered.

11.10.7. U. T. Bödewadt [1] expressed the conjecture that completely monotonic (and thus necessarily analytic) mapping f, i.e., such that $(-1)^{p-1} f^{(p)} \geq 0$ for all $p \in \mathbb{N}$, always possesses completely monotonic iterative roots. This conjecture has been disproved by A. Smajdor [6].

11.10.8. The search for iterative roots of continuous but noninvertible functions has been completely untouched in the present chapter. Actually, the problem is pretty sophisticated. Some partial results are found in Lillo [1], [2]. Lillo exhibits, among others, an example of a continuous map f on a compact interval having no continuous square iterative roots but having such a root with the Darboux property. Sufficient conditions are presented to ensure that for a continuous f any Nth iterative root with the Darboux property will also be continuous.

11.10.9. Further results on convex and/or C^1 iterative roots (see Section 11.4) are due to M. Kuczma [31], M. Kuczma– A. Smajdor [2], [3], A. Smajdor [1], M. C. Zdun [9], [13]; see also Reznick [1] and Bratman [1]. Approximate iterative roots have been investigated by M. Rożnowski [1]. Analytic, in particular, entire square roots of complex functions were considered by I. N. Baker [1], [2]. The so-called *regular iterative roots* are examined in Kuczma–A. Smajdor [3], Kuczma [50] and Zdun [16].

11.10.10. Regarding invariant curves we should emphasize that our treatment was extremely modest not only because of the heavy assumptions on the given transformation but also because of disregarding often quite simple curves omitting the origin. There are a lot of invariant curves, even with respect to linear transformations of the plane, that do not pass through the origin: take $F(x, y) := (\frac{1}{2}x, 2y)$, $(x, y) \in \mathbb{R}^2$ and all hyperbolas $\{(x, y) \in \mathbb{R}^2 : xy = c \neq 0\}$.

11.10.11. Continuous solutions of the equation (of invariant curves)

$$\varphi(f(x, \varphi(x))) = \gamma\varphi(x)$$

have been found by D. Brydak [2], [5], [6] under the assumption that f is continuous and strictly monotonic in either variable (for a related result see also Dhombres [1]).

11.10.12. The existence and uniqueness of Lipschitzian solutions of the pretty general equation

$$\varphi(x) = h(x; \varphi(f_1(x, \varphi(x))), \ldots, \varphi(f_n(x, \varphi(x))))$$

as well as the continuous dependence on given functions have been established (under various assumptions) by S. Czerwik [16] (see also Adamczyk [4]).

11.10.13. The function $\varphi = \text{sgn}$ (sgn $0 := 0$) shows that the assumption of the Darboux property for $\text{id} + \varphi$ in Theorem 11.9.4 is essential. Recently J. Smítal [2] has shown that there exists a nonconstant Darboux solution of Euler's functional equation (11.9.16):

$$\varphi(x + \varphi(x)) = \varphi(x)$$

on the whole real line $\mathbb{R}$. This disproves the natural conjecture that the assumption just mentioned might be replaced by the requirement that φ itself be a Darboux function.

11.10.14. A number of other functional equations containing superpositions of the unknown function have been investigated by various authors. See, in particular, Adamczyk [2]–[4], Bajraktarević [6]–[8], [10], Baron [2], Choczewski–Kuczma [5], Czerwik [16], [17], Dhombres [1], Kuczma [5(a)], Kwapisz [1], Nawrocki [1], Rice [1], Targoński [1]; see also Kuczma [26, Ch. XV, §7] and the references quoted therein.

12

Linear iterative functional inequalities

12.0 Introduction

If we replace an iterative equation by the corresponding inequality, then we certainly may expect a larger set of solutions. As we have seen in previous chapters, the equations themselves often have solutions that depend on an arbitrary function. Thus the results stating the uniqueness of solutions of an inequality are very few. As a rule, such results concern a system of simultaneous inequalities, e.g., one may ask for a solution of an inequality to be nonnegative. Some theorems of this kind are presented in Sections 12.1 and 12.8.

As the existence of solutions in most cases is obvious, in the remaining sections we pay attention mainly to assertions in which solutions of the inequality in question are:

(1) expressed by solutions of simpler inequalities,
(2) estimated by solutions of the corresponding equation,
(3) examined when $x = 0$ or $x \to 0$.

Results of type (1) will be referred to as *representation theorems*, and those of type (2) as *comparison theorems*.

In this chapter we are dealing exclusively with continuous solutions of linear inequalities. We start considering functions u satisfying

$$u(x) \leqslant u(f(x))$$

which are called $\{f\}$-*monotonic* (in the special case $f(x) = x + c$ this notion was introduced by J. Anastassiadis [1]). Inequalities of first order are discussed in Sections 12.2–12.6. The basic reference is the comprehensive work of D. Brydak [8]. Next come a second-order inequality with constant coefficients (Section 12.7) and some inequalities of infinite order (Section 12.8).

Inequalities almost always will be accompanied by associated equations. To make the distinction among their solutions easier, in Sections 12.2–12.6 we accept the following conventions: y denotes solutions of inhomogeneous inequalities, z those of homogeneous ones, whereas φ stands for solutions of equations. Thus we write, e.g.,

$$y(x) \leqslant g(x)y(f(x)) + h(x), \tag{12.0.1}$$

and

$$z(x) \leqslant g(x)z(f(x)),$$

but

$$\varphi(x) = g(x)\varphi(f(x)) + h(x).$$

Moreover, the domain of definition of solutions, both of inequalities and of equations, considered in Sections 12.2–12.7 is always denoted by X, whereas the range is contained in $\mathbb{R}$. We also use the abbreviation 'CS' for 'continuous solution in X'. Therefore, e.g., the expression 'CS y of (12.0.1)' is to be understood as 'continuous solution $y: X \to \mathbb{R}$ of inequality (12.0.1) in X'.

Finally, if real functions φ and ψ are defined on X, then we write '$\varphi = \psi$ in A' iff '$\varphi(x) = \psi(x)$ for $x \in A \subset X$', and analogously for '$<, \leqslant, >, \geqslant$' in place of '$=$'.

12.1 {f}-monotonic functions

The inequality

$$u(x) \leqslant u(f(x)) \tag{12.1.1}$$

is the simplest linear homogeneous iterative inequality.

Let X be an arbitrary set, and $f: X \to X$ an arbitrary function. We accept the following (Kuczma [2]).

Definition 12.1.1. The functions $u: X \to \mathbb{R}$ satisfying (12.1.1) are called {f}-*increasing*. Similarly, the functions $u: X \to \mathbb{R}$ satisfying (12.1.1) but with the inequality sign reversed are called {f}-*decreasing*.

If $X = [0, a|$, $0 < f(x) < x$ in $X \setminus \{0\}$, then every decreasing function is {f}-increasing and every increasing function is {f}-decreasing, but there exist also {f}-monotonic functions which are not monotonic.

In some cases only constant functions are jointly {f_1}-decreasing and {f_2}-increasing. For the case where $f_1(x) = x + p$, $f_2(x) = x + q$, we have (Montel [2], Popoviciu [2]) the following.

Theorem 12.1.1. Let $p, q \in \mathbb{R}$ and assume that p/q is irrational and positive. Then the only functions $w: \mathbb{R} \to \mathbb{R}$ satisfying the inequalities

$$w(x + q) \leqslant w(x) \leqslant w(x + p) \tag{12.1.2}$$

and continuous at a point are $w = \text{const}$.

Proof. Let $w: \mathbb{R} \to \mathbb{R}$ satisfy (12.1.2) and be continuous at $x_0 \in \mathbb{R}$. Take an arbitrary real x and suppose that $w(x) > w(x_0)$. There is a $\delta > 0$ such that $w(x) > w(t)$ for $|t - x_0| < \delta$. We can find $m, n \in \mathbb{N}$ to have $|x - x_0 - mq + np| < \delta$. Hence $w(x) > w(x - mq + np)$, whereas by (12.1.1) we have $w(x - mq + np) \geqslant w(x + np) \geqslant w(x)$, a contradiction. Similarly, $w(x) < w(x_0)$ is also impossible. Thus $w(x) = w(x_0)$ for arbitrary real x, whence w is a constant function. ■

We also have a result (Howroyd [4]) for the more general system

$$w(f_1(x)) \leqslant w(x) \leqslant w(f_2(x)). \tag{12.1.3}$$

Theorem 12.1.2. *Let* $f_1, f_2: [a, b) \to [a, b)$, $-\infty < a < b \leqslant \infty$, *be continuous functions. Assume that for every nonempty open interval* $J \subset [a, b)$ *there exist* $i, j \in \mathbb{N}$ *such that* $f_1^i(a) \in f_2^j(J)$ *and at least one of the following conditions holds:*

(i) *there are* $m, n \in \mathbb{N}$ *such that* $f_2^m(a) \in f_1^n(J)$,
(ii) *the function* f_2 *is a bijection and* $\lim_{n \to \infty} f_2^{-n}(x) = a$ *for* $x \in [a, b)$.

If $w: [a, b) \to \mathbb{R}$ *is a continuous solution of system* (12.1.3) *in* $[a, b)$, *then* w *is a constant function.*

Proof. Without loss of generality it may be assumed that $w(a) = 0$. Iterating the first inequality in (12.1.3) and then setting $x = a$, we get $w(f_1^i(a)) \leqslant 0$ for $i \in \mathbb{N}$. This implies $w(x) \leqslant 0$ everywhere. Indeed, if we had $w(x_0) > 0$ for an $x_0 \in (a, b)$, then w would be positive on an open interval $J \subset [a, b)$. By the inequality $w(f_2(x)) \geqslant w(x)$ we would have $w(x) > 0$ on $f_2^j(J)$, $j \in \mathbb{N}$. But we have assumed that at least one $f_1^i(a)$ lies in at least one of the intervals $f_2^j(J)$, hence a contradiction.

Now, if (i) is assumed, we obtain analogously $w(x) \geqslant 0$, whereas if (ii) holds, then $w(x) \geqslant w(f_2^{-n}(x))$, $n \in \mathbb{N}$, and, on letting $n \to \infty$, we again have $w(x) \geqslant 0$. Thus in both cases $w = 0$. ■

Comments. Some systems more general than (12.1.3) are also considered in Howroyd [4]. For further results on $\{f\}$-monotonic and/or $\{f\}$-convex functions see Brydak [1], [3], Kuczma [9], [26, pp. 126–7], Montel [2].

12.2 Inequalities in the uniqueness case for associated equations

In this section we begin a study of continuous solutions y, resp. z, of the inhomogeneous inequality

$$y(x) \leqslant g(x)y(f(x)) + h(x) \tag{12.2.1}$$

and the homogeneous one

$$z(x) \leqslant g(x)z(f(x)). \tag{12.2.2}$$

12.2A. Comparison theorems

We assume that

(i) $X = [0, a|, 0 < a \leqslant \infty,$

(ii) $f: X \to X$, $g: X \to \mathbb{R}$ and $h: X \to \mathbb{R}$ are continuous, $g(x) > 0$ and $0 < f(x) < x$ in $X \setminus \{0\}$.

Write

$$G_n(x) = \prod_{i=0}^{n-1} g(f^i(x)), \qquad n \in \mathbb{N}_0. \tag{12.2.3}$$

We consider the following cases.

(iii) $G(x) := \lim_{n \to \infty} G_n(x)$ exists, is continuous and positive in X.
(iv) $\lim_{n \to \infty} G_n(x) = 0$ in $X \setminus \{0\}$.

Assumptions (i)–(iii) imply that $g(0) = 1$, whence also $G(0) = 1$. Moreover, the functional equation associated with (12.2.2),

$$\varphi(x) = g(x)\varphi(f(x)), \tag{12.2.4}$$

has a unique one-parameter family of CSs φ given by the formula

$$\varphi(x) = cG(x), \tag{12.2.5}$$

where $c \in \mathbb{R}$ is an arbitrary constant (a parameter) (rewrite (12.2.4) as $\varphi \circ f = \varphi/g$ and apply Theorem 3.1.2). Clearly $c = \varphi(0)$.

Assumptions (i), (ii), (iv) imply that $\varphi = 0$ is the only CS of equation (12.2.4) (see Theorem 3.1.5).

For the homogeneous inequality we obtain the following.

Theorem 12.2.1. *Let hypotheses* (i), (ii) *be satisfied.*
   (a) *If* (iii) *holds, then for every CS z of* (12.2.2) *there exist CSs φ of* (12.2.4) *such that $z \leqslant \varphi$ in X. They are given by* (12.2.5), *with $c \geqslant z(0)$.*
   (b) *If* (iv) *holds, then for every CS z of* (12.2.2) *we have $z \leqslant 0$.*

Proof. By induction we get from (12.2.2)

$$z(x) \leqslant G_k(x)z(f^k(x)), \qquad k \in \mathbb{N}, x \in X.$$

On letting $k \to \infty$ we see that $z(x) \leqslant G(x)z(0)$ in X, if (iii) is assumed; and $z \leqslant 0$ in $X \setminus \{0\}$, if (iv) holds. In the latter case $z(0) \leqslant 0$ by the continuity of z. ∎

The following uniqueness result is a simple consequence of Theorem 12.2.1.

Corollary 12.2.1. *Let hypotheses* (i), (ii) *and* (iii) *or* (iv) *be satisfied and let z be a nonnegative CS of* (12.2.2). *Then $z = 0$ in X, provided $z(0) = 0$ if* (iii) *holds.*

There is an obvious correspondence between solutions of (12.2.1) and (12.2.2) via those of the inhomogeneous equation

$$\varphi(x) = g(x)\varphi(f(x)) + h(x). \tag{12.2.6}$$

Lemma 12.2.1. *Let hypotheses* (i), (ii) *be satisfied, and suppose that equation* (12.2.6) *has a CS* φ. *A function y is a CS of inequality* (12.2.1) *if and only if the function* $z = y - \varphi$ *is a CS of inequality* (12.2.2).

Under our assumptions equation (12.2.6) need not have continuous solutions in X. Assume it has one. Then in case (iii) there is a unique one-parameter family of CSs φ of (12.2.6), viz. (see Theorem 3.1.6)

$$\varphi(x) = cG(x) + \sum_{n=0}^{\infty} G_n(x)h(f^n(x)), \qquad c \in \mathbb{R}. \tag{12.2.7}$$

We have $c = \varphi(0)$, since in view of (12.2.6) $h(0) = 0$ (as $g(0) = G(0) = 1$). On the other hand, in case (iv) the CS φ of equation (12.2.6) is unique (see Theorem 3.1.5).

Lemma 12.2.1 and Theorem 12.2.1 yield the following comparison theorem for the inhomogeneous inequality (12.2.1).

Theorem 12.2.2. *Let hypotheses* (i), (ii) *be satisfied and let y be a CS of inequality* (12.2.1). *Suppose that equation* (12.2.6) *has a CS* φ.

(a) *If* (iii) *holds and* $y(0) \leqslant \varphi(0)$, *then* $y \leqslant \varphi$ *in X*.

(b) *If* (iv) *holds, then* $y \leqslant \varphi$ *in X*.

Remark 12.2.1. In case (a) of Theorem 12.2.1 (resp. 12.2.2), the solution φ of (12.2.6) such that $c = z(0)$ in (12.2.5) (resp. $\varphi(0) = y(0)$) is the minimal one for which $y \leqslant \varphi$ in X.

By contraposition we obtain from Theorem 12.2.2(a) the following.

Theorem 12.2.3. *Let hypotheses* (i)–(iii) *be satisfied, and let y and* φ *be CSs of* (12.2.1) *and* (12.2.6), *respectively. If* $y(x_0) > \varphi(x_0)$ *for an* $x_0 \in X$, *then* $y(0) > \varphi(0)$.

12.2B. Representation theorems

Let conditions (i)–(iii) be fulfilled, and let $z \colon X \to \mathbb{R}$ be any function. Put

$$u(x) := z(x)/G(x).$$

It is easily seen that z is a CS of (12.2.2) if and only if u is a continuous $\{f\}$-increasing function (Definition 12.1.1). Thus we have the following representation theorem.

Theorem 12.2.4. *Let hypotheses* (i)–(iii) *be satisfied. Then the general CS z of inequality* (12.2.2) *is given by*

$$z(x) = u(x)G(x),$$

where u is an arbitrary continuous $\{f\}$-*increasing function.*

Theorem 12.2.4 and Lemma 12.2.1 imply the following representation theorem for continuous solutions of inequality (12.2.1).

Theorem 12.2.5. *Let hypotheses* (i)–(iii) *be satisfied, and suppose that equation* (12.2.6) *has continuous solutions. Then the general CS y of inequality* (12.2.1) *is given by*

$$y(x) = u(x)G(x) + \sum_{n=0}^{\infty} G_n(x)h(f^n(x)),$$

where u is an arbitrary continuous $\{f\}$*-increasing function.*

Remark 12.2.2. In the case where (iv) is assumed we have a representation theorem neither for CSs of (12.2.2) nor for those of (12.2.1).

Comments. The results in this section are essentially due to D. Brydak [8].

12.3 Asymptotic behaviour of nonnegative CSs y of the inhomogeneous inequality

In order to investigate asymptotic properties of CSs y of the inequality

$$y(x) \leqslant g(x)y(f(x)) + h(x) \qquad (12.3.1)$$

we need CSs φ of the equation

$$\varphi(x) = g(x)\varphi(f(x)) + h(x). \qquad (12.3.2)$$

We want to prove that the existence of a CS φ of (12.3.2) satisfying $\varphi(0) = 0$ is ensured by the following conditions.

(i) $X = [0, a|, 0 < a \leqslant \infty$.
(ii) $f: X \to X$, $g: X \to \mathbb{R}$ and $h: X \to \mathbb{R}$ are continuous, $0 < f(x) < x$ and $0 < g(x) < 1$ in $X \setminus \{0\}$; $h(x) \geqslant 0$ in X, and

$$h(x) = o(1 - g(x)), \qquad x \to 0. \qquad (12.3.3)$$

(iii) $\lim_{n \to \infty} G_n(x) = 0$ in $X \setminus \{0\}$, where G_n are given by (12.2.3).
(iv) The function $A: X \setminus \{0\} \to \mathbb{R}$ defined as

$$A(x) := \frac{h(x)}{1 - g(x)}$$

is monotonic in a vicinity $V \subset X$ of zero.

12.3A. Some properties of y

The assumptions we accepted imply that any nonnegative CS y of (12.3.1) satisfies $y(0) = 0$ and $y(x) \leqslant A(x)$ in V. We start by proving these facts.

Lemma 12.3.1. *Let hypotheses* (i)–(iv) *be satisfied. If* y *is a nonnegative CS of* (12.3.1), *then* $y(0) = 0$.

Proof. Suppose that $c = y(0) > 0$. Then we have

$$0 < c_1 < y(x) < c_2 \qquad \text{for } x \in (0, \delta) \tag{12.3.4}$$

with suitable c_1, c_2 and $\delta > 0$. Given an r, $0 < r < 1$, we may assume that δ is sufficiently small that, by (12.3.3),

$$h(x) \leqslant r c_1 (1 - g(x)), \qquad x \in (0, \delta).$$

Using (12.3.1) and this relation we get for $x \in (0, \delta)$

$$\frac{y(x)}{y(f(x))} \leqslant g(x) + \frac{h(x)}{y(f(x))} \leqslant g(x) + c_1^{-1} h(x) \leqslant 1 - (1 - r)(1 - g(x))$$

and, for $n \in \mathbb{N}$,

$$\frac{y(x)}{y(f^n(x))} = \prod_{i=0}^{n-1} \frac{y(f^i(x))}{y(f^{i+1}(x))} \leqslant \prod_{i=0}^{n-1} [1 - (1 - r)(1 - g(f^i(x)))].$$

Hence, on letting $n \to \infty$, we obtain

$$y(x) \leqslant c \prod_{i=0}^{\infty} [1 - (1 - r)(1 - g(f^i(x)))]. \tag{12.3.5}$$

Condition (iii) implies the divergence of the series with the terms $(1 - g(f^i(x)))$, and hence also of that with the terms $(1 - r)(1 - g(f^i(x)))$. Consequently the infinite product in (12.3.5) is zero, which contradicts relation (12.3.4). ∎

Lemma 12.3.3. *Let hypotheses* (i)–(iv) *be satisfied. If* y *is a nonnegative CS of* (12.3.1), *then* $y(x) \leqslant A(x)$ *for* $x \in V$.

Proof. For an indirect proof, suppose that $y(x_0) > A(x_0)$ for an $x_0 \in V$ and write $x_n = f^n(x_0)$, $n \in \mathbb{N}$. The sequence (x_n) is strictly decreasing and tends to zero. Thus $x_n \in V$ for $n \in \mathbb{N}_0$. Since A is increasing in V, as follows from (ii) and (iv), we have

$$A(x_{n+1}) \leqslant A(x_n), \qquad n \in \mathbb{N}_0. \tag{12.3.6}$$

We shall show that

$$y(x_{n+1}) > y(x_n) > A(x_n), \qquad n \in \mathbb{N}_0, \tag{12.3.7}$$

which in the limit yields a contradiction with Lemma 12.3.1:

$$y(0) > y(x_0) > A(x_0) \geqslant 0.$$

To prove (12.3.7), observe first that (12.3.1) and the definition of A imply

$$y(x_{n+1}) - y(x_n) = y(f(x_n)) - y(x_n) \geqslant \frac{y(x_n) - h(x_n)}{g(x_n)} - y(x_n)$$

$$= (y(x_n) - A(x_n)) \frac{1 - g(x_n)}{g(x_n)}.$$

Thus $y(x_{n+1}) > y(x_n)$ whenever $y(x_n) > A(x_n)$. Because of $y(x_0) > A(x_0)$, we get (12.3.7) for $n = 0$. Suppose that (12.3.7) holds for an $n \geqslant 0$. Then, by (12.3.6), $y(x_{n+1}) > y(x_n) > A(x_n) \geqslant A(x_{n+1})$. Consequently, $y(x_{n+2}) > y(x_{n+1})$, and (12.3.7) follows by induction. ∎

12.3B. Unique solution of the associated equation

Now we can prove the announced existence of the continuous solution of (12.3.2). (In this connection see also Theorem 3.6.4 (case (C')).)

Theorem 12.3.1. *Let hypotheses* (i)–(iv) *be satisfied. Then equation* (12.3.2) *has a unique CS* φ. *This solution is given by*

$$\varphi(x) = \sum_{n=0}^{\infty} G_n(x)h(f^n(x)), \tag{12.3.8}$$

and fulfils the condition $\varphi(0) = 0$ *and the inequalities*

$$h(x) \leqslant \varphi(x) \leqslant A(x) \qquad \text{for } x \in V. \tag{12.3.9}$$

Proof. Existence. Write

$$\varphi_k(x) = \sum_{n=0}^{k-1} G_n(x)h(f^n(x)), \qquad k \in \mathbb{N}.$$

Thus $\varphi_k \colon X \to \mathbb{R}$ are nonnegative and continuous, $\varphi_{k+1}(x) \geqslant \varphi_k(x) \geqslant \varphi_1(x) = h(x)$, and they satisfy inequality (12.3.1):

$$h(x) + g(x)\varphi_k(f(x)) = \varphi_{k+1}(x) \geqslant \varphi_k(x). \tag{12.3.10}$$

By Lemma 12.3.2 we have $h(x) \leqslant \varphi_k(x) \leqslant A(x)$ for $x \in V$, $k \in \mathbb{N}$, whence $\lim_{k \to \infty} \varphi_k(x) = \varphi(x)$ exists for $x \in V$ and satisfies (12.3.9). It follows by (12.3.10) that if $\varphi_k(f(x))$ converges to a finite limit then so does $\varphi_k(x)$. Thus series (12.3.8) converges in the whole of X. Since (12.3.3) implies $h(0) = 0$, the terms of the series (12.3.8) at $x = 0$ are all zero, and $\varphi(0) = 0$ follows. Letting $k \to \infty$ in (12.3.10) we see that φ is a solution to (12.3.2).

Continuity. Let $\omega(x)$ denote the oscillation of φ at x. Since $g < 1$, (12.3.2) yields $\omega(f(x)) = \omega(x)/g(x) > \omega(x)$ for $x \in X \setminus \{0\}$. If we had $\omega(x_0) > 0$ for an $x_0 \in X \setminus \{0\}$, then we would get $\omega(f^n(x_0)) > \omega(x_0)$ for $n \in \mathbb{N}$, contrary to $\lim_{n \to \infty} \omega(f^n(x_0)) = \lim_{x \to 0} \omega(x) = 0$ resulting from (12.3.9). Thus $\omega(x) = 0$ in $X \setminus \{0\}$, i.e. φ is continuous in $X \setminus \{0\}$. Relation (12.3.3) implies $\lim_{x \to 0} A(x) = 0$ and the continuity of φ at 0 follows from (12.3.9) and by $\varphi(0) = 0$.

The uniqueness of φ is a consequence of Theorem 3.1.5. ∎

12.3C. Asymptotic behaviour of y

Theorem 12.2.2(b) implies that if y is a nonnegative CS of inequality (12.3.1) then

$$y(x) = O(\varphi(x)), \qquad x \to 0, \tag{12.3.11}$$

where φ is given by (12.3.8). Hence we get in virtue of Theorem 12.3.1 the following.

Theorem 12.3.2. *Let hypotheses* (i)–(iv) *be satisfied. If* y *is a nonnegative CS of inequality* (12.3.1), *then*

$$y(x) = O(A(x)), \qquad x \to 0. \tag{12.3.12}$$

According to (12.3.9), estimate (12.3.11) is better than (12.3.12). In fact, (12.3.11) is sharp. However, since the function A is much simpler than φ, (12.3.12) may be much more convenient to use. For examples see Note 12.9.5. Here we supply a condition under which φ and A are asymptotically equivalent at zero.

Theorem 12.3.3. *Let hypotheses* (i)–(iv) *be satisfied. Moreover, assume that* h *is positive and* g *is decreasing in* V, *and*

$$1 - p(x) = o(1 - g(x)), \qquad x \to 0,[1] \tag{12.3.13}$$

where

$$p(x) := \inf_{t \in (0,x]} \frac{h(f(t))}{h(t)}.$$

The unique CS φ *of equation* (12.3.2) *satisfies*

$$\lim_{x \to 0} \frac{\varphi(x)}{A(x)} = 1. \tag{12.3.14}$$

Proof. The function φ is given by (12.3.8). We have the following estimates of the terms of series (12.3.8) for $x \in V$:

$$\frac{h(f^n(x))}{h(x)} = \prod_{i=0}^{n-1} \frac{h(f^{i+1}(x))}{h(f^i(x))} \geqslant (p(x))^n, \qquad n \in \mathbb{N}_0,$$

and

$$G_n(x) = \prod_{i=0}^{n-1} g(f^i(x)) \geqslant (g(x))^n, \qquad n \in \mathbb{N}_0.$$

Consequently, by (12.3.13), which yields $p(x)g(x) < 1$ (for x small enough)

$$\frac{\varphi(x)}{A(x)} = (1 - g(x)) \sum_{n=0}^{\infty} G_n(x) \frac{h(f^n(x))}{h(x)} \geqslant \frac{1 - g(x)}{1 - p(x)g(x)}.$$

Since (12.3.13) implies $1 - p(x)g(x) = (1 - g(x))(1 + o(1))$, $x \to 0$, we have $\liminf_{x \to 0}[\varphi(x)/A(x)] \geqslant 1$, whereas by (12.3.9) $\limsup_{x \to 0}[\varphi(x)/A(x)] \leqslant 1$, and (12.3.14) follows. ■

Comments. This section is related to Section 3.3 in which asymptotic properties of solution of linear equations are discussed. For the results presented here see Kuczma [47].

---

[1] Condition (12.3.13) is essential; see Note 12.9.6.

12.4 Regular solutions of the homogeneous inequality

Following D. Brydak [7], [8] we consider the inequality

$$z(f(x)) \leq g(x)z(x) \tag{12.4.1}$$

in the case where the continuous solution of the associated functional equation

$$\varphi(f(x)) = g(x)\varphi(x) \tag{12.4.2}$$

depends on an arbitrary function.

We make the following assumptions.

(i) $X = [0, a|, 0 < a \leq \infty$.
(ii) $f: X \to X$, $g: X \to \mathbb{R}$ are continuous, $g(x) > 0$ and $0 < f(x) < x$ in $X \setminus \{0\}$, and f is strictly increasing.
(iii) $\lim_{n \to \infty} G_n(x) = 0$ a.u. in $X \setminus \{0\}$, where G_n are given by (12.2.3).[1]

Condition (iii) guarantees that equation (12.4.2) has a CS φ depending on an arbitrary function (Theorem 3.1.3) and that there exist continuous solutions of (12.4.2) that do not vanish in $X \setminus \{0\}$ (compare also Lemma 3.1.2). Given an $x_0 \in X \setminus \{0\}$, every continuous function $\varphi_0: [f(x_0), x_0] \to \mathbb{R}$, fulfilling the condition

$$\varphi_0(f(x_0)) = g(x_0)\varphi_0(x_0), \tag{12.4.3}$$

can be uniquely extended onto X to a CS φ of (12.4.2). Moreover, $\varphi(0) = 0$ for any CS φ of (12.4.2) (see Corollary 3.1.1).

12.4A. Estimates

Finding upper bounds satisfying (12.4.2), for a CS z of (12.4.1) is simple. First iterate (12.4.1) to get $z(f^n(x)) \leq G_n(x)z(x)$, then let $n \to \infty$ to arrive at the following.

Lemma 12.4.1. *Let hypotheses* (i)–(iii) *be satisfied. If z is a CS of* (12.4.1), *then $z(0) \leq 0$.*

Further, it is clear that if $X = [0, a]$, then to a given CS z of (12.4.1) we can always find a CS φ of (12.4.2) such that $z \leq \varphi$. It is enough to take $x_0 = a$ and an arbitrary continuous function $\varphi_0: [f(a), a] \to \mathbb{R}$ satisfying (12.4.3) and such that $\varphi_0(x) \geq z(x)$ in $[f(a), a]$. It is easily seen that the extension φ to φ_0 to a CS of (12.4.2) in X satisfies the inequality $\varphi(x) \geq z(x)$ in X.

On the other hand, a CS φ of (12.4.2) such that $\varphi(x) \leq z(x)$ in X need not exist. For instance, let $g \leq 1$ in X and $z = \text{const} < 0$. Then z is a CS of (12.4.1), and every CS φ of (12.4.2) satisfies the condition $\varphi(x) > z(x)$ in a neighbourhood of the origin, since $\varphi(0) = 0 > z(0)$.

---

[1] It should be noted that, because of the form (12.4.1) of the inequality, the sequence (G_n) is now the reciprocal of (12.2.3), which we have used in Sections 12.2 and 12.3.

However, if a CS φ of (12.4.2) such that $\varphi \leqslant z$ does exist, then we are able to say something more.

Theorem 12.4.1. *Let hypotheses* (i)–(iii) *be satisfied, let z be a CS of* (12.4.1), *and suppose that there exists a CS $\bar\varphi$ of* (12.4.2) *such that $\bar\varphi \leqq z$. Then the sequence with the terms*

$$Z_n(x) = z(f^n(x))/G_n(x), \qquad x \in X \setminus \{0\}, \; n \in \mathbb{N}, \tag{12.4.4}$$

converges in $X \setminus \{0\}$, and the function

$$\varphi_z(x) = \begin{cases} \lim_{n \to \infty} Z_n(x) & \text{for } x \in X \setminus \{0\}, \\ 0 & \text{for } x = 0, \end{cases} \tag{12.4.5}$$

is a solution of (12.4.2) *satisfying*

$$\varphi_z \leqq z \qquad \text{in } X. \tag{12.4.6}$$

Moreover, φ_z is upper semicontinuous in X and continuous at $x = 0$, and is the maximal solution of (12.4.2) *that estimates z from below.*

Proof. The sequence (12.4.4) is decreasing and bounded below. For (12.4.4) and (12.4.1) yield

$$Z_{n+1}(x) - Z_n(x) = \frac{z(f^{n+1}(x))}{G_{n+1}(x)} - \frac{z(f^n(x))}{G_n(x)}$$

$$= (z(f^{n+1}(x)) - g(f^n(x)))/G_{n+1}(x) \leqq 0,$$

whenever $x \in X \setminus \{0\}$ and $n \in \mathbb{N}$. In turn, as $\bar\varphi \leqq z$,

$$Z_n(x) \geqq \frac{\bar\varphi(f^n(x))}{G_n(x)} = \bar\varphi(x), \tag{12.4.7}$$

by (12.4.2). Thus the sequence (Z_n) converges in $X \setminus \{0\}$.

To prove (12.4.6), note that $\varphi_z \leqq Z_n \leqq Z_1 \leqq z$ in $X \setminus \{0\}$. Further, $\bar\varphi(0) \leqq z(0) \leqq 0$, by the assumption and Lemma 12.4.1. But $\bar\varphi(0) = 0$, as follows from Corollary 3.1.1. Hence $z(0) = 0$, and (12.4.6) is proved.

The relation $g(x)Z_n(x) = Z_{n-1}(f(x))$, resulting from (12.4.4), shows that φ_z satisfies equation (12.4.2) in $X \setminus \{0\}$. Since $\varphi_z(0) = 0$, (12.4.2) holds for $x = 0$ as well.

The function φ_z is upper semicontinuous in $X \setminus \{0\}$ as the limit of a decreasing sequence of continuous functions. The continuity of φ_z at zero results from the estimate $\bar\varphi \leqq \varphi_z \leqq z$ in $X \setminus \{0\}$ (see (12.4.6) and (12.4.7)) and from the condition $\bar\varphi(0) = z(0) = 0$.

Now, if $\hat\varphi$ is another solution of (12.4.2) and $\hat\varphi(x_0) > \varphi_z(x_0)$ for an $x_0 \in X \setminus \{0\}$, then $\hat\varphi(f^n(x_0)) \leqq z(f^n(x_0))$ cannot hold for every $n \in \mathbb{N}$. For it would imply

$$G_n(x_0)\hat\varphi(x_0) = \hat\varphi(f^n(x_0)) \leqq z(f^n(x_0)) = G_n(x_0)Z_n(x_0),$$

whence $\hat{\phi}(x_0) \leq Z_n(x_0)$ for $n \in \mathbb{N}$, and, in the limit, $\hat{\phi}(x_0) \leq \varphi_z(x_0)$, a contradiction. Consequently, there is a $k \in \mathbb{N}$ for which $\hat{\phi}(f^k(x_0)) > z(f^k(x_0))$, i.e. the inequality $\hat{\phi}(x) \leq z(x)$ fails to hold in X. Thus φ_z is maximal. ∎

12.4B. Regular solution

The function φ_z we have just found need not be continuous in X, as may be seen from the following.

Example 12.4.1. Let $X = [0, 1]$, $x_n := 2^{-n}$, $n \in \mathbb{N}_0$. The function

$$z(x) = \begin{cases} x + 2^{n^2 + n}(x - \frac{3}{4}x_n)^n & \text{for } x \in (\frac{3}{4}x_n, x_n], n \in \mathbb{N}_0, \\ x + 2^{n^2 + n - 1}(\frac{3}{4}x_n - x)^n & \text{for } x \in (x_{n+1}, \frac{3}{4}x_n], n \in \mathbb{N}_0, \\ 0 & \text{for } x = 0, \end{cases}$$

is a CS of the inequality

$$z(\tfrac{1}{2}x) \leq \tfrac{1}{2}z(x). \tag{12.4.8}$$

However, the sequence $(Z_n)_{n \in \mathbb{N}}$, $Z_n(x) = 2^n z(2^{-n}x)$, converges to the discontinuous function

$$\varphi_z(x) = \begin{cases} 2x & \text{for } x = x_k, k \in \mathbb{N}_0, \\ x & \text{for remaining } x \in X. \end{cases}$$

Remark 12.4.1. The solution $z(x) = -x^{1/2}$ of (12.4.8) generates the sequence $Z_n(x) = -2^{n/2}x^{1/2}$ which tends in $(0, 1]$ to $-\infty$ as $n \to \infty$. It follows via Theorem 12.4.1 (viz. from the maximality of $\varphi_z = -\infty$) that the inequality $\varphi \leq z$ cannot hold in the whole interval X for any CS φ of the corresponding equation $\varphi(\tfrac{1}{2}x) = \tfrac{1}{2}\varphi(x)$.

We shall adopt the following.

Definition 12.4.1. A CS z of inequality (12.4.1) is called *regular* iff there exists a CS φ of (12.4.2) such that $\varphi \leq z$ in X and the function φ_z defined by (12.4.5) with (12.4.4) is continuous in X.

We are going to show that a regular CS z of (12.4.1) is asymptotically comparable at zero with 'its' φ_z, provided $\varphi_z(x) \neq 0$ in $X \setminus \{0\}$. First we prove the following.

Lemma 12.4.2. *Let the assumptions of Theorem 12.4.1 be fulfilled. If φ is a CS of equation (12.4.2) such that*

$$\varphi(x) > \varphi_z(x) \qquad \text{in } [f(x_0), x_0] \tag{12.4.9}$$

for an $x_0 \in X \setminus \{0\}$, then there exists a $\delta > 0$ such that $\varphi(x) > z(x)$ in $(0, \delta)$.

Proof. We are going to show that there exists a $k \in \mathbb{N}$ such that

$$\varphi(f^k(x)) > z(f^k(x)) \qquad \text{for } x \in [f(x_0), x_0]. \tag{12.4.10}$$

Suppose the contrary. Thus for every $n \in \mathbb{N}$ there is an $x_n \in [f(x_0), x_0]$ such that $\varphi(f^n(x_n)) \leq z(f^n(x_n))$. This inequality remains valid when we replace the nth iterate f^n by f^m with any $m < n$, $m \in \mathbb{N}$. Indeed, if we had $\varphi(f^m(x_n)) > z(f^m(x_n))$ for an $m < n$, then we would get by (12.4.1) and (12.4.2)

$$z(f^n(x_n)) \leq G_{n-m}(x_n)z(f^m(x_n)) < G_{n-m}(x_n)\varphi(f^m(x_n)) = \varphi(f^n(x_n)), \quad (12.4.11)$$

a contradiction. Thus

$$\varphi(f^m(x_n)) \leq z(f^m(x_n)) \qquad \text{for } m \leq n, \, m, \, n \in \mathbb{N}. \quad (12.4.12)$$

Since the interval $[f(x_0), x_0]$ is compact, we can choose from (x_n) a sub-sequence (x_{n_p}) convergent to an element in the interval, say $\bar{x}$. Taking in (12.4.12) $n = n_p$ and letting $p \to \infty$ we obtain in view of the continuity of φ and z that $\varphi(f^m(\bar{x})) \leq z(f^m(\bar{x}))$ for $m \in \mathbb{N}$. But, as has been shown in the proof of Theorem 12.4.1 (maximality of φ_z), this leads to $\varphi(\bar{x}) \leq \varphi_z(\bar{x})$, which contradicts (12.4.9). Relation (12.4.10) has been established.

The argument as in (12.4.11) shows that (12.4.10) holds for k replaced by any $n > k$, $n \in \mathbb{N}$, $x \in [f(x_0), x_0]$. Since f is strictly increasing, this implies that $\varphi(x) > z(x)$ in $(0, f^k(x_0))$. ∎

Now, we are in position to prove the result announced.

Theorem 12.4.2. *Let hypotheses* (i)–(iii) *be satisfied, and let z be a regular CS of* (12.4.1) *such that $\varphi_z(x) \neq 0$ in $X \setminus \{0\}$. Then*

$$\lim_{x \to 0} \frac{z(x)}{\varphi_z(x)} = 1. \quad (12.4.13)$$

Proof. Assume that $\varphi_z > 0$ in $X \setminus \{0\}$. Take an $\varepsilon > 0$ and put $\varphi = (1 + \varepsilon)\varphi_z$. By Lemma 12.4.2 there exists a $\delta > 0$ such that $z < \varphi$ in $(0, \delta)$, whence by (12.4.6) one gets $1 \leq z(x)/\varphi_z(x) \leq 1 + \varepsilon$ in $(0, \delta)$. This implies (12.4.13), as ε can be arbitrarily small. If $\varphi_z < 0$ in $X \setminus \{0\}$, we take $\varphi = (1 - \varepsilon)\varphi_z$, and the proof runs as formerly, with obvious changes. ∎

12.4C. Comparison theorems

Continuous solutions z of inequality (12.4.1) satisfy $z(0) \leq 0$. But any CS φ of equation (12.4.2) satisfies $\varphi(0) = 0$. Thus, in contrast to the situation in Section 12.2 (Theorem 12.2.1), the value $z(0)$ does not suggest any estimate of z by a CS φ of (12.4.2). To get our ends we have to make much stronger assumptions.

Theorem 12.4.3. *Let hypotheses* (i)–(iii) *be satisfied, and let z and φ be CSs of* (12.4.1) *and* (12.4.2), *respectively, $\varphi(x) \neq 0$ in $X \setminus \{0\}$. Suppose that there exists the limit*

$$\lim_{x \to 0} \frac{z(x)}{\varphi(x)} = d. \quad (12.4.14)$$

Then z is regular, $\varphi_z = d\varphi$; and $\varphi \leq z$ whenever either $d \geq 1$ and $\varphi > 0$ in $X \setminus \{0\}$ or $d \leq 1$ and $\varphi < 0$ in $X \setminus \{0\}$.

Proof. By (12.4.4) we have for $x \in X \setminus \{0\}$

$$Z_n(x) = \frac{z(f^n(x))}{\varphi(f^n(x))} \frac{\varphi(f^n(x))}{G_n(x)} = \frac{z(f^n(x))}{\varphi(f^n(x))} \varphi(x).$$

This shows that the limit $\varphi_z(x) = \lim_{n \to \infty} Z_n(x) = d\varphi(x)$ exists and is continuous in $X \setminus \{0\}$, and since $\varphi_z(0) = 0 = \varphi(0)$ (see (12.4.5)), φ_z is continuous in X and $\varphi_z = d\varphi$. Thus z is regular. The inequality $\varphi \leq z$ in the cases specified with $d > 0$ results immediately from (12.4.6). If $d < 0$ and $\varphi < 0$ in $X \setminus \{0\}$, then $\varphi_z = d\varphi > 0 > \varphi$. By this and (12.4.6) we get $\varphi \leq z$. ∎

The next theorem is in a sense converse to Theorem 12.4.2.

Theorem 12.4.4. *Let hypotheses* (i)–(iii) *be satisfied, and let z and φ be CSs of* (12.4.1) *and* (12.4.2), *respectively, $\varphi(x) \neq 0$ in $X \setminus \{0\}$. Further suppose that z is regular, $\varphi_z(x) \neq 0$ in $X \setminus \{0\}$, and*

$$\liminf_{x \to 0} \frac{z(x)}{\varphi(x)} \geq 1 \quad and \quad \varphi > 0 \quad in \ X \setminus \{0\}, \quad (12.4.15)$$

or

$$\limsup_{x \to 0} \frac{z(x)}{\varphi(x)} \leq 1 \quad and \quad \varphi < 0 \quad in \ X \setminus \{0\}. \quad (12.4.16)$$

Then $\varphi \leq z$.

Proof. Assume (12.4.15). Then $\liminf_{x \to 0} [\varphi_z(x)/\varphi(x)] \geq 1$ by Theorem 12.4.2. The function $\omega(x) := \varphi_z(x)/\varphi(x)$ satisfies $\omega(f(x)) = \omega(x)$, whence $\omega(f^n(x)) = \omega(x)$ for $n \in \mathbb{N}$. If we had $\omega(x_0) < 1$ for an $x_0 \in X \setminus \{0\}$, then $\liminf_{x \to 0} \omega(x) \leq \lim_{n \to \infty} \omega(f^n(x_0)) = \omega(x_0) < 1$, a contradiction. Hence $\omega \geq 1$ in $X \setminus \{0\}$, which yields $\varphi_z \geq \varphi$ in X and the theorem results from (12.4.6).

If (12.4.16) holds, the proof is similar. ∎

12.4D. Representation theorems

Regular solutions of inequality (12.4.1) can again be represented as a product of an $\{f\}$-monotonic continuous function u and a CS φ of the associated equation (12.4.2). Two theorems to this effect are presented in this subsection. First we deal with the case where the function φ_z, defined by (12.4.5), does not vanish in $X \setminus \{0\}$.

Theorem 12.4.5. *Let hypotheses* (i)–(iii) *be satisfied. Then the general regular CS z of* (12.4.1) *such that $\varphi_z(x) \neq 0$ in $X \setminus \{0\}$ is given by the formula*

$$z(x) = u(x)\varphi(x), \quad (12.4.17)$$

where φ is an arbitrary CS of (12.4.2) vanishing only at $x = 0$, and u is an arbitrary continuous $\{f\}$- decreasing or $\{f\}$-increasing function (according to the sign of φ), $u(0) = 1$.

Moreover, representation (12.4.17) is unique.

Proof. If z is a regular solution of (12.4.1) and $\varphi_z(x) \neq 0$ in $X \setminus \{0\}$, then φ_z itself is a CS of (12.4.2) that may be used in (12.4.17), and u is given by

$$u(x) = \frac{z(x)}{\varphi_z(x)}, \qquad x \in X \setminus \{0\}; \, u(0) = 1.$$

Theorem 12.4.2 guarantees that u is continuous in X.

Conversely, let z have representation (12.4.17). Then evidently z is a CS of (12.4.1). Moreover, it follows by Theorem 12.4.3 that z is regular and $\varphi_z(x) \neq 0$ for $x \in X \setminus \{0\}$.

The uniqueness of representation (12.4.17) results from Theorem 12.4.3. ∎

The first part of the proof of Theorem 12.4.5 implies the following.

Corollary 12.4.1. *Let hypotheses (i)–(iii) be satisfied. If z is a regular CS of (12.4.1), $\varphi_z(x) \neq 0$ in $X \setminus \{0\}$, then*

$$z(x) = u(x)\varphi_z(x), \qquad x \in X, \tag{12.4.18}$$

where u is a continuous $\{f\}$-decreasing or $\{f\}$-increasing function (according as $\varphi_z > 0$ or $\varphi_z < 0$, in $X \setminus \{0\}$), $u(0) = 1$.

We pass to the case where $\varphi_z = 0$. Let us fix a CS φ_0 of (12.4.2), positive in $X \setminus \{0\}$. (Such a solution does exist in virtue of the remarks at the beginning of the present section.)

Theorem 12.4.6. *Let hypotheses (i)–(iii) be satisfied and let φ_0 be a CS of (12.4.2), $\varphi_0 > 0$ in $X \setminus \{0\}$. Then the general regular CS z of (12.4.1) such that $\varphi_z = 0$ is given by the formula*

$$z(x) = u(x)\varphi_0(x), \tag{12.4.19}$$

where u is an arbitrary continuous $\{f\}$-decreasing function that satisfies $u(0) = 0$.

Proof. Any z representable in form (12.4.19) is clearly a CS of (12.4.1). Moreover, z is regular and $\varphi_z = 0$ by Theorem 12.4.3.

Conversely, suppose that z is a regular CS of (12.4.1) such that $\varphi_z = 0$. Then u defined by (12.4.19) and by $u(0) = 0$ is an $\{f\}$-decreasing function, continuous in $X \setminus \{0\}$. It remains to prove that

$$\lim_{x \to 0} u(x) = \lim_{x \to 0} \frac{z(x)}{\varphi_0(x)} = 0. \tag{12.4.20}$$

The sequence (Z_n) defined by (12.4.4) is decreasing (see the proof of Theorem 12.4.1) and converges to zero in $X\setminus\{0\}$, so the convergence is almost uniform. Fix an $x_0 \in X\setminus\{0\}$ and denote by M the lower bound of φ_0 on $[f(x_0), x_0]$. Given $\varepsilon > 0$, there is an $N \in \mathbb{N}$ such that $|Z_n(x)| < \varepsilon M$ for $x \in [f(x_0), x_0]$ and $n \geq N$. Any point x from $(0, f^N(x_0))$ is an iterate of a point $\bar{x}$ from $[f(x_0), x_0]$, say $x = f^{n(x)}(\bar{x})$, $n(x) \geq N$. Hence

$$\frac{|z(x)|}{\varphi_0(x)} = \frac{|z(f^{n(x)}(\bar{x}))|}{\varphi_0(f^{n(x)}(\bar{x}))} = \frac{|Z_{n(x)}(\bar{x})|}{\varphi_0(\bar{x})} < \varepsilon$$

for $x \in (0, f^N(x_0))$, and (12.4.20) follows. ∎

Remark 12.4.2. It should be stressed that besides the extreme cases covered by Theorems 12.4.5 and 12.4.6 there is also the possibility that φ_z vanishes on a proper subset of $X\setminus\{0\}$.

The following property of regular CSs z of (12.4.1) is a consequence of Corollary 12.4.1 and of the fact that φ_z satisfies (12.4.2).

Theorem 12.4.7. *Let hypotheses* (i)–(iii) *be satisfied. If z is a regular CS of* (12.4.1) *and $\varphi_z(x) \neq 0$ in $X\setminus\{0\}$, then*

$$\lim_{x \to 0} \frac{z(f(x))}{z(x)} = g(0). \tag{12.4.21}$$

A sufficient condition of regularity of z reads as follows.

Theorem 12.4.8. *Let hypotheses* (i)–(iii) *be satisfied. If z is a CS of* (12.4.1), *$z(x) \neq 0$ in $X\setminus\{0\}$, and*

$$\limsup_{x \to 0} \frac{z(f(x))}{z(x)} < g(0), \tag{12.4.22}$$

then z is regular and $\varphi_z = 0$.

Proof. We have, by (12.4.4),

$$Z_n(x) = z(x) \prod_{i=0}^{n-1} \frac{z(f^{i+1}(x))}{z(f^i(x))} \frac{1}{g(f^i(x))} = \prod_{i=0}^{n-1} \tilde{g}(f^i(x)),$$

where $\tilde{g}(x) := z(f(x))/(g(x)z(x)) > 0$. By (12.4.22) there is a $\Theta < 1$ such that $\tilde{g}(x) < \Theta$ in a vicinity of the origin. Thus the sequence (Z_n) tends to zero in $X\setminus\{0\}$, which proves that z is regular and $\varphi_z = 0$. ∎

Remark 12.4.3. Condition (12.4.22) is not necessary for φ_z to vanish identically in X. For instance, the function $z(x) = x \cdot 2^{-1/x}$ is a CS of the inequality

$$z\left(\frac{x}{x+1}\right) \leq \tfrac{1}{2}z(x)$$

in $X = [0, 1]$, and $\varphi_z = 0$ in X. On the other hand, z satisfies (12.4.21). The same example shows that condition (12.4.21) is not sufficient for φ_z to be different from zero in $X \setminus \{0\}$. In fact, (12.4.21) does not even guarantee that z is a regular CS of (12.4.1). The problem will be dealt with again in Section 12.6.

12.5 The inhomogeneous inequality in the nonuniqueness case

Now we consider the inequality

$$y(f(x)) \leqslant g(x)y(x) + h(x) \tag{12.5.1}$$

and the associated equation

$$\varphi(f(x)) = g(x)\varphi(x) + h(x) \tag{12.5.2}$$

in the nonuniqueness case analogous to that we have dealt with in Section 12.4 for the homogeneous inequality (12.4.1). Thus the assumptions we accept are essentially the same as in that section. They are

(i) $X = [0, a|, 0 < a \leqslant \infty$,
(ii) $f: X \to X, g: X \to \mathbb{R}, h: X \to \mathbb{R}$ are continuous, $g(x) > 0$ and $0 < f(x) < x$ in $X \setminus \{0\}$, and f is strictly increasing,
(iii) $\lim_{n \to \infty} G_n(x) = 0$ a.u. in $X \setminus \{0\}$, where G_n is given by (12.2.3).

The results we present here are due to D. Brydak–B. Choczewski [3]. We put

$$h(x; c) := h(x) + c(g(x) - 1).$$

Equation (12.5.2) has CSs φ if and only if there exists a $c \in \mathbb{R}$ such that the sequence $(H_n(x; c))_{n \in \mathbb{N}}$,

$$H_n(x; c) = G_n(x) \sum_{i=1}^{n-1} \frac{h(f^i(x); c)}{G_{i+1}(x)}, \tag{12.5.3}$$

converges to zero a.u. in $X \setminus \{0\}$ (Theorem 3.1.7). If this condition is fulfilled, then the continuous solution of equation (12.5.2) in X depends on an arbitrary function, and $\varphi(0) = c$ for any CS φ of (12.5.2) (see Section 3.1).

12.5A. The best lower bound and the regular solutions

If, for a CS y of (12.5.1), there exists a CS φ of (12.5.2) such that $\varphi \leqslant y$, then we may find among continuous solutions of (12.5.2) the best possible lower bound for y. The proof of the theorem we formulate below differs from that of Theorem 12.4.1 only by details of calculations, and is therefore omitted.

Theorem 12.5.1. *Let hypotheses* (i)–(iii) *be satisfied, let y be a CS of* (12.5.1), *and suppose that there exists a CS φ of* (12.5.2) *such that $\varphi \leqslant y$ in X. Then* the sequence $(Y_n(x))_{n \in \mathbb{N}}$,

$$Y_n(x) = c + (y(f^n(x)) - c - H_n(x; c))/G_n(x), \qquad (12.5.4)$$

$x \in X \setminus \{0\}$, $n \in \mathbb{N}$, where $H(x; c)$ is given by (12.5.3) and $c := \varphi(0)$, converges in $X \setminus \{0\}$, and the function

$$\varphi_y(x) = \begin{cases} \lim_{n \to \infty} Y_n(x) & \text{for } x \in X \setminus \{0\}, \\ c & \text{for } x = 0, \end{cases} \qquad (12.5.5)$$

is a solution of (12.5.2) satisfying

$$\varphi_y \leqslant y \qquad \text{in } X. \qquad (12.5.6)$$

The function (12.5.5) is upper semicontinuous in X and continuous at $x = 0$, and is the maximal solution of (12.5.2) that estimates y from below.

We learn from Subsection 12.4B that φ_y need not be continuous in X. If, for a CS y of inequality (12.5.1) there is a CS φ of (12.5.2) such that $\varphi \leqslant y$ and φ_y is continuous in X, we say that y is *regular* (see Definition 12.4.1).

12.5B. Properties of solutions of the inequality

Theorems 12.4.3, 12.4.4 and Lemma 12.2.1 imply the following two results.

Theorem 12.5.2. *Let hypotheses* (i)–(iii) *be satisfied, and let y be a CS of* (12.5.1), *and φ_0, φ CSs of* (12.5.2), $\varphi(x) \neq \varphi_0(x)$ *in $X \setminus \{0\}$. Suppose that there exists the limit*

$$\lim_{x \to 0} \frac{(y(x) - \varphi_0(x))}{(\varphi(x) - \varphi_0(x))} = d.$$

If $d \geqslant 1$ and $\varphi > \varphi_0$ in $X \setminus \{0\}$, or $d \leqslant 1$ and $\varphi < \varphi_0$ in $X \setminus \{0\}$, then $\varphi \leqslant y$. Moreover, y is regular and its φ_y is $\varphi_0 + d(\varphi - \varphi_0)$.

Theorem 12.5.3. *Let hypotheses* (i)–(iii) *be satisfied, and let y, φ_0, φ be as in Theorem 12.5.2. If y is regular, $\varphi_y(x) \neq \varphi_0(x)$ in $X \setminus \{0\}$, and*

$$\liminf_{x \to 0} \frac{y(x) - \varphi_0(x)}{\varphi(x) - \varphi_0(x)} \geqslant 1, \qquad \varphi > \varphi_0 \text{ in } X \setminus \{0\},$$

or

$$\limsup_{x \to 0} \frac{y(x) - \varphi_0(x)}{\varphi(x) - \varphi_0(x)} \leqslant 1, \qquad \varphi < \varphi_0 \text{ in } X \setminus \{0\},$$

then $\varphi \leqslant y$.

To make sure that Theorem 12.5.3 actually follows from Theorem 12.4.4 and Lemma 12.2.1, we need only to check that if $z = y - \varphi_0$ then $\varphi_z = \varphi_y - \varphi_0$, where φ_z is defined by (12.4.5) and φ_y by (12.5.5). This is so because the sequences generating φ_z (see (12.4.4)) and φ_y (see (12.5.4)) differ just by φ_0.

For first note that

$$\varphi_0(f(x)) - c = g(x)(\varphi_0(x) - c) + h(x; c), \tag{12.5.7}$$

as φ_0 satisfies (12.5.2). Next, iterate (12.5.7) to get

$$\varphi(f^n(x)) - c = G_n(x)(\varphi_0(x) - c) + H_n(x; c), \qquad n \in \mathbb{N}. \tag{12.5.8}$$

By (12.5.8) and (12.5.4) we have

$$z(f^n(x)) = y(f^n(x)) - \varphi_0(f^n(x)) = y(f^n(x)) - c - H_n(x; c) - G_n(x)(\varphi_0(x) - c)$$
$$= G_n(x)(Y_n(x) - \varphi_0(x)),$$

and the equality $Z_n(x) = Y_n(x) - \varphi_0(x)$ follows.

12.5C. Representation theorem

Combining Theorem 12.4.6 with Lemma 12.2.1 we obtain the following result on regular CSs of inequality (12.5.1).

Theorem 12.5.4. *Let hypotheses* (i)–(iii) *be satisfied and let* $\bar{\varphi}_0 > 0$ *in* $X \setminus \{0\}$ *be a fixed CS of the homogeneous equation* (12.4.2). *Then the general regular CS* y *of* (12.5.1) *is given by the formula*

$$y(x) = \varphi(x) + u(x)\bar{\varphi}_0(x), \tag{12.5.9}$$

where φ *is an arbitrary CS of* (12.5.2), *and* u *is an arbitrary continuous* $\{f\}$-*decreasing function,* $u(0) = 0$. *Moreover, for the regular CS* y *of* (12.5.1) *given by* (12.5.9) *we have* $\varphi_y = \varphi$.

12.6 Regular solutions of the homogeneous inequality determined by asymptotic properties

In this section we shall be concerned with continuous solutions of the homogeneous inequality

$$z(f(x)) \leqslant g(x)z(x) \tag{12.6.1}$$

and of the associated equation

$$\varphi(f(x)) = g(x)\varphi(x) \tag{12.6.2}$$

distinguished by a particular asymptotic behaviour at the origin.

The function g is assumed to be a product of functions g_1 and g_2 such that the sequences (G_{n1}) and (G_{n2}), defined by

$$G_{nj}(x) := \prod_{i=0}^{n-1} g_j(f^i(x)), \qquad j = 1, 2, \ n \in \mathbb{N}_0,$$

approach identically zero and nowhere zero limits, respectively.

We assume the following.

(i) $X = [0, a|, \ 0 < a \leqslant \infty$.

(ii) $f: X \to X$, $g_1, g_2: X \to \mathbb{R}$ are continuous on X; $0 < f(x) < x$, $g_1(x) > 0$, $g_2(x) > 0$ in $X \setminus \{0\}$; $g = g_1 g_2$, and f is strictly increasing.

(iii) $\lim_{n \to \infty} G_{n1}(x) = 0$ a.u. in $X \setminus \{0\}$; $\lim_{n \to \infty} G_{n2}(x) = G(x)$ in X, where $G: X \to \mathbb{R}$ is continuous and positive in X.

12.6A. One-parameter family of solutions of the associated equation

We are going to show that regular CSs of inequality (12.6.1) may generate one-parameter families of continuous solutions of equation (12.6.2). Together with (12.6.1) and (12.6.2) we consider the inequality

$$z_1(f(x)) \leqslant g_1(x) z_1(x) \tag{12.6.3}$$

and the equations

$$\varphi_1(f(x)) = g_1(x) \varphi_1(x) \tag{12.6.4}$$
$$\varphi_2(f(x)) = g_2(x) \varphi_2(x). \tag{12.6.5}$$

Theorem 12.6.1. *Let hypotheses* (i)–(iii) *be satisfied, and let* z_1 *be a regular CS of* (12.6.3) *such that* $\varphi_{z_1} > 0$ *in* $X \setminus \{0\}$. *Then for every* $c \in \mathbb{R}$ *there exists exactly one solution* φ *of equation* (12.6.2) *that has the property*

$$\varphi(x) = z_1(x)(c + o(1)), \qquad x \to 0. \tag{12.6.6}$$

This solution is continuous in X, *does not vanish in* $X \setminus \{0\}$ *whenever* $c \neq 0$, *and is given by the formula*

$$\varphi(x) = c \varphi_{z_1}(x) / G(x) \qquad for \ x \in X. \tag{12.6.7}$$

Proof. By Corollary 12.4.1 z_1 has a representation

$$z_1(x) = u_1(x) \varphi_{z_1}(x), \tag{12.6.8}$$

where u_1 is a continuous $\{f\}$-decreasing function such that $u_1(0) = 1$. Function (12.6.7) is a solution of (12.6.2), since φ_{z_1} satisfies equation (12.6.4) and $\varphi_2 = c/G$ satisfies equation (12.6.5) (see Theorem 3.1.2). The function φ is continuous in X and different from zero in $X \setminus \{0\}$, as are φ_{z_1} and G. Thus, by (12.6.8) and $G(0) = 1$ (see Subsection 12.2A),

$$\lim_{x \to 0} \frac{\varphi(x)}{z_1(x)} = \frac{c}{u_1(0) G(0)} = c.$$

We see that φ has property (12.6.6).

In order to prove the uniqueness suppose that we are given a solution φ of (12.6.2) satisfying (12.6.6). Write

$$\hat{\varphi}(x) = \varphi(x) / z_1(x), \qquad x \in X \setminus \{0\}. \tag{12.6.9}$$

Then $\lim_{x \to 0} \hat{\varphi}(x) = c$, and $\hat{\varphi}$ satisfies in $X \setminus \{0\}$ the equation

$$\hat{\varphi}(f(x)) = \hat{g}(x) \hat{\varphi}(x), \tag{12.6.10}$$

where $\hat{g}(x) = g(x) z_1(x) / z_1(f(x)) = g_2(x) u_1(x) / u_1(f(x))$, in view of (12.6.8).

By (12.6.10)

$$\hat{\phi}(f^n(x)) = \hat{\phi}(x) \prod_{i=0}^{n-1} \hat{g}(f^i(x)) = \hat{\phi}(x)u_1(x)G_{n2}(x)/u_1(f^n(x)),$$

for $n \in \mathbb{N}$, whence, on letting $n \to \infty$,

$$\hat{\phi}(x) = \frac{c}{u_1(x)G(x)}. \qquad (12.6.11)$$

Formula (12.6.7) for $x \in X \setminus \{0\}$ results from (12.6.11), (12.6.8) and (12.6.9), and for $x = 0$ $\varphi_{z_1}(0) = 0$ which yields $z_1(0) = 0$ by (12.6.8) and $\varphi(0) = 0$ by (12.6.6). This proves the uniqueness of φ. ∎

12.6B. Regular solutions of the inequality

Since $\varphi_2 = 1/G$ is a CS of (12.6.5), we have $G_{n2}(x) = \varphi_2(f^n(x))/\varphi_2(x)$ and (G_{n2}) converges to G almost uniformly in X (see Theorem 1.2.4). Hence

$$\lim_{n \to \infty} \prod_{i=0}^{n-1} g(f^i(x)) = \lim_{n \to \infty} G_{n1}(x)G_{n2}(x) = 0,$$

a.u. in $X \setminus \{0\}$. Thus hypotheses (i)–(iii) imply hypotheses (i)–(iii) of Section 12.4. Owing to this fact we have the following.

Theorem 12.6.2. *Let hypotheses* (i)–(iii) *be satisfied, and let z_1 be a regular CS of* (12.6.3) *such that $\varphi_{z_1} > 0$ in $X \setminus \{0\}$.*
  (a) *If z is a CS of* (12.6.1) *fulfilling the condition*

$$z(x) = z_1(x)(x + o(1)), \qquad x \to 0, \qquad (12.6.12)$$

with a $c \in \mathbb{R}$, then z is regular, and $\varphi_z(x) \neq 0$ in $X \setminus \{0\}$ if $c \neq 0$, and $\varphi_z = 0$ if $c = 0$.
  (b) *If z is a regular CS of* (12.6.1) *and $\varphi_z(x) \neq 0$ in $X \setminus \{0\}$, then z satisfies* (12.6.6) *with a $c \neq 0$ if and only if φ_z does.*

Proof. (a) By Theorem 12.6.1 equation (12.6.2) has a CS φ such that $\varphi(x) \neq 0$ in $X \setminus \{0\}$ and that (12.6.6) is satisfied with $c = 1$. The latter fact and (12.6.12) imply that $\lim_{x \to 0}[z(x)/\varphi(x)] = c$, whence by Theorem 12.4.3. z is regular and $\varphi_z = c\varphi$.
  (b) The assertion follows from representation (12.4.18) (see Corollary 12.4.1). ∎

12.6C. Special behaviour of given functions

Now we are going to illustrate the applicability of Theorem 12.6.2(a). We replace assumptions (ii) and (iii) by stronger ones.

  (iv) $f: X \to X$ and $g: X \to \mathbb{R}$ are continuous on $X, 0 < f(x) < x$ and $g(x) > 0$ in $X \setminus \{0\}$, and f is strictly increasing.

(v) Either

$$f(x) = sx + O(x^{1+\mu}), \Big\} \qquad x \to 0, \qquad (12.6.13)$$
$$g(x) = b + O(x^p),$$

or

$$f(x) = x - tx^{m+1} + O(x^{m+1+\mu}), \Big\} \qquad x \to 0, \qquad (12.6.14)$$
$$g(x) = 1 - rx^m + O(x^{m+p}),$$

where all the constants occurring are positive and $s < 1$, $b < 1$.

We put

$$p := \begin{cases} \dfrac{\log b}{\log s} & \text{if (12.6.13) holds,} \\[2mm] \dfrac{r}{t} & \text{if (12.6.14) holds.} \end{cases} \qquad (12.6.15)$$

Theorem 12.6.3. *Let hypotheses* (i), (iv), (v) *be satisfied. If z is a CS of* (12.6.1) *fulfilling the condition (with a $c \in \mathbb{R}$)*

$$z(x) = x^p(c + o(1)), \qquad x \to 0, \qquad (12.6.16)$$

then z is regular. Moreover, $\varphi_z(x) \neq 0$ in $X \setminus \{0\}$ if $c \neq 0$, and $\varphi_z = 0$ if $c = 0$.

Proof. Put $g_1(x) = (f(x)/x)^p$ and $g_2(x) = g(x)/g_1(x)$ for $x \in X \setminus \{0\}$, and assign them the values at zero equal to their limits at zero. Then $g = g_1 g_2$ and condition (ii) is fulfilled. Further we have for $x \in X \setminus \{0\}$

$$G_{n1}(x) = \prod_{i=0}^{n-1} (f^{i+1}(x)/f^i(x))^p = (f^n(x)/x)^p$$

so that $\lim_{n \to \infty} G_{n1}(x) = 0$ a.u. in $X \setminus \{0\}$ (Theorem 1.2.4).

It remains to examine the sequence $(G_{n2}(x))$ corresponding to g_2. If (12.6.13) is satisfied, then

$$g_2(x) = 1 + O(x^\sigma), \qquad x \to 0,$$

where $\sigma = \min(\mu, \rho) > 0$, as $s^p = b$ (see (12.6.15)). For every s_1, $s < s_1 < 1$, there is a $\delta > 0$ such that $f(x) \leqslant s_1 x$ in $[0, \delta] \subset X$. Hence it follows that the series $\sum_{n=0}^{\infty} |g_2(f^n(x)) - 1|$ converges a.u. in X, and consequently $G(x) = \prod_{n=0}^{\infty} g_2(f^n(x))$ is continuous and positive in X.

If (12.6.14) is satisfied, then the same properties of (G_{n2}) as shown above are obtained in the proof of Theorem 3.3.2(2) (G_{n2} corresponds to the product $\tilde{G}_n$ there).

Thus in both cases condition (iii) is fulfilled.

Now, $z_1(x) = x^p$ satisfies equation (12.6.4), and hence it is a regular CS of inequality (12.6.3) with $\varphi_{z_1} = z_1 > 0$ in $X \setminus \{0\}$. Thus the theorem results from Theorem 12.6.2(a). ∎

12.6D. Conditions equivalent to regularity

For solutions z of (12.6.1) that behave at the origin like x^q, where q is a nonnegative number, we have a more complete version of Theorems 12.4.7 and 12.4.8. The regularity of solutions of (12.6.1) depends on the relation between q and p, the latter being defined by (12.6.15).

Theorem 12.6.4. *Let hypotheses* (i), (iv), (v) *be satisfied. If z is a CS of* (12.6.1) *having the property*

$$z(x) = x^q(d + o(1)), \qquad x \to 0, \tag{12.6.17}$$

with $q \geqslant 0$, $d \neq 0$, then z is regular if and only if $q \geqslant p$. Moreover, $\varphi_z(x) \neq 0$ in $X \setminus \{0\}$ if $q = p$, whereas $\varphi_z = 0$ if $q > p$.

Proof. If $q \geqslant p$, then z satisfies (12.6.16) with $c = d$ if $q = p$, and with $c = 0$ if $q > p$. Thus the 'if' part of the theorem follows from Theorem 12.6.3.

Conversely, it is enough to prove that if $0 \leqslant q < p$, then z cannot be regular.

First assume (12.6.13) and write $z(x) = x^q \zeta(x)$. In view of (12.6.17), we have $\lim_{x \to 0} \zeta(x) = d \neq 0$, whence

$$\lim_{x \to 0} \frac{z(f(x))}{z(x)} = \lim_{x \to 0} \left(\frac{f(x)}{x} \right)^q \frac{\zeta(f(x))}{\zeta(x)} = s^q. \tag{12.6.18}$$

The situation $0 \leqslant q < p$ may happen only when $d < 0$. For $d > 0$ and the continuity of z imply $z > 0$ in a vicinity of zero, and (12.6.1) yields $s^q \leqslant g(0) = b = s^p$, by (12.6.15). Hence $q \geqslant p$, a contradiction.

Write

$$\hat{g}(x) := \frac{z(f(x))}{g(x) z(x)}, \qquad \text{for } x \in X \setminus \{0\}, \tag{12.6.19}$$

and note that, by (12.6.18), (12.6.13) and (12.6.15), $\lim_{x \to 0} \hat{g}(x) = s^q / b = s^{q-p} > 1$. Now the sequence (12.4.4) generating φ_z becomes

$$Z_n(x) = z(x) \prod_{i=0}^{n-1} \hat{g}(f^i(x)). \tag{12.6.20}$$

Since $d < 0$, we know that $z < 0$ in a vicinity V of zero. Further, $\lim_{i \to \infty} \hat{g}(f^i(x)) > 1$. Therefore sequence (12.6.20) tends (on V) to $-\infty$ as $n \to \infty$ and z is not regular.

Relation (12.6.14) implies the asymptotic property of function (12.6.19)

$$\hat{g}(x) = 1 + (r - qt)x^m + O(x^{m+v}), \qquad x \to 0, \tag{12.6.21}$$

with $v = \min(\mu, m, \rho) > 0$. Since $q < p = r/t$ (see (12.6.15)) we have $r - qt > 0$. Thus (12.6.21) implies $\hat{g}(x) - 1 \geqslant Cx^m$ with a $C > 0$, for $x \in (0, \delta)$, $\delta > 0$. Theorems 1.2.4 and 1.3.5 imply that $f^n(x) \geqslant Dn^{-1/m}$ with a $D > 0$, for n large enough. Hence $\sum_{n=0}^{\infty} (\hat{g}(f^n(x)) - 1) = \infty$, and the sequence $(|Z_n(x)|)$ tends on $(0, \delta)$ to ∞ as n does, since by $d \neq 0$ we may have $z(x) \neq 0$ in $(0, \delta)$.. ∎

Remark 12.6.1. If (12.6.13) is assumed in Theorem 12.6.4, then (12.6.18) shows that there exists the limit $L = \lim_{x \to 0} [z(f(x))/z(x)]$ and $L = s^q$. Thus the assertion of the theorem may be reformulated as follows: z is regular if and only if $L \leqslant g(0)$, and $\varphi_z(x) \neq 0$ in $X \setminus \{0\}$ if $L = g(0)$, whereas $\varphi_z = 0$ if $L < g(0)$. We see that Theorem 12.6.4 is in this case an improved version of Theorems 12.4.7 and 12.4.8.

Remark 12.6.2. In case (12.6.14) we always have $L = 1 = g(0)$, independently of the mutual relation of p and q. Thus (12.4.21) of Theorem 12.4.7 holds independently of whether z is regular or not (and, in the former case, of whether $\varphi_z(x) \neq 0$ in $X \setminus \{0\}$ or $\varphi_z = 0$).

Comments. For all the results discussed in this section see Brydak–Choczewski [1], [4].

12.7 A homogeneous inequality of second order

In this section the functional inequality

$$\varphi(f^2(x)) \leqslant (p + q)\varphi(f(x))) - pq\varphi(x) \tag{12.7.1}$$

is studied after Brydak–Choczewski [2]. Here p and q are some real constants. Our assumptions are the following.

(i) $X = [0, a|, 0 < a \leqslant \infty$,
(ii) $f: X \to X$ is continuous and strictly increasing, $0 < f(x) < x$ in $X \setminus \{0\}$.

The functions φ considered will map X into $\mathbb{R}$, and we shall use the abbreviation 'φ is a CS of ...' for '$\varphi: X \to \mathbb{R}$ is a continuous solution of ... in X'.

12.7A. An equivalent system

Let φ be a CS of (12.7.1). We put

$$z(x) = \varphi(f(x)) - q\varphi(x).$$

Then z is a CS of the inequality

$$z(f(x)) \leqslant pz(x). \tag{12.7.2}$$

Conversely, if z is a CS of inequality (12.7.2), and φ is a CS of the equation

$$\varphi(f(x)) = q\varphi(x) + z(x), \tag{12.7.3}$$

then φ is a CS of inequality (12.7.1). Thus system (12.7.2), (12.7.3) is equivalent to inequality (12.7.1) in the class of functions continuous in X. Whereas we cannot solve system (12.7.2), (12.7.3) in full generality we shall give below some partial results. We are interested in solutions φ of (12.7.1) determining, by (12.7.3), a nonnegative z.

Theorem 3.1.10 implies the following.

Theorem 12.7.1. *Let hypotheses* (i), (ii) *be satisfied, and let* $|q| > 1$. *Then to every nonnegative CS z of* (12.7.2) *there exists exactly one CS φ_0 of* (12.7.1) *satisfying* (12.7.3). *This solution is given by*

$$\varphi_0(x) = -\sum_{n=0}^{\infty} \frac{z(f^n(x))}{q^{n+1}}. \tag{12.7.4}$$

The next theorem concerns the case where $0 < p \leqslant |q| \leqslant 1$.

Theorem 12.7.2. *Let hypotheses* (i), (ii) *be satisfied, and let* $0 < p < |q| \leqslant 1$. *If z is a nonnegative CS of* (12.7.2), *then series* (12.7.4) *converges in X and the φ_0 is a CS of* (12.7.1) *satisfying* (12.7.3). *Moreover, all CSs φ of* (12.7.1) *satisfying* (12.7.3) *are described in Table 12.1.*

Table 12.1

Case	Solution of (12.7.1)	Formula for solutions
$q = -1$	Unique	$\varphi(x) = \varphi_0(x)$
$q = 1$	One-parameter family	$\varphi(x) = c + \varphi_0(x),\ c \in \mathbb{R}$
$|q| < 1$	Depends on an arbitrary function	$\varphi(x) = \varphi_0(x) + \sigma(x)$

The function σ here is an arbitrary CS of the Schröder equation

$$\sigma(f(x)) = q\sigma(x). \tag{12.7.5}$$

Proof. Let $|q| = 1$. Inequality (12.7.2) shows that $0 \leqslant z(0) \leqslant pz(0)$, i.e. $z(0) = 0$, and also

$$0 \leqslant z(f^n(x)) \leqslant p^n z(x), \qquad n \in \mathbb{N}_0. \tag{12.7.6}$$

It follows that series (12.7.4) converges a.u. in X, hence to a continuous function φ_0. Clearly, φ_0 satisfies (12.7.3). This part of the theorem results now from Proposition 3.9.2.

Let $|q| < 1$. We have by (12.7.6)

$$|z(f^n(x))/q^{n+1}| \leqslant (p/|q|)^n(z(x)/|q|),$$

which shows that series (12.7.4) converges a.u. in X. The theorem results now from Theorems 3.1.5 and 3.1.3. ∎

All continuous solutions of inequality (12.7.1) can also be determined in the case where z satisfies a limit condition.

Theorem 12.7.3. *Let hypotheses* (i), (ii) *be satisfied. If z is a CS of* (12.7.2) *such that $z > 0$ in $X \setminus \{0\}$, and*

$$\lim_{x \to 0} \frac{z(f(x))}{z(x)} < |q|, \tag{12.7.7}$$

then there exists the CS φ_0 of (12.7.1) given by (12.7.4). Moreover, all CSs φ of (12.7.1) satisfying (12.7.3) are given by Table 12.1 with the case '$q = -1$' replaced by '$q = -1$ or $|q| > 1$', where σ is any CS of the Schröder equation (12.7.5).

Proof. Now the a.u. convergence of series (12.7.4) in X results from (12.7.7). Namely, for any interval $[0, d] \subset X$ we may find a geometric series with a quotient $\Theta < 1$ majorizing in $[0, d]$ that occurring in (12.7.4). Further the proof is identical to that of Theorems 12.7.1 or 12.7.2. ∎

12.7B. A property of the particular solution

Suppose that we are given a nonnegative CS z of (12.7.2). The CS (12.7.4) of inequality (12.7.1) is distinguished among its CSs satisfying (12.7.3). Assume we are looking for CSs φ of (12.7.1) such that there exists the limit

$$L_\varphi := \lim_{x \to 0} \frac{\varphi(x)}{z(x)}. \qquad (12.7.8)$$

Theorem 12.7.4. *Let hypotheses* (i), (ii) *be satisfied and let*
$$0 < p < |q|.$$
Further, let z be a regular CS of (12.7.2) such that
$$\varphi_z(x) := \lim_{n \to \infty} p^{-n} z(f^n(x)) > 0 \qquad in \ X \setminus \{0\}$$
(see (12.4.4) and (12.4.5)). Function (12.7.4) is the only CS of (12.7.1) satisfying (12.7.3) and such that there exists the finite limit (12.7.8). Moreover

$$L_{\varphi_0} = (p - q)^{-1}. \qquad (12.7.9)$$

Proof. First we prove (12.7.9). Observe that $z \geqslant \varphi_z > 0$, in view of Theorem 12.4.1, and that series (12.7.4) converges in X yielding a CS φ_0 of (12.7.1) (Theorems 12.7.1 and 12.7.2). Moreover, we have by Corollary 12.4.1 $z(x) = u(x)\varphi_z(x)$, where u is a continuous $\{f\}$-decreasing function, $u(0) = 1$. The function φ_z satisfies $\varphi_z(f(x)) = p\varphi_z(x)$, whence $\varphi_z(f^n(x)) = p^n\varphi_z(x)$ for $n \in \mathbb{N}_0$. Thus

$$\varphi_0(x) = -\frac{\varphi_z(x)}{q} \sum_{n=0}^{\infty} \left(\frac{p}{q}\right)^n u(f^n(x)).$$

Clearly this series converges a.u. in X, whence

$$\lim_{x \to 0} \frac{\varphi_0(x)}{\varphi_z(x)} = -\frac{1}{q} \sum_{n=0}^{\infty} \left(\frac{p}{q}\right)^n = \frac{1}{p - q}.$$

On the other hand, $\lim_{x \to 0} [z(x)/\varphi_z(x)] = 1$ by Theorem 12.4.2, and (12.7.9) follows.

Now, let φ be a CS of (12.7.1) satisfying (12.7.3) and such that the limit

(12.7.8) exists. Then the function

$$\psi(x):=\varphi(x)/z(x),\ x\in X\setminus\{0\},\qquad \psi(0):=L_\varphi,$$

is continuous in X and satisfies the equation

$$\psi(f(x))=g(x)\psi(x)+1,$$

where $g(x):=qz(x)/z(f(x)),\ x\in X\setminus\{0\};\ g(0):=q/p$. According to Theorem 12.4.7 g is continuous on X. By virtue of Theorem 3.1.10, ψ is given by

$$\psi(x)=\left(-\frac{1}{q}\right)\sum_{n=0}^{\infty}\prod_{i=0}^{n-1}\frac{z(f^{i+1}(x))}{qz(f^i(x))}=-\frac{1}{z(x)}\sum_{n=0}^{\infty}\frac{z(f^n(x))}{q^{n+1}}.$$

Hence $\varphi=\varphi_0$ in X. ∎

12.7C. Reduction of order

The last theorem in this section shows that if $0<p<q$ one may find CSs of (12.7.1) only among CSs of an inequality of first order.

Theorem 12.7.5. *Let hypotheses* (i), (ii) *be satisfied, and let* $0<p<q$. *Further, let* φ *be a nonnegative CS of inequality* (12.7.1). *Then* φ *satisfies the inequality*

$$\varphi(f(x))\geqslant p\varphi(x). \tag{12.7.10}$$

Proof. Fix an arbitrary $x\in X$ and write

$$a_n:=q^{-n}\varphi(f^n(x)),\qquad n\in\mathbb{N}, \tag{12.7.11}$$

and $s=p/q$. Inequality (12.7.1) yields

$$a_{m+2}\leqslant(1+s)a_{m+1}-sa_m,\qquad m\in\mathbb{N}_0. \tag{12.7.12}$$

Summing up (12.7.12) over m from 0 to $n-1$, $n\in\mathbb{N}$, we obtain

$$a_{n+1}\leqslant sa_n+(a_1-sa_0),\qquad n\in\mathbb{N}. \tag{12.7.13}$$

If we had $a_1-sa_0<0$, (12.7.13) would imply $a_{n+1}<sa_n,\ n\in\mathbb{N}$, whence $\lim_{n\to\infty}a_n=0$. Letting $n\to\infty$ in (12.7.13) we would get $a_1-sa_0\geqslant0$, a contradiction. Thus necessarily $a_1-sa_0\geqslant0$, i.e. since x has been arbitrary, φ satisfies (12.7.10). ∎

Remark 12.7.1. (a) By virtue of Theorem 6.9.2 the sequence (a_n) given by (12.7.11) is bounded and so, by Theorem 6.9.3 applied to inequality (12.7.12), the (Koenigs) limit $\sigma(x)=\lim_{n\to\infty}q^{-n}(f^n(x))$ exists for each $x\in X$. Thus σ satisfies in X the Schröder equation (12.7.5).

(b) Take $0<p<q<1$, $f(x)=qx$, $\varphi(x)=\sqrt{x}$, $x\in\mathbb{R}^+$. Then φ satisfies (12.7.10) but no longer (12.7.1): $\varphi(q^2x)\leqslant(p+q)\varphi(qx)-pq\varphi(x)$. Thus the converse of Theorem 12.7.5 is not true.

Comments. Continuous solutions of higher-order inequalities with constant coefficients are studied by M. Stopa [1]. An analogue of Theorem 12.7.5 for such inequalities may be found in Choczewski–Stopa [1].

12.8 An inequality of infinite order

Here we shall prove some results to the effect that the only nonnegative solution φ of the inequality of infinite order,

$$\varphi(x) \leqq \sum_{n=1}^{\infty} g_n(x)\varphi(f_n(x)), \tag{12.8.1}$$

with a suitable asymptotic property is $\varphi = 0$. Note that the order of (12.8.1) becomes finite if we take $g_n = 0$ from some n on.

The theorem we want to show is useful in proving the uniqueness of solutions of, in general nonlinear, equations in various classes of functions. For instance, one could use it for supplying a direct proof of Theorem 7.2.4.

Assume the following.

(i) X is a subset of a metric space with metric ρ, $\xi \in \text{cl } X$.

(ii) $f_n: X \to X$, $n \in \mathbb{N}$, are arbitrary functions such that for every neighbourhood U of ξ and for every $x \in X$ there exists a $k \in \mathbb{N}$ such that the condition

$$f_{n_1} \circ \cdots \circ f_{n_k}(x) \in U \tag{12.8.2}$$

holds for every $n_1, \ldots, n_k \in \mathbb{N}$. Moreover,

$$\|f_n(x)\| \leqq c_n \|x\|, \qquad n \in \mathbb{N}, \tag{12.8.3}$$

for $x \in X$, with some constants $c_n \in \mathbb{R}^+$, where we put

$$\|x\| := \rho(x, \xi) \qquad \text{for } x \in X.$$

(iii) $g_n: X \to \mathbb{R}^+$, $n \in \mathbb{N}$, satisfy, with some $r \in \mathbb{R}^+$, $\Theta \in (0, 1]$, the condition

$$\sum_{n=1}^{\infty} c_n^r g_n(x) \leqq \Theta \tag{12.8.4}$$

in X if $\Theta = 1$ but in $U_0 \subset X$ if $\Theta < 1$, where U_0 is assumed to have the following property: for every $U_1 \subset U_0$ there is a $U \subset U_1$ such that $f_n(U) \subset U$, $n \in \mathbb{N}_0$, U_0, U_1, U being neighbourhoods of ξ.

Remark 12.8.1. In connection with condition (12.8.2) see Theorem 7.1.3 and the remarks after the proof of Theorem 7.1.1.

Theorem 12.8.1. *Let hypotheses* (i)–(iii) *be satisfied. If a function* $\varphi: X \to \mathbb{R}^+$ *satisfies inequality* (12.8.1), $\varphi(\xi) = 0$ *if* $\xi \in X$, *and the condition*

$$\varphi(x) = o(\|x\|^r), \qquad x \to \xi, \tag{12.8.5}$$

when $\Theta = 1$, *or the condition*

$$\varphi(x) = O(\|x\|^r), \qquad x \to \xi, \tag{12.8.6}$$

when $\Theta < 1$, *then* $\varphi = 0$ *in* X.

To prove the theorem we need an auxiliary result. Let φ be a solution of (12.8.1) satisfying the conditions of Theorem 12.8.1. Write

$$\hat{\varphi}(x) = \|x\|^{-r}\varphi(x), \qquad x \in X \setminus \{\xi\}, \ \hat{\varphi}(\xi) = 0. \tag{12.8.7}$$

In view of (12.8.1) and (12.8.3), $\hat{\varphi}$ satisfies the inequality

$$\hat{\varphi}(x) \leq \sum_{n=1}^{\infty} c_n^r g_n(x)\hat{\varphi}(f_n(x)). \tag{12.8.8}$$

Thus inequality (12.8.1) may first be solved when we have for a $\Theta \in (0, 1]$

$$\sum_{n=1}^{\infty} g_n(x) \leq \Theta. \tag{12.8.9}$$

We do this, putting the problem into a more general setting. Assertion (a) of the subsequent lemma is essentially due to J. Matkowski (unpublished, see Baron [8]).

Lemma 12.8.1. *Let X and $U \subset X$ be arbitrary sets, and let $f_n: X \to X$ be arbitrary functions, $n \in \mathbb{N}$. Further assume that for every $x \in X$ there exists a $k \in \mathbb{N}$ such that (12.8.2) holds for every $n_1, \ldots, n_k \in \mathbb{N}$. Let the functions $g_n: X \to \mathbb{R}^+$ fulfil condition (12.8.9).*
(a) If $\Theta = 1$, (12.8.9) holds in X, and $\varphi(x) \leq c$ for $x \in U$ with a $c \in \mathbb{R}^+$, then $\varphi(x) \leq c$ in the whole of X.
(b) If $\Theta < 1$, $f_n(U) \subset U$, (12.8.9) holds in U, and φ is bounded in U, then $\varphi = 0$ in X.

Proof. We represent X as the sum of its subsets U_m, $m \in \mathbb{N}$, defined by

$$U_1 := U, \qquad U_{m+1} := \bigcap_{n=1}^{\infty} f_n^{-1}(U_m).$$

This representation results from (12.8.2), since $f_n(U_{m+1}) \subset U_m$ for $n, m \in \mathbb{N}$.
In case (a) it is enough to prove that

$$\varphi(x) \leq c \qquad \text{for } x \in U_m, \ m \in \mathbb{N}. \tag{12.8.10}$$

For $m = 1$ this holds by the hypothesis. Assume (12.8.10) true for an $m \in \mathbb{N}$. We have by (12.8.1) and (12.8.9) ($\Theta = 1$), for $x \in U_{m+1}$,

$$\varphi(x) \leq \sum_{n=1}^{\infty} g_n(x)\varphi(f_n(x)) \leq c \sum_{n=1}^{\infty} g_n(x) \leq c, \tag{12.8.11}$$

and induction completes the proof of (12.8.10).
In case (b) we may assume that $c = \sup_U \varphi(x) \geq 0$. By (12.8.1) and (12.8.9), arguing as in (12.8.11), we get $\varphi(x) \leq \Theta c$ for $x \in U$, whence $c \leq \Theta c$, i.e., $c = 0$,

as $\Theta < 1$. Thus $\varphi = 0$ in U, and the argument used in case (a) yields $\varphi = 0$ in the whole of X. ∎

Proof of Theorem 12.8.1. We are going to apply Lemma 12.8.1.

Let $\Theta = 1$ in (12.8.4). For every $\varepsilon > 0$ there is a neighbourhood U of ξ such that $\hat{\phi} \leq \varepsilon$ for $x \in X \cap U$. Lemma 12.8.1(a) applies to (12.8.8), yielding $\hat{\phi} \leq \varepsilon$ in X. Letting $\varepsilon \to 0$, we obtain $\hat{\phi} = 0$. Thus by (12.8.7), $\varphi = 0$ in X.

Let $\Theta < 1$ in (12.8.4). There is a neighbourhood $U_1 \subset U_0$ of ξ such that $\hat{\phi}$ defined by (12.8.7) is bounded on U_1. By (iii), $f_n(U) \subset U$ for a $U \subset U_1$, $n \in \mathbb{N}$, and (12.8.9) holds in U (with $c_n^r g_n(x)$ in place of $g_n(x)$). Lemma 12.8.1(b) implies $\hat{\phi} = 0$ in X, whence $\varphi = 0$ in X. ∎

12.9 Notes

12.9.1. In Theorem 12.1.1 we assume w to be continuous at a point only. Thus this theorem is not a particular case of Theorem 12.1.2, though the assumptions of the latter are satisfied in any interval $[a, \infty)$, $a > -\infty$, as in Theorem 12.1.2 w should be continuous in the whole interval $[a, \infty)$.

12.9.2. Continuous periodic solutions of the equation

$$\varphi(x) = \varphi(2x) + h(x), \qquad x \in \mathbb{R}^+,$$

do not always exist (see Kac [1] and also Fortet [1], Ciesielski [1]). Theorem 12.1.2 implies that the same is true for the inequality

$$y(x) \geq y(cx) + h(x), \qquad x \in \mathbb{R}^+. \tag{12.9.1}$$

Proposition 12.9.1. *If $c > 0$, and $h: \mathbb{R}^+ \to \mathbb{R}^+$ does not vanish identically, then* (12.9.1) *has no continuous periodic solutions.*

(In fact, (12.9.1) implies $y(cx) \leq y(x) = y(x + p)$ for p-periodic y. If $c \geq 1$, Theorem 12.1.2(ii) yields $y = \text{const}$ and (12.9.1) fails to hold. Case $c < 1$ is reduced to the former one. See Howroyd [4].)

12.9.3. The following is a particular case of Theorems 12.2.2 and 12.2.3.

Proposition 12.9.2. *Let hypotheses* (i)–(iii) *of Section 12.2 be satisfied, and let z be a CS of the inequality*

$$z(x) \leq g(x)z(f(x)). \tag{12.9.2}$$

If $z(0) \leq 0$, then $z(x) \leq 0$ for all $x \in X$. If $z(x_0) > 0$ for an $x_0 \in X$, then $z(0) > 0$.

12.9.4. Theorem 12.2.2 contains also the following information.

Proposition 12.9.3. *Under hypotheses* (i)–(iii) *of Section 12.2, if the equation $\varphi(x) = g(x)\varphi(f(x)) + h(x)$ has a CS in X and y is a nonnegative CS of the* *inequality*

$$y(x) \leqslant g(x)y(f(x)) + h(x), \tag{12.9.3}$$

then

$$y(x) = y(0) + O(\varphi_0(x)), \qquad x \to 0, \tag{12.9.4}$$

where

$$\varphi_0(x) = \sum_{n=0}^{\infty} G_n(x)h(f^n(x)).$$

(Results of Section 3.3 may then be used to derive asymptotic properties of y.)

12.9.5. Estimate (12.9.4) ($y(0) = 0$) remains valid in the situation we have considered in Section 12.3 (see (12.3.11)). In this case, however, we have also property (12.3.12), competitive to (12.9.4):

$$y(x) = O(A(x)), \qquad x \to 0, \tag{12.9.5}$$

where $A(x) = h(x)/(1 - g(x))$. We exhibit an example to show that (12.9.5) may yield almost as good an estimate as (12.9.4) does.

Example 12.9.1. (a) Consider the inequality

$$y(x) \leqslant (1 - x)y(x - x^3) + (x^3 + 2x^4 - 2x^5 - x^6 + x^7)$$

in $X = [0, 1)$. Here we have $\varphi_0(x) = x^2$ and $A(x) = x^2 + 2x^3 - 2x^4 - x^5 + x^6$. So A differs from φ_0 only in the terms of higher orders.

(b) In the case of the inequality

$$y(x) \leqslant \left(1 + \frac{1}{\log x}\right) y\left(\frac{x}{x+1}\right) + \frac{x^2 \log x - x}{(x+1) \log x}$$

in $X = [0, e^{-1})$, where at $x = 0$ the functions are assigned their limit values, we have $\varphi_0(x) = x$, whereas

$$A(x) = (x - x^2 \log x)/(x + 1) = x - x^2 \log x - x^2 + x^3 \log x + \cdots.$$

Again A differs from φ_0 only in terms of higher orders.

12.9.6. The situation exemplified in Note 12.9.5 above is a special case of that we have treated in Theorem 12.3.3. An example showing that condition (12.3.13) is essential for the asymptotic equivalence at zero of $A(x)$ and $\varphi_0(x)$ follows.

Example 12.9.2. For the inequality

$$y(x) \leqslant \left(1 + \frac{1}{\log x}\right) y(\tfrac{1}{2}x) + \tfrac{1}{2}x\left(1 - \frac{1}{\log x}\right)$$

in $X = [0, e^{-1})$ we have $\varphi_0(x) = x$ and $A(x) = -\tfrac{1}{2}x(\log x - 1) = O(x \log x)$, $x \to 0$. Thus in this case estimate (12.9.4) is essentially better than (12.9.5).

(In this connection see also Kuczma [47].)

12.9.7. The following result, which may be derived from Theorem 12.4.3, justifies the assumptions of Theorem 12.6.1.

Proposition 12.9.4. *Under assumptions* (i)–(iii) *of Section 12.6, if z_1 is a CS of the inequality $z_1(f(x)) \leqslant g_1(x)z_1(x)$, $z_1 > 0$ in $X \setminus \{0\}$, and the equation*

$$\varphi(f(x)) = g(x)\varphi(x) \qquad (12.9.6)$$

with $g = g_1 g_2$ has a CS φ satisfying $\varphi(x) = z_1(x)(c + o(1))$, $x \to 0$, with a $c \neq 0$, then z_1 is regular and $\varphi_{z_1} \neq 0$ in $X \setminus \{0\}$.

12.9.8. A somewhat weaker condition than we have postulated in Theorem 12.4.3, which also implies the regularity of a CS z of (12.9.6), has recently been found by K. Dankiewicz [1]. Theorem 12.4.3 remains valid if we replace the interval $X \setminus \{0\}$ in (iii) of Section 12.4 by any nonempty, open subinterval of X, and the requirement '$\varphi(x) \neq 0$ in $X \setminus \{0\}$' for the CS φ of equation (12.9.6) by '$\varphi(x) = 0$ if and only if $z(f^i(x)) = 0$ for an $i \in \mathbb{N}_0$'.

(Formula (12.4.14) is then to be understood as the limit of the function z/φ restricted to the set of those $x \in X$ for which $\varphi(x) \neq 0$.)

12.9.9. J. Walorski [1] showed that $\varphi(z) = cz^n$, $c \in \mathbb{C}$, $n \in \mathbb{N}$, are the unique entire solutions of the inequality $|\varphi(rz)| \geqslant r|\varphi(z)|$, $z \in \mathbb{C}$ (r > 1 is fixed).

12.9.10. If y is a solution of the inequality

$$y(x) \geq g(x)y(f(x)) + h(x), \qquad (12.9.7)$$

then (and only then) the function $\bar{y}(x) := -y(x)$ satisfies the inequality

$$\bar{y}(x) \leq g(x)\bar{y}(f(x)) - h(x)$$

of form (12.9.3). Hence one can obtain dual theorems to those we have presented in Sections 12.2–12.6, pertinent to inequality (12.9.7).

12.9.11. The assertion of Theorem 12.8.1 remains valid if we replace (i) and (ii) of Section 12.8, respectively, by the following assumptions.

(i') X is a subset of a metric space, $\xi \in \mathrm{cl}\, X$, and for every $x_0 \in X$ the set $\{\xi\} \cup \{x \in X : \|x\| \leq \|x_0\|\}$ is compact.

(ii') $f_n : X \to X$, $n \in \mathbb{N}$, form a locally equicontinuous family of functions, and $\sup_{\mathbb{N}} \|f_n(x)\| < \|x\|$ for $x \in X \setminus \{\xi\}$, $\|f_n(x)\| \leq c_n \|x\|$, $n \in \mathbb{N}$, $x \in X$, with some constants $c_n \in \mathbb{R}^+$.

(This results directly from Theorems 12.8.1 and 7.1.3.)

12.9.12. Further results on linear inequalities, together with interesting applications to iterative methods of solving equations of various types, can also be found in Kornstaedt [1], Jankowski–Kwapisz [1], Kwapisz [3], Kwapisz–Turo [2]–[4]. Continuous solutions of nonlinear inequalities have been studied by D. Brydak [8]–[10]. For inequalities with superpositions of the unknown function see Brauer [1], [2], Turdza [5]–[10], Czerni [1].

REFERENCES

N. H. ABEL

[1] Détermination d'une fonction au moyen d'une équation qui ne contient qu'une seule variable. *Oeuvres complètes. II*. Christiania, 1881; pp. 36–9.

J. ACZÉL

[1] A remark on involutory functions. *Amer. Math. Monthly* **55** (1948), 638–9.

[2] *Lectures on functional equations and their applications*. Academic Press, New York and London, 1966.

[3] (editor) *Functional equations: history, applications and theory*. D. Reidel Publishing Company, Dordrecht–Boston–Lancaster, 1984.

[4] *A short course on functional equations (Based upon recent applications to the social and behavioural sciences)*. D. Reidel Publishing Company, Dordrecht–Boston–Lancaster–Tokyo, 1987.

J. ACZÉL, C. ALSINA

[1] Characterization of some classes of quasilinear functions with applications to triangular norms and to synthesizing judgements. *Methods of Operation Research* **48** (1984), 3–22.

J. ACZÉL, J. DHOMBRES

[1] *Functional equations containing several variables*. Encyclopaedia of Mathematics and its Applications, Cambridge University Press, 1989.

J. ACZÉL, F. RADÓ

[1] Involutory functions induced by abelian groups. *Studia Univ. Babeş–Bolyai, Ser. Mat. Mech.* **19** (1974), no. 2, 51–5.

J. ACZÉL, T. L. SAATY

[1] Procedures for synthesizing ratio judgements, *J. Mathematical Psychology* **27** (1983), 93–102.

H. ADAMCZYK

[1] On unbounded solution of a system of functional equations [Polish]. *Zeszyty Naukowe Politechniki Warszawskiej, Matematyka* **11** (1968), 159–64.

[2] On a system of functional equations in unbounded domain [Polish]. *Zeszyty Naukowe Politechniki Warszawskiej, Matematyka* **12** (1968), 115–25.

[3] A study of systems of functional equations in a class of unbounded functions

[Polish]. *Demonstratio Math.* **1** (1969), 5–48.

[4]   On the unique solution of some infinite systems of functional equations. *Demonstratio Math.* **8** (1975), 337–45.

P. ALSHOLM

[1]   On integrable solutions to the Baron–Boyarsky functional equation. *C. R. Math. Rep. Acad. Sci. Canada* **8** (1986), 39–41; II. *Ibidem*, 253–4.

J. ANASTASSIADIS

[1]   Fonctions semi-monotones et semi-convexes et solutions d'une équation fonctionnelle. *Bull. Sci. Math.* (2) **76** (1952), 148–60.

[2]   Sur les solutions logarithmiquement convexes ou concaves d'une équation fonctionnelle. *Bull. Sci. Math.* (2) **81** (1957), 78–87.

[3]   Une propriété de la fonction Gamma. *Bull. Sci. Math.* (2) **81** (1957), 116–18.

[4]   Définitions fonctionnelles de la fonction $B(x, y)$. *Bull. Sci. Math.* (2) **83** (1959), 24–32.

[5]   Remarques sur quelques équations fonctionnelles. *C. R. Acad. Sci. Paris* **250** (1960), 2663–5.

[6]   Sur les solutions de l'équation fonctionnelle $f(x + 1) = \varphi(x)f(x)$. *C. R. Acad. Sci. Paris* **253** (1961), 2446–7.

[7]   Définitions de fonctions eulériennes par des équations fonctionnelles. *Mémorial des Sci. Math.* **156**, Paris, 1964.

L. ANCZYK

[1]   On differentiable solutions of Böttcher's functional equation. *Ann. Polon. Math.* **21** (1969), 217–21.

[2]   Special solutions of a linear functional equation. *Zeszyty Nauk. Uniw. Jagiello.* **356**, *Prace Mat.* **16** (1974), 195–205.

G. ANDREOLI

[1]   Le curve tipo Peano–von Koch e certe equazioni funzionali associate. *Ricerca (Napoli)* (2) **14** (1963), September–December, 1–10.

D. V. ANOSOV

[1]   Geodesic flows on closed Riemannian manifolds of a negative curvature [Russian]. *Trudy Mat. Inst. im. V. A. Steklova* **90** (1967).

V. I. ARNOLD, A. AVEZ

[1]   *Ergodic problems of classical mechanics.* Benjamin, New York, 1968.

E. ARTIN

[1]   *Einführung in die Theorie der Gamma-funktion.* Hamb. Math. Einzelschr. 11, Leipzig, 1931. [English translation: *The gamma function.* Holt, Rinehart and Winston, New York–Toronto–London, 1964].

K. B. ATHREYA

[1]   A note on a functional equation arising in Galton–Watson branching processes. *J. Appl. Prob.* **8** (1971), 589–98.

K. B. ATHREYA, P. NEY

[1]   *Branching processes.* Springer Verlag, Berlin–Heidelberg–New York, 1972.

P. C. BAAYEN, W. KUYK, M. A. MAURICE

[1]   On the orbits of the hat-function, and on countable maximal semigroups of continuous mappings of the unit interval into itself. Stichting Mat. Centrum, Afd. Zuiv. Wisk., Report ZW 1962-018, Amsterdam, 1962.

C. BABBAGE

[1] Essay towards the calculus of functions. I, II. *Philosoph. Transact. Roy. Soc. London*: 1815, pp. 389–423; 1816, pp. 179–256.

[2] *Examples of the solutions of functional equations.* Cambridge, 1820.

[3] Des équations fonctionnelles. *Ann. Math. Pures Appl.* **12** (1821/22), 73–103.

K. BADURA

[1] On a linear functional equation of second order. *Ann. Polon. Math.* **24** (1971), 285–93.

J. BAILLIEUL

[1] Green's relations in finite function semigroups. *Aequationes Math.* **7** (1971), 22–7.

M. BAJRAKTAREVIĆ

[1] On solutions of a functional equation [Serbo-Croatian]. *Glasnik Mat.-Fiz. Astr.* (2) **8** (1953), 297–300.

[2] Sur une solution monotone d'une équation fonctionnelle. *Acad. Serbe Sci., Publ. Inst. Math.* **11** (1957), 47–52.

[3] Sur une équation fonctionnelle. *Glasnik Mat.-Fiz. Astr.* (2) **12** (1957), 201–5.

[4] Sur une solution de l'équation fonctionnelle $\varphi(x) + \varphi[f(x)] = F(x)$. *Glasnik Mat.-Fiz. Astr.* (2) **15** (1960), 91–8.

[5] Sur les equations de certaines équations fonctionnelles représentables par des séries. *Glasnik Mat.-Fiz. Astr.* (2) **17** (1962), 27–41.

[6] Sur l'existence de solutions continues monotones de l'équation fonctionnelle $\varphi(x) + \varphi[f(x)] = F(x)$. *Publ. Inst. Math. Beograd* (N.S.) **2 (16)** (1962), 75–80.

[7] Sur l'existence des solutions continues monotones d'une équation fonctionnelle. *Akad. Nauk. Um. Bosne Hercegov., Radovi* **22** (1963), 43–6.

[8] Sur les solutions de certaines équations fonctionnelles et intégrales. *Publ. Inst. Math. Beograd* (N.S.) **4 (18)** (1964), 147–55.

[9] Solution générale de l'équation fonctionnelle $f^N(x) = g(x)$. *Publ. Inst. Math. Beograd* (N.S.) **5 (19)** (1965), 115–24.

[10] On solutions of some functional and integral equations [Russian]. *Math. Sb.* (N.S.) **66 (108)** (1965), 161–9.

I. N. BAKER

[1] The iteration of entire transcendental functions and the solution of the functional equation $f(f(z)) = F(z)$. *Math. Ann.* **120** (1955), 174–80.

[2] Zusammensetzungen ganzer Funktionen. *Math. Zeitschr.* **69** (1958), 121–63.

[4] The existence of fixpoints of entire functions. *Math. Zeitschr.* **73** (1960), 280–4.

[5] Some entire functions with fixpoints of every order. *J. Austral. Math. Soc.* **1** (1960), 203–9.

[6] Solutions of the functional equation $(f(x))^2 - f(x^2) = h(x)$. *Canad. Math. Bull.* **3** (1960), 113–20.

[7] Permutable power series and regular iteration. *J. Austral. Math. Soc.* **2** (1961/62), 265–94.

[8] Permutable entire functions. *Math. Zeitschr.* **79** (1962), 243–9.

[9] Functional iteration near a fixpoint of multiplier 1. *J. Austral. Math. Soc.* **4** (1964), 143–8.

[10] A series associated with the logarithmic function. *J. London Math. Soc.* **42** (1967), 336–8.

[11] Limit functions and sets of non-normality in iteration theory. *Ann. Acad. Sci. Fenn., A II Math.* **467** (1970).

J. A. BAKER

[1] A functional equation from gas-dynamics. *Proceedings of the Nineteenth International Symposium on Functional Equations.* Centre for Information Theory, Faculty of Mathematics, University of Waterloo, Canada, 1981, pp. 10–11.

Ş. BALINT

[1] Über die iterativen Funktionalgleichungen ersten Ranges. *An. Univ. Timişoara, Ser. Şti. Mat.* **10** (1972), 155–66.

[2] Sur les équations fonctionnelles itératives d'ordre *n*. *An. Univ. Timişoara, Ser. Şti. Mat.* **11** (1973), 11–31.

A. D. BARBOUR

[1] The asymptotic behaviour of birth and death and some related processes. *Adv. Appl. Prob.* **7** (1975), 28–43.

B. BARNA

[1] Über die Iteration reeller Funktionen. I, II. *Publ. Math. Debrecen*: **7** (1960), 16–40; **13** (1966), 169–72.

K. BARON

[1] On the continuous solutions of non-linear functional equations of the first order. *Ann. Polon. Math.* **28** (1973), 201–5.

[2] Continuous solutions of a functional equation of *n*-th order. *Aequationes Math.* **9** (1973), 257–9.

[3] On the continuous dependence of continuous solutions of a functional equation of *n*-th order on given functions. *Aequationes Math.* **10** (1974), 78–80.

[4] Note on continuous solutions of a functional equation. *Aequationes Math.* **11** (1974), 267–9.

[5] Note on the existence of continuous solutions of a functional equation of *n*-th order. *Ann. Polon. Math.* **30** (1974), 77–80.

[6] A few observations regarding continuous solutions of a system of functional equations. *Publ. Math. Debrecen* **21** (1974), 185–91.

[7] On extending solutions of a functional equation. *Aequationes Math.* **13** (1975), 285–8.

[8] On approximate solutions of a functional equation. *Bull. Acad. Polon. Sci., Sér. sci. math. astronom. phys.* **23** (1975), 1065–8.

[9] Functional equations of infinite order. *Prace Nauk. Uniw. Śląsk.* **256**, Katowice, 1978.

[10] On approximate solutions of a system of functional equations. *Ann. Polon. Math.* **43** (1983), 305–16.

[11] On integrable solutions of some functional equations. *C. R. Math. Rep. Acad. Sci. Canada* **5** (1983), 265–7.

[12] Elements of the theory of functional equations. Continuous solutions [Polish]. *Prace Nauk. Uniw. Śląsk.* (to be published).

[13] A remark on linear functional equations in the indeterminate case. *Glasnik Mat.* **20** (40) (1985), 373–6.

[14] On a problem of R. Schilling. *Berichte der Math.-Stat. Sektion in der Forschungsgesellschaft Joanneum–Graz*, No. 286 (1988).

[15] On the convergence of sequences of iterates of random-valued functions. *Aequationes Math.* **32** (1987), 240–51.

K. BARON, R. GER

[1] Lyapunov stability of continuous solutions of some functional equations. *An. Şti. Univ. 'Al I. Cuza' Iaşi, Sect. I a Mat.* **22** (1976), 165–72.

[2] An integral version of some theorems occurring in the theory of functional equations. *Mathematica, Cluj* **17** (40) (1975), 5–10.

K. BARON, R. GER, J. MATKOWSKI

[1] Analytic solutions of a system of functional equations. *Publ. Math. Debrecen* **22** (1975), 189–94.

K. BARON, W. JARCZYK

[1] On approximate solutions of functional equations of countable order. *Aequationes Math.* **28** (1985), 22–34.

[2] On a way of division of segments. *Aequationes Math.* **34** (1987), 195–205.

K. BARON, M. KUCZMA

[1] Iteration of random valued functions on the unit interval. *Colloquium Math.* **37** (1977), 263–9.

K. BARON, J. MATKOWSKI

[1] On the solutions fulfilling a Lipschitz condition of a non-linear functional equation of order *n*. *Rev. Roum. Math. Pures Appl.* **17** (1972), 1149–54.

[2] On sequences of contraction mappings. *Rev. Roum. Math. Pures Appl.* **22** (1977), 1041–3.

K. BARON, M. SABLIK

[1] On the uniqueness of continuous solutions of a functional equation of *n*-th order. *Aequationes Math.* **17** (1978), 295–304.

K. BARON, J. WALORSKI

[1] P 235 S1. *Aequationes Math.* **32** (1987), 153.

E. BARVÍNEK

[1] On the distribution of zeroes both of solutions to the linear differential equation $y'' = Q(t)y$ and of their derivatives [Czech]. *Acta Fac. Nat. Univ. Comenian.* **5** (1961), 465–74.

E. F. BECKENBACH

[1] Generalized convex functions. *Bull. Amer. Math. Soc.* **43** (1937), 363–1.

G.P. BELICKIĬ

[1] The local conjugacy of diffeomorphisms [Russian]. *Dokl. Akad. Nauk SSSR* **191** (1970), 515–18. [English translation, *Soviet Math. Doklady* **11** (1970), 300–3].

[2] Functional equations and the conjugacy of local diffeomorphisms of a finite smoothness class [Russian]. *Dokl. Akad. Nauk SSSR* **202** (1972), 255–8. [English translation, *Soviet Math. Doklady* **13** (1972), 56–9].

[3] The conjugacy of local mappings of class C^∞ [Russian]. *Funkc. Anal. Prilož.*

6 (1972), vyp. 1, 66–7. [English translation, *Functional Anal. Appl.* **6** (1972), 57–9].

[4] Functional equations and local conjugacy of mappings of class C^∞ [Russian]. *Mat. Sb. (N.S.)* **91 (133)** (1973), 565–79. [English translation, *Math. USSR–Sb.* **20** (1973), 587–602 (1974)].

[5] Functional equations and conjugacy of local diffeomorphisms of finite smoothness class [Russian]. *Funkc. Anal. Prilož.* **7** (1973), vyp. 4, 17–28. [English translation, *Functional Anal. Appl.* **7** (1973), 268–77 (1974)].

[6] Flat stable germs of C^∞ mappings and their linear approximations [Russian]. *Funkc. Anal. Prilož.* **8** (1974), vyp. 2, 61–2. [English translation, *Functional Anal. Appl.* **8** (1974), 141–4].

[7] On the equivalence of families of local diffeomorphisms [Russian]. *Funkc. Anal. Prilož.* **8** (1974), vyp. 4, 79–80. [English translation, *Functional Anal. Appl.* **8** (1974), 79–80].

[8] On the conjugacy of nonhyperbolic mappings of finite smoothness class [Russian]. *Dokl. Akad. Nauk SSSR* **219** (1974), 275–8. [English translation, *Soviet Math. Doklady* **15** (1974), 1557–61].

[9] Germs of mappings that are co-determined with respect to a given group [Russian]. *Mat. Sb. (N.S.)* **94 (136)** (1974), 452–7, 496. [English translation, *Math. USSR–Sb.* **23** (1974), 425–40 (1975)].

R. BELLMAN

[1] *Introduction to the mathematical theory of control processes.* Vol. 1, Academic Press, New York, 1967.

[2] *Methods of nonlinear analysis.* Vol. II, Academic Press, New York and London, 1973.

L. BERG

[1] The method of reducing the order of linear operator equations. *Demonstratio Math.* **9** (1976), 129–34.

[2] Functional equations in a field. *Math. Nachr.* **74** (1976), 309–12.

A. BIELECKI, J. KISYŃSKI

[1] Sur le problème de E. Goursat relatif à l'équation $\partial^2 z/\partial x\,\partial y = f(x, y)$. *Ann. Univ. M. Curie-Skłodowska, Sect. A* **10** (1956), 99–126.

N. H. BINGHAM, R. A. DONEY

[1] Asymptotic properties of supercritical branching processes. I. The Galton–Watson process. *Adv. Appl. Prob.* **6** (1974), 711–31.

[2] Asymptotic properties of supercritical branching processes. II. Crump–Mode and Jiřina processes. *Adv. Appl. Prob.* **7** (1975), 66–82.

P. BLANCHARD

[1] Complex analytic dynamics on the Riemann sphere. *Bull. Amer. Math. Soc.* **11** (1984), 85–141.

J. BLOCK, J. GUCKENHEIMER, J. MISIUREWICZ, L. S. YOUNG

[1] Periodic points and topological entropy of one dimensional maps. *Global theory of dynamical systems.* Lecture Notes in Mathematics 819, Springer Verlag, Berlin (1980), 18–34.

U. T. BÖDEWADT

[1] Zur Iteration reeller Funktionen. *Math. Zeitschr.* **49** (1944), 497–516.

JU. S. BOGDANOV

[1]   On the functional equation $x^n = t$ [Russian]. *Dokl. Akad. Nauk BSSR* **5** (1961), 235–7.

H. BOHR, J. MOLLERUP

[1]   *Lectures on mathematical analysis. III. Limit processes.* [Danish]. København, 1922.

O. BORŮVKA

[1]   *Linear differential transformations of the second order.* English Universities Press, London, 1971.

[2]   Sur une classe des groupes continus à un paramètre formés des fonctions réelles d'une variable. *Ann. Polon. Math.* **42** (1982), 27–37.

D. BORWEIN

[1]   Convergence criteria for bounded sequences. *Proc. Edin. Math. Soc.* (2) **18** (1972), 99–103.

A. BORZYMOWSKI

[1]   A Goursat problem for some partial differential equations of order $2p$. *Bull. Pol. Sc. Math.* **32** (1984), 577–80.

[2]   Concerning a Goursat problem for some partial differential equations of order $2p$. *Demonstratio Math.* **18** (1985), 253–77.

A. BOSHERNITZAN, L. A. RUBEL

[1]   Coherent families of polynomials. *Analysis* (to appear).

Ł. E. BÖTTCHER

[1]   Beiträge zu der Theorie der Iterationsrechnung. Dissertation. Leipzig, 1898.

[2]   Principles of the iteration calculus [Polish]. I, II. *Prace Mat. Fiz.* **10** (1899), 65–101.

A. BOYARSKY, R. GER

[1]   On the densities of absolutely continuous invariant measures, unpublished.

A. BOYARSKY, G. HADDAD

[1]   All invariant densities of piecewise linear Markov maps are piecewise constant. *Advances in Appl. Math.* **2** (1981), 284–9.

W. M. BOYCE

[1]   Commuting functions with no common fixed point. *Trans. Amer. Math. Soc.* **137** (1969), 77–92.

D. W. BOYD, J. S. WONG

[1]   On nonlinear contractions. *Proc. Amer. Math. Soc.* **20** (1969), 458–69.

S. BRATMAN

[1]   Uniqueness of regular similarity functions. *Ann. Polon. Math.* **31** (1976), 265–7.

G. BRAUER

[1]   A functional inequality. *Amer. Math. Monthly* **68** (1961), 638–42.

[2]   Functional inequalities. *Amer. Math. Monthly* **71** (1964), 1014–17.

F. E. BROWDER

[1]   Nonlinear operators and nonlinear equations of evolution in Banach spaces. *Proc. Symp. Pure Math.* **18** (2) (1976).

J. W. BROWN, M. KUCZMA

[1]   Self-inverse Sheffer sequences. *SIAM J. Math. Anal.* **7** (1976), 723–8.

N. G. DE BRUIJN

[1] *Asymptotic methods in analysis*. Amsterdam, 1958.

D. BRYDAK

[1] On $\{f\}$-semiconvex functions [Polish]. *Rocznik Nauk.-Dydakt. WSP Kraków* **13** (1962), 67–77.

[2] Sur une équation fonctionnelle. I, II. *Ann. Polon. Math.* **15** (1964), 237–51; **21** (1968), 1–13.

[3] A few remarks on $\{f\}$-semiconvex functions [Polish]. *Rocznik Nauk.-Dydakt. WSP Kraków* **25** (1966), 61–70.

[4] On the stability of the functional equation $\varphi[f(x)] = g(x)\varphi(x) + F(x)$. *Proc. Amer. Math. Soc.* **26** (1970), 455–60.

[5] Monotonic solutions of a generalized Euler equation [Polish]. *Rocznik Nauk.-Dydakt. WSP Kraków* **41**, *Prace Mat.* **6** (1970), 5–13.

[6] On a functional equation of invariant curves. *Rocznik Nauk.-Dydakt. WSP Kraków* **51**, *Prace Mat.* **7** (1974), 37–45.

[7] On the homogeneous functional inequality. *Rocznik Nauk.-Dydakt. WSP Kraków*, *Prace Mat.* **9** (1979), 9–13.

[8] On functional inequalities in a single variable. *Dissertationes Math. (Rozprawy Mat.)* **160**, Warsaw, 1979.

[9] Nonlinear functional inequalities in a single variable. *General inequalities 1* (Proceedings of the Symposium, Oberwolfach, 1976), edited by E. F. Beckenbach, ISNM 41, Birkhäuser Verlag, Basle and Stuttgart, 1978, 181–9.

[10] A generalization of theorems concerning a nonlinear functional inequality in a single variable. *General inequalities 2* (Proceedings of the Symposium, Oberwolfach, 1978), edited by E. F. Beckenbach, ISNM 47, Birkhäuser Verlag, Basle and Stuttgart, 1980, 179–84.

[11] On invariant curves under the axial transform. *Ann. Polon. Math.* **40** (1981), 31–8.

D. BRYDAK, B. CHOCZEWSKI

[1] Classification of continuous solutions of a functional inequality. *Zeszyty Naukowe Uniw. Jagiellońskiego* **403**, *Prace Mat.* **17** (1975), 33–40.

[2] Continuous solutions of a functional inequality of second order. *Demonstratio Math.* **9** (1976), 221–8.

[3] On nonhomogeneous functional inequality. *Zeszyty Naukowe Uniw. Jagiellońskiego* **441**, *Prace Mat.* **18** (1977), 89–92.

[4] Application of functional inequalities to determining one-parameter families of solutions of a functional equation. *General inequalities 1* (Proceedings of the Symposium, Oberwolfach, 1976), edited by E. F. Beckenbach, ISNM 41, Birkhäuser Verlag, Basle and Stuttgart, 1978, 191–7.

D. BRYDAK, J. KORDYLEWSKI

[1] On monotonic sequences defined by a recurrence relation. *Publ. Math. Debrecen* **9** (1962), 189–97.

R. C. BUCK

[1] On approximation theory and functional equations. *J. Approximation Theory* **5** (1972), 228–37.

[2]   Approximation theory and functional equations. II. *J. Approximation Theory* **9** (1973), 121–5.

J. BUREK

[1]   Über einseitig beschränkte Lösungen linearer Funktionalgleichung. *Ann. Polon. Math.* **21** (1968), 67–72.

J. BUREK, M. KUCZMA

[1]   Einige Bemerkungen über monotone und konvexe Lösungen gewisser Funktionalgleichungen. *Math. Nachr.* **36** (1968), 121–34.

U. BURKART

[1]   Zur Charakterisierung diskreter dynamischer Systeme. Doctoral Dissertation, Marburg/Lahn, 1981.

[2]   Verallgemeinerte Schrödergleichung und Prä-Schrödergleichungen. *Ann. Polon. Math.* **40** (1983), 109–14.

K. CARATHÉODORY

[1]   *Vorlesungen über reelle Funktionen.* Leipzig–Berlin, 1927.

F. W. CARROLL

[1]   Transcendental transcendence of solutions of Schröder's equation associated with finite Blaschke products. *Michigan Math. J.* **32** (1985), 47–58.

J. W. S. CASSELS

[1]   *An introduction to diophantine approximation.* Cambridge University Press, Cambridge, 1957.

K. T. CHEN

[1]   On local diffeomorphisms about an elementary fixed point. *Bull. Amer. Math. Soc.* **69** (1963), 838–40.

[2]   Normal forms of local diffeomorphisms on the real line. *Duke Math. J.* **35** (1968), 549–55.

B. CHOCZEWSKI

[1]   On continuous solutions of some functional equations of the n-th order. *Ann. Polon. Math.* **11** (1961), 123–32.

[2]   On differentiable solutions of a functional equation. *Ann. Polon. Math.* **13** (1963), 133–8.

[3]   Investigation of the existence and uniqueness of differentiable solutions of a functional equation. *Ann. Polon. Math.* **15** (1964), 117–41.

[4]   Some theorems on the existence of a one-parameter family of C^1 solutions of a linear functional equation. *Mathematica, Cluj* **10 (33)** (1968), 47–52.

[5]   Regular solutions of a linear functional equation in the indeterminate case. *Ann. Polon. Math.* **21** (1969), 257–65.

[6]   On the differentiability of regular solutions of a linear functional equation. *Zeszyty Naukowe Uniw. Jagiellońskiego* **203**, *Prace Mat.* **13** (1969), 19–25.

[7]   Topological methods in the theory of functional equations in a single variable. *Zeszyty Naukowe Uniw. Jagiellońskiego* **223**, *Prace Mat.* **14** (1969), 67–74.

[8]   Some asymptotic properties of continuous solutions of a linear functional equation. *Mathematica, Cluj* **12 (35)** (1970), 227–36.

[9]   Regular solutions of a linear functional equation of the first order. *Publ. Techn. Univ. Miskolc* **30** (1970), 255–62.

[10] Asymptotic behaviour of continuous solutions of some functional equations [Polish]. *Zeszyty Nauk. Akad. Górn.-Hutniczej, Mat.-Fiz.-Chem.* **4** (1970), 66 pp.

[11] Note on the convergence of iterates. *Ann. Polon. Math.* **24** (1971), 229–32.

[12] Asymptotic series expansions of continuous solutions of a linear functional equation. *Bull. Acad. Polon. Sci., Sér. sci. math. astronom. phys.* **21** (1973), 925–30.

[13] On S. Gołąb's characterization of the anharmonic ratio. *Ann. Polon. Math.* **40** (1981), 59–65.

[14] Stability of some iterative functional equations. *General inequalities 4* (Proceedings of the Symposium, Oberwolfach, 1983), edited by W. Walter, ISNM 71, Birkhäuser Verlag, Basle and Stuttgart, 1984, 249–55.

[15] Note on discontinuous solutions of a functional equation. *Ann. Univ. Iagello. Acta Math.* **27** (1988), 281–3.

B. CHOCZEWSKI, M. KUCZMA

[1] On the 'indeterminate case' in the theory of a linear functional equation. *Fund. Math.* **58** (1966), 163–75.

[2] On a problem of Lipiński concerning an integral equation. *Colloquium Math.* **25** (1972), 113–15.

[3] Asymptotic properties of discontinuous solutions of a functional equation. *Zeszyty Naukowe Uniw. Jagiellońskiego* **553**, *Prace Mat.* **22** (1981), 119–23.

[4] Discontinuous solutions of an inhomogeneous linear functional equation. *Ann. Univ. Iagello. Acta Math.* **24** (1984), 155–8.

[5] On a system of functional equations. *Aequationes Math.* **28** (1985), 262–8.

[6] A remark on doubly stochastic measures and functional equations. *C. R. Math. Rep. Acad. Sci. Canada* **11** (1989), 127–32.

B. CHOCZEWSKI, M. STOPA

[1] Some iterative functional inequalities and Schröder's equation. *General inequalities 5* (Proceedings of the Symposium, Oberwolfach, 1986), edited by W. Walter, ISNM 80, Birkhäuser Verlag, Basle and Stuttgart, 1987, 273–5.

B. CHOCZEWSKI, E. TURDZA, R. WĘGRZYK

[1] On the stability of a linear functional equation. *Rocznik Nauk.-Dydakt. WSP Kraków* **69**, *Prace Mat.* **9** (1979), 15–21.

J. D. CHURCH

[1] On infinite composition products of probability generating functions. *Z. Wahrsch. Verw. Gebiete* **19** (1971), 243–56.

Z. CIESIELSKI

[1] On the functional equation $f(t) = g(t) - g(2t)$. *Proc. Amer. Math. Soc.* **13** (1962), 388–93.

W. E. CLARK, A. MUKHERJEA

[1] Comments on a functional equation. *Real Anal. Exchange* **6** (1980–1), 192–9.

R. R. COIFMAN

[1] Sur l'équation d'Abel–Schröder et l'itération continue. *C. R. Acad. Sci. Paris* **258** (1964), 1976–7, 5324–5.

[2] Sur l'unicité des solutions de l'équation d'Abel–Schröder et l'itération continue. *J. Austral. Math. Soc.* **5** (1965), 36–47.

[3]   Pseudo-groupes de transformations et itération continue des fonctions réelles. Thesis, Geneva, 1965.

R. R. COIFMAN, M. KUCZMA

[1]   On asymptotically regular solutions of a linear functional equation. *Aequationes Math.* **2** (1969), 332–6.

P. COLLET, J. P. ECKMANN

[1]   *Iterated maps on the interval as dynamical systems.* Progress in Physics Series, Birkhäuser, Basle, 1980.

R. COOPER

[1]   On a duplication formula. *Math. Gaz.* **41** (1957), 217–18.

E. T. COPSON

[1]   On a generalization of monotonic sequences. *Proc. Edin. Math. Soc.* (2) **17** (1970), 159–74.

H. CREMER

[1]   Über die Häufigkeit der Nichtzentren. *Math. Ann.* **115** (1938), 573–80.

M. CRUM

[1]   On two functional equations which occur in the theory of clock-graduation. *Quart. J. Math., Oxford Ser.* **10** (1939), 155–60.

D. CZAJA-POŚPIECH, M. KUCZMA

[1]   Continuous solutions of some functional equations in the indeterminate case. *Ann. Polon. Math.* **24** (1970), 9–20.

M. CZERNI

[1]   Comparison theorem for a functional inequality. *General inequalities 3* (Proceedings of the Symposium, Oberwolfach, 1981), edited by E. F. Beckenbach and W. Walter, ISNM 64, Birkhäuser Verlag, Basle and Stuttgart, 1981, 253–62.

[2]   Asymptotic set stability for a functional equation of first order. *Rocznik Nauk.-Dydakt. WSP Kraków* **82**, *Prace Mat.* **10** (1982), 15–26.

[3]   Stability of normal regions for linear homogeneous functional equations. *Aequationes Math.* **36** (1988), 176–87.

[4]   Asymptotic interval stability for a linear homogeneous functional equation. *Rocznik Nauk.-Dydakt. WSP Kraków* **2**, *Prace Mat.* **12** (1987), 19–36.

S. CZERWIK

[1]   On a recurrence relation. *Ann. Polon. Math.* **20** (1968), 61–71.

[2]   On sign-preserving solutions of a linear functional equation. *Ann. Polon. Math.* **20** (1968), 73–9.

[3]   On the differentiability of solutions of a functional equation with respect to a parameter. *Ann. Polon. Math.* **24** (1971), 209–17.

[4]   On the continuous dependence of solutions of some functional equations on given functions. *Ann. Polon. Math.* **24** (1971), 247–52.

[5]   On differentiable solutions of a linear functional equation. *Demonstratio Math.* **6** (1973), 87–96.

[6]   On the dependence on a parameter of solutions of a linear functional equation. *Prace Nauk. Uniw. Śląsk.* **30**, *Prace Mat.* **3** (1973), 25–30.

[7]   On some properties of solutions of a functional equation with parameter. *Commentationes Math.* **17** (1974), 335–9.

[8]   Solutions of class C^r with respect to the parameter of a linear functional equation. *Commentationes Math.* **17** (1974), 341–5.

[9] On the existence and uniqueness of convex solutions of a functional equation in the indeterminate case. *Ann. Polon. Math.* **30** (1974), 5–8.

[10] On f-monotonic and f-convex solutions of a functional equation. *Glasnik Mat.* **10 (30)** (1975), 69–72.

[11] Note on the differentiability of solutions of class C^r of a functional equation with respect to a parameter. *Aequationes Math.* **13** (1975), 15–19.

[12] Note on the differentiable solutions of a linear functional equation. *Prace Nauk. Uniw. Śląsk.* **59**, *Prace Mat.* **5** (1975), 7–12.

[13] C-convex solutions of a linear functional equation. *Prace Nauk. Uniw. Śląsk.* **87**, *Prace Mat.* **6** (1975), 49–53.

[14] On the unbounded and continuous solutions of a system of functional equations. *Prace Nauk. Uniw. Śląsk.* **87**, *Prace Mat.* **6** (1975), 63–6.

[15] Special solutions of a functional equation. *Ann. Polon. Math.* **31** (1975), 141–4.

[16] Continuous solutions of a functional equation. *Mat. Vesnik* **12 (27)** (1975), 241–4.

[17] Solutions of a system of functional equations in some class of functions. *Aequationes Math.* **18** (1978), 289–95.

[18] Fixed point theorems and special solutions of functional equations. *Prace Nauk. Uniw. Śląsk.* **428**, Katowice, 1980.

K. DANKIEWICZ

[1] On a homogeneous functional inequality. *Rocznik Nauk.-Dydakt. WSP Kraków* **97**, *Prace Mat.* **11** (1985), 61–9.

[2] On approximate solutions of a functional equation in the class of differentiable functions. *Rocznik Nauk.-Dydakt. WSP Kraków* **115**, *Prace Mat.* **12** (1987).

W. F. DARSOW, M. J. FRANK

[1] Associative functions and Abel–Schröder systems. *Publ. Math. Debrecen* **30** (1983), 253–72.

W. F. DARSOW, M. J. FRANK, H.-H. KAIRIES

[1] Functional equations for a function of van der Waerden type. *Radovi Mat.* **4** (1988), 361–74.

K. DEIMLING

[1] Das Goursat-Problem für $u_{xy} = f(x, y, u)$. *Aequationes Math.* **6** (1971), 206–14.

R. L. DEVANEY

[1] *An introduction to chaotic dynamical systems*. The Benjamin/Cummings Publishing Company, Menlo Park, California, 1986.

J. G. DHOMBRES

[1] Itération linéaire d'ordre 2. *C. R. Acad. Sci. Paris* **280** (1975), A275–7.

[2] On the historical role of functional equations. *Aequationes Math.* **25** (1982), 293–9.

[3] *Some aspects of functional equations*. Lecture Notes. Department of Mathematics, Chulalongkorn University, Bangkok, 1979.

[4] Quelques aspects de l'histoire des équations fonctionnelles liés à l'évolution du concept de fonction. *Archive for History of Exact Sciences* **36** (1986), 151–74.

PH. DIAMOND

[1] Fractional iteration and a generalized Abel's equation. *Aequationes Math.* **6** (1971), 11–23.

[2] A stochastic functional equation. *Aequationes Math.* **15** (1977), 225–33.

[3]   The Schröder and Abel functional equations and nonautonomous difference equations, preprint, University of Queensland, 1981.

L. J. DICKEY, H.-H. KAIRIES, H. S. SHANK

[1]   Analogs for Bernoulli polynomials in fields Z^p. *Aequationes Math.* **14** (1976), 401–4.

K.-D. DIKOF, R. GRAW

[1]   A strictly increasing real functional with no iterative roots of any order. *Proceedings of the 18th International Symposium on Functional Equations* (Waterloo, Canada, 1980).

S. DITOR

[1]   Some binary representations of real numbers and odd solutions to the functional equation $\varphi \circ f = g \circ \varphi$. *Aequationes Math.* **19** (1979), 160–82.

K. DOKSUM

[1]   Empirical probability plots and statistical inference for non-linear models in analysis of variance in the two-sample case. *Ann. Statist.* **2** (1974), 267–77.

R. A. DONEY

[1]   On a functional equation for branching processes. *J. Appl. Prob.* **10** (1973), 198–205.

J. DREWNIAK

[1]   Iteration sequences for some complex functions [Polish]. *Prace Zakładu Systemów Automatyki Kompleksowej PAN* **6** (1972), 27–41.

[2]   Asymptotic properties of iteration sequences [Polish]. *Prace Zakładu Systemów Automatyki Kompleksowej PAN* **22** (1975), 126 pp.

J. DREWNIAK, V. DROBOT

[1]   On the speed of convergence of iterations of functions. *Opuscula Math.* **6** (to appear).

J. DREWNIAK, J. KALINOWSKI

[1]   Les relations entre les équations pré-Schröder. I, II. *Ann. Polon. Math.* **32** (1976), 5–11, 287–92.

J. DREWNIAK, J. KALINOWSKI, B. ULEWICZ

[1]   On the behaviour of continuous real functions in the neighbourhood of a fixed point. *Aequationes Math.* **14** (1976), 123–36.

J. DREWNIAK, M. KUCZMA

[1]   On the asymptotic behaviour of some sequences built of iterates. *Ann. Polon. Math.* **25** (1971), 39–51.

J. DREWNIAK, A. MRÓZEK, B. ULEWICZ

[1]   Discrete iterations of real functions [Polish]. *Prace Zakładu Systemów Automatyki Kompleksowej PAN* **7** (1972), 81 pp.

L. DUBIKAJTIS

[1]   Sur une caractérisation de la fonction sinus. *Ann. Polon. Math.* **16** (1964), 117–20.

[2]   Sur certaines équations fonctionnelles vérifiées par la fonction $\varphi(x) = x^{-1}$. *Ann. Polon. Math.* **22** (1969), 199–205.

S. DUBUC

[1]   Problèmes relatifs à l'itération de fonctions suggérés par les processus en cascade. *Ann. Inst. Fourier de l'Univ. Grenoble* **31** (1971), 172–251.

[2] Etude théorique et numérique de la fonction de Karlin–McGregor. *Journal d'Anal. Math.* **42** (1982/3), 15–37.

[3] Une équation fonctionnelle pour diverses constructions géometriques. *Ann. sc. math. Québec* **9** (1985), 151–74.

J. DUFRESNOY, CH. PISOT

[1] Sur la relation fonctionnelle $f(x + 1) - f(x) = \varphi(x)$. *Bull. Soc. Math. Belg.* **15** (1963), 259–70.

J. DUGUNDJI, A. GRANAS

[1] *Fixed point theory.* Monografie Mat. 61. Polish Scientific Publishers, Warsaw, 1982.

S. D. DURHAM

[1] Limiting distributions for the general branching process with immigration. *J. Appl. Prob.* **11** (1974), 809–13.

R. DURRETT, T. M. LIGETT

[1] Construction of solutions to $\varphi^2(x) = x^4$ (private communication).

CZ. DYJAK

[1] On the dependence of the continuous solutions of a functional equation on an arbitrary function. *Publ. Inst. Math. Beograd (N.S.)* **19 (33)** (1975), 51–3.

[2] On a characterization of trigonometrical and hyperbolic functions by functional equations. *Publ. Inst. Math. Beograd (N.S.)* **32 (46)** (1982), 45–8.

[3] BV-solutions of a linear functional equation. *Publ. Math. Debrecen* **33** (1986), 83–5.

CZ. DYJAK, J. MATKOWSKI

[1] BV-solutions of a nonlinear functional equation. *Glasnik Mat.* **22 (42)** (1987), 335–42.

A. EAGLE

[1] A simple theory of the gamma-function. *Math. Gaz.* **14** (1928), 118–27.

J. ECALLE

[1] Itération fractionnaire des transformations formelles d'une variable complexe. Thèse de 3eme cycle, Orsay, 1970.

[2] Théorie itérative: Introduction à la théorie des invariants holomorphs. *J. Math. Pures Appl.* **54** (1975), 183–258.

[3] Théorie des invariants holomorphs. Publications Mathématiques d'Orsay No. 67-7409, 1974.

R. ENGELKING

[1] *Outline of general topology.* North Holland Publishing Company and Polish Scientific Publishers, Amsterdam, 1968.

P. ERDŐS, E. JABOTINSKY

[1] On analytic iteration. *J. Analyse Math.* **8** (1960/1), 361–76.

L. EULER

[1] *Opera omnia. Series prima. Opera mathematica.* Vol. XXVII. *Commentationes geometricae*, Vol. II. Lausanne, 1954.

P. FATOU

[1] Mémoire sur les équations fonctionnelles. *Bull. Soc. Math. France*: **47** (1919), 161–271; **48** (1920), 33–94, 208–314.

[2] Sur l'équation fonctionnelle d'Abel. *Bull. Soc. Math. France* **47** (1919), 41–2.

[3] Sur les frontières de certains domaines. *Bull. Soc. Math. France* **51** (1923), 16–22.

[4]   Sur l'itération analytique et les substitutions permutables. *J. Math. Pures Appl.*
      (9): **2** (1923), 343–8; **3** (1924), 1–49.

[5]   Sur l'itération des fonctions transcendentes entières. *Acta Math.* **47** (1926),
      337–70.

W. FELLER

[1]   *An introduction to probability theory and its applications.* Vol. 2. John Wiley &
      Sons, New York, 1966.

I. FENYŐ

[1]   Zusammenhang zwischen einer Integrodifferentialgleichung und einer Funk-
      tionalgleichung. *Z. Anal. Anw.* **6 (2)** (1987), 151–7.

N. J. FINE, B. KOSTANT

[1]   The group of formal power series under iteration. *Bull. Amer. Math. Soc.* **61**
      (1955), 36–7.

P. FISCHER

[1]   On Bellman's function equation. *J. Math. Anal. Appl.* **46** (1974), 212–27.

W. B. FITE

[1]   Properties of the solutions of certain functional-differential equations. *Trans.*
      Amer. Math. Soc. **22** (1921), 311–19.

H. G. FORDER

[1]   Duplication formulae. *Math. Gaz.* **41** (1957), 215–18.

W. FÖRG-ROB

[1]   On a problem of R. Schilling, manuscript, University of Innsbruck, 1987.

M. K. FORT, JR.

[1]   Continuous solutions of a functional equation. *Ann. Polon. Math.* **13** (1963),
      205–11.

R. FORTET

[1]   Sur une suite également répartie. *Studia Math.* **9** (1940), 54–70.

G. L. FORTI

[1]   On some conditional Cauchy equation on thin sets. *Boll. Un. Mat. Ital.* (6)
      2-B (1983), 391–402.

J. FRANK

[1]   Conjugacy classes of hat functions (private communication).

G. FUBINI

[1]   Di una nuova successione di numeri. *Period. Mat.* **14** (1899), 147–9.

F. R. GANTMACHER

[1]   *Theory of matrices* [Russian]. Moscow, 1966.

B. GAWEŁ

[1]   On the theorems of Šarkovskiĭ and Štefan on cycles. *Proc. Amer. Math. Soc.*
      (to appear).

A. O. GELFOND

[1]   *Calculus of finite differences* [Russian]. Moscow and Leningrad, 1952.

J. GER

[1]   On convex solutions of the functional equation $\varphi^2(x) = g(x)$. *Zeszyty Naukowe*
      Uniw. Jagiellońskiego **252**, *Prace Mat.* **15** (1971), 61–5.

[2]   On analytic solutions of the functional equation $\varphi(f(x)) = g(x, \varphi(x))$. *Prace*
      Nauk. Uniw. Śląsk. **218**, *Prace Mat.* **8** (1978), 45–59.

[3] On analytic solutions of the equation $\varphi(f(x)) = g(x, \varphi(x))$ (II). *Prace Nauk. Uniw. Śląsk.* **275**, *Prace Mat.* **9** (1979), 74–103.

[4] On analytic solutions of the equation $\varphi(f(x)) = g(x, \varphi(x))$ (III). *Ann. Math. Sil.* **1 (13)** (1985), 93–102.

[5] On analytic solutions of the non-linear functional equations. *Ann. Math. Sil.* **1 (13)** (1985), 103–15.

R. GER, A. SMAJDOR

[1] *p*-convex iteration groups. *Fund. Math.* **91** (1976), 29–38.

N. M. GERSEVANOV

[1] *Iterational calculus and its applications* [Russian]. Moscow, 1950.

M. GHERMĂNESCU

[1] Opérateurs fonctionnelles périodiques. *C. R. Acad. Sci. Paris* **228** (1949), 1190–1.

[2] *Functional equations* [Roumanian]. Editura Acad. R. P. Romine, Bucharest, 1960.

E. GŁOWACKI

[1] On approximate solutions of linear functional equations. *Prace Nauk. Uniw. Śląsk.* **2**, *Prace Mat.* **1** (1969), 41–51.

[2] On approximate solutions of a non-linear functional equation. *Prace Nauk. Uniw. Śląsk.* **12**, *Prace Mat.* **2** (1972), 13–17.

K. GOEBEL, S. REICH

[1] *Uniform convexity, hyperbolic geometry and nonexpansive mappings.* Marcel Dekker, Inc., New York and Basle, 1984.

S. GOŁĄB

[1] Problème 40. *Colloquium Math.* **25** (1948), 240–1.

[2] Sur un système d'équations fonctionnelles lié au rapport anharmonique. *Ann. Polon. Math.* **29** (1974), 273–80.

R. GOLDSTEIN

[1] On certain compositions of functions of a complex variable. *Aequationes Math.* **4** (1970), 103–26.

[2] Some results on factorization of meromorphic functions. *J. London Math. Soc.* (2) **4** (1971), 357–64.

[3] On meromorphic solutions of a functional equation of Ganapathy Iyer. *Aequationes Math.* **8** (1972), 82–94.

[4] On deficient values of meromorphic functions satisfying a certain functional equation. I, II. *Aequationes Math.* **5** (1970), 75–85; **9** (1973), 267–72.

[5] On meromorphic solutions of a functional equation. *Aequationes Math.* **15** (1977), 239–48.

[6] On meromorphic solutions of a functional equation of Kuczma and Smajdor. *Aequationes Math.* **16** (1977), 221–43.

R. GRAW

[1] Über die Orbitstruktur stetiger Abbildungen. Doctoral Dissertation, Marburg/ Lahn, 1981.

R. GRAW, M. KUCZMA

[1] Infinite discrete sets of limit points of iterative sequences of continuous functions. *Aequationes Math.* **22** (1981), 64–9.

A. M. GROBMAN

[1]   On the homeomorphisms of systems of differential equations [Russian]. *Dokl. Akad. Nauk SSSR* **128** (1959), 880–1.

F. GROSS

[1]   On some functional equations of Iyer. *J. Indian Math. Soc.* (2) **32** (1968), 221–7.

A. GULGOWSKA, L. GULGOWSKI

[1]   On problems of B. Choczewski and M. Kuczma concerning an integral equation. *Colloquium Math.* **38** (1977), 91–4.

I. GUMOWSKI, C. MIRA

[1]   *Dynamique chaotique*. Cepadues Editions, Toulouse, 1980.

H. N. GUPTA

[1]   A characterization of cotangent function. Manuscript. University of Regina, 1988.

J. HADAMARD

[1]   Sur l'itération et solutions asymptotiques des équations différentielles. *Bull. Soc. Math. France* **29** (1901), 224–8.

P. I. HAÏDUKOV

[1]   On searching a function from a given iterate [Russian]. *Uč. Zap. Buriatsk. Ped. Inst.* **15** (1958), 3–28.

P. R. HALMOS

[1]   *Measure theory*. Van Nostrand Publ. Co., Toronto–New York–London, 1950.

H. J. HAMILTON

[1]   On monotone and convex solutions of certain difference equations. *Amer. J. Math.* **63** (1941), 427–34.

[2]   Roots of equations by functional iteration. *Duke Math. J.* **13** (1946), 113–21.

G. H. HARDY

[1]   Weierstrass's non-differentiable function. *Trans. Amer. Math. Soc.* **20** (1916), 301–25.

G. H. HARDY, J. E. LITTLEWOOD, G. PÓLYA

[1]   *Inequalities* (2nd edn.). Cambridge University Press, Cambridge, 1952.

T. E. HARRIS

[1]   *The theory of branching processes*. Springer Verlag, Berlin–Göttingen–Heidelberg, 1963.

PH. HARTMAN

[1]   On local homeomorphisms of Euclidean spaces. *Bol. Soc. Mat. Mexicana* (2) **5** (1960), 220–41.

[2]   A lemma in the theory of structural stability of differential equations. *Proc. Amer. Math. Soc.* **11** (1960), 610–20.

[3]   On the local linearization of differential equations. *Proc. Amer. Math. Soc.* **14** (1963), 568–73.

[4]   *Ordinary differential equations*. John Wiley & Sons, New York–London–Sydney, 1964.

H. HARUKI

[1]   On a Laplace integral. *Math. Magazine* **43** (1970), 151–3.

[2]   A functional equation arising from the Joukowski transformation. *Ann. Polon. Math.* **45** (1985), 185–91.

[3]   A new characterization of Euler's gamma function by a functional equation. *Aequationes Math.* **31** (1986), 173–83.

[4]   An application of the cosine duplication formula. *Proceedings of the Eighteenth International Symposium on Functional Equations*, Centre for Information Theory, Faculty of Mathematics, University of Waterloo, Canada, 1980, p. 14.

[5]   A proof of Euler's identity by a functional equation. *Aequationes Math.* **28** (1985), 138–43.

C. R. HEATHCOTE

[1]   A branching process allowing immigration. *J. Roy. Statist. Soc., Ser. B*, 1965, pp. 138–43.

[2]   Corrections and comments on the paper: A branching process allowing immigration. *J. Roy. Statist. Soc., Ser. B*, 1966, pp. 213–17.

C. R. HEATHCOTE, J. W. PITMAN

[1]   An inequality for characteristic functions. *Bull. Austral. Math. Soc.* **6** (1972), 1–9.

C. R. HEATHCOTE, E. SENETA, D. VERE-JONES

[1]   A refinement of two theorems in the theory of branching processes. *Teor. Verojat. Primen.* **12** (1967), 341–6.

M. R. HERMAN

[1]   Sur la conjugaison différentiable des difféomorphismes du cercle à des rotations. *Publ. Math. IHES*, No. 49 (1979).

A. JA. HINČIN

[1]   *Continued fractions* [Russian]. Moscow and Leningrad, 1949. [English translation, The Univ. of Chicago Press, Chicago, Ill., and London, 1964].

M. W. HIRSCH, C. PUGH

[1]   Stable manifolds and hyperbolic sets. *Global analysis*, Proc. Symp. in Pure Math., Vol. 14, Amer. Math. Soc., Providence, RI, 1970, 133–65.

M. W. HIRSCH, C. PUGH, M. SHUB

[1]   Invariant manifolds. *Global analysis*, Proc. Symp. in Pure Math., Vol. 14, Amer. Math. Soc., Providence, RI, 1970, 1015–19.

M. W. HIRSCH, S. SMALE

[1]   *Differential equations and dynamical systems, and linear algebra.* Academic Press, New York and London, 1974.

F. M. HOPPE

[1]   Stationary measures for multitype branching processes. *J. Appl. Prob.* **12** (1975), 219–27.

[2]   The critical Biennaymé–Galton–Watson process. *Stochastic Processes Appl.* **5** (1977), 57–66.

[3]   On a Schröder equation arising in branching processes. *Aequationes Math.* **20** (1980), 33–7.

T. D. HOWROYD

[1]   On certain simultaneous functional equations. *Canad. Math. Bull.* **8** (1965), 77–82.

[2]   On the equivalence of a functional equation in n variables with n functional equations in a single variable. *Aequationes Math.* **4** (1970), 191–7.

[3]   Application of a fixed point theorem to simultaneous functional equations. *Aequationes Math.* **5** (1970), 116–17.

[4]   Functional inequalities. *General inequalities 1* (Proceedings of the Symposium, Oberwolfach, 1976), edited by E. F. Beckenbach, ISNM 41, Birkhäuser Verlag, Basle and Stuttgart, 1978, 153–8.

J. PH. HUNEKE

[1] Two commuting functions from the closed unit interval onto the closed unit interval without a common fixed point. *Topological dynamics* (Symposium, Colorado State Univ., Ft Collins, Colo., 1967). Benjamin, New York, 1968, 291–8.

R. ISAACS

[1] Iterates of fractional order. *Canad. J. Math.* **2** (1950), 409–16.

R. JÄGER

[1] Charakterisierung des Cotanges durch Replikativität. *Manuscripta Math.* **56** (1986), 167–75.

A. M. JAGLOM

[1] Some limit theorems in the theory of branching processes [Russian]. *Dokl. Akad. Nauk SSSR* **56** (1947), 795–8.

M. V. JAKOBSON

[1] Absolutely continuous invariant measures for one-parameter families of one-dimensional maps. *Commun. Math. Phys.* **81** (1981), 39–88.

K. JAKOWSKA-SUWALSKA

[1] On dependence of Lipschitzian solutions of nonlinear functional equation on an arbitrary function. *Annal. Math. Sil.* **1 (13)** (1985), 116–19.

T. JANKOWSKI, M. KWAPISZ

[1] On approximate iterations for systems of dynamic programming equations [Polish]. *Zeszyty Nauk. Politechniki Gdańskiej* **175**, *Matematyka* **6** (1971), 3–22.

W. JARCZYK

[1] A category theorem for linear functional equations in the indeterminate case. *Bull. Acad. Polon. Sci. Ser. sci. math.* **29** (1981), 371–82.

[2] On some infinite systems of inequalities and some fixed point theorem. *Prace Nauk. Uniw. Śląsk.* **425**, *Prace Mat.* **12** (1982), 34–42.

[3] On a set of functional equations having continuous solutions. *Glasnik Mat.* **17 (37)** (1982), 59–64.

[4] On linear functional equations in the indeterminate case. *Glasnik Mat.* **18 (38)** (1983), 91–102.

[5] Generic properties of nonlinear functional equations. *Aequationes Math.* **26** (1983), 40–53.

[6] On linear homogeneous functional equations in the indeterminate case. *Fund. Math.* **120** (1984), 99–104.

[7] On solutions of a certain functional-integral equation. *Ann. Polon. Math.* **44** (1984), 51–65.

[8] Nonlinear functional equations and their Baire category properties. *Aequationes Math.* **31** (1986), 81–100.

[9] On continuous functions which are additive on their graphs. *Berichte der Math.-Stat. Sektion in der Forschungsgesellschaft Joanneum–Graz*, No. 292 (1988).

[10] Nonnegative measurable solutions of a functional equation – a new proof of a result of M. Laczkovich. *Proceedings of the International Conference on Functional Equations and Inequalities*, Szczawnica, June 21–27, 1987. *Rocznik Nauk.-Dydakt. WSP Kraków* **115**, *Prace Mat.* **12** (1987), 188–90.

[11] A recurrent method of solving iterative functional equations. *Prace Nauk. Uniw. Śląsk.* (to appear).

J. JELONEK

[1] Lip $C^r\langle a, b\rangle$ solutions of a linear functional equation. *Demonstratio Math.* **18** (1986), 269–80.

F. JOHN

[1] Special solutions of certain difference equations. *Acta Math.* **71** (1939), 175–89.
[2] Discontinuous convex solutions of difference equations. *Bull. Amer. Math. Soc.* **47** (1941), 175–281.

G. JULIA

[1] Mémoire sur l'itération des fonctions rationnelles. *J. Math. Pures Appl.* (8) **1** (1918), 47–245.
[2] Mémoire sur la permutabilité des fractions rationnelles. *Ann. Sci. Ec. Norm. Sup.* (3) **39** (1922), 131–215.
[3] Mémoire sur la convergence des séries formées avec les itérées successives d'une fraction rationnelle. *Acta Math.* **56** (1930), 149–95. Addition: *Acta Math.* **58** (1932), 407–12.
[4] Fonctions continues sans dérivées formées avec les itérées d'une fraction rationnelle. *Ann. Sci. Ec. Norm. Sup.* (3) **48** (1931), 1–14.

M. KAC

[1] On the distribution of values of sums of type $\sum f(2^k t)$. *Ann. of Math.* **17** (1946), 33–49.

H.-H. KAIRIES

[1] Zur axiomatischen Charakterisierung der Gammafunktion. *J. reine angew. Math.* **236** (1969), 103–11.
[2] Über die logarithmische Ableitung der Gammafunktion. *Math. Ann.* **184** (1970), 157–62.
[3] Multiplikations theoreme. Habilitationsschrift, T. U. Brunswick, 1971.
[4] Definitionen der Bernoulli-Polynome mit Hilfe ihrer Multiplikations theoreme. *Manuscripta Math.* **8** (1973), 363–9.
[5] Ein Existenz- und Eindeutigkeitssatz für die Lösungen einer Funktionalgleichung aus der Nörlundschen Theorie der Differenzengleichungen. *Aequationes Math.* **10** (1974), 278–85.
[6] Convexity in the theory of the Gamma function. *General inequalities 1* (Proceedings of the Symposium, Oberwolfach, 1976), edited by E. F. Beckenbach, ISNN 41, Birkhäuser Verlag, Basle and Stuttgart, 1978, 49–62.
[7] Charakterisierungen von nirgends differenzierbaren Weierstrass-Funktionen durch Replikativität. *Elemente der Math.* (to appear).
[8] Ein System von Funktionalgleichungen vom verallgemeinerten Replikativitäts-Typ. *Per. Math. Hung.* **19 (4)** (1988), 261–71.

J. KALINOWSKI, M. KUCZMA

[1] Some remarks on continuous solutions of a functional equation. *Uniw. Śląsk. Prace Nauk.* **275**, *Prace Mat.* **9** (1979), 64–73.

PL. KANNAPPAN

[1] A characterisation of the cosine. *Studia Sci. Math. Hung.* **5** (1970), 417–19.

J. KARAMATA

[1] Sur une mode de croissance régulière. *Mathematica, Cluj* **4** (1930), 38–53.

[2] On asymptotic behaviour of sequences defined by recurrent relations [Serbo-Croatian]. *Srpska Akad. Nauka, Zbornik Radova* **35**, *Mat. Inst.* **3** (1953), 45–60.

S. KARLIN, J. MCGREGOR

[1] Embedding iterates of analytic functions with two fixed points into continuous groups. *Trans. Amer. Math. Soc.* **132** (1968), 137–45.

J. F. C. KINGMAN

[1] Stationary measures for branching processes. *Proc. Amer. Math. Soc.* **16** (1965), 245–7.

H. KNESER

[1] Reelle analytische Lösungen der Gleichung $\varphi(\varphi(x)) = e^x$ und verwandter Funktionalgleichungen. *J. reine angew. Math.* **187** (1950), 56–67.

G. KOENIGS

[1] Recherches sur les intégrales de certaines équations fonctionnelles. *Ann. Sci. Ec. Norm. Sup.* (3) **1** (1884), Supplement, pp. 3–41.

Z. KOMINEK

[1] C^r solutions of a system of functional equations. *Ann. Polon. Math.* **30** (1974), 191–203.

[2] On the uniqueness and the existence of solutions with some asymptotic properties of a system of functional equations. *Prace Nauk. Uniw. Śląsk.* **158**, *Prace Mat.* **7** (1977), 57–63.

[3] Some theorems on differentiable solutions of systems of functional equations of *n*-th order. *Ann. Polon. Math.* **37** (1980), 71–81.

[4] On the behaviour of C^r solutions of a system of functional equations. *Rev. Roum. Math. Pures Appl.* **23** (1978), 1167–76.

Z. KOMINEK, J. MATKOWSKI

[1] On the existence of a convex solution of the functional equation $\varphi(x) = h(x, \varphi[f(x)])$. *Ann. Polon. Math.* **30** (1974), 1–4.

[2] On the functional equation $\varphi(x) = \alpha\varphi(\alpha x) + (1 - \alpha)\varphi(1 - (1 - \alpha)x)$. *Bull. Math. Soc. Sci. Math., R. S. Roumanie (N.S.)* **30** (1986), 327–34.

N. KOPELL

[1] Commuting diffeomorphisms. *Global analysis* (Proc. Symp. in Pure Math., Vol. 14), Amer. Math. Soc., Providence, RI, 1970, pp. 165–84.

J. KORDYLEWSKI

[1] On the functional equation $F(x, \varphi(x), \varphi[f(x)], \varphi[f^2(x)], \ldots, \varphi[f^n(x)]) = 0$. *Ann. Polon. Math.* **9** (1960), 285–93.

[2] Continuous solutions of the functional equation $\varphi[f(x)] = F(x, \varphi(x))$ with the function $f(x)$ decreasing. *Ann. Polon. Math.* **11** (1961), 115–22.

[3] On continuous solutions of systems of functional equations. *Ann. Polon. Math.* **25** (1971), 53–83.

J. KORDYLEWSKI, M. KUCZMA

[1] On the functional equation $F(x, \varphi(x), \varphi[f(x)]) = 0$. *Ann. Polon. Math.* **7** (1959), 21–32.

[2] On some functional equations [Polish]. *Zeszyty Naukowe Uniw. Jagiellońskiego* **22**, *Prace Mat.* **5** (1959), 23–34.

[3] On the functional equation $F(x, \varphi(x), \varphi[f_1(x)], \ldots, \varphi[f_n(x)]) = 0$. *Ann. Polon. Math.* **8** (1960), 55–60.

[4] On some linear functional equations. I, II. *Ann. Polon. Math.*: **9** (1960), 119–36; **11** (1962), 203–7.

[5] On the continuous dependence of solutions of some functional equations on given functions. I, II. *Ann. Polon. Math.* **10** (1961), 41–8, 167–74.

H.-J. KORNSTAEDT

[1] Funktionalungleichungen und Iterationsverfahren. *Aequationes Math.* **13** (1975), 21–45.

M. A. KRASNOSEL'SKIĬ, G. M. VAĬOTKKO, P. P. ZABREĬKO, JA. B. RUTICKIĬ, V. JA. STECENKO

[1] *Approximate solution of operator equations* [Russian]. Izdat. 'Nauka', Moscow, 1969.

V. G. KRAVČENKO

[1] On solvability of linear functional equations [Russian]. *Mathematical physics* 14, Naukova Dumka, Kiev, 1973, 75–8.

W. KRULL

[1] Bemerkungen zur Differenzengleichung $g(x + 1) - g(x) = \varphi(x)$. I, II. *Math. Nachr.*: **1** (1948), 365–76; **2** (1949), 251–62.

M. KRÜPPEL

[1] Beiträge zur Theorie der vertauschbaren Funktionen. *Math. Nachr.* **56** (1973), 73–100.

Z. KRZESZOWIAK-DYBIEC

[1] Continuous solutions of the functional equation $\varphi(f(x)) = G(x, \varphi(x))$ for vector-valued functions φ. *Ann. Polon. Math.* **22** (1969), 207–16.

[2] On differentiable solutions of the functional equation $\varphi(f(x)) = g(x, \varphi(x))$ for vector-valued functions. *Ann. Polon. Math.* **27** (1973), 121–8.

[3] Existence of differentiable solutions of a system of functional equations of the first order. *Ann. Polon. Math.* **37** (1980), 119–29.

[4] On differentiable solutions of some systems of functional equations of p-th order. *Ann. Polon. Math.* **37** (1980), 135–48.

L. P. KUČKO

[1] On linear functional equations [Ukrainian]. *Dopovidi Akad. Nauk. Ukrain. SSR, Ser. A*, 1973, pp. 320–3.

[2] On linear functional equations [Russian]. *Vestnik Harkovsk. Univ., Ser. Meh.-Mat.* **39** (1974), 3–14.

[3] On local equivalence of functional equations [Russian]. *Ukr. Mat. Žurn.* **37** (1985), 506–9.

M. KUCZMA

[1] On convex solutions of the functional equation $g[\alpha(x)] - g(x) = \varphi(x)$. *Publ. Math. Debrecen* **6** (1959), 40–7.

[2] Remarks on some functional equations. *Ann. Polon. Math.* **8** (1960), 277–84.

[3] On monotonic solutions of a functional equation. I, II. *Ann. Polon. Math.*: **9** (1960), 295–7; **10** (1961), 161–6.

[4] General solution of the functional equation $\varphi(f(x)) = G(x, \varphi(x))$. *Ann. Polon. Math.* **9** (1961), 275–84.

[5]   (a) On some functional equations containing iterations of the unknown function.
      (b) On the functional equation $\varphi^n(x) = g(x)$. *Ann. Polon. Math.* **11** (1961): (a)
      1–5, (b) 161–75.

[6]   Sur une équation fonctionnelle. *Mathematica, Cluj* **3 (26)** (1961), 79–87.

[7]   A uniqueness theorem for a linear functional equation. *Glasnik Mat.-Fiz. Astr.*
      (2) **16** (1961), 177–81.

[8]   On some functional equations whose solutions can be represented with the aid
      of Euler's gamma function [Polish]. *Zeszyty Naukowe WSP w Katowicach,
      Sekcja Mat.* **3** (1962), 71–88.

[9]   On a recurrence relation. *Colloquium Math.* **9** (1962), 105–8.

[10] A characterization of the exponential and logarithmic functions by functional
      equations. *Fund. Math.* **52** (1963), 283–8.

[11] On the Schröder equation. *Rozprawy Mat.* [*Dissertationes Math.*] **34** (1963).

[12] A survey of the theory of functional equations. *Univ. Beograd, Publ. Elektrotehn.
      Fak. Ser. Mat. Fiz.* **130** (1964), 64 pp.

[13] Note on Schröder's functional equation. *J. Austral. Math. Soc.* **4** (1964), 149–51.

[14] Sur une équation aux différences finies et une caractérisation fonctionnelle des
      polynômes. *Fund. Math.* **55** (1964), 77–86.

[15] Sur une équation fonctionnelle qui caractérise la fonction $f(x) = x^{-1}$. *Publ.
      Inst. Math. Beograd (N.S.)* **4 (18)** (1964), 121–4.

[16] On a characterisation of the cosine. *Ann. Polon. Math.* **16** (1964), 53–7.

[17] Remark on a difference equation. *Commentationes Math.* **9** (1965), 2–8.

[18] Third note on the general solution of a functional equation. *Ann. Polon. Math.*
      17 (1965), 179–92.

[19] On convex solutions of Abel's functional equation. *Bull. Acad. Polon. Sci. Sér.
      sci. math. astronom. phys.* **13** (1965), 645–8.

[20] Bemerkungen über die Klassifikation der Funktionalgleichungen. *Roczniki
      PTM, Prace Mat.* **9** (1965), 169–83.

[21] Sur l'équation fonctionnelle de Böttcher. *Mathematica, Cluj* **8 (31)** (1966),
      279–85.

[22] Un théorème d'unicité pour l'équation fonctionnelle de Böttcher. *Mathematica,
      Cluj* **9 (32)** (1967), 285–93.

[23] Une remarque sur les solutions analytiques d'une équation fonctionnelle.
      Colloquium Math. **16** (1967), 93–9.

[24] On the functional characterization of the logarithm. *Funkc. Ekvacioj* **10** (1967),
      67–73.

[25] On a new characterization of the exponential functions. *Ann. Polon. Math.* **21**
      (1968), 39–46.

[26] *Functional equations in a single variable.* Monografie Mat. 46. Polish Scientific
      Publishers, Warsaw, 1968.

[27] On a conjecture of B. Schweizer. *Aequationes Math.* **2** (1968), 122–3.

[28] Some remarks on a functional equation characterizing the root. *Aequationes
      Math.* **2** (1969), 282–6.

[29] Analytic solutions of a linear functional equation. *Ann. Polon. Math.* **21** (1969),
      297–303.

[30] On a functional equation with divergent solutions. *Ann. Polon. Math.* **22** (1969), 173–8.

[31] Fractional iteration of differentiable functions. *Ann. Polon. Math.* **22** (1969), 217–27.

[32] On squares of differentiable functions. *Ann. Polon. Math.* **22** (1969), 229–37.

[33] Fractional iteration of differentiable functions with multiplier zero. *Commentationes Math.* **14** (1970), 35–9.

[34] P 63 R 2. *Aequationes Math.* **5** (1970), 327–8.

[35] On integrable solutions of a functional equation. *Bull. Acad. Polon. Sci., Sér. sci. math. astronom. phys.* **19** (1971), 593–6.

[36] (editor) *Functional equations in the theory of stochastic processes* [Polish]. Silesian University, Katowice, 1972.

[37] On a certain series. *Ann. Polon. Math.* **26** (1972), 199–204.

[38] Special solutions of a functional equation. *Ann. Polon. Math.* **27** (1972), 29–37.

[39] Quelques observations à propos de l'équation pré-Schröder. *Ann. Polon. Math.* **28** (1973), 49–52.

[40] A contribution to the theory of fractional iteration of differentiable functions. *Demonstratio Math.* **6** (1973), 181–9.

[41] Note on linearization. *Ann. Polon. Math.* **29** (1974), 75–81.

[42] Normalizing factors for iterates of random-valued functions. *Prace Nauk. Uniw. Śląsk.* **87**, *Prace Mat.* **6** (1975), 67–72.

[43] Regularly varying solutions of a linear functional equation. *J. Austral. Math. Soc.* **22** (1976), 135–43.

[44] Various aspects of the functional equation of Schröder. *Colloques Internationaux du C.N.R.S. – Transformations ponctuelles et leurs applications* [Toulouse, 10–14 September 1973], 245–57 (1976).

[45] On the number of continuous solutions of a functional equation. *Prace Nauk. Uniw. Śląsk.* **158**, *Prace Mat.* **7** (1977), 51–5.

[46] General continuous solution of a linear homogeneous functional equation. *Ann. Polon. Math.* **35** (1977), 21–5.

[47] Non-negative continuous solutions of a functional inequality. *Ann. Polon. Math.* **36** (1979), 187–99.

[48] Uniqueness of solutions of functional equations in a single variable. *Proceedings of the Symposium on Differential Equations with Deviated Argument* [Kiev, September 1975].

[49] On a theorem of Barna. *Aequationes Math.* **21** (1980), 173–8.

[50] Regular fractional iteration of convex functions. *Ann. Polon. Math.* **38** (1980), 95–100.

[51] An introduction to the theory of functional equations and inequalities. Cauchy's equation and Jensen's inequality. *Prace Nauk. Uniw. Śla* **489**, Polish Scientific Publishers, Warsaw–Cracow–Katowice, 1985, 523 pp.

[52] On some properties of solutions of a functional equation. *Opuscula Math.* **3** (1987), 37–41.

[53] A generic property of a linear functional equation. *Opuscula Math.* **4** (1988), 139–44.

528   References

M. KUCZMA, J. MATKOWSKI

[1] Solutions of a functional equation in a special class of functions. *Ann. Polon. Math.* **26** (1972), 287–93.

M. KUCZMA, A. SMAJDOR

[1] Note on iteration of concave functions. *Amer. Math. Monthly* **74** (1967), 401–2.

[2] Fractional iteration in the class of convex functions. *Bull. Acad. Polon. Sci., Sér. sci. math. astronom. phys.* **16** (1968), 717–20.

[3] Regular fractional iteration. *Bull. Acad. Polon. Sci., Sér. sci. math. astronom. phys.* **19** (1971), 203–7.

M. KUCZMA, W. SMAJDOR

[1] On the problem of uniqueness for solutions of a functional equation. *Bull. Acad. Polon. Sci., Sér. sci. math. astronom. phys.* **19** (1971), 301–4.

[2] On the radius of convergence of series solutions of a functional equation. *Ann. Polon. Math.* **24** (1971), 233–40.

[3] Analytic solutions of some functional equations. *J. London Math. Soc.* (2) **4** (1972), 418–24.

M. KUCZMA, K. SZYMICZEK

[1] On periodic solutions of a functional equation. *Amer. Math. Monthly* **70** (1963), 847–50.

M. KUCZMA, GY. TARGONSKI

[1] On a Pre-Schröder equation. *Bull. Acad. Polon. Sci., Sér. sci. math. astronom. phys.* **18** (1970), 721–4.

M. KUCZMA, P. VOPĚNKA

[1] On the functional equation $\lambda[f(x)]\lambda(x) + A(x)\lambda(x) + B(x) = 0$. *Ann. Univ. Sci. Budapest., Sect. Math.* **3–4** (1960/1), 123–33.

K. KURATOWSKI

[1] Sur une équation fonctionnelle. *Sprawozdania z posiedzeń Tow. Nauk. Warszaw.* **22** (1929), 160–1.

M. KWAPISZ

[1] On a method of successive approximations and qualitative problems for differential-functional and difference equations in Banach space [Polish]. *Zeszyty Nauk. Politechniki Gdańskiej* **79**, *Matematyka* **4** (1965), 3–73.

[2] On the approximate solutions of an abstract equation. *Ann. Polon. Math.* **19** (1967), 47–60.

[3] On a certain functional equation. *Colloquium Math.* **18** (1967), 169–79.

M. KWAPISZ, J. TURO

[1] On the existence and convergence of successive approximations for some functional equations in a Banach space. *J. Diff. Equations* **16** (1974), 298–318.

[2] On the existence and uniqueness of solutions of the Darboux problem for partial differential-functional equations in a Banach space. *Ann. Polon. Math.* **29** (1974), 89–118.

[3] Existence and uniqueness of solutions of non-linear functional equations of r-th order. *Ann. Polon. Math.* **31** (1975), 145–57.

[4] Existence, uniqueness and successive approximation for a class of integral-functional equations. *Aequationes Math.* **14** (1976), 303–23.

M. LACZKOVICH
[1] Nonnegative measurable solutions of a difference equation. *J. London Math. Soc.* (2) **34** (1986), 139–47.

M. LACZKOVICH, SZ. RÉVÉSZ
[1] Decomposition into periodic functions belonging to a given Banach space, manuscript. University of Budapest, 1986.

R. G. LAHA, E. LUKACS
[1] On a functional equation which occurs in a characterization problem. *Aequationes Math.* **16** (1977), 259–74.

R. G. LAHA, E. LUKACS, A. RÉNYI
[1] A generalization of a theorem of E. Vincze. *Magyar Tud. Akad. Mat. Kutató Közl.* **9** (1964), 237–9.

A. LASOTA
[1] Invariant measures and functional equations. *Aequationes Math.* **9** (1973), 193–200.
[2] On mappings isomorphic to *r*-adic transformations. *Ann. Polon. Math.* **35** (1978), 313–22.

A. LASOTA, J. A. YORKE
[1] An example of an Anosov diffeomorphism without a smooth invariant measure. Unpublished.

S. LATTÈS
[1] Sur les équations fonctionnelles qui définissent une courbe ou une surface invariante par une transformation. *Ann. di Mat.* (3) **13** (1906), 1–137.
[2] Sur les courbes invariantes par polaires réciproques. *Nouv. Ann. de Math.* (4) **6** (1906), 308–12.
[3] Nouvelles recherches sur les courbes invariantes par une transformation $(X, Y; x, y, y')$. *Ann. Sci. Ec. Norm. Sup.* (3) **25** (1908), 221–54.
[4] Sur les multiplicités invariantes par une transformation de contact. *Bull. Soc. Math. France* **37** (1909), 137–63.

P. LÉVY
[1] Fonctions à croissance régulière et itération d'ordre fractionnaire. *Ann. Mat. Pura Appl.* (4) **5** (1928), 269–98.

T. Y. LI, J. A. YORKE
[1] Period 3 implies chaos. *Amer. Math. Monthly* **82** (1975), 985–92.

J. C. LILLO
[1] The functional equation $f''(x) = g(x)$. *Arkiv för Mat.* **5** (1965), 357–61.
[2] The functional equation $f^2(x) = g(x)$. *Ann. Polon. Math.* **19** (1967), 123–35.

J. S. LIPIŃSKI
[1] Sur une intégrale. *Colloquium Math.* **7** (1959), 67–74.
[2] Sur la composition commutative des fonctions. *Colloquium Math.* **10** (1963), 271–6.

A. N. LIVŠIC
[1] Some properties of the homology of Y-systems [Russian]. *Mat. Zam.* **10 (5)** (1971), 555–64.

M. LOÈVE
[1] *Probability theory*. 2nd edition. Van Nostrand, Princeton–Toronto–New York–London, 1961.

S. ŁOJASIEWICZ

[1] Solution générale de l'équation fonctionnelle $f(f \cdots f(x) \cdots) = g(x)$. *Ann. Polon. Math.* **24** (1951), 88–91.

A. J. LOTKA

[1] The extinction of families. I, II. *J. Wash. Acad. Sci.* **21** (1931), 377–80; 453–9.

R. D. LUCE, W. EDWARDS

[1] The derivation of subjective scales from just noticeable differences. *Psychological Review* **65** (1958), 222–37.

L. LUCHT

[1] Zur Charakterisierung der Gamma-Funktion. *J. reine angew. Math.* **288** (1976), 77–85.

[2] Zur Gausschen Funktionalgleichung für die Gamma-Funktion auf multiplikativen Zahlenmengen. *Abh. Math. Seminar Univ. Hamburg* **49** (1979), 183–8.

L. LUCHT, D. WOLKE

[1] Trigonometrische Reihen über multiplikativen Zahlenmengen. *Math. Zeitschr.* **149** (1976), 155–67.

[2] Trigonometrische Reihen über multiplikativen Zahlenmengen. II. *Acta Math. Acad. Sci. Hung.* **33** (1979), 263–88.

E. LUKACS

[1] Non-negative definite solutions of certain differential and functional equations. *Aequationes Math.* **2** (1969), 137–43.

E. LUKACS, R. G. LAHA

[1] *Applications of characteristic functions.* Charles Griffin, London, 1964.

A. LUNDBERG

[1] On iterated functions with asymptotic conditions at a fixpoint. *Arkiv för Mat.* **5** (1963), 193–206.

K. LUNDMARK

[1] On the Abel–Schwarzschild functional equation and its astronomical applications. *9. Congr. Math. Scand.* (1939), 323–44.

P. E. LUSH

[1] The stability of Harrod's growth model of an economy. *J. Austral. Math. Soc.* **5** (1965), 207–15.

W. LÜSSY

[1] Lösung der Aufgabe 173. *Elem. Math.* **9** (1954), 40.

W. MAIER, H. KIESEWETTER

[1] *Funktionalgleichungen mit analytischen Lösungen.* VEB Deutscher Verlag der Wissenschaften, Berlin, 1971.

G. MAJCHER

[1] On some functional equations. *Ann. Polon. Math.* **16** (1964), 35–44.

[2] On a new problem for differential equations of hyperbolic type [Polish]. *Politechnika Krakowska, Zeszyt Naukowy* **11** (1965).

[3] On a linear functional equation of order q. *Commentationes Math.* **10** (1966), 21–6.

[4] Sur l'application de la fonction de Riemann et de l'intégrale de Roux à la solution du problème généralisé de Goursat. *Arch. Mech. Stos.* **18** (1966), 181–92.

[5] Applications des équations fonctionnelles dans la théorie des équations différentielles partielles. *Zeszyty Naukowe Uniw. Jagiellońskiego* **223**, *Prace Mat.* **14** (1969), 109–18.

M. MALENICA

[1] On a solution of the functional equation $\varphi(x) + \varphi(f(x)) = F(x)$ [Serbo-Croatian]. *Akad. Nauka Umjet. Bosne Hercegov. Rad. Odjelj. Prirod. Mat. Nauka.* **66** (1980), 103–9.

[2] On some solutions of equations $\varphi(x) + \varphi(f(x)) = F(x)$ under the condition that F satisfies $F(f^p(x)) = F(x)$. *Publ. Inst. Math. Beograd (N.S.)* **29 (43)** (1981), 139–44.

[3] On a solution of the functional equation $\varphi(x) + \varphi(f(x)) = F(x)$ and some special examples of regular transformations [Serbo-Croatian]. *Akad. Nauka Umjet. Bosne Hercegov. Rad. Tehn. Nauka.* **68 (6)** (1981), 127–32.

[4] On some solutions of the equation $\varphi(x) + \varphi(f(x)) = F(x)$ when the function F satisfies the condition $F(f(x)) = F(x)$ [Serbo-Croatian]. *Akad. Nauka Umjet. Bosne Hercegov. Rad. Odjelj. Prirod. Mat. Nauka.* **69 (20)** (1982), 17–21.

[5] On the solutions of the functional equation $\varphi(x) + \varphi(f(x)) = F(x)$ using θ-summability. *Glas. Mat.* **18** (1983), 305–15.

R. H. MARTIN, JR.

[1] *Nonlinear operators and differential equations in Banach spaces*. John Wiley & Sons, New York–London–Sydney–Toronto, 1976.

A. MATKOWSKA

[1] On characterization of Lipschitzian operators of substitution in the class of Hölder's functions. *Zeszyty Nauk. Polit. Łódzkiej* **430**, *Matematyka* **17** (1984), 81–5.

J. MATKOWSKI

[1] On some properties of solutions of a functional equation. *Prace Nauk. Uniw. Śląsk.* **2**, *Prace Mat.* **1** (1969), 79–82.

[2] On meromorphic solutions of a functional equation. I, II. *Ann. Polon. Math.*: **22** (1970), 303–11; **24** (1971), 313–18.

[3] On the continuous dependence of solutions of a functional equation on given functions. *Ann. Polon. Math.* **23** (1970), 37–40.

[4] On the continuous dependence of local analytic solutions of a functional equation on given functions. *Ann. Polon. Math.* **24** (1970), 21–6. Correction: *Ann. Polon. Math.* **26** (1972), 219.

[5] On the uniqueness of differentiable solutions of a functional equation. *Bull. Acad. Polon. Sci., Sér. sci. math. astronom. phys.* **18** (1970), 253–5.

[6] On the existence of differentiable solutions of a functional equation. *Bull. Acad. Polon. Sci., Sér. sci. math. astronom. phys.* **19** (1971), 19–22.

[7] Note on a functional equation. *Zeszyty Naukowe Uniw. Jagiellońskiego* **252**, *Prace Mat.* **15** (1971), 109–11.

[8] On the continuous dependence of local analytic solutions of the functional equation in the non-uniqueness case. *Ann. Polon. Math.* **24** (1971), 319–26.

[9] On the continuous dependence of C^r solutions of a functional equation on given functions. *Aequationes Math.* **6** (1971), 215–27.

[10] The uniqueness of solutions of a system of functional equations in some classes of functions. *Aequationes Math.* **8** (1972), 233–7.

[11] On Lipschitzian solutions of a functional equation. *Ann. Polon. Math.* **28** (1973), 135–9.

[12] Some inequalities and a generalization of Banach's principle. *Bull. Acad. Polon. Sci., Sér. sci. math. astronom. phys.* **21** (1973), 323–5.

[13] Integrable solutions of functional equations. *Dissertationes Math.* **127** (1975).

[14] Functional equations and Nemytskii operators. *Funkcial. Ekvac.* **25** (1982), 127–32.

[15] Form of Lipschitz operators of substitution in Banach spaces of differentiable functions. *Zeszyty Nauk. Polit. Łódzkiej* **430**, *Matematyka* **17** (1984), 5–10.

[16] On a characterization of norms in L^p and functional equations. *Proceedings of the International Conference on Functional Equations and Inequalities, May 27–June 2, 1983, Sielpia* (Poland). *Rocznik Nauk.-Dydakt. WSP Kraków* **97**, *Prace Mat.* **11** (1985), 194–5.

[17] Cauchy functional equation on a restricted domain and commuting functions. *Iteration theory and its functional equations. Proceedings, Schloss Hofen*, edited by R. Liedl, L. Reich and Gy. Targoński, Lecture Notes in Mathematics 1163, Springer Verlag, Berlin–Heidelberg–New York–Tokyo, 1985, 101–6.

[18] Remark on BV solutions of a functional equation connected with invariant measures. *Aequationes Math.* **29** (1985), 210–13.

J. MATKOWSKI, J. MIŚ

[1] On the characterization of Lipschitzian operators of substitution in the space BV$\langle a, b \rangle$. *Math. Nachr.* **117** (1984), 155–9.

J. MATKOWSKI, W. OGIŃSKA

[1] Note on iterations of some entire functions. *Ann. Univ. M. Curie-Skłodowska, Sect. A* **38** (1979), 111–18.

J. MATKOWSKI, R. WĘGRZYK

[1] On equivalence of some fixed point theorems for selfmappings of metrically convex space. *Boll. Un. Mat. Ital.* (5) **15-A** (1978), 359–69.

J. MATKOWSKI, M. C. ZDUN

[1] Solutions of bounded variation of a linear functional equation. *Aequationes Math.* **10** (1974), 223–35.

A. MAYER

[1] Konvexe Lösungen der Funktionalgleichung $1/f(x + 1) = xf(x)$. *Acta Math.* **70** (1939), 57–62.

T. L. MCCOY

[1] On 'cos $F(x) = F(\sin x)$'. *Amer. Math. Monthly* **93** (1986), 111–15.

N. MCSHANE

[1] On the periodicity of homeomorphisms of the real line. *Amer. Math. Monthly* **68** (1961), 562–3.

A. MEIR, E. KEELER

[1] A theorem on contraction mappings. *J. Math. Anal. Appl.* **28** (1969), 326–9.

K. MENGER

[1] An axiomatic theory of functions and fluents. *The axiomatic method*, edited by L. Henkin *et al.*, North Holland, Amsterdam, 1959, pp. 454–73.

P. R. MEYERS

[1] A converse to Banach's contraction theorem. *J. Res. Nat. Bur. Standards-B. Math. and Math. Phys.* **71 B N1** (1967), 73–6.

J. MIŚ

[1] Lipschitzian Nemitskiĭ operators in some Banach spaces and their applications to the theory of functional equations [Polish]. Doctoral Dissertation, Technical University of Łódź, 1983.

M. MISIUREWICZ

[1] On iterates of e^z. *Ergod. Th. & Dynam. Sys.* **1** (1981), 103–6.

E. MOHR

[1] Elementärer Beweis für die Partialbruchzerlegung des Cotangens. *Z. Angew. Math. Mech.* **33** (1953), 247–8.

E. MOLDOVAN

[1] Sur une généralisation de fonctions convexes. *Mathematica, Cluj* **1 (24)** (1959), 49–80.

P. MONTEL

[1] *Leçons sur les récurences et leurs applications.* Gauthier-Villars, Paris, 1957.

[2] Sur les propriétés périodiques des fonctions. *C. R. Acad. Sci. Paris* **251** (1960), 2111–12.

J. MOSER

[1] A rapidly convergent iteration method and non-linear differential equations. I, II. *Ann. Scuola Norm. Super. Pisa, Ser. III* **20** (1966), 265–315, 499–535.

Z. MOSZNER

[1] P 2. *Aequationes Math.* **1** (1968), 150.

[2] P 63 S 1 and P 63 S 2. *Aequationes Math.* **4** (1970), 395.

[3] Sur un problème relatif aux équations de Pré-Schröder. *Ann. Polon. Math.* **27** (1973), 289–92.

[4] Structure de l'automate plein, réduit et inversible. *Aequationes Math.* **9** (1973), 46–59.

[5] The translation equation and its application. *Demonstratio Math.* **6** (1973), 309–27.

B. MUCKENHOUPT

[1] Some results on analytic iteration and conjugacy. *Amer. J. Math.* **84** (1962), 161–9.

M. E. MULDOON

[1] Some characterizations of the Gamma function involving the notion of complete monotonicity. *Aequationes Math.* **8** (1972), 212–15.

[2] Some monotonicity properties and characterizations of the gamma function. *Aequationes Math.* **18** (1978), 54–63.

J. MYCIELSKI

[1] On the conjugates of the function $2|x| - 1$ in $[-1, 1]$. *Bull. London Math. Soc.* **12** (1980), 4–8.

P. J. MYRBERG

[1] Über die Linearisierung der schlichten konformen Abbildungen. *Ann. Acad. Sci. Fenn., Ser. A. I* **145** (1953).

[2] Eine Verallgemeinerung der Abelschen Funktionalgleichung. *Ann. Acad. Sci. Fenn., Ser. A. I* **327** (1962).

S. NABEYA

[1] On the functional equation $f(p + qx + rf(x)) = a + bx + cf(x)$. *Aequationes Math.* **11** (1974), 199–211.

J. NAWROCKI

[1] On the existence of C^r solutions of a functional equation containing a superposition of the unknown function. *Demonstratio Math.* **17** (1984), 897–905.

F. NEUMAN

[1] Sur les équations différentielles linéaires du second ordre dont les solutions ont des racines formant une suite convexe. *Acta Math. Acad. Sci. Hung.* **13** (1962), 281–7.

[2] Construction of second order linear differential equations with solutions of prescribed properties. *Archivum Mathematicum Brno* **1** (1965), 229–46.

[3] A role of Abel's equation in the stability theory of differential equations. *Aequationes Math.* **6** (1971), 66–70.

[4] L^2-solutions of $y'' = q(t)y$ and a functional equation. *Aequationes Math.* **6** (1971), 162–9.

[5] Linear differential equations of the second order and their applications. *Rendiconti di Mat.* (3) **4** (1971), 559–617.

[6] Distribution of zeros of solutions of $y'' = q(t)y$ in relation to their behaviour in large. *Studia Sci. Math. Hung.* **8** (1973), 177–85.

[7] On n-dimensional closed curves and periodic solutions of linear differential equations of the n-th order. *Demonstratio Math.* **6** (1973), 329–37.

[8] On solutions of the vector functional equation $y(\zeta(x)) = f(x)Ay(x)$. *Aequationes Math.* **16** (1977), 245–57.

[9] On transformations of differential equations and systems with deviating argument. *Czechoslovak Math. J.* **31 (106)** (1981), 87–90.

[10] Simultaneous solutions of a system of Abel equations and differential equations with several deviations. *Czechoslovak Math. J.* **32 (10)** (1982), 488–94.

[11] P 235. *Aequationes Math.* **26** (1984), 283.

[12] *Functional equations* [Czech]. Matematicky seminar SNTL, Prague, 1986.

[13] *Ordinary linear differential equations.* Academia, Prague, North Oxford Academic Publishers Ltd., Oxford, 1989.

M. NEWMAN, M. SHEINGORN

[1] Continuous solutions of a homogeneous functional equation. *Aequationes Math.* **13** (1975), 47–59. Corrigendum: *ibidem* **17** (1978), 111.

Z. NITECKI

[1] *Differentiable dynamics – an introduction to the orbit structure of diffeomorphisms.* The MIT Press, Cambridge, Mass., and London, England, 1971.

N. E. NÖRLUND

[1] *Vorlesungen über Differenzenrechnung.* Springer Verlag, Berlin, 1924. [Chelsea, New York, 1954].

W. OGIŃSKA

[1] On the Fatou set for entire functions [Polish]. Doctoral Dissertation, Technical University of Łódź, 1981.

[2] On Fatou's set for some entire functions. *Zeszyty Nauk. Polit. Łódzkiej* **430**, *Matematyka* **17** (1984), 63–8.

A. M. OSTROWSKI

[1] Sur la convergence et l'estimation des erreurs dans quelques procédés de résolution des équations numériques. *Collection of papers in memory of D. A. Grave.* Moscow, 1940, 213–34.

[2] *Solution of equations and systems of equations*. Academic Press, New York and London, 1960.

[3] Über algebräische Lösungen Φ der Funktionalgleichung $\Phi(\varphi(x)) = g(x)\Phi(x)$, für rationale $g(x)$, deren rationale φ entsprechen. (In: Die vierte Tagung über Funktionalgleichungen, Oberwolfach 10.–16. Juli 1966.) *Aequationes Math.* **1** (1968), 135–49.

A. G. PAKES

[1] An asymptotic result for a subcritical branching process with immigration. *Bull. Austral. Math. Soc.* **2** (1970), 223–8.

[2] A branching process with a state dependent immigration component. *Adv. Appl. Prob.* **3** (1971), 301–14.

[3] On supercritical Galton–Watson process allowing immigation. *J. Appl. Prob.* **11** (1974), 814–17.

[4] Some limit theorems for a supercritical branching process allowing immigration. *J. Appl. Prob.* **12** (1975), 17–26.

A. G. PAKES, N. KAPLAN

[1] On the subcritical Bellman–Harris process with immigration. *J. Appl. Prob.* **11** (1974), 652–68.

J. PAVALOIU

[1] La résolution des systèmes d'équations opérationnelles à l'aide des méthodes itératives. *Mathematica, Cluj* **11 (34)** (1969), 137–41.

A. PELCZAR

[1] On the extremal solutions of a functional equation. *Zeszyty Naukowe Uniw. Jagiellońskiego* **55**, *Prace Mat.* **7** (1962), 9–11.

[2] On the extremal solutions of some functional equations. *Zeszyty Naukowe Uniw. Jagiellońskiego* **102**, *Prace Mat.* **10** (1965), 71–8.

[3] On some equations in partially ordered spaces. *Zeszyty Naukowe Uniw. Jagiellońskiego* **167**, *Prace Mat.* **12** (1968), 43–6.

[4] Remarks on some functional equations and inequalities. *Zeszyty Naukowe Uniw. Jagiellońskiego* **167**, *Prace Mat.* **12** (1968), 47–52.

[5] Remarks on commuting mappings in partially ordered spaces. *Zeszyty Naukowe Uniw. Jagiellońskiego* **252**, *Prace Mat.* **15** (1971), 129–32.

G. P. PELYUKH

[1] Existence and uniqueness of the analytic solution of a non-linear functional equation [Russian]. *Aequationes Math.* **16** (1977), 123–37.

G. P. PELYUKH, A. N. ŠARKOVSKIĬ

[1] Characterization of elementary and special functions [Ukrainian]. *Dopovidi Akad. Nauk Ukrain. SSR, Ser. A* **11** (1970), 994–8.

[2] General solution of a system of functional equations in a neighbourhood of a singular point [Russian]. *Functional and differential-difference equations* [Russian]. Izd. Inst. Mat. Akad. Nauk Ukrain. SSR, Kiev, 1974, 100–9.

[3] *Introduction to the theory of functional equations* [Russian]. Izdat. 'Naukova Dumka', Kiev, 1974.

S. PENNER, K. J. SCHROEDER

[1] Abstract characterizations of continuous functions. *Aequationes Math.* **11** (1974), 121–7.

E. PESCHL

[1]   Zur Theorie der schlichten Funktionen. *J. reine angew. Math.* **176** (1937), 69.

E. PESCHL, L. REICH

[1]   Über die globale Lösung der Schröderschen Funktionalgleichung für holomorphe Abbildungen mit Anziehendem Fixpunkt. *Archiv der Math.* **21** (1970), 578–82.

A. PLIŚ, T. WAŻEWSKI

[1]   Functions with all partial derivatives arbitrarily prescribed at a point. *Ann. Polon. Math.* **12** (1962), 155–7.

H. POINCARÉ

[1]   Sur une classe étendue de transcendentes uniformes. *C. R. Acad. Sci. Paris* **103** (1886), 862–4.

[2]   Sur une classe nouvelle de transcendentes uniformes. *J. Math. Pures Appl.* (4) **6** (1890), 313–65.

T. POPOVICIU

[1]   Sur quelques propriétés des fonctions d'une ou de deux variables réelles. *Mathematica, Cluj* **8** (1934), 1–85.

[2]   Remarques sur la définition fonctionnelle d'un polynôme d'une variable réelle. *Mathematica, Cluj* **12** (1936), 5–12.

W. PRANGER

[1]   Series solutions of the functional equation. *J. Austral. Math. Soc.* **5** (1965), 48–55.

S. PREŠIĆ

[1]   Méthode de résolution d'une classe d'équations fonctionnelles linéaires. *C. R. Acad. Sci. Paris* **257** (1963), 2224–6.

[2]   Méthode de résolution d'une classe d'équations fonctionnelles linéaires. *Univ. Beograd, Publ. Elektrotehn. Fak. Ser. Mat. Fiz.* **119** (1963), 21–8.

D. PRZEWORSKA-ROLEWICZ

[1]   Generalized Fox integral equations solved by functional equations. *Selected topics in operation research and mathematical economics. Proceedings* [Karlsruhe, 1983]. Springer Verlag, Berlin–Heidelberg–New York–Tokyo, 1984, 421–9.

R. RACLIS

[1]   Sur la solution méromorphe d'une équation fonctionnelle. *Bull. Math. Soc. Roum. Sci.* **30** (1927), 101–5.

E. RAKOTCH

[1]   A note on contractive mappings. *Proc. Amer. Math. Soc.* **13** (1962), 459–65.

O. RAUSENBERGER

[1]   Theorie der allgemeinen Periodizität. *Math. Ann.* **18** (1881), 379–410.

M. REGHIŞ, L. VUC

[1]   Sur les équations fonctionnelles itératives. I. *Publ. Math. Debrecen* **13** (1966), 25–39.

L. REICH

[1]   Normalformen biholomorpher Abbildungen mit anziehendem Fixpunkt. *Math. Ann.* **180** (1969), 233–55.

B. A. REZNICK

[1]   A uniqueness criterion for fractional iteration. *Ann. Polon. Math.* **30** (1975), 219–24.

G. DE RHAM

[1]   Sur certaines équations fonctionnelles. *Ecole Polytechnique de l'Université de Lausanne. Centenaire 1853–1953*. Lausanne, 1953, 95–7.

[2]   Sur une courbe plane. *J. Math. Pures Appl.* (9) **35** (1956), 35–42.

[3]   Sur quelques courbes définies par des équations fonctionnelles. *Univ. e Politec. Torino, Rend. Sem. Mat.* **16** (1956/7), 101–13.

[4]   Sur un exemple de fonction continue sans dérivée. *Enseignement Math.* (2) **3** (1957), 71–2.

[5]   Sur quelques fonctions différentiables dont toutes les valeurs sont des valeurs critiques. *Celebrazioni Archimedee del Sec. X X (Siracusa, 1961)*, Vol. II, Gubbio, 1962, 61–5.

E. E. RICE

[1]   Iterative square roots of Čebyšev polynomials. *Stochastica* **3** (1979), 1–14.

E. E. RICE, B. SCHWEIZER, A. SKLAR

[1]   When is $f(f(z)) = az^2 + bz + c$? *Amer. Math. Monthly* **87** (1980), 252–63.

E. RIEKSTIŅŠ

[1]   On asymptotic expansion of sequences generated by an iterative process [Russian]. *Latv. Mat. Ežegodnik* **4** (1968), 291–311.

H. E. ROBBINS

[1]   Two properties of the function cos x. *Bull. Amer. Math. Soc.* **50** (1944), 750–2.

M. ROZMUS-CHMURA

[1]   Sur l'équation fonctionnelle $\varphi(x + 1) = f(\varphi(x))$. *Mat. Vesnik* **4** (1967), 75–8.

[2]   Les solutions convexes de l'équation fonctionnelle $g(\alpha(x)) - g(x) = \varphi(x)$. *Publ. Math. Debrecen* **15** (1968), 45–8.

[3]   Regular solutions of some functional equations in the indeterminate case. *Ann. Math. Sil.* **1 (13)** (1985), 120–9.

M. ROŻNOWSKI

[1]   Approximate solutions of the functional equation $\varphi^n(x) = f(x)$ [Polish]. *Prace Nauk. Uniw. Śląsk.* **30**, *Prace Mat.* **3** (1973), 53–73.

D. RUELLE

[1]   Applications conservant une mesure absolument continue par rapport dx sur [0, 1]. *Comm. Math. Phys.* **55** (1977), 47–51.

J. A. RUS

[1]   On fixed points of mappings defined on a Cartesian product. II: metric spaces [Roumanian]. *St. Cerc. Mat.* **24** (1972), 897–904.

D. C. RUSSELL

[1]   On bounded sequences satisfying a linear inequality. *Proc. Edin. Math. Soc.* (2) **19** (1974), 11–16.

H. RÜSSMANN

[1]   Über die Iteration analytischer Funktionen. *J. Math. Mech.* **17** (1967), 523–32.

ST. RUZIEWICZ

[1]   On a continuous function without derivative on an uncountable set [Polish]. *Sprawozdania z posiedzeń Tow. Nauk. Warszaw.* **6** (1913), 282–305.

T. L. SAATY

[1]   *The analytic hierarchy process*. McGraw-Hill, New York, 1980.

M. SABLIK

[1]   On iterates of multi-hat functions. *Prace Nauk. Uniw. Śląsk.* **218**, *Prace Mat.* **8** (1978), 41–4.

[2] Note on a conditional Cauchy equation. *Radovi Matematički* **1** (1985), 241–5.

[3] Differentiable solutions of functional equations in Banach spaces. *Facta Universitatis (Niš), Ser. Math. Inform.* (to appear).

[4] Conditional translation equation. *Berichte der Math.-Stat. Sektion in der Forschungsgesellschaft Joanneum–Graz*, No. 295 (1988).

A. N. ŠARKOVSKIĬ

[1] Necessary and sufficient conditions for the convergence of one-dimensional iterative processes [Russian]. *Ukrain. Mat. Ž.* **12** (1960), No. 4, 484–9.

[2] Coexistence of cycles of the continuous transformation of the real line into itself [Russian]. *Ukrain. Mat. Ž.* **16** (1964), No. 1, 61–71.

[3] On a classification of fixed points. *Ukrain. Mat. Ž.* **17** (1965), No. 5, 80–95.

[4] On attractive and self-attractive sets [Russian]. *Dokl. Akad. Nauk SSSR* **160** (1965), 1036–8.

[5] Continuous transformation on the set of cluster points of an iterative sequence [Russian]. *Ukrain. Mat. Ž.* **18** (1966), No. 5, 127–30.

[6] Characterization of the cosine [Russian]. *Aequationes Math.* **9** (1973), 121–8.

H. SCHMIDT

[1] Eine Bemerkung zum Aufbau der Lehre von der Γ-Funktion. *Archiv der Math.* **9** (1958), 297–9.

E. SCHRÖDER

[1] Über iterierte Funktionen. *Math. Ann.* **3** (1871), 296–322.

C. F. SCHUBERT

[1] Solution of a generalized Schröder equation in two variables. *J. Austral. Math. Soc.* **4** (1964), 410–17.

P. A. SCHWARZMAN

[1] On the convergence of an iterative series [Russian]. *Vestnik Jarosl'av. Univ.* **12** (1975), 155–9.

B. SCHWEIZER, A. SKLAR

[1] The algebra of functions. I, II, III. *Math. Ann.*: **130** (1960), 366–82; **143** (1961), 440–7; **161** (1965), 171–96.

[2] Function systems. *Math. Ann.* **172** (1967), 1–16.

[3] A grammar of functions. I, II. *Aequationes Math.*: **2** (1968), 62–85; **3** (1969), 15–43.

[4] All trapezoid functions are conjugate. *C. R. Math. Rep. Acad. Sci. Canada* **5** (1983), 275–80.

[5] *Probabilistic metric spaces.* North-Holland, New York, 1983.

[6] Continuous functions that conjugate trapezoid functions. *Aequationes Math.* **28** (1985), 303–4.

H. SCHWERDTFEGER

[1] Involutory functions and even functions. *Aequationes Math.* **2** (1968), 50–61.

T. L. SEETHOFF, R. C. SHIFLETT

[1] Doubly stochastic measures with prescribed support. *Z. Wahrscheinlichkeitstheorie verw. Gebiete* **41** (1978), 283–8.

E. SENETA

[1] The Galton–Watson process with mean one. *J. Appl. Prob.* **4** (1967), 489–95.

[2] The stationary distribution of a branching process allowing immigration: a remark on the critical case. *J. Roy. Statist. Soc., Ser. B* 1968, 176–9.

[3] On recent theorems concerning the supercritical Galton–Watson process. *Ann. Math. Statist.* **39** (1968), 2098–102.

[4] Some second-order properties of the Galton–Watson extinction time distribution. *Sankhyā: Indian J. Statist., Ser. A* **31** (1969), 75–8.

[5] On Koenigs' ratios for iterates of real functions. *J. Austral. Math. Soc.* **10** (1969), 207–13.

[6] Functional equations and the Galton–Watson process. *Adv. Appl. Prob.* **1** (1969), 1–42.

[7] On invariant measures for simple branching processes. *J. Appl. Prob.* **8** (1971), 43–51.

[8] The simple branching process with infinite mean. I. *J. Appl. Prob.* **10** (1973), 206–12.

[9] Regularly varying functions in the theory of simple branching processes. *Adv. Appl. Prob.* **6** (1974), 408–20.

[10] On a functional equation with asymptotically unique continuous solution. *Aequationes Math.* **14** (1976), 457–9.

[11] *Regularly varying functions.* Lecture Notes in Math. 508, Springer Verlag, Berlin–Heidelberg–New York, 1976.

E. SENETA, D. VERE-JONES

[1] On the asymptotic behaviour of subcritical branching process with continuum state space. *Z. Wahrscheinlichkeitstheorie verw. Geb.* **10** (1968), 212–25.

[2] On a problem of M. Jiřina concerning continuous state branching process. *Czech. Math. J.* **19 (94)** (1969), 277–83.

W. N. ŠEVELO

[1] *Oscillation of solutions of differential equations with transformed argument* [Russian]. Kiev, 1978.

M. SHERWOOD, M. D. TAYLOR

[1] Doubly stochastic measures with hairpin support. *Probab. Theory Related Fields* (to appear).

A. SIECZKO

[1] Characterization of globally Lipschitzian Nemytskii operator in the Banach space AC^{r-1}. *Math. Nachr.* **141** (1989), 7–11.

C. L. SIEGEL

[1] Iteration of analytic functions. *Ann. of Math.* (2) **43** (1942), 607–12.

[2] *Vorlesungen über Himmelsmechanik.* Springer Verlag, Berlin–Göttingen–Heidelberg, 1956.

W. SIERPIŃSKI

[1] Sur un système d'équations fonctionnelles définissant une fonction avec un ensemble dense d'intervalles d'invariabilité. *Bull. Inter. Acad. Sci. Cracovie, Cl. Sci. Math. Nat. Sér. A* 1911, 577–82. [*Oeuvres choisies*, Tome II, Polish Scientific Publishers, Warsaw, 1975, 44–8.]

[2] Sur une nouvelle courbe continue qui remplit toute une aire plane. *Bull. Inter. Acad. Sci. Cracovie, Cl. Sci. Math. Nat. Sér. A* 1912, 462–78. [*Oeuvres choisies*, Tome II, Polish Scientific Publishers, Warsaw, 1975, 52–66.]

[3]   On an invertible function whose image is everywhere dense on the plane
      [Polish]. *Wektor* **3** (1914), 289–91. French translation: Sur une fonction
      reversible dont l'image est dense dans le plan, *Oeuvres choisies*, Tome II, Polish
      Scientific Publishers, Warsaw, 1975, 84–6.
[4]   Sur deux problèmes de la théorie des fonctions non dérivables. *Bull. Inter. Acad.
      Sci. Cracovie, Cl. Sci. Math. Nat. Sér. A* 1914, 162–82. [*Oeuvres choisies*, Tome
      I, Polish Scientific Publishers, Warsaw, 1974, 258–73.]
[5]   *Infinitesimal operations* [Polish]. Warsaw, 1948.

A. SKLAR

[1]   Canonical decompositions, stable functions and fractional iterates. *Aequationes
      Math.* **3** (1969), 118–29.

F. SKOF

[1]   Intorno a un'equazione funzionale. *Atti Accad. Sci. Torino Cl. Sci. Fis. Mat.
      Natur.* **115**, Suppl. (1982), 71–80.

A. SMAJDOR

[1]   On superpositions of convex functions. *Archiv der Math.* **17** (1966), 333–5.
[2]   On monotonic solutions of a recurrence relation. *Ann. Polon. Math.* **19** (1967),
      169–76.
[3]   On convex iteration groups. *Bull. Acad. Polon. Sci., Sér. sci. math. astronom.
      phys.* **15** (1967), 325–8.
[4]   Remarks on monotonic solutions of a functional equation. *Aequationes Math.*
      1 (1968), 87–93.
[5]   On monotonic solutions of some functional equations. *Dissertationes Math.* **82**
      (1971).
[6]   On some special iteration groups. *Fund. Math.* **82** (1973), 61–8.
[7]   Note on a functional equation. *Ann. Polon. Math.* **30** (1974), 57–61.
[8]   Note on the existence of convex iteration groups. *Fund. Math.* **59** (1975), 213–18.
[9]   Iterations of multi-valued functions. *Prace Nauk. Uniw. Śląsk.* **759**, Katowice,
      1985.

A. SMAJDOR, W. SMAJDOR

[1]   On the existence and uniqueness of analytic solutions of a linear functional
      equation. *Math. Zeitschr.* **98** (1967), 235–42.

W. SMAJDOR

[1]   On the existence and uniqueness of analytic solutions of the functional equation
      $\varphi(z) = h(z, \varphi[f(z)]$. *Ann. Polon. Math.* **19** (1967), 37–45.
[2]   Local analytic solutions of the functional equation $\varphi(z) = h(z, \varphi[f(z)])$ in
      multidimensional spaces. *Aequationes Math.* **1** (1968), 20–36.
[3]   Analytic solutions of the equation $\varphi(z) = h(z, \varphi[f(z)])$ with right side
      contracting. *Aequationes Math.* **2** (1968), 30–8.
[4]   Formal solutions of a functional equation. *Zeszyty Naukowe Uniw.
      Jagiellońskiego* **203**, *Prace Mat.* **13** (1969), 71–8.
[5]   Local analytic solutions of the functional equation $\Phi(z) = H(z, \Phi[f_1(z)], \ldots,
      \Phi[f_n(z)])$. *Ann. Polon. Math.* **24** (1970), 39–43.
[6]   Solutions of the Schröder equation. *Ann. Polon. Math.* **27** (1972), 61–5.
[7]   On continuous solutions of the Schröder equation. *Ann. Polon. Math.* **32** (1976),
      111–18.

D. B. SMALL

[1] On the functional equation $F(u) = F(1 - u) + F\left(\dfrac{u}{1-u}\right)$, $u \in [0, \frac{1}{2}]$. *Aequationes Math.* **6** (1971), 133–40.

H. ŚMIAŁKÓWNA

[1] On some properties of Schroeder's univalent functions. *Bull. Acad. Polon. Sci., Sér. sci. math. astronom. phys.* **23** (1975), 947–9.

J. SMÍTAL

[1] *On functions and functional equations* [Slovak]. Vydavateľstvo Technickej a Ekonomickej Literatúry, Bratislava, 1984. [English translation, Adam Hilger, Bristol and Philadelphia, 1988.]

[2] On Darboux solutions of Euler's equation. *Aequationes Math.* **37** (1989), 279–81.

N. V. ŠRAGIN

[1] Conditions for the measurability of the superposition [Russian]. *Dokl. Akad. Nauk SSSR* **197** (1971), 195–298.

P. ŠTEFAN

[1] A theorem of Šarkovskiĭ on the existence of periodic orbits of continuous endomorphisms of the real line. *Commun. Math. Phys.* **54** (1977), 237–48.

H. STEINHAUS

[1] On a power series [Polish]. *Comment. Math. Prace Mat.* **1** (1955), 276–84.

N. STEINMETZ

[1] On the functional equation $\varphi(x) = \varphi(px) + \varphi(qx + p)$. *C. R. Math. Rep. Acad. Sci. Canada* **4** (1982), 367–71.

N. STEINMETZ, P. VOLKMANN

[1] Funktionalgleichungen für konstante Funktionen. *Aequationes Math.* **27** (1984), 87–96.

S. STERNBERG

[1] On the behaviour of invariant curves near a hyperbolic point of a surface transformation. *Amer. J. Math.* **77** (1955), 526–34.

[2] Local C^n transformations of the real line. *Duke Math. J.* **24** (1957), 97–102.

[3] Local contractions and a theorem of Poincaré. *Amer. J. Math.* **79** (1957), 809–24.

[4] On the structure of local homeomorphisms of euclidean n-space. II. *Amer. J. Math.* **80** (1958), 623–31.

[5] The structure of local homeomorphisms. *Amer. J. Math.* **81** (1959), 578–604.

[6] Infinite Lie groups and the formal aspects of dynamical systems. *J. Math. Mech.* **10** (1961), 451–74.

B. P. STIGUM

[1] A theorem on the Galton–Watson process. *Ann. Math. Statist.* **37** (1966), 695–8.

M. STOPA

[1] On a linear iterative functional inequality of third order. *Opuscula Math.* **3** (1987), 117–26.

G. SZEKERES

[1] Regular iteration of real and complex functions. *Acta Math.* **100** (1958), 203–58.

[2] On a theorem of Paul Lévy. *Magyar Tud. Akad. Mat. Kutató Int. Közl.* **5** (1960), 277–82.

[3] Fractional iteration of exponentially growing functions. *J. Austral. Math. Soc.* **2** (1961/2), 301–20.

[4] Fractional iteration of entire and rational functions. *J. Austral. Math. Soc.* **4** (1964), 129–42.

[5] Scales of infinity and Abel's functional equation. *Math. Chronicle* **13** (1984), 1–27.

J. TABOR

[1] Rational iteration groups. *Rocznik Nauk.-Dydakt. WSP Kraków* **97**, *Prace Mat.* **11** (1985), 153–76.

R. TAMBS LYCHE

[1] Sur l'équation fonctionnelle d'Abel. *Fund. Math.* **5** (1924), 331–3.

GY. TARGOŃSKI

[1] Darstellung von Funktionen durch Kettenreihen. *Publ. Math. Debrecen* **2** (1951/2), 286–9.

[2] Schröder's equation and a generalization of the Chebyshev polynomials. *Trans. New York Acad. Sci.* (2) **27** (1965), 600–6.

[3] Zur Klassifizierung der linearen Operatoren auf Funktionalgebren. *Math. Zeitschr.* **97** (1967), 238–20.

[4] *Seminar on functional operators and equations.* Lecture Notes in Math. 33, Springer Verlag, Berlin–Heidelberg–New York, 1967.

[5] P 63. *Aequationes Math.* **4** (1970), 251.

[6] Orbit properties of functions and 'Pre-Abel' equations. *Ann. Polon. Math.* **33** (1976), 49–55.

[7] An iteration theoretical approach to the concept of time. *Colloques Internationaux du C.N.R.S.* **229**, *Transformations ponctuelles et leurs applications, Toulouse, 10–14 Septembre 1973,* 245–57, CNRS, 1976.

[8] *Topics in iteration theory.* Vandenhoeck and Ruprecht, Göttingen, 1981.

GY. TARGOŃSKI, M. C. ZDUN

[1] Generators and co-generators of substitution semigroups. *Annales Mathematicae Silesianae* **1 (13)** (1985), 169–74.

[2] Substitution operators on L^p-spaces and their semigroups. *Berichte der Math.-Stat. Sektion in der Forschungsgesellschaft Joanneum–Graz,* No. 283, 1987.

N. P. THIELMAN

[1] On the convex solution of a certain functional equation. *Bull. Amer. Math. Soc.* **47** (1941), 118–20.

W. J. THRON

[1] Sequences generated by iteration. *Trans. Amer. Math. Soc.* **96** (1960), 38–53.

H. TÖPFER

[1] Komplexe Iterationsindizes ganzer und rationaler Funktionen. *Math. Ann.* **121** (1949), 191–222.

T. TURD'EV, T. ŠARIFOVA

[1] Linear functional-difference equations [Russian]. *Izv. Akad. Nauk Uzbek. SSR, Ser. fiz.-mat. nauk* **19** (1975), 33–7.

E. TURDZA

[1] On the stability of the functional equation of the first order. *Ann. Polon. Math.* **24** (1970), 35–8.

[2] Some remarks on the stability of the functional equation $\varphi(f(x)) = g(x)\varphi(x) + F(x)$ [Polish]. *Rocznik Nauk.-Dydakt. WSP Kraków* **41**, *Prace Mat.* **6** (1970), 156–64.

[3] On the stability of the functional equation $\varphi(f(x)) = g(x)\varphi(x) + F(x)$. *Proc. Amer. Math. Soc.* **30** (1971), 484–6.

[4] Some remarks on the stability of the non-linear functional equation of the first order. *Demonstratio Math.* **6** (1973), 883–91.

[5] On a functional inequality with the n-th iterate of the unknown function. *Zeszyty Naukowe Uniw. Jagiellońskiego* **356**, *Prace Mat.* **16** (1974), 189–94.

[6] Note on a functional inequality with the n-th iterate of the unknown function. *Zeszyty Naukowe Uniw. Jagiellońskiego* **475**, *Prace Mat.* **19** (1977), 219–21.

[7] Comparison theorems for a functional inequality. *General inequalities* (Proceedings of the Symposium, Oberwolfach, 1976), edited by E. F. Beckenbach, ISNM 41, Birkhäuser Verlag, Basle and Stuttgart, 1978, 199–211.

[8] The solutions of an inequality for the n-th iterate of a function. *Amer. Math. Monthly* **86** (1979), 281–3.

[9] On an iterative functional inequality. *Zeszyty Naukowe Uniw. Jagiellońskiego* **623**, *Prace Mat.* **23** (1982), 105–10.

[10] Set stability for a functional equation of iterative type. *Demonstratio Math.* **15** (1982), 443–8.

S. M. ULAM

[1] *A collection of mathematical problems.* Interscience, New York, 1960.

J. S. ULLIAN

[1] Splinters of recursive functions. *J. Symbolic Logic* **25** (1960), 33–8.

M. URABE

[1] Invariant varieties for finite transformation. *J. Sci. Hiroshima Univ., Ser. A* **16** (1952), 47–55.

D. VAIDA

[1] On the existence of holomorphic solutions of the equation $f(\alpha z) = g(z)f(z)$ [Roumanian]. *Com. Acad. R. P. Romîne* **10** (1960), 403–7.

F. VAJZOVIĆ

[1] On a functional equation. *Glasnik Mat.-Fiz. Astr.* (2) **19** (1964), 217–24.

R. VENTI

[1] Linear normal forms of differential equations. *J. Diff. Equations* **5** (1966), 182–94.

E. VINCZE

[1] Über die Charakterisierung der assoziativen Funktionen von mehreren Veränderlichen. *Publ. Math. Debrecen* **6** (1959), 241–53.

[2] Bemerkung zur Charakterisierung der Gauss'schen Fehlergesetzes. *Magyar Tud. Akad. Mat. Kutató Int. Kösl.* **7** (1962), 357–61.

L. VOLKMANN, P. VOLKMANN

[1] Über die Charakterisierung der Funktion $f(x) = x$ durch Funktionalgleichungen. I, II. *Aequationes Math.*: **28** (1985), 151–5; **30** (1986), 142–50.

P. VOLKMANN

[1] Caractérisation de la fonction $f(x) = x$ par un système de deux équations fonctionnelles. *C. R. Math. Rep. Acad. Sci. Canada* **5** (1983), 27–8.

[2] Charakterisierung der Funktion $1/x$ durch Funktionalgleichungen. *Ann. Polon. Math.* **48** (1988), 91–4.

544   References

R. WAGNER

[1]   Eindeutige Lösungen der Funktionalgleichung $f(x + f(x)) = f(x)$. *Elem. Math.*
      14 (1959), 73–8.

J. WALORSKI

[1]   On a functional inequality. *Aequationes Math.* **32**(1987), 213–15.

W. WALTER

[1]   On a functional equation of Bellman in the theory of dynamic programming.
      Aequationes Math. **14** (1976), 435–44.

[2]   Old and new approaches to Euler's trigonometric expansions. *Amer. Math.
      Monthly* **89** (1982), 225–30.

R. WĘGRZYK

[1]   Continuous solutions of a linear homogeneous functional equation. *Ann. Polon.
      Math.* **35** (1977), 15–20.

[2]   On the extension of integrable solutions of a functional equation of n-th order.
      Ann. Polon. Math. **41** (1983), 229–36.

[3]   Fixed points theorems for multi-valued functions and their applications to
      functional equations. *Dissertationes Math. (Rozprawy Mat.)* **201**, Warsaw, 1982.

J. WEITKÄMPER

[1]   Embeddings in iteration groups and semigroups with nontrivial units.
      Stochastica **7 (3)** (1983), 175–95.

[2]   On conjugate hat functions. *Opuscula Math.* **4** (1988), 307–16.

H. WHITNEY

[1]   Analytic extensions of differentiable functions defined in closed sets. *Trans.
      Amer. Math. Soc.* **36** (1934), 63–89.

G. T. WHYBURN

[1]   Orbit decompositions. *Bull. Amer. Math. Soc. Abstracts* (1941), No. 47-291.

[2]   *Analytic topology.* AMS Coll. Publ. 28 (1942). Fourth printing with corrections
      1971.

R. WOBST

[1]   Correction to a paper by Newman and Sheingorn. *Aequationes Math.* **17** (1978),
      109–10.

W. WUNDERLICH

[1]   Eine überall stetige und nirgends differenzierbare Funktion. *Elem. Math.* **7**
      (1952), 73–9.

M. C. ZDUN

[1]   On the uniqueness of solutions of the functional equation $\varphi(x + f(x)) =
      \varphi(x) + \varphi(f(x))$. *Aequationes Math.* **8** (1972), 229–32.

[2]   On a Böttcher functional equation and its application for evaluation of a
      well-known improper integral. *Prace Nauk. Uniw. Śląsk.* **30**, *Prace Mat.* **3**
      (1973), 79–85.

[3]   On the regular solutions of a linear functional equation. *Ann. Polon. Math.* **30**
      (1974), 89–96.

[4]   Solutions of bounded variation of a linear functional equation and some their
      interpretation for recurrent sequences. *Prace Nauk. Uniw. Śląsk.* **87**, *Prace Mat.*
      6 (1975), 55–62.

[5] Solutions of bounded variation of a linear homogeneous functional equation in the indeterminate case. *Aequationes Math.* **14** (1976), 143–58.

[6] On the orbits of hat-functions. *Colloquium Math.* **36** (1976), 249–54.

[7] Continuous solutions of finite variation of a linear functional equation. *Ann. Polon. Math.* **34** (1977), 179–96.

[8] Some remarks on the continuous solutions of finite variation of a linear functional equation. *Rev. Roum. Math. Pures Appl.* **22** (1977), 863–9.

[9] Differentiable fractional iteration. *Bull. Acad. Polon. Sci., Sér. sci. math. astronom. phys.* **25 (7)** (1977), 643–6.

[10] On the iteration of the hat-functions. *Aequationes Math.* **16** (1977), 181–2.

[11] On integrable solutions of Abel's functional equation. *Glasnik Mat.* **12 (32)** (1977), 49–59.

[12] On differentiable iteration groups. *Publ. Math. Debrecen* **26** (1979), 105–14.

[13] Continuous and differentiable iteration groups. *Prace Nauk. Uniw. Śląsk.* **308**, Katowice, 1979.

[14] Iteration semigroups with restricted domain. *Colloques Internationaux du C.N.R.S.* **332**, *Théorie de l'itération et ses applications, Toulouse, 17–22 Mai 1982,* 75–9, CNRS, 1982.

[15] Note on solutions of bounded variation of a linear non-homogeneous functional equation. *Mathematica, Cluj* **27 (50)** (1985), 79–89.

[16] Regular fractional iterations. *Aequationes Math.* **28** (1985), 73–9.

[17] On embedding of homeomorphisms of the circle in a continuous flow. *Iteration theory and its functional equations* (Proceedings, Schloss Hofen, 1984), edited by R. Liedl, L. Reich and Gy. Targoński, Lecture Notes in Mathematics 1163, Springer Verlag, Berlin –Heidelberg–New York–Tokyo, 1985, 218–31.

[18] Note on commutable functions. *Aequationes Math.* **36** (1988, 153–64.

[19] On simultaneous Abel's equations. *Aequationes Math.* (1989), to appear.

G. ZIMMERMAN-RIGGERT

[1] Iterative roots of Čebyšev polynomials. Abstract in *Aequationes Math.* **17** (1978), 368.

[2] Über die Existenz iterativer Wurzeln von Abbildungen. Doctoral dissertation, Marburg/Lahn, 1978.

AUTHOR INDEX

The numbers in square brackets refer to the reference list.

SUBJECT INDEX